Second Edition

MICROPROCESSORS and MICROCOMPUTER-BASED SYSTEM DESIGN

MOHAMED RAFIQUZZAMAN, PH.D.

Professor
California State Polytechnic University
Pomona, California
and
Adjunct Professor
University of Southern California
Los Angeles, California

CRC Press
Boca Raton New York London Tokyo

Library of Congress Cataloging-in-Publication Data

Rafiquzzaman, Mohamed.
    Microprocessors and microcomputer-based system design / Mohamed Rafiquzzaman. — 2nd ed.
        p.   cm.
    Includes bibliographical references and index.
    ISBN 0-8493-4475-1
    1. Microprocessors.   2. Intel 80xx series microprocessors.   3. Motorola 68000 series microprocessors.
4. Microcomputers.   I. Title.
QA76.5.R27848 1995
004.16—dc20   95-7374
                                                                                                                    CIP

No claim to original U.S. Government works
International Standard Book Number 0-8493-4475-1
Library of Congress Card Number 95-7374
Printed in the United States of America   1 2 3 4 5 6 7 8 9 0
Printed on acid-free paper

Preface

This book is based on the fundamental concepts associated with typical 8-, 16-, and 32-bit microprocessors and microcomputers. These concepts were related in detail to the Intel 8085/8086 and Motorola 68000/68020/68030/88100. A brief coverage of Intel 80386/80486/80960/Pentium, Motorola 68040 and Motorola/IBM/Apple PowerPC is also included.

With the growing popularity of both Intel and Motorola 32-bit microprocessors, it is now necessary to cover these processors at the undergraduate and graduate levels. Therefore, a thorough coverage of these processors is provided.

A detailed treatment of Intel 80386 and Motorola 68020/68030 along with more examples and system design concepts is included. Programming and system design concepts associated with other popular 32-bit microprocessors such as Intel 80486/80960 and Motorola 68040 are also covered in this book. Finally, an overview of Intel Pentium microprocessor is provided. Since the fundamental concept of 8-bit microprocessors, along with the Intel 8085, has proved its worth many times over in the intervening years, the 8085 has been retained in this edition.

This book is divided into ten chapters. Chapter 1 contains the basics of microprocessors, as in the first edition. New topics such as floating-point arithmetic, Program Array Logic (PAL) used for address decoding for 32-bit microprocessors, flash memories and an overview of various 32-bit microprocessors is also included.

Chapter 2 covers details of the 8085 microprocessor.

Chapters 3 through 8 provide detailed descriptions of the architectures, addressing modes, instruction sets, I/O and system design concepts of Intel's 8086, 80386, 80486, and 80960 and Motorola's 68000, 68020, 68030, 68040, and 88100 microprocessors. An overview of Intel 80186, 80286, Pentium and PowerPC microprocessors are also included.

Chapter 9 contains fundamentals of peripheral interfacing.

Chapter 10 includes system design concepts along with the applications of design principles covered in the preceding chapters. Three system design examples using the 8085, 8086, and 68000 are included in detail.

The appendices include materials on the HP 64000 microcomputer development systems, data sheets on various microprocessors and support chips, and a glossary.

The audience of this book can be college students or practicing microprocessor system designers in the industry. It can be used as an undergraduate or graduate text in electrical engineering, computer engineering or computer science. Practitioners of microprocessor system design in the industry will find greater detail and comparison considerations than are found in manufacturers' manuals. The book assumes a familiarity with digital logic and topics such as Boolean Algebra and K-maps.

The author wishes to express his sincere appreciation to his student Frank Lee for making constructive suggestions and typing the manuscript. The author is also grateful to Dr. W.C. Miller of University of Windsor, Canada, and others for their support throughout the writing effort.

Mohamed Rafiquzzaman
Pomona, California

The Author

Mohamed Rafiquzzaman obtained his Ph.D. in Electrical Engineering in Canada in 1974. He worked for Esso/Exxon and Bell Northern Research for approximately 5 years. Dr. Rafiquzzaman is presently a professor of electrical and computer engineering at California State Polytechnic University, Pomona. He was Chair of the department there from 1984 to 1985. Dr. Rafiquzzaman is also an adjunct professor of electrical engineering systems at the University of Southern California, Los Angeles. He consulted for ARCO, Rockwell, Los Angeles County, and Parsons Corporation in the areas of computer applications. He has published six books on computers, which have been translated into Russian, Chinese and Spanish, and has published numerous papers on computers.

Dr. Rafiquzzaman is the founder of Rafi Systems, Inc. a manufacturer of biomedical devices and computer systems consulting firm in California. In 1984, he managed the Olympic Swimming, Diving and Synchronized Swimming teams. He has also managed Swiss timing, scorekeeping, and computer systems.

From 1984 to 1989, he was the instructor for Motorola in Southern California teaching short courses on 68000, 68020, and 68030 for local industries.

Dr. Rafiquzzaman was an advisor to the President of Bangladesh on computers from 1988 to 1990. He is currently involved in research activities in both hardware and software aspects of typical 16- and 32-bit microprocessor-based applications. These activities include image processing, robotics control, and OCR (Optical Character Recognition).

To my parents,
my wife Kusum, son Tito,
and brother Elan

In memory of my brother, Dr. M. Kaisaruzzaman

Table of Contents

1

INTRODUCTION TO MICROPROCESSORS AND MICROCOMPUTER-BASED APPLICATIONS

This chapter provides a brief summary of the features of microprocessors and microcomputer-based applications.

The basic elements of a computer are the Central Processing Unit (CPU), the Memory, and Input/Output (I/O) units. The CPU translates instructions, performs arithmetic or logic operations, and temporarily stores instructions and data in its internal high-speed registers. The memory stores programs and data. The I/O unit interfaces the computer with external devices such as keyboard and display.

With the advent of semiconductor technology, it is possible to integrate the CPU in a single chip. The result is the microprocessor. Metal Oxide Semiconductor (MOS) technology is typically used to fabricate the standard off-the-shelf microprocessors such as those manufactured by Intel and Motorola. Appropriate memory and I/O chips are interfaced to the microprocessor to design a microcomputer. Single-chip microcomputers are also available in which the microprocessor, memory, and I/O are all fabricated in the same chip. These single-chip microcomputers offer limited capabilities. However, they are ideal for certain applications such as peripheral controllers.

Single chip microcomputers are also referred to as "microcontrollers". The microcontrollers are typically used for dedicated applications such as automotive systems, home appliances and home entertainment systems. Typical microcontrollers, therefore, include on-chip timers, A/D (Analog to Digital) and D/A (Digital to Analog) converters. Two popular microcontrollers are Intel 8751 (8-bit)/8096 (16-bit) and Motorola HC11 (8-bit)/HC16 (16-bit). The 16-bit microcontrollers include more on-chip ROM, RAM, and I/O compared to the 8-bit microcontrollers.

The efficient development of microprocessor-based systems necessitates the use of a microcomputer development system. The microcomputer development system is used for the design, debugging, and sometimes the documentation of a microprocessor-based system.

This chapter first covers the evolution of 8-, 16-, and 32-bit microprocessors along with an overview of programming languages, microcomputer hardware, and software. The attributes of typical microcomputer development systems features as well as some specific microprocessor applications are also included.

1.1 Evolution of the Microprocessor

Intel Corporation introduced the first microprocessor, the 4004 (4-bit), in 1971. The 4004 evolved from a development effort while designing a calculator chip set.

Soon after the 4004 appeared in the commercial market, three other microprocessors were introduced. These were the Rockwell International 4-bit PPS-4, the Intel 8-bit 8008, and the National Semiconductor 16-bit IMP-16.

The microprocessors introduced between 1971 and 1973 were the first-generation systems. They were designed using the PMOS (P-type MOS) technology. This technology provided low cost, slow speed, and low output currents and was not compatible with TTL (Transistor Transistor Logic).

After 1973, second-generation microprocessors such as MOS Technology 6502, Motorola 6800 and 6809, Intel 8085, and Zilog Z80 evolved. These 8-bit processors were fabricated using the NMOS (N-type MOS) technology. The NMOS process offers faster speed and higher density than PMOS and is TTL-compatible.

After 1978, the third-generation microprocessors were introduced. These processors are 16 bits wide (16-bit ALU) and include typical processors such as Intel 8086/80186/80286 and Motorola 68000/68010. These microprocessors were designed using the HMOS (high-density MOS) technology. HMOS provides the following advantages over NMOS:

- Speed-Power-Product (SPP) of HMOS is four times better than NMOS:

    NMOS = 4 Picojoules (PJ)

    HMOS = 1 Picojoule (PJ)

    Note that Speed-Power-Product

    = speed * power

    = nanosecond * milliwatt

    = picojoules

- Circuit densities provided by HMOS are approximately twice those of NMOS:

    NMOS = 4128 gates/μm^2

    HMOS = 1852.5 gates/μm^2

    where 1 μm (micrometer) = 10^{-6} meter.

Later, Intel utilized the HMOS technology to fabricate the 8085A. Thus, Intel offers a high-speed version of the 8085A called 8085AH. The price of the 8085AH is higher than the 8085A.

In 1980, fourth-generation microprocessors evolved. Intel introduced the first commercial 32-bit microprocessor, the problematic Intel 432. This processor was eventually discontinued by Intel. Since 1985, more 32-bit microprocessors have been introduced. These include Motorola's MC 68020/68030/68040/PowerPC, Intel's 80386/80486 and the Intel Pentium processor. These processors are fabricated using the low-power version of the HMOS technology called the HCMOS.

The performance offered by the 32-bit microprocessor is more comparable to that of superminicomputers such as Digital Equipment Corporation's VAX11/750 and VAX11/780. Both Intel and Motorola introduced 32-bit RISC (Reduced Instruction Set Computer) microprocessors, namely the Intel 80960 and Motorola MC88100/PowerPC with simplified instruction sets. Note that the purpose of RISC microprocessors is to maximize speed by reducing clock cycles per instruction. Almost all computations can be obtained from a simple instruction set.

RISC microprocessors have hardwired instruction sets like the non-RISC microprocessors such as the Intel 8085 and MOS 6502. This means that for every instruction, there exists actual, physical connections that provide the desired instruction decoding. The microprocessor, therefore, does not use valuable clock cycles (machine cycles) in instruction decoding. Most RISC instructions require a maximum of only two clock cycles to complete. These instructions are restricted to register-to-register operations with load and store for memory access. Since RISC-type microprocessors are hardwired, instructions may be executed simultaneously (as long as the instructions do not share the same register.) This technique is known as pipelining.

Pipelining is the mechanism that actually enables simultaneous processing to occur. Instructions are fetched in sequential order. The processor continues to fetch instructions even though it has not executed the present instruction. This provides an input pipeline to the instruction cache. At this stage, the instructions are intelligently fed into the instruction execution unit for processing. When an instruction is executed, the registers used by the instruction clear respective bits in a scoreboard register. As the microprocessor is executing a given instruction, it checks the scoreboard register to see if the next instructions (residing in a fast read/write memory internal to the microprocessor called the instruction cache) use registers that are currently in use. Instructions not using the same registers can therefore be executed at the same time. The RISC microprocessors can execute as many as five instructions simultaneously.

The trend in microprocessors is implementation of more on-chip functions and for improvement of the speeds of memory and I/O devices. Some manufacturers are speeding up the processors for data crunching type applications. Digital Equipment Corporation's Alpha 21164 with 300 MHz clock, four instruction-per-cycle rate and RISC-based architecture is the fastest microprocessor available today.

1.2 Microprocessor Data Types

This section discusses the data types used by microprocessor. Typical data types include signed and unsigned binary integers, binary coded decimal (BCD), American Standard Code for Information Interchange (ASCII) and floating-point numbers. Note that binary integer numbers do not support fractions in microprocessors. Fractions and mixed numbers (numbers comprised of integers and fractions) use either binary or BCD floating-point formats. Floating-point numbers are often referred to as real numbers.

1.2.1 Unsigned and Signed Binary Integers

An unsigned binary integer has no arithmetic sign. Unsigned binary numbers are therefore always positive. An example is memory address which is always a positive number. An 8-bit unsigned binary integer represents all numbers from 00_{16} through FF_{16} (0_{10} through 255_{10}).

A signed binary integer, on the other hand, includes both positive and negative numbers. It is represented in true form for a positive number and in two's complement form for a negative number. For example, the decimal number +10 can be represented as a true form 8-bit number in a microprocessor as 0000 1010 (binary) or 0A (hexadecimal). The decimal number -10 can be represented in two's complement form as 1111 0110 (binary) or F6 (hexadecimal). The most significant bit of a signed binary number represents the sign of the number. For example, bit 7 of an 8-bit signed number, bit 15 of a 16-bit signed number and bit 31 of a 32-bit signed number represent the signs of the respective numbers. A 32-bit signed binary integer includes all numbers from -2,147,483,648 to 2,147,483,647 with 0 being a positive number.

Finally, note that the hexadecimal number FF_{16} is 255_{10} when represented as an unsigned number. On the other hand, FF_{16} is -1_{10} when represented as a signed number.

1.2.2 BCD (Binary Coded Decimal) Numbers

A BCD digit consists of four bits with a value ranging from 0000_2 to 1001_2 (0 through 9 decimal). A BCD digit greater than 9_{10} can be represented as two or more BCD digits.

Microprocessors store BCD numbers in two forms, packed and unpacked. The unpacked BCD number represents each BCD digit as a byte while the packed BCD number represents two BCD digits in a byte. For example, 23_{10} is represented as $0000\ 0010\ 0000\ 0011_2$ as two unpacked BCD numbers, while it is represented as $0010\ 0011_2$ as a packed BCD number.

Microprocessors normally input data from a keypad in unpacked BCD form. This data can then be converted by writing a program in the microprocessor to packed BCD form for arithmetic operations or for storing in memory. After processing, the packed BCD result is then converted to unpacked BCD form by another microprocessor program written by the user for displays. Typical displays use either unpacked BCD data or unpacked BCD data converted to seven segment code by the microprocessor's program.

1.2.3 ASCII

ASCII (American Standard Code for Information Interchange) is a code that represents alphanumeric (alpha characters and numbers) in a microcomputer's memory. ASCII also represents special symbols such as # and %. It is a 7-bit code. The most significant bit (bit 7) is sometimes used as a parity bit. The parity bit represents the number of ones in the byte. If the number of ones is odd, the parity is odd; otherwise, the parity is even.

Also, note that the hexadecimal numbers 30_{16} through 39_{16} are ASCII codes representing the decimal numbers 0 through 9. A listing of the ASCII codes is included in Appendix F.

1.2.4 Floating-Point Numbers

Floating-Point numbers contain three components. These are a sign, exponent and mantissa. For example, consider the decimal value -2.5×10^{-2}. The sign is negative, the exponent is -2 and the significand or mantissa is 2.5. In floating point numbers, it is possible to store the same number in several different ways. For example, 1 can be represented as 10.0×10^{-1}, 1.0×10^{0}, 0.1×10^{1} and 0.01×10^{2}. To make computations yield the maximum accuracy, the numbers are normalized. This means that the exponent is adjusted so that the mantissa always follows a specific format. A binary floating point number is represented as a normalized binary fraction raised to a power of 2. To convert a binary number to its standard floating-point form, the binary number is converted to a normalized floating-point number. Note that a normalized binary floating-point number is represented as 1.XXXXXX raised to some power of 2 where X can be 0 or 1. The 1. part is implied and is not stored by the microprocessor. First, the binary number is converted to a common 32-bit floating-point format. The most significant bit (bit 31) is the sign bit (S). If S is zero, the number is positive, while if S is one, the number is negative. The next 8 bits contain the biased exponent. This means that an 8-bit number $7F_{16}$ or 127_{10} is added to the exponent. Biased exponent makes numeric comparison (such as less than or greater than) easy. The bias is usually chosen such that the most negative number allowed in the exponent becomes zero and the most positive number becomes the largest value of the representation. For example, with an 8-bit exponent, and $+127_{10}$ bias, the smallest and the largest values of the bias exponent are 0_{10} and 255_{10} respectively. Note that the zero biased exponent is represented as 127_{10}, while the unbiased minimum and maximum values of the 8-bit exponent are -127_{10} and $+128_{10}$ respectively. The remaining 23 bits represent the fractional part of the number. Note that zero is stored as 32 zeros while infinity is stored as 32 ones. As an example, consider converting the decimal number 10 to the standard floating-point format as follows:

1. $10_{10} = 1010_2$
2. Normalize the binary number as $1.XXXXXX \times 2^n = 1.010 \times 2^3$
3. Sign, S = 0 for positive
4. Biased exponent = $7F_{16} + 3 = 82_{16}$

 5. 23-bit Fraction = 0100 0000 0000 0000 0000 000
 6. The floating-point equivalent of 10_{10} is
      S   Exponent   Fraction
      0   1000 0010   0100 0000 0000 0000 0000 0000

A special case of the floating-point format is called NaN (Not a Number). NaNs are results generated by floating-point operations that have no mathematical interpretations. These results may be generated by operations such as multiplication of infinity by infinity.

The BCD floating-point form represents a number in BCD scientific notation. The number is represented as normalized significand raised to some power of 10. Each BCD floating-point number is represented in typical microprocessors as 80 bits. The BCD fraction is 16 digits wide (64-bit) and is stored as packed BCD digits. The whole number portion of the significand is stored as one digit BCD from 0 to 9. The BCD exponent along with the sign is expressed as 12 bits.

Typical floating-point coprocessors such as 80387 (for the Intel 80386 microprocessor) and 68881/68882 (for the Motorola 68020/68030 microprocessor) support several data types. For example, the 80387 coprocessor supports seven data types. These are word integer (16-bits), short integer (32-bit), long integer (64-bit), packed BCD (80-bit), short real (32-bit), long real (64-bit), and temporary real (80-bit).

The 80387 integer data types are represented by the two's complement same as those used by the 80386. The only difference is that the 80386 supports an 8-bit integer while the 80387 supports a 64-bit integer.

The 80387 supports 80-bit packed BCD with 18 decimal digits (bits 0-71), bit 79 as the sign-bit and seven (bits 72-78) unused bits. With 18-digits representation, the COBOL standard (the High level language utilizing BCD) is followed.

The 80387 supports three real data formats. These are short real (32-bit with one sign bit, 8-bit exponent and 23-bit significand), long real (64-bit with one sign-bit, 11-bit for exponent and 62-bit for the significand), temporary real (80-bit with one sign-bit, 15-bit exponent, and 64-bit significand).

The 80387 uses the temporary real format internally. All data types are converted by the 80387 immediately into temporary real. This is done to maximum precision and range of computations.

The 80387 supports five special cases. These are zeros, infinities (both positive and negative), denormals, and NaNs (signalling and quiet).

Denormals represent very small numbers that are not normalized. Normally, numbers are required to be normalized by shifting to left until the most significant is one. Denormals do not have one as the most significant bit of the significand. Denormals permit a gradual underflow. That is the precision is lost gradually rather than abruptly. When the least normalizable number is reached, the next small representation is zero. Denormals provide gradual underflow of numbers that are not normalized. That is, denormals extend the range of very small numbers significantly, but with some loss in precision.

A signaling NaN causes an invalid operation exception when used in an operation. A quiet NaN, on the other hand, does not cause an invalid operation exception.

The floating-point data types supported by the Motorola 68881/68882 floating-point coprocessors are summarized next. Note that the 68881 and 68882 differ in execution speed. They are basically identical. The 68882 is an enhanced version of the 68881 in that it executes several floating-point instructions concurrently with the 68020/68030. The 68881/68882 supports integers, binary floating-point numbers, and packed floating-point BCD. Data are represented externally by using these formats. The 68881/68882 utilizes an 80-bit binary floating point form to represent all data internally.

The 68881/68882 supports these signed integer formats. These are 8-bit byte, 16-bit word, and 32-bit long word.

Binary floating-point format is also called binary real form. The 68881/68882 supports binary floating-point form which contains three fields. These are a sign, biased exponent and a significand. The 68881/68882 operates on these sizes, namely 32-bit single-precision, 64-bit double precision, and 96-bit extended precision.

For single precision (bit 31: sign bit, bits 23-30: 8-bit exponent, bits 0-22: 23-bit significand), 64-bit double precision (bit 63: sign bit, bits 62-52: 11-bit exponent, bits 0-51: 52-bit significand), and 96-bit extended-precision (bit 95: sign bit, bits 80-94: 15-bit exponent, bits 64-79: zero; sixteen unused bits, bits 0-63: 64-bit significand).

The biased exponent is used. The Single Precision adds a bias of 127_{10} ($7F_{16}$), double Precision uses a bias of 1023_{10} ($3FF_{16}$), and the extended-Precision uses a bias of 16383_{10} ($3FFFF_{16}$). The bias is added to the exponent before it is stored in this format and subtracted to convert to a true exponent when the number is interpreted.

A few special cases that do not conform to the floating-point form are also handled by the 68881/68882. For example, a zero is represented with all bits of the exponent and significand as zeros. The sign bit may be a one or a zero representing +0 or -0. The infinity, on the other hand is represented by all bits in the exponent and significand set to ones. The sign bit may be zero or one representing positive or negative infinity. The 68881/68882 supports BCD floating point form which represent each number as normalized significand raised to a power of 10. This format stores a number as 96 bits. The least significant 64 bits (8 bytes) contain the 16-digit BCD fraction. The next byte contains the whole number portion of the significand (0-9). The most significant bit (bit 95) contains the sign of significand while bit 94 includes the sign of the exponent. The exponent is represented by three digit BCD packed exponent (000-999) in bits 80–91. Similar to the 80387, the 68881/68882 also represents NaN's and also provides exceptions for signaling NaN's.

IEEE has established the standard for floating-point arithmetic specified by ANSI-IEEE754-1985. Typical 32-bit microprocessors use this standard.

1.3 Microcomputer Hardware

In this section, some unique features associated with various microcomputer components will be described.

The microcomputer contains a microprocessor, a memory unit, and an input/output unit. These elements are explained in the following in detail. Figure 1.1 shows a simplified block diagram of a microcomputer.

1.3.1 The System Bus

The system bus contains three buses. These are the address bus, the data bus, and the control bus. These buses connect the microprocessor to each of the memory and I/O elements so that information transfer between the microprocessor and any of the other elements can take place.

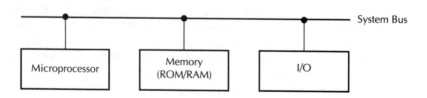

FIGURE 1.1   Simplified block diagram of a microprocessor.

On the address bus, information transfer normally takes place only in one direction, from the microprocessor to the memory or I/O elements. Therefore, this is called a unidirectional bus. This bus is usually 16 to 32 bits wide. The number of unique addresses that the microprocessor can generate on this bus depends on the width of this bus. For example, for a 16-bit address bus, the microprocessor can generate $2^{16} = 65,536$ different possible addresses. A different memory location or an I/O element can be represented by each one of these addresses.

The data bus is a bidirectional bus, that is, information can flow in both directions, to or from the microprocessor. This bus is normally 8, 16, or 32 bits wide.

The control bus is used to transmit signals that are used to synchronize the operation of the individual microcomputer elements. Typical control signals include READ, WRITE, and RESET. Some signals on the control bus such as interrupt signals are unidirectional, while some others such as RESET may be bidirectional.

1.3.2 The Microprocessor

The commercial microprocessor, fabricated using the MOS technology, is normally contained in a single chip. The microprocessor is comprised of a register section, one or more ALUs (Arithmetic Logic Units), and a control unit. Depending on the register section, the microprocessor can be classified either as an accumulator-based or a general-purpose register-based machine.

In an accumulator-based microprocessor such as the Intel 8085 and Motorola 6809, one of the operands is assumed to be held in a special register called the "accumulator". All arithmetic and logic operations are performed using this register as one of the data sources. The result after the operation is stored in the accumulator. One-address instructions are very predominant in this organization. Eight-bit microprocessors are usually accumulator-based.

The general-purpose register-based microprocessor is usually popular with 16- and 32-bit microprocessors, such as Intel 8086/80386/80486 and Motorola 68000/68020/68030/68040, and is called general-purpose, since its registers can be used to hold data, memory addresses, or the results of arithmetic or logic operations. The number, size, and types of registers vary from one microprocessor to another. Most registers are general-purpose registers, while some are provided with dedicated functions.

Typical dedicated registers include the Program Counter (PC), the Instruction Register (IR), Status Register (SR), the Stack Pointer (SP) and the Index Register. The 32-bit microprocessors include special on-chip combinational network called the Barrel Shifter.

The PC normally contains the address of the next instruction to be executed. Upon activating the microprocessor chip's RESET input, the PC is normally initialized with the address of the first instruction. For example, the 80486, upon hardware reset, reads the first instruction from the 32-bit address $FFFFFFF0_{16}$. In order to execute the instruction, the microprocessor normally places the PC contents on the address bus and reads (fetches) the first instruction from external memory. The program counter contents are then automatically incremented by the ALU. The microcomputer thus executes a program sequentially unless it encounters a jump or branch instruction. The size of the PC varies from one microprocessor to another depending on the address size. For example, the 68000 has a 24-bit PC, while the 68040 contains a 32-bit PC.

The instruction register (IR) contains the instruction to be executed. After fetching an instruction from memory, the microprocessor places it in the IR for translation.

The status register contains individual bits with each bit having a special meaning. The bits in the status register are called flags. Each flag is usually set or reset by an ALU operation. The flags are used by the Conditional Branch instructions. Typical flags include carry, sign, zero, and overflow.

The carry (C) flag is used to reflect whether or not an arithmetic operation such as ADD generates a carry. The carry is generated out of the 8th bit (bit 7) for byte operations, 16th bit (bit 15) for 16-bit, or 32nd bit (bit 31) for 32-bit operations. The carry is used as the borrow flag for subtraction. In multiple word arithmetic operations, any carry from a low-order word must be reflected in the high-order word for correct results.

The zero (Z) flag is used to indicate whether the result of an arithmetic or logic operation is zero. Z = 1 for a zero result and Z = 0 for a non-zero result. The sign flag (sometimes also called the negative flag) indicates whether a number is positive or negative. S = 1 indicates a negative number if the most significant bit of the number is one; S = 0 indicates a positive number if the most significant bit of the number is zero.

The overflow (V) flag is set to one if the result of an arithmetic operation on signed (two's complement) numbers is too large for the microprocessor's maximum word size; the C flag is overflow for unsigned numbers. The overflow flag for signed 8-bit numbers can be shown as $V = C7 \oplus C6$, where C7 is the final carry and C6 is the previous bit's carry. The $\oplus$ symbol indicates exclusive-OR operation. This can be illustrated by the numerical examples shown below:

$$
\begin{array}{r}
0000\ 0100 \\
+\ 0000\ 0010 \\
\hline
C7=\ 0 \leftarrow 0000\ 0110 \\
\end{array}
\qquad
\begin{array}{r}
04_{10} \\
+\ 02_{10} \\
\hline
06_{10} \\
\end{array}
$$

$$C6=\ 0 \leftarrow 0 \qquad 6_{16}$$

From the above, the result is correct when C6 and C7 have the same values (0 in this case). When C6 and C7 are different, an overflow occurs. For example, consider the following:

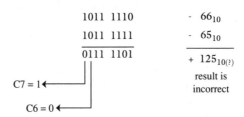

$$
\begin{array}{r}
1011\ 1110 \\
1011\ 1111 \\
\hline
0111\ 1101 \\
\end{array}
\qquad
\begin{array}{r}
-\ 66_{10} \\
-\ 65_{10} \\
\hline
+\ 125_{10(?)} \\
\text{result is} \\
\text{incorrect}
\end{array}
$$

C7 = 1

C6 = 0

The result is incorrect. Since $V = C6 \oplus C7 = 0 \oplus 1 = 1$, the overflow flag is set. Note that this applies to signed two's complemented numbers only.

The stack pointer (SP) register addresses the stack. A stack is Last-In First-Out (LIFO) read/write memory in the sense that items that go in last will come out first. This is because stacks perform all read (POP) and write (PUSH) operations from one end.

The stack is addressed by a register called the stack pointer (SP). The size of the SP is dependent on the microprocessor's address size. The stack is normally used by subroutines or interrupts for saving certain registers such as the program counter.

Two instructions, PUSH (stack write) and POP (stack read), can usually be performed by the programmer to manipulate the stack. If the stack is accessed from the top, the stack pointer is decremented before a PUSH and incremented after a POP. On the other hand, if the stack is accessed from the bottom, the SP is incremented before a PUSH and decremented after a POP. Typical microprocessors access the stack from the top. Depending on the microproces-

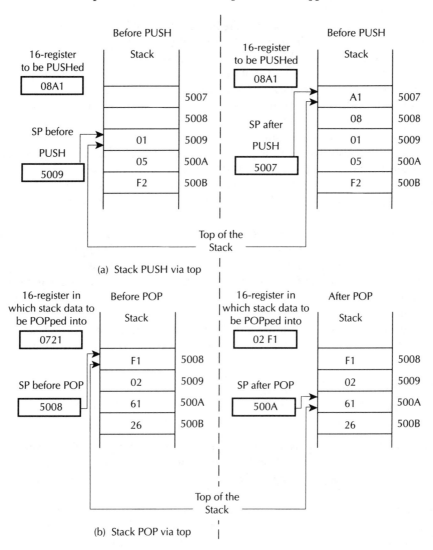

FIGURE 1.2 PUSH and POP operations via top.

sor, an 8-, 16-, or 32-bit register can be pushed onto or popped from the stack. The value by which the SP is incremented or decremented after POP or PUSH operations depends on the register size. For example, values of one for an 8-bit register, two for 16-bit registers, and four for 32-bit registers are used. Figure 1.2 shows the stack data when accessed from the top before and after PUSHing a 16-bit register onto the stack or POPping 16 bits from the stack into the 16-bit register. Note that stack items PUSHed must be POPped in reverse order. The item pushed last must be popped first.

Consider the PUSH operation in Figure 1.2a when the stack is accessed from the top. The SP is decremented by 2 after the PUSH. The SP is decremented since it is accessed from the top. A decrement value of 2 is used since the register to be pushed is 16 bits wide.

The POP operation shown in Figure 1.2b is the reverse of the PUSH. The SP is incremented after POP. The contents of locations 5008_{16} and 5009_{16} are assumed to be empty.

An index register is typically used as a counter for an instruction or for general storage functions. The index register is useful with instructions where tables or arrays of data are accessed. The general-purpose register-based microprocessor can use any general-purpose register as the index register.

Typical 32-bit microprocessors such as the Intel 80386/80486 and Motorola 68020/68030/68040 include a special type of shifter called barrel shifter for performing fast shift operations.

The barrel shifter is an on-chip combinational network for 32-bit microprocessors and provides fast shift operations. For example, the 80386 barrel shifter can shift a number from 0 through 64 positions in one clock period (clock rate is 16.67 MHz).

The ALU in the microprocessor performs all arithmetic and logic operations on data. The size of the ALU defines the size of the microprocessor. For example, Intel 8086 (or Motorola 68000) is a 16-bit microprocessor since its ALU is 16 bits wide. The Intel 8088 (or Motorola 68008) is also a 16-bit microprocessor since its ALU is 16 bits wide, even though its data bus is 8 bits wide. Motorola 68040 (or Intel 80486) is a 32-bit microprocessor since its ALU is 32 bits wide. The ALU usually performs operations such as binary addition and subtraction. The 32-bit microprocessors include multiple ALUs for parallel operations and thus achieve fast speed.

The control unit of the microprocessor performs instruction interpreting and sequencing. In the fetch phase, the control unit reads instructions from memory using the PC as a pointer. It then recognizes the instruction type, gets the necessary operands, and routes them to the appropriate functional units of the execution unit. Necessary signals are issued to the execution unit to perform the desired operations, and the results are routed to the specified destination.

In the sequencing phase, the control unit determines the address of the next instruction to be executed and loads it into the PC. The control unit is typically designed using one of three techniques:

- Hardwired control
- Microprogramming
- Nanoprogramming

The hardwired control unit is designed by physically connecting typical components such as gates and flip-flops. Typical 32-bit RISC microprocessors such as the Intel 80960 and Motorola 88100 are designed using hardwired control. The microprogrammed control unit includes a control ROM for translating the instructions. Intel 8086 is a microprogrammed microprocessor. Nanoprogramming includes two ROMs inside the control unit. The first ROM (microROM) stores all the addresses of the second ROM (nanoROM). If the microinstructions (which is the case with the 68000/68020/68030/68040) repeat many times in a microprogram, use of two-level ROMs provides tremendous memory savings. This is the reason that the control units of the 68000, 68020, 68030, and 68040 are nanoprogrammed.

1.3.3　Memory Organization

1.3.3.a　Introduction

A memory unit is an integral part of any microcomputer system and its primary purpose is to hold programs and data.

In a broad sense, a microcomputer memory system can be logically divided into three groups:

- Processor memory
- Primary or main memory
- Secondary memory

Processor memory refers to the microprocessor registers. These registers are used to hold temporary results when a computation is in progress. Also, there is no speed disparity between these registers and the microprocessor because they are fabricated using the same technology.

However, the cost involved in this approach forces a microcomputer architect to include only a few registers (usually 8 or 16) in the microprocessor.

Primary or main memory is the storage area in which all programs are executed. The microprocessor can directly access only those items that are stored in primary memory. Therefore, all programs and data must be within the primary memory prior to execution.

Secondary memory refers to the storage medium comprising slow devices such as magnetic tapes and disks. These devices are used to hold large data files and huge programs such as compilers and data base management systems which are not needed by the processor frequently. Sometimes secondary memories are also referred to as auxiliary or backup store or virtual memory.

Secondary memory stores programs and data in excess of the main memory. The microcomputer cannot directly execute programs stored in the secondary memory. In order to execute these programs, the microcomputer must transfer them to its main memory by a system program called the operating system. This topic is covered later in the chapter.

Data in disk memories are stored in tracks. A track is a concentric ring of data stored on a surface of a disk. Each track is further subdivided into several sectors. Each sector typically stores 512 or 1024 bytes of data. All disk memories use magnetic media except the optical disk memory which stores data on a plastic disk. Data is read or sometimes written on the optical disk with a laser beam. There are two types of optical disks. These are the CD-ROM (Compact Disk Read Only Memory) and the WORM (Write Once Read Many). The CD-ROM is inexpensive compared to the WORM drive. However it suffers from lack of speed and has limited software applications at the present time. The WORM drive is typically used in huge data storing applications such as insurance and banking since data can be written only once. The optical disk memory is currently becoming popular with microcomputer systems. One of the commonly used disk memories with microcomputer systems is the floppy or flexible disk. The floppy disk is a flat, round piece of plastic coated with magnetically sensitive oxide material. The disk is provided with a protective jacket to prevent fingerprints or foreign matter from contaminating the disk's surface. The floppy disk is available in three sizes. These are the 8 inch, 5.25 inch, and 3.5 inch. The 8 inch floppy disk is not used in present systems. These days, the 5.25 inch and 3.5 inch are very popular. Also, the 3.5 inch floppy is replacing the 5.25 inch floppy in newer systems since it is smaller in size and does not bend easily. All floppy disks are provided with an off-center index hole that allows the electronic system reading the disk to find the start of a track and the first sector.

Hard disk memory is also frequently used with microcomputer systems. The hard disk, also known as the fixed disk, is not removable like the floppy disk.

A comparison of the some of the features associated with the hard disk and floppy disk is provided below:

Characteristic	Hard Disk	Floppy Disk
Size	5 Mbytes to several Gbytes	1.2 Mbytes typical for 5.25 inch floppy. 1.44 Mbytes typical for 3.5 inch floppy.
Rotational Speed	3600 rpm	300 rpm
Number of heads	May have up to 8 disk surfaces with up to two heads per surface	Two heads; One head for the upper surface and the other head for the lower surface

Primary memory normally includes ROM (Read-only Memory) and RAM (Random Access Memory). As the name implies, a ROM permits only a read access. Some ROMs are custom made, that is, their contents are programmed by the manufacturer. Such ROMs are called mask programmable ROMs. Sometimes a user may have to program a ROM in the field. For instance, in a fusible-link ROM, programmable read-only memory (PROM) is available. The main disadvantage of a PROM is that it cannot be reprogrammed.

Some ROMs can be reprogrammed, these are called Erasable Programmable Read-Only Memories (EPROMs).

In an EPROM, programs are entered using electrical impulses and the stored information is erased by using ultraviolet rays. Usually an EPROM is programmed by inserting the EPROM chip into the socket of a PROM programmer and providing program addresses and voltage pulses at the appropriate pins of the chip. Typical erase times vary between 10 and 30 minutes.

With advances in IC technology, it is possible to achieve an electrical means of erasure. These new ROMs are called Electrically Alterable ROMs (EAROMs) or Electrically Erasable PROMs (EEPROMs or E2PROMs) and these ROM chips can be programmed even when they are in the circuit board. These memories are also called Read Mostly Memories (RMMs), since they have much slower write times than read times. Random Access Memories (RAMs) are read/write memories.

Information stored in random access memories will be lost if the power is turned off. This property is known as volatility and, hence, RAMs are usually called volatile memories. RAMs can be backed up by batteries for a certain period of time and are sometimes called nonvolatile RAMs. Stored information in a magnetic tape or magnetic disk is not lost when the power is turned off. Therefore, these storage devices are called nonvolatile memories. Note that a ROM is a nonvolatile memory.

Some RAMs are constructed using bipolar transistors, and the information is stored in the form of voltage levels in flip-flops. These voltage levels do not usually drift away, or decay. Such memories are called static RAMs because the stored information remains constant for some period of time.

On the other hand, in RAMs that are designed using MOS transistors, the information is held in the form of electrical charges in capacitors. Here, the stored charge has the tendency to decay. Therefore, a stored 1 would become a 0 if no precautions were taken. These memories are referred to as dynamic RAMs. In order to prevent any information loss, dynamic RAMs have to be refreshed at regular intervals. Refreshing means boosting the signal level and writing it back. This activity is performed by a hardware unit called "refresh logic" which can either be a separate chip or is contained in the microprocessor chip.

Since the static RAM maintains information in active circuits, power is required even when the chip is inactive or in standby mode. Therefore, static RAMs require large power supplies. Also, each static RAM cell is about four times larger in area than an equivalent dynamic cell; a dynamic RAM chip contains about four times as many bits as a static RAM chip using the same or comparable semiconductor technology. Figure 1.3 shows the subcategories of ROMs, RAMs, and their associated technologies.

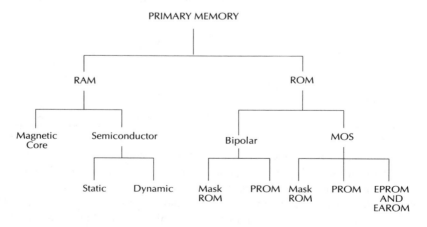

FIGURE 1.3   Subcategories of RAMs and ROMs.

Today, one megabit of data can be stored in an ordinary dynamic RAM chip. The data can be accessed in 80 nanoseconds or less. The RAM chip costs $5. In contrast, it takes 150 nanoseconds to access a one-megabit EEPROM which costs $150. Sixteen megabit DRAMs are very popular these days at a price of approximately 0.3 millicent per bit. Recently, IBM, Hitachi, Toshiba and others have introduced 64 mega-bit DRAMs. It is expected that giga-bit DRAMs will not be introduced until the next century.

In the mid 1980s, Toshiba Semiconductor invented flash memory. About the same time, Intel and Seeq Semiconductor were also working on flash memories. While each manufacturer implemented its flash memory differently, they operate in a similar way.

Like EPROMs and EEPROMs or EAROMs, flash memory is nonvolatile and reprogrammable. Flash memory is fabricated by using ETOX II (EPROM Tunnel Oxide) technology which is a combination of EPROM and EEPROM technologies. Flash memory is relatively inexpensive compared to EEPROM. A one megabit flash memory costs about $15. Flash memory can be reprogrammed electrically while embedded in the board. However, one can only change a sector or a block (consisting of multiple bytes) at a time.

Flash memory cells contain a single transistor like the EPROM cell. In contrast, a DRAM cell typically contains a transistor and a capacitor, an EEPROM cell contains two transistors while a static RAM cell requires four or six transistors.

The non-volatility and DRAM-like speed of flash memory are ideal for solid-state "disk" drives. Flash based disks do not have any disks or moving parts. Flash disks are very fast compared to most available disk drives.

Data can be accessed in 120 nanoseconds in flash memories while it takes 15 to 30 milliseconds to access data stored in today's typical hard disk. However, flash disks are limited to up to 40 megabytes in capacity whereas hard disk drives can store from 5 megabytes to several gigabytes.

A flash disk can be built from one or more flash-memory IC chips and some controlling logic devices. For example, to build a 512Kbyte flash disk, four one-megabit flash memory chips can be connected on a small card. An example of such a flash memory system is the Intel iMC004FLKA 4 Megabyte flash memory card. In addition to flash-disk hardware, software to manage files on a flash disk is required. The file system software handles creating and deleting files, changing the file sizes and formatting the flash disk. Microsoft offers flash file system software for the MS-DOS operating system.

The most severe limitation of flash disks has been its cost.However, the cost of flash ROM is significantly decreasing. In the future, high density flash memory is expected to be available at an inexpensive cost.

Flash memory can be programmed using either 5V or 12V. The 5V feature becomes more desirable for portable equipment where no 12V power is available. The speed, rugged construction, and lower power consumption of flash disks is ideal for laptop and notebook computers.

In summary, due to the high cost of flash disks, desktop computers will continue to use hard disk drives. Since flash memory combines the advantages of an EPROM's low cost with an EEPROM's ease of reprogramming, flash memories are being extensively used these days as a microcomputer's main non-volatile memory. An example of flash memory is the Intel 28F020 256K x 8 flash memory. By 1994, the cost of a megabyte of flash memory is expected to move from its current level of $120 to about $15. At that time, flash disks will be able to replace hard disks in many applications.

1.3.3.b Main Memory Array Design

In many applications, a memory of large capacity is often realized by interconnecting several small-size memory blocks. In this section, design of a large main memory using small-size memories as building blocks is presented. The memory map defining all memory addresses is determined. Note that the microprocessor's reset vector must be included in the memory map.

There are three types of techniques used for designing the main memory. These are linear decoding, full decoding/partial decoding and memory decoding using PALs. We will illustrate the concepts associated with these techniques in the following.

First, consider the block diagram of a typical static RAM chip shown in Figure 1.4.

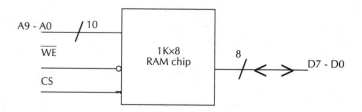

FIGURE 1.4   Typical Static RAM Chip.

The capacity of this chip is 8192 bits and these bits are organized as 1024 words with 8 bits/word. Each word has a unique address and this is specified on 10-bit address lines A9—A0 (note that $2^{10} = 1024$). The inputs and outputs are routed through the 8-bit bidirectional data lines D7 through D0. The operation of this chip is governed by the two control inputs: $\overline{WE}$ (write Enable) and CS (chip select). The truth table that describes the operation of this chip is shown in Table 1.1.

TABLE 1.1   Truth Table for 1K × 8 Static RAM

CS	$\overline{WE}$	MODE	Status of D7—D0	Power
L	X	Not selected	High impedance	Standby
H	L	Write	Acts as an input bus	Active
H	H	Read	Acts as an output bus	Active

Note: H — high, L — low, X — don't care.

From this table, it is easy to see that when CS input is low, the chip is not selected and thus the lines D7 through D0 are driven to the high impedance state. When CS = 1 and $\overline{WE}$ is LOW, data on lines D7—D0 are written into the word addressed by A0 through A9. Similarly, when CS = 1 and WE is high, the contents of the memory word (whose address is specified on address lines A9 through A0) will appear on lines D7 through D0. Note that when the chip select input CS goes to low, the device is disabled and the chip automatically reduces its power requirements and remains in this low-power standby mode as long as CS remains low. This feature results in system power savings as high as 85% in larger systems, where the majority of devices are disabled.

1.3.3.b.i Linear Decoding.   This technique uses the unused address lines of the microprocessor as chip selects for the memory chips. This method is used for small systems.

A simple way to connect an 8-bit microprocessor to a 6K RAM system using linear decoding is shown in Figure 1.5. In this approach, the address lines A9 through A0 of the microprocessor are used as a common input to each 1K × 8 RAM chip. The remaining 6 high-order lines are used to select one of the 6 RAM chips. For example, if A15A14A13A12A11A10 = 000010, then the RAM chip 1 is selected. The address map realized by this arrangement is summarized in Figure 1.6. This method is known as the linear select decoding technique. The principal advantage of this method is that it does not require any decoding hardware. However, this approach has some disadvantages:

- Although with a 16-bit address bus we have 64K bytes of RAM space, we are able to interface only 6K bytes of RAM. This means that this idea wastes address space.
- The address map is not contiguous; rather, it is sparsely distributed.
- If both A11 and A10 are high at the same time, both RAM chips 0 and 1 are selected and thus a bus conflict occurs. This can be avoided by proper programming to select the desired memory chip and deselect the others.
- Also, if all unused address lines are not utilized as chip selects for memory, then these unused pins become don't cares (can be 0 or 1). This results in foldback, meaning that a memory location will have its image in the memory map. For example, if A15 is don't care in design and if A14 to A0 address lines are used, then address 0000_{16} and address 8000_{16} are the same locations. This is called foldback and it wastes memory space.

1.3.3.b.ii Full/Partial Decoding. Difficulties such as the bus conflict and sparse address distribution are eliminated by the use of the full/partial decoded addressing technique. To see this, consider the organization shown in Figure 1.7. In this setup, we use a 2-to-4 decoder and interface the 8-bit microprocessor with 4K bytes of RAM. In particular, the four combinations of the lines A11 and A10 select the RAM chips as follows::

A11	A10	Device Selected
0	0	RAM chip 0
0	1	RAM chip 1
1	0	RAM chip 2
1	1	RAM chip 3

Also observe that this hardware makes sure that the memory system is enabled only when the lines A15 through A12 are zero. The complete address map corresponding to this organization is summarized in Figure 1.8.

1.3.3.b.iii   ***Memory Decoding by using Programmable Array Logic (PAL).***   A Programmable Array Logic (PAL) is similar to a ROM in concept except that it does not provide full decoding of the input lines. Instead, a PAL provides a partial sum of products which can be obtained via programming and saves a lot of space on the board. The PAL chip contains a fused programmable AND array and a fixed OR array. Note that in PLA (Programmable Logic Array) both AND and OR arrays are programmable. The AND and OR gates are fabricated inside the PAL without interconnections. The specific functions desired are implemented during programming via software. Programming of the PAL provides connections of the inputs of the AND gates and the outputs of the AND gates to the inputs of the OR gates. Therefore, the PAL implements sum of products of the inputs. PALs are used extensively these days with 32-bit microprocessors such as the Intel 80386/80486 and Motorola 68030/68040 for performing the memory decode function. PALs connect these microprocessors to memory, I/O and other chips without the use of any additional logic gates or circuits.

Each input has both true and complemented forms. A look at the NOR gate output $\emptyset_1$ indicates that there are two X connections at the inputs of this NOR gate. The three X inputs are wire-ANDed together with programming the PAL so that $\emptyset = \left(I_0 \cdot \bar{I_2}\right) + I_1$.

The PAL chips are usually identified by a two-digit number followed by a letter and then a digit. The two-digit number specifies the number of input lines while the last digit defines the number of output lines. The fixed number of AND gates are connected to either an OR or a NOR gate. The letter 'H' indicates that the output gates are OR gates. The letter 'L' is used when the outputs are NOR gates.

As an example, the 10H8 provides eight OR gate outputs driven by ten AND gate inputs. The 10L8, on the other hand, is the same as the 10H8 except that the eight output gates are NOR gates.

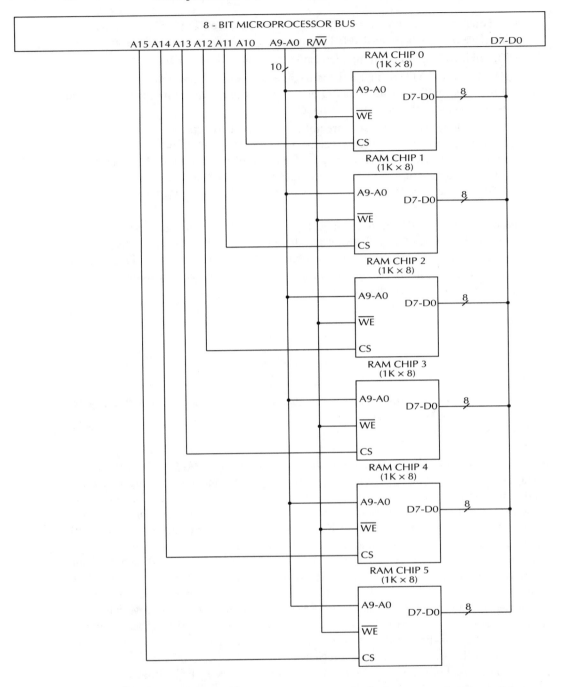

FIGURE 1.5   An 8-bit microprocessor interfaced to a 6K RAM system using the linear select decoding technique.

Some PALs provide additional features. The 16L8 includes tristate outputs. Each of the eight NOR gate outputs is driven internally by six AND gates. A seventh AND gate provides the enable signal for the tristate buffers.

The 16L8 is a popular PAL used with 32-bit microprocessors. The 16L8 is a 20-pin PAL manufactured using bipolar transistors. It has ten input pins (labeled I), two outputs (labeled O) and six programmable Input/Output (labeled I/O) lines. Using the programmable I/O lines, the number of input lines can be increased to a maximum of 16 and the number of output lines can be increased to 8.

Binary Address Pattern A15 A14 A13 A12 A11 A10 A9 A7 A6 A5 A4 A3 A2 A1 A0	Device Selected	Address Assignment in Hex
0 0 0 0 0 1 0 0 0 0 0 0 0 0 0 · · · · · · · · · · · · · · · 0 0 0 0 0 1 1 1 1 1 1 1 1 1 1	RAM CHIP 0	0400 to 07FF
0 0 0 0 1 0 0 0 0 0 0 0 0 0 0 · · · · · · · · · · · · · · · 0 0 0 0 1 0 1 1 1 1 1 1 1 1 1	RAM CHIP 1	0800 to 0BFF
0 0 0 1 0 0 0 0 0 0 0 0 0 0 0 · · · · · · · · · · · · · · · 0 0 0 1 0 0 1 1 1 1 1 1 1 1 1	RAM CHIP 2	1000 to 13FF
0 0 1 0 0 0 0 0 0 0 0 0 0 0 0 · · · · · · · · · · · · · · · 0 0 1 0 0 0 1 1 1 1 1 1 1 1 1	RAM CHIP 3	2000 to 23FF
0 1 0 0 0 0 0 0 0 0 0 0 0 0 0 · · · · · · · · · · · · · · · 0 1 0 0 0 0 1 1 1 1 1 1 1 1 1	RAM CHIP 4	4000 to 43FF
1 0 0 0 0 0 0 0 0 0 0 0 0 0 0 · · · · · · · · · · · · · · · 1 0 0 0 0 0 1 1 1 1 1 1 1 1 1	RAM CHIP 5	8000 to 83FF

FIGURE 1.6   Address map realized by the system shown in Figure 1.5.

Programming PALs can be accomplished by first creating a file by using a text editor on a personal computer. The file should include information such as the pin assignments of the PAL and the boolean equation for the outputs. By inserting the PAL into the programming module included with the personal computer, the PAL can then be programmed with the PAL programming software provided with the personal computer.Note that PAL programming hardware and software are sold separately and not usually included with a personal computer.

1.3.3.c Memory Management Concepts

Due to the massive amount of information that must be saved in most systems, the mass storage is often a disk. If each access is to a disk (even a hard disk), then system throughput will be reduced to unacceptable levels.

An obvious solution is to use a large and fast locally accessed semiconductor memory. Unfortunately the storage cost per bit for this solution is very high. A combination of both off-board disk (secondary memory) and on-board semiconductor main memory must be designed into a system. This requires a mechanism to manage the two-way flow of information between the primary (semiconductor) and secondary (disk) media. This mechanism must be able to transfer blocks of data efficiently, keep track of block usage, and replace them in a nonarbitrary way. The primary memory system must therefore be able to dynamically allocate memory space.

An operating system must have resource protection from corruption or abuse by users. Users must be able to protect areas of code from each other, while maintaining the ability to communicate and share other areas of code. All these requirements indicate the need for a device, located between the microprocessor and memory, to control accesses, perform address mappings, and act as an interface between the logical (programmer's memory) and microprocessor physical (memory) address spaces. Since this device must manage memory use, it is appropriately called the memory management unit (MMU). Typical 32-bit microprocessors such as Motorola 68030 and Intel 80386 include on-chip MMU.

The MMU reduces the burden of the memory management function on the operating system.

The basic functions provided by the MMU are address translation and protection. The MMU translates logical program addresses to physical memory addresses. The addresses in a

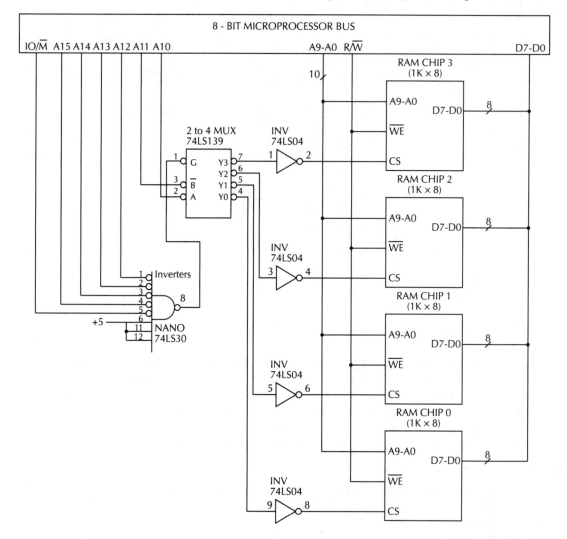

FIGURE 1.7   An 8-bit microprocessor interfaced to a 4K RAM system using a full/partial decoded addressing technique

Binary Address Pattern	Device Selected	Address Assignment in Hex
A15 A14 A13 A12 A11 A10 A9 A7 A6 A5 A4 A3 A2 A1 A0		
0 0 0 0 0 0 0 0 0 0 0 0 0 0 0 · · · · · · · · · · · · · · · 0 0 0 0 0 0 1 1 1 1 1 1 1 1 1	RAM CHIP 0	0000 to 03FF
0 0 0 0 0 1 0 0 0 0 0 0 0 0 0 · · · · · · · · · · · · · · · 0 0 0 0 0 1 1 1 1 1 1 1 1 1 1	RAM CHIP 1	0400 to 07FF
0 0 0 0 1 0 0 0 0 0 0 0 0 0 0 · · · · · · · · · · · · · · · 0 0 0 0 1 1 1 1 1 1 1 1 1 1 1	RAM CHIP 2	0800 to 0BFF
0 0 0 0 1 1 0 0 0 0 0 0 0 0 0 · · · · · · · · · · · · · · · 0 0 0 0 1 1 1 1 1 1 1 1 1 1 1	RAM CHIP 3	0C00 to 0FFF

FIGURE 1.8   Address map corresponding to the organization shown in Figure 1.7.

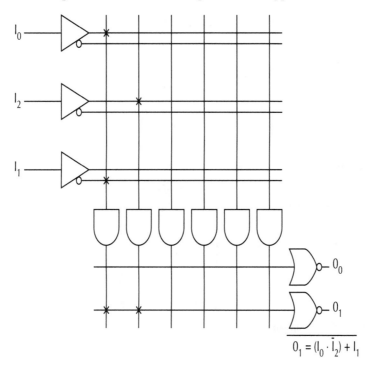

$$O_1 = \overline{(I_0 \cdot \overline{I_2}) + I_1}$$

FIGURE 1.9 A typical PAL.

program are called logical addresses since they indicate the logical positions of instructions and data. The MMU translates these logical addresses to physical addresses provided by the memory chips. The MMU can perform address translation in one of two ways:

1. By using the substitution technique as shown in Figure 1.10a
2. By adding an offset to each logical address to obtain the corresponding physical address as shown in Figure 1.10b

Address translation using substitution is faster than the offset method. However, the offset method has the advantage of mapping a logical address to any physical address as determined by the offset value.

Memory is usually divided into small manageable units. The terms "page" and "segment" are frequently used to describe these units. Paging divides the memory into equal-sized pages, while segmentation divides the memory into variable-sized segments.

It is relatively easier to implement the address translation table if the logical and main memory spaces are divided into pages. The term "page" is associated with logical address space, while the term "block" usually refers to a page in main memory space.

There are three ways to map logical addresses to physical addresses. These are paging, segmentation, and combined paging/segmentation.

In a paged system, a user has access to a larger address space than physical memory provides. The virtual memory system is managed by both hardware and software. Note that memory in excess of the main memory such as floppy disk storage is called virtual memory. The hardware included in the memory management unit handles address translation. The memory management software in the operating system performs all functions including page replacement policies in order to provide efficient memory utilization. The memory management software performs functions such as removal of the desired page from main memory to accommodate a new page, transferring a new page from secondary to main memory at the right instant of time, and placing the page at the right location in memory.

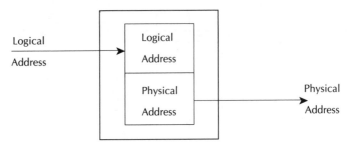

FIGURE 1.10a   Address translation using substitution technique.

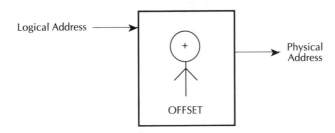

FIGURE 1.10b   Address translation by offset technique.

If the main memory is full during transfer from secondary to main memory, it is necessary to remove a page from main memory to accommodate the new page. Two popular page replacement policies are first-in first-out (FIFO) and least recently used (LRU). The FIFO policy removes the page from main memory that has been resident in memory for the longest amount of time. The FIFO replacement policy is easy to implement. One of the main disadvantages of the FIFO policy is that it is likely to replace heavily used pages. Note that heavily used pages are resident in main memory for the longest amount of time. Sometimes this replacement policy might be a poor choice. For example, in a time-shared system, several users normally share a copy of the text editor in order to type and correct programs. The FIFO policy on such a system might replace a heavily used editor program page to make room for a new page. This program page might be recalled to main memory immediately. The FIFO, in this case, would be a poor choice.

The LRU policy, on the other hand, replaces that page which has not been used for the longest amount of time.

In the segmentation method, the MMU utilizes the segment selector to obtain a descriptor from a table in memory containing several descriptors. A descriptor contains the physical base address for a segment, the segment's privilege level, and some control bits. When the MMU obtains a logical address from the microprocessor, it first determines whether the segment is already in the physical memory. If it is, the MMU adds an offset component to the segment base component of the address obtained from the segment descriptor table to provide the physical address. The MMU then generates the physical address on the address bus for accessing the memory. On the other hand, if the MMU does not find the logical address in physical memory, it interrupts the microprocessor. The microprocessor executes a service routine to bring the desired program from a secondary memory such as disk to the physical memory. The MMU determines the physical address using the segment offset and descriptor as above and then generates the physical address on the address bus for memory. A segment will usually consist of an integral number of pages, say, each 256 bytes long. With different-sized segments being swapped in and out, areas of valuable primary memory can become unusable. Memory is unusable for segmentation when it is sandwiched between already allocated segments and if it is not large enough to hold the latest segment that needs to be loaded. This is called external fragmentation and is handled by MMUs using special techniques. An example of external fragmentation is given in Figure 1.11. The advantages of segmented memory management are that few descriptors are required for large programs or data spaces, and internal fragmentation (to be discussed later) is minimized. The disadvantages include external fragmentation, involved algorithms for placing data are required, possible restrictions on starting address, and longer data swap times are required to support virtual memory.

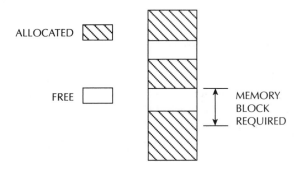

FIGURE 1.11   Memory fragmentation (external).

Address translation using descriptor tables offers a protection feature. A segment or a page can be protected from access by a program section of a lower privilege level. For example, the selector component of each logical address includes one or two bits indicating the privilege level of the program requesting access to a segment. Each segment descriptor also includes one or two bits providing the privilege level of that segment. When an executing program tries to access a segment, the MMU can compare the selector privilege level with the descriptor privilege level. If the segment selector has the same or higher privilege level, then the MMU permits the access. If the privilege level of the selector is lower than the descriptor, the MMU can interrupt the microprocessor informing of a privilege level violation. Therefore, the indirect technique of generating physical address provides a mechanism of protecting critical program sections in the operating system.

Paging divides the memory into equal-sized pages, it avoids the major problem of segmentation-external fragmentation. Since the pages are of the same size, when a new page is requested and an old one swapped out, the new one will always fit into the vacated space. However, a problem common to both techniques remains — internal fragmentation. Internal fragmentation is a condition where memory is unused but allocated due to memory block size implementation restrictions. This occurs when a module needs, say, 300 bytes and page is 1K bytes, as shown in Figure 1.12.

In the paged-segmentation method, each segment contains a number of pages. The logical address is divided into three components: segment, page, and word. The segment component defines a segment number, the page component defines the page within the segment, and the word component provides the particular word within the page. A page component of n bits can provide up to 2^n pages. A segment can be assigned with one or more pages up to a maximum of 2^n pages; therefore, a segment size depends on the number of pages assigned to it.

Protection mechanisms can operate either on physical address or logical address. Physical memory protection can be accomplished by using one or more protection bits with each block to define the access type permitted on the block. This means that each time a page is transferred from one block to another, the block protection bits must be updated. A more efficient approach is to provide a protection feature in logical address space by including protection bits in the descriptors of the segment table in the MMU.

PAGES ≡≡≡ 1 K
IF 300 BYTES NEEDED 1 K BYTES ARE ALLOCATED

MEMORY UNUSED BUT ALLOCATED BECAUSE OF   1 K
IMPLEMENTATION RESTRICTIONS ON BLOCK SIZES   PAGE

USED
UNUSED BUT ALLOCATED

FIGURE 1.12 Memory fragmentation (internal).

1.3.3.d Cache Memory Organization

The performance of a microcomputer system can be significantly improved by introducing a small, expensive, but fast memory between the microprocessor and main memory. This memory is called cache memory and this idea was first introduced in the IBM 360/85 computer. Later on, this concept was also implemented in minicomputers such as PDP-11/70. With the advent of VLSI technology, the cache memory technique is gaining acceptance in the microprocessor world. For example, an on-chip cache memory is implemented in Intel's 32-bit microprocessor, the 80486, and Motorola's 32-bit microprocessor, the MC 68020/68030/68040. The 80386 does not have on-chip cache but external cache memory can be interfaced to it. Studies have shown that typical programs spend most of their execution times in loops. This means that the addresses generated by a microprocessor have a tendency to cluster around a small region in the main memory. This phenomenon is known as locality of reference. The 32-bit microprocessor can execute the same instructions in a loop from the on-chip cache rather than reading them repeatedly from the external main memory. Thus the performance offered by 32-bit microprocessors is greatly improved.

The block diagram representation of a microprocessor system that employs an on-chip cache memory is shown in Figure 1.13. Usually, a cache memory is very small in size and its access time is less than that of the main memory by a factor of 5.

Cache hit means that the reference is found in the cache and the data pertaining to the microprocessor reference is transferred to the microprocessor from the cache. However, if the reference is not found in the cache, we call it a cache miss. When there is a cache miss, the main memory is accessed by the microprocessor and the data are then transferred to the microprocessor from the main memory. At the same time, a block of data containing the desired data needed by the microprocessor is transferred from the main memory to the cache. The block normally contains 4 to 16 bytes, and this block is placed in the cache using the standard replacement policies such as FIFO (First-In First-Out) or LRU (Least Recently Used). This block transfer is done with a hope that all future references made by the microprocessor will be confined to the fast cache.

The relationship between the cache and main memory blocks is established using mapping techniques. Three widely used mapping techniques are

- Direct mapping
- Fully-associative mapping
- Set-associative mapping

In direct mapping, the main memory address is divided into two fields: an index field and a tag field. The number of bits in the index field is equal to the number of address bits required to access the cache memory.

Assume that the main memory address is m bits wide and the cache memory address is n bits wide. Then the index field requires n bits and the tag field is $(m - n)$ bits wide. The n-

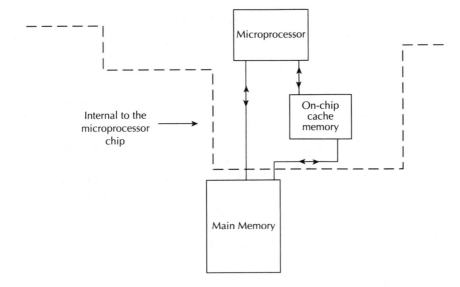

FIGURE 1.13   Memory organization of a computer system that employs a cache memory.

bit address accesses the code. Each word in the cache includes the data word and its associated tag. When the microprocessor generates an address for main memory, the index field is used as the address to access the cache. The tag field of the main memory is compared with the tag field in the word read from cache. A hit occurs if the tags match. This means that the desired data word is in cache. A miss occurs if there is no match, and then the required word is read from main memory. It is written in the cache along with the new tag. A random access memory is used as the cache memory.

One of the main drawbacks of direct mapping is that numerous misses may occur if two or more words with addresses having the same index but different tags are accessed several times. This can be minimized by incorporating a larger cache.

The fastest cache memory utilizes an associative memory. The method is known as fully associative mapping. Each associative memory content contains main memory address and its content (data). When the microprocessor generates a main memory address, it is compared associatively (simultaneously) with all addresses in the associative memory. If there is a match, the corresponding data word is read from the associative cache memory and sent to the microprocessor. If a miss occurs, the main memory is accessed, and the address and its corresponding data are written to the associative cache memory. If the cache is full, certain policies such as FIFO (first-in first-out) are used as replacement algorithm for the cache. The associative cache is expensive but provides fast operation.

The set-associative mapping is a combination of direct and associative mapping. Each cache word stores two or more main memory words using the same index address. Each main memory word consists of a tag and its data word. An index with two or more tags and data words forms a set. When the microprocessor generates a memory request, the index of the main memory address is used as the cache address. The tag field of the main memory address is then compared associatively (simultaneously) with all tags stored under the index. If a match occurs, the desired data word is read. If a miss occurs, the data word, along with its tag, is read from main memory and also written into the cache. The hit ratio improves as the set size increases. This is because more words with the same index but different tags can be stored in cache.

There are two ways of writing into cache: the write-back and write-through methods. In the write-back method, whenever the microprocessor writes something into a cache word, a dirty bit is assigned to the cache word. When a dirty word is to be replaced with a new word, the dirty word is first copied into the main memory before it is overwritten by the incoming new word. The advantage of this method is that it avoids unnecessary writing into main memory.

In the write-through method, whenever the microprocessor alters cache data, the same alteration is made in the main memory copy of the altered cache data. This policy can be easily implemented and also it insures that the contents of the main memory are always valid. This feature is desirable in a multiprocessor system where the main memory is shared by several processors. However, this approach may lead to several unnecessary writes to main memory.

One of the important aspects of cache memory organization is to devise a method that insures proper utilization of the cache. Usually, the tag directory contains an extra bit for each entry. This additional bit is called a valid bit. When the power is turned on, the valid bit corresponding to each cache block entry of the tag directory is reset to zero. This is done in order to indicate that the cache block holds invalid data. When a block of data is first transferred from the main memory to a cache block, the valid bit corresponding to this cache block is set to 1. In this arrangement, whenever the valid is a zero, it implies that a new incoming block can overwrite the existing cache block. Thus, there is no need to copy the contents of the cache block being replaced into the main memory.

1.3.4 Input/Output (I/O)

This section describes the basic input and output techniques used by microcomputers to transfer data between the microcomputer and external devices. The general characteristics of I/O are described. One communicates with a microcomputer system via the I/O devices interfaced to it. The user can enter programs and data using the keyboard on a terminal and execute the programs to obtain results. Therefore, the I/O devices connected to a microcomputer system provide an efficient means of communication between the computer and the outside world. These I/O devices are commonly called peripherals and include keyboards, CRT displays, printers, and disks.

The characteristics of the I/O devices are normally different from those of the microcomputer. For example, the speed of operation of the peripherals is usually slower compared to the microcomputer, and the word length of the microcomputer may be different from the data format of the peripheral device. To make the characteristics of the I/O devices compatible with those of the microcomputer, interface hardware circuitry between the microcomputer and I/O devices is necessary.

In a typical microcomputer system, the user gets involved with two types of I/O devices: physical I/O and logical I/O. When the microcomputer has no operating system, the user must work directly with physical I/O devices and perform detailed I/O design.

There are three ways of transferring data between the microcomputer and a physical I/O device:

- Programmed I/O
- Interrupt driven I/O
- Direct memory access (DMA)

The microcomputer executes a program to communicate with an external device via a register called the I/O port for programmed I/O.

An external device requests the microcomputer to transfer data by activating a signal on the microcomputer's interrupt line during interrupt I/O. In response, the microcomputer executes a program called the interrupt-service routine to carry out the function desired by the external device, again by way of one or more I/O ports.

Data transfer between the microcomputer's memory and an external device occurs without microprocessor involvement with direct memory access.

For a microcomputer with an operating system, the user works with virtual I/O devices. The user does not have to be familiar with the characteristics of the physical I/O devices. Instead, the user performs data transfers between the microcomputer and the physical I/O devices indirectly by calling the I/O routines provided by the operating system using virtual I/O instructions. This is called logical I/O.

1.3.4.a Programmed I/O

As described earlier, the microcomputer communicates with an external device via one or more registers called I/O ports using programmed I/O. These I/O ports are occasionally fabricated by the manufacturer in the same chip as the memory chip to achieve minimum chip count for small system applications. For example, the Intel 8355/8755 contains 2K bytes of ROM/EPROM with two I/O ports. The Motorola 6846 has 2K bytes of ROM and an 8-bit I/O port.

I/O ports are usually of two types. For one type, each bit in the port can be individually configured as either input or output. For the other type, all bits in the port can be set up as either all parallel input or output bits. Each port can be configured as an input or output port by another register called the command, or data-direction register. The port data register contains the actual input or output data. The data-direction register is an output register and can be used to configure the bits in the port as inputs or outputs.

Each bit in the port can usually be set up as an input or output by respectively writing a 0 or a 1 in the corresponding bit of the data-direction register (DDR). A bidirectional buffer (one input buffer and one output buffer) is connected at each bit of the port. A '1' written to a particular bit in DDR enables the output buffer while a '0' enables the input buffer connected at the corresponding bit of the port. As an example, if an 8-bit data-direction register contains 34_{16}, then the corresponding port is defined as follows:

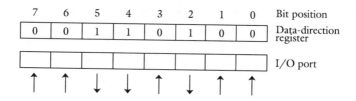

In the preceding example, since 34_{16} ($0011\ 0100_2$) is sent as an output into the data-direction register, bits 0, 1, 3, 6, and 7 of the port are set up as inputs, and bits 2, 4, and 5 of the port are defined as outputs. The microcomputer can then send outputs to external devices, such as LEDs, connected to bits 2, 4, and 5 through a proper interface. Similarly, the microcomputer can input the status of external devices, such as switches, through bits 0, 1, 3, 6, and 7. To input data from the input switches, the 8-bit microcomputer assumed here inputs the complete byte, including the bits to which LEDs are connected. While receiving input data from an I/O port, however, the microcomputer places a value, probably 0, at the bits configured as outputs and the program must interpret them as "don't cares". At the same time, the microcomputer's outputs to bits configured as inputs are disregarded.

For parallel I/O, there is only one register, known as the command register, for all ports. A particular bit in the command register configures all bits in a port as either inputs or outputs.

Consider two I/O ports in an I/O chip along with one command register. Assume that a 0 or a 1 in a particular bit position defines all bits of ports A or B as inputs or outputs.

For example,

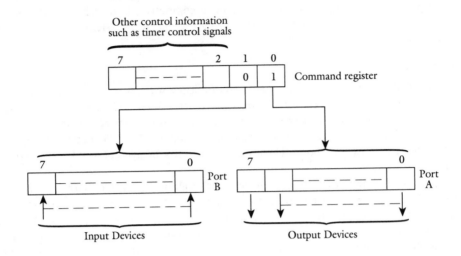

Some I/O ports are called handshake ports. Data transfer occurs via these ports through exchanging of control signals between the I/O controller and an external device.

1.3.4.b Standard I/O Versus Memory-Mapped I/O

I/O ports are addressed using either standard I/O or memory-mapped I/O techniques. The standard I/O, also called isolated I/O, uses the $\overline{\text{IO/M}}$ control pin on the microprocessor chip. The processor outputs a HIGH on this pin to indicate to memory and the I/O chips that an I/O operation is taking place. A LOW output from the processor to this pin indicates a memory operation. Execution of IN or OUT instructions makes the $\overline{\text{IO/M}}$ HIGH, whereas memory-oriented instructions, such as LDA and STA, drive the IO/M to LOW. In standard I/O, the processor uses the $\overline{\text{IO/M}}$ pin to distinguish between I/O and memory. For 8-bit microprocessors, an 8-bit address is typically used for each I/O port. This is because 8 bits are the basic data unit for these processors. Eight-bit processors are usually capable of directly addressing 64K bytes of memory using 16 address lines. With an 8-bit I/O port address, these processors are capable of addressing 256 ports. However, in a typical application, there are usually four or five I/O ports required. Some of the address bits of the microprocessor are normally decoded to obtain the I/O port addresses. With memory-mapped I/O, the processor does not differentiate between I/O and memory and, therefore, does not use the microprocessor's $\overline{\text{IO/M}}$ control pin. The microprocessor uses the memory addresses (which may not exist in the microcomputer's physical memory) to represent I/O ports. The I/O ports are mapped into the microprocessor's main memory and, hence, are called *memory-mapped I/O.*

In memory-mapped I/O, the most significant bit (MSB) of the address may be used to distinguish between I/O and memory. If the MSB of address is 1, an I/O port is selected. If the MSB of address is 0, a memory location is accessed. This reduces the microprocessor's addressing memory (main memory) by 50%. Sixteen and thirty-two bit microprocessors provide special control signals for performing memory-mapped I/O. Thus, these processors do not use MSB of the address lines. Intel microprocessors can use either standard or memory-mapped I/O while Motorola microprocessors use only memory-mapped I/O. For example, with standard I/O, Intel 8086 uses IN AL, PortA and OUT PortA, AL for inputing and outputing data. On the other hand, the 8086 uses memory-oriented instructions such as MOV AL, START, and MOV START, AL for inputing and outputing data. Note that START is an

I/O port mapped as a memory address. Motorola, on the other hand, does not have any IN or OUT instructions and uses memory-oriented instructions for I/O operation.

1.3.4.c Unconditional and Conditional Programmed I/O

The microprocessor can send data to an external device at any time during unconditional I/O. The external device must always be ready for data transfer. A typical example is when the processor outputs a 7-bit code through an I/O port to drive a seven-segment display connected to this port.

In conditional I/O, the microprocessor outputs data to an external device via handshaking. Data transfer occurs by the exchanging of control signals between the microprocessor and an external device. The microprocessor inputs the status of the external device to determine whether the device is ready for data transfer. Data transfer takes place when the device is ready.

The concept of conditional I/O will now be demonstrated by means of data transfer between a microprocessor and an analog-to-digital (A/D) converter. Consider, for example, the A/D converter shown in the accompanying figure.

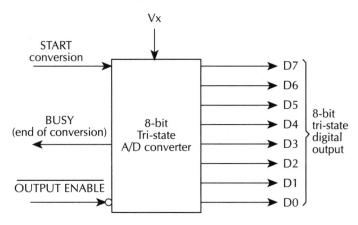

The A/D converter just shown transforms an analog voltage Vx into an 8-bit binary output at pins D7—D0. A pulse at the START conversion pin initiates the conversion. This drives the BUSY signal LOW. The signal stays LOW during the conversion process. The BUSY signal goes HIGH as soon as the conversion ends. Since the A/D converter's output is tristated, a LOW on the $\overline{\text{(OUTPUT ENABLE)}}$ transfers the converter's output. A HIGH on the $\overline{\text{(OUTPUT ENABLE)}}$ drives the converter's output to a high impedance state.

The concept of conditional I/O can be demonstrated by interfacing the A/D converter to an 8-bit processor. Figure 1.14 shows such an interfacing example.

The user writes a program to carry out the conversion process. When this program is executed, the processor sends a pulse to the START pin of the converter via bit 2 of port A. The microprocessor then checks the BUSY signal by bit 1 of port A to determine if the conversion is completed. If the BUSY signal is HIGH (indicating the end of conversion), the microprocessor sends a LOW to the $\overline{\text{(OUTPUT ENABLE)}}$ pin of the A/D converter. The microprocessor then inputs the converter's D0—D7 outputs via port B. If the conversion is not completed, the microprocessor waits in a loop checking for the BUSY signal to go HIGH.

1.3.4.d Typical Microcomputer Output Circuit

The microcomputer designer is often concerned with the output circuit because of the microcomputer's small output current drive capability. The tristate output circuit shown in Figure 1.15 typically is utilized by a microcomputer as the output circuit. This circuit uses totem pole-type output called a PUSH-PULL circuit providing low-output currents. Therefore, a current amplifier (buffer) is required to drive devices such as LEDs.

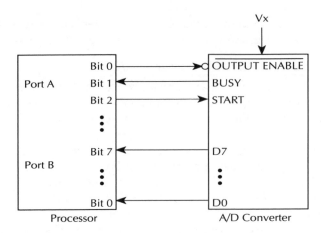

FIGURE 1.14   Interfacing an A/D converter to an 8-bit processor.

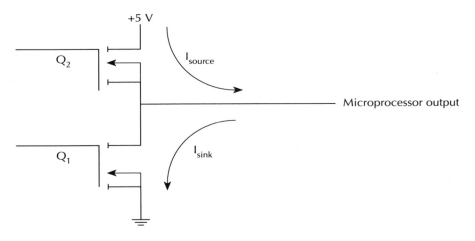

FIGURE 1.15   Digital microprocessor output circuit.

In the preceding figure, when Q_1 is ON, Q_2 is OFF and I_{sink} will flow from the external device into Q_1. Also, when Q_2 is ON, Q_1 is OFF and I_{source} will flow from Q_2 into the output device. Figure 1.16 shows a hardware interface circuit using the push-pull circuit for driving an external device such as LED.

Assume I_{source} to be 400 μA (usually represented by a negative sign such as $I_{OH} = -400$ μA; the negative sign indicates that the chip is losing current) with a minimum voltage V_A of 2.4 at point A, and that the LED requires 10 mA at 1.7 V. Therefore, a buffer such as a transistor is required at the output circuit to increase the current drive capability to drive the LED. In order to design the interface, the values of R_1, R_2 and minimum β of the transistor will be determined in the following:

$$R_1 = \frac{V_A - V_{BE}(Q_3)}{400\ \mu A} = \frac{2.4 - 0.7}{400\ \mu A} = 4.25\ k\Omega$$

Since $I_{LED} = 10$ μA at 1.7 V and assuming that V_{CE} (saturation) = 0 V,

$$R_2 = \frac{5 - 1.7 - V_{CE}(Q_3)}{10\ mA} = \frac{3.3}{10\ mA} = 330\ \Omega$$

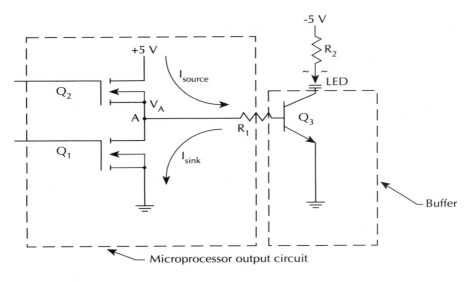

FIGURE 1.16   Microprocessor I/O interfaces for driving an LED.

Since $I_{source} = 400\ \mu A = I_B(Q_3)$, β for transistor Q_3 is

$$\beta = \frac{I_C(Q_3)}{I_B(Q_3)} = \frac{10\ mA}{400\ \mu A} = \frac{10 \times 10^{-3}}{400 \times 10^{-6}} = 25$$

Therefore, the interface design is complete, and a transistor with a mimimum of saturation β of 25 and $R_1 = 4.25\ k\Omega$ and $R_2 = 330\ \Omega$ is required. Note that a MOS outputs more sink current than source current. If sink current is used, the LED in Figure 1.16 can directly be connected to the microcomputer's output through an appropriate resistance. However, if the resistor value is not large enough, it may damage the transistor Q1.

1.3.4.e Interrupt Driven I/O

A disadvantage of conditional programmed I/O is that the microcomputer needs to check the status bit (BUSY signal for the A/D converter) by waiting in a loop. This type of I/O transfer is dependent on the speed of the external device. For a slow device, this waiting may slow down the capability of the microprocessor to process other data. The polled I/O and interrupt I/O techniques are efficient in this type of situation.

Interrupt I/O is a device-initiated I/O transfer. The external device is connected to a pin called the *interrupt* (INT) pin on the microprocessor chip. When the device needs an I/O transfer with the microcomputer, it activates the interrupt pin of the microprocessor chip. The microcomputer usually completes the current instruction and saves at least the contents of the current program counter on the stack.

The microcomputer then automatically loads an address into the program counter to branch to a subroutine-like program called the interrupt-service routine. This program is written by the user. The external device wants the microcomputer to execute this program to transfer data. The last instruction of the service routine is a RETURN, which is typically the same instruction used at the end of a subroutine. This instruction normally loads the return address (saved in the stack before going to the service routine) in the program counter. Then, the microcomputer continues executing the main program. Note that subroutines and interrupts are handled in a similar way: subroutines are initiated via execution of an instruction such as subroutine CALL while interrupts are initiated via activation of the microprocessor's interrupt pin by the interrupting device.

1.3.4.e.i ***Interrupt Types.***   There are typically three types of interrupts: external interrupts, traps or internal interrupts, and software interrupts.

External interrupts are initiated through the microcomputer's interrupt pins by external devices such as A/D converters. External interrupts can further be divided into two types: maskable and nonmaskable. A maskable interrupt is enabled or disabled by executing instructions such as EI or DI. If the microcomputer's interrupt is disabled, the microcomputer ignores the maskable interrupt. Some microprocessors, such as the Intel 8086, have an interrupt-flag bit in the processor status register. When the interrupt is disabled, the interrupt-flag bit is 1, so no maskable interrupts are recognized by the processor. The interrupt-flag bit resets to zero when the interrupt is enabled.

The nonmaskable interrupt has higher priority than the maskable interrupt. If both maskable and nonmaskable interrupts are activated at the same time, the processor will service the nonmaskable interrupt first. The nonmaskable interrupt is typically used as a power failure interrupt. Microprocessors normally use +5 V DC, which is transformed from 110 V AC. If the power falls below 90 V AC, the DC voltage of +5 V cannot be maintained. However, it will take a few milliseconds before the AC power can drop this low (below 90 V AC). In these few milliseconds, the power failure-sensing circuitry can interrupt the microprocessor. An interrupt service routine can be written to store critical data in nonvolatile memory such as battery-backed CMOS RAM. The interrupted program can continue without any loss of data when the power returns.

Some microprocessors are provided with a maskable handshake interrupt. This interrupt is usually implemented by using two pins — INTR and $\overline{\text{INTA}}$. When the INTR pin is activated by an external device, the processor completes the current instruction, saves at least the current program counter onto stack, and generates an interrupt acknowledge $\left(\overline{\text{INTA}}\right)$. In response to the $\overline{\text{INTA}}$, the external device provides an instruction, such as CALL, using external hardware on the data bus of the microcomputer. This instruction is then read and executed by the microcomputer to branch to the desired service routine.

Internal interrupts, or traps, are activated internally by exceptional conditions such as overflow, division by zero, or execution of an illegal op-code. Traps are handled the same way as external interrupts. The user writes a service routine to take corrective measures and provide an indication to inform the user that an exceptional condition has occurred.

Many microprocessors include software interrupts, or system calls. When one of these instructions is executed, the microprocessor is interrupted and serviced similarly to external or internal interrupts. Software interrupt instructions are normally used to call the operating system. These instructions are shorter than subroutine calls, and no calling program is needed to know the operating system's address in memory. Software interrupt instructions allow the user to switch from user to supervisor mode. For some microprocessors, a software interrupt is the only way to call the operating system, since a subroutine call to an address in the operating system is not allowed.

1.3.4.e.ii   ***Interrupt Address Vector.***   The technique used to find the starting address of the service routine (commonly known as the *interrupt address vector*) varies from one microprocessor to another. With some microprocessors, the manufacturers define the fixed starting address for each interrupt. Other manufacturers use an indirect approach by defining fixed locations where the interrupt address vector is stored.

1.3.4.e.iii   ***Saving the Microprocessor Registers.***   When a microprocessor is interrupted, it saves at least the program counter on the stack so the microprocessor can return to the main program after executing the service routine. Some microprocessors save only one or two registers, such as the program counter and status register. Some other microprocessors save all microprocessor registers before going to the service routine. The user should know the specific registers the microprocessor saves prior to executing the service routine. This will enable the user to use the appropriate return instruction at the end of the service routine to restore the original conditions upon return to the main program.

1.3.4.e.iv Interrupt Priorities.   A microprocessor is typically provided with one or more interrupt pins on the chip. Therefore, a special mechanism is necessary to handle interrupts from several devices that share one of these interrupt lines. There are two ways of servicing multiple interrupts: polled and daisy chain techniques.

Polled interrupts are handled by software and therefore are slower when compared with daisy chaining. The processor responds to an interrupt by executing one general service routine for all devices. The priorities of devices are determined by the order in which the routine polls each device. The processor checks the status of each device in the general service routine, starting with the highest-priority device to service an interrupt. Once the processor determines the source of the interrupt, it branches to the service routine for the device.

In a daisy chain priority system, devices are connected in a daisy chain fashion to set up a priority system. Suppose one or more devices interrupt the processor. In response, the processor pushes at least the PC and generates an interrupt acknowledge ($\overline{\text{INTA}}$) signal to the highest-priority device. If this device has generated the interrupt, it will accept the $\overline{\text{INTA}}$. Otherwise it will pass the $\overline{\text{INTA}}$ onto the next device until $\overline{\text{INTA}}$ is accepted. Once accepted, the device provides a means for the processor to find an interrupt address vector by using external hardware. The daisy chain priority scheme is based on mostly hardware and is therefore faster than the polled interrupt.

1.3.4.f Direct Memory Access (DMA)

Direct Memory Access (DMA) is a technique that transfers data between a microcomputer's memory and an I/O device without involving the microprocessor. DMA is widely used in transferring large blocks of data between a peripheral device and the microcomputer's memory. The DMA technique uses a DMA controller chip for the data-transfer operation. The main functions of a typical DMA controller are summarized as follows:

- The I/O devices request DMA operation via the DMA request lines of the controller chip.
- The controller chip activates the microprocessor HOLD pin, requesting the CPU to release the bus.
- The processor sends HLDA (hold acknowledge) back to the DMA controller, indicating that the bus is disabled. The DMA controller places the current value of its internal registers, such as the address register and counter, on the system bus and sends a DMA acknowledge to the peripheral device. The DMA controller completes the DMA transfer and releases the buses.

There are three basic types of DMA: block transfer, cycle stealing, and interleaved DMA.

For block-transfer DMA, the DMA controller chip takes the bus from the microcomputer to transfer data between the memory and I/O device. The microprocessor has no access to the bus until the transfer is completed. During this time, the microprocessor can perform internal operations that do not need the bus. This method is popular with microprocessors. Using this technique, blocks of data can be transferred.

Data transfer between the microcomputer memory and an I/O device occurs on a word-by-word basis with cycle stealing. Typically, the microprocessor clock is enabled by ANDing an $\overline{\text{INHIBIT}}$ signal with the system clock. The system clock has the same frequency as the microprocessor clock.

The DMA controller controls the $\overline{\text{INHIBIT}}$ line. During normal operation, the $\overline{\text{INHIBIT}}$ line is HIGH, providing the microprocessor clock. When DMA operation is desired, the controller makes the INHIBIT line LOW for one clock cycle. The microprocessor is then stopped completely for one cycle. Data transfer between the memory and I/O takes place during this cycle. This method is called cycle stealing because the DMA controller takes away or steals a cycle without microprocessor recognition. Data transfer takes place over a period of time.

With interleaved DMA, the DMA controller chip takes over the system bus when the microprocessor is not using it. For example, the microprocessor does not use the bus while incrementing the program counter or performing an ALU operation. The DMA controller chip identifies these cycles and allows transfer of data between the memory and I/O device. Data transfer takes place over a period of time for this method.

The DMA controller chip usually has at least three registers normally selected by the controller's register select (RS) line: an address register, a terminal count register, and a status register. Both the address and terminal count registers are initialized by the microprocessor. The address register contains the starting of the data to be transferred, and the terminal count register contains the desired block to be transferred. The status register contains information such as completion of DMA transfer.

1.3.4.g Summary of Microcomputer I/O Methods

Figure 1.17 summarizes the I/O structure (explained so far) of typical microcomputers.

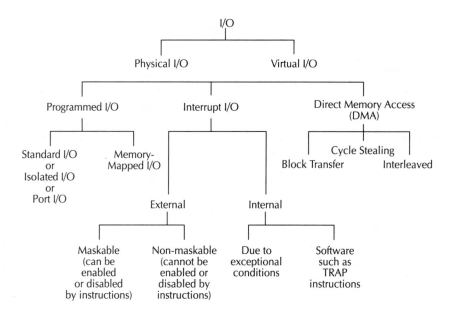

FIGURE 1.17   I/O structure of a typical microcomputer.

It should be mentioned that while using either block transfer or cycle stealing DMA in systems with dynamic RAMs, circuitry must be included for refreshing the dynamic RAM's during DMA transfer.

1.3.4.h Coprocessors

In typical 8-bit microprocessors such as the Intel 8085, technology places a limit on the chip area. As a consequence, these microprocessors include no hardware or firmware for performing scientific computations such as floating-point arithmetic, matrix manipulation, and graphic-data processing. Therefore, users of these systems must write these programs. Unfortunately, this approach is unacceptable in high-speed applications, since program execution takes a significant amount of time. To eliminate this problem, coprocessors are used.

In this approach, a single chip is built for performing scientific computations at high speed. However, the chip is regarded as a companion to the original or host microprocessor. Typically, each special operation is encoded as an instruction that can be interpreted only by the companion processor. When the companion processor encounters one of these special instructions, it assumes the processing functions independent of the host microprocessor. The companion processor that operates in this manner is called the coprocessor. Therefore, this concept not only extends the capabilities of the host microprocessor, but also increases the processing rate of the system. The coprocessor concept is widely used with typical 32-bit microprocessors such as the Motorola 68020 and Intel 80386.

It is important to make the distinction between standard peripheral hardware and a coprocessor. A coprocessor is a device that has the capability of communicating with the main processor through the protocol defined as the coprocessor interface. As mentioned before, the coprocessor also adds additional instructions, registers, and data types that are not directly supported by the main microprocessor. The coprocessor provides capabilities to the user without appearing to be hardware external to the main microprocessor.

Standard peripheral hardware, on the other hand, is generally accessed through the use of interface registers mapped into the memory space of the main processor. The programmer uses standard processor instructions to access the peripheral interface registers and thus utilize the services provided by the peripheral. It should be pointed out that even though a peripheral can provide capabilities equivalent to a coprocessor for many applications, the programmer must implement the communication protocol between the main microprocessor and the peripheral necessary to use the peripheral hardware. Two main techniques may be used to pass commands to a coprocessor. These are intelligent monitor interface and coprocessors using special signals.

In the *intelligent monitor interface,* the coprocessor monitors the instruction stream by obtaining commands directly from the bus at the same time as the main microprocessor. The Intel 80387 floating-point coprocessor is of this type, as it monitors the instruction stream simultaneously with a main microprocessor such as the Intel 80386. This has the obvious advantage of requiring no additional bus cycles to pass the content of the instruction word to the coprocessor. One of the main disadvantages of this approach is that each coprocessor in the system must duplicate the bus monitoring circuitry and instruction queue, tracking all branches, wait states, operand fetches, and instruction fetches.

In the second type, the coprocessor may be explicitly addressed by certain instructions, which initiate a special sequence of microinstructions in the main microprocessor to effect command and operand transfer.

In this approach, when the main microprocessor executes a coprocessor instruction, it decodes the instruction and writes a command in the command register (one of the interface registers) specifying the operation required by the coprocessor. In response, the coprocessor writes data back in a register called the *response register* (one of the interface registers). The main microprocessor can read these data, and it tells the main microprocessor certain information such as whether additional information is required by the coprocessor to carry out the operation. If such data are required, the main microprocessor provides this; otherwise, the coprocessor carries out the operation concurrently with the main microprocessor and provides the result.

An advantage of this approach is that no special signals are required for the coprocessor interface.

One of the main disadvantages of this method is that once the main processor detects a coprocessor instruction, the main has to use bus bandwidth and timing to transmit the command to the appropriate coprocessor. The Motorola 68881 (floating-point coprocessor) is of this type.

Note that state-of-the-art 32-bit microprocessors such as the Intel 80486 and the Motorola 68040 implement coprocessor hardware such as floating point hardware and MMU on the microprocessor chip.

1.4 Microcomputer System Software and Programming Concepts

In this section, the basic concepts associated with system software and programming will be discussed.

1.4.1 System Software

Typical microcomputer system software provided in microcomputer development systems such as the HP64000 includes editors, assemblers, compilers, interpreters, debuggers, and an operating system.

The editor is used to create and change source programs. Source programs can be written in assembly language or a high-level language such as Pascal. The editor has commands to change, delete, or insert lines or characters. The text editor is a special type of editor that is used to enter and edit text in a general-purpose computer, whether the text is a report, a letter, or a program.

An assembler translates a source text that was created using the editor into a target machine language in object code (binary).

High-level languages contain English-like commands that are readily understandable by the programmer. High-level languages normally combine a number of assembly-level statements into a single high-level statement. A compiler is used to translate the high-level languages such as Pascal into machine language. The advantage of high-level languages over assembly language are ease of readability and maintainability.

Like a compiler, an interpreter usually processes a high-level language program. Unlike a compiler, an interpreter actually executes the high-level language program one statement at a time, rather than translating the whole program into a sequence of machine instructions to be run later.

The debugger provides an interactive method of executing and debugging the user's software, one or a few instructions at a time, allowing the user to see the effects of small pieces of the program and thereby isolate programming errors.

An operating system performs resource management and human-to-machine translation functions. A resource may be the microprocessor, memory, or an I/O device. Basically, an operating system is another program that tells the machine what to do under a variety of conditions. Major operating system functions include efficient sharing of memory, I/O peripherals, and the microprocessor among several users. An operating system is

1. The interface between hardware and users
2. The manager of system resources in accordance with system policy to achieve system objectives

Operating systems for microcomputers became available when microcomputers moved from process control applications to the general-purpose computer applications. It was appropriate to write a process control program in assembly language because the microcomputer was required to perform dedicated real-time control functions. But when the microcomputers evolved to the point of controlling several I/O devices (disks, printers), an organized operating system was needed.

Note that many laboratory trainers such as the Intel SDK-86 include primitive operating systems called monitors to provide functions such as keyboard/display interfaces and program debugging features.

1.4.2 Programming Concepts

In general, programs are developed using assembly and high-level languages.

1.4.2.a Assembly Language Programming

Program designers realized the importance of symbols in programming to improve readability and expedite the program development process. As a first step, they came up with the idea of giving a symbolic name for each instruction. These names are called mnemonics, and a program written using such mnemonics is called an assembly language program. Given below is a typical 8086 assembly language program:

```
MOV    AL,5 ; Load AL reg with 5
ADD    AL,3 ; (AL) ← (AL) + 3
HLT           ; Halt processing
```

From this example, it is clear that the usage of mnemonics (in our example MOV, ADD, HLT are the mnemonics) improves the readability of our program significantly.

An assembly language program cannot be executed by a machine directly, as it is not in binary form. Usually, we refer to a symbolic program as a source program. An assembler is needed in order to translate an assembly language (source) program into the machine language (object) executable by the machine. This is illustrated in Figure 1.18.

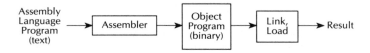

FIGURE 1.18   Assembly Process.

1.4.2.b High-Level Language Programming

Typical examples of high-level languages include FORTRAN, BASIC, Pascal, C and C++. The program shown below is written in FORTRAN, in order to obtain the sum of the first N natural numbers:

```
        READ (5,10) N
10      FORMAT (I4)
        NSUM = N*(N+1)/2
        WRITE (2,10) NSUM
        STOP
        END
```

A translator called a compiler is needed to translate a program written in a high-level language into binary form.

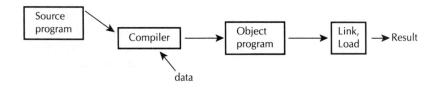

FIGURE 1.19   Compilation process.

1.4.2.c Which Programming Language to Choose?

Compilers normally provide inefficient machine codes because of the general guidelines which must be followed for designing them. It was found that a compiled high level language generates many more lines of machine code than an equivalent assembly language program. Therefore, the machine code program generated by the assembler will take up less memory space and also will execute much faster.

Therefore, although the C language includes I/O instructions, applications involving I/O are normally written in assembly language. One of the main uses of assembly language is in writing programs for real-time applications. Real time applications means that the task required by the application must be completed before any other input to the program can occur which will change its operation.

High level languages are normally used for applications which require extensive mathematical computations. These applications include optimizations and other non-real time control system applications.

1.5 Typical Microcomputer Addressing Modes and Instructions

In this chapter, some important characteristics and properties of microcomputer instruction sets are discussed. Topics include addressing modes and instruction types.

1.5.1 Introduction

An instruction manipulates the stored data, and a sequence of instructions constitutes a program. In general, an instruction has two components:

- Op-code field
- Operand field(s)

The op-code field specifies how data are to be manipulated.The op-code field may contain data or a microprocessor register or a memory address. Consider the following instruction:

```
ADD              R1, R0
op-code field  operand fields
```

Assume that this microcomputer uses R1 as the source register and R0 as the destination register. The preceding instruction then adds the contents of registers R0 and R1 and saves the sum in register R0.

Depending on the number of addresses specified, one can have the following instruction formats:

- Three-operand
- Two-operand
- One-operand
- Zero-operand

The 8-bit microprocessors include mostly one-operand instructions along with some zero- and two-operand instructions. In 16-bit microprocessors, two-operand instructions are predominant although some zero- and one-operand instructions are present. The 32-bit microprocessors include all four instruction formats.

1.5.2 Addressing Modes

The sequence of operations that a microprocessor has to carry out while executing an instruction is called its *instruction cycle*. One of the activities in an instruction cycle is the determination of the addresses of the operands involved in that instruction.

The way in which a microprocessor accomplishes this task is by recognizing the *addressing mode* used in the instruction. Typical addressing modes supported by the instruction sets of popular processors will be examined.

An instruction is said to have an *inherent* addressing mode if it is a zero-operand instruction. As an example, consider the zero-operand instruction CLC which clears the carry flag to zero.

Whenever an instruction contains data in the operand field, it is called an *immediate mode* instruction. For example, consider the following instruction:

$$\text{ADD \#25, R1; R1} \leftarrow \text{R1} + 25$$

In this instruction, the symbol # indicates that it is an immediate-mode instruction. This convention is adopted in the assemblers for processors such as MC68000.

An instruction is said to have an *absolute addressing mode* if it contains the address of the operand. For example, consider the following move instruction:

$$\text{MOVE 5000, R2; R2} \leftarrow [5000]$$

This instruction copies the contents of memory location 5000 in the register R2.

An instruction is said to have a *register mode* if it contains a register in the operand field. For example, consider the following register mode intruction.

$$\text{ADD R2, R3; R3} \leftarrow \text{R2} + \text{R3}$$

This instruction uses register mode for both source and destination operands.

Whenever an instruction specifies a register that holds the address of an operand, the resulting addressing mode is known as the *register indirect* mode. From this definition, it follows that the Effective Address (EA) of an operand in the register-indirect mode is the contents of the register R. More formally, this result is written as follows:

$$\text{EA} = [\text{R}]$$

To illustrate this idea clearly, consider the following instruction:

$$\text{MOVE (R2), (R3); [R3]} \leftarrow [\text{R2}]$$

Assume that the following configuration exists:

$$[\text{R2}] = 5000_{16}$$
$$[\text{R3}] = 4000_{16}$$
$$[5000] = 1256_{16}$$
$$[4000] = 4629_{16}$$

This instruction copies the contents of the memory location, whose address is specified by the register R2, into the location whose address is specified by the register R3. Thus, after the execution of this instruction, the memory location 4000 will contain the value 1256_{16}.

1.5.3 Instruction Types

In general, instructions available in a processor may be broadly classified into five groups:

- Data transfer instructions
- Arithmetic instructions
- Logical instructions
- Program control instructions
- I/O instructions

Data transfer instructions are primarily concerned with data transfers between the microprocessor registers or between register and memory. An example is MOVE R0, R1 which transfers the contents of register R0 to register R1.

Typical arithmetic instructions include ADD and SUBTRACT instructions. For example, ADD R0, R1 adds the contents of R0 to R1 and stores the result in R1.

Logical instructions perform Boolean AND, OR, NOT, and EXCLUSIVE-OR operations on a bit-by-bit basis. An example is OR R0, R1 which logically ORs the contents of R0 with R1 and places the result in R1.

Typical program control instructions include unconditional and conditional branch and subroutine CALL instructions. For example, JMP 2035H unconditionally branches to the 16-bit address 2035H.

I/O instructions perform input and output operations. An example is IN PORTA which inputs the contents of an I/O port called Port A into a microprocessor register such as the accumulator.

1.6 Basic Features of Microcomputer Development Systems

A microcomputer development system is a tool that allows the designer to develop, debug, and integrate error-free application software in microprocessor systems.

Development systems fall into one of two categories: systems supplied by the device manufacturer (non-universal systems) and systems built by after-market manufacturers (universal systems). The main difference between the two categories is the range of microprocessors that a system will accommodate. Non-universal systems are supplied by the microprocessor manufacturer (Intel, Motorola, RCA) and are limited to use for the particular microprocessor manufactured by the supplier. In this manner, an Intel development system may not be used to develop a Motorola-based system. The universal development systems (Hewlett-Packard, Tektronix) can develop hardware and software for several microprocessors.

Within both categories of development systems, there are basically three types available: single-user systems, time-shared systems, and networked systems. A single-user system consists of one development station that can be used by one user at a time. Single-user systems are low in cost and may be sufficient for small systems development. Time-shared systems usually consist of a "dumb"-type terminal connected by data lines to a centralized microcomputer-based system that controls all operations. A networked system usually consists of a number of smart Cathode Ray Tubes (CRTs) capable of performing most of the development work and can be connected over data lines to a central microcomputer. The central microcomputer in a network system usually is in charge of allocating disk storage space and will download some programs into the user's work station microcomputer. A microcomputer development system is a combination of the hardware necessary for microprocessor design and the software to control the hardware. The basic components of the hardware are the central processor, the CRT terminal, mass storage device (floppy or hard disk), and usually an In-Circuit Emulator (ICE).

In a single-user system, the central processor executes the operating system software, handles the Input/Output (I/O) facilities, executes the development programs (editor, assembler, linker), and allocates storage space for the programs in execution. In a large multiuser networked system the central processor may be responsible for mass storage allocation, while a local processor may be responsible for the I/O facilities and execution of development programs.

The CRT terminal provides the interface between the user and the operating system or program under execution. The user enters commands or data via the CRT keyboard and the program under execution displays data to the user via the CRT screen.

Each program (whether system software or user program) is stored in an ordered format on disk. Each separate entry on the disk is called a *file*. The operating system software contains the routines necessary to interface between the user and the mass storage unit. When the user requests a file by a specific *file name*, the operating system finds the program stored on disk by the file name and loads it into main memory. More advanced development systems contain *memory management* software that protects a user's files from unauthorized modification by another user. This is accomplished via a unique user identification code called USER ID. A user can only access files that have the user's unique code.

The equipment listed above comprises a basic development system, but most systems have other devices such as printers and PROM programmers attached. A printer is needed to provide the user with a hard copy record of the program under development.

After the application system software has been completely developed and debugged, it needs to be permanently stored for execution in the target hardware. The EPROM programmer takes the machine code and programs it into an EPROM. Erasable/Programmable Read Only Memories (EPROMs) are more generally used in system development as they may be erased and reprogrammed if the program changes. EPROM programmers usually interface to circuits particularly designed to program a specific PROM. These interface boards are called person-ality cards and are available for all the popular PROM configurations.

Most development systems support one or more in-circuit emulators (ICEs). The ICE is one of the most advanced tools for microprocessor hardware development. To use an ICE, the microprocessor chip is removed from the system under development (called the target processor) and the emulator plugged into the microprocessor socket. The ICE will function-ally and electrically act identically to the target processor with the exception that the ICE is under the control of development system software. In this manner the development system may exercise the hardware that is being designed and monitor all status information avail-able about the operation of the target processor. Using an ICE, processor register contents may be displayed on the CRT and operation of the hardware observed in a single-stepping mode. In-circuit emulators can find hardware and software bugs quickly that might take many hours using conventional hardware testing methods.

Architectures for development systems can be generally divided into two categories: the master/slave configuration and the single-processor configuration. In a master/slave configu-ration, the master (host) processor controls all development functions such as editing, assem-bling, and so on. The master processor controls the mass storage device and processes all I/O (CRT, printer).

The software for the development systems is written for the master processor which is usually not the same as the slave (target) processor. The slave microprocessor is typically connected to the user prototype via a 40-pin connector (the number varies with the processor) which links the slave processor to the master processor.

Some development systems such as the HP 64000 completely separate the system bus from the emulation bus and therefore use a separate block of memory for emulation. This separa-tion allows passive monitoring of the software executing on the target processor without stopping the emulation process. A benefit of the separate emulation facilities allows the master processor to be used for editing, assembling, and so on, while the slave processor continues the emulation. A designer may therefore start an emulation running, exit the emulator program, and at some future time return to the emulation program.

Another advantage of the separate bus architecture is that an operating system needs to be written only once for the master processor and will be used no matter what type of slave processor is being emulated. When a new slave processor is to be emulated, only the emulator probe needs to be changed.

A disadvantage of the master/slave architecture is that it is expensive. In single-processor architecture, only one processor is used for system operation and target emulation. The single processor does both jobs of executing system software as well as acting as the target processor. Since there is only one processor involved, the system software must be rewritten for each type of processor that is to be emulated. Since the system software must reside in the same memory used by the emulator, not all memory will be available to the emulation process, which may be a disadvantage when large prototypes are being developed. The single processor systems are inexpensive.

The programs provided for microprocessor development are the operating system, editor, assembler, linker, compiler, and debugger.

The operating system is responsible for executing the user's commands. The operating system (such as UNIX) handles I/O functions, memory management, and loading of programs from mass storage into RAM for execution.

The editor allows the user to enter the source code (either assembly language or some high-level language) into the development system.

Almost all current microprocessor development systems use the character-oriented editor, more commonly referred to as the screen editor. The editor is called a screen editor because the text is dynamically displayed on the screen and the display automatically updates any edits made by the user.

The screen editor uses the pointer concept to point to the character(s) that need editing. The pointer in a screen editor is called the cursor and special commands allow the user to position the cursor to any location displayed on the screen. When the cursor is positioned, the user may insert characters, delete characters, or simply type over the existing characters.

Complete lines may be added or deleted using special editor commands. By placing the editor in the *insert* mode, any text typed will be inserted at the cursor position when the cursor is positioned between two existing lines. If the cursor is positioned on a line to be deleted, a single command will remove the entire line from the file.

Screen editors implement the editor commands in different fashions. Some editors use dedicated keys to provide some cursor movements. The cursor keys are usually marked with arrows to show the direction of cursor movement.

More advanced editors (such as the HP 64000) use *soft keys*. A soft key is an unmarked key located on the keyboard directly below the bottom of the CRT screen. The mode of the editor decides what functions the keys are to perform. The function of each key is displayed on the screen directly above the appropriate key. The soft key approach is valuable because it frees the user from the problem of memorizing many different special control keys. The soft key approach also allows the editor to reassign a key to a new function when necessary.

The source code generated on the editor is stored as ASCII or text characters and cannot be executed by a microprocessor. Before the code can be executed, it must be converted to a form acceptable by the microprocessor. An assembler is the program used to translate the assembly language source code generated with an editor into object code (machine code) which may be executed by a microprocessor.

Assemblers recognize four *fields* on each line of source code. The fields consist of a variable number of characters and are identified by their position in the line. The fields, from left to right on a line, are the label field, the mnemonic or op-code field, the operand field, and the comment field. Fields are separated by characters called *delimiters* which serve as a flag to the assembler that one field is done and the next one is to start. Typical delimiters and their uses are

space	used to separate fields
TAB	used to separate fields
,	used between addresses or data in the operand field
;	used before a comment statement
:	used after a label

A few typical lines of 8085 source code are

LABEL FIELD	MNEMONIC FIELD	OPERAND FIELD	COMMENT FIELD
	MOV	A,5	; LOAD A5 INTO A
	ADD	AL,2	; ADD 5 AND 2, STORE RESULT IN AL

As can be seen in the above example, tab keys are used instead of spaces to separate the fields to give a more *spread out* line which is easier to read during debugging.

In order for the assembler to differentiate between numbers and labels, specific rules are set up which apply to all assemblers. A label must start with a letter. After the letter, a combination of letters and numbers (called alphanumerics) may be used. For example, when grouping lines of code by function, a common alphabetic string may be used followed by a unique number for the label: L00P01, L00P02, L00P10, and so on.

A numeric quantity must start with a number, even though the number may be in hex (which may start with a letter). Most assemblers assume that a number is expressed in the decimal system and if another base is desired, a special code letter is used immediately following the number. The usual letter codes used are

> B   binary
> C   octal
> H   hex (Motorola uses $ before the number)

To avoid confusion when hex quantities are used, a leading zero is inserted to tell the assembler that the quantity is a number and not a label (for example, the quantity FA in hex would be represented by 0FAH in the source code).

Assembler Directives

Assembler directives are instructions entered into the source code along with the assembly language. These directives do not get translated into object code but are used as special instructions to the assembler to perform some special functions. The assembler will recognize directives that assign memory space, assign addresses to labels, format the pages of the source code, and so on.

The directive is usually placed in the op-code field. If any labels or data are required by the directive, they are placed in the label or operand field as necessary.

Some common directives will now be discussed in detail.

a. ORIGIN (ORG).   The ORG statement is used by the programmer when it is necessary to place the program in a particular location in memory. As the assembler is translating the source code, it keeps an internal counter (similar to the microprocessor program counter) that keeps track of the address for the machine code. The counter is incremented automatically and sequentially by the assembler. If the programmer wishes to alter the locations where the machine code is going to be located, the ORG statement is used.

For example, if it is desired to have a subroutine at a particular location in memory, such as 2000H, the statement ORG 2000H, would be placed immediately before the subroutine to direct the assembler to alter the internal program counter.

Most assemblers will assume a starting address of zero if no ORG statement is given in the source code.

b. EQUATE (EQU).   The EQU instruction is used to assign the data value or address in the operand field to the label in the label field. The EQU instruction is valuable because it allows the programmer to write the source code in symbolic form and not be concerned with the numeric value needed. In some cases, the programmer is developing a program without knowing what addresses or data may be required by the hardware. The program may be written and debugged in symbolic form and the actual data added at a later time. Using the EQU instruction is also helpful when a data value is used several times in a program. If, for example, a counter value was loaded at ten different locations in the program, a symbolic label (such as COUNT) could be used and the label count defined at the end of the program. By using this technique, if it is found during debugging that the value in COUNT must be changed, it need only be changed at the EQU instruction and not at each of the ten locations where it is used in the program.

c. *DEFINE BYTE (DEFB or DB).*   The DB instruction is used to set a memory location to a specific data value. The DB instruction is usually used to create data tables or to preset a flag value used in a program. As the name implies, the DB instruction is used for creating an 8-bit value.

For example, if a table of four values, 45H, 34H, 25H, and 0D3H, had to be created at address 2000H, the following code could be written:

```
       ORG  2000H           ; SET TABLE ADDRESS
  TABLE DB  44H,34H,25H,0D3H; PRESET TABLE VALUES
```

The commas are necessary for the assembler to be able to differentiate between data values. When the code is assembled, the machine code would appear as follows:

```
            . . .
           2000   45
           2001   34
           2002   25
           2003   D3
            . . .
```

d. *DEFINE WORD (DEFW or DW).*   Similarly to DB, DW defines memory locations to specific values. As the name implies, the memory allotted is in word lengths which are usually 16 bits wide. When assigning a 16-bit value to memory locations, two 8-bit memory locations must be used. By convention, most assemblers store the least significant byte of the 16-bit value in the first memory location and the most significant byte of the 16-bit value in the next memory location. This technique is sometimes referred to as *Intel style,* because the first microprocessors were developed by Intel, and this storage method is how the Intel processors store 16-bit words.

Data tables may be created with the DW instruction, but care must be taken to remember the order in which the 16-bit words are stored. For example, consider the following table:

```
       ORG              2500H
  DATA DW    4000H,  2300H,  4BCAH
```

The machine code generated for this table would appear as follows:

```
            . . .
           2500   00
           2501   40
           2502   00
           2503   23
           2504   CA
           2505   4B
            . . .
```

e. TITLE. TITLE is a formatting instruction that allows the user to name the program and have the name appear on the source code listing. Consider the following line:

TITLE 'MULTIPLICATION ROUTINE'

When the assembler generates the program listing, each time it starts a new page the title MULTIPLICATION ROUTINE appears at the top of each page.

Several types of assemblers are available, the most common types are discussed below.

a. One-Pass Assembler. The one-pass assembler was the first type to be developed and is therefore the most primitive. Very few systems use a one-pass assembler because of the inherent problem that only *backward references* may be used.

In a one-pass assembler the source code is processed only once. As the source code is processed, any labels encountered are given an address and stored in a table. Therefore, when the label is encountered again, the assembler may look backward to find the address of the label. If the label has not been defined yet (for example, a jump instruction that jumps forward), the assembler issues an error message.

b. Two-Pass Assembler. In the two-pass assembles, the source code is passed twice through the assembler. The first pass made through the source code is specifically for the purpose of assigning an address to all labels. When all labels have been stored in a table with the appropriate addresses, a second pass is made to actually translate the source code into machine code.

The two-pass style assembler is the most popular type of assembler currently in use.

c. Macroassembler. A macroassembler is a type of two-pass assembler that allows the programmer to write the source code in *macros*. A macro is a sequence of instructions that the programmer gives a name. Whenever the programmer wishes to duplicate the sequence of instructions, the macro name is inserted into the source code.

d. Cross Assemblers. A cross assembler may be of any of the types already mentioned. The distinguishing feature of a cross assembler is that it is not written in the same language used by the microprocessor that will execute the machine code generated by the assembler.

Cross assemblers are usually written in a high-level language such as FORTRAN which will make them machine independent. For example, an 8086 assembler may be written in FORTRAN and then the assembler may be executed on another machine such as the Motorola 6800.

e. Metaassembler. The most powerful assembler is the metaassembler because it will support many different microprocessors. The programmer merely specifies at the start of the source code which microprocessor assembly language will be used and the metaassembler will translate the source code to the correct machine code.

The output file from most development system assemblers is an object file. The object file is usually relocatable code that may be configured to execute at any address. The function of the linker is to convert the object file to an *absolute* file which consists of the actual machine code at the correct address for execution. The absolute files thus created are used for debugging and finally for programming EPROMs.

Debugging a microprocessor-based system may be divided into two categories: software debugging and hardware debugging. Both debug processes are usually carried out separately from each other because software debugging can be carried out on an Out-Of-Circuit-Emulator (OCE) without having the final system hardware.

The usual software development tools provided with the development system are

- Single-step facility
- Breakpoint facility

A single-stepper simply allows the user to execute the program being debugged one instruction at a time. By examining the register and memory contents during each step, the debugger can detect such program faults as incorrect jumps, incorrect addressing, erroneous op codes, and so on.

A breakpoint allows the user to execute an entire section of a program being debugged.

There are two types of breakpoint systems: hardware and software. The hardware breakpoint uses hardware to monitor the system address bus and detect when the program is executing the desired breakpoint location. When the breakpoint is detected, the hardware uses the processor control lines to either halt the processor for inspection or cause the processor to execute an interrupt to a breakpoint routine. Hardware breakpoints can be used to debug both ROM- and RAM-based programs. Software breakpoint routines may only operate on a system with the program in RAM because the breakpoint instruction must be inserted into the program that is to be executed.

Single-stepper and breakpoint methods complement each other. The user may insert a breakpoint at the desired point and let the program execute up to that point. When the program stops at the breakpoint the user may use a single-stepper to examine the program one instruction at a time. Thus, the user can pinpoint the error in a program.

There are two main hardware debugging tools: the logic analyzer and the in-circuit emulator.

Logic analyzers are usually used to debug hardware faults in a system. The logic analyzer is the digital version of an oscilloscope because it allows the user to view logic levels in the hardware.

In-circuit emulators can be used to debug and integrate software and hardware.

PC-based workstations are extensively used as development systems.

1.7 System Development Flowchart

The total development of a microprocessor-based system typically involves three phases: software design, hardware design, and program diagnostic design. A systems programmer will be assigned the task of writing the application software, a logic designer will be assigned the task of designing the hardware, and typically both designers will be assigned the task of developing diagnostics to test the system. For small systems, one engineer may do all three phases, while on large systems several engineers may be assigned to each phase. Figure 1.20 shows a flowchart for the total development of a system. Notice that software and hardware development may occur in parallel to save time.

1.7.1 Software Development

The first step in developing the software is to take the system specifications and write a flowchart to accomplish the desired tasks that will implement the specifications.

The assembly language or high-level source code may now be written from the system flowchart.

The complete source code is then assembled. The assembler will check for syntax errors and print error messages to help in the correction of errors.

The normal output of an assembler is the object code and a program listing. The object code will be used later by the linker. The program listing may be sent to a disk file for use in debugging or it may be directed to the printer.

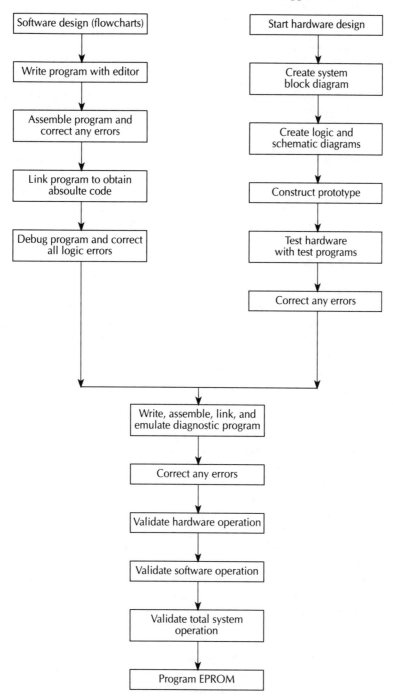

FIGURE 1.20   Microprocessor system development flowchart.

The linker can now take the object code generated by the assembler and create the final absolute code that will be executed on the target system. The emulation phase will take the absolute code and load it into the development system RAM. From here, the program may be debugged using breakpoints or single-stepping.

1.7.2 Hardware Development

Working from the system specifications, a block diagram of the hardware must be developed.
The logic diagram and schematics may now be drawn using the block diagram as a guide.
A prototype may now be constructed and tested for wiring errors.

When the prototype has been constructed it may be debugged for correct operation using standard electronic testing equipment such as oscilloscopes, meters, logic probes, and logic analyzers, all with test programs created for this purpose.

After the prototype has been debugged electrically, the development system in-circuit emulator may be used to check it functionally. The ICE will verify memory map, correct I/O operation, and so on.

The next step in the system development is to validate the complete system by running operational checks on the prototype with the finalized application software installed. The EPROM is then programmed with the error-free programs.

1.8 Typical Microprocessors

Intel and Motorola established the standard for future microprocessors.

Both Intel and Motorola introduced popular 8-bit microprocessors, namely, Intel 8080/8085, and Motorola 6800/6809 in the mid 1970's. These microprocessors are capable of directly addressing 64k of memory and were popular for several years. They are still being utilized in many inexpensive low-speed applications.

In 1978, Intel introduced its first 16-bit microprocessor, the Intel 8086. Several variations of the Intel 16-bit microprocessor family include the 8088, 80186/80188 and 80286. The 8088 is similar to the 8086 except that the 8088 has an 8-bit data bus. Both the 8086 and 8088 can run at 5 Mhz. They are packaged in 40 pins and can directly address one megabyte of memory. In 1982, Intel introduced an enhanced version of the 8086/8088 which is the Intel 80186/80188. The 80186 has a 16-bit data bus while the 80188 has an 8-bit data bus. The 80186/80188 can operate at either 8 Mhz (80186) or 6 Mhz (80186-6). The 80186 integrates the 8086 and several new functional units into a single chip. The new on-chip components include a clock generator, two DMA channels, interrupt controller, address decoding, and three 16-bit programmable timers. Like the 8086, the 80186 can directly address one megabyte of memory. The 80186 is housed in a 68-pin leadless package.

The 80286 is Intel's other 16-bit microprocessor with on-chip memory protection capability primarily designed for multiuser/multitasking systems and introduced virtual memory.

Intel's 32-bit microprocessor family includes the 80386, 80486, and Pentium. The 80386 includes an on-chip memory management unit while the 80486 (DX and DX2) contains all the features of the 80386 along with floating-point hardware in the same chip. The Pentium, on the other hand, is a superscalar processor. This means that it includes dual pipelining and executes more than one instruction per cycle.

Soon after introduction of the Intel 8086, Motorola released its first 16-bit microprocessor, the 68000. It is housed in 64-pin or 68-pin package with a direct addressing capability of 16 megabytes. The 68008 is similar to the 68000 except that the 68000 has a 16-bit data bus while the 68008 has an 8-bit data bus. Also, the 68008 can directly address one megabyte of memory. The 68010 and 68012 are 16-bit microprocessors similar to the Intel 80286 and provide on-chip memory management support. Motorola CPU 32 is an enhanced 68000 along with many features of the 16-bit 68010 and the 32-bit 68020. The CPU 32 can support high level languages and is suited for controller applications. For example, in applications where power consumption is a consideration, the CPU32 forces the device into a low-power standby mode when immediate processing is not required upon execution of the LPSTOP instruction. Also, to maximize throughput for real-time applications, reference data is often precalculated and stored in memory for quick access upon execution of the TBL instruction.

Motorola's 32-bit microprocessor family includes the 68020, 68030, and 68040.

Since 1988, Intel, Morotola, and others have been introducing the RISC microprocessors. Some of these are the Intel 80960 family, the Motorola MC88100, the Apple/IBM/Motorola PowerPC, and Digital Equipment Corporation's Alpha 21164. The 80960 and 88100 are 32-bit microprocessors. They can directly address 4 gigabytes of memory. The PowerPC family, on the other hand, includes both 32-bit and 64-bit microprocessors. The 80960 and 88100 include a 32-bit data bus while the PowerPC contains a 64-bit data bus. The PowerPc is based on IBM PowerPC architecture and Motorola's 88100 bus interface design. These RISC microprocessors find extensive applications in embedded controls such as laser printers.

The PowerPC 601 outperforms and under-prices the Pentium but to be successful must build up its selection of software applications. Note that the Pentium had a flaw in its division algorithm caused by a problem with a Lookup table used in the division. Intel recently corrected this problem.

Tables 1.2a and 1.2b provide a brief description of Intel and Motorola's typical 16- and 32-bit microprocessors.

1.9 Typical Practical Applications

Microprocessors are being extensively used in a wide variety of applications. Typical applications include dedicated controllers, personal workstations, and real-time robotics control. Some of these applications are described in the following.

1.9.1 Personal Workstations

Personal workstations are designed using the high-performance 16- and 32-bit microprocessors. A dedicated single user (rather than multiple users sharing resources of a single microcomputer) can obtain significant computing power from these workstations.

The state-of-the-art workstations use 32-bit microprocessors to provide certain sophisticated functions such as IC layout, 3D graphics, and stress analysis.

1.9.2 Fault-Tolerant Systems

In many applications such as control of life-critical systems, control of nuclear waste, and unattended remote system operation, the reliability of the hardware is of utmost importance. The need for such reliable systems resulted in fault-tolerant systems. These systems use redundant computing units to provide reliable operation. However, the cost of fault-tolerant

Table 1.2a   Intel 16-bit microprocessors

	8086	8088	80286
Introduced	June 1978	June 1978	Feb. 1982
Maximum clock speed	10 Mhz	10 Mhz	20 Mhz
On-chip cache	No	No	No
MIPS (Millions of Instructions Per Second)	0.33	0.33	1.2
Addressing Modes	12	12	8
Transistors	29,000	29,000	134,000
Data Bus	16-bit	8-bit	16-bit
Address Bus	20-bit	20-bit	24-bit
Directly Addressable memory	One megabyte	One megabyte	Sixteen megabytes
Number of pins	40	40	68
Virtual Memory	No	No	Yes; One gigabyte per task
On-Chip memory management and protection	No	No	Yes
Math Coprocessor Interface	8087	8087	80287XL/XLT

Table 1.2a   Intel 32-bit microprocessors

	80386DX	80386SX	486DX	486SX	486DX2	Pentium
Introduced	October 1985	June 1988	April 1989	April 1991	March 1992	May 1993
Maximum clock speed	40 Mhz	33 Mhz	50 Mhz	25 Mhz	66 Mhz	100 Mhz
MIPS	6	2.5	20	16.5	40	112
Transistor	275,000	275,000	1.2 Millions	1.185 Millions	1.2 Millions	3.1 Millions
On-chip cache memory	Support chips avaliable	Support chips avaliable	Yes	Yes	Yes	Yes
Data Bus	32-bit	16-bit	32-bit	32-bit	32-bit	64-bit
Address Bus	32-bit	24-bit	32-bit	32-bit	32-bit	32-bit
Directly Addressable Memory	4 gigabytes	16 megabytes	4 gigabytes	4 gigabytes	4 gigabytes	4 gigabytes
Pins	132	100	168	168	168	273
Virtual Memory	Yes	Yes	Yes	Yes	Yes	Yes
On-chip Memory Management and Protection	Yes	Yes	Yes	Yes	Yes	Yes
Addressing Modes	11	11	11	11	11	11
Floating-point	387DX	387SX	*	487SX	*	*

* Indicates on-chip floating point.

Note that the 80386SL is also a 32-bit microprocessor with a 16-but data bus like the 80386SX (not listed in the table above). The 80386SL can run at a speed of up to 25 MHz and has a direct addressing capability of 32 megabytes. The 80386SL provides virtual memory support along with on-chip memory management and protection. It can be interfaced to the 80387SX to provide floating-point support. The 80386SL includes an on-chip disk controller hardware.

Table 1.2b   Motorola 16-bit Microprocessors

	MC68000	MC68010	CPU32
Maximum Clock Speed	25 Mhz	25 Mhz	16.67 Mhz
Pins	64, 68	64,68	64, 68
Data Bus	16-bit	16-bit	16-bit
Virtual Memory	No	Yes	Yes
Directly Addressable Memory	16 Megabytes	16 Megabytes	16 Megabytes
Control Registers	None	3	3
Stack Pointers	2	2	2
Cache	No	No	No
Addressing Modes	14	14	16
Coprocessor Interface	No on-chip hardware	No on-chip hardware	No on-chip hardware

Table 1.2b   Motorola 32-bit Microprocessors

	MC68020	MC68030	MC68040
Maximum Clock Speed	33 Mhz (8 Mhz min.)	33 Mhz (8 Mhz min.)	33 Mhz (8 Mhz min.)
Pins	114	118	179
Address Bus	32-bit	32-bit	32-bit
Addressing Modes	18	18	18
Maximum Addressable Memory	4 Gigabytes	4 Gigabytes	4 Gigabytes
Memory Management	By interfacing the 68851 MMU chip	On-chip MMU	On-chip MMU
Cache (on-chip)	Instruction cache	Instruction and data cache	Instruction and data cache
Floating Point	By interfacing 68881/68882 floating-point coprocessor chip	By interfacing 68881/68882 floating-point coprocessor chip	On-chip floating-point hardware

systems can be very high if the performance requirements of the application need high-performance VAX-type computers. Since the performance levels of 32-bit microprocessors are comparable to the VAX-type computer, multiple 32-bit microprocessors in a redundant configuration outperform the VAXs. Thus, the 32-bit microprocessors provide efficient fault-tolerant systems.

1.9.3 Real-Time Controllers

Real-time controllers such as flight-control systems for aircraft, flight simulators, and automobile engine control require high-performance computers. Some of these applications were handled in the past by using mainframe computers which resulted in high cost and the controllers occupied large spaces.

The state-of-the-art flight simulators use multiple 32-bit microprocessors to perform graphic manipulation, data gathering, and high-speed communications. Obviously, an application such as real-time automobile engine control using mainframe computers is not practical since these systems are not small enough to fit under a car hood. These controllers are currently being designed using the small-sized 32-bit microprocessors to perform high-speed data manipulation and calculation.

1.9.4 Robotics

The processing requirements of complex robots attempting to emulate human activities exceed the capabilities of 8- and 16-bit microprocessors. With 32-bit microprocessors, it is now feasible to design these controllers at low cost. In many cases, the microprocessor is used as the brain of the robot. In a typical application, the microprocessor will input the actual arm angle measurement from a sensor, compare it with the desired arm angle, and will then send outputs to a motor to position the arm.

Mitsubishi manufactured the first 68020-based system robot control system.

1.9.5 Embedded Control

Embedded Control microprocessors, also called embedded controllers, are designed to be programmed to manage specific tasks. Once programmed, the embedded controllers can manage the functions of a wide variety of electronic products. Since the microprocessors are embedded in the host system, their presence and operation are basically hidden from the host system. Typical embedded control applications include office automation products such as copiers, laser printers, fax machines, and consumer electronics like VCRs, microwave ovens, and automative systems.

There are two types of embedded control applications, event control (real-time) and data control. Even control applications require distributed embedded controllers dedicated to single functions.

These applications include motor engine or instrument control system. In these applications, the controllers offer real-time response to support programs with small amounts of data. Typically, 8- and 16-bit single chip microcontrollers such as the Intel 8751/8096 and Motorola HC11/HC16 are used.

The 80186/80188 based microcontrollers are used for data control applications.

RISC microprocessor-based embedded controllers perform many different functions while handling large programs using large amounts of data. Applications such as laser printers require a high performance microprocessor with on-chip floating-point hardware. The RISC microprocessors are ideal for these types of applications.

QUESTIONS AND PROBLEMS

1.1   What is the basic difference between the microprocessor and the microcomputer?

1.2   What is meant by an 8-bit microprocessor and a 16-bit microprocessor?

1.3   Interpret FE_{16} as unsigned, signed, and floating-point numbers.

1.4   Determine the carry, sign, overflow, and zero flags for the following operation:

ADD 82A1H, 231FH

1.5   What are the basic difference between EPROM, EAROM, and flash ROM?

1.6   What is the difference between static and dynamic RAM?

1.7   Assume the following ROM chip and the microprocessor:

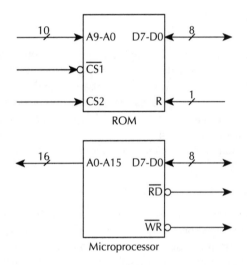

Connect the microprocessor to the ROM to obtain the following memory map: 0000_{16} through $03FF_{16}$. Use only the signals shown. Draw a neat logic diagram and analyze the memory map.

1.8   Assume the following microprocessor and the RAM chip:

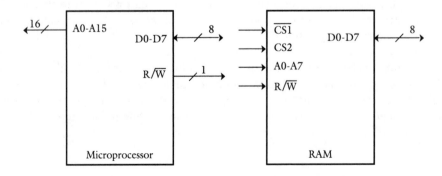

Draw a neat logic diagram showing connections between the above microprocessor and the RAM chip using only the signals shown to include the memory map 0200_{16} through $02FF_{16}$. Use linear decoding.

1.9   Use the following chips to design a microcomputer memory to include the map 3000_{16} through $33FF_{16}$.

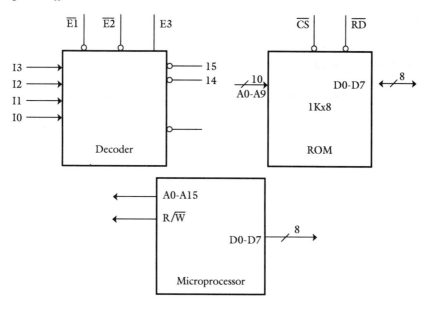

Draw a neat logic diagram.

1.10   What is meant by foldback in linear decoding?

1.11   Define the three basic types of I/O.

1.12   What is the difference between:

i)   Standard and memory-mapped programmed I/O
ii)   Maskable and nonmaskable interrupts
iii)   Internal and external interrupts
iv)   Block transfer and interleaved DMA
v)   Polled and daisy chain
vi)   One-pass and two-pass assemblers

1.13   Comment on the importance of the following architectural features in an operating system implementation:

a)   Address translation
b)   Protection
c)   Program relocation

1.14   Explain clearly the differences between segmentation and paging. Can you think of a situation where it would be advantageous to define a virtual memory that is smaller than available physical memory?

1.15   What is the purpose of cache memory? Discuss briefly the various types of cache.

1.16   Discuss the main features of typical coprocessors.

1.17   What is the difference between software breakpoint and hardware breakpoint?

1.18   Discuss the basic features of microcomputer development systems.

1.19   Compare the typical features of the 16- and 32-bit microprocessors by Intel and Motorola.

1.20   What types of applications are the RISC microprocessors used for?

1.21   Discuss floating-point data formats supported by both Intel 80387 and Motorola 68881/6882.

2

INTEL 8085

This chapter describes hardware, software, and interfacing aspects of the Intel 8085. Topics include 8085 register architecture, addressing modes, instruction set, input/output, and system design.

2.1 Introduction

The Intel 8085 is an 8-bit microprocessor. The 8085 is designed using NMOS in 40-in DIP (Dual In-line Package). The 8085 can be operated from either 3.03 MHz maximum (8085A) or 5 MHz maximum (8085A-2) internal clock frequency.

The 8085 has three enhanced versions, namely, the 8085AH, 8085AH-2, and 8085AH-1. These enhanced processors are designed using the HMOS (High-density MOS) technology. Each is packaged in a 40-pin DIP like the 8085. These enhanced microprocessors consume 20% lower power than the 8085A. The internal clock frequencies of the 8085AH, 8085AH-2, and 8085AH-1 are 3, 5, 6 MHz, respectively. These HMOS 8-bit microprocessors are expensive compared to the NMOS 8-bit 8085A.

Figure 2.1 shows a simplified block diagram of the 8085 microprocessor. The accumulator connects to the data bus and the Arithmetic and Logic Unit (ALU). The ALU performs all data manipulation, such as incrementing a number or adding two numbers.

The temporary register feeds the ALU's other input. This register is invisible to the programmer and is controlled automatically by the microprocessor's control circuitry.

The flags are a collection of flip-flops that indicate certain characteristics of the result of the most recent operation performed by the ALU. For example, the zero flag is set if the result of an operation is zero. The zero flag is tested by the JZ instruction.

The instruction register, instruction decoder, program counter, and control and timing logic are used for fetching instructions from memory and directing their execution.

2.2 Register Architecture

The 8085 registers and status flags are shown in Figure 2.2.

The accumulator (A) is an 8-bit register. Most arithmetic and logic operations are performed using the accumulator. All I/O data transfers between the 8085 and the I/O devices are performed via the accumulator. Also, there are a number of instructions that move data between the accumulator and memory.

The B, C, D, E, H, and L are each 8 bits long. Registers H and L are the memory address register or data counter. This means that these two registers are used to store the 16-bit address

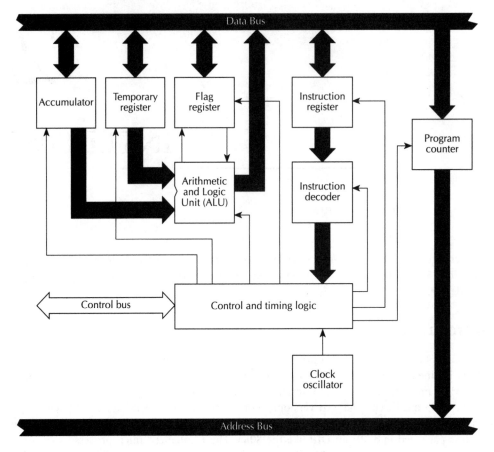

FIGURE 2.1   Simplified 8085 block diagram.

of 8-bit data being accessed from memory. This is the implied or register indirect addressing mode. There are a number of instructions, such as MOV reg, M, and MOV M, reg, which move data between any register and memory location addressed by H and L. However, using any other memory reference instruction, data transfer takes place between a memory location and the only 8085 register, the accumulator. The instruction LDAX B is a typical example.

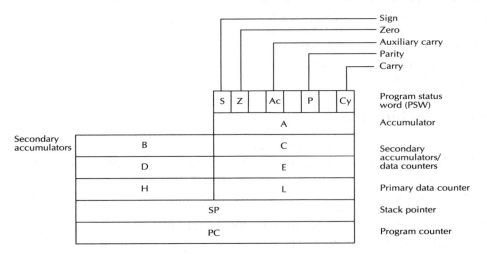

FIGURE 2.2   8085 microprocessor registers and status flags.

Registers B, C, D, and E are secondary accumulators or data counters. There are a number of instructions to move data between any two registers. There are also a few instructions that combine registers B and C or D and E, as a 16-bit data counter with high byte of a pair contained in the first register and low byte in the second. These instructions typically include LDAX B, LDAX D, STAX B, and STAX D, which transfer data between memory and the accumulator.

Each of these 8-bit registers can be incremented and decremented by a single byte instruction. There are a number of instructions which combine two of these 8-bit registers to form 16-bit register pairs as follows:

$$
\begin{array}{ccc}
A & \text{and} & PSW \\
B & \text{and} & C \\
D & \text{and} & E \\
H & \text{and} & L \\
\end{array}
$$

high-order byte   low-order byte

The 16-bit register pair obtained by combining the accumulator and the program status word (PSW) is used only for stack operations. Sixteen bit arithmetic operations use B and C, D and E, or H and L as 16-bit data registers.

The **program status word** consists of five status flags. These are described below.

The **carry flag** (Cy) reflects the final carry out of the most significant bit of any arithmetic operation. Any logic instruction resets or clears the carry flag. This flag is also used by the shift and rotate instructions. The 8085 does not have any CLEAR CARRY instruction. One way of clearing the carry will be by ORing or ANDing the accumulator with itself.

The **parity status flag** (P) is set to 1 if an arithmetic or logic instruction generates an answer with even parity, that is, containing an even number of 1 bits. This flag is 0 if the arithmetic or logic instruction generates an answer with odd parity, that is, containing an odd number of 1s.

The **auxiliary carry flag** (Ac) reflects any carry from bit 3 to bit 4 (assuming 8-bit data with bit 0 as the least significant bit and bit 7 as the most significant bit) due to an arithmetic operation. This flag is useful for BCD operations.

The **zero flag** (Z) is set to 1 whenever an arithmetic or logic operation produces a result of 0. The zero flag is cleared to zero for a nonzero result due to arithmetic or logic operation.

The **sign status flag** (S) is set to the value of the most significant bit of the result in the accumulator after an arithmetic or logic operation. This provides a range of -128_{10} to $+127_{10}$ (with 0 being considered positive) as the 8085's data-handling capacity.

The 8085 does not have an overflow flag. Note that execution of arithmetic or logic instructions in the 8085 affects the flags. All conditional instructions in the 8085 instruction set use one of the status flags as the required condition.

The **stack pointer** (SP) is 16 bits long. All stack operations with the 8085 use 16-bit register pairs. The stack pointer contains the address of the last data byte written into the stack. It is decremented by 2 each time 2 bytes of data are written or pushed onto the stack and is incremented by 2 each time 2 bytes of data are read from or pulled (popped) off the stack, that is, the top of the stack has the lowest address in the stack that grows downward.

The **program counter** (PC) is 16 bits long to address up to 64K of memory. It usually addresses the next instruction to be executed.

2.3 Memory Addressing

When addressing a memory location, the 8085 uses either register indirect or direct memory addressing. With register indirect addressing, the H and L registers perform the function of the memory address register or data counter; that is, the H, L pair holds the address of the data.

With this mode, data transfer may occur between the addressed memory location and any one of the registers A, B, C, D, E, H, or L.

Also, some instructions, such as LDAX B, LDAX D, STAX B, and STAX D, use registers B and C or D and E to hold the address of data. These instructions transfer data between the accumulator and the memory location addressed by registers B and C or D and E using the register indirect mode.

There are also a few instructions, such as the STA ppqq, which use the direct-memory addressing mode to move data between the accumulator and the memory. These instructions use 3 bytes, with the first byte as the OP code followed by 2 bytes of address.

The stack is basically a part of the RAM. Therefore, PUSH and POP instructions are memory reference instructions.

All 8085 JUMP instructions use direct or absolute addressing and are 3 bytes long. The first byte of this instruction is the OP code followed by a 2-byte address. This address specifies the memory location to which the program would branch.

2.4 8085 Addressing Modes

The 8085 has five addressing modes:

1. **Direct** — Instructions using this mode specify the effective address as a part of the instruction. These instructions contain 3 bytes, with the first byte as the OP code followed by 2 bytes of address of data (the low-order byte of the address in byte 2, the high-order byte of the address in byte 3). Consider LDA 2035H. This instruction loads accumulator with the contents of memory location 2035_{16}. This mode is also called the absolute mode.
2. **Register** — This mode specifies the register or register pair that contains data. For example, MOV B, C moves the contents of register C to register B.
3. **Register Indirect** — This mode contains a register pair which stores the address of data (the high-order byte of the address in the first register of the pair, and the low-order byte in the second). As an example, LDAX B loads the accumulator with the contents of a memory location addressed by B, C register pair.
4. **Implied or Inherent** — The instructions using this mode have no operands. Examples include STC (Set the Carry Flag).
5. **Immediate** — For an 8-bit datum, this mode uses 2 bytes, with the first byte as the OP code, followed by 1 byte of data. On the other hand, for 16-bit data, this instruction contains 3 bytes, with the first byte as the OP code followed by 2 bytes of data. For example, MVI B, 05 loads register B with the value 5, and LXI H, 2050H loads H with 20H and L with 50H.

A JUMP instruction interprets the address that it would branch to in the following ways:

1. **Direct** — The JUMP instructions, such as JZ ppqq, use direct addressing and contain 3 bytes. The first byte is the OP code, followed by 2 bytes of the 16-bit address where it would branch to unconditionally or based on a condition if satisfied. For example, JMP 2020 unconditionally branches to location 2020H.
2. **Implied or Inherent Addressing** — This JUMP instruction using this mode is 1 byte long. A 16-bit register pair contains the address of the next instruction to be executed. The instruction PCHL unconditionally branches to a location addressed by the H, L pair.

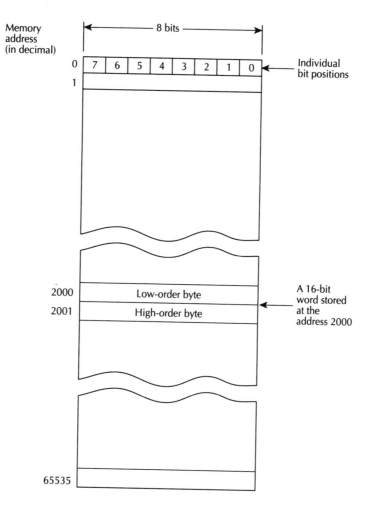

FIGURE 2.3   8085 addressing structures.

2.5 8085 Instruction Set

As mentioned before, the 8085 uses a 16-bit address. Since the 8085 is a byte-addressable machine, it follows that it can directly address 65,536 (2^{16}) distinct memory locations. The addressing structure of the 8085 processor is shown in Figure 2.3.

From this figure, we notice that two consecutive memory locations may be used to represent a 16-bit data item. However, according to the Intel convention, the high-order byte of a 16-bit quantity is always assigned to the high memory address.

The 8085 instructions are 1 to 3 bytes long and these formats are shown in Figure 2.4. The 8085 instruction set contains 74 basic instructions and supports conventional addressing modes such as immediate, register, absolute, and register indirect addressing modes.

Table 2.1 lists the 8085 instructions in alphabetical order; the object codes and instruction cycles are also included. When two instruction cycles are shown, the first is for "condition not met", while the second is for "condition met". Table 2.2 provides the 8085 instructions affecting the status flags. Note that not all 8085 instructions affect the status flags. The 8085 arithmetic and logic instructions normally affect the status flags.

In describing the 8085 instruction set, we will use the symbols in Table 2.3.

The 8085 move instruction transfers 8-bit data from one register to another, register to memory, and vice versa. A complete summary of these instructions is presented in Table 2.4.

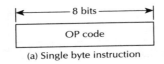

(a) Single byte instruction

(b) Two byte instructions

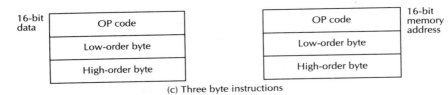

(c) Three byte instructions

FIGURE 2.4   8085 instructions format.

Table 2.1   Summary of 8085 Instruction Set

Instruction	OP Code	Bytes	Cycles	Operations performed
ACI DATA	CE	2	7	[A] ← [A] + second instruction byte + [Cy]
ADC A	8F	1	4	[A] ← [A] + [A] + [Cy]
ADC B	88	1	4	[A] ← [A] + [B] + [Cy]
ADC C	89	1	4	[A] ← [A] + [C] + [Cy]
ADC D	8A	1	4	[A] ← [A] + [D] + [Cy]
ADC E	8B	1	4	[A] ← [A] + [E] + [Cy]
ADC H	8C	1	4	[A] ← [A] + [H] + [Cy]
ADC L	8D	1	4	[A] ← [A] + [L] + [Cy]
ADC M	8E	1	7	[A] ← [A] + [[H L]] + [Cy]
ADD A	87	1	4	[A] ← [A] + [A]
ADD B	80	1	4	[A] ← [A] + [B]
ADD C	81	1	4	[A] ← [A] + [C]
ADD D	82	1	4	[A] ← [A] + [D]
ADD E	83	1	4	[A] ← [A] + [E]
ADD H	84	1	4	[A] ← [A] + [H]
ADD L	85	1	4	[A] ← [A] + [L]
ADD M	86	1	7	[A] ← [A] + [[H L]]
ADI DATA	C6	2	7	[A] ← [A] + second instruction byte
ANA A	A7	1	4	[A] ← [A] ∧ [A]
ANA B	A0	1	4	[A] ← [A] ∧ [B]
ANA C	A1	1	4	[A] ← [A] ∧ [C]
ANA D	A2	1	4	[A] ← [A] ∧ [D]
ANA E	A3	1	4	[A] ← [A] ∧ [E]
ANA H	A4	1	4	[A] ← [A] ∧ [H]
ANA L	A5	1	4	[A] ← [A] ∧ [L]
ANA M	A6	1	4	[A] ← [A] ∧ [[H L]]
ANI DATA	E6	2	7	[A] ← [A] ∧ second instruction byte
CALL ppqq	CD	3	18	Call A subroutine addressed by ppqq
CC ppqq	DC	3	9/18	Call a subroutine addressed by ppqq if Cy = 1
CM ppqq	FC	3	9/18	Call a subroutine addressed by ppqq if S = 1
CMA	2F	1	4	[A] ← 1's complement of [A]
CMC	3F	1	4	[Cy] ← 1's complement of [Cy]
CMP A	BF	1	4	[A] − [A] and affects flags
CMP B	B8	1	4	[A] − [B] and affects flags

Table 2.1 Summary of 8085 Instruction Set (*continued*)

Instruction	OP Code	Bytes	Cycles	Operations performed
CMP C	B9	1	4	[A] − [C] and affects flags
CMP D	BA	1	4	[A] − [D] and affects flags
CMP E	BB	1	4	[A] − [E] and affects flags
CMP H	BC	1	4	[A] − [H] and affects flags
CMP L	BD	1	4	[A] − [L] and affects flags
CMP M	BE	1	7	[A] − [[H L]] and affects flags
CNC ppqq	D4	3	9/18	Call a subroutine addressed by ppqq if Cy = 0
CNZ ppqq	C4	3	9/18	Call a subroutine addressed by ppqq if Z = 0
CP ppqq	F4	3	9/18	Call a subroutine addressed by ppqq if S = 0
CPE ppqq	EC	3	9/18	Call a subroutine addressed by ppqq if P = 1
CPI DATA	FE	2	7	[A] − second instruction byte and affects flags
CPO ppqq	E4	3	9/18	Call a subroutine addressed by ppqq if P = 0
CZ ppqq	CC	3	9/18	Call a subroutine addressed by ppqq if Z = 1
DAA	27	1	4	Decimal adjust accumulator
DAD B	09	1	10	[HL] ← [HL] + [BC]
DAD D	19	1	10	[HL] ← [HL] + [DE]
DAD H	29	1	10	[HL] ← [HL] + [HL]
DAD SP	39	1	10	[HL] ← [HL] + [SP]
DCR A	3D	1	4	[A] ← [A] − 1
DCR B	05	1	4	[B] ← [B] − 1
DCR C	0D	1	4	[C] ← [C] − 1
DCR D	15	1	4	[D] ← [D] − 1
DCR E	1D	1	4	[E] ← [E] − 1
DCR H	25	1	4	[H] ← [H] − 1
DCR L	2D	1	4	[L] ← [L] − 1
DCR M	35	1	4	[[HL]] ← [[HL]] − 1
DCX B	0B	1	6	[BC] ← [BC] − 1
DCX D	1B	1	6	[DE] ← [DE] − 1
DCX H	2B	1	6	[HL] ← [HL] − 1
DCX SP	3B	1	6	[SP] ← [SP] − 1
DI	F3	1	4	Disable interrupts
EI	FB	1	4	Enable interrupts
HLT	76	1	5	Halt
IN PORT	DB	2	10	[A] ← [specified port]
INR A	3C	1	4	[A] ← [A] + 1
INR B	04	1	4	[B] ← [B] + 1
INR C	0C	1	4	[C] ← [C] + 1
INR D	14	1	4	[D] ← [D] + 1
INR E	1C	1	4	[E] ← [E] + 1
INR H	24	1	4	[H] ← [H] + 1
INR L	2C	1	4	[L] ← [L] + 1
INR M	34	1	4	[[HL]] ← [[HL]] +1
INX B	03	1	6	[BC] ← [BC] +1
INX D	13	1	6	[DE] ← [DE] + 1
INX H	23	1	6	[HL] ← [HL] + 1
INX SP	33	1	6	[SP] ← [SP] + 1
JC ppqq	DA	3	7/10	Jump to ppqq if Cy = 1
JM ppqq	FA	3	7/10	Jump to ppqq if S = 1
JMP ppqq	C3	3	10	Jump to ppqq
JNC ppqq	D2	3	7/10	Jump to ppqq if Cy = 0
JNZ ppqq	C2	3	7/10	Jump to ppqq if Z = 0
JP ppqq	F2	3	7/10	Jump to ppqq if S = 0
JPE ppqq	EA	3	7/10	Jump to ppqq if P = 1
JPO ppqq	E2	3	7/10	Jump to ppqq if P = 0
JZ ppqq	CA	3	7/10	Jump to ppqq if Z = 1
LDA ppqq	3A	3	13	[A] ← [ppqq]
LDAX B	0A	1	7	[A] ← [[BC]]
LDAX D	1A	1	7	[A] ← [[DE]]
LHLD ppqq	2A	3	16	[L] ← [ppqq],[H] ← [ppqq + 1]

Table 2.1   Summary of 8085 Instruction Set (*continued*)

Instruction	OP Code	Bytes	Cycles	Operations performed
LXI B	01	3	10	[BC] ← second and third instruction bytes
LXI D	11	3	10	[DE] ← second and third instruction bytes
LXI H	21	3	10	[HL] ← second and third instruction bytes
LXI SP	31	3	10	[SP] ← second and third instruction bytes
MOV A,A	7F	1	4	[A] ← [A]
MOV A,B	78	1	4	[A] ← [B]
MOV A,C	79	1	4	[A] ← [C]
MOV A,D	7A	1	4	[A] ← [D]
MOV A,E	7B	1	4	[A] ← [E]
MOV A,H	7C	1	4	[A] ← [H]
MOV A,L	7D	1	4	[A] ← [L]
MOV A,M	7E	1	7	[A] ← [[HL]]
MOV B,A	47	1	4	[B] ← [A]
MOV B,B	40	1	4	[B] ← [B]
MOV B,C	41	1	4	[B] ← [C]
MOV B,D	42	1	4	[B] ← [D]
MOV B,E	43	1	4	[B] ← [E]
MOV B,H	44	1	4	[B] ← [H]
MOV B,L	45	1	4	[B] ← [L]
MOV B,M	46	1	7	[B] ← [[HL]]
MOV C,A	4F	1	4	[C] ← [A]
MOV C,B	48	1	4	[C] ← [B]
MOV C,C	49	1	4	[C] ← [C]
MOV C,D	4A	1	4	[C] ← [D]
MOV C,E	4B	1	4	[C] ← [E]
MOV C,H	4C	1	4	[C] ← [H]
MOV C,L	4D	1	4	[C] ← [L]
MOV C,M	4E	1	7	[C] ← [[HL]]
MOV D,A	57	1	4	[D] ← [A]
MOV D,B	50	1	4	[D] ← [B]
MOV D,C	51	1	4	[D] ← [C]
MOV D,D	52	1	4	[D] ← [D]
MOV D,E	53	1	4	[D] ← [E]
MOV D,H	54	1	4	[D] ← [H]
MOV D,L	55	1	4	[D] ← [L]
MOV D,M	56	1	7	[D] ← [[HL]]
MOV E,A	5F	1	4	[E] ← [A]
MOV E,B	58	1	5	[E] ← [B]
MOV E,C	59	1	4	[E] ← [C]
MOV E,D	5A	1	4	[E] ← [D]
MOV E,E	5B	1	4	[E] ← [E]
MOV E,H	5C	1	4	[E] ← [H]
MOV E,L	5D	1	4	[E] ← [L]
MOV E,M	5E	1	7	[E] ← [[HL]]
MOV H,A	67	1	4	[H] ← [B]
MOV H,B	60	1	4	[H] ← [A]
MOV H,C	61	1	4	[H] ← [C]
MOV H,D	62	1	4	[H] ← [D]
MOV H,E	63	1	4	[H] ← [E]
MOV H,H	64	1	4	[H] ← [H]
MOV H,L	65	1	4	[H] ← [L]
MOV H,M	66	1	7	[H] ← [[HL]]
MOV L,A	6F	1	4	[L] ← [A]
MOV L,B	68	1	4	[L] ← [B]
MOV L,C	69	1	4	[L] ← [C]
MOV L,D	6A	1	4	[L] ← [D]
MOV L,E	6B	1	4	[L] ← [E]
MOV L,H	6C	1	4	[L] ← [H]
MOV L,L	6D	1	4	[L] ← [L]

Table 2.1 Summary of 8085 Instruction Set (*continued*)

Instruction	OP Code	Bytes	Cycles	Operations performed
MOV L,M	6E	1	7	[L] ← [[HL]]
MOV M,A	77	1	7	[[HL]] ← [A]
MOV M,B	70	1	7	[[HL]] ← [B]
MOV M,C	71	1	7	[[HL]] ← [C]
MOV M,D	72	1	7	[[HL]] ← [D]
MOV M,E	73	1	7	[[HL]] ← [E]
MOV M,H	74	1	7	[[HL]] ← [H]
MOV M,L	75	1	7	[[HL]] ← [L]
MVI A, DATA	3E	2	7	[A] ← second instruction byte
MVI B, DATA	06	2	7	[B] ← second instruction byte
MVI C, DATA	0E	2	7	[C] ← second instruction byte
MVI D, DATA	16	2	7	[D] ← second instruction byte
MVI E, DATA	1E	2	7	[E] ← second instruction byte
MVI H, DATA	26	2	7	[H] ← second instruction byte
MVI L, DATA	2E	2	7	[L] ← second instruction byte
MVI M, DATA	36	2	10	[[HL]] ← second instruction byte
NOP	00	1	4	No operation
ORA A	B7	1	4	[A] ← [A] ∨ [A]
ORA B	B0	1	4	[A] ← [A] ∨ [B]
ORA C	B1	1	4	[A] ← [A] ∨ [C]
ORA D	B2	1	4	[A] ← [A] ∨ [D]
ORA E	B3	1	4	[A] ← [A] ∨ [E]
ORA H	B4	1	4	[A] ← [A] ∨ [H]
ORA L	B5	1	4	[A] ← [A] ∨ [L]
ORA M	B6	1	7	[A] ← [A] ∨ [[HL]]
ORI DATA	F6	2	7	[A] ← [A] ∨ second instruction byte
OUT PORT	D3	2	10	[specified port] ← [A]
PCHL	E9	1	6	[PCH][a] ← [H], [PCL][a] ← [L]
POP B	C1	1	10	[C] ← [[SP]], [SP] ← [SP] + 2 [B] ← [[SP] + 1]
POP D	D1	1	10	[E] ← [[SP]], [SP] ← [SP] + 2 [D] ← [[SP] + 1]
POP H	E1	1	10	[L] ← [[SP]], [SP] ← [SP] + 2 [H] ← [[SP] + 1]
POP PSW	F1	1	10	[A] ← [[SP] + 1], [PSW] ← [[SP]], [SP] ← [SP] + 2 [[SP] − 1] ← [B], [SP] ← [SP] − 2
PUSH B	C5	1	12	[[SP] − 2] ← [C]
PUSH D	D5	1	12	[[SP] − 1] ← [D], [[SP] − 2] ← [E] [SP] ← [SP] − 2
PUSH H	E5	1	12	[[SP] − 1] ← [H], [SP] ← [SP] − 2 [[SP] − 2] ← [L]
PUSH PSW	F5	1	12	[[SP] − 1] ← [A], [SP] ← [SP] − 2 [[SP] − 2] ← [PSW]
RAL	17	1	4	(rotate accumulator left through carry)
RAR	1F	1	4	(rotate accumulator right through carry)
RC	D8	1	6/12	Return if carry; [PC] ← [SP]]
RET	C9	1	10	[PCL][a] ← [[SP]], [SP] ← [SP] + 2 [PCH][a] ← [[SP] + 1]
RIM	20	1	4	Read interrupt mask
RLC	07	1	4	(rotate accumulator left)
RM	F8	1	6/12	Return if minus; [PC] ← [[SP]]

Table 2.1 Summary of 8085 Instruction Set (*continued*)

Instruction	OP Code	Bytes	Cycles	Operations performed
RNC	D0	1	6/12	Return if no carry; [PC] ← [[SP]]
RNZ	C0	1	6/12	Return if result not zero; [PC] ← [[SP]]
RP	F0	1	6/12	Return if positive; [PC] ← [[SP]], [SP] ← [SP] + 2
RPE	E8	1	6/12	Return if parity even; [PC] ← [[SP]], [SP] ← [SP] + 2
RPO	E0	1	6/12	Return if parity odd; [PC] ← [[SP]], [SP] ← [SP] + 2
RRC	0F	1	4	A → [ ][ ][ ][ ][ ][ ][ ][ ] → Cy
RST0	C7	1	12	Restart
RST1	CF	1	12	Restart
RST2	D7	1	12	Restart
RST3	DF	1	12	Restart
RST4	E7	1	12	Restart
RST5	EF	1	12	Restart
RST6	F7	1	12	Restart
RST7	FF	1	12	Restart
RZ	C8	1	6/12	Return if zero; [PC] ← [[SP]]
SBB A	9F	1	4	[A] ← [A] − [A] − [Cy]
SBB B	98	1	4	[A] ← [A] − [B] − [Cy]
SBB C	99	1	4	[A] ← [A] − [C] − [Cy]
SBB D	9A	1	4	[A] ← [A] − [D] − [Cy]
SBB E	9B	1	4	[A] ← [A] − [E] − [Cy]
SBB H	9C	1	4	[A] ← [A] − [H] − [Cy]
SBB L	9D	1	4	[A] ← [A] − [L] − [Cy]
SBB M	9E	1	7	[A] ← [A] − [[HL]] − [Cy]
SBI DATA	DE	2	7	[A] ← [A] − second instruction byte − [Cy]
SHLD ppqq	22	3	16	[ppqq] ← [L], [ppqq + 1] ← [H]
SIM	30	1	4	Set interrupt mask
SPHL	F9	1	6	[SP] ← [HL]
STA ppqq	32	3	13	[ppqq] ← [A]
STAX B	02	1	7	[[BC]] ← [A]
STAX D	12	1	7	[[DE]] ← [A]
STC	37	1	4	[Cy] ← 1
SUB A	97	1	4	[A] ← [A] − [A]
SUB B	90	1	4	[A] ← [A] − [B]
SUB C	91	1	4	[A] ← [A] − [C]
SUB D	92	1	4	[A] ← [A] − [D]
SUB E	93	1	4	[A] ← [A] − [E]
SUB H	94	1	4	[A] ← [A] − [H]
SUB L	95	1	4	[A] ← [A] − [L]
SUB M	96	1	7	[A] ← [A] − [[HL]]
SUI DATA	D6	2	7	[A] ← [A] − second instruction byte
XCHG	EB	1	4	[D] ↔ [H], [E] ↔ [L]
XRA A	AF	1	4	[A] ← [A] ⊕ [A]
XRA B	A8	1	4	[A] ← [A] ⊕ [B]
XRA C	A9	1	4	[A] ← [A] ⊕ [C]
XRA D	AA	1	4	[A] ← [A] ⊕ [D]
XRA E	AB	1	4	[A] ← [A] ⊕ [E]
XRA H	AC	1	4	[A] ← [A] ⊕ [H]
XRA L	AD	1	4	[A] ← [A] ⊕ [L]
XRA M	AE	1	7	[A] ← [A] ⊕ [[HL]]
XRI DATA	EE	2	7	[A] ← [A] ⊕ second instruction byte
XTHL	E3	1	16	[[SP]] ↔ [L], [[SP] + 1]] ↔ [H]

[a]PCL — program counter low byte; PCH — program counter high byte.

All mnemonics copyright Intel Corporation 1976.

Table 2.2   8085 Instructions Affecting the Status Flags

Instructions[a]	Status flags[b]				
	Cy	Ac	Z	S	P
ACI DATA	+	+	+	+	+
ADC reg	+	+	+	+	+
ADC M	+	+	+	+	+
ADD reg	+	+	+	+	+
ADD M	+	+	+	+	+
ADI DATA	+	+	+	+	+
ANA reg	0	1	+	+	+
ANA M	0	1	+	+	+
ANI DATA	0	1	+	+	+
CMC	+				
CMP reg	+	+	+	+	+
CMP M	+	+	+	+	+
CPI DATA	+	+	+	+	+
DAA	+	+	+	+	+
DAD rp	+				
DCR reg		+	+	+	+
DCR M		+	+	+	+
INR reg		+	+	+	+
INR M		+	+	+	+
ORA reg	0	0	+	+	+
ORA M	0	0	+	+	+
ORI DATA	0	0	+	+	+
RAL	+				
RAR	+				
RLC	+				
RRC	+				
SBB reg	+	+	+	+	+
SBB M	+	+	+	+	+
SBI DATA	+	+	+	+	+
STC	+				
SUB reg	+	+	+	+	+
SUB M	+	+	+	+	+
SUI DATA	+	+	+	+	+
XRA reg	0	0	+	+	+
XRA M	0	0	+	+	+
XRI DATA	0	0	+	+	+

[a]reg — 8-bit register; M — memory; rp — 16-bit register pair.
[b]Note that instructions which are not shown in the table do not affect the flags; + indicates that the particular flag is affected; 0 or 1 indicates that these flags are always 0 or 1 after the corresponding instructions are executed.

All mnemonics copyright Intel Corporation 1976.

Table 2.3   Symbols to be Used in 8085 Instruction Set

Symbol	Interpretation
r1, r2	8-bit register
rp	Register pair
data8	8-bit data
data16	16-bit data
M	Memory location indirectly addressed through the register pair H,L
addr16	16-bit memory address

Table 2.4 8085 MOVE Instructions

| Instruction | Symbolic description | Addressing mode | | Example | Illustration |
		Source	Destination		Comments
MOV r1, r2	(r1) ← (r2)	Register	Register	MOV A, B	Copy the contents of the register B into the register A
MOV r, M	(r) ← M((HL))	Register indirect	Register	MOV A, M	Copy the contents of the memory location whose address is specified in the register pair H,L into the A register
MVI r, data8	(r) ← data8	Immediate	Register	MVI A, 08	Initialize the A register with the value 08
MOV M, r	M((HL)) ← (r)	Register	Register indirect	MOV M, B	Copy the contents of the B register into the memory location addressed by H,L pair
MVI M, data8	M((HL)) ← data8	Immediate	Register indirect	MVI M, 07	Initialize the memory location whose address is specified in the register pair H,L with the value 07

Table 2.5 8085 Load and Store Instructions

Instruction	Symbolic description	Restriction	Addressing mode	Example	Illustration Comments
LDA addr16	$(A) \leftarrow M(addr16)$	The destination is always the accumulator register	Absolute	LDA 2000H	Load the accumulator with the contents of the memory location whose address is 2000H
LHLD addr16	$(L) \leftarrow M(addr16)$ $(H) \leftarrow M(addr16+1)$	The destination is always the register pair H,L	Absolute	LHLD 2000H	Load the H and L registers with contents of the memory locations 2001H and 2000H, respectively
LXI rp, data16	$(rp) \leftarrow data16$	rp may be HL, DE, BC, or SP	Immediate	LXI H, 2024H	$H \leftarrow 20_{16}\ L \leftarrow 24_{16}$
LDAX rp	$(A) \leftarrow M((rp))$	Destination is always the accumulator; also rp may be either B,C or D,E	Register indirect	LDAX B	Load the accumulator with the contents of the memory location whose address is specified with the register pair B,C
STA addr16	$M(addr16) \leftarrow (A)$	Source is always the accumulator register	Absolute	STA 2001H	Save the contents of the accumulator into the memory location whose address is 2001H
SHLD addr16	$M(addr16) \leftarrow (L)$ $M(addr16+1) \leftarrow (H)$	The source is always the register pair H,L	Absolute	SHLD 2000	$M(2000) \leftarrow (L)$ $M(2001) \leftarrow (H)$
STAX rp	$M((rp)) \leftarrow (A)$	The source is always the accumulator and the register pair may be B,C or D,E	Register indirect	STAX D	Save the contents of the accumulator register into the memory location whose address is specified with the register pair D,E
XCHG	$[D] \leftrightarrow [H], [E] \leftrightarrow [L]$		Inherent	XCHG	Exchanges [DE] with [HL]

Table 2.6   8085 Arithmetic Instructions

Operation	Instruction	Interpretation	Addressing mode	Illustration Example	Comments
8-bit addition	ADD r	(A) ← (A) + (r)	Register	ADD B	(A) ← (A) + (B)
	ADI data8	(A) ← (A) + data8	Immediate	ADI 05	(A) ← (A) + 05
	ADD M	(A) ← (A) + M((HL))	Register indirect	—	—
8-bit addition with	ADC r	(A) ← (A) + (r) + Cy	Register	ADC C	(A) ← (A) + (C) + Cy
a carry	ACI data8	(A) ← (A) + data8 + Cy	Immediate	ACI 07	(A) ← (A) + 07 + Cy
	ADC M	(A) ← (A) + M((HL)) + Cy	Register indirect	—	—
8-bit subtraction	SUB r	(A) ← (A) – (r)	Register	SUB C	(A) ← (A) – (C)
	SUI data8	(A) ← (A) – data8	Immediate	SUI 03	(A) ← (A) – 03
	SUB M	(A) ← (A) – M((HL))	Register indirect	—	—
8-bit subtraction	SBB r	(A) ← (A) – (r) – Cy	Register	SBB D	(A) ← (A) – (D) – Cy
with a borrow	SBI data8	(A) ← (A) – data8 – Cy	Immediate	SBI 04	(A) ← (A) – 04 – Cy
	SBB M	(A) ← (A) – M((HL)) – Cy	Register indirect	—	—
16-bit addition	DAD rp	(HL) ← (HL) + (rp)	Register	DAD B	(HL) ← (HL) + (BC)
Decimal adjust	DAA	Convert the 8-bit number stored in the accumulator into BCD	Inherent	—	—
8-bit increment	INR r	(r) ← (r) + 1	Register	INR B	(B) ← (B) + 1
	INR M	M((HL)) ← M((HL)) + 1	Register indirect	—	—
16-bit increment[a]	INX rp	(rp) ← (rp) + 1	Register	INX D	(DE) ← (DE) + 1
8-bit decrement	DCR r	(r) ← (r) – 1	Register	DCR B	(B) ← (B) – 1
	DCR M	M((HL)) ← M((HL)) – 1	Register indirect	—	—
16-bit decrement[a]	DCX rp	(rp) ← (rp) – 1	Register	—	

[a]rp = BC, DE, HL, or SP.

The 8085 instruction set also accomplishes the 8- and 16-bit data transfers using the load and store instructions. These instructions are summarized in Table 2.5.

From Table 2.5 notice that we adopt the following convention when we specify a register pair in the instruction.

Symbol	Register pairs used
B	B,C
D	D,E
H	H,L

Also, observe that in Table 2.5 the 8085 processor does not provide LDAX H instruction. This is because the same result can be obtained by using the MOV A,M instruction. The 8085 includes a one-byte exchange instruction, namely, XCHG. The XCHG exchanges the contents of DE with HL. That is, it performs the following operation:

$$[D] \leftrightarrow [H]$$
$$[E] \leftrightarrow [L]$$

The arithmetic instructions provided by the 8085 processor allow one to add (or subtract) two 8-bit data with or without carry (or borrow). The subtraction operation is realized by adding the two's complement of the subtrahend to the minuend. During the subtraction operation, the carry flag will be treated as the borrow flag. Table 2.6 lists these instructions. For some instructions such as ADD M, examples and comments are not included. This is due to limited space in the table.

As far as logical operations are concerned, the 8085 includes some instructions to perform traditional Boolean operations such as AND, OR, EXCLUSIVE-OR. In addition, instructions are available to complement the accumulator and to set the carry flag. The 8085 COMPARE instructions subtract the specified destination from the contents of the

Table 2.7 8085 Logical Instructions

Operation	Instruction	Interpretation	Addressing mode	Illustration Example	Comments
Boolean AND	ANA r	$(A) \leftarrow (A) \wedge (r)$	Register	ANA B	$(A) \leftarrow (A) \wedge (B)$
	ANI data8	$(A) \leftarrow (A) \wedge$ data8	Immediate	ANI 0FH	$(A) \leftarrow (A) \wedge 00001111_2$
	ANA M	$(A) \leftarrow (A) \wedge M((HL))$	Register indirect	—	—
Boolean OR	ORA r	$(A) \leftarrow (A) \vee (r)$	Register	ORA C	$(A) \leftarrow (A) \vee (C)$
	ORI data8	$(A) \leftarrow (A) \vee$ data8	Immediate	ORI 08H	$(A) \leftarrow (A) \vee 00001000_2$
	ORA M	$(A) \leftarrow (A) \vee M((HL))$	Register indirect	—	—
Boolean EXCLUSIVE-OR	XRA r	$(A) \leftarrow (A) \oplus (r)$	Register	XRA A	$(A) \leftarrow (A) \oplus (A)$
	XRI data8	$(A) \leftarrow (A) \oplus$ data8	Immediate	XRI 03H	$(A) \leftarrow (A) \oplus 00000011_2$
	XRA M	$(A) \leftarrow (A) \oplus M((HL))$	Register indirect	—	—
Compare	CMP r	$(A) - (r)$ and affect flags	Register	CMP D	Compare (A) register with (D) register
	CPI data8	$(A) -$ data8 and affect flags	Immediate	CPI 05	Compare (A) with 05
	CMP M	$(A) - M((HL))$ and affect flags	Register indirect	—	—
Complement	CMA	$(A) \leftarrow (\bar{A})$	Inherent	—	—
Bit manipulation	STC	$Cy \leftarrow 1$ (set carry to 1)	Inherent	—	—
	CMC	$Cy \leftarrow Cy'$ (complement carry flag)	Inherent	—	—

accumulator and affect the status flags according to the result. However, in this case the result of the subtraction is not provided in the accumulator. All 8085 logical instructions are specified in Table 2.7. For some instructions in this figure, examples and comments are not provided. This is due to limited space in the table.

The AND instruction can be used to perform a masking operation. If the bit value in a particular bit position is desired in a word, the word can be logically ANDed with appropriate masking data to accomplish this. For example, the bit value at bit 3 of the word $1011 X011_2$ can be determined as follows:

$$
\begin{array}{lll}
 & 1011 \quad x011 & \text{Word} \\
\text{AND} & \underline{0000 \quad 1000} & \text{Masking Data} \\
 & 0000 \quad x000 & \text{Result}
\end{array}
$$

If the bit value X at bit 3 is 1, then the result is nonzero $(Z = 0)$; otherwise the result is zero $(Z = 1)$. The Z-flag can be tested using JZ (jump if $Z = 1$) or JNZ (jump if $Z = 0$) to determine whether $X = 0$ or 1. The AND instruction can also be used to determine whether a binary number is odd or even by checking the least significant bit (LSB) of the number (LSB = 0 for even and LSB = 1 for odd). XRA instruction can be used to find ones compliment of a binary number by exclusive-ORing the number with all ones as follows:

$$
\begin{array}{lll}
 & 1010 \quad 1001 & \text{Original number} \\
\text{XOR} & \underline{1111 \quad 1111} & \\
 & 0101 \quad x110 & \text{Ones complement}
\end{array}
$$

One of the applications of the compare instruction is to find a match in an array. The number to be matched can be loaded in the accumulator and compared with each element in the array. The JZ (jump if $Z = 1$) can then be executed to find the match. If the subtraction instruction is used in place of the COMPARE, the number to be matched in the accumulator

TABLE 2.8   8085 Rotate Instructions

Instruction	Interpretation	Illustration
RLC	Rotate left accumulator by one position without the carry flag Cy	
RRC	Rotate right accumulator by one position without the carry flag Cy	
RAR	Rotate right accumulator by one position through the carry flag	
RAL	Rotate left accumulator by one position through the carry flag	

will be lost after each subtraction, and therefore the number needs to be loaded for the next subtraction.

The 8085 instruction set includes rotating the contents of the A register to the left or right without or through the carry flag. These instructions are listed in Table 2.8.

In the 8085, only the absolute mode branch instruction is of the form JMP addr16. There is also a one-byte implied unconditional jump instruction, namely, PCHL. The PCHL loads [H] into PC high byte and [L] into PC low byte. That is, PCHL performs an unconditional JUMP to a location addressed by the contents of H and L. The general format of an 8085 conditional branch instruction is J <condition code> addr16 where the condition code may represent one of the following conditions:

Conditional jumps	Condition	Comment
JZ	$Z = 1$	Z flag is set (result equal to zero)
JNZ	$Z = 0$	Z flag is reset (result not equal to zero)
JC	$Cy = 1$	Cy flag is set
JNC	$Cy = 0$	Cy flag is reset
JPO	$P = 0$	The parity is odd
JPE	$P = 1$	The parity is even
JP	$S = 0$	S flag is reset (or the number is positive)
JM	$S = 1$	S flag is set (or the number is negative, or minus)

For example, the following instruction sequence causes a branch to the memory address 2000_{16} only if the contents of the A and B registers are equal:

CMP B
JZ 2000H

In the 8085, the subroutine call instruction is of the form

CALL addr16

The instruction RET transfers the control to the caller, and it should be the last instruction of the subroutine. Both instructions use the PC and SP for subroutine linkage.

For example, consider the following:

	Main program		Subroutine
	—	**SUB**	—
	—		—
	—		—
	CALL SUB		—
START	—		—
	—		—
	—		**RET**
	—		

The call SUB instruction pushes or saves the current PC contents (START which is the address of the next instruction) onto the stack and loads PC with the starting address of the Subroutine (SUB) specified with the CALL instruction. The RET instruction at the end of the subroutine pops or reads the address START (saved onto the stack by the CALL instruction) into PC and transfers control to the right place in the main program.

There are a number of conditional call instructions. These include:

```
CC  addr (call if Cy = 1)
CNC addr (call if Cy = 0)
CZ  addr (call if Z = 1)
CNZ addr (call if Z = 0)
CM  addr (call if S = 1)
CP  addr (call if S = 0)
CPE addr (call if P = 1)
CPO addr (call if P = 0)
```

Also, there are a number of conditional return instructions. These include:

```
RC  (Return if Cy = 1)
RNC (Return if Cy = 0)
RZ  (Return if Z = 1)
RNZ (Return if Z = 0)
RPE (Return if P = 1)
RPO (Return if P = 0)
RM  (Return if S = 1)
RP  (Return if S = 0)
```

There are eight one-byte call instructions (RST 0 to 7) which have predefined addresses. The format for these instructions is

```
11  XXX  111  =  000 for RST0
              =  001 for RST1
              =  010 for RST2
              =  011 for RST3
              =  100 for RST4
              =  101 for RST5
              =  110 for RST6
              =  111 for RST7
```

RSTs are one-byte call instructions used mainly with interrupts. Each RST has a predefined address. However, RST0 and the hardware reset vector have the same address 0000_{16}. Therefore, use of RST0 is not usually recommended. The RSTs cause the 8085 to push the PC onto the stack. The 8085 then loads the PC with a predefined address based on the particular RST being used.

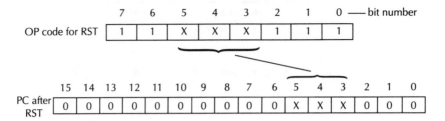

FIGURE 2.5   Execution of the RST instruction.

Table 2.9   RST0-RST7 Vector Addresses

Instruction	OP code (hexadecimal)	Vector address (hexadecimal)
RST0	C7	0000
RST1	CF	0008
RST2	D7	0010
RST3	DF	0018
RST4	E7	0020
RST5	EF	0028
RST6	F7	0030
RST7	FF	0038

A 3-bit code in the OP code for a particular RST determines the address to which the program would branch. This is shown in Figure 2.5.

The vector addresses for the RSTs are listed in Table 2.9. Note that a limited number of locations are available for each RST instruction. This may not provide enough locations for writing a complete subroutine. Therefore, one may place an unconditional JUMP at the vector address to JUMP to an address in RAM where the actual subroutine is written.

The 8085 stack manipulation instructions allow one to save and retrieve the contents of the register pairs into and from the stack, respectively. For example, the following instruction saves the register pair B,C into the stack:

```
PUSH B ; (SP) ← (SP) - 1
       ; M((SP)) ← (B)
       ; (SP) ← (SP) - 1
       ; M((SP)) ← (C)
```

Similarly, the instruction POP D retrieves the top two words of the stack and places them into the registers E and D in that order as follows:

```
POP D ; (E) ← M((SP))
      ; (SP) ← (SP) + 1
      ; (D) ← M((SP))
      ; (SP) ← (SP) + 1
```

This means that all 8085 registers can be saved onto the stack using the following instruction sequence:

```
PUSH PSW ; save the A and flags register
PUSH B   ; save the D,E pair
PUSH D   ; save the D,E pair
PUSH H   ; save the H,L pair
```

Similarly, the saved status can be restored by using the following sequence of POP instructions:

```
POP  H   ; restore H,L pair
POP  D   ; restore D,E pair
POP  B   ; restore B,C pair
POP  PSW ; restore A and flags register
```

There are two other stack instructions: SPHL and XTHL. SPHL is a one-byte instruction. It moves the [L] to SP high byte and [H] to SP low byte. XTHL is also a one-byte instruction. It exchanges the [L] with the top of the stack addressed by SP and [H] with the next stack addressed by SP + 1. That is, XTHL performs the following:

$$[[SP]] \leftrightarrow [L]$$
$$[[SP+1]] \leftrightarrow [H]$$

The 8085 can use either standard or memory-mapped I/O. Using standard I/O, the input and output instructions have the following format:

```
IN  (8-bit port address)  ;       input instruction
OUT (8-bit port address)  ;       output instruction
```

For example, the instruction IN 02H transfers the contents of the input port with address 02_{16} into the accumulator. Similarly, the instruction OUT 00H transfers the contents of the accumulator to the output port with address 00_{16}. Using memory-mapped I/O, LDA addr and STA addr can be used as input and output instructions, respectively.

The 8085 HLT instruction forces the 8085 to enter into the halt state. Similarly, the dummy instruction NOP neither achieves any result nor affects any CPU registers. This is a useful instruction for producing software delay routines and to insert diagnostic messages.

The 8085 8-bit increment (INR) and decrement (DCR) instructions affect the status flags. However, the 16-bit increment (INX) and decrement (DCX) instructions do not affect the flags. Therefore, while using these instructions in a loop counter value greater than 256_{10}, some other instructions must be used with DCX or INX to affect the flags after their execution. For example, the following instruction sequence will affect the flags for DCX:

```
        LXI B,16-BIT DATA ;  Load
Loop      —               ;  initial
          —               ;  16-bit loop
          —               ;  count to BC
          —
          —
          —               ;  Decrement
        DCX B             ;  counter
        MOV A,B           ;  Move B to A to
                             test for zero
        ORA C             ;  Logically or with A
        JNZ Loop          ;  Jump if not zero
          —
          —
          —
```

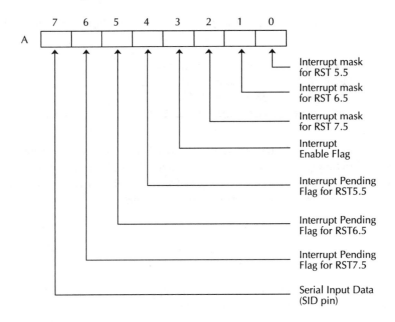

FIGURE 2.6   Accumulator data format after execution of RIM.

There are four one-byte 8085 interrupt instructions. These are DI, EI, RIM, and SIM.

DI disables the 8085's maskable interrupt capability. EI, on the other hand, enables the 8085 maskable interrupt capability.

RIM is a one-byte instruction. It loads the accumulator with 8 bits of data as shown in Figure 2.6.

Bits 0, 1, and 2 provide the values of the RST 5.5, RST 6.5, and RST 7.5 mask bits, respectively. If the mask bit corresponding to a particular RST is one, the RST is disabled; a zero in a specific RST (bits 0, 1, and 2) means that RST is enabled.

If the interrupt enable bit (bit 3) is 0, the 8085's maskable interrupt capability is disabled; the interrupt is enabled if this bit is one.

A "one" in a particular interrupt pending bit indicates that an interrupt is being requested on the identified RST line; if this bit is zero, no interrupt is waiting to be serviced. The serial input data (bit 7) indicate the value of the SID pin.

The SIM instruction outputs the contents of the accumulator to define interrupt mask bits and the serial output data line. The bits in the accumulator before execution of the SIM are defined as shown in Figure 2.7.

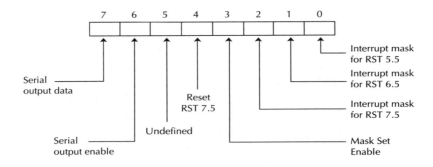

FIGURE 2.7   Accumulator data format before execution of the SIM.

If the mask set enable bit is set to one, interrupt mask bits for RST 7.5, RST 6.5, and RST 5.5 are sent out; a zero value at the mask set enable does not affect the interrupt mask bits. A one at a particular interrupt mask disables that interrupt and a zero enables it.

The RESET RST 7.5, if set to one, resets as internal flip-flop to zero in order to clear the 7.5 interrupt.

If the serial output enable is one, the serial output data are sent to the SOD pin.

The interrupt instructions will be covered in detail during discussion of the 8085 interrupts.

Example 2.1

Write an 8085 assembly language program to add a 16-bit number in locations 5000H (high byte) and 5001H (low byte) with another 16-bit number stored in 5002H (high byte) and 5003H (low byte). Store result in BC.

Solution

```
        ORG     2000H
        LDA     MEM2      ;   Load low byte of
        LXI     H,5003H   ;   number 1
        ADD     M         ;   Add with low
                          ;   byte of number 2
        MOV     C, A      ;   Store result in C
        LDA     MEM1      ;   Load high byte
        DCX     H         ;   of number 1
        ADC     M         ;   Add with
                          ;   high byte and Cy
        MOV     B, A      ;   Store result
                          ;   in B
        HLT               ;   stop
        ORG     5000H
        MEM1    DB DATA1
        MEM2    DB DATA2
        MEM3    DB DATA3
        MEM4    DB DATA4
```

Example 2.2

Write an 8085 assembly language program to perform a parity check on an 8-bit word in location 5000_{16}. If the parity is odd, store DD16 in location 5000_{16}. However, if the parity is even, store EE16 in location 5000_{16}.

Solution

```
        ORG 3000H
        LDA PARITY   ;   Load 8-bit
                     ;   data into A
        ADI 00H      ;   Add with
                     ;   00H to affect
        MVI A, 0DDH  ;   flags
        JPO ODD      ;   Check for
                     ;   odd parity
        MVI A,0EEH   ;   If parity
                     ;   even, store
                     ;   EEH
```

```
ODD      STA PARITY   ;  in 5000H
         HLT          ;  Stop
         ORG 5000H
PARITY   DB DATA
```

Note that address 3000H at ORG 3000H is arbitrarily chosen.

Example 2.3

Write a program in 8085 assembly language to perform an unsigned 8-bit by 8-bit multiplication via repeated addition. Assume that the multiplicand is in "B" register and the multiplier is in "C" register. Store the product in HL.

Solution

```
         ORG 5000H
         LXI H,0000H  ;  Initialize the
                      ;  16-bit product to zero
         MVI C, MULT  ;  Move multiplier to C
         MOV E,B      ;  Move multiplicand
                      ;  to register E
         MVI D,OOH    ;  Convert multiplicand
                      ;  to unsigned 16-bit
START    DAD D        ;  Perform 16-bit addition
         DCR C        ;  Decrement multiplier
         JNZ START    ;  Jump if multiplier not zero
         HLT          ;  Stop. Product
                      ;  is in HL
```

Example 2.4

Write a subroutine in 8085 assembly language to check whether an 8-bit number in location 3000_{16} is odd or even. If the number is odd, store DD16 in location 3000_{16}; if the number is even, store EE16 in location 3000_{16}.

Also, write the main program to initialize stack pointer 5020_{16}, load the 8-bit number to be checked for odd or even into A, store DD16 or EE16 depending on the result, and stop.

Solution

Main Program

```
         ORG 4000H
         LXI SP, 5020H ;  Initialize SP
         LDA START     ;  Load number
         CALL CHECK    ;  Call subroutine
         STA START     ;  Store DD16 or EE16
         HLT           ;  stop
         ORG 3000H
START    DB DATA
```

Subroutine

```
         ORG 7000H
CHECK    RAR           ;  Rotate 'A' for checking
         MVI A,0EEH    ;  Jump if even
```

```
        JNC RETURN      ; Store DD16 if
        MVI A,0DDH      ; odd and return
                        ; Store EE16
RETURN  RET             ; if even and return
```

Example 2.5

Write an 8085 assembly language program to clear 100_{10} consecutive bytes starting at 2050_{16}.

Solution

```
        ORG 2000H
        MVI A, 64H    ; Load 'A' with number of
                      ; bytes to be cleared
        LXI H,2050H   ; Load HL with 2050H
LOOP    MVI M, 00H    ; Clear memory location
        INX H         ; Increment address
        DCR A         ; Decrement loop counter
        JNZ LOOP      ; Jump to loop if
                      ; loop counter zero
        HLT           ; Stop
```

2.6 Timing Methods

Timing concepts are very important in microprocessor applications. Typically, in sequential process control the microprocessor is required to provide time delays for on-off devices such as pumps or motor-operated valves. DELAY routines are used to provide such time delays.

A delay program typically has an input register that contains the initial count. The register pair D,E is used for this purpose. A typical delay subroutine is given below:

```
DELAY   DCX D       ; Decrement the D,E contents
        MOV A,D     ;
        ORA E       ; Are the contents zero?
        JNZ DELAY   ; Jump if not zero
        RET         ; Return
```

We now calculate the total time required by the DELAY routine using the following data:

Instruction	Number of cycles
CALL	18
DCX D	6
MOV A,D	4
ORA E	4
JNZ	7/10
RET	10

Note that in the above, if the JNZ condition is met ($Z = 0$), ten cycles are required and the program branches back to the DCX D instruction. However, if the JNZ condition is not met ($Z = 1$), the seven cycles are required, and the program executes the next instruction, that is, the RET instruction. Also, note that the CALL instruction is used in the main program written by the user, and the 3-byte instruction CALL DELAY is used.

For each iteration in which the JNZ condition is met ($Z = 0$), the number of cycles is equal to cycles for DCX D + cycles for MOV A,D + cycles for ORA E + cycles for JNZ = $6 + 4 + 4 + 10 = 24$ cycles.

These 24 cycles will be performed $(y - 1)$ times, where y is the initial contents of D,E. For the final iteration in which no jump is performed and the JNZ condition is not satisfied ($Z \neq 0$), the number of cycles is equal to cycles for DCX D + cycles for MOV A,D + cycles for ORA E + cycles for JNZ + cycles for RET = 6 + 4 + 4 + 7 + 10 = 31 cycles. Therefore, the time used, including a CALL instruction, is

$$18 + 31 + 24(y - 1) = 49 + 24(y - 1)\text{ clock cycles}$$

Suppose that in a program a delay time of 1/3 ms is desired. The DELAY routine can be used to accomplish this in the following way. Each cycle of the 8085 clock is 1/3 ms (3 MHz). The number of cycles required in the DELAY routine is

$$\frac{1/3 \text{ ms}}{1/3 \text{ μs}} = \frac{10^{-3}}{10^{-6}} = 1000 \text{ cycles}$$

Therefore, the initial counter value y of the D,E register pair can be calculated:

$$49 + 24\,(y - 1) = 1000$$
$$24\,(y - 1) = 951$$
$$y = \frac{951}{24} + 1 \cong 40_{10} = 28_{16}$$

Therefore, in the program the D,E register pair can be loaded with 0028_{16} and the DELAY routine can be called to obtain 1/3 ms of time delay. Table 2.10 shows initial counts for various time delays.

Table 2.10   Time Intervals along with Initial Counts

3-MHz clock milliseconds	Initial count (hexadecimal)
1/3	0028
1	007C
2	00F9
10	04E1
100	30D3

The following program produces a delay of 10 ms:

```
LXI SP, 5000H  ;  Set stack pointer
LXI D, 04E1H   ;  Load D,E with initial count value of 04E1
               ;  to provide 10 ms of delay
CALL DELAY     ;  Call DELAY routine
HLT            ;  STOP
```

The delay times can be increased by using a counter. Suppose that a delay of 5 s is desired in a program. From Table 2.10 an initial count of $30D3_{16}$ produces a 100-ms delay. We can use a counter along with the 100-ms delay to obtain the 5-s delay as follows

$$(100 \text{ ms}) \times X = 5 \text{ s}$$

where X is the value of the counter. Then

$$X = \frac{5}{100 \times 10^{-3}} = \frac{5}{10^{-1}} = 50$$

Therefore, a counter of 50_{10} or 32_{16} is required. Now the program for the 5-s delay can be written as follows:

```
        LXI SP, 5000H   ;   Set stack pointer
        MVI C, 32H      ;   Do DELAY loop 50₁₀ times by loading C
                        ;   with count 32₁₆
START   LXI D, 30D3H    ;   Load initial count
        CALL DELAY      ;   Call DELAY loop
        DCR C           ;   Decrement C and check if zero: if
                        ;   not, do another delay
        JNZ START       ;   Loop back
        HLT             ;   STOP
```

In the above, since execution times of DCR C and JNZ START are very small compared to 5 s, they are not considered in computing the delay.

2.7 8085 Pins and Signals

The 8085 is housed in a 40-pin dual in-line package (DIP). Figure 2.8 shows the 8085 pins and signals.

The low-order address byte and data lines AD0 to AD7 are multiplexed. These lines are bidirectional. The beginning of an instruction is indicated by the rising edge of the ALE signal. At the falling edge of ALE, the low byte of the address is automatically latched by some of the 8085 support chips such as 8155 and 8355: AD0 to AD7 lines can then be used as data lines. Note that ALE is an input to these support chips. However, if the support chips do not latch AD0 to AD7, then external latches are required to generate eight separate address lines A7 to A0 at the falling edge of ALE.

Pins A8 to A15 are unidirectional and contain the high byte of the address.

Table 2.11 lists the 8085 pins along with a brief description of each.

The RD pin signal is output LOW by the 8085 during a memory or I/O READ operation. Similarly, the WR pin signal is output LOW during a memory or I/O WRITE.

Next, we explain the purpose of $IO/\overline{M}$, S0, and S1 signals. The $IO/\overline{M}$ signal is output HIGH by the 8085 to indicate execution of an I/O instruction such as IN or OUT. This pin is output LOW during execution of a memory instruction such as LDA 2050H.

The $IO/\overline{M}$, S0, and S1 are output by the 8085 during its internal operations, which can be interpreted as follows:

$IO/\overline{M}$	S1	S0	Operation performed by the 8085
0	0	1	Memory WRITE
0	1	0	Memory READ
1	0	1	I/O WRITE
1	1	0	I/O READ
0	1	1	OP code fetch
1	1	1	Interrupt acknowledge

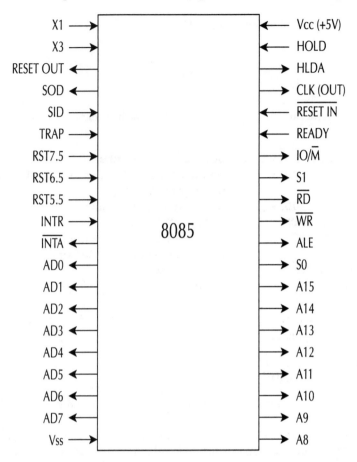

FIGURE 2.8   8085 microprocessor signals and pin assignments.

Table 2.11   8085 Signal Description Summary

Pin name	Description	Type
AD0–AD7	Address/data bus	Bidirectional, tristate
A8–A15	Address bus	Output, tristate
ALE	Address latch enable	Output, tristate
RD	Read control	Output, tristate
WR	Write control	Output, tristate
IO / M	I/O or memory indicator	Output, tristate
S0, S1	Bus state indicators	Output
READY	Wait state request	Input
SID	Serial data input	Input
SOD	Serial data output	Output
HOLD	Hold request	Input
HLDA	Hold acknowledge	Output
INTR	Interrupt request	Input
TRAP	Nonmaskable interrupt request	Input
RST5.5	Hardware vectored	Input
RST6.5	Hardware vectored interrupt request	Input
RST7.5	Hardware vectored	Input
INTA	Interrupt acknowledge	Output
RESET IN	System reset	Input
RESET OUT	Peripherals reset	Output
X1, X2	Crystal or RC connection	Input
CLK (OUT)	Clock signal	Output
Vcc, Vss	Power, ground	

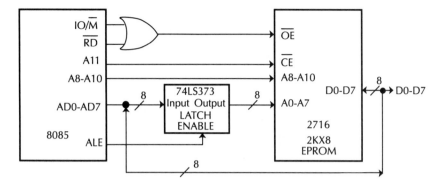

(a) 8155 - 2716 interface using internal latches.

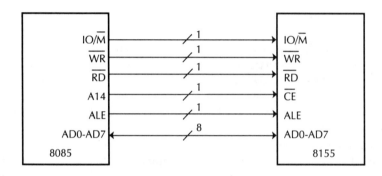

(b) 8085 - 8155 interface using ALE and AD0-AD7

FIGURE 2.9   8085's interface to external device using ALE and the multiplexed AD0 to AD7 pins.

Figure 2.9 illustrates the utilization of ALE and AD0 to AD7 signals for interfacing an EPROM and a RAM.

The 2716 is a $2K \times 8$ EPROM with separate address and data lines without any built-in latches. This means that a separate latch such as the 74LS373 must be used to isolate the 8085 low byte address and D0-D7 data lines at the falling edge of ALE (Figure 2.9a).

The 8155 contains 256-byte static RAM, three user ports, and a 14-bit timer. The 8155 is designed for 8085 in the sense that it has built-in latches with ALE as input along with multiplexed address (low byte) and data lines, AD0 to AD7. Therefore, as shown in Figure 2.9b, external latches are not required.

The READY input can be used by the slower external devices for obtaining extra time in order to communicate with the 8085. The READY signal (when LOW) can be utilized to provide wait-state clock periods in the 8085 machine. If READY is HIGH during a read or write cycle, it indicates that the memory or peripheral is ready to send or receive data. If not used, it must be tied high.

The Serial Input Data (SID) and Serial Output Data (SOD) lines are associated with the 8085 serial I/O transfer. The SOD line can be used to output the most significant bit of the accumulator. The SID signal can be input into the most significant bit of the accumulator.

The HOLD and HLDA signals are used for the Direct Memory Access (DMA) type of data transfer. The external devices place a HIGH on HOLD line in order to take control of the system bus. The HOLD function is acknowledged by the 8085 by placing a HIGH output on the HLDA pin and driving the tristate outputs high impedance.

The signals on the TRAP, RST7.5, RST6.5, RST5.5, INTR, and INTA are related to the 8085 interrupt signals. TRAP is a nonmaskable interrupt; that is, it cannot be enabled or disabled

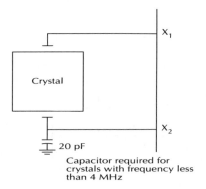

FIGURE 2.10   Crystal connection to X1 and X2 pins.

by an instruction. The TRAP has the highest priority. RST7.5, RST6.5, and RST5.5 are maskable interrupts used by the devices whose vector addresses are generated automatically. INTA is an interrupt acknowledge signal which is pulsed LOW by the 8085 in response to the interrupt INTR request. In order to service INTR, one of the eight OP codes (RST0 to RST7) has to be provided on the 8085 AD0-AD7 bus by external logic. The 8085 then executes this instruction and vectors to the appropriate address to service the interrupt.

All unused control pins such as interrupts and HOLD must be disabled by grounding them. (READY must be tied high).

The 8085 has the clock generation circuit on the chip and, therefore, no external oscillators need to be designed. The 8085A can operate with a maximum clock frequency of 3.03 MHz and the 8085A-2 can be driven with a maximum of 5 MHz clock. The 8085 clock frequency can be generated by a crystal, an LC tuned circuit, or an external clock circuit. The frequency at X1X2 is divided by 2 internally. This means that in order to obtain 3.03 MHz, a clock source of 6.06 MHz must be connected to X1X2. For crystals of less than 4 MHz, a capacitor of 20 pF should be connected between X2 and a ground to ensure the starting up of the crystal at the right frequency (Figure 2.10).

There is a TTL signal which is output on pin 37, called the CLK (OUT) signal. This signal can be used by other external microprocessors or support chips.

The RESET IN signal, when pulsed LOW then high, causes the 8085 to execute the first instruction at the 0000_{16} location. In addition, the 8085 resets instruction register, interrupt mask (RST5.5, RST6.5, and RST7.5) bits, and other registers. The RESET IN must be held LOW for at least three clock periods. A typical 8085 reset circuit is shown in Figure 2.11. In this circuit, when the switch is activated, RESET IN is driven to LOW with a large time constant providing adequate time to reset the system.

The 8085 requires a minimum operating voltage of 4.75 V. Upon applying power, the 8085A attains this voltage after 500 μs. The reset circuit of Figure 2.11 resets the 8085 upon activation

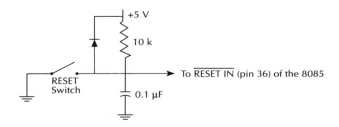

FIGURE 2.11   8085 reset circuit.

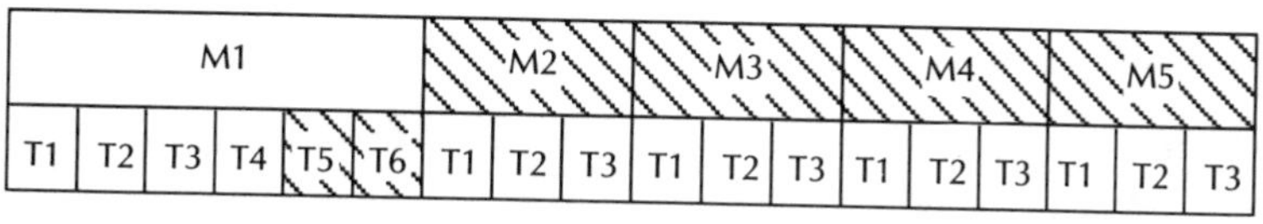

FIGURE 2.12   8085 machine cycles.

of the switch. The voltage across the 0.1-μF capacitor is zero on power-up. The capacitor then charges Vcc after a definite time determined by the time constant RC. The chosen values of RC in the figure will drive the $\overline{\text{RESET IN}}$ pin to low for at least three clock periods. In this case, after activating the switch, $\overline{\text{RESET IN}}$ will be low (assuming capacitance charge time is equal to the discharge time) for $10K*0.1\ \mu F = 1$ ms, which is greater than three clock periods ($3*1/3\ \mu s = 1\ \mu s$) of the 3-MHz 8085A. During normal operation of the 8085, activation of the switch will short the capacitor to ground and will discharge it. When the switch is opened, the capacitor charges and the $\overline{\text{RESET IN}}$ pin becomes HIGH. Upon hardware reset, the 8085 clears PC, IR, HALT flip-flop, and some other registers; the 8085 registers PSW, A, B, C, D, E, H, and L are unaffected. Upon activation of the $\overline{\text{RESET IN}}$ to low, the 8085 outputs HIGH at the RESET OUT pin which can be used to reset the memory and I/O chips connected to the 8085. Note that since hardware reset initializes PC to 0, the 8085 fetches the first instruction for address 0000_{16} after reset.

2.8 8085 Instruction Timing and Execution

An 8085's instruction execution consists of a number of machine cycles (MCs). These cycles vary from one to five (M1 to M5) depending on the instruction. Each machine cycle contains a number of 320-ns clock periods. The first machine cycle will be executed in either four or six clock periods, and the machine cycles that follow will have three clock periods. This is shown in Figure 2.12.

The shaded MCs indicate that these machine cycles are required by certain instructions. Similarly, the shaded clock periods (T5 and T6) mean that they are needed in M1 by some instructions.

The clock periods within a machine cycle can be illustrated as shown in Figure 2.13. Note that the beginning of a new machine cycle is indicated on the 8085 by outputting the Address Latch Enable (ALE) signal HIGH. During this time, lines AD0 to AD7 are used for placing the low byte of the address.

When the ALE signal goes LOW, the low byte of the address is latched so that the AD0 to AD7 lines can be used for transferring data.

We now discuss the timing diagrams for instruction fetch, READ, and WRITE.

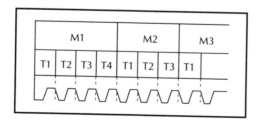

FIGURE 2.13   Clock period within a machine cycle.

2.8.1 Basic System Timing

Figure 2.14 shows the 8085 basic system timing. An instruction execution includes two operations: OP code fetch and execution.

The OP code fetch cycle requires either four (for one-byte instructions such as MOV A,B) or six cycles (for 3 byte instructions such as LDA 2030H). The machine cycles that follow will need three clock periods.

The purpose of an instruction fetch is to read the contents of a memory location containing an instruction addressed by the program counter and to place it in the instruction register. The 8085 instruction fetch timing diagram shown in Figure 2.14 can be explained in the following way:

1. The 8085 puts a LOW on the $IO/\overline{M}$ line of the system bus, indicating a memory operation.
2. The 8085 sets S0 = 1 and S1 = 1 on the system bus, indicating the memory fetch operation.
3. The 8085 places the program counter high byte on the A8 to A15 lines and the program counter low byte on the AD0 to AD7 lines of the system bus. The 8085 also sets the ALE signal to HIGH. As soon as the ALE signal goes to LOW, the program counter low byte on the AD0 to AD7 is latched automatically by some 8085 support chips such as 8155 (if 8085 support chips are not used, these lines must be latched using external latches), since these lines will be used as data lines for reading the OP code.
4. At the beginning of T2 in M1, the 8085 puts the $\overline{RD}$ line to LOW indicating a READ operation. After some time, the 8085 loads the OP code (the contents of the memory location addressed by the program counter) into the instruction register.
5. During the T4 clock period in M1, the 8085 decodes the instruction.

The Machine Cycle M2 of Figure 2.14 shows a memory (or I/O) READ operation as seen by the external logic, and the status of the S0 and S1 signals indicates whether the operation is instruction fetch or memory READ; for example, S1 = 1, S0 = 1 during instruction fetch and S1 = 1, S0 = 0 during memory READ provided $IO/\overline{M}$ = 0.

The purpose of the memory READ is to read the contents of a memory location addressed by a register pair, such as the H,L pair, or a memory location specified with the instruction and the data placed in a microprocessor register such as the accumulator. In contrast, the purpose of the memory fetch is to read the contents of a memory location addressed by PC into IR. The machine cycle M3 of Figure 2.14 indicates a memory (or I/O) write operation. In this case, S1 = 0 and S0 = 1 indicate a memory write operation when $IO/\overline{M}$ = 0 and an I/O write operation when $IO/\overline{M}$ = 1.

2.8.2 8085 Memory READ ($IO/\overline{M}$ = 0, $\overline{RD}$ = 0) and I/O READ ($IO/\overline{M}$ = 1, $\overline{RD}$ = 0)

Figure 2.15a shows an 8085A clock timing diagram. The machine cycle of M2 of Figure 2.14 shows a memory READ timing diagram.

The purpose of the memory READ is to read the contents of a memory location addressed by a register pair, such as HL. Let us explain the 8085 memory READ timing diagram of Figure 2.15b along with the READ timing signals of Figure 2.14:

1. The 8085 uses machine cycle M1 to fetch and decode the instruction. It then performs the memory READ operation in M2.
2. The 8085 continues to maintain $IO/\overline{M}$ at LOW in M2 indicating a memory READ operation (or $IO/\overline{M}$ = 1 for I/O READ).
3. The 8085 puts S1 = 1, S0 = 0, indicating a READ operation.
4. The 8085 places the contents of the high byte of the memory address register, such as the contents of the H register, on lines A8 to A15.

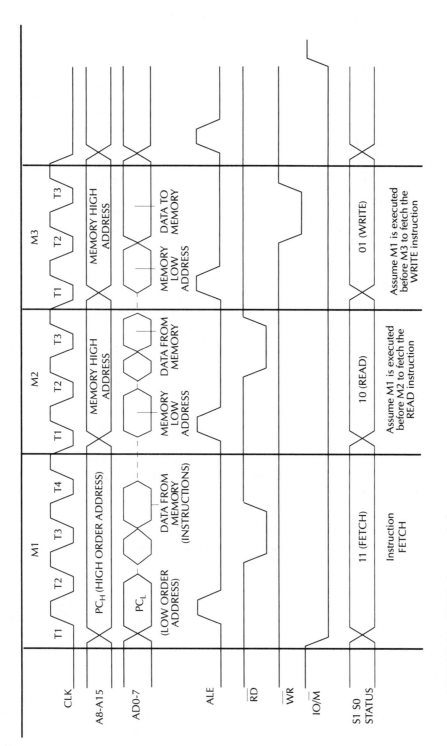

FIGURE 2.14 8085 Basic system timing.

5. The 8085 places the contents of the low byte of the memory address register, such as the contents of the L register, in lines AD0 to AD7.
6. The 8085 sets ALE to high, indicating the beginning of M2. As soon as ALE goes to low, the memory or support chip must latch the low byte of the address lines, since the same lines are going to be used as data lines.
7. The 8085 puts the $\overline{RD}$ signal to LOW, indicating a READ operation.
8. The 8085 gets the data from the memory location addressed by the memory address register, such as the H,L pair, and places the data into a register such as the accumulator. In case of I/O, the 8085 inputs data from the I/O port into the accumulator.

2.8.3 8085 Memory WRITE (IO/$\overline{M}$ = 0, $\overline{WR}$ = 0) and I/O WRITE (IO/$\overline{M}$ = 1, $\overline{WR}$ = 0)

The machine cycle M3 of Figure 2.14 shows a memory WRITE timing diagram. As seen by the external logic, the signals S1 = 0, S0 = 1, and $\overline{WR}$ = 0 indicate a memory WRITE operation.

The purpose of a memory WRITE is to store the contents of the 8085 register, such as the accumulator, into a memory location addressed by a pair, such as H,L.

The WRITE timing diagram of Figure 2.14 can be explained as follows:

1. The 8085 uses machine cycle M1 to fetch and decode the instruction. It then executes the memory WRITE instruction in M3.
2. The 8085 continues to maintain IO/$\overline{M}$ at LOW, indicating a memory operation (or IO/$\overline{M}$ = 1 for I/O WRITE).
3. The 8085 puts S1 = 0, S0 = 1, indicating a WRITE operation.
4. The 8085 places the Memory Address Register high byte, such as the contents of the H register, on lines A8 to A15.
5. The 8085 places the Memory Address Register low byte, such as the contents of L register, on lines AD0 to AD7.
6. The 8085 sets ALE to HIGH, indicating the beginning of M3. As soon as ALE goes to LOW, the memory or support chip must latch the low byte of the address lines, since the same lines are going to be used as data lines.
7. The 8085 puts the $\overline{WR}$ signal to LOW, indicating a WRITE operation.
8. It also places the contents of the register, say, accumulator, on data lines AD0 to AD7.
9. The external logic gets data from the lines AD0 to AD7 and stores the data in the memory location addressed by the Memory Address Register, such as the H,L pair. In case of I/O, the 8085 outputs [A] to an I/O port.

Figures 2.15a through c show the 8085A clock and read and write timing diagrams.

2.9 8085 Input/Output (I/O)

The 8085 I/O transfer techniques are discussed. The 8355/8755 and 8155/8156 I/O ports and 8085 SID and SOD lines are also included.

2.9.1 8085 Programmed I/O

There are two I/O instructions in the 8085, namely, IN and OUT. These instructions are 2 bytes long. The first byte defines the OP code of the instruction and the second byte specifies the I/O port number. Execution of the IN PORT instruction causes the 8085 to receive one byte of data into the accumulator from a specified I/O port. On the other hand, the OUT PORT instruction, when executed, causes the 8085 to send one byte of data from the accumulator into a specified I/O port.

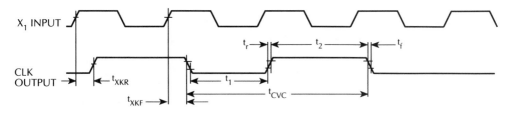

FIGURE 2.15a   8085A clock.

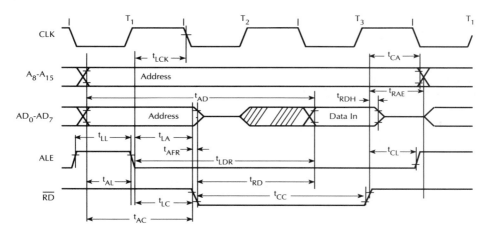

FIGURE 2.15b   8085 Read Timing diagram.

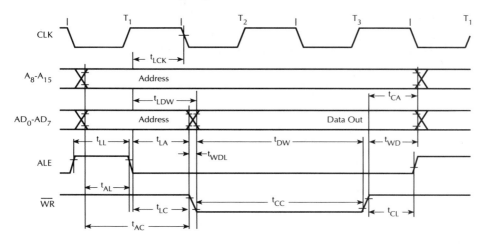

FIGURE 2.15c   8085 Write Timing diagram.

The 8085 can access I/O ports using either standard I/O or memory-mapped I/O.

In standard I/O, the 8085 inputs or outputs data using IN or OUT instructions.

In memory-mapped I/O, the 8085 maps I/O ports as memory addresses. Hence, LDA addr or STA addr instructions are used to input or output data to or from the 8085. The 8085's programmed I/O capabilities are obtained via the support chips, namely, 8355/8755 and 8155/8156. The 8355/8755 contains a 2K-byte ROM/EPROM and two 8-bit I/O ports (ports A and B).

The 8155/8156 contains 256-byte RAM, two 8-bit and one 6-bit I/O ports, and a 14-bit programmable timer. The only difference between the 8155 and 8156 is that chip enable is LOW on the 8155 and HIGH on the 8156.

2.9.1.a 8355/8755 I/O Ports

Two 8-bit ports are included in the 8355/8755. These are ports A and B. Another 8-bit port, called the data direction register, is associated with each one of these ports. These registers (DDRA and DDRB) can be used to configure each bit in ports A or B as either input or output. For example, a "0" written into a bit position of the data direction register sets up the corresponding bit in the I/O port as input. On the other hand, a "1" written in a particular bit position in the data direction register sets up the corresponding bit in the I/O port as output. For example, consider the following instruction sequence:

```
MVI   A,  05H
OUT   DDRA
```

The above instruction sequence assumes DDRA as the data direction register for port A. The bits of port A are configured as follows:

	7	6	5	4	3	2	1	0
DDRA	0	0	0	0	0	1	0	1

	7	6	5	4	3	2	1	0
Port A	IN	IN	IN	IN	IN	OUT	IN	OUT

The 8355/8755 uses the $\overline{IO/M}$ pin on the chip in order to distinguish between standard and memory-mapped I/O. This pin is controlled by the 8085 as shown in Figure 2.16.

The 8085 outputs a HIGH on the IO/M pin when it executes either an IN or OUT instruction. This means that $IO/\overline{M}$ in the 8355/8755 becomes HIGH during execution of IN or OUT. This, in turn, tells the 8355/8755 to decode the AD1 and AD0 lines in order to obtain the 8-bit address of various ports in the chip as follows:

AD1	AD0	
0	0	Port A
0	1	Port B
1	0	Data Direction Register A
1	1	Data Direction Register B

The other 6 bits of each 8-bit port address are don't care conditions. This means that these bits can be either one or zero. The 8085/8355/8755 standard I/O is illustrated in Figure 2.17.

Since the 8355/8755 is provided with 2K bytes of memory, 11 address lines A0 to A10 are required for memory addressing. Since the 8085 has 16 address pins A0 to A15, A11 to A15 will not be used for memory addressing. Note that the 8355/8755 includes two chip enables $\overline{CE}$ and CE. In Figure 2.17 these two chip enables are connected to Vcc and A11, respectively. It should be pointed out that the 8085 duplicates low and high bytes of the 16-bit address lines with the port address when it executes an IN or OUT instruction. This means that if the 8085 executes IN 01 instruction, it puts 0101_{16} on the 16 address lines. Note that in Figure 2.17, A11 = 0 for both memory and port addressing. This is because A11 = 0 enables this chip. When the 8085 executes an LDA addr or STA addr instruction, $IO/\overline{M}$ becomes LOW. This tells the 8355/8755 to interpret A0 to A10 as memory addresses. On the other hand, when the 8085 executes an IN PORT or OUT PORT instruction, the 8085 drives $IO/\overline{M}$ to HIGH. This tells the 8355/8755 to decode AD1 and AD0 for I/O port addresses. The port addresses are as follows:

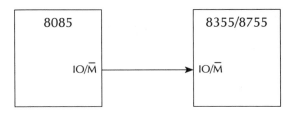

FIGURE 2.16 Interfacing the 8085 with 8355/8755 via the $IO/\overline{M}$ pin.

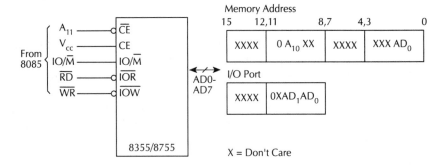

FIGURE 2.17 8355/8755 standard I/O.

| | A15 | A14 | A13 | A12 | A11 | A10 | A9 | A8 | |
	AD7	AD6	AD5	AD4	AD3	AD2	AD1	AD0	Address
Port A =	X	X	X	X	X	0	X	0	$= 00_{16}$
Port B =	X	X	X	X	0	X	0	1	$= 01_{16}$
DDRA =	X	X	X	X	0	X	1	0	$= 02_{16}$
DDRB =	X	X	X	X	0	X	1	1	$= 03_{16}$

X is don't care. Assume X is zero in the above.

Let us now discuss 8355/8755 memory-mapped I/O. Figure 2.18 provides such an example. In Figure 2.18, A11 must be zero for selecting the 8355/8755 and A15 is connected to $IO/\overline{M}$ of the 8355/8755. When A15 = 1, $IO/\overline{M}$ becomes HIGH. This tells the 8355/8755 chip to decode AD1 and AD0 for obtaining I/O port addresses. For example, if we assume all don't cares in the I/O port address are 1, then the I/O port addresses will be mapped into memory locations as follows:

Port name	16-bit memory address
I/O port A	$F7FC_{16}$
I/O port B	$F7FD_{16}$
DDRA	$F7FE_{16}$
DDRB	$F7FF_{16}$

Note that in Figure 2.18 memory addresses are mapped as 7000_{16} through $77FF_{16}$ assuming all don't cares to be ones.

Note that the above port addresses may not physically exist in memory. However, input or output operations with these ports can be accomplished by generating the necessary signals by executing LDA or STA instructions with the above addresses. For example, outputting to DDRA or DDRB can be accomplished via storing to locations $F7FE_{16}$ or $F7FF_{16}$, respectively. The instructions STA F7FEH or STA F7FFH will generate all of the required signals for OUT

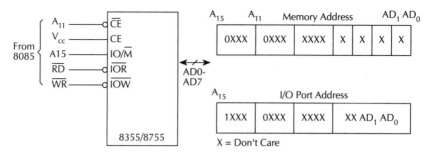

FIGURE 2.18   8355/8755 memory-mapped I/O.

DDRA or OUT DDRB, respectively. For example, upon execution of the STA F7FEH, the 8085 sends a LOW to the $\overline{WR}$ pin and places F7FEH on the address bus. This will make A15 = 1, A11 = 0, AD1 = 1, AD0 = 0, and thus will make the IO / M = 1, $\overline{CE}$ = 0, $\overline{IOW}$ = 0 on the 8355/ 8755 in Figure 2.18.

When ALE goes to LOW, the 8085 places the contents of the accumulator on the AD7 to AD0 pins. The 8355/8755 then takes this data and writes into DDRA. Therefore, STA F7FEH is equivalent to OUT DDRA instruction.

2.9.1.b 8155/8156 I/O Ports

The 8155 or 8156 includes 256 bytes of static RAM and three parallel I/O ports. These ports are port A (8-bit), port B (8-bit), and port C (6-bit). By parallel it is meant that all bits of the port are configured as either all input or all output. Bit-by-bit configuration like the 8355/8755 is not permitted. The only difference between the 8155 and 8156 is that the 8155 has LOW chip enable ($\overline{CE}$), while the 8156 includes a HIGH chip enable (CE). The 8155/8156 ports are configured by another port called the command status register (CSR). When data are output to the CSR via the accumulator, each bit is interpreted as a command bit to set up ports and control timer as shown in Figure 2.19. Port C can be used as a 6-bit parallel port or as a control port to support data transfer between the 8085 and an external device via ports A and B using handshaking. Note that handshaking means data transfer via exchange of control signals. Two bits (bits 2 and 3) are required in CSR to configure port C. Note that port A interrupt and port B interrupt are associated with handshaking and are different from the 8085 interrupts. For example, port A interrupt is HIGH when data are ready to be transferred using handshaking signals such as port A buffer full and port A strobe. The port A interrupt (PC0 in ALT3) can be connected to an 8085 interrupt pin and data can be transferred to or from the 8085 via port A by executing appropriate instructions in the interrupt service routine.

When the 8085 reads the CSR, it accesses the status register and information such as status of handshaking signals and timer interrupt is obtained.

Three bits are used to decode the 8155 six ports (CSR, port A, port B, port C, timer high port, timer low port) as follows:

AD2	AD1	AD0	Port selected
0	0	0	CSR
0	0	1	Port A
0	1	0	Port B
0	1	1	Port C
1	0	0	Timer-low port
1	0	1	Timer-high port

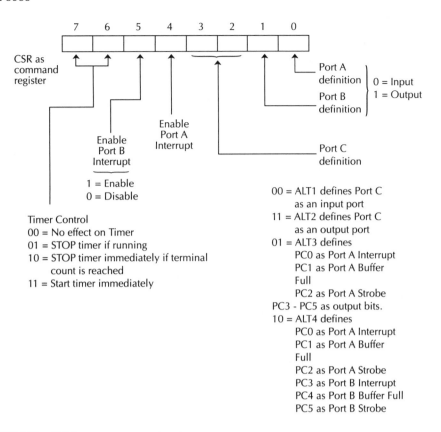

FIGURE 2.19   CSR format as command register.

A typical interface between the 8085 and 8155 is shown in Figure 2.20. Consider Figure 2.20a. Since the 8085 duplicates low byte address bus with the high byte address (i.e., AD7 to AD0 same as A15 to A8 for 8085 standard I/O), the address pins AD2 to AD0 will be the same as A10 to A8. This means that the pins A10 to A8 must not be used as chip enables since they will be used for decoding of port addresses. Therefore, A14 is used as chip enable in the figure. The unused address lines A11 to A13 and A15 are don't cares and are assumed to be zero in the following. Therefore, the port addresses are

A15	A14	A13	A12	A11	A10	A9	A8	
AD7	AD6	AD5	AD4	AD3	AD2	AD1	AD0	Address
0	0	0	0	0	0	0	0	= 00H
								= CSR
0	0	0	0	0	0	0	1	= 01H
								= Port A
0	0	0	0	0	0	1	0	= 02H
								= Port B
0	0	0	0	0	0	1	1	= 03H
								= Port C

For memory-mapped I/O, consider Figure 2.20b. In this case, the 8085 low byte and high byte address bus are not duplicated. The ports will have 16-bit addresses as follows. Assume the unused address pins A8 to AD3 to be zeros.

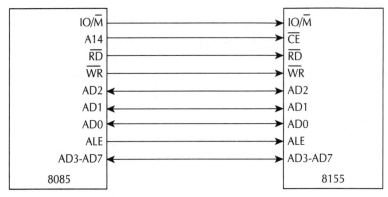

(a) 8085 – 8155 Interface using standard I/O

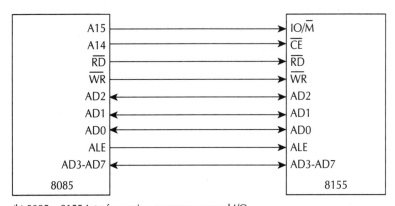

(b) 8085 – 8155 Interface using memory-mapped I/O

FIGURE 2.20   8085-8155 interface for I/O ports.

A15	A14	A13	A12	A11	A10	A9	A8	AD7	AD6	AD5	AD4	AD3	AD2	AD1	AD0	Port address
1	0	0	0	0	0	0	0	0	0	0	0	0	0	0	0	$= 8000_{16}$ CSR
1	0	0	0	0	0	0	0	0	0	0	0	0	0	0	1	$= 8001_{16}$ (Port A)
1	0	0	0	0	0	0	0	0	0	0	0	0	0	1	0	$= 8002_{16}$ (Port B)
1	0	0	0	0	0	0	0	0	0	0	0	0	0	1	1	$= 8003_{16}$ (Port C)

Like the 8355/8755 memory-mapped I/O, the above port addresses may not physically exist. However, read or write operations with them will generate the necessary signals for input or output transfer with the ports. Assuming all don't cares to be zeros, the memory map of either configuration of Figure 2.20a or b includes addresses 0000H through 00FFH.

Example 2.6

An 8085-8355-based microcomputer is required to drive an LED connected to bit 0 of port A based on the input conditions set by a switch on bit 1 of port A. The input/output conditions are as follows: if the input to bit 1 of port A is HIGH then the LED will be turned ON; otherwise the LED will be turned OFF. Assume that a HIGH will turn the LED ON and a LOW will turn it OFF. Write an 8085 assembly language program starting at 5000H.

Solution

```
        ORG  5000H
PORT A  EQU  00H
DDRA    EQU  02H
        MVI  A, 01H   ;  Configure Port A
        OUT  DDRA
START   IN PORT A     ;  Input Port A
        RAR           ;  Rotate switch to LED position
        OUT PORT A    ;  Output to LED
        JMP START     ;  Endless loop
```

Example 2.7

An 8085-8155-based microcomputer is required to drive an LED connected to bit 0 of port A based on two switch inputs connected to bits 6 and 7 of port A. If both switches are either HIGH or LOW, turn the LED on; otherwise turn it OFF. Assume that a HIGH will turn the LED ON and a LOW will turn it OFF. Use port addresses of CSR and port A as 20_{16} and 21_{16}, respectively.

Write an 8085 assembly language program to accomplish this starting at address 3000H.

Solution

```
        ORG  3000H
CSR     EQU  20H
PORT A  EQU  21H
START   MVI  A, 00H   ;  Configure Port A
        OUT  CSR      ;  as input
        IN   PORT A   ;  Input Port A
        ANI  0C0H     ;  Retain bits 6 and 7
                      ;  turn LED ON
        MVI  A, 01H   ;  Otherwise, configure
        OUT  CSR      ;  Port A as output
        JPE  LEDON
        MVI  A, 00H   ;  and turn LED OFF.
LEDON   OUT  PORT A   ;
        JMP  START    ;  Jump to START
```

Example 2.8

Write an 8085 assembly language program starting at address 5000H to turn on an LED connected to bit 4 of the 8155 I/O port B. Use address of port B as 22_{16}.

Solution

```
        ORG  5000H
Port B  EQU  22H
        MVI  A,02H    ;  Configure
        OUT  CSR      ;  Port B as output
        MVI  A,10H    ;  Output HIGH
        OUT  PORT B   ;  to LED
        HLT
```

Example 2.9

An 8085-8355-based microcomputer is required to drive a common anode seven-segment display connected to port A as follows:

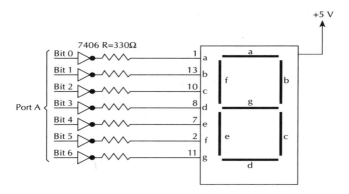

Write an 8085 assembly language program starting at address at 5000H to display a single hexidecimal digit (0 to F) from location 4000_{16}. Use a look-up table. Note that the 7406 shown in the figure contains six inverting buffers on one 7406. Therefore, two 7406 or one 7406 and one transistor are required for the seven segments. Use port addresses of DDRA and port A as 42H and 40H, respectively.

Solution

The decode table can be obtained as follows:

Hex digit	Bits								Decode byte
	7	6	5	4	3	2	1	0	
		g	f	e	d	c	b	a	
0	0	0	1	1	1	1	1	1	3F
1	0	0	0	0	0	1	1	0	06
2	0	1	0	1	1	0	1	1	5B
3	0	1	0	0	1	1	1	1	4F
4	0	1	1	0	0	1	1	0	66
5	0	1	1	0	1	1	0	1	6D
6	0	1	1	1	1	1	0	1	7D
7	0	0	0	0	0	1	1	1	07
8	0	1	1	1	1	1	1	1	7F
9	0	1	1	0	0	1	1	1	67
A	0	1	1	1	0	1	1	1	77
B	0	1	1	1	1	1	0	0	7C
C	0	0	1	1	1	0	0	1	39
D	0	1	0	1	1	1	1	0	5E
E	0	1	1	1	1	0	0	1	79
F	0	1	1	1	0	0	0	1	71

```
            ORG      5000H
DDRA        EQU      42H
Port  A     EQU      40H
MVI         A,7FH
OUT         DDRA     ;  Port  A
LXI         H,4015H  ;  Load  HL  with
                     ;  starting  address  of  table
LXI         D,0000H  ;  Load  0000H  to  DE
```

```
        LDA     DIGIT    ;  Load digit to be
                         ;  Displayed into A
        MOV     E,A      ;  Move digit to E
        DAD     D        ;  Determine digit address
        MOV     A,M      ;  Load decode byte to A
        OUT     Port A   ;  Outport decode byte
        HLT              ;  to display
        ORG     4015H
TABLE   DB      3FH, 06H, 5BH, 4FH
        DB      66H, 6DH, 7DH, 07H
        DB      7FH, 67H, 77H, 7CH
        DB      39H, 5EH, 79H, 71H
        ORG     4000H
DIGIT   DB      DATA
```

2.9.2 8085 Interrupt System

The 8085 chip has five interrupt pins, namely, TRAP, RST7.5, RST6.5, RST5.5, and INTR. If the signals on these interrupt pins go to HIGH simultaneously, then TRAP will be serviced first (i.e., highest priority) followed by RST7.5, RST6.5, RST5.5, and INTR. Note that once an interrupt is serviced, all the interrupts except TRAP are disabled. They can also be enabled or disabled simultaneously by executing the EI or DI instruction respectively. The 8085 interrupts are

1. TRAP — TRAP is a nonmaskable interrupt. That is, it cannot be enabled or disabled by an instruction. In order for the 8085 to service this interrupt, the signal on the TRAP pin must have a sustained HIGH level with a low to high transition. If this condition occurs, then the 8085 completes execution of the current instruction, pushes the program counter onto the stack, and branches to location 0024_{16} (interrupt address vector for the TRAP). Note that the TRAP interrupt is cleared by the falling edge of the signal on the pin.
2. RST7.5 — RST7.5 is a maskable interrupt. This means that it can be enabled or disabled using the SIM or EI/DI instruction. The 8085 responds to the RST7.5 interrupt when the signal on the RST7.5 pin has a low to high transition. In order to service RST7.5, the 8085 completes execution of the current instruction, pushes the program counter onto the stack, and branches to $003C_{16}$. The 8085 remembers the RST7.5 interrupt by setting an internal D flip-flop by the leading edge.
3. RST6.5 — RST6.5 is a maskable interrupt. It can be enabled or disabled using the SIM or EI/DI instruction. RST6.5 is HIGH level sensitive. In order to service this interrupt, the 8085 completes execution of the current instruction, saves the program counter onto the stack, and branches to location 0034_{16}.
4. RST5.5 — RST5.5 is a maskable interrupt. It can be enabled or disabled by the SIM or EI/DI instruction. RST5.5 is HIGH level sensitive. In order to service this interrupt, the 8085 completes execution of the current instruction, saves the program counter onto the stack, and branches to $002C_{16}$.
5. INTR — INTR is a maskable interrupt. It can be enabled or disabled by EI or DI instruction. This is also called the handshake interrupt. INTR is HIGH level sensitive. When no other interrupts are active and the signal on the INTR pin is HIGH, the 8085 completes execution of the current instruction, and generates an interrupt acknowledge, INTA, LOW pulse on the control bus. The 8085 then expects either a 1-byte CALL (RST0 through RST7) or a 3-byte CALL on the data lines. This instruction must be provided by external hardware. In other words, the INTA can be used to enable a tristate

buffer. The output of this buffer can be connected to the 8085 data lines. The buffer can be designed to provide the appropriate op code on the data lines. Note that the occurrence of INTA turns off the 8085 interrupt system in order to avoid multiple interrupts from a single device. Also note that there are eight RST instructions (RST0 through RST7). Each of these RST instructions has a vector address. These were shown in Table 2.9.

In response to a HIGH on the INTR, the 8085 proceeds with the sequence of events described below. If INTR is the only interrupt and if the 8085 system interrupt is enabled by executing the EI instruction, the 8085 will turn off the system interrupt and then make the $\overline{\text{INTA}}$ LOW for about two cycles. This $\overline{\text{INTA}}$ signal can be used to enable an external hardware to provide an op code on the data bus. The 8085 can then read this op code. Typically, the 1-byte RST or 3-byte CALL instruction can be used as the op code. If the 3-byte CALL is used, then the 8085 will generate two additional $\overline{\text{INTA}}$ cycles in order to fetch all 3 bytes of the instruction. However, on the other hand, if RST is used, then no additional $\overline{\text{INTA}}$ is required. The call op code is normally placed on the data bus by the 8259 programmable interrupt controller. At this point, only the op code for the CALL (CD16) is fetched by the 8085. The 8085 executes this instruction and determines that it needs two more bytes (the address portion of the 3-byte instruction). The 8085 then generates a second $\overline{\text{INTA}}$ cycle followed by a third $\overline{\text{INTA}}$ cycle in order to fetch the address portion of the CALL instruction from the 8259. The 8085 executes the CALL instruction and branches to the interrupt service routine located at an address specified in the CALL instruction. Note that the recognition of any maskable interrupt (RST7.5, RST6.5, RST5.5, and INTR) disables all maskable interrupts to avoid multiple interrupts from the same device.

Therefore, in order that the 8085 can accept another interrupt, the last two instructions of the interrupt service routine will be EI followed by RET.

One can produce a single RST instruction, say RST7 (op code FFH in HEX), using 74LS244. The inputs I0 to I7 of the 74LS244 (Figure 2.21) are connected to HIGH and its enable line ($\overline{\text{OE}}$) is tied to an $\overline{\text{INTA}}$. In response to INTA LOW, the 74LS244 places FF in hex (RST7) on the data bus. Figure 2.21 shows a typical circuit. Figure 2.22 provides a schematic for eight priority interrupts.

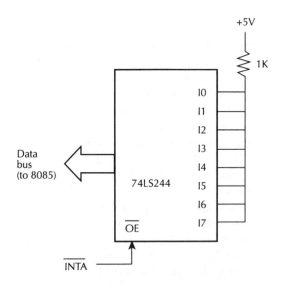

FIGURE 2.21 Using an octal buffer to provide RST7 instruction.

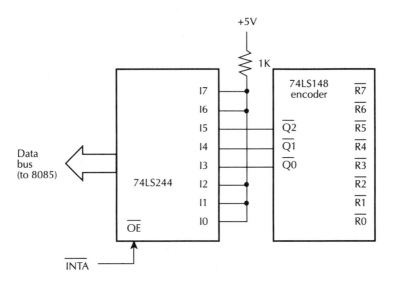

FIGURE 2.22 Forming eight RST instructions with a priority encoder.

Let us elaborate on Figure 2.23. Execution of the EI instruction sets the RS flip-flop of Figure 2.23 and makes one of the inputs to the AND gates #1 through #4 HIGH. Hence, in order for all the interrupts (except TRAP) to work, the interrupt system must be enabled. Execution of Disable Interrupts (DI) clears the RS flip-flop and disables all interrupts except TRAP. The SIM instruction outputs the contents of the accumulator which can be interpreted as shown in Figure 2.24.

The interrupt mask function is only executed if the mask set enable bit is 1. Suppose that if 06_{16} is stored in the accumulator and the SIM instruction is executed. A 1 will be sent to the interrupt mask for RST7.5 and RST6.5, and 0 will be sent to RST5.5. That is, in Figure 2.23, a 1 will be sent to the inputs of the AND gates #1 and #2, and a 0 will be sent to the AND gate #3, then inverted at the AND gate inputs (shown by circles), giving two LOW outputs disabling RST7.5, RST6.5, and a HIGH input to AND gate #3. Therefore, in order to enable RST7.5, RST6.5, or RST5.5, the interrupt system must be enabled by executing EI, the appropriate interrupt mask bit must be LOW by executing SIM, and the appropriate interrupt signal (leading edge or high level) at the respective pins must be available. For example, consider the RST7.5 interrupt. When the EI and SIM instructions are executed, the interrupt system can be enabled and also the interrupt mask bit for RST7.5 can be set to LOW, making the two inputs to AND gate #1 HIGH. The third input to this AND gate can be set to a HIGH by a leading edge at the RST7.5 pin. This sets the D flip-flop, thus making the output of the AND gate #1 HIGH, enabling RST7.5. The 8085 branches to location 003C16 where a 3-byte JMP instruction takes the program to the service routine. The RST5.5 and RST6.5 can similarly be explained from Figure 2.23.

The RIM instruction can be used to check whether one or more of the RST7.5, RST6.5, and RST5.5 interrupts are waiting to be serviced. The RIM instruction also provides the status of the mask bits for RST7.5, RST6.5, and RST5.5 and the status of the SID pin (HIGH or LOW). After execution of the RIM instruction, the 8085 loads the accumulator with 8 bits of data which can be interpreted as shown in Figure 2.25.

The bits are interpreted as follows:

Mask bits — Bits 0, 1, and 2 are the status of the mask bits for RST5.5, RST6.5, and RST7.5.
Interrupt Enable bit — This bit indicates whether one maskable interrupt capability is
   enabled (if this bit = 1) or disabled (if this bit = 0).

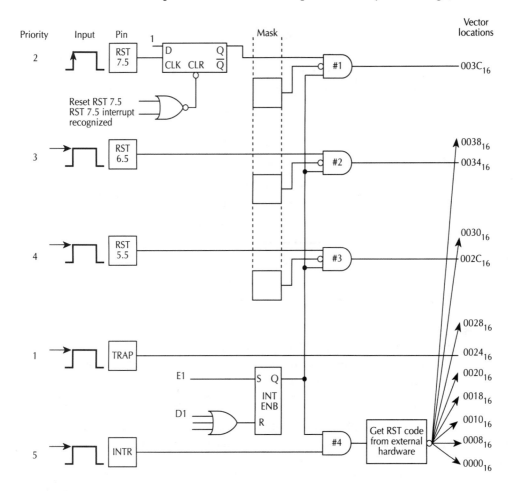

FIGURE 2.23   8085 interrupt structure.

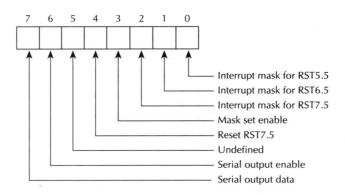

FIGURE 2.24   Interpretation of data output by the SIM instruction.

Interrupt Pending bits — A one in a particular bit position indicates that the particular RST is waiting to be serviced, while a 0 indicates that no interrupt is pending.

Serial Input data — These data indicate the status of the SID pin.

The RIM instruction can be used to read the interrupt pending bits in the accumulator. These bits can then be checked by software to determine whether any higher priority interrupts

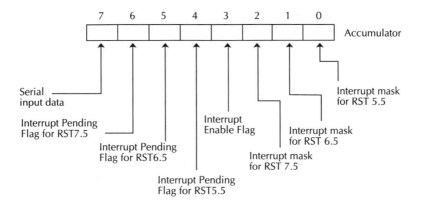

FIGURE 2.25 Execution of RIM instruction.

are pending. For example, suppose that RST6.5 and RST5.5 interrupts are used in a system. If RST5.5 interrupt first occurs, assuming no other higher level interrupts, the 8085 will execute the RST5.5 service routine. Note that while in this service routine, the RST6.5 interrupt (higher priority than RST5.5 interrupt) may occur. Since the 8085's maskable interrupt capability is disabled, the user can check whether the RST6.5 is pending while executing the RST5.5 service routine in one of two ways:

- By executing EI at the beginning of the RST5.5 service routine provided that the level at RST5.5 pin is LOW (RST5.5 disabled). This will service the RST6.5 immediately if pending and will then complete RST5.5 service routine (nested interrupts).
- By executing RIM and using rotate instruction to check whether RST6.5 is pending. The RIM instruction can be executed at several places in the RST5.5 service routine. If the RST6.5 is found to be pending, the interrupt can be enabled by executing EI in the RST5.5 service routine.

2.9.3 8085 DMA

The Intel 8257 DMA controller chip is a 40-pin DIP and is programmable. It is compatible with the 8085 microprocessor. The 8257 is a four-channel DMA controller with priority logic built into the chip. This means that the 8257 provides for DMA transfers for a maximum of up to four devices via the DMA request lines DRQ0 to DRQ3 (DRQ0 has the highest priority and DRQ3 the lowest). Associated with each DRQ is a DMA acknowledge (DACK0 to DACK3 for four DMA requests DRQ0 to DRQ3). Note that the DACK signals are active LOW. The 8257 uses the 8085 HOLD pin in order to take over the system bus. After initializing the 8257 by the 8085, the 8257 performs the DMA operation in order to transfer a block of data of up to 16,384 bytes between the memory and a peripheral without involving the microprocessor. A typical 8085-8257 interface is shown in Figure 2.26. An I/O device, when enabled by the 8085, can request a DMA transfer by raising the DMA request (DRQ) line of one of the channels of the 8257. In response, the 8257 will send a HOLD request (HRQ) to the 8085. The 8257 waits for the HOLD acknowledge (HLDA) from the 8085. On receipt of HLDA from the 8085, the 8257 generates a LOW on the DACK lines for the I/O device. Note that DACK is used as a chip select bit for the I/O device. The 8257 sends the READ or WRITE control signals, and data are transferred between the I/O and memory. On completion of the data transfer, the DACK0 is set to HIGH, and the HRQ line is reset to LOW in order to transfer control of the bus to the 8085. The 8257 utilizes four clock cycles in order to transfer 8 bits of data.

The 8257 has three main registers. These are a 16-bit DMA address register, a terminal count register, and a status register. Both address and terminal count registers must be initialized

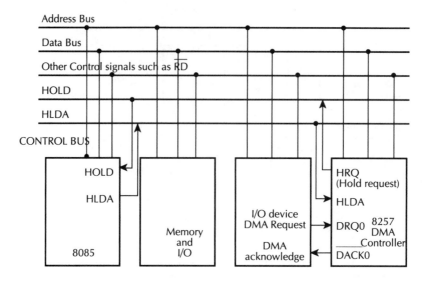

FIGURE 2.26 An 8085-8257 interface.

before a DMA operation. The DMA address register is initialized with the starting address of the memory to be written into or read from. The low-order 14 bits of the terminal count register are initialized with the value $(n - 1)$, where n is the desired number of DMA cycles. A Terminal Count (TC) pin on the 8257 is set to HIGH in order to indicate to the peripheral device that the present DMA cycle is the last cycle. An 8-bit status register in the 8257 is used to indicate which channels have attained a terminal count.

2.9.4 8085 SID and SOD Lines

Serial I/O is extensively used for data transfer between a peripheral device and the microprocessor. Since microprocessors perform internal operations in parallel, conversion of data from parallel to serial and vice versa is required to provide communication between the microprocessor and the serial I/O. The 8085 provides serial I/O capabilities via SID (Serial Input Data) and SOD (Serial Output Data) lines.

One can transfer data to or from the SID or SOD lines using the instruction RIM and SIM. After executing the RIM instruction, the bits in the accumulator are interpreted as follows:

1. Serial input bit is bit 7 of the accumulator.
2. Bits 0 to 6 are interrupt masks, the interrupt enable bit, and pending interrupts.

The SIM instruction sends the contents of the accumulator to the interrupt mask register and serial output line. Therefore, before executing the SIM, the accumulator must be loaded with proper data. The contents of the accumulator are interpreted as follows:

1. Bit 7 of the accumulator is the serial output bit.
2. The SOD enable bit is bit 6 of the accumulator. This bit must be 1 in order to output bit 7 of the accumulator to the SOD line.
3. Bits 0 to 5 are interrupt masks, enables, and resets.

Example 2.10

An 8085/8155-based microcomputer is required to input a switch via the SID line and output the switch status to an LED connected to the SOD line. Write an 8085 assembly language program to accomplish this.

Solution

```
      ORG   5000H
START RIM           ; Bit 7 of 'A' is SID
      ORI   40H     ; Set SOD Enable to one
      SIM           ; output to LED
      JMP   START   ; REPEAT
```

Example 2.11

Write an 8085 assembly language program to implement the following requirements. i) if V1 > V2, the 8085/8155-based microprocessor system will read the switch input from port B. If switch is open (input high), turn the LED off. If switch is closed (input low), turn the LED on. ii) Repeat i) by using a) TRAP, b) RST6.5 with SID/SOD, c) INTR.

Assume CSR
                = 00H
        Port A = 01H
        Port B = 02H
        Port C = 03H

Write all programs starting at 2000H and service routines at 3000H. Intialize SP to 20C0H.

i) *Solution*

```
            ORG   2000H
      CSR   EQU   00H     ; Define CSR address.
      PORTA EQU   01H     ; Define Port A address.
      PORTB EQU   02H     ; Define Port B address.
      PORTC EQU   03H     ; Define Port C address.
            MVI   A,0CH   ; Set Port A/Port B input and Port
                          ; C output.
            OUT   CSR     ; Write to CSR.
            OUT   PORTC   ; Initialize bit 5 of port C. Turn
                          ; LED off.
      SCAN  IN    PORTA   ; Get the data from Port A.
            RAR           ; Move the bit0 of A to Cy.
            JNC   SCAN    ; Go to SCAN if V1 < V2. Go to
                          ; next if V1>V2.
            IN    PORTB   ; Get the switch input from Port B.
            XRI   02H     ; Invert the input data.
            RAL           ; Move the data in bit 1 of Port A
                          ; to bit 5.
            RAL           ;
            RAL           ;
            RAL           ;
            OUT   PORTC   ; Write to Port C.

            JMP   SCAN    ; Loop
```

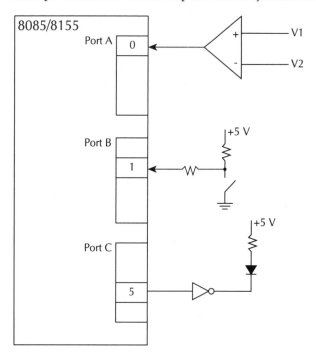

ii) *Solution*

```
    a)       ORG    0024H     ; Trap vector address entry.
             JMP    3000H     ; Go to trap service routine.
             ORG    2000h
    CSR      EQU    00H       ; Define CSR address.
    PORTB    EQU    02H       ; Define Port B address.
    PORTC    EQU    03H       ; Define Port C address.
             LXI    SP,20C0H  ; Initialize SP.
             MVI    A,0CH     ; Set Port B input and Port C
                               output.
             OUT    CSR       ; Write to CSR.
             OUT    PORTC     ; Initialize bit 5 of Port C. Turn
                               LED off.
    LOOP     NOP             ; Wait for TRAP request.
             JMP    Loop
             ORG    3000H     ; TRAP service routine entry.
             IN     PORTB     ; Get the switch input from Port
                               B.
             XRI    02H       ; Invert the input data.
             RAL             ; Move the data in bit 1 of A to
                               bit 5.
             RAL             ;
             RAL             ;
             RAL             ;
             OUT    PORTC     ; Write to Port C.
             RET
```

ii) *Solution*

```
    b)       ORG    0034H     ; RST6.5 vector address entry.
```

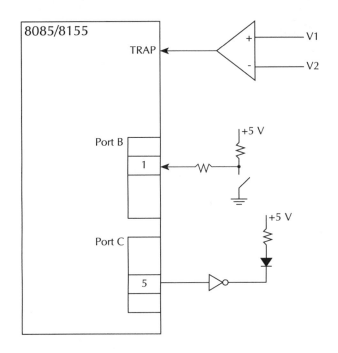

```
        JMP    3000H       ; Go to RST6.5 service routine.
        ORG    2000h
        LXI    SP,20C0H    ; Initialize SP.
        MVI    A,40h       ; Initialize SOD. Turn LED off.
        SIM                ;
                           ;
        MVI    A,0DH       ; Set data for RST6.5.
        SIM                ; Enable RST6.5.
        EI                 ; Enable interrupt.
Loop    NOP
        JMP    LOOP        ; Wait for RST6.5 request.
        ORG    3000H       ; RST6.5 service routine entry.
        RIM                ; Get the data from SID>
        XRI    80H         ; Invert the input data in bit
                             7 of A.
        ORI    40H         ; Enable SOD.
        SIM                ; Write to SOD.
        RET
```

ii) *Solution*

```
   c)  ORG    0030H       ; RST6 vector address entry
        JMP    3000H       ; Go to RST6 service routine
        ORG    2000H
CSR     EQU    00H         ; Define CSR address.
PORTB   EQU    02H         ; Define Port B address.
PORTC   EQU    03H         ; Define Port C address.
        LXI    SP,20C0H    ; Initialize SP.
        MVI    A,0CH       ; Set Port B input and Port C
                             output.
        OUT    CSR         ; Write to CSR.
```

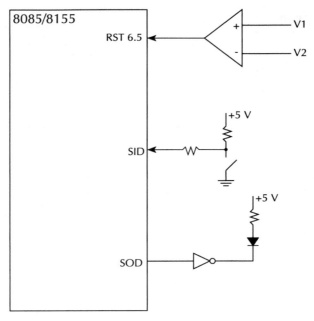

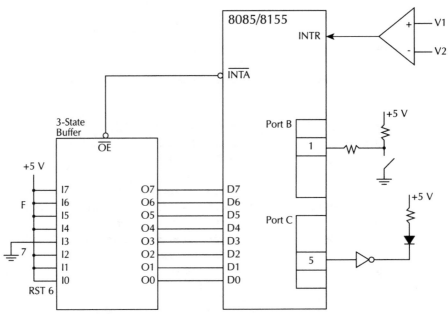

```
         OUT   PORTC      ; Initialize bit 5 of port C. Turn
                            LED off.
         EI               ; Enable interrupt.
Loop     NOP              ; Wait for INTR request.
         JMP   LOOP
         ORG   3000H      ; RST6 service routine entry.
         IN    PORTB      ; Get the switch input from Port
                            B.
         XRI   02H        ; Invert the input data.
         RAL              ; Move the data in bit 1 of A to
                            bit 5.
```

```
RAL                 ;
RAL                 ;
RAL                 ;
OUT    PORTC        ; Write to Port C.
RET
```

2.10 8085-Based System Design

In order to illustrate the concepts associated with 8085-based system design, a microcomputer with 2K EPROM (2716), 256 byte RAM, and 3 ports (8155) is designed. A hardware schematic is included. Also, an 8085 assembly language program is provided to multiply 4-bit unsigned numbers entered via DIP switches connected to port A. The 8-bit product is displayed on two seven-segment displays interfaced via port B. Repeated addition will be used for multiplication. Figure 2.27 shows a schematic of the hardware design. Full decoding using the 74LS138 decoder is utilized. Texas Instruments TIL 311's displays with on-chip decoder are used. The memory and I/O maps are given in the following:

1. Memory map

    2716

A15	A14	A13	A12	A11	A10	A9	A8	AD7	AD6	AD5	AD4	AD3	AD2	AD1	AD0
0	0	0	0	0	{- - - - - - - - - - - - - - - - - all zeros to ones - - - - - - - - - - - -}

    Result 0000H-07FFH

    8155

A15	A14	A13	A12	A11	A10	A9	A8	AD7	AD6	AD5	AD4	AD3	AD2	AD1	AD0
0	0	0	0	1	0	0	0	{- - - - - - - - - - - all zeros to ones - - - - - - - -}

    Result 0800H-08FFH

2. I/O map using standard I/O

    Ports

A15 AD7	A14 AD6	A13 AD5	A12 AD4	A11 AD3	A10 AD2	A9 AD1	A8 AD0		
0	0	0	0	1	0	0	0	CSR	08H
0	0	0	0	1	0	0	1	Port A	09H
0	0	0	0	1	0	1	0	Port B	0AH
0	0	0	0	1	0	1	1	Port C	0BH

A listing of the 8085 assembly language program for performing 4 bit × 4 bit multiplication is provided below:

```
CSR EQU   08H
PORT A    EQU   09H
PORT B    EQU   0AH
REPEAT    MVI   A,02H   ; Configure Port A as input
    OUT   CSR           ; and Port B as output
    MVI   L,00H         ; Initialize product to zero
    IN    Port A        ; Input multiplier and
                          multiplicand
```

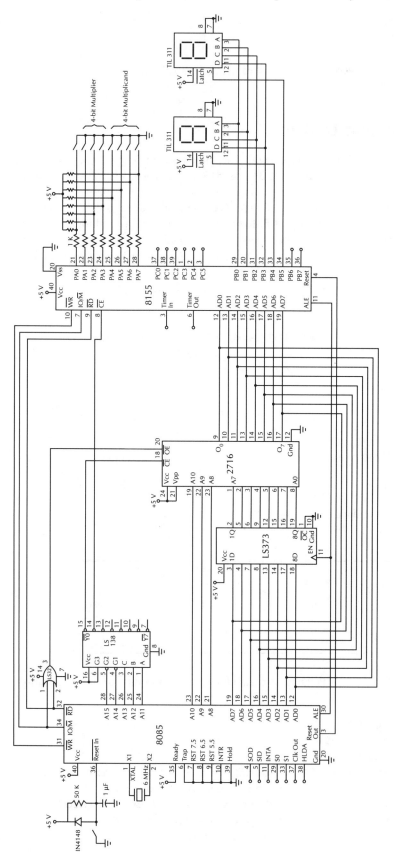

FIGURE 2.27   8085-based system design.

```
          MOV   B,A           ; Save multiplicand and
                              ; multiplier in B
          ANI   OFH           ; Mask multiplicand and retain
                              ; multiplier
          MOV   C,A           ; Save 4-bit multiplier in C
          MOV   A,B           ; Move multiplicand and
                              ; multiplier to A
          RAR                 ; Move 4-bit multiplicand
          RAR                 ; (upper nibble of A)
          RAR                 ; into LOW nibble
          RAR                 ; of
                              ; accumulator
          ANI   OFH           ; Mask high nibble and retain
          MOV   D,A           ; multiplicand
START     ADD   D             ; Perform repeated addition
          DCR   C             ; Decrement multiplier value
          MOV   L,A           ; Save sum
          JNZ   START         ; if Z = 0, repeat addition
                              ; else product in L
          MOV   A,L           ; Move product to A
          ANI   OFH           ; Retain product LOW nibble and
          MOV   H,A           ; Save in H
          MOV   A,L           ; Move product to A
          RAR                 ; Move
          RAR                 ; high nibble
          RAR                 ; of product
          RAR                 ; to LOW nibble
                              ; of A
          ANI   2FH           ; Retain high nibble of
                              ; product and enable latch
          OUT   PORTB         ; of high hex display and
                              ; disable low latch
                              ; Display high nibble of product
          MOV   A,H           ; Move LOW product nibble to A
          ANI   1FH           ; Enable LOW latch and disable
                              ; high latch
          OUT   PORTB         ; Display LOW nibble of product
          JMP   REPEAT        ; On low display and continue
```

The above program can be assembled. The 2716 can then be programmed by using an EPROM programmer with the machine code of the preceding program starting at location 000H. Then upon activation of the switch at the 8085 reset input, the dip switch low and high nibbles will be multiplied and the result of the multiplication will be displayed on the two TIL311s. By changing DIP switch inputs at port A, new results can be displayed.

Questions and Problems

2.1 Compare the main differences of 8085AH, 8085AH-2, and 8085AH-1 with the 8085.

2.2 What is the primary purpose of the 8085 H-L pair with respect to external memory? List two of its main functions.

2.3 Identify the addressing modes of the following instructions:
   i) **MOV A, M**
   ii) **RAR**
   iii) **STAX D**
   iv) **LDA START**

2.4 Assume register pair BC contains $7F02_{16}$. For the following 8085 assembly language program, determine the carry, zero, parity, and sign flags after execution of the MVI A,05H instruction:

```
ORG    4000H
LXI    SP, 2050H
PUSH   B
MVI    A,02H
ADI    03H
POP    PSW
MVI    A, 05H
HLT
```

2.5 What function is performed by each of the following 8085 instructions:
   i) **XOR A**
   ii) **MOV D,D**
   iii) **DAD H**

2.6 i) What is the function of the 8085 ALE pin?
   ii) If a crystal of 4 MHz is connected to the 8085 X1X2 pins, what is the 8085 internal clock period?

2.7 Write an 8085 assembly language program to divide an 8-bit unsigned number in accumulator by 4. Neglect remainder. Use minimum number of instructions. Store result in A.

2.8 Using the simplest possible algorithm, write an 8085 assembly language program with a minimum number of instructions to divide a 16-bit unsigned number in DE by 16. Neglect remainder. Store result in DE.

2.9 Assume the contents of DE are 2050H. Write an 8085 assembly language program to unconditionally branch to 2050H. Do not use any conditional or unconditional Jump or CALL or return instructions. Use three instructions maximum.

2.10 Write an 8085 assembly language program using a minimum number of instructions to add the 16-bit numbers in BC, DE, and HL. Store the 16-bit result in DE.

2.11 Write an 8085 assembly language program to add two 24-bit numbers located in memory address 2000_{16} through 2005_{16} with the most significant byte of the first number in 2000_{16} and the most significant byte of the second number in 2003_{16}. Store the 24-bit result in three consecutive bytes starting at address 2000_{16}.

2.12 Write a subroutine in 8085 assembly language to divide a 16-bit unsigned number in BC by 32_{10}. Neglect the remainder. Use the simplest possible algorithm. Also, write the main program in 8085 assembly language to perform all the initializations. The main program will then call the subroutine, store the 16-bit quotient in DE, and stop.

2.13 Write an 8085 assembly language program to shift the contents of the DE register twice to left without using any ROTATE instructions. After shifting, if the contents of DE are nonzero, store $FFFF_{16}$ in the HL pair. On the other hand, if the contents of DE pair are zero, then store 0000_{16} in the HL pair.

2.14 Write an 8085 assembly language program to check the parity of an 8-bit number in "A" without using any instructions involving the parity flag. Store EE16 in location 3000_{16} if the parity is even; otherwise, store DD16 in location 3000_{16}.

2.15 Write an 8085 assembly language program to move a block of data of length 100_{10} from the source block starting at 2000_{16} to the destination block starting at 3000_{16}.

2.16 Write a subroutine in 8085 assembly language to divide an 8-bit unsigned number Xi by 2. Also, write the main program in 8085 assembly language which will call the subroutine to compute

$$\sum_{i=1}^{3} \frac{Xi}{2}$$

Store the result in location 5000_{16}. Use a minimum number of instructions.

2.17 Assume an 8085/8355-based microcomputer. Suppose that four switches are connected to bits 0 through 3 of port A, an LED to bit 4 of port A, and another LED to bit 2 of port B. If the number of low switches is even, turn the port A LED ON and port B LED OFF. If the number of low switches is odd, turn port A LED OFF and port B LED ON. Write an 8085 assembly language program to accomplish the above.

2.18 Draw a simplified diagram using an 8085, one 8156, and one 8355 to include the following memory map and I/O ports:
- i)   *8355*   *8156*
  2000_{16} thru $27FF_{16}$ 4000_{16} thru $40FF_{16}$
- ii)   Using memory-mapped I/O, configure ports A and B of the 8355 at addresses 8000_{16} and 8001_{16}. Show only connections for the lines which are pertinent.

2.19 It is desired to interface a pump to a 8085/8355/8156-based microcomputer. The pump will be started by the microcomputer at the trailing edge of a start pulse via the SOD pin. When the pump runs, a HIGH status signal from the pump called "Pump Running" will be used to interrupt the microcomputer via its INTR interrupt. In response, the microcomputer will turn an LED on connected to bit 7 of the 8355 port A. Assume DDRA as the data direction register for this port.
- i)   Draw a simplified schematic for accomplishing the above.
- ii)   Write the main and interrupt service routines using the 8085 assembly language programs.

2.20 Will the circuit shown work? If so, determine the memory and I/O maps in hex. If not, justify briefly, modify the circuit and then determine the memory and I/O maps. Use only the pins and signals as shown. Also, assume all don't cares to be zeros.

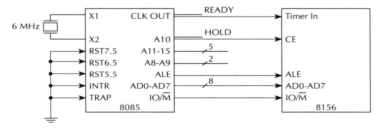

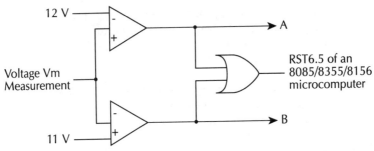

In the above, if Vm > 12 V, turn an LED ON connected at bit 3 of port A. On the other hand, if Vm < 11 V, turn the LED OFF. Use ports, registers, and memory locations as needed.

  i) Draw a hardware block diagram showing the microcomputer and the connections of the above diagram to its ports.
  ii) Write main program and service routines in 8085 assembly language.

2.22 Interface the following 8156 RAM chip to obtain the following memory map: 0400H thru 04FFH. Show only the connections for the pins which are shown. Assume all unused address lines to be zero.

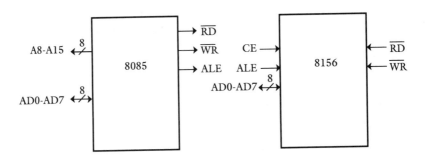

2.23

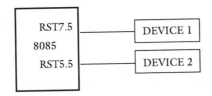

If device 2 is presently being serviced by the 8085 and the device 1 interrupt occurs, explain briefly what the user needs to do in the service routine of device 2 in order that device 1 will be serviced before device 2.

2.24 Assume an 8085/8355 microcomputer. Suppose that two switches are connected to bit 1 of Port A and the SID line. Also, an LED is connected to the SOD line. It is desired to turn the LED ON if both switches are HIGH; otherwise the LED is to be turned off. Write an 8085 assembly language program to input both switches and turn the LED ON or OFF based on the above conditions.

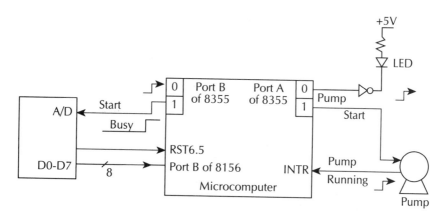

2.25 It is desired to interface a pump and an A/D converter to an 8085/8355/8156-based microcomputer as follows: The pump can be started by a HIGH at the pump start signal. If the pump starts, the pump-running signal goes to HIGH; otherwise the pump running stays LOW. The A/D converter can be started by a HIGH and the conversion is completed when the Busy signal is HIGH. Start the pump and A/D converter at nearly the same time. If the pump runs, the LED is to be turned on. On the other hand, if the Busy signal goes to HIGH, the A/D converter's output is to be read.

   i)   Draw a simplified hardware schematic to accomplish the above.
   ii)   Write main program and service routines in 8085 assembly language. Include the interrupt priority concept in the programs.

2.26 Assume an 8085/8355-based microcomputer. Write an 8085 assembly language program to turn an LED (LED 1) ON connected to bit 3 of port A if a 16-bit number in DE register pair is negative after shifting once to the left; if the number is positive after shifting, turn the LED (LED 1) OFF. Also, after shifting, if there is a sign change of the 16-bit number, turn another LED (LED 2) ON connected to bit 1 of port A; otherwise, turn the LED (LED 2) OFF.

2.27 Design an 8085/2716/8155-based microcomputer to include the following:
   i)   4K EPROM, 512 bytes static RAM, four 8-bit and two 6-bit ports. Use standard I/O and linear decoding.
   ii)   Repeat i) except use fully decoding using a 3 × 8 decoder.
   iii) Repeat i) except use memory-mapped I/O.
   iv)   Repeat i) except use memory-mapped I/O and full decoding.

Draw a neat schematic of each design and also determine memory and I/O maps in each case.

3

INTEL 8086

This chapter describes the internal architecture, addressing modes, instruction set, and I/O techniques associated with the 8086 microprocessor. Interfacing capabilities to typical memory and I/O chips such as the 2716, 6116, and 8255 are included.

A design technique is presented showing interconnection of the 8086 to 2716 EPROM, 6116 RAM, and 8255 I/O chips. The memory and I/O maps are then determined.

3.1 Introduction

The 8086 is Intel's first 16-bit microprocessor.

The 8086 is designed using the HMOS technology and contains approximately 29,000 transistors. The 80C86A is the low power version of the 8086 designed using HCMOS technology. The 8086 is packaged in a 40-pin CERDIP or plastic package and requires a single 5V power supply. The 8086 can be operated at three different clock speeds. The standard 8086 runs at 5 MHz internal clock frequency, whereas the 8086-1 and 8086-2 run at internal clock frequencies of 10 and 8 MHz, respectively. An external clock generator/driver chip such as the Intel 8284 is needed to generate the 8086 clock input signal. The 8284 divides the external crystal input internally by three. This means that for a 5-MHz 8086 internal clock, the 8284's X1 and X2 pins must be connected to a 15-MHz crystal. The 8284 will then generate a 5 MHz clock at its CLK pin which should be connected to the 8086 CLK input.

The 8086 has a 20-bit address and, hence, it can directly address up to one megabyte (2^{20}) of memory. The 8086 uses a segmented memory. An interesting feature of the 8086 is that it prefetches up to six instruction bytes from memory and queues them in order to speed up instruction execution.

There are some advantages of working with the segmented memory. First of all, after initializing the 16-bit segment registers, the 8086 has to deal with only 16-bit effective addresses. That is, the 8086 has to manipulate and store 16-bit address components. Secondly, because of memory segmentation, the 8086 can be effectively used in time-shared systems. For example, in a time-shared system, several users share a microprocessor. The microprocessor works with one user's program for say, 10 milliseconds. After spending 10 milliseconds with each of the other users, the microprocessor returns to execute the first user's program. Each time the microprocessor switches from one user's program to the next, it must execute a new section of code and new sections of data. Segmentation makes it easy to switch from one user program (and data) to another.

The memory of an 8086-based microcomputer is organized as bytes. Each byte can be uniquely addressed with 20-bit addresses of 00000_{16}, 00001_{16}, 00002_{16} ..., $FFFFF_{16}$. An 8086 16-

bit word consists of any two consecutive bytes; the low-addressed byte is the low byte of the word and the high-addressed byte contains the high byte as follows:

Low byte of the word	High byte of the word
07_{16}	26_{16}
Address 00520_{16}	Address 00521_{16}

The 16-bit word stored at the even address 00520_{16} is 2607_{16}.

Next consider a word stored at an odd address as follows:

Low byte of the word	High byte of the word
05_{16}	$3F_{16}$
Address 01257_{16}	Address 01258_{16}

The 16-bit word stored at the odd address 01257_{16} is $3F05_{16}$. Note that for word addresses, the programmer uses the low-order address (odd or even) to specify the whole 16-bit word.

The 8086 always accesses a 16-bit word to or from memory. The 8086 can read a 16-bit word in one operation if the first byte of the word is at an even address. On the other hand, the 8086 must perform two memory accesses to two consecutive memory even addresses, if the first byte of the word is at an odd address. In this case, the 8086 discards the unwanted bytes of each. For example, consider MOV BX, [ADDR]. Note that the X or H (or L) following the 8086 register name in an instruction indicates whether the transfer is 16-bit (for X) or 8-bit (for H or L). The instruction, MOV BX, [ADDR] moves the contents of a memory location addressed by ADDR into the 8086 16-bit register BX. Now, if ADDR along with the data segment register provides a 20-bit even address such as 30024_{16}, then this MOV instruction loads the low (BL) and high (BH) bytes of the 8086 16-bit register BX with the contents of memory locations 30024_{16} and 30025_{16}, respectively, in a single access. Now, if ADDR is an odd address such as 40005_{16}, then the MOV BX, [ADDR] instruction loads BL and BH with the contents of memory locations 40005_{16} and 40006_{16}, respectively, in two accesses. Note that the 8086 accesses locations 40004_{16} and 40005_{16} in the first operation but discards the contents of 40004_{16}, and in the second operation accesses 40006_{16} and 40007_{16} but ignores the contents of 40007_{16}.

Next, consider a byte move such as MOV BH, [ADDR]. If ADDR is an even address such as 50002_{16}, then this MOV instruction accesses both 50002_{16} and 50003_{16}, but loads BH with the contents of 50002_{16} and ignores the contents of 50003_{16}. However, if ADDR is an odd address such as 50003_{16}, then this MOV loads BH with the contents of 50003_{16} and ignores the contents of 50002_{16}.

The 8086 family consists of two types of 16-bit microprocessors — the 8086 and 8088. The 8088 has an 8-bit external data path to memory and I/O, while the 8086 has a 16-bit external data path. This means that the 8088 will have to do two read operations to read a 16-bit word into memory. In most other respects, the processors are identical. Note that the 8088 accesses memory in bytes. No alterations are needed to run software written for one microprocessor on the other. Because of similarities, only the 8086 will be considered here. The 8088 was used in designing the original IBM Personal computer.

An 8086 can be configured as a small uniprocessor system (minimum mode if the MN/MX pin is tied to HIGH) or as a multiprocessor system (maximum mode when MN/MX pin is tied to LOW). In a given system, the MN/MX pin is permanently tied to either HIGH or LOW. Some of the 8086 pins have dual functions depending on the selection of the MN/MX pin level. In the minimum mode (MN/MX pin high), these pins transfer control signals directly to memory and input/output devices. In the maximum mode ($\overline{MN/MX}$ pin low), these same pins have different functions which facilitate multiprocessor systems. In the maximum mode, the control functions normally present in minimum mode are performed by a support chip, the 8288 bus controller.

Due to technological advances, Intel introduced the high performance 80186 and 80188 which are enhanced versions of the 8086 and 8088, respectively. The 8-MHz 80186/80188 provides two times greater throughput than the standard 5-MHz 8086/8088. Both have integrated several new peripheral functional units such as a DMA controller, a 16-bit timer unit, and an interrupt controller unit into a single chip. Just like the 8086 and 8088, the 80186 has a 16-bit data bus and the 80188 has an 8-bit data bus; otherwise, the architecture and instruction set of the 80186 and 80188 are identical. The 80186/80188 has an on-chip clock generator so that only an external crystal is required to generate the clock. The 80186/80188 can operate at either 6 or 8 MHz. Like the 8085, the crystal frequency is divided by 2 internally. In other words, external crystals of 12 or 16 MHz must be connected to generate the 6- or 8-MHz internal clock frequency. The 80186/80188 is fabricated in a 68-pin package. Both processors have on-chip priority interrupt controller circuits to provide five interrupt pins. Like the 8086/8088, the 80186/80188 can directly address one megabyte of memory. The 80186/80188 is provided with 10 new instructions beyond the 8086/8088 instruction set. Examples of these instructions include INS and OUTS for inputting and outputting string byte or string word. The 80286, on the other hand, has added memory protection and management capabilities to the basic 8086 architecture. An 8-MHz 80286 provides up to six times greater throughput than the 5-MHz 8086. The 80286 is fabricated in a 68-pin package. The 80286 can be operated at 6, 8, 10 or 12.5 MHz clock frequency. The 80286 is typically used in a multiuser or multitasking system. The 80286 was used as the CPU of the IBM PC/AT Personal computer. Intel's 32-bit microprocessor family includes 80386, 80486 and Pentium microprocessors which will be covered later in this book.

3.2 8086 Architecture

Figure 3.1 shows a block diagram of the 8086 internal architecture. As shown in the figure, the 8086 microprocessor is internally divided into two separate functional units. These are the Bus Interface Unit (BIU) and the Execution Unit (EU). The BIU fetches instructions, reads data from memory and ports, and writes data to memory and I/O ports. The EU executes instructions that have already been fetched by the BIU. The BIU and EU function independently. The BIU interfaces the 8086 to the outside world. The BIU contains segment registers, instruction pointer, instruction queue, and address generation/bus control circuitry to provide functions such as fetching and queuing of instructions, and bus control.

The BIU's instruction queue is a First-In-First-Out (FIFO) group of registers in which up to six bytes of instruction code are prefetched from memory ahead of time. This is done in order to speed up program execution by overlapping instruction fetch with execution. This mechanism is known as pipelining.

The bus control logic of the BIU generates all the bus control signals such as read and write signals for memory and I/O. The 8086 contains the on-chip logical address to physical address mapping hardware. The programmer works with the logical address which includes the 16-bit contents of a segment register and a 16-bit displacement or offset value. The 8086 on-chip mapping hardware translates this logical address to 20-bit physical address which it then generates on its twenty addressing pins.

The BIU has four 16-bit segment registers. These are the Code Segment (CS), the Data Segment (DS), the Stack Segment (SS), and the Extra Segment (ES). The 8086's one megabyte memory is divided into segments of up to 64K bytes each. The 8086 can directly address four segments (256K byte within the 1 Mbyte memory) at a particular time. Programs obtain access to code and data in the segments by changing the segment register contents to point to the

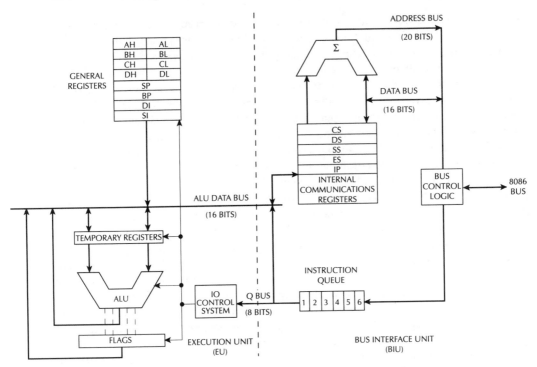

FIGURE 3.1 Internal architecture of the 8086.

desired segments. All program instructions must be located in main memory pointed to by the 16-bit CS register with a 16-bit offset in the segment contained in the 16-bit instruction pointer (IP). The BIU computes the 20-bit physical address internally using the programmer-provided logical address (16-bit contents of CS and IP) by logically shifting the contents of CS four bits to left and then adding the 16-bit contents of IP. For example, if $[CS] = 456A_{16}$ and $[IP] = 1620_{16}$, then the 20-bit physical address is generated by the BIU as follows:

$$\text{Four times log ically shifted } [CS] \text{ to left } = 456A0_{16}$$

$$+ [IP] \text{ as offset} \qquad\qquad = 1620_{16}$$

$$20\text{ - bit physical address} \qquad\qquad = 46CC0_{16}$$

The SS register points to the current stack. The 20-bit physical stack address is calculated from SS and SP for stack instructions such as PUSH and POP. The programmer can use the BP register instead of SP for accessing the stack using the based addressing mode. In this case, the 20-bit physical stack address is calculated from BP and SS.

The DS register points to the current data segment; operands for most instructions are fetched from this segment. A 16-bit offset (Effective Address, EA) along with the 16-bit contents of DS are used for computing the 20-bit physical address.

The ES register points to the extra segment in which data (in excess of 64K pointed to by DS) is stored. String instructions use ES and DI to determine the 20-bit physical address for the destination, and DS and SI for the source address.

The segments can be contiguous, partially overlapped, fully overlapped, or disjoint. An example of how five segments (segment 0 through segment 4) may be stored in physical memory are shown below:

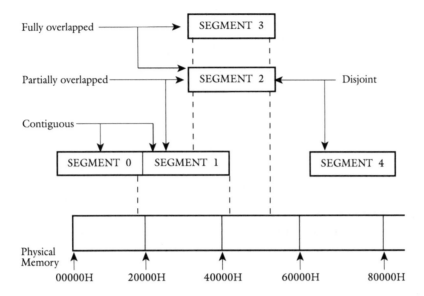

In the above, SEGMENTs 0 and 1 are contiguous (adjacent), SEGMENTs 1 and 2 are partially overlapped, SEGMENTs 2 and 3 are fully overlapped, and SEGMENTs 2 and 4 are disjoint. Every segment must start on 16-byte memory boundaries.

Typical examples of values of segments should then be selected based on physical addresses starting at 00000_{16}, 00010_{16}, 00020_{16}, 00030_{16} ..., $FFFF0_{16}$. A physical memory location may be mapped into (contained in) one or more logical segments. Many applications can be written to simply initialize the segment registers and then forget them. One can then work with a 64K memory as with the 8085.

The EU decodes and executes instructions. A decoder in the EU control system translates instructions. The EU has a 16-bit ALU for performing arithmetic and logic operations.

The EU has eight 16-bit general registers. These are AX, BX, CX, DX, SP, BP, SI, and DI. The 16-bit registers AX, BX, CX, and DX can each be used as two 8-bit registers (AH, AL, BH, BL, CH, CL, DH, DL). For example, the 16-bit register DX can be considered as two 8-bit registers DH (high byte of DX) and DL (low byte of DX). The general-purpose registers AX, BX, CX, and DX are named after special functions carried out by each one of them. For example, the AX is called the 16-bit accumulator while the AL is the 8-bit accumulator. The use of accumulator registers is assumed by some instructions. The Input/Output (IN or OUT) instructions always use AX or AL for inputting/outputting 16- or 8-bit data to or from an I/O port.

Multiplication and division instructions also use AX or AL. The AL register is the same as the 8085 A register.

The BX register is called the base register. This is the only general-purpose register, the contents of which can be used for addressing 8086 memory. All memory references utilizing these register contents for addressing use DS as the default segment register. The BX register is similar to 8085 HL register. In other words, 8086 BH and BL are equivalent to 8085 H and L registers, respectively.

The CX register is known as the counter register. This is because some instructions such as shift, rotate, and loop instructions use the contents of CX as a counter. For example, the instruction LOOP START will automatically decrement CX by 1 without affecting flags and will check if [CX] = 0. If it is zero, the 8086 executes the next instruction; otherwise the 8086 branches to the label START.

The data register DX is used to hold high 16-bit result (data) in 16×16 multiplication or high 16-bit dividend (data) before a $32 \div 16$ division and the 16-bit remainder after the division.

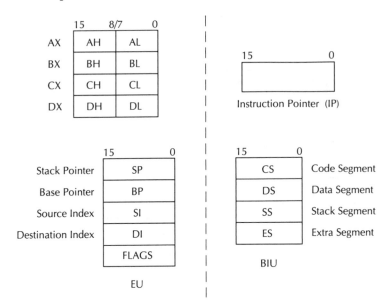

FIGURE 3.2 8086 registers.

The two pointer registers, SP (stack pointer) and BP (base pointer), are used to access data in stack segment. The SP is used as an offset from the current SS during execution of instructions that involve stack segment in external memory. The SP contents are automatically updated (incremented or decremented) due to execution of POP or PUSH instruction.

The base pointer contains an offset address in the current SS. This offset is used by the instructions utilizing the based addressing mode.

The FLAG register in the EU holds the status flags after an ALU operation. Figure 3.2 shows the 8086 registers.

The 8086 has six one-bit flags. Figure 3.3 shows the flag register. The AF (Auxiliary carry Flag) is used by BCD arithmetic instructions. AF = 1 if there is a carry from the low nibble (4-bit) into the high nibble or a borrow from the high nibble into the low nibble of the low-order 8-bit of a 16-bit number. The CF (Carry Flag) is set if there is a carry from addition or borrow from subtraction. The OF (Overflow Flag) is set if there is an arithmetic overflow, that is, if the size of the result exceeds the capacity of the destination location. An interrupt on overflow instruction is available which will generate an interrupt in this situation. The SF (Sign Flag) is set if the most significant bit of the result is one (negative) and is cleared to zero for non-negative result. The PF (Parity Flag) is set if the result has even parity; PF is zero for odd parity of the result. The ZF (Zero Flag) is set if result is zero; ZF is zero for nonzero result.

The 8086 has three control bits in the flag register which can be set or reset by the programmer: Setting DF (Direction Flag) to one causes string instructions to autodecrement the appropriate index register(s), and clearing DF to zero causes string instructions to autoincrement. Setting IF (Interrupt Flag) to one causes the 8086 to recognize external maskable interrupts; clearing IF to zero disables these interrupts. Setting TF (Trace Flag) to one places the 8086 in the single-step mode. In this mode, the 8086 generates an internal interrupt

15		12	11	10	9	8	7	6	5	4	3	2	1	0
	. . .		OF	DF	IF	TF	SF	ZF		AF		PF		CF

FIGURE 3.3 8086 flag register.

after execution of each instruction. The user can write a service routine at the interrupt address vector to display the desired registers and memory locations. The user can thus debug a program.

3.3 8086 Addressing Modes

The 8086 has 12 basic addressing modes. The various 8086 addressing modes can be classified into five groups:

1. Addressing modes for accessing immediate and register data (register and immediate modes)
2. Addressing modes for accessing data in memory (memory modes)
3. Addressing modes for accessing I/O ports (I/O modes)
4. Relative addressing mode
5. Implied addressing mode

The assembler directives for the Microsoft 8086 asembler written for IBM PC will be used to illustrate the above 8086 addressing modes.

3.3.1 Addressing Modes for Accessing Immediate and Register Data (Register and Immediate Modes)

3.3.1.a Register Addressing Mode

This mode specifies the source operand, destination operand, or both to be contained in an 8086 register. An example is MOV DX, CX which moves the 16-bit contents of CX into DX. Note that in the above both source and destination operands are in register mode. Another example is MOV CL, DL which moves 8-bit contents of DL into CL. MOV BX, CH is an illegal instruction; the register sizes must be the same.

3.3.1.b Immediate Addressing Mode

In immediate mode, 8- or 16-bit data can be specified as part of the instruction. For example, MOV CL, 03H moves the 8-bit data 03H into CL. On the other hand, MOV DX, 0502H moves the 16-bit data 0502H into DX. Note that in both of the above MOV instructions, the source operand is in immediate mode and the destination operand is in register mode.

A constant such as "VALUE" can be defined by the assembler EQUATE directive such as VALUE EQU 35H. An 8086 instruction with immediate mode such as MOV BH, VALUE can then be used to load 35H into BH. Note that even though the immediate mode specifies data with the instruction, these immediate data must be part of the program located in the code segment. That is, the memory must be addressed by the 8086 CS and IP registers. This is because these data are considered part of the instruction.

3.3.2 Addressing Modes for Accessing Data in Memory (Memory Modes)

The 8086 must use a segment register whenever it accesses the memory. Also, every memory addressing instruction uses an Intel-defined standard default segment register which is DS in most cases. However, a segment override prefix can be placed before most of the memory addressing instructions whose default segment register is to be overridden. For example, INC BYTE PTR [START] will increment the 8-bit content of memory location in DS with offset START by one. However, segment DS can be overridden by ES as follows: INC ES: BYTE PTR [START]; segments cannot be overridden for some cases such as stack reference instructions (such as PUSH and POP). There are six modes in this category. These are

a. Direct addressing mode
b. Register indirect addressing mode
c. Based addressing mode
d. Indexed addressing mode
e. Based indexed addressing mode
f. String addressing mode

3.3.2.a Direct Addressing Mode

In this mode, the 16-bit effective address (EA) is directly included with the instruction. As an example, consider MOV CX, DS:START. This instruction moves the 16-bit contents of a 20-bit physical memory location computed from START and DS into CX.

Note that in the above instruction, the source is in direct addressing mode. If the 16-bit value assigned to the offset START by the programmer using an assembler directive such as EQU is 0040_{16} and [DS] = 3050_{16}, then the BIU generates the 20-bit physical address 30540_{16} on the 8086 address pins and then initiates a memory read cycle to read the 16-bit data from memory location starting at 30540_{16} location. The memory logic places the 16-bit contents of locations 30540_{16} and 30541_{16} on the 8086 data pins. The BIU transfers these data to the EU; the EU then moves these data to CX; $[30540_{16}]$ to CL and $[30541_{16}]$ to CH.

3.3.2.b Register Indirect Addressing Mode

In this mode, the EA is specified in either a pointer register or an index register. The pointer register can be either base register BX or base pointer register BP and index register can be either the Source Index (SI) register or the Destination Index (DI) register. The 20-bit physical address is computed using DS and EA. For example, consider MOV [DI], BX. The destination operand of the above instruction is in register indirect mode, while the source operand is in register mode. The instruction moves the 16-bit content of BX into a memory location offset by the value of EA specified in DI from the current contents in DS*16. Now, if [DS] = 5004_{16}, [DI] = 0020_{16}, and [BX] = 2456_{16}, then after MOV [DI], BX, contents of BL (56_{16}) and BH (24_{16}) are moved to memory locations 50060_{16} and 50061_{16}, respectively.

Using this mode, one instruction can operate on many different memory locations if the value in the base or index register is updated.

3.3.2.c Based Addressing Mode

In this mode, EA is obtained by adding a displacement (signed 8-bit or unsigned 16-bit) value to the contents of BX or BP. The segment registers used are DS and SS. When memory is accessed using BX, the 20-bit physical address is computed from BX and DS. On the other hand, when the user stack is accessed using BP, the 20-bit physical address is computed from BP and SS. This allows the programmer to access the stack without changing the SP contents. Note that SP is called the system stack pointer since some 8086 stack instructions such as PUSH and POP automatically use SP as the stack pointer while BP is called the user stack pointer. As an example of this mode, consider MOV AL, START [BX]. Note that some assemblers use MOV AL, [BX + START] rather than MOV AL, START [BX]. The source operand in the above instruction is in based mode. EA is obtained by adding the value of START and [BX]. The 20-bit physical address is produced from DS and EA. The 8-bit content of this memory location is moved to AL. The displacement START can be either unsigned 16-bit or signed 8-bit. However, a byte is saved for the machine code representation of the instruction if 8-bit displacement is used. The 8086 sign-extends the 8-bit displacement and then adds it to [BX] in the above MOV instruction for determining EA. On the other hand, the 8086 adds an unsigned 16-bit displacement directly with [BX] for determining EA.

Based addressing provides a convenient way to address a structure which may be stored at different places in memory:

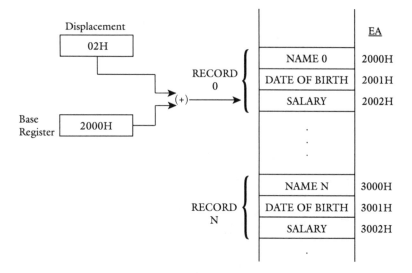

For example, the element salary in record 0 of the employee NAME 0 can be loaded into an 8086 internal register such as AL using the instruction MOV AL, ALPHA [BX], where ALPHA is the 8-bit displacement 02H and BX contains the starting address of the record 0. Now, in order to access the salary of RECORD N, the programmer simply changes the contents of the base register to 3000H.

If BP is specified as a base register in an instruction, the 8086 automatically obtains the operand from the current SS (unless a segment override prefix is present). This makes based addressing with BP a very convenient way to access stack data. BP can be used as a stack pointer in SS to access local variables. Consider the following instruction sequence:

```
PUSH BP              ;   Save BP
MOV BP, SP           ;   Establish BP
PUSH DX              ;   Save
PUSH AX              ;   registers
SUB SP, 4            ;   Allocate 2 words of
  :                  ;   stack for accessing stack
MOV [BP - 6], AX     ;   Arbitrary instructions for
MOV [BP - 8], BX     ;   accessing stack data using BP
ADD SP, 4            ;   Deallocate storage
POP AX               ;   Restore
POP DX               ;   all registers
POP BP               ;   that were pushed before
```

This instruction sequence is arbitrarily chosen to illustrate the use of BP for accessing the stack.

Figure 3.4 shows the 8086 stack during various stages. Figure 3.4A shows the stack before execution of the instruction sequence.

The instruction sequence from PUSH BP to SUB SP, 4 pushes BP, DX, and AX and then subtracts 4 from SP, and this allocates 2 words of the stack. The stack at this point is shown in Figure 3.4B. Note that in 8086, SP is decremented by 2 for PUSH and incremented by 2 for POP. The [BP] is not affected by PUSH or POP. The instruction sequence MOV [BP – 6], AX saves AX in the stack location addressed by [BP – 6] in SS. The instruction MOV [BP – 8], BX writes the [BX] into the stack location [BP – 8] in SS. These instructions are arbitrarily chosen to illustrate how BP can be used to access the stack. These two local variables can be accessed by the subroutine, using BP. The instruction ADD SP, 4 releases two words of the allocated

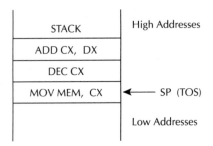

A) Stack before executing the instruction sequence.

(Assumed)

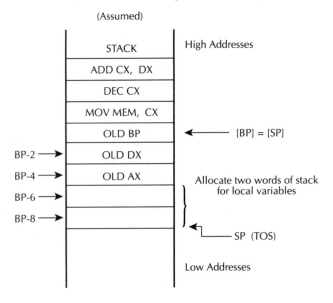

B) Stack after execution of the instruction sequence
from PUSH BP to SUB SP, 4

FIGURE 3.4　Accessing stack using BP.

stack. The stack at this point is shown in Figure 3.4C. The last three POP instructions restore the contents of AX, DX, and BP to their original values and return the stack as it was before the instruction sequence was executed. This is shown in Figure 3.4D.

3.3.2.d Indexed Addressing Mode

In this mode, the effective address is calculated by adding the unsigned 16-bit or sign-extended 8-bit displacement and the contents of SI or DI.

As an example, MOV BH, ARRAY[SI] moves the contents of the 20-bit address computed from the displacement ARRAY, SI and DS into BH. The 8-bit displacement is provided by the programmer using an assembler directive such as EQU. For 16-bit displacement, the EU adds this to SI to determine EA. On the other hand, for 8-bit displacement the EU sign-extends it to 16 bits and then adds to SI for determining EA.

The Indexed addressing mode can be used to access a single table. The displacement can be the starting address of the table. The content of SI or DI can then be used as an index from the starting address to access a particular element in the table.

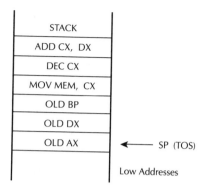

C) Stack after execution of ADD SP, 4

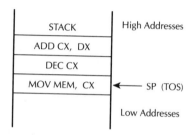

D) Stack before execution of POP BP.

FIGURE 3.4 (continued).

3.3.2.e Based Indexed Addressing Mode

In this mode, the EA is computed by adding a base register (BX or BP), an index register (SI or DI), and a displacement (unsigned 16-bit or sign-extended 8-bit). As an example, consider MOV ALPHA [SI] [BX], CL if [BX] = 0200H, value of ALPHA = 08H, [SI] = 1000H, and [DS] = 3000 H, then 8-bit content of CL is moved to 20-bit physical address 31208_{16}.

Based indexed addressing mode provides a convenient way for a subroutine to address an array allocated on a stack. Register BP can be loaded with the offset in segment SS (top of the stack after the subroutine has saved registers and allocated local storage). The displacement can be the value which is the difference between the top of the stack and the beginning of the array. An index register can then be used to access individual array elements as shown in Figure 3.5.

In the following, [BP] = top of the stack = 2005H; displacement = difference between the top of the stack and start of the array = 04H; [SI or DI] = N = 16-bit number (0, 2, 4, 6 in the example). As an example, the instruction MOV DX, 4 [SI] [BP] with [SI] = 6 will read the array (3) which is the content of 200FH in SS into DX. Since in the based indexed mode, the contents of two registers such as BX and SI can be varied, two-dimensional arrays such as matrices can also be accessed.

3.3.2.f String Addressing Mode

This mode uses index registers. The string instructions automatically assume SI points to the first byte or word of the source operand and DI points to the first byte or word of the destination operand. The contents of SI and DI are automatically incremented (by clearing DF to 0 by CLD instruction) or decremented (by setting DF to 1 by STD instruction) to point to the next byte or word. The segment register for the source is DS and may be overridden.

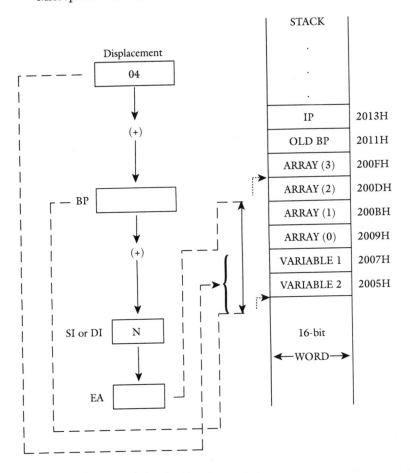

FIGURE 3.5 An example of Based Indexed Addressing.

The segment register for the destination must be ES and cannot be overridden. As an example, consider MOVS BYTE. If $[DF] = 0$, $[DS] = 2000_{16}$, $[SI] = 0500_{16}$, $[ES] = 4000_{16}$, $[DI] = 0300_{16}$, $[20500]_{16} = 38_{16}$, and $[40300_{16}] = 45_{16}$, then after execution of the MOVS BYTE, $[40300_{16}] = 38_{16}$, $[SI] = 0501_{16}$, and $[DI] = 0301_{16}$. The contents of other registers and memory locations are unchanged. Note that SI and DI can be used in either the source or destination operand of a two-operand instruction, except for string instructions in which SI points to the source (may be overridden) and DI must point to the destination.

3.3.3 Addressing Modes for Accessing I/O Ports (I/O Modes)

Standard I/O uses port addressing modes. For memory-mapped I/O, memory addressing modes are used. There are two types of port addressing modes: direct and indirect.

In direct port mode, the port number is an 8-bit immediate operand. This allows fixed access to ports numbered 0 to 255. For example, OUT 05H, AL outputs [AL] to 8-bit port 05H. In indirect port mode, the port number is taken from DX allowing 64K 8-bit ports or 32K 16-bit ports. For example, if $[DX] = 5040_{16}$, then IN AL, DX inputs the 8-bit content of port 5040_{16} into AL. On the other hand, IN AX, DX inputs the 8-bit contents of ports 5040_{16} and 5041_{16} into AL and AH, respectively. Note that 8-bit and 16-bit I/O transfers must take place via AL and AX, respectively.

3.3.4 Relative Addressing Mode

Instructions using this mode specify the operand as a signed 8-bit displacement relative to PC. An example is JNC START. This instruction means that if carry = 0, then PC is loaded with current PC contents (next instruction address) plus the 8-bit signed value of START; otherwise the next instruction is executed. Relative mode with signed 8-bit displacement provides a range of -128_{10} to $+127_{10}$ (0 being positive). If branching beyond this range is necessary, one must use the unconditional 8086 JUMP instruction which uses direct mode.

3.3.5 Implied Addressing Mode

Instructions using this mode have no operands. An example is CLC which clears the carry flag to zero.

3.4 8086 Instruction Set

The 8086 instruction set includes equivalents of the 8085 instructions plus many new ones. The new instructions contain operations such as signed and unsigned multiplication and division, bit manipulation instructions, string instructions, and interrupt instructions.

The 8086 has approximately 117 different instructions with about 300 op codes. The 8086 instruction set contains no operand, single operand, and two operand instructions. Except for string instructions which involve array operations, the 8086 instructions do not permit memory-to-memory operations. Table 3.1 lists a summary of 8086 instructions in alphabetical order. Tables supplied in Appendix E provide a detailed description of the 8086 instructions including the instruction execution times.

Examples of some of the 8086 instructions are given in the following.

1. Data transfer instructions

- MOV CX, DX copies the 16-bit content of DX into CX. MOV AX, 0205H moves immediate data 0205H into 16-bit register AX. MOV CH, [BX] moves the 8-bit content of memory location addressed by BX and segment register DS into CH. If [BX] = 0050H, [DS] = 2000H, [20050H] = 08H, then after MOV CH, [BX], the content of CH will be 08H.

  MOV START [BP], CX moves the 16-bit (CL to first location and then CH) content of CX into two memory locations addressed by the sum of the displacement START and BP, and segment register SS. For example, if [CX] = 5009H, [BP] = 0030H, [SS] = 3000H, START = 06H, then after MOV START [BP], CX physical memory location [30036H] = 09H and [30037H] = 50H. Note that the segment register SS can be overridden by CS using MOV CS: START [BP], CX.

- PUSH START [BX] pushes the 16-bit contents of two memory locations starting at the 20-bit physical address computed from START, BX, and DS after decrementing SP by 2.

- POP ES pops the top stack word into ES and then increments SP by 2.

- XCHG START [BX], AX exchanges the 16-bit word in AX with the contents of two consecutive memory locations starting at 20-bit physical address computed from START, BX, and DS. [AL] is exchanged with the content of the first location and [AH] is exchanged with the content of the next location.

- XLAT can be used to convert a code such as ASCII into another code such as EBCDIC. This instruction is equivalent to MOV AL, [AL][BX].

Table 3.1   Summary of 8086 Instructions

Instructions	Interpretation	Comments
AAA	ASCII adjust [AL] after addition	This instruction has implied addressing mode; this instruction is used to adjust the content of AL after addition of two ASCII characters
AAD	ASCII adjust for division	This instruction has implied addressing mode; converts two unpacked BCD digits in AX into equivalent binary numbers in AL; AAD must be used before dividing two unpacked BCD digits by an unpacked BCD byte
AAM	ASCII adjust after multiplication	This instruction has implied addressing mode; after multiplying two unpacked BCD numbers, adjust the product in AX to become an unpacked BCD result; ZF, SF, and PF affected
AAS	ASCII adjust [AL] after subtraction	This instruction has implied addressing mode used to adjust [AL] after subtraction of two ASCII characters
ADC mem/reg 1, mem/reg 2	[mem/reg 1] ← [mem/reg 1] + [mem/reg 2] + CY	Memory or register can be 8- or 16-bit; all flags are affected; no segment registers are allowed; no memory-to-memory ADC is permitted
ADC mem, data	[mem] ← [mem] + data + CY	Data can be 8- or 16-bit; mem uses DS as the segment register; all flags are affected
ADC reg, data	[reg] ← [reg] + data + CY	Data can be 8- or 16-bit; register cannot be segment register; all flags are affected; reg is not usually used as AX or AL
ADD mem/reg 2, mem/reg 1	[mem/reg 1] ← [mem/reg 2] + [mem/reg 1]	Add two 8- or 16-bit data; no memory-to-memory ADD is permitted; all flags are affected; mem uses DS as the segment register; reg 1 or reg 2 cannot be segment register
ADD mem, data	[mem] ← [mem] + data	Mem uses DS as the segment register; data can be 8- or 16-bit; all flags are affected
ADD reg, data	[reg] ← [reg] + data	Data can be 8- or 16-bit; no segment registers are allowed; all flags are affected; this instruction should not be used to add AL or AX with 8- or 16-bit immediate data
AND mem/reg 1, mem/reg 2	[mem/reg 1] ← [mem/reg 1] ∧ [mem/reg 2]	This instruction logically ANDs 8- or 16-bit data in [mem/reg 1] with 8- or 16-bit data in [mem/reg 2]; all flags are affected; OF and CF are cleared to zero; no segment registers are allowed; no memory-to-memory operation is allowed; mem uses DS as the segment register
AND mem, data	[mem] ← [mem] ∧ data	Data can be 8- or 16-bit; mem uses DS as the segment register; all flags are affected with OF and CF always cleared to zero
AND reg, data	[reg] ← [reg] ∧ data	Data can be 8- or 16-bit; reg cannot be segment register; all flags are affected with OF and CF cleared to zero
CALL PROC (NEAR)	Call a subroutine in the same segment with signed 16-bit displacement (to CALL a subroutine in ±32K)	NEAR in the statement BEGIN PROC NEAR indicates that the subroutine 'BEGIN' is in the same segment and BEGIN is 16-bit signed; CALL BEGIN instruction decrements SP by 2 and then pushes IP onto the stack and then adds the signed 16-bit value of BEGIN to IP and CS is unchanged; thus, a subroutine is called in the same segment (intrasegment direct)
CALL reg 16	CALL a subroutine in the same segment addressed by the contents of a 16-bit general register	The 8086 decrements SP by 2 and then pushes IP onto the stack, then specified 16-bit register contents (such as BX, SI, and DI) provide the new value for IP; CS is unchanged (intrasegment indirect)

Table 3.1 Summary of 8086 Instructions (*continued*)

Instructions	Interpretation	Comments
CALL mem 16	CALL a subroutine addressed by the content of a memory location pointed to by 8086 16-bit register such as BX, SI, and DI	The 8086 decrements SP by 2 and pushes IP onto the stack; the 8086 then loads the contents of a memory location addressed by the content of a 16-bit register such as BX, SI, and DI into IP; [CS] is unchanged (intrasegment indirect)
CALL PROC (FAR)	CALL a subroutine in another segment	FAR in the statement BEGIN PROC FAR indicates that the subroutine 'BEGIN' is in another segment and the value of BEGIN is 32 bit wide
		The 8086 decrements SP by 2 and pushes CS onto the stack and moves the low 16-bit value of the specified 32-bit number such as 'BEGIN' in CALL BEGIN into CS; SP is again decremented by 2; IP is pushed onto the stack; IP is then loaded with high 16-bit value of BEGIN; thus, this instruction CALLS a subroutine in another code segment (intersegment direct)
CALL DWORDPTR [reg 16]	CALL a subroutine in another segment	This instruction decrements SP by 2, and pushes CS onto the stack; CS is then loaded with the contents of memory locations addressed by [reg 16 + 2] and [reg 16 + 3] in DS; the SP is again decremented by 2; IP is pushed onto the stack; IP is then loaded with the contents of memory locations addressed by [reg 16] and [reg 16 + 1] in DS; typical 8086 registers used for reg 16 are BX, SI, and DI (intersegment indirect)
CBW	Convert a byte to a word	Extend the sign bit (bit 7) of AL register into AH
CLC	CF ← 0	Clear carry to zero
CLD	DF ← 0	Clear direction flag to zero
CLI	IF ← 0	Clear interrupt enable flag to zero to disable maskable interrupts
CMC	CF ← CF'	One's complement carry
CMP mem/reg 1, mem/reg 2	[mem/reg 1] − [mem/reg 2], flags are affected	mem/reg can be 8- or 16-bit; no memory-to-memory comparison allowed; result of subtraction is not provided; all flags are affected
CMP mem/reg, data	[mem/reg] − data, flags are affected	Subtracts 8- or 16-bit data from [mem or reg] and affects flags; no result is provided
CMPS BYTE or CMPSB	FOR BYTE [[SI]] − [[DI]], flags are affected [SI] ← [SI] ± 1 [DI] ← [DI] ± 1	8- or 16-bit data addressed by [DI] in ES is subtracted from 8- or 16-bit data addressed by SI in DS and flags are affected without providing any result; if DF = 0, then SI and DI are incremented by one for byte and two for word; if DF = 1, then SI and DI are decremented by one for byte and two for word; the segment register ES in destination cannot be overridden
CMPS WORD or CPSW	FOR WORD [[SI]] − [[DI]], flags are affected [SI] ← [SI] + 2 [DI] ← [DI] + 2	
CWD	Convert a word to 32 bits	Extend the sign bit of AX (bit 15) into DX
DAA	Decimal adjust [AL] after addition	This instruction uses implied addressing mode; this instruction converts [AL] into BCD; DAA should be used after BCD addition
DAS	Decimal adjust [AL] after subtraction	This instruction uses implied addressing mode; converts [AL] into BCD; DAS should be used after BCD subtraction
DEC reg 16	[reg 16] ← [reg 16] − 1	This is a one-byte instruction; used to decrement a 16-bit register except segment register; does not affect the carry flag
DEC mem/reg 8	[mem] ← [mem] − 1 or [reg 8] ← [reg 8] − 1	Used to decrement a byte or a word in memory or an 8-bit register content; segment register cannot be decremented by this instruction; does not affect carry flag

Table 3.1   Summary of 8086 Instructions (*continued*)

Instructions	Interpretation	Comments
DIV mem/reg	16/8 bit divide: $\dfrac{[AX]}{[mem\,8\,/\,reg\,8]}$ [AH] ← Remainder [AL] ← Quotient 32/16 bit divide: $\dfrac{[DX]\,[AX]}{[mem\,16\,/\,reg\,16]}$ [DX] ← Remainder [AX] ← Quotient	Mem/reg is 8-bit for 16-bit by 8-bit divide and 16-bit for 32-bit by 16-bit divide; this is an unsigned division; no flags are affected; division by zero automatically generates an internal interrupt
ESC external OP code, source	ESCAPE to external processes	This instruction is used to pass instructions to a coprocessor such as the 8087 floating point coprocessor which simultaneously monitors the system bus with the 8086; the coprocessor OP codes are 6-bit wide; the coprocessor treats normal 8086 instructions as NOP's; the 8086 fetches all instructions from memory; when the 8086 encounters an ESC instruction, it usually treats it as NOP; the coprocessor decodes this instruction and carries out the operation using the 6-bit OP code independent of the 8086; for ESC OP code, memory, the 8086 accesses data in memory for the coprocessor; for ESC data, register, the coprocessor operates on 8086 registers; the 8086 treats this as an NOP
HLT	HALT	Halt
IDIV mem/reg	Same as DIV mem/reg	Signed division. No flags are affected.
IMUL mem/reg	For 8 × 8 [AX] ← [AL] * [mem 8/reg 8] FOR 16 × 16 [DX] [AX] ← [AX]* [mem 16/reg 16]	Mem/reg can be 8- or 16-bit; only CF and OF are affected; signed multiplication
IN AL, DX	[AL] ← [PORT DX]	Input AL with the 8-bit content of a port addressed by DX; this is a one-byte instruction
IN AX, DX	[AX] ← [PORT DX]	Input AX with the 16-bit content of a port addressed by DX and DX + 1; this is a one-byte instruction
IN AL, PORT	[AL] ← [PORT]	Input AL with the 8-bit content of a port addressed by the second byte of the instruction
IN AX, PORT	[AX] ← [PORT]	Input AX with the 16-bit content of a port addressed by the 8-bit address in the second byte of the instruction
INC reg 16	[reg 16] ← [reg 16] + 1	This is a one-byte instruction; used to increment a 16-bit register except the segment register; does not affect the carry flag
INC mem/reg 8	[mem] ← [mem] + 1 or [reg 8] ← [reg 8] + 1	This is a two-byte instruction; can be used to increment a byte or word in memory or an 8-bit register content; segment registers cannot be incremented by this instruction; does not affect the carry flag
INT n (n can be zero thru 255)	[SP] ← [SP] – 2 [[SP]] ← Flags IF ← 0 TF ← 0 [SP] ← [SP] – 2 [[SP]] ← [CS] [CS] ← 4n + 2 [SP] ← [SP] –2	Software interrupts can be used as supervisor calls; that is, request for service from an operating system; a different interrupt type can be used for each type of service that the operating system could supply for an application or program; software interrupt instructions can also be used for checking interrupt service routines written for hardware-initiated interrupts

Table 3.1 Summary of 8086 Instructions (*continued*)

Instructions	Interpretation	Comments
	$[[SP]] \leftarrow [IP]$	
	$[IP] \leftarrow 4n$	
INTO	Interrupt on Overflow	Generates an internal interrupt if OF = 1; executes INT 4; can be used after an arithmetic operation to activate a service routine if OF = 1; when INTO is executed and if OF = 1, operations similar to INT n take place
IRET	Interrupt Return	POPS IP, CS and Flags from stack; IRET is used as return instruction at the end of a service routine for both hardware and software interrupts
JA/JNBE disp 8	Jump if above/jump if not below or equal	Jump if above/jump if not below or equal with 8-bit signed displacement; that is, the displacement can be from -128_{10} to $+127_{10}$, zero being positive; JA and JNBE are the mnemonic which represent the same instruction; Jump if both CF and ZF are zero; used for unsigned comparison
JAE/JNB/JNC disp 8	Jump if above or equal/jump if not below/jump if no carry	Same as JA/JNBE except that the 8086 Jumps if CF = 0; used for unsigned comparison
JB/JC/JNAE disp 8	Jump if below/jump if carry/jump if not above or equal	Same as JA/JNBE except that the jump is taken CF = 1, used for unsigned comparison
JBE/JNA disp 8	Jump if below or equal/jump if not above	Same as JA/JNBE except that the jump is taken if CF = 1 or ZF = 0; used for unsigned comparison
JCXZ disp 8	Jump if CX = 0	Jump if CX = 0; this instruction is useful at the beginning of a loop to bypass the loop if CX = 0
JE/JZ disp 8	Jump if equal/jump if zero	Same as JA/JNBE except that the jump is taken if ZF = 1; used for both signed and unsigned comparison
JG/JNLE disp 8	Jump if greater/jump if not less or equal	Same as JA/JNBE except that the jump is taken if $((SF \oplus OF) \text{ or } ZF) = 0$; used for signed comparison
JGE/JNL disp 8	Jump if greater or equal/ jump if not less	Same as JA/JNBE except that the jump is taken if $(SF \oplus OF) = 0$; used for signed comparison
JL/JNGE disp 8	Jump if less/Jump if not greater nor equal	Same as JA/JNBE except that the jump is taken if $(SF \oplus OF) = 1$; used for signed comparison
JLE/JNG disp 8	Jump if less or equal/ jump if not greater	Same as JA/JNBE except that the jump is taken if $((SF \oplus OF) \text{ or } ZF) = 1$; used for signed comparison
JMP Label	Unconditional Jump with a signed 8-bit (SHORT) or signed 16-bit (NEAR) displacement in the same segment	The label START can be signed 8-bit (called SHORT jump) or signed 16-bit (called NEAR jump) displacement; the assembler usually determines the displacement value; if the assembler finds the displacement value to be signed 8-bit (−128 to +127, 0 being positive), then the assembler uses two bytes for the instruction: one byte for the OP code followed by a byte for the displacement; the assembler sign extends the 8-bit displacement and then adds it to IP; [CS] is unchanged; on the other hand, if the assembler finds the displacement to be signed 16-bit (±32 K), then the assembler uses three bytes for the instruction: one byte for the OP code followed by 2 bytes for the displacement; the assembler adds the signed 16-bit displacement to IP; [CS] is unchanged; therefore, this JMP provides a jump in the same segment (intrasegment direct jump)
JMP reg 16	$[IP] \leftarrow [reg \ 16]$ [CS] is unchanged	Jump to an address specified by the contents of a 16-bit register such as BX, SI, and DI in the same code segment; in the example JMP BX, [BX] is loaded into IP and [CS] is unchanged (intrasegment memory indirect jump)
JMP mem 16	$[IP] \leftarrow [mem]$ [CS] is unchanged	Jump to an address specified by the contents of a 16-bit memory location addressed by 16-bit register

Table 3.1   Summary of 8086 Instructions (*continued*)

Instructions	Interpretation	Comments
		such as BX, SI, and DI; in the example, JMP [BX] copies the content of a memory location addressed by BX in DS into IP; CS is unchanged (intrasegment memory indirect jump)
JMP Label (FAR)	Unconditionally jump to another segment	This is a 5-byte instruction: the first byte is the OP code followed by four bytes of 32-bit immediate data; bytes 2 and 3 are loaded into IP; bytes 4 and 5 are loaded into CS to JUMP unconditionally to another segment (intersegment direct)
JMP DWORDPTR [reg 16]	Unconditionally jump to another segment	This instruction loads the contents of memory locations addressed by [reg 16] and [reg 16 + 1] in DS into IP; it then loads the contents of memory locations addressed by [reg 16 + 2] and [reg 16 + 3] in DS into CS; typical 8086 registers used for reg 16 are BX, SI, and DI (intersegment indirect)
JNE/JNZ disp 8	Jump if not equal/jump if not zero	Same as JA/JNBE except that the jump is taken if ZF = 0; used for both signed and unsigned comparison
JNO disp 8	Jump if not overflow	Same as JA/JNBE except that the jump is taken if OF = 0
JNP/JPO disp 8	Jump if no parity/jump if parity odd	Same as JA/JNBE except that the jump is taken if PF = 0
JNS disp 8	Jump if not sign	Same as JA/JNBE except that the jump is taken if SF = 0
JO disp 8	Jump if overflow	Same as JA/JNBE except that the jump is taken if OF = 1
JP/JPE disp 8	Jump if parity/jump if parity even	Same as JA/JNBE except that the jump is taken if PF = 1
JS disp 8	Jump if sign	Same as JA/JNBE except that the jump is taken if SF = 1
LAHF	[AH] ← Flag low-byte	This instruction has implied addressing mode; it loads AH with the low byte of the flag register; no flags are affected
LDS reg, mem	[reg] ← [mem] [DS] ← [mem + 2]	Load a 16-bit register (AX, BX, CX, DX, SP, BP, SI, DI) with the content of specified memory and load DS with the content of the location that follows; no flags are affected; DS is used as the segment register for mem
LEA reg, mem	[reg] ← [offset portion of address]	LEA (load effective address) loads the value of the source operand rather than its content to register (such as SI, DI, BX) which are allowed to contain offset for accessing memory; no flags are affected
LES reg, mem	[reg] ← [mem] [ES] ← [mem + 2]	DS is used as the segment register for mem; in the example LES DX, [BX], DX is loaded with 16-bit value from a memory location addressed by 20-bit physical address computed from DS and BX; the 16-bit content of the next memory is loaded into ES; no flags are affected
LOCK	LOCK bus during next instruction	Lock is a one-byte prefix that causes the 8086 (configured in maximum mode) to assert its bus LOCK signal while following instruction is executed; this signal is used in multiprocessing; the LOCK pin of the 8086 can be used to LOCK other processors off the system bus during execution of an instruction; in this way, the 8086 can be assured of uninterrupted access to common system resources such as shared RAM

Table 3.1 Summary of 8086 Instructions (*continued*)

Instructions	Interpretation	Comments
LODS BYTE or LODSB	FOR BYTE [AL] ← [[SI]] [SI] ← [SI] ± 1	Load 8-bit data into AL or 16-bit data into AX from a memory location addressed by SI in segment DS; if DF = 0, then SI is incremented by 1 for byte or
LODS WORD or LODSW	FOR WORD [AX] ← [[SI]] [SI] ← [SI] ± 2	incremented by 2 for word after the load; if DF = 1, then SI is decremented by 1 for byte or decremented by 2 for word; LODS affects no flags
LOOP disp 8	Loop if CX not equal to zero	Decrement CX by one, without affecting flags and loop with signed 8-bit displacement (from −128 to +127, zero being positive) if CX is not equal to zero
LOOPE/LOOPZ disp 8	Loop while equal/loop while zero	Decrement CX by one without affecting flags and loop with signed 8-bit displacement if CX is equal to zero, and if ZF = 1 which results from execution of the previous instruction
LOOPNE/LOOPNZ disp 8	Loop while not equal/loop while not zero	Decrement CX by one without affecting flags and loop with signed 8-bit displacement if CX is not equal to zero and ZF = 0 which results from execution of previous instruction
MOV mem/reg 2, mem/reg 1	[mem/reg 2] [mem/reg 1]	mem uses DS as the segment register; no memory-to-memory operation allowed; that is, MOV mem, mem is not permitted; segment register cannot be specified as reg or reg; no flags are affected; not usually used to load or store 'A' from or to memory
MOV mem, data	[mem] ← data	mem uses DS as the segment register; 8- or 16-bit data specifies whether memory location is 8- or 16-bit; no flags are affected
MOV reg, data	[reg] ← data	Segment register cannot be specified as reg; data can be 8- or 16-bit; no flags are affected
MOV segreg, mem/reg	[segreg] ← [mem/reg]	mem uses DS as segment register; used for initializing CS, DS, ES, and SS; no flags are affected
MOV mem/reg, segreg	[mem/reg] ← [segreg]	mem uses DS as segment register; no flags are affected
MOVS BYTE or MOVSB	FOR BYTE [[DI]] ← [[SI]] [SI] ← [SI] ± 1	Move 8-bit or 16-bit data from the memory location addressed by SI in segment DS location addressed by DI in ES; segment DS can be overridden by a prefix but destination segment must be ES and
MOVS WORD or MOVSW	FOR WORD [[DI]] ← [[SI]] [SI] ← [SI] ± 2	cannot be overridden; if DF = 0, then SI is incremented by one for byte or incremented by two for word; if DF = 1, then SI is decremented by one for byte or by two for word
MUL mem/reg	FOR 8 × 8 [AX] ← [AL]* [mem/reg] FOR 16 × 16 [DX] [AX] ← [AX]* [mem/reg]	mem/reg can be 8- or 16-bit; only CF and OF are affected; unsigned multiplication
NEG mem/reg	[mem/reg] ← [mem/reg]′ + 1	mem/reg can be 8- or 16-bit; performs two's complement subtraction of the specified operand from zero, that is, two's complement of a number is formed; all flags are affected except CF = 0 if [mem/reg] is zero; otherwise CF = 1
NOP	No Operation	8086 does nothing
NOT reg	[reg] ← [reg]′	mem and reg can be 8- or 16-bit; segment registers are not allowed; no flags are affected; ones complement reg
NOT mem	[mem] ← [mem]′	mem uses DS as the segment register; no flags are affected; ones complement mem
OR Mem/reg 1, Mem/reg 2	[mem/reg 1] ← [mem/reg 1] ∨ [mem/reg 2]	No memory-to-memory operation is allowed; [mem] or [reg 1] or [reg 2] can be 8- or 16-bit; all flags are affected with OF and CF cleared to zero; no

Table 3.1   Summary of 8086 Instructions (*continued*)

Instructions	Interpretation	Comments
		segment registers are allowed; mem uses DS as segment register
OR mem, data	[mem] ← [mem] ∨ data	mem and data can be 8- or 16-bit; mem uses DS as segment register; all flags are affected with CF and OF cleared to zero
OR reg, data	[reg] ← [reg] ∨ data	reg and data can be 8- or 16-bit; no segment registers are allowed; all flags are affected with CF and OF cleared to zero
OUT DX, AL	[PORT] ← [AL] DX	Output the 8-bit contents of AL into an I/O Port addressed by the 16-bit content of DX; this is a one-byte instruction
OUT DX, AX	[PORT] ← [AX] DX	Output the 16-bit contents of AX into an I/O Port addressed by the 16-bit content of DX; this is a one-byte instruction
OUT PORT, AL	[PORT] ← [AL]	Output the 8-bit contents of AL into the Port specified in the second byte of the instruction
OUT PORT, AX	[PORT] ← [AX]	Output the 16-bit contents of AX into the Port specified in the second byte of the instruction
POP mem	[mem] ← [[SP]] [SP] ← [SP] + 2	mem uses DS as the segment register; no flags are affected
POP reg	[reg] ← [[SP]] [SP] ← [SP] + 2	Cannot be used to POP segment registers or flag register
POP segreg	[segreg] ← [[SP]] [SP] ← [SP] + 2	POP CS is illegal
POPF	[Flags] ← [[SP]] [SP] ← [SP] + 2	This instruction pops the top two stack bytes in the 16-bit flag register
PUSH mem	[SP] ← [SP] − 2 [[SP]] ← [mem]	mem uses DS as segment register; no flags are affected; pushes 16-bit memory contents
PUSH reg	[SP] ← [SP] − 2 [[SP]] ← [reg]	reg must be a 16-bit register; cannot be used to PUSH segment register or Flag register
PUSH segreg	[SP] ← [SP] − 2 [[SP]] ← [segreg]	PUSH CS is illegal
PUSHF	[SP] ← [SP] − 2 [[SP]] ← [Flags]	This instruction pushes the 16-bit Flag register onto the stack
RCL mem/reg, 1	ROTATE through carry left once byte or word in mem/reg	FOR BYTE FOR WORD
RCL mem/reg, CL	ROTATE through carry left byte or word in mem/reg by [CL]	Operation same as RCL mem/reg, 1 except the number of rotates is specified in CL for rotates up to 255; zero or negative rotates are illegal
RCR mem/reg, 1	ROTATE through carry right once byte or word in mem/reg	FOR BYTE FOR WORD
RCR mem/reg, CL	ROTATE through carry right byte or word in mem/reg by [CL]	Operation same as RCR mem/reg, 1 except the number of rotates is specified in CL for rotates up to 255; zero or negative rotates are illegal

Table 3.1 Summary of 8086 Instructions (*continued*)

Instructions	Interpretation	Comments
RET	.POPS IP for intrasegment CALLS .POPS IP and CS for intersegment CALLS	The assembler generates an intrasegment return if the programmer has defined the subroutine as NEAR; for intrasegment return, the following operations take place: $[IP] \leftarrow [[SP]]$, $[SP] \leftarrow [SP] + 2$; on the other hand, the assembler generates an intersegment return if the subroutine has been defined as FAR; in this case, the following operations take place: $[IP] \leftarrow [[SP]]$, $[SP] \leftarrow [SP] + 2$, $[CS] \leftarrow [[SP]]$, $[SP] \leftarrow [SP] + 2$; an optional 16-bit displacement 'START' can be specified with the intersegment return such as RET START; in this case, the 16-bit displacement is added to the SP value; this feature may be used to discard parameter pushed onto the stack before the execution of the CALL instruction
ROL mem/reg, 1	ROTATE left once byte or word in mem/reg	FOR BYTE FOR WORD
ROL mem/reg, CL	ROTATE left byte or word by the content of CL	[CL] contains rotate count up to 255; zero and negative shifts are illegal; CL is used to rotate count when the rotate is greater than once; mem uses DS as the segment register
ROR mem/reg, 1	ROTATE right once byte or word in mem/reg	FOR BYTE FOR WORD
ROR mem/reg, CL	ROTATE right byte or word in mem/reg by [CL]	Operation same as ROR mem/reg, 1; [CL] specifics the number of rotates for up to 255; zero and negative rotates are illegal; mem uses DS as the segment register
SAHF	[Flags, low-byte] $\leftarrow$ [AH]	This instruction has the implied addressing mode; the content of the AH register is stored into the low-byte of the flag register; no flags are affected
SAL mem/reg, 1	Shift arithmetic left once byte or word in mem or reg	FOR BYTE FOR WORD Mem uses DS as the segment register; reg cannot be segment registers; OF and CF are affected; if sign bit is changed during or after shifting, the OF is set to one

Table 3.1 Summary of 8086 Instructions (*continued*)

Instructions	Interpretation	Comments
SAL mem/reg, CL	Shift arithmetic left byte or word by shift count on CL	Operation same as SAL mem/reg, 1; CL contains shift count for up to 255; zero and negative shifts are illegal; [CL] is used as shift count when shift is greater than one; OF and SF are affected; if sign bit of [mem] is changed during or after shifting, the OF is set to one; mem uses DS as segment register
SAR mem/reg, 1	SHIFT arithmetic right once byte or word in mem/reg	FOR BYTE FOR WORD
SAR mem/reg, CL	SHIFT arithmetic right byte or word in mem/reg by [CL]	Operation same as SAR mem/reg, 1; however, shift count is specified in CL for shifts up to 255; zero and negative shifts are illegal
SBB mem/reg 1, mem/reg 2	[mem/reg 1] ← [mem/reg 1] − [mem/reg 2] − CY	Same as SUB mem/reg 1, mem/reg 2 except this is a subtraction with borrow
SBB mem, data	[mem] ← [mem] − data − CY	Same as SUB mem, data except this is a subtraction with borrow
SBB reg, data	[reg] ← [reg] − data − CY	Same as SUB reg, data except this is a subtraction with borrow
SBB A, data	[A] ← [A] − data − CY	Same as SUB A, data except this is a subtraction with borrow
SCAS BYTE or SCASB	FOR BYTE [AL] − [[DI]], flags are affected, [DI] [DI] ± 1	8- or 16-bit data addressed by [DI] in ES is subtracted from 8- or 16-bit data in AL or AX and flags are affected without affecting [AL] or [AX] or string data; ES cannot be overridden; if DF = 0, then DI is incremented by one for byte and two for word; if DF = 1, then DI is decremented by one for byte or decremented by two for word
SCAS WORD or SCASW	FOR WORD [AX] − [[DI]], flags are affected, [DI] ← [DI] ± 2	
SHL mem/reg, 1	SHIFT logical left once byte or word in mem/reg	Same as SAL mem/reg, 1
SHL mem/reg, CL	SHIFT logical left byte or word in mem/reg by the shift count in CL	Same as SAL mem/reg, CL except overflow is cleared to zero
SHR mem/reg, 1	SHIFT right logical once byte or word in mem/reg	FOR BYTE FOR WORD
SHR mem/reg, CL	SHIFT right logical byte or word in mem/reg by [CL]	Operation same as SHR mem/reg, 1; however, shift count is specified in CL for shifts up to 255; zero and negative shifts are illegal
STC	CF ← 1	Set carry to one
STD	DF ← 1	Set direction flag to one
STI	IF ← 1	Set interrupt enable flag to one to enable maskable interrupts
STOS BYTE or STOSB	FOR BYTE [[DI]] ← [AL] [DI] ← [DI] ± 1	Store 8-bit data from AL or 16-bit data from AX into a memory location addressed by DI in segment ES; segment register ES cannot be overridden; if DF = 0, then DI is incremented by one for byte or incremented by two for word after the store
STOS WORD or STOSW	FOR WORD [[DI]] ← [AX] [DI] ← [DI] ± 2	

Table 3.1   Summary of 8086 Instructions (*continued*)

Instructions	Interpretation	Comments
SUB mem/reg 1, mem/reg 2	[mem/reg 1] ← [mem/reg 1] − [mem/reg 2]	No memory-to-memory SUB permitted; all flags are affected; mem uses DS as the segment register
SUB mem, data	[mem] ← [mem] − data	Data can be 8- or 16-bit; mem uses DS as the segment register; all flags are affected
SUB reg, data	[reg] ← [reg] − data	Data can be 8- or 16-bit; all flags are affected
TEST mem/reg 1, mem/reg 2	[mem/reg 1] ∧ [mem/reg 2], no result; flags are affected	No memory-to-memory TEST is allowed; no result is provided; all flags are affected with CF and OF cleared to zero; [mem], [reg 1] or [reg 2] can be 8- or 16-bit; no segment registers are allowed; mem uses DS as the segment register
TEST mem, data	[mem] ∧ data, no result; flags are affected	Mem and data can be 8- or 16-bit; no result is provided; all flags are affected with CF and OD cleared to zero; mem uses DS as the segment register
TEST reg, data	[reg] ∧ data, no result; flags are affected	Reg and data can be 8- or 16-bit; no result is provided; all flags are affected with CF and OF cleared to zero; reg cannot be segment register;
WAIT	8086 enters wait state	Causes CPU to enter wait state if the 8086 TEST pin is high; while in wait state, the 8086 continues to check TEST pin for low; if TEST pin goes back to zero, the 8086 executes the next instruction; this feature can be used to synchronize the operation of 8086 to an event in external hardware
XCHG mem, reg	[mem] ↔ [reg]	reg and mem can be both 8- or 16-bit; mem uses DS as the sement register; reg cannot be segment register; no flags are affected
XCHG reg, reg	[reg] ↔ [reg]	reg not used to exchange reg with AX; reg can be 8- or 16-bit; reg cannot be segment register; no flags are affected
XLAT	[AL] ← [AL] + [BX]	This instruction is useful for translating characters from one code such as ASCII to another such as EBCDIC; this is a no-operand instruction and is called an instruction with implied addressing mode; the instruction loads AL with the contents of a 20-bit physical address computed from DS, BX, and AL; this instruction can be used to read the elements in a table where BX can be loaded with a 16-bit value to point to the starting address (offset from DS) and AL can be loaded with the element number (0 being the first element number); no flags are affected; the XLAT instruction is equivalent to MOV AL, [AL] [BX]
XOR mem/reg 1, mem/reg 2	[mem/reg 1] ← [mem/reg 1] ⊕ [mem/reg 2]	No memory-to-memory operation is allowed; [mem] or [reg 1] or [reg 2] can be 8- or 16-bit; all flags are affected with CF and OF cleared to zero; mem uses DS as the segment register
XOR mem, data	[reg] ← [mem] ⊕ data	Data and mem can be 8- or 16-bit; mem uses DS as the segment register; mem cannot be segment register; all flags are affected with CF and OF cleared to zero
XOR Reg, data	[reg] ⊕ [reg] ≈ data	Same as XOR mem, data; should not be used for XORing AL or AX with immediate data

Suppose that an 8086-based microcomputer is interfaced to an ASCII keyboard and an IBM printer (EBCDIC code). In such a system, any number entered into the microcomputer will be in ASCII code which must be converted to EBCDIC code before outputting to the printer. If the number entered is 4, then the ASCII code for 4 (34H) must be translated to the EBCDIC code for 4 (F4H). A look-up table containing EBCDIC code for all the decimal numbers (0 to 9) can be stored in memory at the 16-bit starting address in the data segment of memory; with code for 0 stored at 3030H, code for 1 at 3031H, and so on. Now, if AL is loaded with the ASCII code 34H and BX is loaded with 3000H, then by using the XLAT instruction, content of memory location 3034H in the data segment containing the EBCDIC code for 4 (F4H) from the look-up table is read into AL, thus replacing the ASCII code for 4 with the EBCDIC code for 4.

2. *Examples of input/output instructions*

- Consider fixed port addressing in which the 8-bit port address is directly specified as part of the instruction. IN AL, 38H inputs 8-bit data from port 38H into AL. IN AX, 38H inputs 16-bit data from ports 38H and 39H into AX. OUT 38H, AL outputs the contents of AL to port 38H. OUT 38H, AX, on the other hand, outputs the 16-bit contents of AX to ports 38H and 39H.

- For the variable port addressing, the port address is 16-bit and is contained in the DX register. Consider ports addressed by 16-bit address contained in DX. Assume [DX] = 3124_{16} in all the following examples:

  IN AL, DX inputs 8-bit data from 8-bit port 3124_{16} into AL.

  IN AX, DX inputs 16-bit data from ports 3124_{16} and 3125_{16} into AL and AH respectively.

  OUT DX, AL outputs 8-bit data from AL into port 3124_{16} .

  OUT DX, AX outputs 16-bit data from AL and AH into ports 3124_{16} and 3125_{16} respectively.

Variable port addressing allows up to 65,536 ports with addresses from 0000H to FFFFH. The port addresses in the variable port addressing can be calculated dynamically in a program. For example, assume that an 8086-based microcomputer is connected to three printers via three separate ports. Now, in order to output to each one of the printers, separate programs are required if fixed port addressing is used. However, with variable port addressing one can write a general subroutine to output to the printers and then supply the address of the port for a particular printer for which data output is desired to register DX in the subroutine.

3. *Examples of address initialization instructions*

- LEA reg, mem loads an offset (mem) directly into the specified register. This instruction is useful when address computation is required. LEA BX, 5000H and MOV BX, 5000H accomplish the same task. That is, both of these instructions load 5000H into BX. On the other hand, LEA DI, [SI] [BX] loads the 16-bit value computed from BX and SI into DI whereas MOV DI, [SI] [BX] loads the 16 bit contents of a memory location computed from SI and BX into DI. LEA can be used to load these addresses. For example, LEA can be used to load the address of the table used by the XLAT instructions.

- LDS reg, mem can be used to initialize SI and DS to point to the start of the source string before using one of the string instructions. For example, LDS SI, [BX] loads the 16-bit contents of memory offset by [BX] in the data segment into SI and the 16-bit contents of memory offset by [BX + 2] in the data segment into DS.

- LES reg, mem can be used to point to the start of the destination string before using one of the string instructions. For example, LES DI, [BX] loads the 16-bit contents of memory offset by [BX] in the data segment to DI and then initializes ES with the 16-bit contents of memory offset by [BX + 2] in DS.

4. *Examples of flag register instructions*

- PUSHF pushes the 16-bit flag register onto the stack.
- LAHF loads AH with the condition codes from the low byte of the flag register.

5. *Explanation of arithmetic instructions*

- Numerical data received by an 8086-based microcomputer from a terminal is usually in ASCII code. The ASCII codes for numbers 0 to 9 are 30H through 39H. Two 8-bit data can be entered into an 8086-based microcomputer via a terminal. The ASCII codes for these data (with 3 as the upper middle for each type) can be added. AAA instruction can then be used to provide the correct unpacked BCD. Suppose that ASCII codes for 2 (32_{16}) and 5 (35_{16}) are entered into an 8086-based microcomputer via a terminal. These ASCII codes can be added and then the result can be adjusted to provide the correct unpacked BCD using AAA instructions as follows:

```
ADD CL, DL  ;  [CL] = 32₁₆ = ACSII for 2
            ;  [DL] = 35₁₆ = ASCII for 5
            ;  Result [CL] = 67₁₆
MOV AL, CL  ;  Move ASCII result
            ;  into AL since AAA
            ;  adjust only [AL]
AAA         ;  [AL] = 07, unpacked
            ;  BCD for 7
```

Note that in order to send the unpacked BCD result 07_{16} back to the terminal, [AL] = 07 can be ORed with 30H to provide 37H, the ASCII code for 7.

- DAA is used to adjust the result of adding two packed BCD numbers in AL to provide a valid BCD number. If after the addition, the low 4-bit of the result in AL is greater than 9 (or if AF = 1), then the DAA adds 6 to the low 4 bits of AL. Then, if the high 4 bits of the result in AL is greater than 9 (or if CF = 1), then DAA adds 60H to AL. As an example, consider adding two packed BCD digits 55 with 18 as follows:

```
ADD AL, DL  ;  [AL] = 55 BCD
            ;  [DL] = 18 BCD
            ;  Result = [AL] = 6DH
DAA         ;  Since low nibble
            ;  D = 1101₂ > 9, add i.e. 1101₂ + 0110₂ →
```

$$1 \qquad \underbrace{0011}_{}{}_2$$

$$\uparrow \qquad\qquad \uparrow$$

$$\text{carry} \qquad \text{3BCD}$$

- The ASCII codes for two 8-bit numbers in an 8086-based microcomputer can be subtracted. Assume

```
[AL] = 35H = ASCII for 5
[DL] = 37H = ASCII for 7
```

The following instruction sequence provides the correct subtraction result:

```
SUB AL, DL  ;  [AL] = 1111 1110₂ = subtraction result
            ;  in 2's complement
            ;  CF = 1
AAS         ;  [AL] = BCD02
            ;  CF = 1 means
            ;  borrow to be
            ;  used in multibyte BCD subtraction
```

AAS adjusts [AL] and leaves zeros in the upper nibble. To output to the ASCII terminal from the microcomputer, BCD data can be ORed with 30H to produce the correct ASCII code.

- DAS can be used to adjust the result of subtraction in AL of two packed BCD numbers to provide the correct packed BCD. If low 4-bit in AL is greater than 9 (or if AF = 1), then DAS subtracts 6 from the low 4-bit of AL. Then, if the upper 4-bit of the result in AL is greater than 9 (or if CF = 1), DAS subtracts 60 from AL. While performing these subtractions, any borrows from LOW and HIGH nibbles are ignored. For example, consider subtracting BCD 55 in DL from BCD 94 in AL.

```
SUB AL, DL  ;  [AL] = 3FH  →  Low nibble = 1111
DAS         ;  CF = 0                   -6 = 1010
                                       1    1001
            ;  [AL] = 39 BCD            ↑
                                      ignore
```

- IMUL mem/reg provides signed 8 × 8 or signed 16 × 16 multiplication. As an example, if [CL] = FDH = -3_{10}, [AL] = FEH = -2_{10}, then after IMUL CL, register AX contains 0006H.
- Consider 16 × 16 unsigned multiplication, MUL WORDPTR [BX]. If [BX] = 0050H, [DS] = 3000H, [30050H] = 0002H, [AX] = 0006H, then after MUL WORDPTR [BX], [DX] = 0000H, [AX] = 000CH.
- Consider DIV BL. If [AX] = 0009H, [BL] = 02H, then after DIV BL,

```
[AH] = Remainder = 01H
[AL] = Quotient  = 04H
```

- Consider IDIV WORDPTR [BX]. If [BX] = 0020H, [DS] = 2000H, [20020H] = 0004H, [DX] [AX] = 00000011H, then after IDIV. WORDPTR [BX],

```
[DX] = Remainder = 0001H
[AX] = Quotient  = 0004H
```

- AAD converts two unpacked BCD digits in AH and AL to an equivalent binary number in AL. AAD must be used before dividing two unpacked BCD digits in AX by an unpacked BCD byte. For example, consider dividing [AX] = unpacked BCD 0508 (58 decimal) by [DH] = 07H. [AX] must first be converted to binary by using AAD. The register AX will then contain 003AH = 58 decimal. After DIV DH, [AL] = quotient = 08 unpacked BCD, [AH] = remainder = 02 unpacked BCD.

- Consider CBW. This instruction extends the sign from the AL register to AH register. For example, if [AL] = E2H, then after CBW, AH will contain FFH since the most significant bit of E2H is one. Note that sign extension is useful when one wants to perform an arithmetic operation on two signed numbers of different sizes. For example, the 8-bit signed number 02H can be subtracted from 16-bit signed number 2005H as follows:

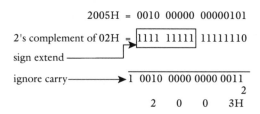

$$2005H = 0010\ 00000\ 00000101$$

Another example of sign extension is that in order to multiply a signed 8-bit number by a signed 16-bit number, one must first sign-extend the signed 8-bit into a signed 16-bit number and then the instruction IMUL can be used for 16×16 signed multiplication.

- AAM adjusts the product of two unpacked BCD digits in AX. If [AL] = BCD3 = 00000011_2 and [CH] = BCD8 = $0000\ 1000_2$, then after MUL CH, [AX] = 0000000000011000_2 = 0018H, and after using AAM, [AX] = $00000010\ 00000100_2$ = unpacked BCD 24. The following instruction sequence accomplishes this:

MUL CH
AAM

Note that the 8086 does not allow multiplication of two ASCII codes. Therefore, before multiplying two ASCII bytes received from a terminal, one must mask the upper 4-bits of each one of these bytes and then multiply them as two unpacked BCD digits and then use AAM for adjustment. In order to convert the unpacked BCD product back to ASCII, for sending back to the terminal, one must OR the product with 3030H.

6. *Examples of logical, shift and rotate instructions*

- TEST BL, 3 logically ANDs the contents of BL with 00000011_2 but does not store the result in BL. All flags are affected.
- All shift and rotate instructions include two operands. The destination operand specifies the register or memory to be shifted or rotated while the source operand specifies the number of times the register or memory contents are to be shifted. For example, SHL DX, 1 logically shifts the 16-bit contents of DX once to the left. On the other hand, SHL DX,CL, with CL = 5, logically shifts the 16-bit contents of DX five times to the left. For all 8086 shift and rotate instructions, a shift count of one must be represented in the source operand by '1' while shift counts from 2_{10} to 255_{10} must be represented in the source operand by the contents of CL.

7. *Example of string instructions*

- LODS can be represented in four forms:

For Byte   LODS BYTE
or
LODSB

For Word   LODS WORD
or
LODSW

If [SI] = 0020H, [DS] = 3000H, [30020H] = 05H, DF = 0, then after LODS BYTE or LODSB, [AL] = 05H, [SI] = 0021H.

- If [DS] = 2000H, [ES] = 3000H, [SI] = 0020H, [DI] = 0050H, DF = 0, [20020H] = 0205H, [30050H] 4071H, then after execution of MOVSW or MOVS WORD, memory location 30050H will contain 0205H. Since DF = 0, register SI will contain 0022H and register DI will contain 0052H.

- REP, a one-byte prefix, can be used with MOVS to cause the instruction MOVS to continue executing until CX = 0. Each time the instructions such as REP MOVSB or REP MOVSW are executed, CX is automatically decremented by 1, and if [CX] ≠ 0, MOVS is reexecuted and then CX is decremented by 1 until CX = 0; the next instruction is then executed. REP MOVSB or REP MOVSW can be used to move string bytes until the string length (loaded into CX before the instructions REP MOVSB or REP MOVSW) is decremented for zero.

- REPE/REPZ or REPNE/REPNZ prefix can be used with CMPS or SCAS to cause one of these instructions to continue executing until ZF = 0 (for REPNE/REPNZ prefix) or CX = 0. Note that REPE and REPZ are two mnemonics for the same prefix byte. Similarly, REPNE and REPNZ also provide the same purpose.

- If CMPS is prefixed with REPE or REPZ, the operation is interpreted as "compare while not end-of-string (CX not zero) and strings are equal (ZF = 1)". If CMPS is preceded by REPNE or REPNZ, the operation is interpreted as "compare while not end-of-string (CX not zero) and strings not equal (ZF = 0)". Thus, repeated CMPS can be used to find matching or differing string elements.

- If SCAS is prefixed with REPE or REPZ, the operation is interpreted as "scan while not end-of-string (CX not 0) and string-element = scan-value (ZF = 1)". This form may be used to scan for departure from a given value. If SCAS is prefixed with REPNE or REPNZ, the operation is interpreted as "scan while not end-of-string (CX not 0) and string-element is not equal to scan-value (ZF = 0)". This form may be used to locate a value in a string.

- Repeated string instructions are interruptible; the processor recognizes the interrupt before processing the next string element. Upon return from the interrupt, the repeated operation is resumed from the point of interruption. When multiple prefixes (such as LOCK and segment override) are specified in addition to any of the repeat prefixes, program execution does not resume properly upon return from interrupt.

  The processor remembers only one prefix in effect at the time of the interrupt, the prefix that immediately precedes the string instructions. Upon return from interrupt, program execution resumes at this point but any additional prefixes specified are not recognized. If multiple prefix must be used with a string instruction, maskable interrupts should be disabled for the duration of the repeated execution. However, this will not prevent a nonmaskable interrupt from being recognized.

- Note that the segment register for destination for all string instructions is always ES and cannot be overridden. However, DS can be overridden for the source using a prefix. For example, ES: MOVSB instruction uses the segment register as ES for both source and destination.

8. *Examples of unconditional transfers*

- There are two types of Jump instructions. These are intersegment Jumps (both IP and CS change; Jump to a different code segment) and intrasegment Jumps (IP changes and CS is fixed; Jump in the same code segment). For example, JMP FAR BEGIN (or some 8086 assemblers use JMP FAR PTR BEGIN) unconditionally branches to a label BEGIN in a different code segment. JMP START jumpts to a label START in the same code segment.

- CALL instructions can be intersegment and intrasegment. For example, CALL DWORD PTR [BX] pushes CS and IP onto the stack and loads IP and CS with the contents of four consecutive locations pointed to by BX. CALL BX, on the other hand, pushes IP onto the stack, the new value of IP is loaded from BX and CS in unchanged.

9. *Examples of conditional transfers*

All 8086 conditional branch instructions use 8-bit signed displacement. That is, the displacement covers a branch range of -128_{10} to $+127_{10}$ with 0 being positive. In order to branch out of this range, the 8086 unconditional jump instructions (having direct mode) must be used.

Conditional jumps are typically used with compare instructions to find the relationship (equal to, greater than, or less than) between two numbers. The use of conditional instructions depends on whether the numbers to be compared are signed or unsigned. The 8-bit number $1111\ 1110_2$, when considered as signed, has a value of -2_{10}; the same number has a value of $+254_{10}$ when considered as unsigned. This number, when considered signed, will be smaller than zero, and when unsigned will be greater than zero. Some new terms are used to differentiate between signed and unsigned conditional transfers. For the unsigned numbers, the terms used are "below and above", while for signed numbers, the terms "less than" and "greater than" are used. The number $0111\ 1110_2$ when considered signed is greater than $0000\ 0000_2$, while the same number $0111\ 1110_2$ when considered unsigned is above $0000\ 0000_2$. The conditional transfer instructions for equality of two numbers are the same for both signed and unsigned numbers. This is because when two numbers are compared for equality irrespective of whether they are signed or unsigned, they will provide a zero result (ZF = 1) if equal or a nonzero result (ZF = 0) if not equal. Therefore, the same instructions apply for both signed and unsigned numbers for "equal to" or "not equal to" conditions, and the various signed and unsigned conditional branch instructions for determining the relationship between two numbers are as follows:

Signed		Unsigned	
Name	Alternate name	Name	Alternate name
JE disp8 (JUMP if equal)	JZ disp8 (JUMP if result zero)	JE disp8 (JUMP if equal)	JZ disp8 (JUMP if zero)
JNE disp8 (JUMP if not equal)	JNZ disp8 (JUMP if not zero)	JNE disp8 (JUMP if not equal)	JNZ disp8 (JUMP if not zero)
JG disp8 (JUMP if greater)	JNLE disp8 (JUMP if not less or or equal)	JA disp8 (JUMP if above)	JNBE disp8 (JUMP if not below or equal)
JGE disp8 (JUMP if greater or equal)	JNL disp8 (JUMP if not less)	JAE disp8 (JUMP if above or equal)	JNB disp8 (JUMP if not below)
JL disp8 (JUMP if less than)	JNGE disp8 (JUMP if not greater or equal)	JB disp8 (JUMP if below)	JNAE disp8 (JUMP if not above or equal)
JLE disp8 (JUMP if less or equal)	JNG disp8 (JUMP if not greater)	JBE disp8 (JUMP if below or equal)	JNA disp8 (JUMP if not above)

There are also conditional transfer instructions that are concerned with the setting of status flags rather than relationship between two numbers. The table below lists these instructions:

JC disp8	JUMP if carry, i.e., CF = 1
JNC disp8	JUMP if no carry, i.e., CF = 0
JP disp8	JUMP if parity, i.e., PF = 1
JNP disp8	JUMP if no parity, i.e., PF = 0
JO disp8	JUMP if overflow, i.e., OF = 1
JNO disp8	JUMP if no overflow, i.e., OF = 0
JS disp8	JUMP if sign, i.e., SF = 1
JNS disp8	JUMP if no sign, i.e., SF = 0
JZ disp8	JUMP if result zero, i.e., Z = 1
JNZ disp8	JUMP if result not zero, i.e., Z = 0

10. *Examples of LOOP instructions*

- LOOP BEGIN automatically decrements CX by 1 without affecting the flags and jumps to the label BEGIN if CX ≠ 0; goes to the next instruction if CX = 0 after autodecrement.
- LOOPNE NEXT automatically decrements CX by 1 and jumps to the label NEXT if CX ≠ 0 and ZF = 0. However, if CX = 0 after autodecrement or ZF = 1, execution will go on to the next instruction. Note that LOOPNE can be read as 'Loop while not equal and CX not 0.'

11. *Examples of interrupt instructions*

- INT n is a software interrupt instruction. This instruction pushes Flags, CS, and IP onto the stack and loads CS from the memory location 4n and IP from the memory location 4n+2. An interrupt service routine can be written at this location.
- IRET pops IP, CS, and flags from the stack.

12. *Examples of processor or control instructions*

- STD sets direction flag to one.
- CLI clears the IF flag in the status register to zero to disable maskable interrupts.

3.5 8086 Assembler-Dependent Instructions

Some 8086 instructions do not define whether an 8-bit or 16-bit operation is to be executed. Instructions with one of the 8086 registers as an operand typically define the operation as 8-bit or 16-bit based on the register size. An example is MOV CL, [BX] which moves an 8-bit number with the offset defined by [BX] in the data segment into register CL; MOV CX, [BX], on the other hand, moves the number from offsets [BX] and [BX + 1] in the data segment into CX.

The string instructions may define this in two ways. Typical examples are MOVSB or MOVS BYTE for 8-bit and MOVSW or MOVS WORD for 16-bit. Memory offsets can also be specified by including BYTEPTR for 8-bit and WORDPTR for 16-bit with the instruction. Typical examples are INC BYTEPTR [BX] and INC WORDPTR [BX].

3.6 ASM-86 Assembler Directives

The ASM-86 is the assembler written by Intel for the 8086 microprocessor. Other 8086 assemblers include the Microsoft 8086 assembler written for the IBM personal computer and the Hewllet Packard 8086 assembler written for the HP 64000 microcomputer development system. These assemblers allow the programmer to assign the values of CS, DS, SS, and ES.

These assemblers can be used to declare a variable's type as byte (8-bit), word (16-bit), or double word (4 bytes or 2 words) as follows:

```
START DB 0  ;  START is declared
            ;  as a byte offset
            ;  and its content is initialized to zero.

BEGIN DW 0  ;  BEGIN is declared
            ;  as a word offset
            ;  and its content is initialized to zero.

NAME DD 0   ;  NAME is declared
            ;  as a double word
            ;  (4 bytes) offset
            ;  and initialized to zero.
```

The EQU directive can be used to assign a name to constants.

In the following, typical ASM-86 assembler directives such as SEGMENT, ENDS, ASSUME, and DUP are discussed.

3.6.1 SEGMENT and ENDS Directives

A section of a program or a data array can be defined by the SEGMENT and ENDS directives as follows:

```
             ORG      5000H
JOHN     SEGMENT
X1       DB       0
X2       DB       0
X3       DB       0
JOHN     ENDS
```

The segment name is JOHN. The assembler will assign a numeric value 5000H to JOHN. The programmer must use the 8086 instructions to load JOHN into DS as follows:

```
MOV BX,  JOHN
MOV DS,  BX
```

Note that the segment registers (except CS) must be loaded via a 16-bit register (AX, BX, CX, or DX).

3.6.2 Assume Directive

As mentioned before, the 8086, at any time, can directly address four physical segments which include a code segment, a data segment, a stack segment, and an extra segment. An 8086 program may contain a number of logical segments containing code, data, and stack. The ASSUME pseudoinstruction assigns a logical segment to a physical segment at any given time. That is, the ASSUME directive tells the assembler what addresses will be in the segment registers at execution time.

For example, the statement ASSUME CS: PROGRAM, DS: DATA, SS: STACK directs the assembler to use the logical code segment PROGRAM as CS containing the instructions, the logical data segment DATA as DS containing data, and the logical STACK segment STACK as

SS containing the stack. But, it is the responsibility of the programmer to set these segment registers in the program.

3.6.3 DUP Directive

The DUP directive can be used to initialize several locations to zero. For example, the statement START DW 4 DUP (0) reserves four words starting at the offset START in DS and initializes them to zero. The DUP directive can also be used to reserve several locations which need not be initialized. A question mark must be used with DUP in this case. For example, the statement BEGIN DB 100 DUP (?) reserves 100 bytes of uninitialized data space to an offset BEGIN in the data segment. Note that BEGIN should be typed in the label field, DB in the op field, and 100 DUP (?) in the operand field.

A typical example illustrating the use of these directives is given below:

```
DATA          SEGMENT
ADDR          DW  3005H
ADDR          DW  2003FH
DATA          ENDS
STACK         SEGMENT
              DW  60  DUP  (0)   ;  Assign 60₁₀ words.
                                 ;  of stack with zeros.
STACK_TOP     LABEL WORD         ;  Initialize Stack_Top
                                 ;  to the next
STACK         ENDS               ;  location after the
                                 ;  top of the stack.

CODE          SEGMENT
              ASSUME CS: CODE,  DS: DATA,  SS: STACK
              MOV AX,  STACK
              MOV SS,  AX
              LEA SP,  STACK_TOP
              MOV AX,  DATA
              MOV DS,  AX
              LEA SI,  ADDR
              LEA DI,  ADDRR
                 -
                 -          ⎤  Main Program
                 -          ⎬      Body
                 -          ⎦
CODE          ENDS
```

Note that LABEL is a directive used to initialize STACK_TOP to the next location after the top of the stack. The statement STACK_TOP LABEL WORD gives the name STACK_TOP to the next address after the 60 words are set aside for the stack. The WORD in this statement indicates that PUSH into and POP from the stack are done as words.

When the assembler translates an assembly language program, it computes the displacement, or offset, of each instruction code byte from the start of a logical segment that contains it. For example, in the above program the CS: CODE in the ASSUME statement directs the assembler to compute the offsets or displacements of the following instructions from the start of the logical segment CODE. This means that when the program is run, the CS will contain the 16-bit value where the logical segment CODE was located in memory. The assembler keeps track of the instruction byte displacements which are loaded into IP. The 20-bit physical address generated from CS and IP are used to fetch each instruction.

Another example to store data bytes in a data segment and to allocate stack is given in the following:

```
DSEG        SEGMENT
ARRAY       DB 02H, F1H, A2H    ;   store 3 bytes
                                ;   of data in an
DSEG        ENDS                ;   address defined
            -                   ;   by DSEG as DS
            -                   ;   and ARRAY as
            -                   ;   offset
SSEG        SEGMENT
            DW 10 DUP (0)       ;   Allocate
                                ;   10 word stack
STACK_TOP LABEL WORD            ;   Label initial
                                ;   TOS
SSEG        ENDS
            -
            -
            -
            -
            MOV AX, DSEG        ;   Initialize
            MOV DS, AX          ;   DS
            MOV AX, SSEG        ;   Initialize
            MOV SS, AX          ;   SS
            MOV SP, STACK_TOP   ;   Initialize SP
            -
            -
            -
```

Note that typical 8086 assemblers on the Microsoft software and HP64000 system use the ORG directive to load CS and IP. For example, CS and IP can be initialized with 0500H and 0020H as follows:

For Microsoft 8086 assembler: ORG 05000020H
For HP64000 8086 assembler: ORG 0500H:0020H

Example 3.1

Determine the effect of each one of the following 8086 instructions:

i)   **PUSH [BX]**
ii) **DIV DH**
iii) **CWD**
iv) **MOVSB**
v)   **MOV START [BX], AL**

Assume the following data prior to execution of each one of the above instructions independently. Assume all numbers in hexadecimal.

[DS]	= 3000H	[SI]	= 0400H
[ES]	= 5000H	[DI]	= 0500H
[DX]	= 0400H	DF	= 0
[SP]	= 5000H	[BX]	= 6000H
[SS]	= 6000H	Value of START	= 05H
[AX]	= 00A9H		

[36000H] = 02H, [36001H] = 03H
[50500H] = 05H
[30400H] = 02H, [30401H] = 03H

Solution

i) 20-bit physical memory addressed by DS and BX = 36000H. 20-bit physical location pointed to by SP and SS = 65000H. PUSH [BX] pushes [36001H] and [36000H] into stack locations 64FFFH and 64FFEH, respectively. The SP is then decremented by 2 to contain the 20-bit physical address 64FFEH (SS = 6000H, SP = 4FFEH). Therefore, [64FFFH] = 03H and [64FFEH] = 02H.

ii) Before unsigned division, [DX] = 0400H, DH contains 04H, and [AX] = 00A9H = 169_{10}. After DIV DH, [AH] = Remainder = 01H and [AL] = Quotient = 2AH = 42_{10}.

iii) CWD sign extends AX register into the DX register. Since the sign bit of [AX] = 0, after CWD, [DX AX] = 000000A9H.

iv) MOVSB moves the content of memory addressed by the source [DS] and [SI] to the destination addressed by [ES] and [DI] and then it increments SI and DI by 1 for byte move. Since DF = 0, [DS] = 3000H, [SI] = 0400H, [ES] = 5000H, [DI] = 0500H, and the content of physical memory location 30400H is moved to physical memory 50500H. Since [30400H] = 02H, the location 50500H will also contain 02H. Since DF = 0 after MOVSB, [SI] = 0401H, [DI] = 0501H.

v) Since [BX] = 6000H, [DS] = 3000H, START = 05H, and the physical memory for destination = 36005H. After MOV START [BX], AL, memory location 36005H will contain A9H.

Example 3.2

Write 8086 assembly program to clear 100_{10} consecutive bytes.

Solution

```
0000                    CODE  SEGMENT  AT  1000H        ;  Program to
                                                        ;  clear 100 bytes
                        ASSUME  CS:CODE,  DS:DATA
0000  B8 07D0           MOV AX,  2000                   ;  Initialize
0003  8E D8             MOV DS,  AX                      ;  Data Segment
0005  7D 1E 0000  R     LEA BX,  ADDR                    ;  Initialize BX
0009  B9 0064           MOV CX,  100                     ;  Initialize loop
                                                        ;  count
000C  C6 07 00  START:MOV BYTEPTR[BX],  0              ;  Clear memory byte
000F  43                INC BX                          ;  Update pointer
0010  E2 FA             LOOP  START                     ;  Decrement CX and
                                                        ;    loop
0012  F4                HLT                             ;  Halt
0013                    CODE  ENDS
0000                    DATA  SEGMENT  AT  2000H
0000  3000          ADDR DW 3000H                       ;  Store the initial
                                                        ;  address
0002                    DATA  ENDS                      ;  End program
                        END
```

```
Microsoft (R) Macro Assembler Version 5.10 5/13/92 13:12:12
                                              Symbols-1
```

Segments and Groups:

Name	Length	Align	Combine Class
CODE	0013	AT	1000
DATA	0002	AT	2000

Symbols:

Name	Type	Value	Attr
ADDR	L WORD	0000	DATA
START	L NEAR	000C	CODE
@CPU	TEXT	0101h	
@FILENAME	TEXT	ex32	
@VERSION	TEXT	510	

```
   28  Source Lines
   28  Total Lines
    9  Symbols

47886 + 445965 Bytes symbol space free

    0  Warning Errors
    0  Severe Errors
```

Example 3.3

Write 8086 assembly program to computer $\sum_{i=1}^{N} X_i Y_i$ where Xi and Yi are signed 8-bit numbers. N = 100. Assume no overflow.

Solution

```
0000                    CODE SEGMENT AT 1000H
                        ASSUME CS:CODE, DS:DATA
0000  B8 2000           MOV AX, 2000H          ;  Initialize
0003  8E D8             MOV DS, AX             ;  Data Segment
0005  B9 0064           MOV CX, 100            ;  Initialize loop
                                               ;  count
0008  8D 1E 0000 R      LEA BX, ADDR1          ;  Load ADDR1
                                               ;  into BX
000C  8D 36 0002 R      LEA SI, ADDR2          ;  Load ADDR2
                                               ;  into SI
0010  BA 0000           MOV DX, 0000H          ;  Initialize sum
                                               ;  to zero
0013  8A 07      START:MOV AL, [BX]            ;  Load data into
                                               ;  AL
0015  F6 2C             IMUL BYTEPTR [SI]       ;  Signed 8x8
                                               ;  multiplication
```

```
0017   03 D0           ADD DX, AX           ;  Sum XiYi
0019   43              INC BX               ;  Update pointer
001A   46              INC SI               ;  Update pointer
001B   E2 F6           LOOP START           ;  Decrement CX
                                            ;  and loop
001E                   CODE ENDS
0000                   DATA SEGMENT AT 2000H
0000   0000            ADDR1 DW 0000H       ;  Location of Xi
0002   1000            ADDR2 DW 1000H       ;  Location of Yi
0004                   DATA ENDS
                       END                  ;  End program
```

Microsoft (R) Macro Assembler Version 5.10   5/13/92   13:26:33

Segments and Groups:

Name	Length	Align	Combine Class
CODE	001E	AT	1000
DATA	0004	AT	2000

Symbols:

Name	Type	Value	Attr
ADDR1	L WORD	0000	DATA
ADDR2	L WORD	0002	DATA
START	L NEAR	0013	CODE
@CPU	TEXT	0101h	
@FILENAME	TEXT	ex33	
@VERSION	TEXT	510	

```
    39  Source Lines
    39  Total Lines
    10  Symbols

47848 + 445003 Bytes symbol space free

     0  Warning Errors
     0  Severe Errors
```

Example 3.4

Write 8086 assembly language program to add two words; each word contains four packed BCD digits. The first word is stored in two consecutive locations with the low byte at the offset pointed by SI at 0500H, while the second word is stored in two consecutive locations with the low byte pointed by BX at the offset 1000H. Store the result in memory pointed to by BX.

Solution

```
0000                   CODE SEGMENT AT 1000H
                       ASSUME CS:CODE, DS:DATA
0000   BE 2000         MOV AX, 2000H        ;  Initialize
```

```
0003  8E D8           MOV  DS, AX         ;  Data Segment
0005  B9 0002         MOV  CX, 2          ;  Initialize loop
                                          ;  count
0008  BE 0500         MOV  SI, 0500H      ;  Initialize SI
000B  BB 1000         MOV  BX, 1000H      ;  Initialize BX
000E  F8              CLC                 ;  Clear carry
000F  8A 04      START:MOV  AL, [SI]      ;  Move data
0011  12 07           ADC  AL, [BX]       ;  Perform addition
0013  27              DAA                 ;  BCD adjust
0014  88 07           MOV  [BX], AL       ;  Store result
0016  46              INC  SI             ;  Update pointer
0017  43              INC  BX             ;  Update pointer
0018  E2 F5           LOOP START          ;  Decrement CX and
                                          ;  loop
001A  F4              HLT                 ;  Halt
001B            CODE  ENDS
0000            DATA  SEGMENT AT 2000H
0000            DATA  ENDS
                      ENDS                ;  End program
```

Microsoft (R) Macro Assembler Version 5.10

Segments and Groups:

Name	Length	Align	Combine Class
CODE	001B	AT	1000
DATA	0000	AT	2000

Symbols:

Name	Type	Value	Attr
START	L NEAR	000F	CODE
@CPU	TEXT	0101h	
@FILENAME	TEXT	ex34	
@VERSION	TEXT	510	

```
   36 Source Lines
   36 Total Lines
    8 Symbols

47890 + 445961 Bytes symbol space free

    0 Warning Errors
    0 Severe Errors
```

Example 3.5

Write an 8086 assembly language program to add two words; each contains two ASCII digits. The first word is stored in two consecutive locations with the low byte pointed to by SI at offset 0300H, while the second byte is stored in two consecutive locations with the low byte pointed to by DI at offset 0700H. Store the result in DI.

Solution

```
0000                    CODE SEGMENT AT 1000H
                        ASSUME CS:CODE, DS:DATA
0000  B8 2000           MOV AX, 2000H            ;  Initialize
0003  8E D8             MOV DS, AX               ;  Data Segment
0005  B9 0002           MOV CX, 2                ;  Initialize loop
                                                 ;  count
0008  BE 0300           MOV SI, 0300H            ;  Initialize SI
000B  BF 0700           MOV DI, 0700H            ;  Initialize DI
000E  F8                CLC                      ;  Clear carry
000F  8A 04      START:MOV AL, [SI]              ;  Move data
0011  12 05             ADC AL, [SI]             ;  Perform addition
0013  37                AAA                      ;  ASCII adjust
0014  88 05             MOV [DI], AL             ;  Store result
0016  46                INC SI                   ;  Update pointer
0017  47                INC DI                   ;  Update pointer
0018  E2 F5             LOOP START               ;  Decrement CX and
                                                 ;  loop
001A                    HLT                      ;  Halt
001B              CODE ENDS
0000              DATA SEGMENT AT 2000H
0000              DATA ENDS
                  END                            ;  End program
```

Microsoft (R) Macro Assembler Version 5.10 10/05/93 22:25:10

                                            Symbols-1

Segments and Groups:

Name	Length	Align	Combine Class
CODE	001B	AT	1000
DATA	0000	AT	2000

Symbols:

Name	Type	Value	Attr
START	L NEAR	000F	CODE
@CPU	TEXT	0101h	
@FILENAME	TEXT	ex35	
@VERSION	TEXT	510	

```
    36 Source Lines
    36 Total Lines
     8 Symbols

47890 + 445961 Bytes symbol space free

     0 Warning Errors
     0 Severe Errors
```

Example 3.6

Write an 8086 assembly language program to compare a source string of 50_{10} words pointed to by an offset of 2000H in the data segment with a destination string pointed to by an offset 3000H in another segment. The program should be halted as soon as a match is found or the end of string is reached. Assume CS and DS are initialized.

Solution

```
0000                    CODE  SEGMENT  AT  1000H
                        ASSUME  CS:CODE,DS:DATA,ES:DATAA
0000   B8  2000         MOV  AX,2000H            ;  initialize
0003   8E  D8           MOV  DS,AX               ;  Data segment
0005   B8  4000         MOV  AX,4000H            ;  Initialize
0008   8E  C0           MOV  ES,AX               ;  ES
000A   BE  2000         MOV  SI,2000H            ;  Initialize SI
000D   BF  3000         MOV  DI,3000H            ;  Initialize DI
0010   B9  0032         MOV  CX,50               ;  Initialize CX
0013   FC               CLD                      ;  DF is cleared so
                                                 ;  that SI and DI
                                                 ;  will autoincrement
                                                 ;  after compare
0014   F2/A7            REPNE  CMPSW             ;  Repeat CMPSW until
                                                 ;  CX=0 or until
                                                 ;  compared words are
                                                 ;  equal
0016   F4               HLT                      ;  Halt
0017             CODE  ENDS
0000             DATA  SEGMENT  AT  2000H
0000             DATA  ENDS
0000             DATAA  SEGMENT  AT  4000H
                 END                             ;  End program
```

Microsoft (R) Macro Assembler Version 5.10 10/05/92 11:45:41

Segments and Groups:

Name	Length	Align	Combine Class
CODE	0017	AT	1000
DATA	0000	AT	2000
DATAA	0000	AT	4000

Symbols:

Name	Type	Value	Attr
@CPU	TEXT	0101h	
@FILENAME	TEXT	TEST1	
@VERSION	TEXT	510	

     20 Source Lines

```
            20 Total Lines
             8 Symbols

      47884 + 433567 Bytes symbol space free

             0 Warning Errors
             0 Severe Errors
```

Example 3.7

Write a subroutine in 8086 assembly language which can be called by a main program in a different code segment. The subroutine will multiply a signed 16-bit number in CX by a signed 8-bit number in AL. The main program will call this subroutine, store the result in two consecutive memory words, and stop. Assume SI and DI contain pointers to the signed 8-bit and 16-bit data, respectively.

Solution

```
0000                    CODE SEGMENT AT 1000H
                        ASSUME CS:CODE,DS:DATA,SS:STACK
0000  B8 5000           MOV AX, 5000H          ;  Initialize
0003  8E D8             MOV DS, AX             ;  Data Segment
0005  B8 6000           MOV AX, 6000H          ;  Initialize
0008  8E D0             MOV SS, AX             ;  Stack Segment
000A  BC 0020           MOV SP, 0020H          ;  Initialize SP
000D  BB 2000           MOV BX, 2000H          ;  Initialize BX
0010  8A 04             MOV AL, [SI]           ;  Move 8-bit data
0012  8B 0D             MOV CX, [DI]           ;  Move 16-bit data
0014  9A 0000---R       CALL FAR PTR MULTI     ;  Call MULTI
                                               ;  subroutine
0019  89 17             MOV [BX], DX           ;  Store high word of
                                               ;  result
001B  89 47 02          MOV [BX+2], AX         ;  Store low word of
                                               ;  result
001E  F4                HLT                    ;  Halt
001F               CODE ENDS
0000               SUBR SEMENT AT 7000H
                   ASSUME CS:SUBR
0000               MULTI PROC FAR              ;  Must be called from
0000  51               PUSH CX                 ;  another code
                                               ;  segment
0001  50               PUSH AX                 ;
0002  98               CBW                     ;  Sign extend AL
0003  F7 E9            IMUL CX                 ;  [DX][AX]←[AX]*[CX]
0005  58               POP AX                  ;
0006  59               POP CX                  ;
0007  CB               RET                     ;  Return
0008               MULTI ENDP                  ;  End of procedure
0004               SUBR ENDS                   ;  End subroutine
0000               DATA SEGMENT AT 5000H
0000               DATA ENDS
0000               STACK SEGMENT AT 6000H
0000               STACK ENDS
                   END                         ;  End program
```

Microsoft (R) Macro Assembler Version 5.10 10/05/93 22:40:32
Symbols-1

Segments and Groups:

Name	Length	Align	Combine Class
CODE	001F	AT	1000
DATA	0000	AT	5000
STACK	0000	AT	6000
SUBR	0004	AT	7000

Symbols:

Name	Type	Value	Attr
MULTI	F PROC	0000	SUBR Length=0004
@CPU	TEXT	0101h	
@FILENAME	TEXT	EX37	
@VERSION	TEXT	510	

```
    52 Source Lines
    52 Total Lines
    10 Symbols

47852 + 443951 Bytes symbol space free

     0 Warning Errors
     0 Severe Errors
```

Example 3.8

Write an 8086 assembly language program to subtract two 64-bit numbers. Assume SI and DI contain the starting address of the numbers. Store the result in memory pointed to by [DI].

Solution

```
0000                 CODE SEGMENT AT 4000H
3000                      ORG 3000H
3000   05712FA2   DATAA DD 05712FA2H           ; DATAA LOW
3004   0248A201         DD 0248A201H           ; DATAA HIGH
3008   561A2604   DATAB DD 561A2604H           ; DATAB LOW
300C   72A2B270         DD 72A2B270H           ; DATAB HIGH
3010              DATA ENDS
0000              PROG SEGMENT
                      ASSUME CS:PROG,DS:DATA
0000   B8 4000         MOV AX, 4000H           ; Initialize
0003   8E D8           MOV DS, AX              ; DS
0005   BA 0004         MOV DX, 4               ; Load 4 into DX
0008   8D 36 3000 R    LEA SI, DATAA           ; Initialize SI
000C   8D 3E 30008 R   LEA DI, DATAB           ; Initialize DI
0010   F8              CLC                     ; Clear carry
0011   8B 04      START: MOV AX, [SI]          ; Load DATAA
0013   19 05           SBB [DI], AX            ; Perform subtraction
```

```
0015    83 C6 02            ADD SI, 2                    ;  Update
0018    83 C7 02            ADD DI, 2                    ;  pointers
001B    4A                  DEC DX
001C    75 F3               JNZ START
001E    F4                  HLT
001F                PROG ENDS
                    END
```

```
Microsoft (R) Macro Assembler Version 5.10  10/10/93 22:14:0
                                            Symbols-1
```

Segments and Groups:

Name	Length	Align	Combine	Class
DATA	3010	AT	4000	
PROG	001F	PARA	NONE	

Symbols:

Name	Type	Value	Attr
DATAA	L DWORD	3000	DATA
DATAB	L DWORD	3008	DATA
START	L NEAR	0011	PROG
@CPU	TEXT	0101h	
@FILENAME	TEXT	examp38	
@VERSION	TEXT	510	

```
    25 Source Lines
    25 Total Lines
    10 Symbols

47518 + 428045 Bytes symbol space free

     0 Warning Errors
     0 Severe Errors
```

Example 3.9

Write an 8086 assembly language program that will perform the following operation:

$$5 * AL - 6 * BH + (BH/8) \rightarrow CX$$

Assume that the stack pointer is already initialized.

Solution

```
0000                PROG SEGMENT AT 5000H
0500                    ORG 0500H
                        ASSUME CS:PROG
0500    B3 05           MOV BL,5                    ;  Compute
0502    F6 E3           MUL BL                      ;  AX←5*AL
0504    8B C8           MOV CX,AX                   ;  Store in CX
```

0506	B3 FA	MOV BL,-6	;
0508	8A C7	MOV AL,BH	; Compute
050A	F6 E3	MUL BL	; AX←6*BH
050C	03 C8	ADD CX,AX	; store in CX
050E	51	PUSH CX	
050F	B1 03	MOV CL,3	
0511	D2 EF	SHR BH,CL	; Compute BH/8
0513	8A DF	MOV BL,BH	; Convert BH/8 to
0515	B7 00	MOV BH,0	; 16-bit unsigned
			; number in BX
0517	59	POP CX	
0518	03 CB	ADD CX,BX	; Store final result
			; in CX
051A	F4	HLT	
051B		PROG ENDS	
		END	

```
Microsoft (R) Macro Assembler Version 5.10  1/16/80 12:07:44
                                          Symbols-1
Segments and Groups:
```

Name   **Length**   **Align**   **Combine Class**

PROG ....   051B   AT   5000

Symbols:

Name   **Type**   **Value**   **Attr**

@CPU	TEXT	0101h	
@FILENAME		TEXT	TEST1
@VERSION		TEXT	510

```
   20 Source Lines
   20 Total Lines
    6 Symbols

47808 + 433643 Bytes symbol space free

    0 Warning Errors
    0 Severe Errors
```

Example 3.10

Write an 8086 assembly program that converts a number from Fahrenheit degrees to Celsius degrees. Each number is only one byte. The source byte is assumed to reside at offset 1000H in the data segment, and the destination at an offset of 2000H in the data segment.

$$\text{use } C = \frac{(F - 32)}{9} \times 5$$

Solution

```
0000                    CODE  SEGMENT
                        ASSUME  CS:CODE,DS:DATA
0000  BE 1000           MOV SI, 1000H          ;   Initialize source
                                               ;   pointer
0003  BF 2000           MOV DI, 2000H          ;   Initialize
                                               ;   destination pointer
                                               ;
0006  B4 00             MOV AH, 0              ;   Clear AX high byte
0008  8A 04             MOV AL, [SI]           ;   Get degrees F
000A  98                CBW                    ;   Sign extend
000B  2D 0020           SUB AX, 32             ;   Subtract 32
                                               ;
000E  B9 0005           MOV CX, 5              ;   Get multiplier
0011  F7 E9             IMUL CX                ;   Multiply by 5
                                               ;
0013  B9 0009           MOV CX, 9              ;   Get divisor
0016  F7 F9             IDIV CX                ;   divide by 9 to
                                               ;   get Celsius
                                               ;
0017  88 05             MOV [DI], AL           ;   Put result in
                                               ;   destination
                                               ;
001A                    CODE ENDS              ;   End segment
                                               ;
                        END                    ;
```

Microsoft (R) Macro Assembler Version 5.10   10/10/93 22:27:4
Symbols-1

Segments and Groups:

Name	Length	Align	Combine Class
CODE	0019	PARA	NONE

Symbols:

Name	Type	Value	Attr
GO	L NEAR	0000	CODE
@CPU	TEXT	0101h	
@FILENAME	TEXT	examp	310
@VERSION		TEXT	510

    22 Source Lines
    22 Total Lines
    11 Symbols

47464 + 428099 Bytes symbol space free

     0 Warning Errors
     0 Severe Errors

3.7 System Design Using the 8086

This section covers the basic concepts associated with interfacing the 8086 to its support chips such as memory and I/O. Topics such as timing diagrams and 8086 pins and signals will also be included.

3.7.1 Pins and Signals

The 8086 pins and signals are shown in Figure 3.6. Unless otherwise indicated, all 8086 pins are TTL compatible. As mentioned before, the 8086 can operate in two modes. These are minimum mode (uniprocessor system — single 8086) and maximum mode (multiprocessor system — more than one 8086). MN/MX is an input pin used to select one of these modes. When MN/MX is HIGH, the 8086 operates in the minimum mode. In this mode, the 8086 is configured (that is, pins are defined) to support small, single processor systems using a few devices that use the system bus.

When MN/MX is LOW, the 8086 is configured (that is, pins are defined in the maximum mode) to support multiprocessor systems. In this case, the Intel 8288 bus controller is added to the 8086 to provide bus controls and compatibility with the multibus architecture. Note that in a particular application, the MN/MX must be tied to either HIGH or LOW.

The AD0-AD15 lines are a 16-bit multiplexed address/data bus. During the first clock cycle AD0-AD15 are the low order 16 bits of address. The 8086 has a total of 20 address lines. The upper four lines are multiplexed with the status signals for the 8086. These are the A16/S3, A17/S4, A18/S5, and A19/S6. During the first clock period of a bus cycle (read or write cycle), the entire 20-bit address is available on these lines. During all other clock cycles for memory and I/O operations, AD15-AD0 contain the 16-bit data, and S3, S4, S5, and S6 become status lines. S3 and S4 lines are decoded as follows:

A17/S4	A16/S3	Function
0	0	Extra segment
0	1	Stack segment
1	0	Code or no segment
1	1	Data segment

Therefore, after the first clock cycle of an instruction execution, the A17/S4 and A16/S3 pins specify which segment register generates the segment portion of the 8086 address. Thus, by decoding these lines and then using the decoder outputs as chip selects for memory chips, up to 4 megabytes (one megabyte per segment) can be provided. This provides a degree of protection by preventing erroneous write operations to one segment from overlapping into another segment and destroying information in that segment. A18/S5 and A19/S6 are used as A18 and A19, respectively, during the first clock period of an instruction execution. If an I/O instruction is executed, they stay low during the first clock period. During all other cycles, A18/S5 indicates the status of the 8086 interrupt enable flag and a low A19/S6 pin indicates that the 8086 is on the bus. During a "Hold Acknowledge" clock period, the 8086 tristates the A19/S6 pin and thus allows another bus master to take control of the system bus.

The 8086 tristates AD0-AD15 during Interrupt Acknowledge or Hold Acknowledge cycles.

BHE/S7 is used as BHE (Bus High Enable) during the first clock cycle of an instruction execution. The 8086 outputs a low on this pin during read, write, and interrupt acknowledge cycles in which data are to be transferred in a high-order byte (AD15-AD8) of the data bus. BHE can be used in conjunction with AD0 to select memory banks. A thorough discussion is provided later. During all other cycles BHE/S7 is used as S7 and the 8086 maintains the output level (BHE) of the first clock cycle on this pin.

Common Signals

Name	Function	Type
AD15-AD0	Address/Data Bus	Bidirectional, 3-State
A19/S6-A16/S3	Address/Status	Output, 3-State
BHE/S7	Bus High Enable/Status	Output, 3-State
MN/MX	Minimum/Maximum Mode Control	Input
RD	Read Control	Output, 3-State
TEST	Wait On Test Control	Input
READY	Wait State Control	Input
RESET	System REset	Input
NMI	Non-Maskable Interrupt Request	Input
INTR	Interrupt Request	Input
CLK	System Clock	Input
Vcc	+5V	Input
GND	Ground	

Mimimum Mode Signals (MN/MX = Vcc)

Name	Function	Type
HOLD	Hold Request	Input
HLDA	Hold Acknowledge	Output
WR	Write Control	Output, 3-State
M/IO	Memory/IO Control	Output, 3-State
DT/R	Data Transmit/Receive	Output, 3-State
DEN	Data Enable	Output, 3-State
ALE	Address Latch Enable	Output
INTA	Interrupt Acknowledge	Output

Maximum Mode Signals (MN/MX = GND)

Name	Function	Type
RQ/GT1, 0	Request/Grant Bus Access Control	Bidirectional
LOCK	Bus Priority Lock Control	Output, 3-State
S2-S0	Bus Cycle Status	Output, 3-State
QS1, QS0	Instruction Queue Status	Output

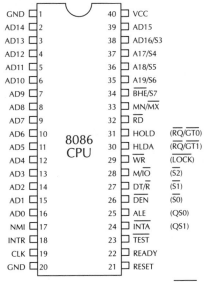

Maximum mode pin functions (e.g., LOCK) are shown in parenthesis

FIGURE 3.6   8086 pins and signals.

RD is LOW whenever the 8086 is reading data from memory or an I/O location.

TEST is an input pin and is only used by the WAIT instruction. The 8086 enters a wait state after execution of the WAIT instruction until a LOW is seen on the TEST pin. This input is synchronized internally during each clock cycle on the leading edge of the CLK pin.

INTR is the maskable interrupt input. This line is not latched and, therefore, INTR must be held at a HIGH level until recognized to generate an interrupt.

NMI is the nonmaskable interrupt input activated by a leading edge.

RESET is the system reset input signal. This signal must be high for at least four clock cycles to be recognized, except after power-on which requires a 50-ms reset pulse. It causes the 8086 to initialize registers DS, SS, ES, IP, and flags to all zeros. It also initializes CS to FFFFH. Upon removal of the RESET signal from the RESET pin, the 8086 will fetch its next instruction from 20-bit physical address FFFF0H (CS = FFFFH, IP = 0000H).

The reset signal to the 8086 can be generated by the 8284. The 8284 has a Schmitt Trigger input (RES) for generating reset from a low active external reset.

Since the reset vector is located at the physical address FFFF0H, there may not be enough locations available to write programs. Typical assemblers such as the Microsoft 8086 assembler use the following to jump to a different code segment to write programs:

```
ORG      0FFFF0H ;reset vector
         JMP      FAR BEGIN
         ORG      30000500H
BEGIN    —
         —      } user
         —      } program
         —
```

The above instruction sequence will allow the 8086 to jump to the offset BEGIN (offset 0500H) in code segment 3000H upon hardware reset where the user can write his/her programs.

To guarantee reset from power-up, the 8284 reset input must remain below 1.05 volts for 50 microseconds after Vcc has reached the minimum supply voltage of 4.5V. The $\overline{\text{RES}}$ input of the 8284 can be driven by a simple RC circuit as shown in Figure 3.7.

The values of R and C can be selected as follows:

$$Vc\ (t) = V\ (1 -\exp - (t/RC))$$

where t = 50 microseconds, V = 4.5 V, Vc = 1.05V, and RC = 188 microseconds. For example, if C is chosen arbitrarily to be 0.1 µF, then R = 1.88 KΩ.

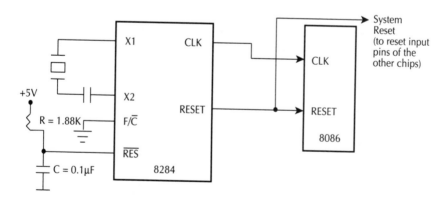

FIGURE 3.7   8086 reset and system resst.

As mentioned before, the 8086 can be configured in either minimum or maximum mode using the MN/MX input pin. In minimum mode, the 8086 itself generates all bus control signals. These signals are

- DT/R̄ (Data Transmit/Receive). DT/R̄ is an output signal required in minimum system that uses an 8286/8287 data bus transceiver. It is used to control direction of data flow through the transceiver.
- DEN (Data Enable) is provided as an output enable for the 8286/8287 in a minimum system which uses the transceiver.
- DEN is active LOW during each memory and I/O access and for INTA cycles.
- ALE (Address Latch Enable) is an output signal provided by the 8086 and can be used to demultiplex the AD0-AD15 into A0-A15 and D0-D15 at the falling edge of ALE. The 8086 ALE signal is similar to the 8085 ALE.
- M/IO. This 8086 output signal is similar to the 8085 IO/M̄. It is used to distinguish a memory access (M/IO = HIGH) from an I/O access (M/IO = LOW). When the 8086 executes an I/O instruction such as IN or OUT, it outputs a LOW on this pin. On the other hand, the 8086 outputs HIGH on this pin when it executes a memory reference instruction such as MOVE AX, [SI].
- WR̄. The 8086 outputs LOW on this pin to indicate that the processor is performing a write memory or write I/O operation, depending on the M/IO signal.
- INTA. The 8086 INTA is similar to the 8085 INTA. For Interrupt Acknowledge cycles (for INTR pin), the 8086 outputs LOW on this pin.
- HOLD (input), HLDA (output). These pins have the same purpose as the 8085 HOLD/ HLDA pins and are used for DMA. A HIGH on the HOLD pin indicates that another master is requesting to take over the system bus. The processor receiving the HOLD request will output HLDA high as an acknowledgment. At the same time, the processor tristates the system bus. Upon receipt of LOW on the HOLD pin, the processor places LOW on the HLDA pin. HOLD is not an asynchronous input. External synchronization should be provided if the system cannot otherwise guarantee the setup time.
- CLK (input) provides the basic timing for the 8086.

The maximum clock frequencies of the 8086-1, 8086, and 8086-2 are 5 MHz, 10 MHz, and 8 MHz, respectively. Since the design of these processors incorporates dynamic cells, a minimum frequency of 2 MHz is required to retain the state of the machine. The 8086, 8086-1, and 8086-2 will be referred to as 8086 in the following. Since the 8086 does not have on-chip clock generation circuitry, an 8284 clock generator chip must be connected to the 8086 CLK pin as shown in Figure 3.8.

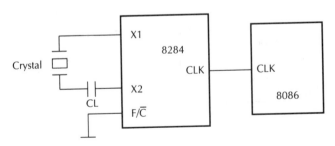

FIGURE 3.8   8284 clock generator connections to the 8086.

The crystal must have a frequency three times the 8086 internal frequency. That is, the 8284 divides the crystal clock frequency by 3. In other words, to generate a 5 MHz 8086 internal clock, the crystal clock must be 15 MHz. To select the crystal inputs of the 8284 as the frequency source for clock generation, the F/C input must be strapped to ground. This strapping option allows either the crystal or an external frequency input as the source for clock generation. When selecting a crystal for use with the 8284, the crystal series resistance should be as low as possible. Two crystal manufacturers recommended by Intel are Crystle Corp. model CY 15A (15 MHz) and CTS Knight Inc. model CY 24 A (24 MHz).

Note that the 8284 can be used to generate the 8086 READY input signal based on inputs from slow memory and I/O devices which are not capable of transferring information at the 8086 rate.

In the maximum mode, some of the 8086 pins in the minimum mode are redefined. For example, pins HOLD, HLDA, WR, M/IO, DT/R, DEN, ALE, and INTA in the minimum mode are redefined as RQ/GT0, RQ/GT1, LOCK, S2, S1, S0, QS0, and QS1, respectively. In maximum mode, the 8288 bus controller decodes the status information from S0, S1, S2 to generate bus timing and control signals required for a bus cycle. S2, S1, S0 are 8086 outputs and are decoded as follows:

$\overline{S2}$	$\overline{S1}$	$\overline{S0}$	
0	0	0	Interrupt Acknowledge
0	0	1	Read I/O port
0	1	0	Write I/O port
0	1	1	Halt
1	0	0	Code access
1	0	1	Read memory
1	1	0	Write memory
1	1	1	Inactive

- $\overline{RQ}/\overline{GT0}$, $\overline{RQ}/\overline{GT1}$. These request/grant pins are used by other local bus masters to force the processor to release the local bus at the end of the processor's current bus cycle. Each pin is bidirectional, with $\overline{RQ}/\overline{GT0}$ having higher priority than $\overline{RQ}/\overline{GT1}$. These pins have internal pull-up resistors so that they may be left unconnected. The request/ grant function of the 8086 works as follows:

  1. A pulse (one clock wide) from another local bus master ($\overline{RQ}/\overline{GT0}$ or $\overline{RQ}/\overline{GT1}$ pins) indicates a local bus request to the 8086.
  2. At the end of 8086 current bus cycle, a pulse (one clock wide) from the 8086 to the requesting master indicates that the 8086 has relinquished the system bus and tristated the outputs. Then the new bus master subsequently relinquishes control of the system bus by sending a LOW on $\overline{RQ}/\overline{GT0}$ or $\overline{RQ}/\overline{GT1}$ pins. The 8086 then regains bus control.

- $\overline{LOCK}$. The 8086 outputs LOW on the $\overline{LOCK}$ pin to prevent other bus masters from gaining control of the system bus. The $\overline{LOCK}$ signal is activated by the 'LOCK' prefix instruction and remains active until the completion of the instruction that follows.

- QS1, QS0. The 8086 outputs to QS1 and QS0 pins to provide status to allow external tracking of the internal 8086 instruction queue as follows:

QS1	QS0	
0	0	No operation
0	1	First byte of op code from queue
1	0	Empty the queue
1	1	Subsequent byte from queue

QS0 and QS1 are valid during the clock period following any queue operation. The 8086 can be operated from a +5V to +10V power supply. There are two ground pins on the chip to distribute power for noise reduction.

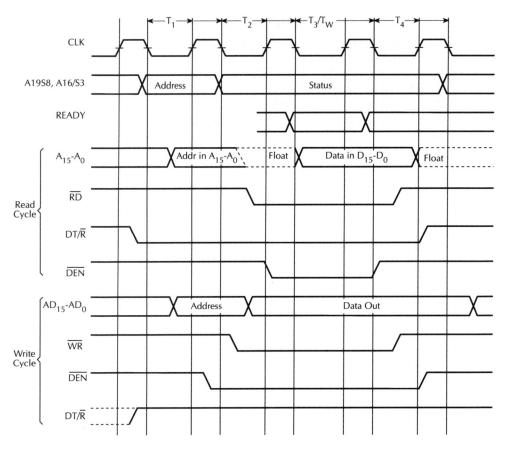

FIGURE 3.9   Basic 8086 bus cycle.

3.7.2 8086 Basic System Concepts

This section describes basic concepts associated with 8086 bus cycles, address and data bus, system data bus, and multiprocessor environment.

3.7.2.a 8086 Bus Cycle

In order to communicate with external devices via the system bus for transferring data or fetching instructions, the 8086 executes a bus cycle. The 8086 basic bus cycle timing diagram is shown in Figure 3.9. The minimum bus cycle contains four CPU clock periods called T States. The bus cycle timing diagram depicted in Figure 3.9 can be described as follows:

1. During the first T State (T1), the 8086 outputs the 20-bit address computed from a segment register and an offset on the multiplexed address/data/status bus.
2. For the second T State (T2), the 8086 removes the address from the bus and either tristates or activates the AD15-AD0 lines in preparation for reading data via AD15-AD0 lines during the T3 cycle. In case of a write bus cycle, the 8086 outputs data on AD15-AD0 lines. Also, during T2, the upper four multiplexed bus lines switch from address (A19-A16) to bus cycle status (S6, S5, S4, S3). The 8086 outputs LOW on RD (for read cycle) or WR (for write cycle) during portions of T2, T3, and T4.
3. During T3, the 8086 continues to output status information on the four A19-A16/S6-S3 lines and will either continue to output write data or input read data to or from AD15-AD0 lines. If the selected memory or I/O device is not fast enough to transfer data

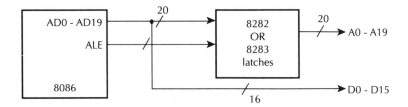

FIGURE 3.10   Separate address and data buses.

with the 8086, the memory or I/O device activates the 8086's READY input line low by the start of T3. This will force the 8086 to insert additional clock cycles (wait states TW) after T3. Bus activity during TW is the same as T3. When the selected device has had sufficient time to complete the transfer, it must activate the 8086 READY PIN HIGH. As soon as the TW clock periods end, the 8086 executes the last bus cycle, T4. The 8086 will latch data on AD15-AD0 lines during the last wait state or during T3 if no wait states are requested.

4. During T4, the 8086 disables the command lines and the selected memory and I/O devices from the bus. Thus, the bus cycle is terminated in T4. The bus cycle appears to devices in the system as an asynchronous event consisting of an address to select the device, a register or memory location within the device, a read strobe, or a write strobe along with data.

5. DEN and DT/R pins are used by the 8286/8287 transceiver in a minimum system. During the read cycle, the 8086 outputs DEN LOW during part of T2 and all of T3 cycles. This signal can be used to enable the 8286/8287 transceiver. The 8086 outputs LOW on the DT/R pin from the start of T1 and part of T4 cycles. The 8086 uses this signal to receive (read) data from the receiver during T3-T4. During a write cycle, the 8086 outputs DEN LOW during part of T1, all of T2 and T3, and part of T4 cycles. The signal can be used to enable the transceiver. The 8086 outputs HIGH on DT/R throughout the four bus cycles to transmit (write) data to the transceiver during T3-T4.

3.7.2.b   8086 Address and Data Bus Concepts

The majority of memory and I/O chips capable of interfacing to the 8086 require a stable address for the duration of the bus cycle. Therefore, the address on the 8086 multiplexed address/data bus during T1 should be latched. The latched address is then used to select the desired I/O or memory location. Note that the 8086 has a 16-bit multiplexed address and data bus, while the 8085's 8-bit data lines and LOW address byte are multiplexed. Hence, the multiplexed bus components of the 8085 family are not applicable to the 8086. To demultiplex the bus, the 8086 provides an ALE (Address Latch Enable) signal to capture the address in either the 8282 (noninverting) or 8283 (inverting) 8-bit bistable latches. These latches propagate the address through to the outputs while ALE is HIGH and latch the address in the falling edge of ALE. This only delays address access and chip select decoding by the propagation delay of the latch. Figure 3.10 shows how the 8086 demultiplexes the address and data buses.

The programmer views the 8086 memory address space as a sequence of one million bytes in which any byte may contain an eight-bit data element and any two consecutive bytes may contain a 16-bit data element. There is no constraint on byte or word addresses (boundaries). The address space is physically implemented on a 16-bit data bus by dividing the address space into two banks of up to 512K bytes as shown in Figure 3.11. These banks can be selected by BHE (Bus High Enable) and A0 as follows:

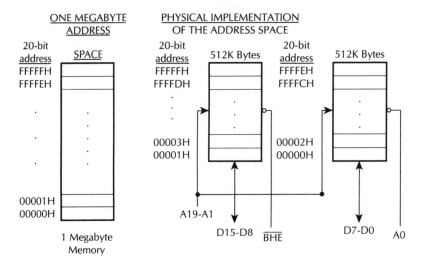

FIGURE 3.11   8086 memory.

BHE	A0	Byte transferred
0	0	Both bytes
0	1	Upper byte to/from odd address
1	0	Lower byte to/from even address
1	1	None

One bank is connected to D7-D0 and contains all even addressed bytes (A0 = 0). The other bank is connected to D15-D8 and contains odd-addressed bytes (A0 = 1). A particular byte in each bank is addressed by A19-A1. The even-addressed bank is enabled by LOW A0 and data bytes transferred over D7-D0 lines. The 8086 outputs HIGH on BHE (Bus High Enable) and thus disables the odd-addressed bank. The 8086 outputs LOW on BHE to select the odd-addressed bank and HIGH on A0 to disable the even-addressed bank. This directs the data transfer to the appropriate half of the data bus. Activation of A0 and BHE is performed by the 8086 depending on odd or even addresses and is transparent to the programmer. As an example, consider execution of the instruction MOV DH, [BX]. Suppose the 20-bit address computed by BX and DS is even. The 8086 outputs LOW on A0 and HIGH on BHE. This will select the even-addressed bank. The content of the selected memory is placed on the D7-D0 lines by the memory chip. The 8086 reads this data via D7-D0 and automatically places it in DH. Next, consider accessing a 16-bit word by the 8086 with low byte at an even address as shown in Figure 3.12.

For example, suppose that the 8086 executes the instruction MOV [BX], CX. Assume [BX] = 0004H, [DS] = 2000H. The 20-bit physical address for the word is 20004H. The 8086 outputs

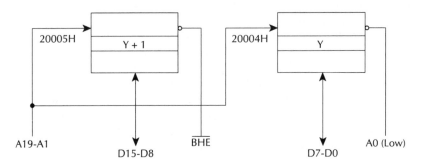

FIGURE 3.12   Even-addressed word transfer.

LOW on both A0 and $\overline{BHE}$, enabling both banks simultaneously. The 8086 outputs [CL] to D7-D0 lines and [CH] to D15-D8 lines with WR LOW and M/IO HIGH. The enabled memory banks obtain the 16-bit data and write [CL] to location 20004H and [CH] to location 20005H. Next, consider accessing an odd-addressed 16-bit word by the 8086. For example, suppose the 20-bit physical address computed by the 8086 is 20005H. The 8086 accomplishes this transfer in two bus cycles (A19-A1 = 20004H in the first cycle, and 200006H in the second cycle). In the first bus cycle, the 8086 outputs HIGH on A0, LOW on $\overline{BHE}$, and thus enables the odd-addressed bank and disables the even-addressed bank. The 8086 also outputs LOW on RD and HIGH on M/IO pins. In this bus cycle, the odd memory bank places [20005H] on D15-D8 lines. The 8086 reads this data into CL. In the second bus cycle, the 8086 outputs LOW on A0, HIGH on $\overline{BHE}$, and thus enables the even-addressed bank and disables the odd-addressed bank. The 8086 also outputs LOW on RD and HIGH on M/IO pins. The selected even-addressed memory bank places [20006H] on D7-D0 lines. The 8086 reads this data into CH.

During a byte read, the 8086 floats the entire D15-D0 lines during portions of T2 cycle even though data are expected on the upper or lower half of the data bus. As will be shown later, this action simplifies the chip select decoding requirements for ROMs and EPROMs. During a byte write, the 8086 will drive the entire 16-bit data bus. The information on the half of the data bus not transferring data is indeterminate. These concepts also apply to I/O transfers.

If memory of I/O devices are directly connected to the multiplexed bus, the designer must guarantee that the devices do not corrupt the address on the bus during T1. To avoid this, the memory or I/O devices should have an output enable controlled by the 8086 read signal.

3.7.3 Interfacing with Memories

Figure 3.13 shows a general block diagram of an 8086 memory array. In Figure 3.13, the 16-bit word memory is partitioned into high and low 8-bit banks on the upper and lower halves of the data bus selected by $\overline{BHE}$ and A0.

3.7.3.a ROM and EPROM

ROMs and EPROMs are the simplest memory chips to interface to the 8086. Since ROMs and EPROMs are read-only devices, A0 and $\overline{BHE}$ are not required to be part of the chip enable/select decoding (chip enable is similar to chip select except chip enable also provides whether

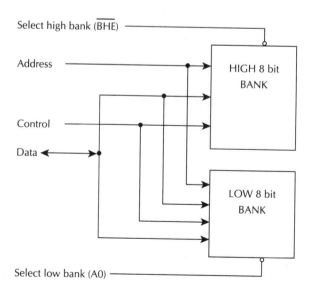

FIGURE 3.13   8086 memory array.

the chip is in active or standby power mode). The 8086 address lines must be connected to the address pins of the ROM/EPROM chips starting with A1 and higher to all the address lines of the ROM/EPROM chips. The 8086 unused address lines can be used as chip enable/select decoding. To interface the ROMs/RAMs directly to the 8086 multiplexed bus, they must have output enable signals.

Byte accesses are obtained by reading the full 16-bit word onto the bus with the 8086 discarding the unwanted byte and accepting the desired byte. If RD, WR, and M/IO are not decoded to generate separate memory and I/O commands for memory and I/O chips and the I/O space overlaps with the memory space of ROM/EPROM, then M/IO must be a condition of chip select decode.

3.7.3.b Static RAMs

Since static RAMs are read/write memories, both A0 and $\overline{BHE}$ must be included in the chip select/chip enable decoding of the devices and write timing must be considered in the compatibility analysis.

For each static RAM (containing odd or even addresses), the memory data lines must be connected to either the upper half AD15-AD8 for static RAM with odd addresses or lower half AD7-AD0 of the 8086 data lines for static RAM with even addresses.

3.7.3.c Dynamic RAMs

Dynamic RAMs store information as charges in capacitors. Since capacitors can hold charges for a few milliseconds, refresh circuitry is necessary in dynamic RAMs for retaining these charges. Therefore, dynamic RAMs are complex devices to design into a system. To relieve the designer of most of these complicated interfacing tasks, Intel provides the 8202 dynamic RAM controller as part of the 8086 family of peripheral devices. The 8202 can be interfaced with the 8086 to build a dynamic memory system. A thorough discussion on this topic can be found in the Intel manuals.

3.7.4 8086 Programmed I/O

The 8086 can be interfaced to 8- and 16-bit I/O devices using either standard or memory-mapped I/O. The standard I/O uses the instructions IN and OUT and is capable of providing 64K bytes of I/O ports. Using standard I/O, the 8086 can transfer 8- or 16-bit data to or from a peripheral device. The 64K byte I/O locations can then be configured as 64K 8-bit ports or 32K 16-bit ports. All I/O transfer between the 8086 and the peripheral devices take place via AL for 8-bit ports (AH is not involved) and AX for 16-bit ports. The I/O port addressing can be done either directly or indirectly as follows:

DIRECT
- IN AL, PORTA or IN AX, PORTB inputs 8-bit contents of port A into AL or 16-bit contents of port B into AX, respectively.

  Port A and port B are assumed as 8- and 16-bit ports, respectively.
- OUT PORTA, AL or OUT PORT B, AX outputs 8-bit contents of AL into port A or 16-bit contents of AX into port B, respectively.

INDIRECT
- IN AX, DX or IN AL, DX inputs 16-bit data addressed by DX into AX or 8-bit data addressed by DX into AL, respectively.
- OUT DX, AX or OUT DX, AL outputs 16-bit contents of AX into the port addressed by DX or 8-bit contents of AL into the port addressed by DX, respectively. In indirect addressing, register DX is used to hold the port address.

Data transfer using the memory-mapped I/O is accomplished by using memory-oriented instructions such as MOV reg 8 or reg 16, [BX] and MOV [BX], reg 8 or reg 16 for inputting and outputting 8- or 16-bit data from or to an 8-bit register or a 16-bit register addressed by the 20-bit memory-mapped port location computed from DS and BX.

Note that the indirect I/O transfer method is desirable for service routines that handle more than one device such as multiple printers by allowing the desired device (a specific printer) to be passed to the procedure as a parameter.

Devices with I/O ports can be connected to either the upper or lower half of the data bus. If the I/O port chip is connected to the 8086 lower half of the data lines (AD0-AD7), the port addresses will be even (A0 = 0). On the other hand, the port addresses will be odd ($\overline{BHE}$ = 0) if the I/O port chip is connected to the upper half of the 8086 data lines (AD8-AD15).

3.8 8086-Based Microcomputer

In this section, an 8086 will be interfaced in the minimum mode to provide 2K × 16 EPROM, 2K × 16 static RAM, and six 8-bit I/O ports. 2716 EPROM, 6116 static RAM, and 8255 I/O chips will be used for this purpose. Memory and I/O maps will also be determined. Figure 3.14 shows a hardware schematic for accomplishing this.

Three 74LS373 latches are used. The 2716 is a 2K × 8 ultraviolet EPROM with eleven address pins A0-A10 and eight data pins O0-7. Two 2716s are used. The 8086 A1-A11pins are connected to the A0-A10 pins of these chips. The 2716 even EPROM's O0-7 pins are connected to the 8086 D0-D7 pins. This is because the 8086 reads data via the D0-D7 pins for even addresses. On the other hand, the O0-7 pins of the odd 2716 are connected to the 8086 D8-D15 pins. The 8086 reads data via D8-D15 pins for odd addresses.

The 6116 is a 2KX8 SRAM. Two 6116s are used.

Table 3.2 shows memory and I/O maps of the 8086-based microcomputer of Figure 3.14. Note that the reset vector FFFF0H (CS = FFFFH, IP = 0000H) is included in the 2716.

For I/O ports, two 8255 chips are used. The 8255 is a general-purpose programmable I/O chip. The 8255 has three 8-bit I/O ports: ports A, B, and C. Ports A and B are latched 8-bit ports for both input and output. Port C is also an 8-bit port with latched output but the inputs are not latched. Port C can be used in two ways. It can either be used as a simple I/O port or as a control port for data transfer using handshaking via ports A and B.

The 8086 can configure the three ports by outputting appropriate data to the 8-bit control register. The ports can be decoded by two 8255 input pins A0 and A1 as follows:

A1	A0	
0	0	Port A
0	1	Port B
1	0	Port C
1	1	Control register

The structure of the control register is given below:

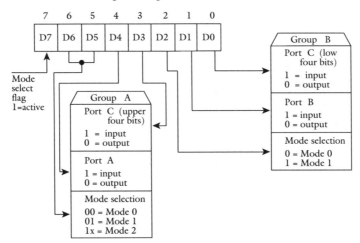

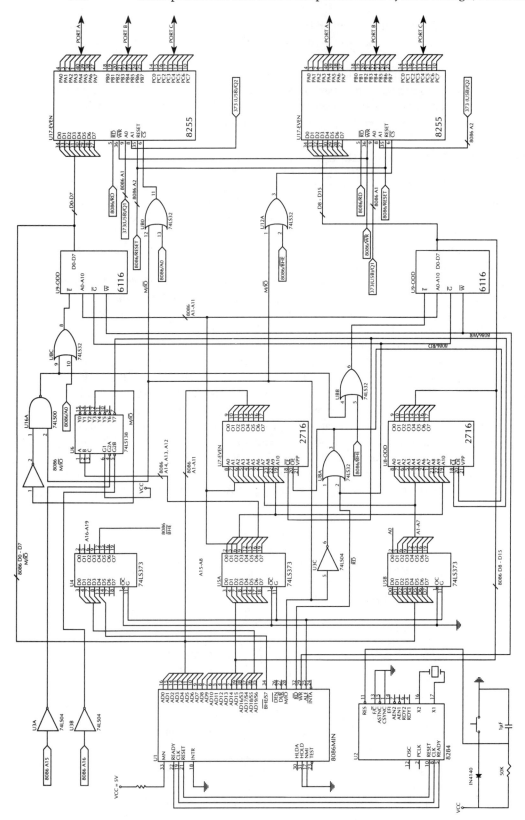

FIGURE 3.14 8086-Based microcomputer.

TABLE 3.2 Memory and I/O Maps

(a) Physical Memory Map

								2716 (U7) EVEN # EPROM											
A19	A18	A17	A16	A15	A14	A13	A12	A11	A10	A9	A8	A7	A6	A5	A4	A3	A2	A1	A0
1	1	1	1	1	1	1	1	0	0	0	0	0	0	0	0	0	0	0	0
TO																			
1	1	1	1	1	1	1	1	1	1	1	1	1	1	1	1	1	1	1	0
don't cares asumed high			to enable the 74LS138		output line 7 of the 74LS138														
FF000H, FF002H, … FFFFEH																			

								2716 (U8) ODD # EPROM											
A19	A18	A17	A16	A15	A14	A13	A12	A11	A10	A9	A8	A7	A6	A5	A4	A3	A2	A1	A0
1	1	1	1	1	1	1	1	0	0	0	0	0	0	0	0	0	0	0	1
TO																			
1	1	1	1	1	1	1	1	1	1	1	1	1	1	1	1	1	1	1	1
don't cares asumed high			to enable the 74LS138		output line 7 of the 74LS138														
FF001H, FF003H, … FFFFFH																			

								6116 (U9) EVEN # RAM											
A19	A18	A17	A16	A15	A14	A13	A12	A11	A10	A9	A8	A7	A6	A5	A4	A3	A2	A1	A0
1	1	1	1	1	0	1	1	0	0	0	0	0	0	0	0	0	0	0	0
TO																			
1	1	1	1	1	0	1	1	1	1	1	1	1	1	1	1	1	1	1	0
don't cares asumed high			to enable the 74LS138		output line 3 of the 74LS138														
FB800H, FB802H, … FBFFEH																			

								6116 (U10) ODD # RAM											
A19	A18	A17	A16	A15	A14	A13	A12	A11	A10	A9	A8	A7	A6	A5	A4	A3	A2	A1	A0
1	1	1	1	1	0	1	1	0	0	0	0	0	0	0	0	0	0	0	1
TO																			
1	1	1	1	1	0	1	1	1	1	1	1	1	1	1	1	1	1	1	1
don't cares asumed high			to enable the 74LS138		output line 3 of the 74LS138														
FB801H, FB803H, … FBFFFH																			

TABLE 3.2 Memory and I/O Maps (*continued*)

(b) Logical Memory Map

Chip	Segment Value	Offset
Even 2716	FF00H	0000H, 0002H, 0004H, ..., 0FFEH
Odd 2716	FF00H	0001H, 0003H, 0005H, ..., 0FFFH
Eben 6116	FB00H	0000H, 0002H, 0004H, ..., 0FFEH
Odd 6116	FB00H	0001H, 0003H, 0005H, ..., 0FFFH

I/O Map	
Even 8255	Port A = F8H
	Port B = FAH
	Port C = FCH
	CSR = FEH
Odd 8255	Port A = F9H
	Port B = FBH
	Port C = FDH
	CSR = FFH

The bit 7 (D7) of the control register must be one to send the above definitions for bits 0 through 6 (D0-D6).

In this format, bits D0-D6 are divided into two groups: groups A and B. Group A configures all 8 bits of port A and upper 4 bits of port C, while Group B defines all 8 bits of port B and lower 4 bits of Port C. All bits in a port can be configured as a parallel input port by writing a 1 at the appropriate bit in the control register by the 8086 OUT instructions, and a 0 to a particular bit position will configure the appropriate port as a parallel output port. Group A has three modes of operation. These are modes 0, 1, and 2. Group B has two modes: modes 0 and 1. Mode 0 for both groups A and B provides simple I/O operation for each of the three ports. No handshaking is required. Mode 1 for both groups A and B is the strobed I/O mode used for transferring I/O data to or from a specified port in conjuction with strobes or handshaking signals. Ports A and B use the lines on port C to generate or accept these handshaking signals.

The mode 2 of group A is the strobed bidirectional bus I/O and may be used for communicating with a peripheral device on a single 8-bit data bus for both transmitting and receiving data (bidirectional bus I/O). Handshaking signals are required. Interrupt generation and enable/disable functions are also available.

When D7 = 0, the bit set/reset control word format is used for the control register as follows:

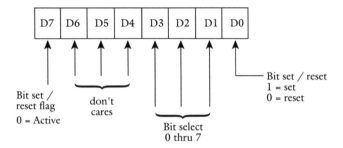

This format is used to set or reset the output on a pin of port C or when enabling the interrupt output signals for handshake data transfer. For example, the following 8-bits

$$
\begin{array}{ccccc}
0 & X\ X\ X & 1\ 1\ 0 & & 0 \\
\uparrow & \text{don't} & \text{bit 6} & & \uparrow \\
\text{bit set/} & \text{care} & & & \text{clear} \\
\text{reset mode} & & & &
\end{array}
$$

will clear bit 6 of port C to zero. Note that the control word format can be output to the 8255 control register by using the 8086 OUT instruction.

Next, only mode 0 of the 8255 will only be considered in the following discussion to illustrate 8086's programmed I/O capability.

Now, let us define the control word format for mode 0 more precisely by means of numerical example. Consider that the control word format is 1 000 0 010$_2$. With this data in the control register, all 8 bits of port A are configured as outputs, 8 bits of port C are also configured as outputs, but all 8 bits of port B are defined as inputs. On the other hand, outputting 1 0 011 011$_2$ into the control register will configure all 3 8-bit ports (ports A, B, and C) as inputs.

Let us decode the I/O port addresses. One 8255 will contain the odd-addressed ports since it is enabled by $\overline{\text{BHE}}$, while the other 8255 will include the even-addressed ports since it is enabled by A0.

Since the 8086 A1 and A2 pins are utilized in addressing the ports, bits A3-A7 are don't cares and are assumed to be ones here. Note that A0 = 1 for odd-addressed ports, while A0 = 0 for even-addressed ports.

For 8255-1 (Odd-Addressed Ports)

Port name	A7	A6	A5	A4	A3	A2	A1	A0	
Port A	1	1	1	1	1	0	0	1	= F9H
Port B	1	1	1	1	1	0	1	1	= FBH
Port C	1	1	1	1	1	1	0	1	= FDH
Control Register	1	1	1	1	1	1	1	1	= FFH

For 8255-2 (Even-Addressed Ports)

Port name	A7	A6	A5	A4	A3	A2	A1	A0	
Port A	1	1	1	1	1	0	0	0	= F8H
Port B	1	1	1	1	1	0	1	0	= FAH
Port C	1	1	1	1	1	1	0	0	= FCH
Control Register	1	1	1	1	1	1	1	0	= FEH

In the above, standard I/O technique is used. The 8255s can also be interfaced to the 8086 using memory-mapped I/O. In this case the 8086 M/IO pin will not be used. The 20-bit physical addresses for the ports can be determined in a similar way by considering any unused 8086 address bits (A3-A19) as don't cares.

From the above discussion, the following points can be summarized:

1. For ROMs and EPROMs, $\overline{BHE}$ and A0 are not required to be part of chip enable/select decoding.
2. For RAMs and I/O chips, both $\overline{BHE}$ and A0 must be used in chip select logic.
3. For ROMs/EPROMs and RAMs, both odd and even addressed chips are required. However, for I/O chips, either an odd-addressed I/O chip or even-addressed I/O chip or both can be used depending on the number of ports required in an application.
4. For enterfacing ROMs/EPROMs and RAMs to the 8086, the same chip select logic must be used for both even and its corresponding odd memory bank. For example, the even memory chip containing address 00000H, 00002H, 00004H, ..., must have the same chip select logic as its odd counterpart containing address 00001H, 00003H, 00005H,.... The memory map of the two memory banks will be distinguished by the value of A0 (A0 = 0 for even addresses and A0 = 1 for odd addresses).

Example 3.11

Assume 8086/8255 based microcomputer. Write an 8086 assembly language program to drive an LED connected to bit 2 of Port B based on two switch inputs at bits 6 and 7 of Port A. If both switch inputs are HIGH or LOW, turn the LED "ON"; otherwise turn the LED "OFF".

Solution

```
              TITLE    PROGRAM
1000          ORG      1000H
0000          PROG     SEGMENT
                       ASSUME  CS:PROG
=0001         PORTA    EQU  01H
```

```
=0003            PORTB    EQU  03H
=0007            CNTRL    EQU  07H
0000  B0  90              MOV  AL,90H
0002  E6  07              OUT  CNTRL,AL
0004  E4  01     BEGIN:   IN   AL,PORTA
0006  24  C0              AND  AL,0C0H
0008  7A  06              JPE  LEDON
000A  B0  00              MOV  AL,00H
000C  E6  03              OUT  PORTB,AL
000E  EB  F4              JMP  BEGIN
0010  B0  04     LEDON:   MOV  AL,04H
0012  E6  03              OUT  PORTB,AL
0014  EB  EE              JMP  BEGIN
0016             PROG     ENDS
                          END
```

Microsoft (R) Macro Assembler Version 5.10 10/23/93 21:58:39
PROGRAM   Symbols-1

Segments and Groups:

Name	Length	Align	Combine Class
PROG	0016	PARA	NONE

Symbols:

Name	Type	Value	Attr
BEGIN	L NEAR	0004	PROG
CNTRL	NUMBER	0007	
LEDON	L NEAR	0010	PROG
PORTA	NUMBER	0001	
PORTB	NUMBER	0003	
@CPU	TEXT	0101h	
@FILENAME		TEXT	test1
@VERSION		TEXT	510

```
   19  Source  Lines
   19  Total  Lines
   11  Symbols
```

47884 + 433567 Bytes symbol space free

```
    0  Warning  Errors
    0  Severe  Errors
```

Example 3.12

Assume an 8086/8255 configuration with the following port addresses:

$$
\begin{array}{ll}
\text{Port A} & = \text{F9H} \\
\text{Port B} & = \text{FBH} \\
\text{Control Register} & = \text{FFH}
\end{array}
$$

Port A has three switches connected to bits 0, 1, and 2 and port B has an LED connected to bit 2 as follows.

Write an 8086 assembly language program to turn the LED ON if port A has an odd number of HIGH switch inputs; otherwise turn the LED OFF. Do not use any instructions involving the parity flag.

Solution

```
1000              ORG    1000H
0000              CODE   SEGMENT
                         ASSUME CS:CODE
=00F9             PORTA  EQU 0F9H
=OOFB             PORTB  EQU 0FBH
=00FF             CNTRL  EQU 0FFH
0000 B0 90               MOV AL, 90H      ;  Configure Port A as
                                          ;  input
0002 E6 FF               OUT CNTRL, AL    ;  Port B as output
0004 E4 F9        BEGIN: IN AL, PORTA     ;  Input switches
0006 24 07               AND AL, 07H      ;  Mask high five bits
0008 3C 07               CMP AL, 07H      ;  Are all three inputs
                                          ;  HIGH?
000A 74 12               JZ ODD           ;  If so, turn LEDON
000C 3C 01               CMP AL, 01H      ;  Is only input 0 HIGH?
000E 74 0E               JZ ODD           ;  If so, turn LED ON
0010 3C 02               CMP AL, 02H      ;  Is only input 1 HIGH?
0012 74 0A               JZ ODD           ;  If so, turn LED ON
0014 3C 04               CMP AL, 04H      ;  Is only input 2 HIGH?
0016 74 06               JZ ODD           ;  If so, turn LED ON
0018 B0 00               MOV AL, 00H      ;  Else turn LED
001A E6 FB               OUT PORTB, AL    ;  OFF
001C EB E6               JMP BEGIN        ;  REPEAT
001E B0 04        ODD:   MOV AL, 04H      ;  Turn LED
0020 E6 FB               OUT PORTB, AL    ;  ON
0022 EB E0               JMP BEGIN        ;  REPEAT
0024              CODE   ENDS
                         END
```

Microsoft (R) Macro Assembler Version 5.10   11/14/93 16:40:4
                                             Symbols-1

Segments and Groups:

Name	Length	Align	Combine Class
CODE	0024	PARA	NONE

Symbols:

Name	Type	Value	Attr
BEGIN	L NEAR	0004	CODE
CNTRL	NUMBER	00FF	
ODD	L NEAR	001E	CODE
PORTA	NUMBER	00F9	

```
PORTB  . . . . . . . . . . .     NUMBER    00FB
@CPU  . . . . . . . . . . . .    TEXT      0101h
@FILENAME  . . . . . . .          TEXT      examp17
@VERSION  . . . . . . . .         TEXT      510

    26 Source Lines
    26 Total Lines
    11 Symbols

47576 + 427923 Bytes symbol space free

     0 Warning Errors
     0 Severe Errors
```

3.9 8086 INTERRUPT SYSTEM

The 8086 interrupts can be classified into three types. These are

1. Predefined interrupts
2. User-defined software interrupts
3. User-defined hardware interrupts

The interrupt vector addresses for all the 8086 interrupts are determined from a table stored in locations 00000H through 003FEH. The starting addresses for the service routines for the interrupts are obtained by the 8086 using this table. Four bytes of the table are assigned to each interrupt: two bytes for IP and two bytes for CS. The table may contain up to 256 32-bit vectors. If fewer than 256 interrupts are defined in the system, the user is required to provide enough memory for the interrupt pointer table for obtaining the defined interrupts.

The 8086 assigns every interrupt a type code for identifying the interrupt. There are 256 type codes associated with the 256 table entries. Each entry consists of two word addresses, one for storing the IP contents and the other for storing the CS contents. Each 8086 interrupt physical address vector is 20 bits wide and is computed from the 16-bit contents of IP and CS.

For obtaining an interrupt address vector, the 8086 uses the two addresses in the pointer table where IP and CS are stored for a particular interrupt type.

For example, for the interrupt type nn (instruction INT nn), the table address for IP = 4 * nn and the table address for CS = (4 * nn) + 2. For servicing the 8086's nonmaskable interrupt (NMI pin), the 8086 assigns the type code 2 to this interrupt. The 8086 automatically executes the INT2 instruction internally to obtain the interrupt address vector as follows:

$$\text{Address for IP} = 4 * 2 = 00008H$$
$$\text{Address for CS} = (4 * 2) + 2 = 0000AH$$

The 8086 loads the values of IP and CS from the 20-bit physical addresses 00008H and 0000AH in the pointer table. The user must store the desired 16-bit values of IP and CS in these locations. Similarly, the IP and CS values for other interrupts are calculated. The 8086 interrupt pointer table layout is shown in Table 3.3.

In response to an interrupt, the 8086 pushes flags, CS, and IP onto the stack, clears TF and IF flags, and then loads IP and CS from the pointer table using the type code.

Interrupt service routines must be terminated with the IRET (Interrupt Return) instruction which pops the top three stack words into IP, CS, and flags, thus returning to the right place in the main program. The 256 interrupt type codes are assigned as follows:

TABLE 3.3   8086 Interrupt Pointer Table

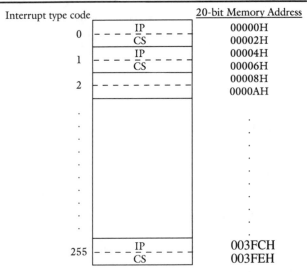

- Types 0 to 4 are for the predefined interrupts.
- Types 5 to 31 are reserved by Intel for future use.
- Types 32 to 255 are available for maskable interrupts.

3.9.1 Predefined Interrupts (0 to 4)

The predefined interrupts include DIVISION BY ZERO (type 0), SINGLE STEP (type 1) NONMASKABLE INTERRUPT pin (type 2), BREAKPOINT-INTERRUPT (type 3), and INTERRUPT ON OVERFLOW (type 4). The user must provide the desired IP and CS values in the interrupt pointer table. The user may also imitate these interrupts through hardware or software. If a predefined interrupt is not used in a system, the user may assign some other function to the associated type.

The 8086 is automatically interrupted whenever a division by zero is attempted. This interrupt is nonmaskable and is implemented by Intel as part of the execution of the divide instruction. When the TF (TRAP flag) is set by an instruction, the 8086 goes into the single step mode. The TF can be set to one as follows:

```
PUSHF                 ;  Save flags
MOV BP, SP            ;  Move [SP] to [BP]
OR [BP + 0], 0100H    ;  Set TF
POPF                  ;  Pop flags
```

Note that in the above [BP + 0] rather than [BP] is used since BP cannot be used without displacement.

Once TF is set to one, the 8086 automatically generates a TYPE 1 interrupt after execution of each instruction. The user can write a service routine at the interrupt address vector to display memory locations and/or register to debug a program. Single step is nonmaskable and cannot be enabled by STI (enable interrupt) or CLI (disable interrupt) instruction.

The nonmaskable interrupt is initiated via the 8086 NMI pin. It is edge triggered (LOW to HIGH) and must be active for two clock cycles to guarantee recognition. It is normally used for catastrophic failures such as power failure. The 8086 obtains the interrupt vector address by automatically executing the INT2 (type 2) instruction internally.

Type 3 interrupt is used for breakpoint and is nonmaskable. The user inserts the one-byte instruction INT3 into a program by replacing an instruction. Breakpoints are useful for program debugging.

The INTERRUPT ON OVERFLOW is a type 4 interrupt. This interrupt occurs if the overflow flag (OF) is set and the INTO instruction is executed. The overflow flag is affected, for example, after execution of signed arithmetic such as IMUL (signed multiplication) instruction. The user can execute the INTO instruction after the IMUL. If there is an overflow, an error service routine written by the user at the type 4 interrupt address vector is executed.

3.9.2 User-Defined Software Interrupts

The user can generate an interrupt by executing a two-byte interrupt instruction INT nn. The INT nn instruction is not maskable by the interrupt enable flag (IF). The INT nn instruction can be used to test an interrupt service routine for external interrupts. Type codes 0 to 255 can be used. If predefined interrupt is not used in a system, the associated type code can be utilized with the INT nn instruction to generate software (internal) interrupts.

3.9.3 User-Defined Hardware (Maskable Interrupts, Type Codes 32_{10}–255_{10})

The 8086 maskable interrupts are initiated via the INTR pin. These interrupts can be enabled or disabled by STI (IF = 1) or CLI (IF = 0), respectively. If IF = 1 and INTR is active (HIGH) without occurrence of any other interrupts, the 8086, after completing the current instruction, generates INTA LOW twice, each time for about 2 cycles.

The state of the INTR pin is sampled during the last clock cycle of each instruction. In some instances, the 8086 samples the INTR pin at a later time. An example is execution of POP to a segment register. In this case, the interrupts are sampled until completion of the following instruction. This allows a 32-bit pointer to be loaded to SS and SP without the danger of an interrupt occurring between the two loads.

INTA is only generated by the 8086 in response to INTR, as shown in Figure 3.15. The interrupt acknowledge sequence includes two INTA cycles separated by two idle clock cycles. ALE is also generated by the 8086 and will load the address latches with indeterminate

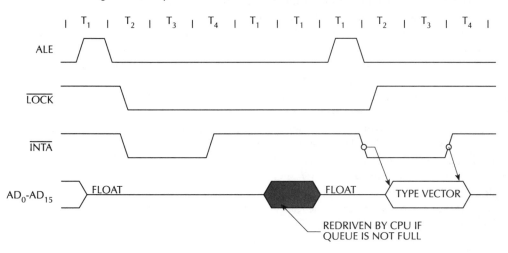

FIGURE 3.15 Interrupt acknowledge sequence.

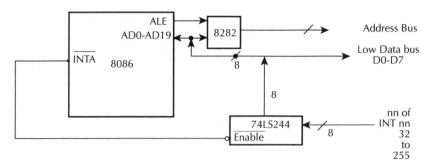

FIGURE 3.16   Servicing the INTR in the minimum mode.

information. The ALE is useful in maximum systems with multiple 8259A priority interrupt controllers. During the INTA bus cycles, DT/R and DEN are LOW (see 8086 minimum mode bus cycle). The first INTA bus cycle indicates that an interrupt acknowledge cycle is in progress and allows the system to be ready to place the interrupt type code on the next INTA bus cycle. The 8086 does not obtain the information from the bus during the first cycle. The external hardware must place the type code on the lower half of the 16-bit data bus during the second cycle.

In the minimum mode, the M/IO is low indicating I/O operation during the INTA bus cycles. The 8086 internal LOCK signal is also low from T2 of the first bus cycle until T2 of the second bus cycle to avoid the BIU from accepting a hold request between the two INTA cycles. Figure 3.16 shows a simplified interconnection between the 8086 and 74LS244 for servicing the INTR. INTA enables 74LS244 to place the type code nn on the 8086 data bus.

In the maximum mode, the status lines S0 to S2 will enable the INTA output for each cycle via the 8288. The 8086 LOCK output will be active from T2 of the first cycle until T2 of the second to prevent the 8086 from accepting a hold request on either RQ/GT input and to prevent bus arbitration logic from releasing the bus between INTAs in multimaster systems. The LOCK output can be used in external logic to lock other devices off the system bus, thus ensuring the INTA sequence to be completed without intervention.

Once the 8086 has the interrupt-type code (via the bus for hardware interrupts, from software interrupt instructions INT nn or from the predefined interrupts), the type code is multiplied by four to obtain the corresponding interrupt vector in the interrupt vector table. The four bytes of the interrupt vector are least significant byte of the instruction pointer, most significant byte of the pointer, least significant byte of the code segment register, and most significant byte of the code segment register. During the transfer of control, the 8086 pushes the flags and current code segment register and instruction pointer into the stack. Flags TF and IF are then cleared to zero. The CS and IP values are read by the 8086 from the interrupt vector table. No segment registers are used when accessing the interrupt pointer table. S4, S3 has the value 10_2 to indicate no segment register selection.

As far as the 8086 interrupt priorities are concerned, single-step interrupt has the highest priority, followed by NMI, followed by the software interrupts (all interrupts except single step, NMI, and INTR interrupts). This means that a simultaneous NMI and single step will cause the NMI service routine to follow single step; a simultaneous software interrupt and single step will cause the software interrupt service routine to follow single step and a simultaneous NMI and software interrupt will cause the NMI service routine to be executed prior to the software interrupt service routine. An exception to this priority scheme occurs if all three nonmaskable interrupts (single step, software, and NMI) are pending. For this case, software interrupt service routine will be executed first followed by the NMI service routine, and single stepping will not be serviced. However, if software interrupt and single stepping are pending, single stepping resumes upon execution of the instruction causing the software interrupt (the next instruction in the routine being single stepped).

The INTR is maskable and has the lowest priority. If the user does not wish to single step before INTR is serviced, the single-step routine must disable interrupts during execution of the program being single stepped, and reenable interrupts on entry to the single-step routine. To avoid single stepping before the NMI service routine, the single-step routine must check the return address on the stack for the NMI service routine address and return control to that routine without single step enabled.

A priority interrupt controller such as the 8259A can be used with the 8086 INTR to provide eight levels of hardware interrupts. The 8259A has built-in features for expansion of up to 64 levels with additional 8259As. The 8259A is programmable and can be readily used with the 8086 to obtain multiple interrupts from the single 8086 INTR pin.

Example 3.13

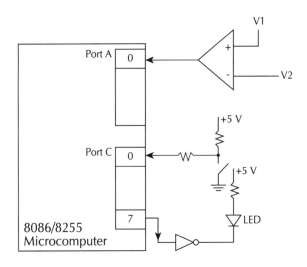

i)   In the above, the 8086/8255 microcomputer is required to perform the following:

   If V1 > V2, turn the LED "ON" if the switch is open or turn the LED "OFF" if the switch is closed. If V1 ≤ V2, loop and test again.

Write an 8086 assembly language program to accomplish the above.

ii)   Repeat part i) using NMI and INTR interrupts. Write service routine at IP = 40000H, CS = 3000H. Assume that the stack pointer and stack segment are already initialized.

Solution

i)

```
0000            JOHN    SEGMENT AT 1000H
                        ASSUME CS: JOHN
=0001           PORTA EQU 01H
=0005           PORTC EQU 05H
=0007           CNTRL EQU 07H
0000 B0 91              MOV AL, 91H        ;  Configure
0002 E6 04              OUT CNTRL,AL       ;  Port A and Port C
```

```
0004 E4 01    START:    IN AL, PORTA      ;  Input comparator
0006 24 01              AND AL, 01H       ;  Check if high
0008 74 FA              JZ START          ;  Jump back if low
000A E4 03              IN AL, PORTC      ;  Input switch
000C B1 02              MOV CL, 2
000E D2 D8              RCR AL, CL
0010 E6 03              OUT PORTC, AL     ;  Output to LED
0012 F4                 HLT
0013          JOHN      ENDS
                        END
```

Microsoft (R) Macro Assembler Version 5.10   11/17/93 22:35:4

Symbols-1

Segments and Groups:

Name	Length	Align	Combine	Class
JOHN	0013	AT	1000	

Symbols:

Name	Type	Value	Attr
CNTRL	NUMBER	0007	
PORTA	NUMBER	0001	
PORTC	NUMBER	0005	
START	L NEAR	0004	JOHN
@CPU	TEXT	0101h	
@FILENAME	TEXT	EXP318A	
@VERSION	TEXT	510	

```
    17 Source Lines
    17 Total Lines
    10 Symbols

47576 + 427939 Bytes symbol space free

    0 Warning Errors
    0 Severe Errors
```

ii)
  Using NMI
    Main program

```
0000          JOHN      SEGMENT AT 2000H
                        ASSUME CS: JOHN
=0001         PORTA EQU 01H
=0005         PORTC EQU 05H
=0007         CNTRL EQU 07H
0000 B0 91              MOV AL, 91H       ;  Configure
0002 E6 07              OUT CNTRL,AL      ;  Port A and Port C
0004 9B       WAIT JMP WAT
0005 F4                 HLT               ;  Wait for interrupt
0006          JOHN ENDS
                        END
```

Service routine

```
0008                           ORG           00000008H
0008  40003000                 DD            40003000H
30004000                       ORG           30004000H
30004000 E4 03   START:   IN AL, PORTC    ;  Input switch
30004002 B1 02            MOV CL, 2
30004004 D2 D8            RCR AL, CL
30004006 E6 03            OUT PORTC, AL   ;  Output to LED
30004008 CF               IRET
```

Microsoft (R) Macro Assembler Version 5.10 11/17/93 22:18:20
                                              Symbols-1

Segments and Groups:

Name	Length	Align	Combine	Class
JOHN	0006	AT	2000	

Symbols:

Name	Type	Value	Attr
CNTRL	NUMBER	0007	
PORTA	NUMBER	0001	
PORTC	NUMBER	0005	
@CPU	TEXT	0101h	
@FILENAME		TEXT	test
@VERSION		TEXT	510

```
    11  Source Lines
    11  Total Lines
     9  Symbols

47884 + 433567 Bytes symbol space free

     0  Warning Errors
     0  Severe Errors
```

Using INTR (vector 256_{10})

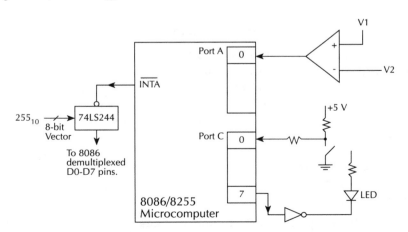

Main program

```
0000            JOHN     SEGMENT AT 3000H
                         ASSUME CS: JOHN
=0001           PORTA EQU 01H
=0005           PORTC EQU 05H
=0007           CNTRL EQU 07H
0000 FB                  STI                    ;  Enable maskable
                                                ;  interrupt
0001 B0 91               MOV AL, 91H            ;  Configure Port A and
0003 E6 04               OUT CNTRL, AL          ;  Port C
            WAIT:JMP WAIT
0005 F4                  HLT                    ;  Wait for interrupt
0006            JOHN     ENDS
                         END
```

Service routine

```
03FC                            ORG           000003FCH
03FC 40003000                   DD            40003000H
                BOB             SEGMENT
                                ASSUME CS:BOB
30004000                        ORG           30004000H
30004000 E4 03                  IN AL, PORTC    ;  Input switch
30004002 B1 02                  MOV CL, 2
30004004 D2 D8                  RCR AL, CL
30004006 E6 03                  OUT PORTC, AL   ;  Output to LED
30004008 F4     BOB             ENDS
                                END
```

Microsoft (R) Macro Assembler Version 5.10   11/17/93 22:55:4
                                              Symbols-1

Segments and Groups:

Name	Length	Align	Combine Class
JOHN	0006	AT	3000

Symbols:

Name	Type	Value	Attr
CNTRL	NUMBER	0007	
PORTA	NUMBER	0001	
PORTC	NUMBER	0005	
@CPU	TEXT	0101h	
@FILENAME		TEXT	EXP318C
@VERSION		TEXT	510

```
    11 Source Lines
    11 Total Lines
     9 Symbols

47576 + 427939 Bytes symbol space free
```

0 **Warning Errors**
0 **Severe Errors**

3.10 8086 DMA

When configured in the minimum mode (MN/ $\overline{\text{MX}}$ pin HIGH), the 8086 provides HOLD
(DMA request) and HLDA (DMA acknowledge) signals to take over the system bus for DMA
applications. The Intel DMA controller chips 8257 and 8237 can be used with the 8086. The
8257 or 8237 can request DMA transfer between the 8086 memory and I/O device by activating
the 8086 HOLD pin. The 8086 will complete the current bus cycle (if there is one presently in
progress) and then output HLDA, relinquishing the system bus to the DMA controller. The
8086 will not try to use the system bus until the HOLD pin is negated.

As mentioned before, the 8086 memory addresses are organized in two separate banks —
one containing even-addressed bytes and the other containing odd-addressed bytes. An 8-bit
DMA controller must alternately select these two banks to access logically adjacent bytes in
memory.

QUESTIONS AND PROBLEMS

3.1 What is the basic difference between the 8086 and 8088 microprocessors? Name one
reason why these two microprocessors are included in the i APX 86 family by Intel.

3.2 List the 8086 minimum and maximum mode signals. How are these modes selected?

3.3 What are the functions provided by 8288 bus controller in a maximum mode 8086
system?

3.4 Which bit of the 8086 FLAG register is used by the string instructions? How? Illustrate
this by using the 8086 MOVSB instruction.

3.5 What is the relationship between the 8086 and 8284 input clocks? Does the 8086 have an
on-chip clock circuitry? Comment.

3.6 What is the purpose of the TF bit in the FLAG register?

3.7 If $[BL] = 36_{16}$ (ASCII code for 6) and $[CL] = 33_{16}$ (ASCII code for 3), write an 8086
assembly program which will add the contents of BL and CL, and then provide the result in
decimal. Store result in CL.

3.8 What happens to the contents of the AX register after execution of the following 8086
instruction sequence:

```
MOV AX,  0F180H
CBW
CWD
```

3.9 Determine the addressing modes for the following instructions:

(a) **MOV CH, 8**
(b) **MOV AX, DS:START**
(c) **MOV [SI], AL**

(d) **MOV SI, BYTEPTR[BP+2][DI]**

3.10   Consider MOV BX, DS: BEGIN. How many memory accesses are required by the 8086 to execute the above instruction if BEGIN = 0401H.

3.11   Write an 8086 assembly program to implement the following Pascal segment:

```
SUM: =0; for i: = 0 to 15 do
SUM: = SUM + A(i)
```

Assume CS and DS are already initialized. A(i)s are 16-bit numbers.

3.12   Write and 8086 assembly program to add two 128-bit numbers stored in memory in consecutive locations.

3.13   What are the remainder, quotient, and registers containing them after execution of the following instruction sequence?

```
MOV  DX,  0
MOV  AX,  −5
MOV  BX,  2
IDIV BX
```

3.14   Write an 8086 assembly language program to divide A5721624H by F271H. Store the remainder and quotient onto the stack. Assume that the numbers are signed and stored in the stack as follows:

Assume that the stack segment and stack pointer are already initialized.

3.15   Write an 8086 assembly language program to compute X = Y + Z − 12FFH where X, Y, Z are 64-bit variables. The lower 32 bits of Y and Z are stored respectively at offset 5000H and 5008H followed by the upper 32 bits. Store the lower 32 bits of the 64-bit result at offset 6000H followed by the upper 32 bits.

3.16   Assume that registers AL, BX and DX CX contain a signed byte, a signed word, and a signed 32-bit number respectively. Write an 8086 assembly language program that will compute the signed 32-bit result:

$$AL + BX - DXCX \rightarrow DXCX$$

3.17   Write an 8086 assembly language program to compute X = 5 * Y + (Z/W) where offsets 5000H, 5002H, and 5004H respectively contain the 16-bit signed integers Y, Z, and W. Store the 32-bit result in memory starting at offset 5006H. Discard the remainder of Z/W.

3.18   Write an 8086 instruction sequence to clear the trap flag to zero.

3.19   Write an 8086 subroutine to compute

$$Y = \sum_{i=1}^{N} \frac{Xi^2}{N}$$

where Xis are signed 8-bit integers and N = 100. The numbers are stored in consecutive locations. Assume SI points to Xis and SP, DS, SS are already initialized.

3.20   Write an 8086 assembly program to move a block of data bytes of length 100_{10} from the source block starting at location 2000H in ES = 1000H to the destination block starting at location 3000H in the same extra segment.

3.21   Write an 8086 assembly program to logically shift a 128-bit number stored in location starting at 4000H in DS twice to the right. Store the result in memory location starting at 5000H in the data segment. Assume DS is initialized.

3.22   Connect one 2732, and one 8255 to an 8086 to obtain even 2732 locations and odd addresses for the 8255's Port A, Port B, Port C, and the Control register. Show only the connections for the pins shown in the figure below. Assume all unused address lines to be zeros. Use latches and gates as required.

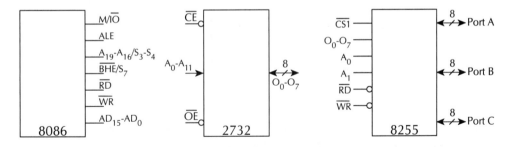

3.23   Determine the number of 2732 4K × 8 EPROMs and 6116 2K × 8 static RAMs to provide a 4K ×16 EPROM and a 2K ×16 word RAM. Draw a neat schematic connecting these chips to the 8086 and determine the map.

3.24   Assume the memory and I/O maps of Table 3.2. Interface the following A/D to the 8086/2716/6116/8255 of Figure 3.14:

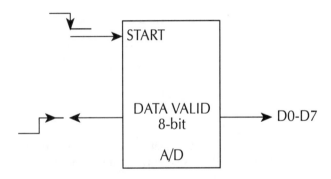

Write an 8086 assembly program to input the A/D converter and turn an LED ON connected to bit 5 of port A of the even 8255 if the number read from the A/D is odd; otherwise turn the LED OFF. Assume that the LED is turned ON by a HIGH and turned OFF by a LOW.

3.25 Repeat problem 3.24 using the 8086 INTR interrupt.

3.26 Write an 8086 assembly language program to add a 16-bit number stored in DX (bits 0 to 7 containing the high-order byte of the number and bits 8 to 15 containing the low-order byte) with another 16-bit number stored in BX (bits 0 to 7 containing the low-order 8 bits of the number and bits 8 through 15 containing the high-order 8 bits). Store the result in CX.

3.27 Assume an 8086/8255 based system. Write an 8086 assembly language program to input 16-bit data via Port B and Port C, and then divide this by the 8-bit input data at Port A. Assume all numbers to be signed.

3.28 An 8086/8255 based microcomputer is required to drive a common anode seven-segment display connected to Port C as follows:

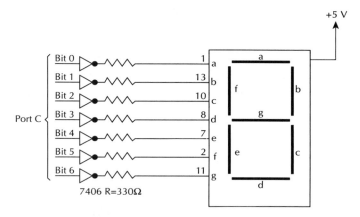

Write an 8086 assembly language program to display a single hexadecimal digit (0 to F) from offset 3000H at DS = 1000H. Use a look-up-table.

3.29 Assume an 8086/8255 based microcomputer. Suppose that four switches are connected to bits 0 through 3 of Port A, an LED is connected at bit 4 of Port B and another LED is connected at bit 2 of Port C. If the number of low switches is even, turn the Port B LED "ON" and Port C LED "OFF". If the number of low switches is ODD turn Port B LED "OFF" and Port C LED "ON". Write an 8086 assembly language program to accomplish the above.

3.30

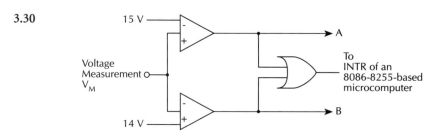

In the above figure, if $V_M > 15V$, turn an LED "ON" connected at bit 2 of Port C. On the other hand if $V_M < 14V$, turn the LED "OFF". Use registers and memory locations of your choice.

Draw a block diagram showing the microcomputer and the connections of the figure to its ports. Also, write 8086 assembly language programs to accomplish the above using:

a)   Polled I/O by inputting one or more outputs in the figure.
b)   INTR
c)   NMI

4

INTEL
80186/80286/80386

This chapter describes the internal architecture, addressing modes, instruction set, and I/O techniques associated with the 80186, 80286, and 80386 microprocessors. Interfacing capabilities to typical memory and I/O chips are also included. Finally, virtual memory concepts associated with the 80286 and 80386 are covered.

4.1 Intel 80186 and 80286

This section covers the two enhanced versions of the 8086 microprocessor: Intel 80186 and 80286. The Intel 80186 includes the Intel 8086 and six separate functional units in a single chip, while the 80286 has integrated memory protection and management into the basic 8086 architecture.

4.1.1 Intel 80186

The Intel 80186 family is fabricated using HMOS technology and contains two microprocessors: Intel 80186 and 80188. The only difference between them is that the 80186 has a 16-bit data bus, while the 80188 includes an 8-bit data bus. The 80186 is packaged in a 68-pin leadless package. The 80186 can be operated at three different clock speeds: 8 MHz, 10 MHz and 12.5 MHz. The 80C186 and 80C188 are the Low-Power (HCMOS) versions of the 80186 and 80188 respectively. The 80C186 like the 80186 can be operated at 8-, 10-, or 12.5-MHz while the 80C188, like the 80188 can be operated at 8- or 10-MHz. The 80186 can directly address one megabyte of memory. It contains the 8086 microprocessor and several additional functional units. The major on-chip circuits include a clock generator, two independent DMA channels, a programmable interrupt controller, three programmable 16-bit timers, and a chip select unit.

The 80186 provides double the performance of the standard 8086. The 80186 includes 10 new instructions beyond the 8086. The 80186 is completely object code compatible with the 8086. It contains all the 8086 registers and generates the 20-bit physical address from a 16-bit segment register and a 16-bit offset in the same way as the 8086. The 80186 does not have the 8086's MN/$\overline{\text{MX}}$ pin. The 80186 has enough pins to generate the minimum mode-type pins. S0-S3 status signals can be connected to external bus controller chips such as 8288 for generating the maximum mode type signals. Like the 8086, the 80186 fetches the first instruction from physical address FFFF0H upon hardware reset.

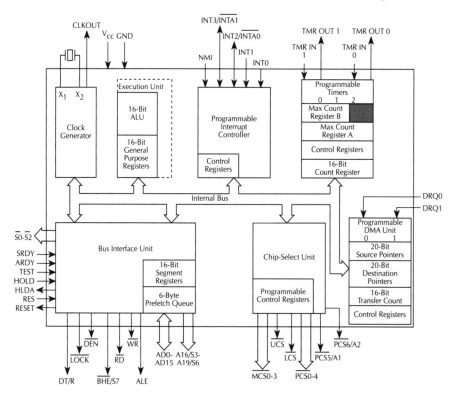

FIGURE 4.1   80186 functional block diagram.

Figure 4.1 shows the 80186 functional block diagram. The DMA unit provides two DMA request input signals, DRQ0 and DRQ1. One of these input signals can be used by external devices such as disk controller to request a data transfer between the memory and disk via direct memory access technique. Each 80186 DMA channel contains two 20-bit registers and a 16-bit counter. One of the 20-bit registers holds the destination address of the DMA transfers while the 16-bit counter stores the number of words or bytes to be transferred. DMA transfers can take place between memory and I/O or from memory to memory or from I/O to I/O. The DMA channels may be programmed such that one channel has priority over the other. DMA transfers will take place whenever the ST/STOP bit in the control register is set to one.

The 80186 contains three independent 16-bit timers/counters namely counter 0, 1, and 2. Counters 0 and 1 can be programmed to count external events. The inputs and outputs of these two counters are available on the 80186 pins. The third timer, counter 2, is not connected to any external pins. This timer only counts the 80186 clock cycles. Counter 2 is decremented every four 80186 clocks. One can connect the output of counter 2 to a DMA unit or to an interrupt input or to the input of counter 1 and/or 0 by setting or clearing the appropriate bits in a control word. Counter 2 can, therefore, be used to interrupt the 80186 after a programmed amount of time or to provide a pulse to the DMA unit after a specific amount of time.

The priorities of the four interrupt pins, INT0, INT1, INT2/ INTA0 and INT3/ INTA1 are programmable. If these four interrupt inputs are programmed in their internal mode, then the activation of one of them by an external signal will cause the 80186 to push the return address on the stack and vector directly to the interrupt address vector for that interrupt. The INT2/ INTA0 and INT3/ INTA1 pins have dual functions. They can be programmed as interrupt inputs or as interrupt acknowledge output signals (INTA0 for INT0 and INTA1 for INT1). These interrupt pins can be used to connect the 80186 to an external priority interrupt controller such as the 8259A. For example, suppose that the interrupt request line from an external 8259A is connected to the 80186 INT1 input pin and the 80186 INT3/ INTA1 pin is connected to the 8259A interrupt acknowledge input pin. When the 8259A receives an interrupt request, it activates the 80186 INT1 pin. In response, the 80186 sends the interrupt acknowledge signal via its INT3/ INTA1 pin. The 8259A then places the desired interrupt type code on the 80186 data bus. The 80186 obtains the CS and IP values based on the type code and branches to the service routine.

The 80186 interrupt controller allows the 80186 to receive interrupts from internal or external sources. Internal interrupt sources (timers and DMA channels) can be disabled by their own control registers or by mask bits within the interrupt controller. The 80186 is provided with five dedicated pins for external interrupts. These pins are NMI, INT0, INT1, INT2/ INTA0, and INT3/ INTA1. NMI is the only nonmaskable interrupt.

In the master mode, the interrupt controller provides three modes of operation. These are fully nested mode, cascade mode, and special fully nested mode. In the fully nested mode, all four maskable interrupt pins are used as direct interrupt requests. The interrupt vectors are obtained by the 80186 internally.

Upon acceptance of an interrupt (hardware, INT instructions, or instruction exceptions such as divide by 0), the 80186 pushes CS, IP, and the status word onto the stack just like the 8086. Also, similar to the 8086, an interrupt pointer table with 256 entries provides interrupt address vectors (IP and CS) for each interrupt type. This type identifies the appropriate table entry. Nonmaskable interrupts use an internally supplied type, while the types for maskable interrupts are provided by the user via external hardware. The types for INT instructions and instruction exceptions are generated internally by the 80186.

The 80186 includes an on-chip clock generator/crystal oscillator circuit. Like the 8085, a crystal connected at the 80186 X1 X2 pins is divided by 2 internally. The built-in chip select unit is an address decoder. This unit can be programmed to generate six memory chip selects (LCS, UCS, and MCS0 — 3 pins) and seven I/O or peripheral chip selects (PCS0-4, PCS5/ A1, and PCS6/A2 pins). This unit can be programmed to generate an active low chip select when a memory or port address in a particular range is sent out. For example, the 80186 outputs low on the LCS (lower chip select) pin when it accesses an address between 00000H and a higher address (in the range of 1K to 256K) programmable by the user via a control word. On the other hand, the 80186 outputs low on the UCS (upper chip select) pin when it accesses an address between a user programmable lower address (by placing some bits in a control word via an instruction) and upper fixed address FFFFFH. The four middle chip select pins (MCS0-3) are activated low by the 80186 when it accesses an address in the mid range. Both the starting address and the size of the four blocks can be specified via the control word. The specified size of blocks can be from 2K to 128K. The memory areas assigned to different chip selects must not overlap; otherwise, two chip selects will be asserted and bus contention will occur. The built-in decoder allows the 80186 to select large memory blocks. For peripheral chip selects, a base address can be programmed via a control word. The 80186 sends low on the PCS0 when it accesses a port address located in a block from this base address to up to 128 bytes. The 80186 sends low on the other chip selects PCS1-6 when one of six contiguous 128-byte blocks above the block for PCS0 is selected. Like the 8086, memory for the 80186 is set up as odd (BHE = 0) and even (A0 = 0) memory banks.

The 80186 provides eight addressing modes. These include register, immediate, direct, register indirect (SI, DI, BX, or BP), based (BX or BP), indexed (SI or DI), based indexed, and based indexed with displacement modes.

Typical data types provided by the 80186 include signed integer, ordinal (unsigned binary number), pointer, string, ASCII, unpacked/packed BCD, and floating point. The 80186 instruction set is divided into seven types. These are data transfer, arithmetic, shift/rotate, string, control transfer, high level instructions (for example, the BOUND instruction detects values outside prescribed range), and processor control. As mentioned before, the 80186 includes 10 new instructions beyond the 8086. These 10 additional instructions are listed below:

Data Transfer

PUSHA	— Push all registers onto stack
POPA	— Pop all registers from stack
PUSH immediate	— Push immediate numbers onto stack

Arithmetic

IMUL destination register, source, immediate data means immediate data*source → destination register.

Logical

SHIFT/ROTATE destination, immediate data or CL shifts/rotates register or memory contents by the number of times specified in immediate data or by the contents of CL.

String Instructions

INSB or INSW	— Input string byte or string word
OUTSB or OUTSW	— Output string byte or string word

High Level Instructions

ENTER	— Format stack for procedure entry
LEAVE	— Restore stack for procedure exit
BOUND	— Detect values outside predefined range

Let us explain some of these instructions.

- IMUL destination register, source, immediate data. This is a signed multiplication. This instruction multiplies signed 8- or 16-bit immediate data with 8- or 16-bit data in a specified source register or memory location and places the result in a general-purpose destination register. As an example, IMUL DX, CX, –3 multiplies the contents of CX by –3 and places the lower 16-bit result in DX. Note that the immediate 8-bit data of –3 are sign extended to 16-bit prior to multiplication. A 32-bit result is obtained but only the lower 16-bit is saved by this instruction.

- ROL/ROR/SAL/SAR destination, immediate data or CL. Shift count can be specified by immediate data (one) or in CL up to a maximum of 32_{10}.

- INSB DX or INSW DX respectively inputs a byte or a word from a port addressed by DX to a memory location in ES pointed to by DI. If DF = 0, DI will automatically be incremented (by 1 for byte and 2 for word) after execution of this instruction. On the other hand, if DF = 1, DI is automatically decremented (by 1 for byte and 2 for word) after execution of this instruction. The instructions INSB for byte and INSW for word are used. A typical example of inputting 50 bytes of I/O data via a port into a memory location is given below (assume ES is already initialized):

```
        STD                 ;  Set DF to 1.
        LEA DI, ADDR        ;  Initialize DI.
        MOV DX, 0E124H      ;  Load port address.
        MOV CX, 50          ;  Initialize count.
        REP INSB DX         ;  Input port until CX = 0.
STOP    JMP STOP            ;  Halt.
```

- OUTSB DX or OUTSW DX respectively provides outputting to a port addressed by DX from a source string in DS with offset in SI.

- The ENTER instruction is used at the beginning of an assembly language subroutine which is to be called by a high level language program such as Pascal. The main purpose of ENTER is to reserve space on the stack for variables used in the subroutine.

The ENTER instruction has two immediate operands:

ENTER imm16, imm8

The first operand imm16 specifies the total memory area allocated to the local variables, which is 16 bits wide (0 to 64K bytes). The second operand imm8, on the other hand, is 8 bits wide and specifies the number of nested subroutines.

For the main subroutine, imm8 = 0. Note that nested subroutines mean a subroutine calling another subroutine. For example, if there are three subroutines SUB1, SUB2, and SUB3 such that the main program M calls SUB1, SUB1 calls SUB2 and SUB2 calls SUB3, then imm8 = 0 for SUB1, 1 for SUB2, and 2 for SUB3. ENTER can be used to allocate temporary stack space for local variables for each subroutine.

In the second operand, if imm8 = 0, the ENTER instruction pushes the frame printer BP onto the stack. ENTER then subtracts the first operand imm16 from the stack pointer and sets the frame pointer, BP, to the current stack pointer value.

The LEAVE instruction is used at the end of each subroutine (usually before the RET instruction). The LEAVE does not have any operand. The LEAVE instruction should be used with the ENTER instruction. The ENTER allocates space in stack for variables used in the subroutine, while the LEAVE instruction deallocates this space and ensures that SP and BP have the original values that they had prior to execution of the ENTER. The RET instruction then returns to the appropriate address in the main program.

As an example of application of ENTER and LEAVE instructions, suppose that a subroutine requires 16 bytes of stack for local variables. The instructions ENTER 16, 0 at the subroutine's entry point and a LEAVE before the RET instruction will accomplish this. The 16 local bytes may be accessed.

When the 80186 accesses an array, the BOUND instruction can be used to ensure that data outside the array are not accessed. When the BOUND is executed, the 80186 compares the content of a general-purpose register (initialized by the user with the offset of the array element currently being accessed) with the lower and upper bounds of the array (loaded by the user prior to BOUND). The format for BOUND is BOUND reg16, memory32. The first operand is the register containing the array index and the second operand is a memory location containing the array bounds. If the index value violates the array bounds, an exception (maskable interrupt 5) takes place. A service routine can be executed by the user to indicate that the array element being accessed is out of bounds. As an example, consider BOUND SI, ADDR. The lower bound of the array is contained in address ADDR and the upper bound is in address ADDR + 2. Both bounds are 16 bits wide. For a valid access content of SI must be greater than or equal to the content of the memory location with offset ADDR and less than or equal to the contents of the memory location with offset ADDR + 2; otherwise interrupt 5 occurs. The BOUND instruction is normally placed just before the array itself, making the array addressable via a constant from the start of the array.

The BOUND instruction is normally placed following the computation of an offset value to ensure that the limits of the array boundaries are not violated. This permits checking whether or not the offset of an array being accessed is within the boundaries when the based addressing mode is used to access an element in the array. For example, the instruction segment shown below will allow accessing of an array with base address in BX and array length of 50_{10} bytes:

```
MOV     ADDR, 1000;
MOV     ADD+2, 1049;
BOUND   BX, ADDR;
MOV     CL, [BX];
```

In the above, it is assumed that the offset of the lowest array element is 1000. With an array length of 50, the offset of the highest array element is 1049. It is assumed that BX contains the offset of the array element currently being worked on. The BOUND instruction checks whether the contents of BX is between 1000 and 1049. If it is, the MOV instruction accesses the desired array element into CL; otherwise type 5 interrupt is generated.

The 80186/80188 is used in embedded control. In these applications, the microcomputer performs a dedicated control function. Embedded control applications are divided into two types. These are event control and data control.

In embedded control applications involving event control, the microprocessor initiates a timed sequence of events. An example of such an application is the industrial process control.

In embedded control applications involving data control, the microprocessor transfers volumes of data to be processed from secondary memory such as disk to main memory.

The 80186/80188 is, therefore, highly integrated to satisfy the requirements of data control applications. The 80186/80188 is also provided with added features such as string I/O instructions and DMA channels to better handle fast movement of data.

4.1.2 Intel 80286

The Intel 80286 is a high-performance 16-bit microprocessor with on-chip memory protection capabilities primarily designed for multiuser/multitasking systems. The IBM PC/AT and its clones capable of multitasking operations use the 80286 as their CPU. The 80286 can address 16 megabytes (2^{24}) of physical memory and 1 gigabyte (2^{30}) of virtual memory per task. The 80286 can be operated at three different clock speeds. These are 8 MHz (80286-4), 10 MHz (80286-6), and 12.5 MHz (80286).

The 80286 has two modes of operations. These are real address mode and protected virtual address mode (PVAM). In the real address mode, the 80286 is object code compatible with the Intel 8086/8088/80186/80188. In protected virtual address mode, the 80286 is source code compatible with the iAPX 86/88 family and may require some software modification to use virtual address features of the 80286. Note that the protected virtual address mode is not used by PCDOS.

The 80286 includes special instructions to support operating systems. For example, one instruction can end a current task execution, save its state, switch to a new task, load its state, and begin executing the new task.

The 80286's performance is up to six times faster than the standard 5-MHz 8086. The 80286 is housed in a 68-pin leadless flat package. Figure 4.2 shows a functional diagram of the 80286. It contains four separate processing units. These are the Bus Unit (BU), the Instruction Unit (IU), the Address Unit (AU), and the Execution Unit (EU). The BU provides all memory and I/O read and write operations. The BU also performs data transfer between the 80286 and

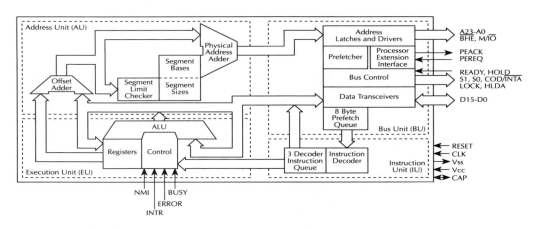

FIGURE 4.2   80286 internal block diagram.

coprocessors such as the 80287. The prefetcher in the BU prefetches instructions of up to 6 bytes and places them in a queue.

The Instruction Unit (IU) translates or decodes up to 3 prefetched instructions and places them in a queue for execution by the execution unit.

The Execution Unit (EU) executes instructions from the IU sequentially. The EU contains a 16-bit ALU, an 8086 flag register, general-purpose registers, pointer registers, index registers, and one 16-bit additional register called the machine status word (MSW) register. The lower four bits of the MSW are used. One bit places the 80286 into PVAM mode while the other three bits control the processor extension (coprocessor) interface. The LMSW and SMSW instructions can load and store MSW in real address mode.

The Address Unit (AU) calculates a 20-bit physical address based on the 16-bit contents of a segment register and a 16-bit offset just like the 8086. In this mode, the 80286 addresses one megabyte of physical memory. The 80286 has 24 address pins. However, in the real address mode, pins A23-A20 are ignored and A19-A0 pins are used. In the protected virtual address mode (PVAM), the AU operates as a memory management unit (MMU) and utilizes all 24 address lines to provide 16 megabytes of physical memory. The BU outputs memory or I/O addresses to devices connected to the 80286 after receiving them from the AU.

The 80286 does not have on-chip clock generator circuitry. Therefore, an external 82284 chip is required. The 80286 has a single CLK pin for a single-phase clock input. The 80286 divides its input clock by 2 internally and then provides the processor clock. The 82284 also provides the 80286 RESET and READY signals.

The 80286 external memory is configured as odd (BHE = 0) and even (A0 = 0) memory banks just like the 8086. The 80286 operates in a mode similar to the 8086 maximum mode. Some of the 80286 pins such as M/IO, S0, S1, HOLD, HLDA, READY, and LOCK have identical functions as the 8086. Two external interrupt pins (NMI and INTR) are provided. The nonmaskable interrupt NMI is serviced in the same way as the 8086. The INTR and COD/INTA are used together to provide an interrupt-type code on the data bus via external hardware. Note that the 80286 INTA is multiplexed with another function called the COD (code). This pin distinguishes instruction fetch cycles from memory data read cycles. Also, it distinguishes interrupt acknowledge cycles from I/O cycles. M/IO = HIGH and COD/INTA = HIGH define instruction fetch cycle. On the other hand, M/IO = LOW and COD/INTA = LOW specify interrupt acknowledge cycle.

A new pin called the CAP is provided on the chip. The 80286 MOS substrate must be applied with a negative voltage for maximum speed. The negative voltage is obtained from the +5V. An external capacitor must be connected to the CAP pin for filtering this bias voltage.

Four pins are provided to interface the 80286 with a coprocessor. These are PEREQ, PEACK, BUSY, and ERROR. The PEREQ (Processor Extension Request) input pin can be activated by the coprocessor to tell the 80286 to perform data transfer to or from memory for it. When the 80286 is ready, it activates the PEACK (Processor Extension Acknowledge) signal to inform the coprocessor of the start of the transfer. The 80286 BUSY signal input, when activated LOW by the coprocessor, stops 80286 program execution on WAIT and some ESC instructions until BUSY is HIGH. If a coprocessor finds some error during processing, it will activate the 80286 ERROR input pin. This will generate an interrupt. A service routine for this interrupt can be written to provide an indication to inform the user of the error.

Upon hardware reset, the 80286 operates in real (physical) address mode and starts executing programs at physical address FFFF0H like the 8086. In this mode, 20-bit physical addresses are generated by adding a 16-bit offset to the shifted (4 times to the left) segment register just like the 8086. Note that after hardware reset, the 80286 sets A23-A20 to all ones, CS = F000H, DS = 0000H, ES = 0000H, and SS = 0000H.

The 80286 on-chip MMU is disabled in the real address mode. In this mode, the 80286 acts functionally as a high-performance 8086. This mode averages $2 \frac{1}{2}$ times the performance of an 8086 running at the same clock frequency. All instructions of the 8086, 80186, plus a few more such as LMSW (Load Machine Status Word) and SMSW (Store Machine Status Word) are available with the 80286 in the real address mode. In this mode, the 80286 supports only the 8086 data types and can directly execute 8086 machine code programs with minor modifications. When interfaced with an 80287 floating point coprocessor, the 80286 supports 8087 floating point data types also. Upon hardware reset, the 80286 operates in real address mode unless the user sets a bit in the Machine States Word (MSW) register by using the LMSW instruction to change the 80286 mode to Protected Virtual Address Mode (PVAM). Note that before changing to PVAM, descriptor tables must be in memory. The 80286, in real address mode, can run 8086 or 8088 software. Now, to change the 80286 mode from real address to PVAM, the user should read the contents of MSW, set just the Protection Enable (PE) bit to 1 without changing the other bits, and then load the new data into the MSW. The following instruction sequence will accomplish this:

```
SMSW CX     ;   Store MSW into a general register such as CX
OR CX, 1    ;   Set only the PE bit (bit 0 in MSW)
LMSW CX     ;   Load the new value back to MSW.
```

After the above instruction sequence is executed, the 80286 operates in PVAM with memory management capabilities. In the PVAM, the 80286 is compatible with the 8086/8088 at the source code level but not at the machine code level. This means that most 8086/8088 programs must be recompiled or reassembled. Note that the real address mode is normally used to initialize peripheral devices, transfer the main portion of the operating system from disk to main memory, initialize some registers, enable interrupts and place the 80286 into PVAM.

When the 80286 is in the protected mode, the on-chip MMU is enabled which expects several address-mapping tables to exist in memory. The 80286, in this mode, will automatically access these tables for translating the logical addresses used by the user to physical addresses. Once the 80286 is in PVAM, the only way to get back to the real mode is via hardware reset. This is intentionally done so that a malicious programmer cannot switch the mode from PVAM to real mode and thus the protection feature in PVAM is maintained.

The 80286 supports the following data types:

- 8-bit or 16-bit signed binary numbers (integers)
- Unsigned 8- or 16-bit numbers (ordinal)
- A 32-bit pointer comprised of a 16-bit segment selector and 16-bit offset
- A contiguous sequence of bytes or words (strings)
- ASCII
- Packed and unpacked BCD
- Floating point

The 80286 provides 8 addressing modes. These include register, immediate, direct, register indirect, based, indexed, based index, and based indexed with displacement modes. The new 80286 instructions are for supporting the PVAM of the 80286 via an operating system. These instructions are listed in the following and are used by the operating system:

CTST	Clear task switch flag to zero located in the MSW register
LGDT	Load global descriptor table register from memory
SGDT	Store global descriptor table register into memory
LIDT	Load interrupt descriptor table register from memory
LLDT	Load selector and associated descriptor into LDTR (local descriptor table register)
SLDT	Store selector from LDTR in specified register or memory
LTR	Load task register and descriptor for TSS (task state segment)

STR	Store selector from task register in register or memory
LMSW	Load MSW register from register or memory
SMSW	Store MSW register in register or memory
LAR	Load access rights byte of descriptor into register or memory
LSL	Load segment limit from descriptor into register or memory
ARPL	Adjust register privilege Level of selector
VERR	Determine if segment addressed by a selector is readable
VERW	Determine if segment pointed to the selector is writable

Next, the 80286 will be considered from an operating systems point of view. In this context, the memory management capabilities protection and task switching features of the 80286 will be covered. Using these on-chip hardware features, a multitasking operating system can be implemented in the 80286-based microcomputer system.

The 80286 memory management features provide the operating system with the following capabilities:

- An operating system which can separate tasks from each other. This avoids an 80286 system failure due to task errors.
- As tasks begin and end, the operating system can optimize memory usage by moving them around, a process referred to as dynamic relocation. This is because a program can be executed in different parts of memory without being reassembled or recompiled.
- Use of virtual memory becomes easy. Note that virtual memory is a method for executing programs larger than the main memory by automatically transferring parts of the programs between main memory and disk.
- Controlled sharing of information between tasks.

The 80286 protection features allow:

- An operating system to protect itself from malicious users in a multiuser environment
- Critical subsystems such as disk I/O from being destroyed by program bugs under development

The 80286 task switching provides:

- Fast task switching due to 80286's hardware implementation for accomplishing this feature. This permits the 80286 to spend more time on task execution than switching. Real-time systems can thus be supported by the 80286 since they may require fast task switching. Note that an exception in a running task or an interrupt from a peripheral device requires task switching.

4.1.2.a 80286 Memory Management

The 80286 logical segments may be called virtual segments because all of them may not be resident in physical memory at the same time. A 80286 logical segment can be of any length from 1 byte to 64K bytes. During creation of each segment, a size or limit value is defined. This makes it easier for the 80286 to determine if a memory access is within bounds of a segment. The segments presently used by a task are stored in physical memory. In PVAM, all 24 address pins are used and therefore, the directly addressable (physical) memory is 16 megabytes. While writing 80286 programs in PVAM, one can refer to the current segment by the assigned names. For example, if a segment called JOHN is to be used as the current data segment, then the following instructions should be used to load JOHN into DS:

```
MOV BX, JOHN
MOV DS, BX.
```

When the 80286 executes the above instruction segment, it will use JOHN as the current data segment. If the segment JOHN is not currently resident in Physical memory, the 80286 will generate an interrupt. A service routine is written to load this segment from disk to physical memory and then resume execution of the main program.

In PVAM, when a program is assembled, a descriptor (8-byte wide) is assigned to each segment. The descriptor contains information such as segment length in bytes, 24-bit base address where the segment is located in physical memory and the privilege level.

The descriptors are held in tables in main memory and are read into the 80286 as needed. There are two main types of descriptor tables. These are the global and the local descriptor tables. A system can contain only one global descriptor table. The global descriptor table includes information such as the descriptors for the operating system segments and the descriptors for segments which are accessed by user tasks. A local descriptor is created in the system for each task. All tasks share a global descriptor table with the memory areas specified by its descriptors. Also, each task contains its own local descriptor table with the memory areas specified by its descriptors.

In PVAM, all programs are written using segments. Each segment is assigned with an 8-byte descriptor which includes the length, starting address, and access rights for that segment. Segment descriptors for programs are stored in memory either in the global descriptor table or in a local descriptor table.

The 80286 memory management is based on address translation. That is, the 80286 translates logical addresses (addresses used in programs) to physical addresses (addressing required by memory hardware). The 80286 memory pointer includes two 16-bit words: one word for a segment selector and the other as an offset into the selected segment. The real and virtual modes compute physical addresses from these selector and offset values in different ways.

In the real address mode, the 80286 computes the physical address from a 16-bit (selector) content and a 16-bit offset just like the 8086/80186. It shifts the 16-bit selector four times to the left and then adds the 16-bit offset to determine the 20-bit physical address. As mentioned before, even though the 80286 has 24 address pins (A0-A23), in the real address mode pins A0-A19 are used. Also, A20-A23 pins are only used at reset for CS and are zero otherwise.

In the Protected Virtual Address Mode (PVAM or virtual mode for short), the 32-bit address is called a virtual address. Just like the logical address, the virtual address includes a 16-bit selector and a 16-bit offset. The 80286 determines the 24-bit physical address by first obtaining a 24-bit value from a table in memory using the segment value (selector) as an index (rather than shifting the segment value 4 times to left as in the real mode) and then adding the 16-bit offset. Figure 4.3 shows the 80286 virtual address translation scheme.

The 16-bit selector is divided into a 13-bit index, one-bit Table Indicator (TI), and two-bit Requested Privilege Level (RPL). The 13-bit index is used as a displacement to access the selected table. Each entry in the table is termed a descriptor. An index can start from a value of 0 to a higher value. The index value refers to a descriptor in the table. For example, index value K refers to the descriptor K. Each descriptor is 8 bytes wide and contains the 24-bit base address required for physical address calculation. This address only occupies three bytes of the 8-byte descriptor. The meaning of the other bytes will be explained later. The single-bit TI tells the 80286 to select one of two tables: Global Descriptor Table (GDT) and Local Descriptor Table (LDT). All tasks in the 80286 share a common single table called the GDT, while each task has its own LDT. Therefore, the 24-bit base addresses required in physical address calculation for segments to be shared by all tasks are stored in the GDT and the base addresses for segments dedicated to a particular task are stored in its LDT. When TI = 0, the GDT is used as the look-up table, and when TI = 1, the LDT is used as the look-up table. The 2-bit RPL is used by the operating system for implementing the 80286's protection features. RPL is not used in physical address calculation.

The 24-bit physical address is then generated by the 80286 by adding the 24-bit base address of the selected descriptor and the 16-bit offset.

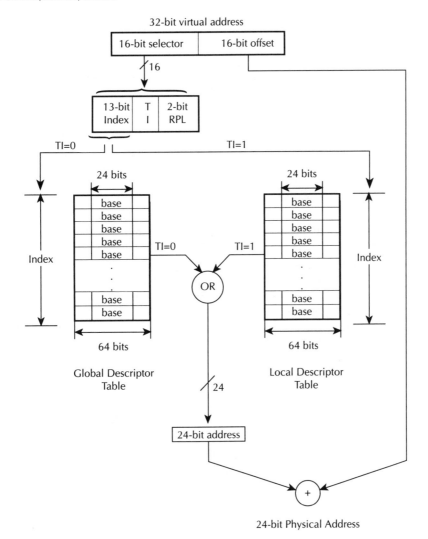

FIGURE 4.3 80286 virtual address translation.

Note that in the above, when TI = 0 (GDT selector) and Index = 0, a null selector is selected. The selector does not correspond to the 0th GDT descriptor. Null selectors can be loaded into a segment register, but use of null selectors in virtual address translation would generate an 80286 exception.

Figure 4.4 shows the 80286 address translation registers. The segment registers CS, DS, SS, and ES have already been discussed before. The GDT and LDT registers are only used in the 80286 virtual mode address translation. The GDT register stores the 24-bit base address and the length of the GDT (in bytes) minus one. During system initialization, the GDT is loaded and is usually kept unchanged after this. The 80286 generates an exception when indexing beyond the GDT limit is attempted.

The LDT register stores a 16-bit selector for a descriptor in the GDT which defines the location of the present LDT. The 80286 task switch operation is invoked by executing an inter-segment JMP or CALL instruction which refers to a task state segment (TSS) or task gate descriptor in the GDT or LDT. Each task must have a TSS associated with it. The current TSS is identified by a special register in the 80286 called the Task Register (TR). The TR register makes task switching automatic and very fast. For example, the 80286 local address space can be modified during task switching by updating the LDT register. Figure 4.5 shows flowchart for accessing memory.

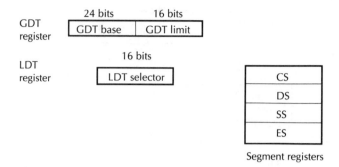

FIGURE 4.4   80286 address translation registers.

The 80286 contains a number of special registers called shadow registers which are provided to speed up memory references. The shadow registers are internal and cannot be accessed by instructions. Whenever a selector is moved into a segment register, the associated shadow register is updated with its descriptor automatically. Therefore, any memory accessed with respect to the segment register does not require referral to a look-up table, since the descriptor loaded into the shadow register contains the base address of the selected segment.

Note that in the virtual mode, the 80286 descriptor table can store a maximum of 2^{13} (13-bit index) descriptors and each segment can specify a segment of up to 2^{16} bytes. Therefore, a task can have its own LDT address space of up to $2^{13} \times 2^{16} = 2^{29}$ bytes and can share 2^{29} bytes (GDT) with all other tasks. Therefore, an address space of 1 gigabyte (2^{30} bytes) can be assigned to a task.

The following 80286 memory management instructions include loading and storing the address translation registers and checking the contents of descriptors:

LGDT	Load GDT register
SGDT	Store GDT register
LLDT	Load LDT register
SLDT	Store LDT register
LAR	Load Access Rights
LSL	Load Segment Limit

4.1.2.b Protection

In PVAM, the 80286 has built-in features for the following protection schemes:

1. Protecting system software such as the operating system from user programs.
2. Protecting one user task from another.
3. Protecting portions of memory from accidental access.

The 80286 protection mechanism is implemented by using the contents of the descriptors. Any access to memory is validated by checking the information in segment descriptors. The 80286 generates an interrupt if the memory access is invalid.

The 80286 provides protection mechanism for supporting multitasking and virtual memory features. The 80286 includes certain basic protection features such as segment limit and segment usage checking. These basic protections are useful even though multitasking and virtual memory may not be available in a system. The basic protection mechanism also allows assignment of privilege levels to virtual memory space in a hierarchical manner.

The 80286 privilege level mechanism uses certain rules to define the hierarchical order. This allows protection of the operating system independent of the user. The 80286 includes special descriptor table entries named call gates to permit CALLS to higher privilege code segments

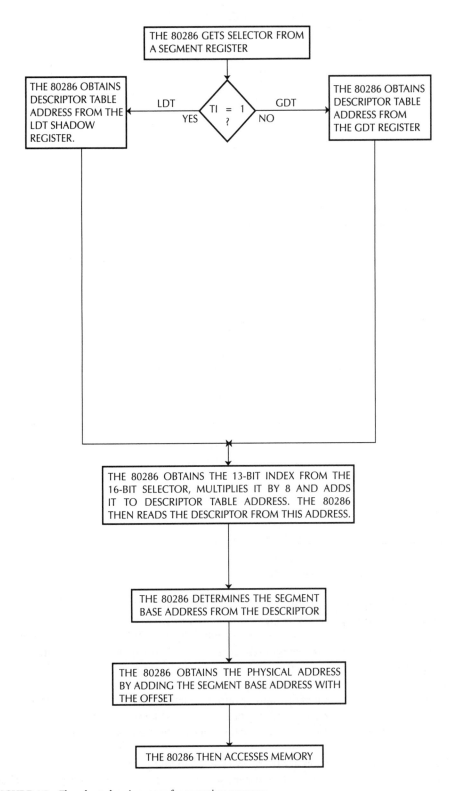

FIGURE 4.5 Flowchart showing steps for accessing memory.

Code or Data Segment Descriptor

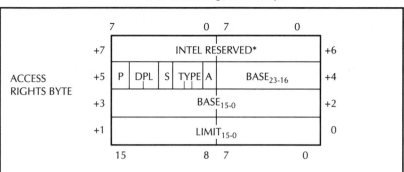

*Must be set to 0 for compatibility with IAPX 386.

Access Rights Byte Definition

Bit Position	Name	Function	
7	Present (P)	P = 1	Segment is mapped into physical Memory.
6-5	Decriptor Privilege Level (DPL)	P = 0	No mapping to physical memory eexists, base and limit are not used. Segment privilege attribute used in privilege test.
4	Segment Descriptor (S)	S = 1	Code or data (includes stacks) segment descriptor
		S = 0	System Segment Descriptor or Gate Descriptor
3	Executable (E)	E = 0	Data segment descriptor type is:
2	Expansion Direction (ED)	ED = 0	Expand up segment, offsets must be ≤ limit.
		ED = 1	Expand down segment, offsets must be > limit.
1	Writeable (W)	W = 0	Data segment may not be written into.
		W = 1	Data segment may be written into.
3	Executable (E)	E = 1	Code Segment Descriptor type is:
2	Conforming (C)	C = 1	Code segment may only be executed when CPL ≥ DPL and CPL remains unchanged.
1	Readable (R)	R = 0	Code segment may not be read.
		R = 1	Code segment may be read.
0	Accessed (A)	A = 0	Segment has not been accessed.
		A = 1	Segment selector has been loaded into segment register or used by selector test instructions.

(Bracket notes in table: "If Data Segment (S =1, E = 0)" for the executable/expansion/writeable rows; "If Code Segment (S =1, E = 0)" for the executable/conforming/readable rows.)

FIGURE 4.6 Code and data segment descriptor contents.

at specific valid entry points. In order to protect the operating system, the 80286 does not allow accessing a higher privilege entry point without its access gate.

The 80286 provides some instructions controlled by its IOPL (Input/Output Privilege Level) feature to protect shared system resources. The I/O instructions are permitted at the highest privilege level.

The 80286 on-chip protection features handle basic violations such as trying to access code segment instead of stack segment by trapping into the operating systems's appropriate routine. Thus, the operating system recovers a faulty task by taking whatever actions are necessary. The 80286 protection hardware provides information such as stack status to inform the operating system of the fault type.

The 80286 on-chip protection hardware provides up to four privilege levels in a hierarchical manner which can be used to protect system software such as the operating system from unauthorized access. The 80286 can read specific bits in its segment descriptor to obtain privilege levels of each code and data segment. Figure 4.6 shows the format for a segment descriptor for a code or data segment.

The descriptors include four words. Bit 3 of the 3-bit-type field of the access rights byte is called the Executable (E) bit. E = 1 identifies a code segment descriptor, while E = 0 identifies the segment as a data segment descriptor. The two-bit DPL (Descriptor Privilege Level) provides the privilege level of the descriptor. The DPL field specifies a hierarchical privilege system of four levels 0 thru 3.

Level 0 is the highest level while level 3 is the lowest. Privilege level provides protection within a task. Operating system routines, interrupt handlers, and other system software can be included and protected within the virtual address space of each task using the four levels of privileges. Each task in the system has a separate stack for each of its privilege levels. Typical examples of privilege levels are summarized as follows:

- A single privilege level can be assigned to all the code of a dedicated 80286. In this case, all load/store and I/O instructions are available. The 80286 can be initialized by setting the PE bit in MSW to one and then loading the GDTR with appropriate values to address a valid global descriptor table.

- Privilege levels can be assigned to user and supervisor mode types of applications. In this case, all system software can be defined as level 0 (highest level) and all other programs at some lower privilege level.

- For large applications, the system software can be divided into critical and noncritical. All critical software can be defined as a kernel with the highest privilege level (level 0), the noncritical portion of the system software is defined with levels 1 and 2, while all user application programs are defined with the lowest level (level 3).

Upon enabling the protected mode bit, the 80286 basic protection features are available irrespective of a single 80286-type application or user/supervisor configuration or a large application. The descriptor's limit field (the maximum offset from the base) and the access rights byte provides the 80286's basic protection features.

Segment limit checking ensures that all memory accesses are physically available in the segment. For all read and write operations with memory, the 80286 in the protected mode automatically checks the offset of an effective address with the descriptor limit (predefined). This limit checking feature ensures that a software fault in a segment does not interfere with any other segments in the system.

The descriptor access rights byte complements the limit checking. It differentiates code segments from data segments. The access right byte along with limit checking ensures proper usage of the segments. At least three types of segments can be defined using the access byte. Data segments can be defined as read/write or read-only. Code segments can be designated as execute-only and can be defined as conforming segments. A code segment in a particular privilege level can be accessed by using 80286 CALL or JMP instruction without a privilege level transition. A segment of equal or lower privilege level than another segment (defined as conforming via the access byte) can access that segment.

The hierarchical protection levels have four logical rules. These are summarized below:

- The Current Privilege Level (CPL) at any instant of time represents the level of the code segment presently being executed. This is provided by the privilege level in the access rights byte of the descriptor. The 80286 gives the value of CPL in its code register.

- Since every privilege level has its own stack, a stack segment rule is implemented in the 80286 to ensure using the proper stack. According to this rule, the stack segment (stack addressed by the stack segment register) and the current code segment must have the identical privilege level.

- As far as the data segments are concerned, the DPL (Descriptor Privilege Level) of an accessed data segment must be lower than or equal to the CPL. This rule allows protection of privileged data segments from unprivileged codes.

- The 80286 is provided with a rule which pertains to accessing data segments. The 80286 can access data segments of equal or lower privilege with respect to the CPL. For example, if CPL is 1, the 80286 codes in current code segment can access data segments with privilege levels of 1, 2, or 3, but not 0.
- The 80286 allows CALLing a subroutine in a code segment with higher privilege level by using call gates, and returns to code with lower privilege code segments. This is called the flow control rule and it protects higher privilege code segments. For example, if the CPL is 2, all code segments with levels 2 and 3 can be accessed by the 80286; code segments of levels 0 or 1 cannot be accessed directly. However, lower or equal privilege level accesses can be done directly. Also, higher privilege level accesses can be controlled by special descriptor table entries known as call gates. A call gate is 8 bytes wide and is stored like a descriptor in a descriptor table.

The main difference between a descriptor and a call gate is that a descriptor's contents refer to a segment in memory. On the other hand, a gate refers to another descriptor.

A descriptor includes a 24-bit physical base address, while a gate contains a 32-bit virtual address. When the effective address of an intersegment CALL references a call gate, the 80286 redirects control to the destination address defined within the gate. The 32-bit virtual address (selector and offset) of the gate can be used by the 80286 to access a higher privilege code segment.

The 80286 controls the use of I/O instructions. The user may choose the level at which these I/O instructions can be used. This level is called the IOPL (Input Output Privilege Level).

IOPL is a two-bit flag whose value varies from 0 to 3. In a user/supervisor configuration in which all supervisor code is at level 0 (highest) and all user code is at lower levels, the IOPL should be 0. This zero value of IOPL allows the supervisor code to carry out I/O operations but ensures that the user code cannot execute these I/O instructions.

Protection of a task from unauthorized access by another is provided by the 80286 both in virtual and physical memory spaces. The 80286 provides a multitasking feature via its virtual memory capabilities. As mentioned before, the virtual memory space consists of two spaces: global and local. The local space is unique to the present task being executed. This uniqueness of the local spaces provides intertask protection in the virtual memory space. The 80286's limit checking feature in the physical memory space avoids illegal accesses of segments beyond the defined segment limits and thus provides protection.

The 80286 is especially designed to execute several different tasks simultaneously (appears to be simultaneous). This is called multitasking. If the present task needs to wait for some external data, the 80286 can be programmed to switch to another task until such data are available. This mechanism of switching from one task to another is called task switching. The 80286 automatically performs all the necessary steps in order to properly switch from one task to another. When a task switching takes place, the 80286 stores the state of the present task (typically most of the 80286 registers), loads the state of the new task, and starts executing the new task. If execution of the outgoing task is desired after completion of the incoming one, the 80286 can automatically go back to the right place where the task switch took place.

Task switching may occur due to hardware or software reasons. For example, task switching may take place due to 80286 external interrupt requests (hardware reason) or due to the operating system's desire to time-share the 80286 among multiple user tasks (software reason). The task to be executed due to interrupts is termed interrupt-scheduled, while the task to be executed due to time-sharing by the operating system is called software-scheduled.

As soon as an interrupt-scheduled or a software-scheduled task is ready to be run by the 80286, it becomes the currently active (incoming) task. All inactive tasks (outgoing) have code and data segments saved in memory or disk by the 80286. Each outgoing task has a Task State Segment (TSS) associated with it. The TSS holds the task register state of an inactive task.

The TSS includes a special access right byte in its descriptor in the GDT in order that the 80286 can identify it as code or data segments. TSSs are referenced by 16-bit selectors (each task has a unique selector) that identify a TSS descriptor in the GDT. The 80286 stores the TSS selector of the presently active task in its Task State Segment register (TR). The first 44 bytes of a TSS store the complete state of a task. Information such as selectors and 80286 registers is saved.

The 80286 provides protection for portions of memory from accidental access in the following ways. When a segment selector is to be loaded into a segment register, the 80286 automatically verifies whether the descriptor table indexed by the selector contains a valid descriptor for that selector. An interrupt is automatically generated by the 80286 if a valid descriptor is not present. However, if the descriptor is valid, the shadow registers are loaded with the base, limit, and access rights byte of the descriptor. The 80286 then verifies whether the segment for that descriptor is resident in physical memory. An interrupt is generated by the 80286 if it is not present (as indicated by the P-bit of the access rights byte of the descriptor). An interrupt service routine can be written to move the desired segment into physical memory and return execution to the main program. The 80286 also verifies whether the segment descriptor is of the right type to be loaded into the appropriate segment register. For example, the descriptor for a read-only data segment cannot be moved into a stack segment register since the stack is a read/write memory. After a segment selector and descriptor are loaded into a segment register, checks are made each time an address in the actual segment is accessed. For example, an attempt to write to a read-only data segment will generate an error. Also, the limit value in the segment descriptor is used to ensure that an address generated by program instructions is within the limit specified for the segment.

Example 4.1

Discuss the 80286's performance impact on memory management while executing the following program:

```
          MOV DS, Segmentselector    ;  Load data selector
          MOV BX, Displ              ;  Load offset
          MOV CX, Count              ;  Load loop count
BEGIN     MOV DX, data               ;  Move 16-bit data to
                                     ;  DX
          CMP DX, WORDPTR[BX]        ;  Find match
          JZ DONE                    ;  If match
          JMP BEGIN                  ;  found, stop
                                     ;  else compare
DONE      HLT
```

Solution

The 80286 memory management capabilities are only utilized when loading the selector value (first instruction in the above program). By loading the selector value into DS, the 80286 chooses a descriptor from a descriptor table and then automatically determines the physical address of that segment using the descriptor. Thus, by executing the first instruction MOV DS, segmentselector, the 80286 automatically determines the segment's physical address (transparent to the user).

After determining the segment's physical address, the 80286 executes all other instructions that follow. The 80286 reads data from the data segment every time it goes through the BEGIN loop. The 80286 does not refer to the descriptor table because of its on-chip cache.

Like any other memory management system, the 80286 will have some overhead for performing the virtual address to the physical address translation. This is transparent to the user's application programs and is automatically carried out by the 80286. The memory management overhead (virtual to physical address translation) is minimized due to on-chip cache. One of the main characteristics of on-chip MMU is that after the descriptor is read into the on-chip cache memory, no overhead occurs. Therefore, the overhead is kept at a minimum, thus providing good performance.

4.1.2.c 80286 Exceptions

In the real address mode, the 80286 exception mechanism is similar to the 8086 except a few more exceptions such as the 'invalid opcode exception' is included in the 80286. External interrupts such as the maskable (INTR) and nonmaskable (NMI) interrupts in the 80286 are serviced in the real address mode by using interrupt vector table with 256 vectors similar to the 8086.

In the protected mode, the 80286 detects exceptions essential to its protection model, and its support for multitasking and virtual memory. Some examples of the non-real mode exceptions include 'exception for code, data, or extra segment not present' and 'Privilege violation'. The protected mode pointers are actually gates, since gates in protected mode serve as a redirection mechanism. The interrupt table can contain task gates, interrupt gates, and trap gates. The interrupt table, in protected mode, is known as the IDT (Interrupt Descriptor Table). A special register called the IDTR (Interrupt Descriptor Table register) is used by the 80286 to locate the interrupt table both in real and protected modes. The IDTR is typically initialized by the system programmer once, to point at the desired area of physical memory.

4.2 INTEL 80386

In 1985, Intel introduced its first 32-bit microprocessor, the 80386DX. It initially ran at a clock frequency of 16 MHz and contained 275,000 transistors. In 1988, Intel introduced the 80386SX which was a 16-bit external data bus version of the 80386DX (32-bit data bus). The 80386DX is a full 32-bit microprocessor and is available in 16-, 20-, 25- and 33-Mhz speed versions. It can directly address a maximum of 4 gigabytes of memory.

The 80386SX, on the other hand, can operate at 16- and 20-Mhz. The 80386SL is similar to the 80386SX with 16-bit data bus and can operate at 20- and 25-Mhz. The 80386SX and 80386SL can directly address up to 16 megabytes and 32 megabytes of memory respectively. Compaq was the first major OEM (Original Equipment Manufacturer) to use the 80386DX in 1986 in its PC. The 80386SL, on the other hand, is used in notebook computers with built-in power management options.

The 80386DX will be covered in detail in this section. The 80386 family of microprocessors will be referred to as the 80386 in the following.

The 80386 is a logical extension of the Intel 80286.

The 80386 provides multitasking support, memory management, pipelined architecture, address translation caches, and a high-speed bus interface in a single chip.

The 80386 is software compatible at the object code level with the Intel 8086, 80186, and 80286. The 80386 includes separate 32-bit internal and external data paths along with eight general-purpose 32-bit registers. The processor can handle 8-, 16-, and 32-bit data types. It has separate 32-bit data and generates a 32-bit physical address. The chip has 132 pins and is housed in a Pin Grid Array (PGA) package. The 80386 is designed using high-speed CHMOS III technology.

The 80386 is highly pipelined and can perform instruction fetching, decoding, execution, and memory management functions in parallel. The on-chip memory management and protection hardware translates logical addresses to physical addresses and provides the protec-

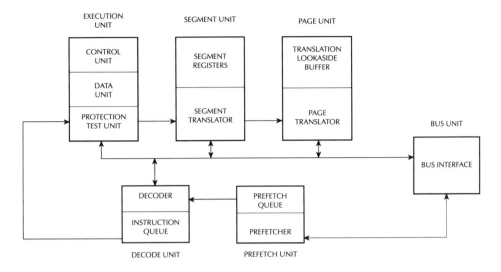

FIGURE 4.7 80386 functional units.

tion rules required in a multitasking environment. The 80386 includes special hardware for task switching. A single instruction or an interrupt is required for the 80386 to perform complete task switching. A 16-MHz 80386 can save the state of one task (all registers), load the state of another task (all registers, segment, and paging registers if needed), and resume execution in less than 16 microseconds. The 80386 contains a total of 129 instructions.

The 80386 protection mechanism, paging, and the instructions to support them are not present in the 8086. Also, the semantics of all instructions that affect segment registers (PUSH, POP, MOV, LES, LDS) and those affecting program flow (CALL, INTO, INT, IRET, JMP, RET) are quite different than the 8086 on the 80386 in protected mode.

The main differences between the 80286 and the 80386 are the 32-bit addresses and data types and paging and memory management. To provide these features and other applications, several new instructions are added in the 80386 instruction set beyond those of the 80286.

The internal architecture of the 80386 includes six functional units (Figure 4.7) that operate in parallel. The parallel operation is known as pipelined processing. Fetching, decoding, execution, memory management, and bus access for several instructions are performed simultaneously. The six functional units of the 80386 are

- Bus interface unit
- Code prefetch unit
- Decode unit
- Execution unit
- Segmentation unit
- Paging unit

The bus interface unit connects the 80386 with memory and I/O. Based on internal requests for fetching instructions and transferring data from the code prefetch unit, the 80386 generates the address, data, and control signals for the current bus cycles.

The code prefetch unit prefetches instructions when the bus interface unit is not executing bus cycles. It then stores them in a 16-byte instruction queue for decoding by the instruction decode unit.

The instruction decode unit translates instructions from the prefetch queue into microcodes. The decoded instructions are then stored in an instruction queue (FIFO) for processing by the execution unit.

The execution unit processes the instructions from the instruction queue. It contains a control unit, a data unit, and a protection test unit.

The control unit contains microcode and parallel hardware for fast multiply, divide, and effective address calculation.

The data unit includes a 32-bit ALU, 8 general-purpose registers, and a 64-bit barrel shifter for performing multiple bit shifts in one clock. The data unit carries out data operations requested by the control unit. The protection test unit checks for segmentation violations under the control of the microcode.

The segmentation unit translates logical addresses into linear addresses at the request of the execution unit.

The translated linear address is sent to the paging unit. Upon enabling of the paging mechanism, the 80386 translates these linear addresses into physical addresses. If paging is not enabled, the physical address is identical to the linear addresses and no translation is necessary.

Figure 4.8 shows a typical 80386 system block diagram.

The 80287 or 80387 numeric coprocessor can be interfaced to the 80386 to extend the 80386 instruction set to include instructions such as floating point operations. These instructions are executed in parallel by the 80287 or 80387 with the 80386 and thus off-load the 80386 of these functions.

The 82384 clock generator provides system clock and reset signals. The 82384 generates both the 80386 clock (CLK2) and a half-frequency clock (CLK) to drive the 80286-compatible devices that may be included in the system. It also generates the 80386 RESET signal. The internal frequency of the 80386 is 1/2 the frequency of CLK2.

The 8259A interrupt controller provides interrupt control and management functions. Interrupts from as many as eight external sources are accepted by one 8259A and up to 64 interrupt requests can be handled by connecting several 8259A chips. The 8259A manages priorities between several interrupts, then interrupts the 80386 and sends a code to the 80386 to identify the source of the interrupt.

The 82258 Advanced DMA (ADMA) controller performs DMA transfers between the main memory and the I/O device such as a hard disk or floppy disk without involving the 80386. It provides four channels and all signals necessary to perform DMA transfers.

The 80386 has three processing modes: protected mode, real-address mode, and virtual 8086 mode.

Protected mode is the normal 32-bit application of the 80386. All instructions and features of the 80386 are available in this mode.

Real-address mode (also known as the "real mode") is the mode of operation of the processor upon hardware RESET. This mode appears to programmers as a fast 8086 with a few new instructions. This mode is utilized by most applications for initialization purposes only.

Virtual 8086 mode (also called V86 mode) is a mode in which the 80386 can go back and forth repeatedly between V86 mode and protected mode at a fast speed. The 80386, when entering into the V86 mode, can execute an 8086 program. The processor can then leave V86 mode and enter protected mode to execute an 80386 program.

As mentioned before, the 80386 enters real address mode upon hardware reset. In this mode, the Protection Enable (PE) bit in a control register called the Control Register 0 (CR0) is cleared to zero. Setting the PE bit in CR0 places the 80386 in protected mode. When in protected mode, setting the VM (Virtual Machine) bit in the flag register (called the EFLAGS register) will place the 80386 in V86 mode. Details of these modes are discussed later.

4.2.1 Basic 80386 Programming Model

The 80386 basic programming model includes the following aspects:

a) Memory organization and segmentation
b) Data types

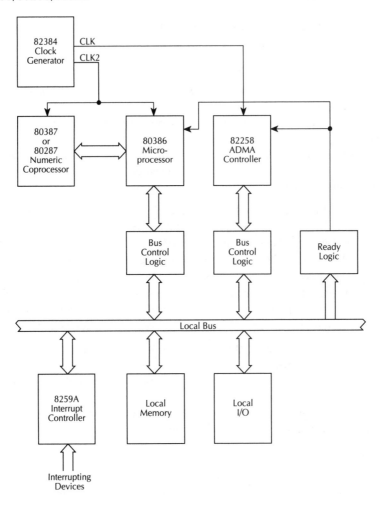

FIGURE 4.8   80386 system block diagram.

Component	Description
80386 Microprocessor	32-bit high-performance microprocessor with on-chip memory management and protection
80287 or 80387 Numeric Coprocessor	Performs numeric instruction in parallel with 80386; expands instruction set
82384 Clock Generator	Generates system clock and RESET signal
8259A Programmable Interrupt Controller	Provides interrupt control and management
82258 Advanced DMA	Performs direct memory controller access (DMA)

  c) Registers
  d) Addressing modes

I/O is not included as part of the basic programming model. This is because systems designers may select to use I/O instructions for application programs or may select to reserve them for the operating system. Therefore, 80386 I/O capabilities will be covered during the discussion of systems programming.

4.2.1.a Memory Organization and Segmentation

The 4-gigabyte physical memory of the 80386 is structured as 8-bit bytes. Each byte can be uniquely accessed by a 32-bit address.

The programmer can write assembly language programs without a knowledge of physical address space.

The memory organization model available to applications programmers is determined by the system software designers. The memory organization model available to the programmer for each task can vary between the following possibilities:

- A "flat" address space includes a single array of up to 4 gigabytes. Even though the physical address space can be up to 4 gigabytes, in reality it is much smaller. The 80386 maps the 4-gigabyte flat space into the physical address space automatically by using an address translation scheme transparent to the applications programmers.

- A segmented address space includes up to 16,383 linear address spaces of up to 4 gigabytes each. In a segmented model, the address space is called the logical address space and can be up to 2^{46} bytes (64 tetrabytes). The processor maps this address space onto the physical address space (up to 4 gigabytes) by an address translation technique.

To applications programmers, the logical address space appears as up to 16,383 one-dimensional subspaces, each with a specified length. Each of these linear subspaces is called a segment. A segment is a unit of contiguous address space with sizes varying from one byte up to a maximum of 4 gigabytes.

A pointer in the logical address space consists of a 16-bit segment selector identifying a segment and a 32-bit offset addressing a byte within a segment.

4.2.1.b Data Types

Data types can be byte (8-bit), word (16-bit with low byte address n and high byte by address n + 1), and double word (32-bit with byte 0 addressed by address n and byte 3 by address n + 3). All three data types can start at any byte address. Therefore, the words are not required to be aligned at even-numbered addresses and double words need not be aligned at addresses evenly divisible by 4. However, for maximum speed performance, data structures (including stacks) should be designed in such a way that, whenever possible, word operands are aligned at even addresses and double-word operands are aligned at addresses evenly divisible by 4.

Depending on the instruction referring to the operand, the following additional data types are available: integer (signed 8-, 16-, or 32-bit), ordinal (unsigned 8-, 16-, or 32-bit), near pointer (a 32-bit logical address which is an offset within a segment), far pointer (a 48-bit logical address consisting of a 16-bit selector and a 32-bit offset), string (8-, 16-, or 32-bit data from 0 bytes to $2^{32} - 1$ bytes), bit field (a contiguous sequence of bits starting at any bit position of any byte and may contain up to 32 bits), bit string (a contiguous sequence of bits starting at any position of any byte and may contain up to $2^{32} - 1$ bits), and packed/unpacked BCD and ASCII-type data. When the 80386 is interfaced to a coprocessor such as the 80287 or 80387, then floating point numbers (signed 32-, 64-, or 80-bit real numbers) are supported.

4.2.1.c 80386 Registers

Figure 4.9 shows 80386 registers. The 80386 has 16 registers classified as general, segment, status, and instruction.

The eight general registers are the 32-bit registers EAX, EBX, ECX, EDX, EBP, ESP, ESI, and EDI. The low-order word of each of these eight registers has the 8086/80186/80286 register

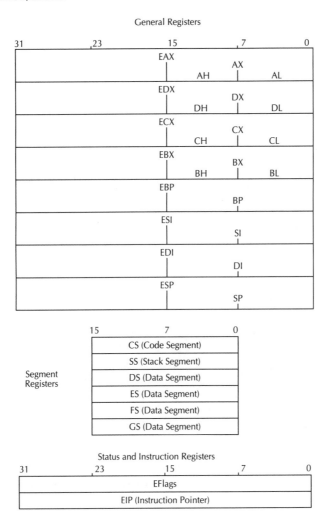

General Registers

FIGURE 4.9   80386 applications register set.

names AX (AH or AL), BX (BH or BL), CX (CH or CL), DX (DH or DL), BP, SP, SI, and DI. They are useful for making the 80386 compatible with the 8086, 80186, and 80286 processors.

The six 16-bit segment registers (CS, SS, DS, ES, FS, and GS) allow systems software designers to select either a flat or segmented model of memory organization. The purpose of CS, SS, DS, and ES is obvious. Two additional data segment registers FS and GS are included in the 80386. The four data segment registers (DS, ES, FS, GS) can access four separate data areas and allow programs to access different types of data structures. For example, one data segment register can point to the data structures of the current module, another to the exported data of a higher level module, another to a dynamically created data structure, and another to data shared with another task.

The flag register is a 32-bit register named EFLAGS. Figure 4.10 shows the meaning of each bit in this register. The low-order 16 bits of EFLAGS is named FLAGS and can be treated as

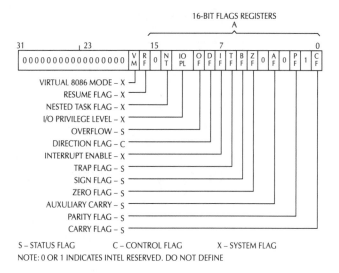

FIGURE 4.10   EFLAGS register.

a unit. This is useful when executing 8086/80186/80286 code, because this part of EFLAGS is the same as the FLAGS register of the 80286/80186/80286. The 80386 flags are grouped into three types: the status flags, the control flags, and the system flags.

The status flags include CF, PF, AF, ZF, SF, and OF as in the 8086/80186/80286. The control flag DF is used by strings as in the 8086/80186/80286. The system flags control I/O, maskable interrupts, debugging, task switching, and enabling of virtual 8086 execution in a protected, multitasking environment. The purpose of IF and TF is identical to the 8086/80186/80286. Let us explain the other flags:

- IOPL (I/O Privilege Level) — This is a 2-bit field and supports the 80386 protection feature. The IOPL field defines the privilege level needed to execute I/O instructions. If the present privilege level is less than or equal to IOPL (privilege level is specified by numbers), the 80386 can execute I/O instructions; otherwise it takes a protection exception.

- NT (Nested Task) — The NT bit controls the IRET operation. If NT = 0, a usual return from interrupt is taken by the 80386 by popping EFLAGS, CS, and EIP from the stack. If NT = 1, the 80386 returns from an interrupt via task switching.

- RF (Resume Flag) — If RF = 1, the 80386 ignores debug faults and does not take another exception so that an instruction can be restarted after a normal debug exception. If RF = 0, the 80386 takes another debug exception to service debug faults.

- VM (Virtual 8086 Mode) — When VM bit is set to one, the 80386 executes 8086 programs. When VM bit is zero, the 80386 operates in the protected mode.

The RF, NT, DF, and TF can be set or reset by an 80386 program executing at any privilege level. The VM and IOPL bits can be modified by a program running at only privilege level 0 (the highest privilege level). An 80386 with I/O privilege level can only modify the IF bit. The IRET instruction or a task switch can set or reset the RF and VM bits. The other control bits can also be modified by the POPF instruction.

The instruction pointer register (EIP) contains the offset address relative to the start of the current code segment of the next sequential instruction to be executed. The EIP is not directly accessible by the programmer; it is controlled implicitly by control-transfer instructions, interrupts, and exceptions. The low-order 16 bits of EIP is called IP and is useful when the 80386 executes 8086/80186/80286 instructions.

4.2.1.d 80386 Addressing Modes

The 80386 has 11 addressing modes which are classified into register/immediate and memory addressing modes. Register/immediate type includes two addressing modes, while the memory addressing type contains the other nine modes.

1. Register/Immediate Modes — Instructions using these register or immediate modes operate on either register or immediate operands.
   i) Register Mode — The operand is contained in one of the 8, 16, or 32-bit general registers. An example is DEC ECX which decrements the 32-bit register ECX by one.
   ii) Immediate Mode — The operand is included as part of the instruction. An example is MOV EDX, 5167812FH which moves the 32-bit data $5167812F_{16}$ to EDX register. Note that the source operand in this case is in immediate mode.
2. Memory Addressing Modes — The other 9 addressing modes specify the effective memory address of an operand. These modes are used when accessing memory. An 80386 address consists of two parts: a segment base address and an effective address. The effective address is computed by adding any combination of the following four components:
   - Displacement: 8- or 32-bit immediate data following the instruction; 16-bit displacements can be used by inserting an address prefix before the instruction.
   - Base: The contents of any general-purpose register can be used as base. Compilers normally use these base registers to point to the beginning of the local variable area.
   - Index: The contents of any general-purpose register except ESP can be used as an index register. The elements of an array or a string of characters can be accessed via the index register.
   - Scale: The index register's contents can be multiplied (scaled) by a factor of 1, 2, 4, or 8. Scaled index mode is efficient for accessing arrays or structures.

The nine memory addressing modes are a combination of the above four elements. Of these nine modes, eight of them are executed with the same number of clock cycles, since the Effective Address calculation is pipelined with the execution of other instructions; the mode containing base, index, and displacement components requires additional clocks.

As shown in Figure 4.11, the Effective Address (EA) of an operand is computed according to the following formula:

$$EA = Base\ reg + (Index\ Reg * Scaling) + Displacement$$

1. Direct Mode — The operand's effective address is included as part of the instruction as an 8-, 16-, or 32-bit displacement. An example is DEC WORDPTR [4000H].
2. Register Indirect Mode — A base or index register contains the operand's effective address. An example is MOV EBX, [ECX].
3. Based Mode — The contents of a base register are added to a displacement to obtain the operand's effective address. An example is MOV [EDX + 16], EBX.
4. Index Mode — The contents of an index register are added to a displacement to obtain the operand's effective address. An example is ADD START [EDI], EBX.
5. Scaled Index Mode — The contents of an index register are multiplied by a scaling factor (1, 2, 4, or 8) which is added to a displacement to obtain the operand's effective address. An example is MOV START [EBX * 8], ECX.
6. Based Index Mode — The contents of a base register are added to the contents of an index register to obtain the operand's effective address. An example is MOV ECX, [ESI] [EAX].
7. Based Scaled Index Mode — The contents of an index register are multiplied by a scaling factor (1, 2, 4, or 8) and the result is added to the contents of a base register to determine the operand's effective address. An example is MOV [ECX * 4] [EDX], EAX.

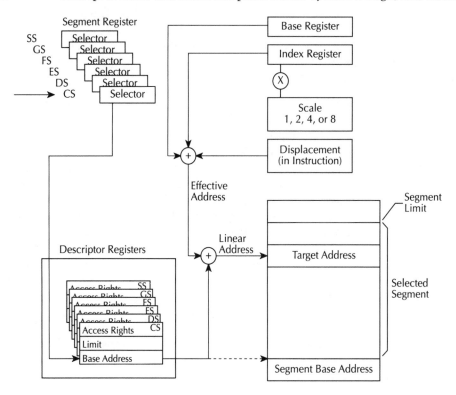

FIGURE 4.11   Addressing mode calculations.

8. Based Index Mode with Displacement — The operand's effective address is obtained by adding the contents of a base register and an index register with a displacement. An example is MOV [EBX] [EBP + 0F24782AH], ECX.

9. Based Scaled Index Mode with Displacement — The contents of an index register are multiplied by a scaling factor, and the result is added to the contents of a base register and a displacement to obtain the operand's effective address. An example is MOV [ESI*8] [EBP + 60H], ECX.

The 80386 can execute 8086/80186/80286 16-bit instructions in real and protected modes. This is provided in order to make the 80386 software compatible with the 80286, 80186, and the 8086. The 80386 uses the D bit in the segment descriptor register (8 bytes wide) to determine whether the instruction size is 16 or 32 bits wide. If D = 0, the 80386 uses all operand lengths and effective addresses as 16 bits long. On the other hand, if D = 1, then the default length for operands and addresses is 32 bits. Note that in the protected mode, the operating system can set or reset the D bit using proper instructions. In real mode, the default size for operands and addresses is 16 bits. Note that real address mode does not use descriptors.

Irrespective of the D-bit definition, the 80386 can execute either 16- or 32-bit instructions via the use of two override prefixes such as operand size prefix and address length prefix. These prefixes override the D bit on an individual instruction basis. These prefixes are automatically included by Intel assemblers. For example, if D = 1 and the 80386 wants to execute INC WORD PTR [BX] to increment a 16-bit memory location, the assembler automatically adds the operand length prefix to specify only a 16-bit value.

The 80386 uses either 8- or 32-bit displacements and any register as base or index register while executing a 32-bit code. However, the 80386 uses either 8- or 16-bit displacements with the base and index registers conforming to the 80286 while executing 16-bit code. The base and index registers utilized by the 80386 for 16- and 32-bit addresses are given in the following:

	16-bit addressing	32-bit addressing
Base Register	BX, BP	Any 32-bit general-purpose register
Index Register	SI, DI	Any 32-bit general-purpose register except ESP
Scale Factor	None	1, 2, 4, 8
Displacement	0, 8, 16 bits	0, 8, 32 bits

4.2.2 80386 Instruction Set

The 80386 extends the 8086/80186/80286 instruction set in two ways: 32-bit forms of all 16-bit instructions are included to support the 32-bit data types and 32-bit addressing modes are provided for all memory reference instructions. The 32-bit extension of the 8086/80186/80286 instruction set is accomplished by the 80386 via the default bit (D) in the code segment descriptor and by having 2 prefixes to the instruction set.

The 80386 instruction set is divided into nine types:

Data transfer
Arithmetic
String
Logical
Bit manipulation
Program Control
High-level language
Protection Model
Processor control

These instructions are listed in Table 4.1.

TABLE 4.1 80386 Instructions

	Data Transfer Instructions
General purpose	
MOV	Move operand
PUSH	Push operand onto stack
POP	Pop operand off stack
PUSHA	Push all registers on stack
POPA	Pop all registers off stack
XCHG	Exchange operand, register
XLAT	Translate
Conversion	
MOVZX	Move Byte or Word, Dword, with zero extension
MOVSX	Move Byte or Word, Dword, sign extended
CBW	Convert Byte to Word, or Word to Dword
CDW	Convert Word to Dword
CDWE	Convert Word to Dword extended
CDQ	Convert Dword to Qword
Input/output	
IN	Input operand from I/O space
OUT	Output operand to I/O space
Address object	
LEA	Load effective address
LDS	Load pointer into D segment register
LES	Load pointer into E segment register
LFS	Load pointer into F segment register
LGS	Load pointer into G segment register
LSS	Load pointer into S (stack) segment register
Flag manipulation	
LAHF	Load A register from flags

TABLE 4.1   80386 Instructions (*continued*)

	Data Transfer Instructions
SAHF	Store A register in flags
PUSHF	Push flags onto stack
POPF	Pop flags off stack
PUSHFD	Push Eflags onto stack
POPFD	Pop Eflags off stack
CLC	Clear carry flag
CLD	Clear direction flag
CMC	Complement carry flag
STC	Set carry flag
STD	Set direction flag

	Arithmetic Instructions
Addition	
ADD	Add operand
ADC	Add with carry
INC	Increment operand by 1
AAA	ASCII adjust for addition
DAA	Decimal adjust for addition
Subtraction	
SUB	Subtract operand
SBB	Subtract with borrow
DEC	Decrement operand by 1
NEG	Negate operand
CMP	Compare operands
AAS	ASCII adjust for subtraction
Multiplication	
MUL	Multiply double/single precision
IMUL	Integer multiply
AAM	ASCII adjust after multiply
Division	
DIV	Divide unsigned
IDIV	Integer divide
AAD	ASCII adjust after division

	String Instructions
MOVS	Move Byte or Word, Dword string
INS	Input string from I/O space
OUTS	Output string to I/O space
CMPS	Compare Byte or Word, Dword string
SCAS	Scan Byte or Word, Dword string
LODS	Load Byte or Word, Dword string
STOS	Store Byte or Word, Dword string
REP	Repeat
REPE/REPZ	Repeat while equal/zero
RENE/REPNZ	Repeat while not equal/not zero

	Logical Instructions
Logicals	
NOT	"NOT" operand
AND	"AND" operand
OR	"Inclusive OR" operand
XOR	"Exclusive OR" operand
TEST	"Test" operand
Shifts	
SHL/SHR	Shift logical left or right
SAL/SAR	Shift arithmetic left or right
SHLD/SHRD	Double shift left or right

TABLE 4.1 80386 Instructions (*continued*)

Rotates	
ROL/ROR	Rotate left/right
RCL/RCR	Rotate through carry left/right

Bit Manipulation Instructions

Single bit instructions	
BT	Bit test
BTS	Bit test and set
BTR	Bit test and reset
BTC	Bit test and complement
BSF	Bit scan forward
BSR	Bit scan reverse
Bit string instructions	
IBTS	Insert bit string
XBTS	Exact bit string

Program Control Instructions

Conditional transfers	
SETCC	Set byte equal to condition code
JA/JNBE	Jump if above/not below nor equal
JAE/JNB	Jump if above or equal/not below
JB/JNAE	Jump if below/not above nor equal
JBE/JNA	Jump if below or equal/not above
JC	Jump if carry
JE/JZ	Jump if equal/zero
JG/JNLE	Jump if greater/not less nor equal
JGE/JNL	Jump if greater or equal/not less
JL/JNGE	Jump if less/not greater nor equal
JLE/JNG	Jump if less or equal/not greater
JNC	Jump if not carry
JNE/JNZ	Jump if not equal/not zero
JNO	Jump if not overflow
JNP/JPO	Jump if not parity/parity odd
JNS	Jump if not sign
JO	Jump if overflow
JP/JPE	Jump if parity/parity even
JS	Jump if sign
Unconditional transfers	
CALL	Call procedure/task
RET	Return from procedure/task
JMP	Jump
Iteration controls	
LOOP	Loop
LOOPE/LOOPZ	Loop if equal/zero
LOOPNE/	Loop if not equal/not zero
LLOPNZ	
JCXZ	JUMP if register CX = 0
Interrupts	
INT	Interrupt
INTO	Interrupt if overflow
IRET	Return from interrupt
CLI	Clear interrupt enable
SLI	Set interrupt enable

High Level Language Instructions

BOUND	Check array bounds

TABLE 4.1   80386 Instructions (*continued*)

ENTER	Setup parameter block for entering procedure
LEAVE	Leave procedure

Protection Model	
SGDT	Store global descriptor table
SIDT	Store interrupt descriptor table
STR	Store task register
SLDT	Store local descriptor table
LGDT	Load global dedcriptor table
LIDT	Load interrupt descriptor table
LTR	Load task register
LLDT	Load local descriptor table
ARPL	Adjust requested privilege level
LAR	Load access rights
LSL	Load segment limit
VERR/VERW	Verify segment for reading or writing
LMSW	Load machine status word (lower 16 bits of CR0)
SMSW	Store machine status word

Processor Control Instructions	
HLT	Halt
WAIT	Wait until BUSY # negated
ESC	Escape
LOCK	Lock bus

The 80386 instructions include zero-operand, single-operand, two-operand, and three-operand instructions. Most zero-operand instructions such as STC occupy only one byte. Single operand instructions are usually two bytes wide. The two-operand instructions usually allow the following types of operations:

Register-to-register
Memory-to-register
Immediate-to-register
Memory-to-memory
Register-to-memory
Immediate-to-memory

The operands can be either 8, 16, or 32 bits wide. In general, operands are 8 or 32 bits long when the 80386 executes the 32-bit code. On the other hand, operands are 8 or 16 bits wide when the 80386 executes the existing 80286 or 8086 code (16-bit code). Prefixes can be added to all instructions which override the default length of the operands. That is, 32-bit operands for 16-bit code or 16-bit operands for 32-bit code can be used.

The 80386 various instructions affecting the status flags are summarized in Table 4.2.

Table 4.3 lists the various conditions referring to the relation between two numbers (signed and unsigned) for Jcond and SETcond instructions.

All new 80386 instructions along with those which have minor variations from the 80286 are listed in alphabetical order below in Table 4.4.

A detailed description of most of the new 80386 instructions is given in the following.

TABLE 4.2 Status Flag Summary

Status Flag Functions

Bit	Name	Function
0	CF	Carry flag — Set on high-order bit carry or borrow; cleared otherwise
2	PF	Parity flag — Set if low-order eight bits of result contain an even number of 1 bits; cleared otherwise
4	AF	Adjust flag — Set on carry from or borrow to the low-order four bits of AL; cleared otherwise; used for decimal arithmetic
6	ZF	Zero flag — Set if result is zero; cleared otherwise
7	SF	Sign flag — Set equal to high-order bit of result (0 is positive, 1 is negative)
11	OF	Overflow flag — Set if result is too large a positive number or too small a negative number (excluding sign-bit) to fit in destination operand; cleared otherwise

Key to Codes

T	Instruction tests flag
M	Instruction modifies flag (either sets or resets depending on operands)
0	Instruction resets flag
—	Instruction's effect on flag is undefined
Blank	Instruction does not affect flag

Instruction	OF	SF	ZF	AF	PF	CF
AAS	—	—	—	TM	—	M
AAD	—	M	M	—	M	—
AAM	—	M	M	—	M	—
DAA	—	M	M	TM	M	TM
DAS	—	M	M	TM	M	TM
ADC	M	M	M	M	M	TM
ADD	M	M	M	M	M	M
SBB	M	M	M	M	M	TM
SUB	M	M	M	M	M	M
CMP	M	M	M	M	M	M
CMPS	M	M	M	M	M	M
SCAS	M	M	M	M	M	M
NEG	M	M	M	M	M	M
DEC	M	M	M	M	M	
INC	M	M	M	M	M	
IMUL	M	—	—	—	—	M
MUL	M	—	—	—	—	M
RCL/RCR 1	M					TM
RCL/RCR count	—					TM
ROL/ROR 1	M					M
ROL/ROR count	—					M
SAL/SAR/SHL/SHR 1	M	M	M	—	M	M
SAL/SAR/SHL/SHR count	—	M	M	—	M	M
SHLD/SHRD	—	M	M	—	M	M
BSF/BSR	—	—	M	—	—	—
BT/BTS/BTR/BTC	—	—	—	—	—	M
AND	0	M	M	—	M	0
OR	0	M	M	—	M	0
TEST	0	M	M	—	M	0
XOR	0	M	M	—	M	0

TABLE 4.3   Condition Codes
(For Conditional Instructions Jcond, and SETcond)

Mnemonic	Meaning	Condition tested
O	Overflow	OF = 1
NO	No overflow	OF = 0
B	Below	
NAE	Neither above nor equal	CF = 1
NB	Not below	CF = 0
AE	Above or equal	
E	Equal	ZF = 1
Z	Zero	
NE	Not equal	ZF = 0
NZ	Not zero	
BE	Below or equal	(CF or ZF) = 1
NA	Not above	
NBE	Neither below nor equal	(CF or ZF) = 0
A	Above	
S	Sign	SF = 1
NS	No sign	SF = 0
P	Parity	PF = 1
PE	Parity even	
NP	No parity	PF = 0
PO	Parity odd	
L	Less	$(SF \oplus OF) = 1$
NGE	Neither greater nor equal	
NL	Not less	$(SF \oplus OF) = 0$
GE	Greater or equal	
LE	Less or equal	$((SF \oplus OF)$ or ZF$) = 1$
NG	Not greater	
NLE	Neither less nor equal	$((SF \oplus OF)$ or ZF$) = 0$
G	Greater	

Note: The terms "above" and "below" refer to the relation between two unsigned values (neither SF nor OF is tested). The terms "greater" and "less" refer to the relation between two signed values (SF and OF are tested).

TABLE 4.4   80386 Instructions (New Instructions beyond Those of 8086/80186/80286)

Instruction	Comment
ADC EAX, imm32	Add CF with sign-extended immediate byte and 32-bit data in reg32
ADC reg32/mem32, imm32	or mem32
ADC reg32/mem32, imm8	
ADC reg32/mem32, reg32	
ADC reg32, reg32/mem32	
ADD EAX, imm32	
ADD reg32/mem32, imm32	Immediate data byte is sign-extended before addition
ADD reg32/mem32, imm8	
ADD reg32/mem32, reg32	
ADD reg32, reg32/mem32	
AND EAX, imm32	
AND reg32/mem32, imm32	
AND reg32/mem32, imm8	
AND reg32/mem32, reg32	
AND reg32, reg32/mem32	
BOUND reg32, mem64	Check if reg32 is within bounds specified in mem64; the first 32 bits of mem64 contain the lower bound and the second 32 bits contain the upper bound
BSF	Bit scan forward
BSR	Bit scan reverse

TABLE 4.4   80386 Instructions (New Instructions beyond Those of 8086/80186/80286) (*continued*)

Instruction	Comment
BT	Bit test
BTC	Bit test and complement
BTR	Bit test and reset
BTS	Bit test and set
CALL label	There are several variations of the label such as disp16, disp32, reg16, reg32, mem16, mem32, ptr16:16, ptr16:32, mem16:16, and mem16:32; NEAR CALLS use reg16/mem16, reg32/mem32, and disp16/disp32 and the CALL is in the same segment with CS unchanged; CALL disp16 or disp32 adds a signed 16- or 32-bit offset to the address of the next instruction (current EIP) and the result is stored in EIP; when disp16 is used, the upper 16 bits of EIP are cleared to zero; CALL reg16/mem16 or reg32/mem32 specifies a register or memory location from which the offset is obtained; near-return instruction should be used for these CALL instructions; the far calls CALL ptr16:16 or ptr16:32 uses a four-byte or six-byte operand as a long pointer to the procedure called; the CALL mem16:16 or mem16:32 reads the long pointer from the memory location specified (indirection); in real-address mode or virtual 86 mode, the long pointer provides 16 bits for CS and 16 or 32 bits for EIP depending on operand size; the far-call instruction pushes CS and IP (or EIP) and return addresses; in protected mode, both long pointers have different meaning; consult Intel manuals for details
CDQ	Convert doubleword to quadword
	EDX : EAX ← sign extend EAX
CMP reg32/mem32, imm32	$\begin{bmatrix} \text{reg32} \\ \text{or} \\ \text{mem32} \end{bmatrix} - \text{imm32} \Rightarrow$ affects flags
CMP EAX, imm32	
CMP reg32/mem32, imm8	Compare sign extended 8-bit data with reg32 or mem32
CMP reg32mem32, reg32	
CMP reg32, reg32/mem32	
CMPSD	
CMPS mem32, mem31	
CWDE	EAX ← sign extend AX
DEC reg32/mem32	
DIV EAX, reg32/mem32	Unsigned divide EDX:EAX by reg32 or mem32 (EAX = quotient, EDX = remainder)
ENTER imm16, imm8	Create a stack frame before entering a procedure
IDIV EAX,	reg32/mem32 Signed divide; everything else is same as DIV
IMUL reg32/mem32	Signed multiply EDX: EAX ← EAX * reg32 or mem32
IMUL reg32, reg32/mem32	reg32 ← reg32 *reg32 or mem32; upper 32 bits of the product are discarded
IMUL reg16, reg16/mem16	reg16 ← reg16* reg16/mem16; result is low 16 bits of the product
IMUL reg16, reg16/mem16,	reg16 ← reg16/mem16 * (sign extended imm8); result is low 16 bits of the product
IMUL reg32, reg32/mem32,	reg32 ← reg32/mem32* (sign extended imm8); result is low 32 bits of the product
IMUL reg16, imm8	In all IMUL instructions, size of the result is defined by the size of the
IMUL reg32, imm8	destination (first) operand; the other
IMUL reg16, reg16/mem16,	
IMUL reg32, reg32/mem32,	
IMUL reg16, imm16	
IMUL reg32, imm32	
IN EAX, imm8	Input 32 bits from immediate port into EAX
IN EAX, DX	Input 32 bits from port DX into EAX
INC reg32/mem32	
INS reg32/mem32, DX or INSD	

TABLE 4.4   80386 Instructions (New Instructions beyond Those of 8086/80186/80286) (*continued*)

Instruction	Comment
IRET/IRETD	In real-address mode, IRET pops IP, CS, and FLAGS, and IRETD POPS EIP, CS, and EFLAGS; for protection mode, consult Intel manuals
Jcc label	cc can be any of the 31 conditions, including the flag settings and less, greater, above, equal, etc.; label can be disp8 as with the 80286; in 80386, however, one can have disp16 and disp32 as label; all these are signed displacements; the 80386 includes the 80286 JCXZ disp8 instruction; furthermore, the 80386 includes a new instruction called JECXZ disp8; (Jump if ECX = 0)
JMP label	The label can be specified in the same way as the CALL label instruction; see CALL label explanation for near-jump and far-jump explanations
LAR reg32, reg32/mem32	Load access right byte; reg32 ← reg32 or mem32 masked by 00FXFF00H
LEA reg32, m	Calculate the effective address (offset part) m and store it in reg32
LEAVE	LEAVE releases the stack space used by a procedure for its local variables; LEAVE reverses the action of ENTER instruction; LEAVE sets ESP to EBP and then POPS EBP
LGS/LSS/LDS/LES/LFS	Load full pointer; explained later
LODS mem32 or LODSD	To be explained later
LOOP disp8	To be explained later
LOOPcond disp8	To be explained later
MOV reg32/mem32, reg32	
MOV reg32, reg32/mem32	
MOV EAX, mem32	Move 32 bits from mem32 to EAX
MOV mem32, EAX	
MOV reg32, imm32	
MOV reg32/mem32, imm32	
MOV reg32, CR0/CR2/CR3	CRs are control registers
MOV CR0/CR2/CR3, reg32	
MOV reg32, DR0/DR1/DR2/D3	DRs are debug registers
MOV reg32, DR6/DR7	
MOV DR0/DR1/DR2/DR3, reg32	
MOV DR6/DR7, reg32	
MOV reg32, TR6/TR7	TRs are test registers
MOV TR6/TR7, reg32	
MOVS mem32, mem32 or MOVSD	To be explained later
MOVSX	Move with sign extend; to be discussed later
MOVZX	Move with zero extend; to be discussed later
MUL EAX, reg32/mem32	Unsigned multiply; EDX: EAX ← EAX * (reg32 or mem32)
NEG reg32/mem32	Two's complement. Negate 32 bits in reg32 or mem32
NOT reg32/mem32	Ones complement
OR d,s	The definitions for d and s are same as AND
OUT imm8, EAX	Output 32-bit EAX to immediate port number
OUT DX, reg32/mem32 or OUTSD	To be explained later
POP reg32	To be explained later
POP mem32	To be explained later
POP FS	To be explained later
POP GS	To be explained later
POPAD	To be explained later
POPFD	To be explained later
PUSH reg32	To be explained later
PUSH mem32	To be explained later
PUSH FS	To be explained later
PUSH GS	To be explained later
PUSHAD	To be explained later
PUSHFD	To be explained later
RCL reg32/mem32, 1	Rotate reg32 or mem32 thru CF once to left

TABLE 4.4   80386 Instructions (New Instructions beyond Those of 8086/80186/80286) (*continued*)

Instruction	Comment
RCL reg32/mem32, CL	Rotate reg32 or mem32 thru CF to left CL times
RCL reg32/mem32, imm8	
RCR reg32/mem32, 1	
RCR reg32/mem32, CL	
RCR reg32/mem32, imm8	
ROL reg32/mem32, 1	
ROL reg32/mem32, CL	
ROL reg32/mem32, imm8	
ROR reg32/mem32, 1	
ROR reg32/mem32, CL	
ROR reg32/mem32, imm8	
SAL/SAR/SHL/SHR d, n	d and n have same definitions as RCL/RCR/ROL/ROR
SBB d,s	d and s have same definitions as ADD d,s
SCAS mem32 or SCASD	To be explained later
SET cc	To be explained later
SHLD	To be explained later
SHRD	To be explained later
STOS mem32 or STOSD	To be explained later
SUB d,s	d and s have same definitions as ADD
TEST EAX, imm32	
TEST reg32/mem32, imm32	
TEST reg32/mem32, reg32	
XCHG reg32, EAX	Exchange 32-bit register contents with EAX
XCHG EAX, reg32	
XCHG reg32/mem32, reg32	
XCHG reg32, reg32/mem32	
XLATB	Set AL to memory byte DS: [EBX + unsigned AL]. DS cannot be overridden. In XLAT, DS can be overridden.
XOR d,s	d and s have same definitions as AND

4.2.2.a Arithmetic Instructions

There are two new instructions beyond those of 80286. These are CWDE and CDQ. CWDE sign extends the 16-bit contents of AX to a 32-bit doubleword in EAX. CDQ instruction sign extends a doubleword (32 bits) in EAX to a quadword (64 bits) in EDX: EAX.

4.2.2.b Bit Manipulation Instructions

The following lists the 80386 six-bit manipulation instructions:

```
BSF   Bit scan forward
BSR   Bit scan reverse
BT    Bit test
BTC   Bit test and complement
BTR   Bit test and reset
BTS   Bit test and set
```

The above instructions are discussed in the following:

   • BSF (Bit Scan Forward)

```
BSF   d , s
      reg16,  reg16
      reg16,  mem16
      reg32,  reg32
      reg32,  mem32
```

The 16-bit (word) or 32-bit (doubleword) number defined by s is scanned (checked) from right to left (bit 0 to bit 15 or bit 31). The bit number of the first one found is stored in d. If the whole 16-bit or 32-bit number is zero, the zero flag is set to one; if a one is found, the zero flag is reset to zero. For example, consider BSF EBX, EDX. If [EDX] = 01241240_{16}, then [EBX] = 00000006_{16} and ZF = 0. This is because the bit number 6 in EDX (contained in second nibble of EDX) is the first one when EDX is scanned from the right.

- BSR (Bit Scan Reverse)

```
BSR   d ,  s
      reg16,  reg16
      reg16,  mem16
      reg32,  reg32
      reg32,  mem32
```

BSR scans or checks a 16-bit or 32-bit number specified by s from the most significant bit (bit 15 or bit 31) to the least significant bit (bit 0). The destination operand d is loaded with the bit index (bit number) of the first set bit. If the bits in the number are all zero, the ZF is set to one and operand d is undefined; ZF is reset to zero if a one is found.

- BT (Bit Test)

```
BT    d ,  s
      reg16,  reg16
      mem16,  reg16
      reg16,  imm8
      mem16,  imm8
      reg32,  reg32
      mem32,  reg32
      reg32,  imm8
      mem32,  imm8
```

BT assigns the bit value of operand d (base) specified by operand s (the bit offset) to the carry flag. Only the CF is affected. If operand s is an immediate data, only eight bits are allowed in the instruction. This operand is taken modulo 32, so the range of immediate bit offset is from 0 to 31. This permits any bit within a register to be selected. If d is a register, the bit value assigned to CF is defined by the value of the bit number defined by s taken modulo the register size (16 or 32). If d is a memory bit string, the desired 16-bit or 32-bit can be determined by adding s (bit index) divided by operand size (16 or 32) to the memory address of d. The bit within this 16- or 32-bit word is defined by d modulo the operand size (16 or 32). If d is a memory operand, the 80386 may access four bytes in memory starting at Effective address + (4* [bit offset divided by 32]). As an example, consider BT CX, DX. If [CX] = $081F_{16}$, [DX] = 0021_{16}, then since the content of DX is 33_{10}, the bit number one (remainder of 33/16 = 1) of CX (value 1) is reflected in the CF and therefore CF = 1.

- BTC (Bit Test and Complement)

```
BTC  d ,  s
```

d and s have the same definitions as the BT instruction. The bit of d defined by s is reflected in the CF. After CF is assigned, the same bit of d defined by s is ones complemented. The 80386 determines the bit number from s (whether s is immediate data or register) and d (whether d is register or memory bit string) in the same way as the BT instruction.

- BTR (Bit Test and Reset)

$$\text{BTR} \quad \text{d} \quad , \quad \text{s}$$

d and s have the same definitions as for the BT instruction. The bit of d defined by s is reflected in CF. After CF is assigned, the same bit of d defined by s is reset to zero. Everything else that is applicable to the BT instruction also applies to BTR.

- BTS (Bit Test and Set)

$$\text{BTS} \quad \text{d} \quad , \quad \text{s}$$

Same as BTR except the specified bit in d is set to one after the bit value of d defined by s is reflected into CF. Everything else applicable to the BT instruction also applies to BTS.

4.2.2.c Byte-Set-On Condition Instructions

These instructions set a byte to one or reset a byte to zero depending on any of the 16 conditions defined by the status flags. The byte may be located in memory or in a one-byte general register. These instructions are very useful in implementing Boolean expressions in high level languages such as Pascal. The general structure of this instruction is SETcc (set byte on condition cc) which sets a byte to one if condition cc is true; or else, reset the byte to zero. The following is a list of these instructions:

Instruction	Condition codes	Description
SETA/SETNBE reg 18/mem8	CF = 0 and ZF = 0	Set byte if above or not below/equal
SETAE/SETNB/SETNC reg8/mem8	CF = 0	Set if above/equal, set if not below, or set if not carry
SETB/SETNAE/SETC reg8/mem8	CF = 1	Set if below, set if not above/equal, or set if carry
SETBE/SETNA reg8/mem8	CF = 1 or ZF = 1	Set if below/equal or set if not above
SETE/SETZ reg8/mem8	ZF = 1	Set if equal or set if zero
SETG/SETNLE reg8/mem8	ZF = 0 or SF = OF	Set if greater or set if not less/equal
SETGE/SETNL reg8/mem8	SF = OF	Set if greater/equal or set if not less
SETL/SETNGE reg8/mem8	SF ≠ OF	Set if less or set if not greater/equal
SETLE/SETNG reg8/mem8	ZF = 1 and SF ≠ OF	Set if less/equal or set if not greater
SETNE/SETNZ reg8/mem8	ZF = 0	Set if not equal or set if not zero
SETNO reg8/mem8	OF = 0	Set if no overflow
SETNP/SETPO reg8/mem8	PF = 0	Set if no parity or set if parity odd
SETNS reg8/mem8	SF = 0	Set if not sign
SETO reg8/mem8	OF = 1	Set if overflow
SETP/SETPE reg8/mem8	PF = 1	Set if parity or set if parity even
SETS reg8/mem8	SF = 1	Set if sign

As an example, consider SETB BL. If [BL] = 52_{16} and CF = 1, then after this instruction is executed [BL] = 01_{16} and CF remains at 1; all other flags (OF, SF, ZF, AF, PF) are undefined. On the other hand, if CF = 0, then after execution of SETB BL, BL contains 00_{16}, CF = 0 and ZF = 1; all other flags are undefined. Similarly, the other SETcc instructions can be explained.

4.2.2.d Conditional Jumps and Loops

JECXZ disp8 jumps if ECX is zero. disp8 means a relative address range from 128 bytes before the end of the instruction. JECXZ tests the contents of ECX register for zero and not the flags.

If [ECX] = 0, then after execution of JECXZ instruction, the program branches with signed 8-bit relative offset ($+127_{10}$ to -128_{10} with 0 being positive) defined by disp8.

JECXZ instruction is useful at the beginning of a conditional loop that terminates with a conditional loop instruction such as LOOPNE label. The JECXZ prevents entering the loop with ECX = 0, which would cause the loop to execute up to 2^{32} times instead of zero times.

LOOP Instructions

Instruction	Description
LOOP disp8	Decrement CX/ECX by one and jump if CX/ECX ≠ 0
LOOPE/LOOPZ disp8	Decrement CX/ECX by one and jump if CX/ECX ≠ 0 and ZF = 1
LOOPNE/LOOPNZ disp8	Decrement CX/ECX by one and jump if CX/ECX ≠ 0 and ZF = 0

The 80386 LOOP instructions are similar to those of 8086/80186/80286, except that if the counter is more than 16 bits, ECX rather than CX register is used as the counter.

4.2.2.e Data Transfer

- Move instructions description

```
MOVSX   d,s  Move  and  sign  extend
MOVZX   d,s  Move  and  zero  extend
```

The d and s operands are defined as follows:

```
MOVSX   d,s
or
MOVZX   reg16,  reg8
        reg16,  mem8
        reg32,  reg8
        reg32,  mem8
        reg32,  reg16
        reg32,  mem16
```

MOVSX reads the contents of the effective address or register as a byte or a word from the source and sign-extends the value to the operand size of the destination (16 or 32 bits) and stores the result in the destination. No flags are affected. MOVZX, on the other hand, reads the contents of the effective address or register as a byte or a word and zero-extends the value to the operand size of the destination (16 or 32 bits) and stores the result in the destination. No flags are affected. For example, consider MOVSX BX, CL. If CL = 81_{16} and [BX] = $21AF_{16}$, then after execution of MOVSX BX, CL, register BX will contain $FF81_{16}$ and CL contents do not change. Also, consider MOVZX CX, DH. If CX = $F237_{16}$ and [DH] = 85_{16}, then after execution of this MOVZX, CX register will contain 0085 and DH contents do not change.

- PUSHAD and POPAD Instructions — There are two new PUSH and POP instructions in the 80386 beyond those of 80286. These are PUSHAD and POPAD. PUSHAD saves all 32-bit general registers (the order is EAX, ECX, EDX, EBX, original ESP, EBP, ESI, and EDI) onto the 80386 stack. PUSHAD decrements the stack pointer (ESP) by 32_{10} to hold the eight 32-bit values. No flags are affected. POPAD reverses a previous PUSHAD. It pops the eight 32-bit registers (the order is EDI, ESI, EBP, ESP, EBX, EDX, ECS, and EAX). The ESP value is discarded instead of loading onto ESP. No flags are affected. Note that ESP is actually popped but thrown away, so that [ESP], after popping all the registers, will be incremented by 32_{10}.

• Load Pointer Instruction — There are five instructions in this category. These are LDS, LES, LFS, LGS, and LSS. The first two instructions LDS and LES are available in the 80286. However, the 80286 loads 32 bits from a specified location (16-bit offset and DS) into a specified 16-bit register such as BX and the other into DS for LDS or ES for LES. The 80386, on the other hand, can have four versions of these instructions as follows:

```
LDS    reg16, mem16:  mem16
LDS    reg32, mem16:  mem32
LES    reg16, mem16:  mem16
LES    reg32, mem16:  mem32
```

Note that mem16: mem16 or mem16: mem32 defines a memory operand containing four pointers composed of two numbers. The number to the left of the colon corresponds to the pointer's segment selector. The number to the right corresponds to the offset. These instructions read a full pointer from memory and store it in the selected segment register: specified register. The instruction loads 16 bits into DS (for LDS) or into ES (for LES). The other register loaded is 32 bits for 32-bit operand size and 16 bits for 16-bit operand size. The 16- and 32-bit registers to be loaded are determined by reg16 or reg32 register specified.

The three new instructions LFS, LGS, and LSS associated with segment registers FS, GS, and SS can similarly be explained.

4.2.2.f Flag Control

There are two new 80386 instructions beyond those of the 80286. These are PUSHFD and POPFD. PUSHFD decrements the stack pointer by 4 and saves the 80386 EFLAGS register to the new top of the stack. No flags are affected. POPFD, on the other hand, pops the 32-bit (doubleword) from the stack-top and stores the value in EFLAGS. All flags except VM and RF are affected.

4.2.2.g Logical

There are two new 80386 logical instructions beyond those of 80286. These are SHLD and SHRD.

Instruction	Description
SHLD d,s, count	Shift left double
SHRD d,s, count	Shift right double

The operands are defined as follows:

d	s	count
reg16	reg16	imm8
mem16	reg16	imm8
reg16	reg16	CL
mem16	reg16	CL
reg32	reg32	CL
mem32	reg32	imm8
reg32	reg32	CL
mem32	reg32	CL

For both SHLD and SHRD, the shift count is defined by the low five bits and, therefore, shifts up to 0 to 31 can be obtained.

SHLD shifts by the specified shift count the contents of d:s with the result stored back into d; d is shifted to the left by the shift count with the low-order bits of d being filled from the high-order bits of s. The bits in s are not altered after shifting. The carry flag becomes the value of the last bit shifted out of the most significant bit of d.

If the shift count is zero, the instruction works as a NOP. For a specified shift count, the SF, ZF, and PF flags are set according to the result in d. CF is set to the value of the last bit shifted out. OF and AF are undefined.

SHRD, on the other hand, shifts the contents of d:s by the specified shift count to the right with the result being stored back into d. The bits in d are shifted right by the shift count with the high-order bits being filled from the low-order bits of s. The bits in s are not altered after shifting.

If the shift count is zero, this instruction operates as a NOP. For the specified shift count, the SF, ZF, and PF flags are set according to the value of the result. CF is set to the value of the last bit shifted out. OF and AF are undefined.

As an example, consider SHLD BX, DX, 2. If $[BX] = 183F_{16}$, $[DX] = 01F1_{16}$, then after this SHLD, $[BX] = 60FC_{16}$, $[DX] = 01F1_{16}$, CF = 0, SF = 0, ZF = 0, and PF = 1.

Similarly, the SHRD instruction can be illustrated.

4.2.2.h String

- **Compare String** — There is a new instruction CMPS mem32, mem32 (or CMPSD) beyond the compare string instruction available with the 80286. This instruction compares 32-bit words ES:EDI (second operand) with DS:ESI and affects the flags. The direction of subtraction of CMPS is [[ESI]] − [[EDI]]. The left operand ESI is the source and the right operand EDI is the destination. This is a reverse of the normal Intel convention in which the left operand is the destination and the right operand is the source. This is true for byte (CMPSB) or word (CMPSW) compare instructions. The result of subtraction is not stored; only the flags are affected. For the first operand (ESI), the DS is used as segment unless a segment override byte is present, while the second operand (EDI) must use ES as the segment register and cannot be overridden. ESI and EDI are incremented by 4 if DF = 0, while they are decremented by 4 if DF = 1. CMPSD can be preceded by the REPE or REPNE prefix for block comparison. All flags are affected.

- **Load and Move Strings** — There are two new 80386 instructions beyond those of 80286. These are LODS mem32 (or LODSD) and MOVS mem32, mem32 (or MOVSD). LODSD loads the doubleword (32-bit) from a memory location specified by DS:ESI into EAX. After the load, ESI is automatically incremented by 4 if DF = 0, while ESI is automatically decremented by 4 if DF = 1. No flags are affected. LODS can be preceded by REP prefix. LODS is typically used within a loop structure because further processing of the data moved into EAX is normally required. MOVSD copies the doubleword (32-bit) at memory location addressed by DS:ESI to the memory location at ES:EDI. DS is used as the segment register for the source and may be overridden. ES must be used as the segment register and cannot be overridden. After the move, ESI and EDI are incremented by four if DF = 0, while they are decremented by 4 if DF = 1. MOVS can be preceded by the REP prefix for block movement of ECX doublewords. No flags are affected.

- **String I/O Instructions** — There are two new 80386 string I/O instructions beyond those of the 80286. These are INS mem32, DX (or INSD) and OUTS DX, mem32 (or OUTSD). INSD inputs 32-bit data from a port addressed by the content of DX into a memory location specified by ES:EDI. ES cannot be overridden. After data transfer, EDI is automatically incremented by 4 if DF = 0, while it is decremented by 4 if DF = 1. INSD can be preceded by the REP prefix for block input of ECX doublewords. No flags are affected. OUTSD instruction outputs 32-bit data from a memory location addressed by DS:ESI to a port addressed by the content of DX. DS can be overridden. After data transfer, ESI is incremented by 4 if DF = 0 and decremented by 4 if DF = 1. OUTSD can be preceded by the REP prefix for block output of ECX doublewords.

- **Store and Scan Strings** — There is a new 80386 STOS mem32 (or STOSD) instruction. STOS stores the contents of the EAX register to a doubleword addressed by ES and EDI. ES cannot be overridden. After storing, EDI is automatically incremented by 4 if DF = 0 and decremented by 4 if DF = 1. No flags are affected. STOS can be preceded by the REP prefix for a block fill of ECX doublewords. There is a new scan instruction called the SCAS mem32 (or SCASD) in the 80386. SCASD performs the 32-bit subtraction [EAX] – [memory addressed by ES and EDI]. The result of subtraction is not stored, and the flags are affected. SCASD can be preceded by the REPE or REPNE prefix for block search of ECX doublewords. All flags are affected.

4.2.2.i Table Look-Up Translation Instruction

There is a modified version of the 80286 XLAT instruction available in the 80386.

XLAT mem8 (or XLATB) replaces the AL register from the table index to the table entry. AL should be the unsigned index into a table addressed by DS:BX for 16-bit address (available in 80286 and 80386) and DS:EBX for 32-bit address (available only in 80386). DS can be overridden. No flags are affected. If DF = 0, EDI is incremented by 4 and if DF = 1, EDI is decremented by 4.

4.2.2.j High-Level Language Instructions

The three instructions ENTER, LEAVE, and BOUND (also available with 80186/80286) in this category have been enhanced in the 80386.

Before a subroutine is called by a main program, it is required quite often to pass some parameters to the subroutine by the main program. Normally these parameters are pushed onto the stack before calling the subroutine and then they are used by the subroutine by popping them from stack during its execution. In the 80386, a portion of the stack called the stack frame is used to store these parameters. Two 80386 instructions, namely, ENTER and LEAVE, are included for allocating and deallocating stack frames.

The ENTER imm16, imm8 instruction creates a stack frame. The data imm8 defines the nesting depth (also called the lexical level) of the subroutine and can be from 0 to 31. The value 0 specifies the first subroutine only. The data imm16 defines the number of stack frame pointers copied into the new stack frame from the preceding frame.

After the instruction is executed, the 80386 uses EBP as the current frame pointer and ESP as the current stack pointer. The data imm16 specifies the number of bytes of local variables for which the stack space is to be allocated.

ENTER can be used for either nested or non-nested subroutines or procedures. For example, if the lexical level imm8 is zero, the non-nested form is used. If imm8 is zero, ENTER pushes the frame pointer EBP onto the stack; ENTER then subtracts the first operand imm16 from the ESP and sets EBP to the current ESP.

For example, a procedure with 28 bytes of local variables would have an ENTER 28, 0 instruction at its entry point and a LEAVE instruction before every RET. The 28 local bytes would be addressed as offset from EBP. Note that the LEAVE instruction sets ESP to EBP and then pops EBP. For the 80186 and 80286, ENTER and LEAVE instructions use BP and SP instead of EBP and ESP. The 80386 uses BP (low 16 bits of EBP) and SP (low 16 bits of ESP) for 16-bit operands, and EBP and ESP for 32-bit operands.

The formal definition of the ENTER instruction is given in the following:

```
LEVEL denotes the value of the second operand, imm8:
Push EBP
Set a temporary value FRAME-PTR: = ESP
If LEVEL > 0 then
      Repeat (LEVEL-1) times:
              EBP: = EBP-4
              Push all EBP for the previous subroutines
                  and then EBP for the present subroutine.
      End repeat
      Push FRAME-PTR
End if
EBP:  = FRAME-PTR
ESP:  = ESP - first operand, imm16
```

The LEAVE instruction performs the following:

- (ESP) ← (EBP)
- POP into EBP.

In order to illustrate the Enter and Leave instructions, consider the following:

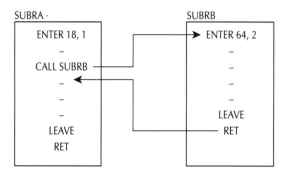

In the above, subroutine A (SUBRA) calls subroutine B. It is assumed that the nesting depth for these subroutines are 1 and 2 respectively.

The stack frames created after execution of the ENTER instructions in the two subroutines are shown below:

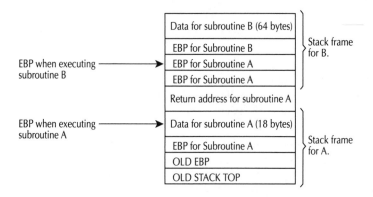

As the ENTER instruction in subroutine A is executed, the old EBP from the subroutine that called SUBRA is pushed onto the stack. EBP is loaded from ESP to point to the location of the old EBP. The lexical level for the ENTER for subroutine A is pushed onto the stack. Finally, in order to allocate 18 bytes, 12H is subtracted from the current ESP.

After entering into the subroutine B, a second ENTER instruction ENTER 64, 2 is executed. Since the lexical level in this case is 2, the Old EBP (EBP for subroutine A) is first pushed onto the stack. The EBP for subroutine A previsouly stored on the stack frame is pushed onto the stack. Finally, the current EBP for subroutine B is pushed onto the stack. This mechanism provides access to the stack frame for subroutine A from subroutine B. Next, local storage of 64 bytes as specified in the ENTER instruction are allocated for local storage.

The BOUND instruction checks to determine if the contents of a register called the array index lie within the minimum (lower bound) and maximum (upper bound) limits of an array.

The 80386 provides two forms of the BOUND instruction:

```
BOUND    reg16,  mem32
BOUND    reg32,  mem64
```

The first form is for 16-bit operands and is also available with the 80186 and 80286. The second form is for 32-bit operands and is included in the 80386 instruction set. For example, consider BOUND EDI, ADDR. Suppose [ADDR] = 32-bit lower bound, d_L and [ADDR + 4] = 32-bit upper bound d_U. If, after execution of this instruction, [EDI] < d_L and > du, the 80386 traps to interrupt 5; otherwise the array is accessed.

The BOUND instruction is usually placed following the computation of an index value to ensure that the limits of the index value are not violated. This allows checking whether or not the address of an array being accessed is within the array boundaries when the address register indirect with index mode is used to access an element in the array. For example, the following instruction segment will allow accessing an array with base address in ESI, the index value in EDI, and an array length of 200_{10} bytes. Assume the 32-bit contents of memory location 52070422_{16} and 52070426_{16} are 0 and 199_{10}, respectively:

```
           —
           —
           —
BOUND    EDI,  52070422H
MOV      EAX,  [EDI][ESI]
```

In the above, if the contents of EDI are not within the array boundaries of 0 and 199_{10}, the 80386 will trap to interrupts and the MOV instruction will not be executed.

In the following 80386 programming examples, the 80386 is assumed to be real mode. The 80386 assembler directives are not included. These directives are similar to those of the 8086.

Example 4.2

Determine the effect of each one of the following 80386 instructions:

i) **CDQ**
ii) **BTC CX, BX**
iii) **MOVSX ECX, E7H**

Assume [EAX] = FFFFFFFFH, [ECX] = F1257124H, [EDX] = EEEEEEEEH, [BX] = 0004H, [CX] = 0FA1H, prior to execution of each of the above instructions.

i) After CDQ,
    [EAX] = FFFF FFFFH
    [EDX] = FFFF FFFFH
ii) After BTC CX, BX, bit 4 of register CX is reflected in the ZF and
    then ones complemented in CX.

Before

$$[CX] = \begin{matrix} 15 & 14 & 13 & 12 & & 11 & 10 & 9 & 8 & & 7 & 6 & 5 & 4 & & 3 & 2 & 1 & 0 \end{matrix} \quad \text{Bit no.}$$

$$\begin{matrix} 0 & 0 & 0 & 0 & & 1 & 1 & 1 & 1 & & 1 & 0 & 1 & 0 & & 0 & 0 & 0 & 1_2 \end{matrix}$$

CF=0

1's complement

After BTC CX, BX,

$$[CX] = \begin{matrix} 0 & 0 & 0 & 0 & & 1 & 1 & 1 & 1 & & 1 & 0 & 1 & 1 & & 0 & 0 & 0 & 1_2 \end{matrix}$$

$$\begin{matrix} 0 & & F & & B & & 1_{16} \end{matrix}$$

Hence, [CX] = 0FB1H and [BX] = 0004H

ii) MOVSX ECX, E7H copies the 8-bit data E7H into low byte of ECX and then sign-extends to 32 bits. Therefore, after MOVSX ECX, E7H, [ECX] = FFFFFFE7H.

Example 4.3

Write an 80386 assembly language program to multiply a signed 8-bit number by a signed 32-bit number in ECX. Store result in EAX. Assume that the segment registers are already initialized and also that the result fits within 32 bits.

Solution

```
IMUL ECX, data8  ;  Perform signed
                 ;  multiplication
MOV EAX, ECX     ;  Store result
HLT              ;  in EAX and stop
```

Example 4.4

Write an 80386 assembly program to move two columns of 10,000 32-bit numbers from A (i) to B (i). In other words, move A (1) to B (1), A (2) to B (2), and so on. Initialize DS to 2000H, ES to 3000H, ESI to 0100H and EDI to 0200H.

Solution

```
MOV ECX, 10,000  ;  Initialize counter
MOV BX, 2000H    ;  Initialize DS
MOV DS, BX       ;  register
MOV BX, 3000H    ;  Initialize ES
MOV ES, BX       ;  register
MOV ESI, 0100H   ;  Initialize ESI
MOV EDI, 0200H   ;  Initialize EDI
CLD              ;  Clear DF to Autoincrement
REPMOVSD         ;  MOV A (i) to
                 ;  B (i) until ECX = 0
HLT
```

Example 4.5

Identify the addressing modes and the size of the operation for the following instructions:

   i) **MOV EAX, [EBX*4]**
  ii) **ADD AH, [EBX+35][ESI]**

Solution

   i)   <u>Source</u>   <u>Destination</u>
       Scaled Index   Register
         32-bit
  ii)   <u>Source</u>   <u>Destination</u>
      Based Indexed   Register
         8-bit

Example 4.6

Compute the physical address for the specified operands of the following instructions. Assume $DS = 0300_{16}$, $ESI = 00005000_{16}$, $EBX = 00000200_{16}$.

   i) **MOV BH, [SI]**
  ii) **ADD [BX+50H], CX**

Solution

  i) Physical address for the source $= 03000_{16} + 5000_{16} = 08000_{16}$
    Physical address for the source $= 03000_{16} + 0250_{16}$
                                $= 03250_{16}$

Example 4.7

Write an 80386 instruction sequence to compute the following:

$$\text{INTEGER} = (\text{INTEGER1} \oplus \text{EDX}) \vee (\text{ECX} \cdot \text{EDX})$$

Assume that the contents of locations INTEGER and INTEGER1 are 32-bit wide.

Solution

```
MOV EAX, EDX
NOT EDX              ;   EDX ← EDX
XOR EDX, [INTEGER1]  ;   EDX ← (INTEGER1) ⊕ EDX
AND ECX, EAX         ;   ECX ← ECX · EDX
OR ECX, EDX          ;   ECX ← EDX ∨ ECX
MOV [INTEGER], ECX   ;   [INTEGER] ← ECX
```

Example 4.8

Write an 80386 assembly language program to add two 64-bit binary numbers. The numbers are pointed to by SI and DI, respectively. Store result in location pointed to by DI. Assume data are already loaded in memory locations.

Solutions

```
MOV EAX, [SI+0]   ;  Load first 64-bit
MOV EBX, [SI+4]   ;  number
ADD [DI], EAX     ;  Add with
ADC [DI+4], EBX   ;  second 64-bit number
HLT
```

Example 4.9

Write an 80386 assembly language program to check whether the 64-bit number stored in memory pointed to by SI is zero. If it is zero, store 0 in AL; otherwise store 1 in AL.

Solution

```
        MOV       EBX, [SI+0]  ;  Move the upper 32-bit into
                               ;  EBX
        OR        EBX, [SI+4]  ;  Check whether both halves
                               ;  are zero.
        JNZ       ZERO
        MOV       AL, 1        ;  The number is not zero
        HLT
ZERO:   MOV AL, 0              ;  The number is zero
        HLT
```

Example 4.10

Assume the content of physical memory location 21010_{16} is $2F_{16}$, [ADDR] = 2000_{16} and double word stored at location ADDR+2 is 02340110_{16}. Find the contents of EAX register after execution of the following 80386 instruction sequence:

```
LDS    EAX,  [ADDR]
MOV    EBX,  00001000H
XLAT
```

Solution

LDS instruction in the above loads DS with 2000_{16} and EAX with 02340110_{16}. MOV loads EBX with $00001000H_{16}$. The XLAT instruction loads AL with the contents of memory location addressed by DS + BX + AL which is 20000_{16} + 1000_{16} + 10_{16} = 21010_{16}. Therefore, [EAX] = $0234012F_{16}$.

Example 4.11

Write an 80386 assembly language program to multiply an unsigned 32-bit number in EBX by an unsigned 16-bit number in CX. Store result in EDX:EAX.

Solution

```
MOVZX ECX, CX   ;  Zero Extend CX.
MOV EAX, EBX    ;  Move to AX for multiplication
MUL EAX, ECX    ;  Unsigned multiplication
HLT
```

Example 4.12

Write an 80386 assembly language instruction sequence to extract the bit field EAX[31:24] and store it in EBX[7:0], so that EBX[31:8] is zero and the original EAX is not affected.

Solution

```
MOV   EBX,  EAX
MOV   CL,  8
ROL   EBX,  CL
AND   EBX,  000000FFH
```

Example 4.13

Write 80386 assembly language program to compute the following: $X = Y + Z - 1F20_{16}$ where X, Y, and Z are 64-bit variables. The lower 32 bits of Y and Z are stored starting at locations NUMBER and NUMBER + 8, each followed by the upper 32 bits. Store the lower 32-bit of the 64-bit result in EAX followed by the upper 32 bits in ECX.

Solution

```
MOV   EAX,  [NUMBER]      ;  Load EAX with low 32-bit of Y
MOV   EBX,  [NUMBER+4]    ;  Load EBX with high 32-bit of Y
ADD   EAX,  [NUMBER+8]    ;  Add low 32 bits
MOV   ECX,  [NUMBER+12]   ;  Load ECX with high 32-bit of Z
ADC   ECX,  EBX           ;  Add high 32 bits and carry
SUB   EAX,  1F20H         ;  Subtract 1F20H from 32-bit of
                         ;  result
MOV   EBX,  0             ;  If borrow,
SBB   ECX,  EBX           ;  Subtract from high 32-bit of result
HLT
```

The 80386 assembly language programs can be assembled by using 80386 assemblers on IBM PC's.

A typical program structure is given below:

```
        PAGE 55, 132            ;  Set page dimensions
        .386
STACK   SEGMENT 'STACK' STACK
        DW 50 DUP(?)
STACK   ENDS
PROG    SEGMENT PARA 'CODE' PUBLIC USE16
        ASSUME CS:PROG, DS:DATA, SS:STACK
BEGIN:  MOV BX, CS
        MOV DS, BX
        JMP START
(Specify the constants and variables here)
START:  (Enter program here))
        MOV AX, 4C00H          ;  RETURN
        INT 21H                ;  to DOS
PROG    ENDS
        END BEGIN
```

In the above, the PAGE directive in the first line specifies the number of lines on a page and the width of each line for the assembler. The notation .386 in the second line indicates to the assembler that the program includes the 80386 instructions. The next three lines define the stack segment with 50 bytes.

The code segment where the actual program starts is defined next. The logical name of the code segment in the above is PROG. The code segment is public and tells the assembler to use 16-bit registers.

The 80386 uses a new directive USE16 or USE32. Programs that are developed to run on the 8086/80286 must use the USE16 to specify that all operand and address sizes are 16 bits. This automatically limits segment size to 64K. With the USE32, programs are assembled with an operand and address sizes of 32 bits. This allows access to up to 4 gigabytes of memory.

INT 21H with 4C00H in AX returns control to DOS operating system. The 80386 assembly language programs can be assembled with an assembler such as the PHAR LAP '386' assembler. All examples in this chapter are written without the complete 80386 directives. The main purpose is to illustrate use of 80386 instructions for writing assembly language programs.

4.2.3 Memory Organization

Memory on the 80386 is structured as 8-bit (byte), 16-bit (word), or 32-bit (doubleword) quantities. Words are stored in two consecutive bytes with low byte at the lower address and high byte at the higher address. The byte address of the low byte addresses the word. Doublewords are stored in four consecutive bytes in memory with byte 0 at the lowest address and byte 3 at the highest address. The byte address of byte 0 addresses the doubleword.

Memory on the 80386 can also be organized as pages or segments. The entire memory can be divided into one or more variable length segments which can be shared between programs or swapped to disks. Memory can also be divided into one or more 4K-byte pages. Segmentation and paging can also be combined in a system. The 80386 includes three types of address spaces. These are logical or virtual, linear, and physical. A logical or virtual address contains a selector (contents of a segment register) and offset (effective address) obtained by adding the base, index, and displacement components discussed earlier. Since each task on the 80386 can have a maximum of 16K selectors and offsets can be 4 gigabytes (2^{32} bits), the programmer views a virtual address space of 2^{46} or 64 tetrabytes of logical address space per task.

The 80386 on-chip segment unit translates the logical address space to 32-bit linear address space. If the paging unit is disabled, then the 32-bit linear address corresponds to the physical address. On the other hand, if the paging unit is enabled, the paging unit translates the linear address space to the physical address space. Note that the physical addresses are generated by the 80386 on its address pins.

The main difference between real and protected modes is how the 80386 segment unit translates logical addresses to linear addresses. In real mode, the segment unit shifts the selector to the left four times and adds the result to the offset to obtain the linear address. In protected mode, every selector has a linear base address. The linear base address is stored in one of two operating system tables (local descriptor table or global descriptor table). The selector's linear base address is summed with the offset to obtain the linear address. Figure 4.12 shows the 80386 address translation mechanism.

There are three main types of 80386 segments. These are code, data, and stack segments. These segments are of variable size and can be from 1 byte to 4 gigabytes (2^{32} bits) in length.

Instructions do not explicitly define the segment type. This is done in order to obtain compact instruction encoding. A default segment register is automatically selected by the 80386 according to Table 4.5.

In general, DS, SS, and CS use the selectors for data, stack, and code. Segment override prefixes can be used to override a given segment register as per Table 4.5.

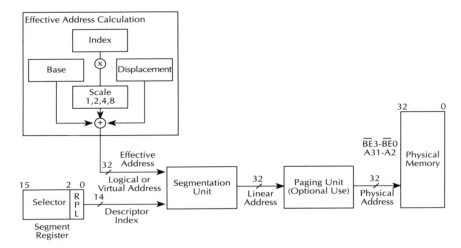

FIGURE 4.12   Address translation mechanism.

4.2.4 I/O Space

The 80386 supports both standard and memory-mapped I/O. The I/O space contains 64K 8-bit ports, 32K 16-bit ports, 16K 32-bit ports, or any combination of ports up to 64K bytes. I/O instructions do not go through the segment or paging units. Therefore, the I/O space refers to physical memory. The M/IO pin distinguishes between the memory and I/O.

The 80386 includes IN and OUT instructions to access I/O ports with port address provided by DL, DX, or EDX registers. All 8- and 16-bit port addresses are zero-extended on the upper address lines. The IN and OUT instructions drive the 80386 M/IO pin to low.

I/O port addresses 00F8H through 00FFH are reserved by Intel. The coprocessors in the I/O space are at locations 800000F8H through 800000FFH.

4.2.5 80386 Interrupts

Earlier interrupts and exceptions which are of interest to application programmers were discussed. In this section, details of these interrupts and exceptions are covered. The difference between interrupts and exceptions is that interrupts are used to handle asynchronous external events and exceptions handle instruction faults. The 80386 also treats software interrupts such as INT n as exceptions.

TABLE 4.5   Segment Register Selection Rules

Type of memory reference		Implied (default) segment use	Segment override prefixes possible
Code Fetch		CS	None
Destination of PUSH, PUSHA instructions		SS	None
Source of POP, POPA instructions		SS	None
Other data references, with effective address			
using base register of:	[EAX]	DS	CS,SS,ES,FS,GS
	[EBX]	DS	CS,SS,ES,FS,GS
	[ECX]	DS	CS,SS,ES,FS,GS
	[EDX]	DS	CS,SS,ES,FS,GS
	[EBX]	DS	CS,SS,ES,FS,GS
	[ESI]	DS	CS,SS,ES,FS,GS
	[EDI]	DS	CS,SS,ES,FS,GS
	[EBP]	SS	CS,DS,ES,FS,GS
	[ESP]	SS	CS,DS,ES,FS,GS

The 80386 interrupts and exceptions are similar to those of the 8086.

There are three types of interrupts/exceptions. These are hardware interrupts, exceptions, and software interrupts.

Hardware interrupts can be of two types. The 80386 provides the NMI pin for the nonmaskable interrupt. When the NMI pin encounters a LOW to HIGH transition by an external device such as an A/D converter, the 80386 services the interrupt via the internally supplied instruction INT2. The INT2 instruction does not need to be provided via external hardware.

The 80386 services a maskable interrupt when its INTR pin is activated HIGH and the IF bit is set to one. An 8-bit vector can be supplied by the user via external hardware which identifies the interrupt source.

The IF bit in the EFLAG registers is reset when an interrupt is being serviced. This, in turn, disables servicing additional interrupts during an interrupt service routine.

When an interrupt occurs, the 80386 completes execution of the current instruction. The 80386 then pushes the EIP, CS, and flags onto the stack. Next, the 80386 obtains an 8-bit vector via either external hardware (maskable) or internally (nonmaskable) which identifies the appropriate entry in the interrupt table. The table contains the starting address of the interrupt service routine. At the end of the interrupt service routine, IRET can be placed to resume the program at the appropriate place in the main program.

The software interrupt due to execution of INT n has the same effect as the hardware interrupt. A special case of the software interrupt INT n is the INT3 or breakpoint interrupt. Like the 8086, the single-step interrupt is enabled by setting the TF bit. The TF bit is set by altering the stack image and executing a POPF or IRET instruction. The single step uses INT1. Exceptions are classified as faults, traps, or aborts depending on the way they are reported and whether or not the instruction causing the exception is restarted. Faults are exceptions that are detected and serviced before the execution of the faulting instruction. A fault can occur in a virtual memory system when the 80386 references a page or a segment not present in the main memory. The operating system can execute a service routine at the fault's interrupt address vector to fetch the page or segment from disk. Then the 80386 restarts the instruction traps and immediately reports the cause of the problem via the execution of the instruction. Typical examples of traps are user-defined interrupts. Aborts are exceptions which do not allow the exact location of the instruction causing the exception to be determined. Aborts are used to report severe errors such as a hardware error or illegal values in system tables.

Therefore, upon completion of the interrupt service routine, the 80386 resumes program execution at the instruction following the interrupted instruction. On the other hand, the return address from an exception fault routine will always point to the instruction causing the exception. Table 4.6 lists the 80386 interrupts along with to where the return address points.

The 80386 can handle up to 256 different interrupts/exceptions. For servicing the interrupts, a table containing up to 256 interrupt vectors must be defined by the user. These interrupt vectors are pointers to the interrupt service routine. In real mode, the vectors contain two 16-bit words: the code segment and a 16-bit offset. In protected mode, the interrupt vectors are 8-byte quantities, which are stored in an interrupt descriptor table. Of the 256 possible interrupts, 32 are reserved by Intel and the remaining 224 are available to be used by the system designer.

If there are several interrupts/exceptions occurring at the same time, the 80386 handles them according to the following priorities:

Priority	Interrupt/exception
1. (Highest)	Exception faults
2.	TRAP instructions
3.	Debug traps for this instruction
4.	Debug faults for next instruction
5.	NMI
6. (Lowest)	INTR

TABLE 4.6 Interrupt Vector Assignments

Function	Interrupt number	Instruction which can cause exception	Return address points to faulting instruction	Type
Divide error	0	DIV, IDIV	Yes	Fault
Debug exception	1	Any instruction	Yes	Trap[a]
NMI interrupt	2	INT 2 or NMI	No	NMI
One-byte interrupt	3	INT	No	Trap
Interrupt on overflow	4	INTO	No	Trap
Array bounds check	5	BOUND	Yes	Fault
Invalid OP-code	6	Any illegal instruction	Yes	Fault
Device not available	7	ESC, WAIT	Yes	Fault
Double fault	8	Any instruction that can generate an exception		Abort
Coprocessor segment	9	Coprocessor tries to access data past the end of a segment	No	Trap[b]
Invalid TSS	10	JMP, CALL, IRET, INT	Yes	Fault
Segment not present	11	Segment register instructions	Yes	Fault
Stack fault	12	Stack references	Yes	Fault
General protection fault	13	Any memory reference	Yes	Fault
Page fault	14	Any memory access or code fetch	Yes	Fault
Coprocessor error	16	ESC, WAIT	Yes	Fault
Intel reserved	17—32			
Two-byte interrupt	0—255	INT n	No	Trap

[a]Some debug exceptions may report both traps on the previous instruction and faults on the next instruction.
[b]Exception 9 no longer occurs on the 80386 due to the improved interface between the 80386 and its coprocessors.

As an example, suppose an instruction causes both a debug exception (interrupt no. 1) and page fault (interrupt vector 14). According to the built-in priority mechanism, the 80386 will first service the page fault by executing the exception 14 handler. The exception 14 handler will be interrupted by the debug exception handler (1). An address in the page fault handler will be pushed onto the stack and the service routine for the debug handler (1) will be completed. After this, the exception 14 handler will be executed. This permits the system designer to debug the exception handler.

In real mode, the 80386 obtains the values of Ip and CS similar to the 8086 by using a table called the interrupt pointer table. In protected mode, this table is called Interrupt descriptor table. The 80386 real-mode interrupt pointer table is shown on the following page.

4.2.6 80386 Reset and Initialization

When the 80386 is initialized and reset, the 80386 will start executing instructions near the top of physical memory, at location FFFFFFF0H. When the first Intersegment Jump or call is executed, address lines A20-A31 will drop low, and the 80386 will only execute instructions in the lower one megabyte of physical memory. This allows the system designer to use a ROM at the top of physical memory to intialize the system and take care of Resets. Driving the RESET input pin HIGH for at least 78 CLK2 periods resets the 80386.

Upon hardware reset, the 80386 registers contain the values as shown in Table 4.7.

		Physical Memory Address
Vector Number 255	CS	003FEH
	IP	003FCH
	⋮	
32	CS	00082H
	IP	00080H
	⋮	
Coprocessor not present	CS	0001EH
	IP	0001CH
Invalid opcode 6	CS	0001AH
	IP	00018H
Bound check 5	CS	00016H
	IP	00014H
OVERFLOW 4	CS	00012H
	IP	00010H
BREAKPOINT #	CS	0000EH
	IP	0000BH
NMI 2	CS	0000AH
	IP	00008H
DEBUG 1	CS	00006H
	IP	00004H
DIVIDE BY 0 ERROR 0	CS	00002H
	IP	00000H

4.2.7 Testability

The 80386 provides capability to perform self-test. The self-test checks all of the control ROM and the associate nonrandom logic inside the 80386. The self-test feature is performed when the 80386 RESET pin goes from HIGH to LOW and the BUSY # pin is LOW. The self-test takes above 30 milliseconds with a 16-MHz clock. After self-test, the 80386 performs reset and begins program execution. If the self-test is successful, the contents of both EAX and EDX are zero; otherwise the contents of EAX and EDX are not zero, indicating a faulty chip.

4.2.8 Debugging

In addition to the software breakpoint and single-stepping features, the 80386 also includes six program-accessible 32-bit registers for specifying up to four district breakpoints. Unlike the INT3 which only allows instruction breakpointing, the 80386 debug registers permit breakpoints to be set for data accesses. Therefore, a breakpoint can be set up if a variable is accidentally being overwritten. Thus, the 80386 can stop executing the program whenever the variable's contents are being changed.

4.2.9 80386 Pins and Signals

As mentioned before, the 80386 is a 132-pin ceramic Pin Grid Array (PGA). Pins are arranged 0.1 inch (2.54 mm) center-to-center, in a 14 × 14 matrix, three rows around.

A number of sockets are available for low insertion force or zero insertion force mountings. Three types of terminals include soldertail, surface mount, or wire wrap. These application

TABLE 4.7   Register Values after Reset

Flag word(EFLAGS)	00000002H	Note 1
Machine status word (CR0)	UUUUUUU0H	Note 2
Instruction pointer (IP)	0000FFF0H	
Code segment	F000H	Note 3
Data segment	0000H	
Stack segment	0000H	
Extra segment (ES)	0000H	
Extra segment (FS)	0000H	
Extra segment (GS)	0000H	
All other registers	Undefined	

Note: U means undefined.

Note 1: The upper 14 bits of the EFLAGS registers are zero, VM (Bit 17) and RF (Bit 16) and other defined flag bits are zero.

Note 2: All of the defined fields in the CR0 are 0 (PG Bit 31, TS Bit 3, EM Bit 2, MP Bit 1, and PE Bit 0) except for ET Bit 4 (processor Extension type). The ET Bit is set during Reset according to the type of coprocessor in the system. If the coprocessor is an 80387 then ET will be 1, if the coprocessor is an 80287 or no coprocessor is present then ET will be 0. All other bits are undefined.

Note 3: The code segment Register (CS) will have its base address set to FFF00000H and Limit set to 0FFFFH. All undefined bits are reserved and should not be used.

sockets are manufactured by Amp, Inc. of Harrisburg, PA, Advanced Interconnections of Warwick, RI, and Textool Products of Irving, TX.

Figure 4.13 shows the 80386 pinout as viewed from the pin side of the chip. Table 4.8 provides the 80386 pinout functional grouping description.

Figure 4.14 shows functional grouping of the 80386 pins. A brief description of the 80386 pins and signals is provided in the following. The # symbol at the end of the signal name or the — symbol above a signal name indicates the active or asserted state when it is low. When the symbol # is absent after the signal name or the symbol — is absent above a signal name, the signal is asserted when high.

The 80386 has 20 Vcc and 21 GND pins for power distribution. These multiple power and ground pins reduce noise. Preferably, the circuit board should contain Vcc and GND planes.

CLK2 pin provides the basic timing for the 80386. This clock is divided by 2 internally to provide the internal clock used for instruction execution. The CLK2 signal is specified at CMOS-compatible voltage levels and not at TTL levels.

There are two phases (phase one and phase two) of the internal clock. Each CLK2 period defines a phase of the internal clock. Figure 4.15 shows the relationship. The 80386 is reset by activating the RESET pin for at least 15 CLK2 periods. The RESET signal is level-sensitive. When the RESET pin is asserted, the 80386 ignores all input pins and drives all other pins to idle bus state. The 82384 clock generator provides system clock and reset signals.

D0-D31 provides the 32-bit data bus. The 80386 can transfer 16- or 32-bit data via the data bus.

The address pins A2-A31 along with the byte enable signals BE0# thru BE3# to generate physical memory or I/O port addresses. Using these pins, the 80386 can directly address 4

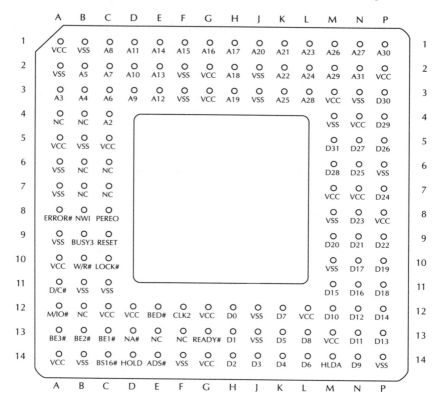

FIGURE 4.13   80386 PGA pinout view from pin side.

gigabytes by physical memory (00000000H thru FFFFFFFFH) and 64 kilobytes of I/O addresses (00000000H thru 0000FFFFH). The coprocessor addresses range from 800000F8H thru 800000FFH. Therefore, coprocessor select signal is generated by the 80386 when M/IO # is LOW and A31 is HIGH.

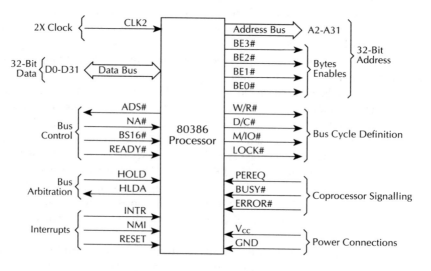

FIGURE 4.14   Functional signal groups.

TABLE 4.8 80386 PGA Pinout Functional Grouping

Pin	signal	Pin	signal	Pin	signal	Pin	signal
NS	A31	M5	D31	A1	Vcc	A2	Vss
P1	A30	P3	D30	A5	Vcc	A6	Vss
M2	A29	P4	D29	A7	Vcc	A9	Vss
L3	A28	M6	D28	A10	Vcc	B1	Vss
N1	A27	N5	D27	A14	Vcc	B5	Vss
M1	A26	P5	D26	C5	Vcc	B11	Vss
K3	A25	N6	D25	C12	Vcc	B14	Vss
L2	A24	P7	D24	D12	Vcc	C11	Vss
L1	A23	N8	D23	G2	Vcc	FS	Vss
K2	A22	P9	D22	G3	Vcc	F3	Vss
K1	A21	N9	D21	G12	Vcc	F14	Vss
J1	A20	M9	D20	G14	Vcc	J2	Vss
H3	A19	P10	D19	L12	Vcc	J3	Vss
H2	A18	P11	D18	M3	Vcc	J12	Vss
H1	A17	N10	D17	M7	Vcc	J13	Vss
G1	A16	N11	D16	M13	Vcc	M4	Vss
F1	A15	M11	D15	N4	Vcc	M8	Vss
E1	A14	P12	D14	N7	Vcc	M10	Vss
E2	A13	P13	D13	P2	Vcc	N3	Vss
E3	A12	N12	D12	P8	Vcc	P6	Vss
D1	A11	N13	D11			P14	Vss
D2	A10	M12	D10				
D3	A9	N14	D9	F12	CLK2	A4	N.C.
C1	A8	L13	D8			D4	N.C.
C2	A7	K12	D7	E14	ADS#	B6	N.C.
C3	A6	L14	D6			B12	N.C.
B2	A5	K13	D5	B10	W/R#	C6	N.C.
B3	A4	K14	D4	A11	D/C#	C7	N.C.
A3	A3	J14	D3	A12	M/IO#	E13	N.C.
C4	A2	H14	D2	C10	LOCK#	F13	N.C.
A13	BE3#	H13	D1				
B13	BE2#	H12	D0	D13	NA#	C8	PEREQ
C13	BE1#			C14	BS16#	B9	BUSY#
E12	BE0#			G13	READY#	A8	ERROR#
		D14	HOLD				
C9	RESET	M14	HLDA	B7	INTR	B8	NMI

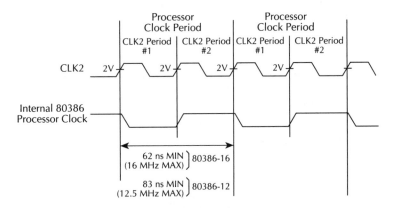

FIGURE 4.15 CLK2 signal and internal processor clock.

The byte enable outputs, BE0# thru BE3# by the 80386, define which bytes of D0-D31 are utilized in the current data transfer. These definitions are given below:

> BE0# is low when data is transferred via D0-D7
> BE1# is low when data is transferred via D8-D15
> BE2# is low when data is transferred via D16-D23
> BE3# is low when data is transferred via D24-D31

The 80386 asserts one or more byte enables depending on the physical size of the operand being transferred (1, 2, 3, or 4 bytes).

When the 80386 performs a word memory write or word I/O write cycle via D16-D31 pins, it duplicates this data on D0-D15.

W/R#, D/C#, M/IO#, and LOCK# output pins specify the type of bus cycle being performed by the 80386. W/R# pin, when HIGH, identifies write cycle and, when LOW, indicates read cycle. D/C# pin, when HIGH, identifies data cycle and, when LOW, indicates control cycle. M/IO# differentiates between memory and I/O cycles. LOCK# distinguishes between locked and unlocked bus cycles. W/R#, D/C#, and M/IO# pins define the primary bus cycle. This is because these signals are valid when ADS# (address status output) is asserted. LOCK# output is valid as soon as the bus cycle begins, but due to address pipelining LOCK# may be valid later when ADS# is asserted. Table 4.9 defines the bus cycle definitions.

The 80386 bus control signals include ADS# (address status), READY# (transfer acknowledge), NA# (next address request), and BS16# (bus size 16).

The 80386 outputs LOW on the ADS# pin to indicate a valid bus cycle (W/R#, D/C#, M/IO#) and address (BE0#-BE3#, A2-A31) signals.

When READY# input is asserted during a read cycle or an interrupt acknowledge cycle, the 80386 latches the input data on the data pins and ends the cycle. When READY# is low during a write cycle, the 80386 ends the bus cycle.

The NA# input pin is activated low by external hardware to request address pipelining. A low on this pin means that the system is ready to receive new values of BE0#-BE3#, A2-A31, W/R#, D/C#, and M/IO# from the 80386 even if the completion of the present cycle is not acknowledged on the READY# pin.

BS16# input pin permits the 80386 to interface to 32- and 16-bit data buses. When the BS16# input pin is asserted low by an external device, the 80386 uses the low-order half (D0-D15) of the data bus corresponding to BE0# and BE1# for data transfer. If the 80386 asserts BE2# or BE3# during a bus cycle, then assertion of BS16# by an external device in this cycle

TABLE 4.9   Bus Cycle Definition

M/IO#	D/C#	W/R#	Bus cycle type		Locked?
Low	Low	Low	INTERRUPT ACKNOWLEDGE		Yes
Low	Low	High	Does not occur		—
Low	High	Low	I/O DATA READ		No
Low	High	High	I/O DATA WRITE		No
High	Low	Low	MEMORY CODE READ		No
High	Low	High	HALT:	SHUTDOWN:	No
			Address = 2	Address = 0	
			(BE0# High	(BE0# Low	
			BE1# High	BE1# High	
			BE2# Low	BE2# High	
			BE3# High	BE3# High	
			A2-A31 Low)	A2-A31 Low)	
High	High	Low	MEMORY DATA READ		Some cycles
High	High	High	MEMORY DATA WRITE		Some cycles

will automatically cause the 80386 to transfer the upper byte(s) via only D0-D15. For 32-bit data operands with BS16# asserted, the 80386 will automatically execute two consecutive 16-bit bus cycles to accomplish this.

HOLD (input) and HLDA (output) pins are 80386 bus arbitration signals. These signals are used for DMA transfers. PEREQ, BUSY#, and ERROR# pins are used for interfacing coprocessors such as 80287 or 80387 to the 80386. A HIGH on PEREQ (coprocessor request) input pin indicates that a coprocessor is requesting the 80386 to transfer data to or from memory. The 80386 thus transfers data between the coprocessor and memory. This signal is level-sensitive. A LOW on the BUSY# (coprocessor Busy) input pin means that the coprocessor is still executing an instruction and is not capable of accepting another instruction. The BUSY# pin avoids interference with a previous coprocessor instruction.

ERROR# (coprocessor error) input pin, when asserted LOW by the coprocessor, indicates that the previous coprocessor instruction generated a coprocessor error of a type not masked by the coprocessor's control register. This input pin is automatically sampled by the 80386 when a coprocessor instruction is encountered and, if asserted, the 80386 generates exception 7 for executing the error-handling routine.

There are two interrupt pins on the 80386. These are INTR (maskable) and NMI (nonmaskable) pins. INTR is level-sensitive. When INTR is asserted and if the IF bit in the EFLAGS is 1, the 80386 (when ready) responds to the INTR by performing two interrupt acknowledge cycles and at the end of the second cycle latches an 8-bit vector on D0-D7 to identify the source of interrupt. To ensure INTR recognition, it must be asserted until the first interrupt acknowledge cycle starts.

NMI is leading-edge sensitive. It must be negated for at least 8 CLK2 periods and then be asserted for at least 8 CLK2 periods to assure recognition by the 80386. The servicing of NMI was discussed earlier.

Table 4.10 summarizes the characteristics of all 80386 signals.

TABLE 4.10   80386 Signal Summary

Signal name	Signal function	Active state	Input/output	Input synch or asynch to CLK2	Output high impedance during HDLA?
CLK2	Clock	—	I	—	—
D0-D31	Data bus	High	I/O	S	Yes
BE0#-BE3#	Byte enables	Low	O	—	Yes
A2-A31	Address bus	High	O	—	Yes
W/R#	Write-read indications	High	O	—	Yes
D/C#	Data-control indication	High	O	—	Yes
M/IO#	Memory-I/O indication	High	O	—	Yes
LOCK#	Bus lock indication	Low	O	—	Yes
ADS#	Address status	Low	O	—	Yes
NA#	Next address request	Low	I	S	—
BS16#	Bus size 16	Low	I	S	—
READY#	Transfer acknowledge	Low	I	S	—
HOLD	Bus hold request	High	I	S	—
HLDA	Bus hold acknowledge	High	O	—	No
PEREQ	Coprocessor request	High	I	A	—
BUSY#	Coprocessor busy	Low	I	A	—
ERROR#	Coprocessor error	Low	I	A	—
INTR	Maskable interrupt request	High	I	A	—
NMI	Nonmaskable interrupt request	High	I	A	—
RESET	Reset	High	I	A (note)	—

Note: If the phase of the internal processor clock must be synchronized to external circuitry, RESET falling edge must meet setup and hold times t_{25} and t_{26}.

4.2.10 80386 Bus Transfer Technique

The 80386 uses one or more bus cycles to perform all data transfers.

The 32-bit address is generated by the 80386 from BE0#-BE3# and A2-A32 as follows:

80386 address signals							
Physical base address			80386 address pins and BE0#-BE3# signals				
A31—A2	A1	A0	A31—A2	BE3#	BE2#	BE1#	BE0#
A31—A2	0	0	A31—A2	X	X	X	Low
A31—A2	0	1	A31—A2	X	X	Low	High
A31—A2	1	0	A31—A2	X	Low	High	High
A31—A2	1	1	A31—A2	Low	High	High	High

Dynamic bus sizing feature connects the 80386 with 32-bit or 16-bit data buses for memory or I/O. A single 80386 can be connected to both 16- and 32-bit buses. During each bus cycle, the 80386 dynamically determines bus width and then transfers data to or from 32- or 16-bit devices. During each bus cycle, the 80386 BS16# pin can be asserted for 16-bit ports or negate BS16# for 32-bit ports by the external device. With BS16# asserted all transfers are performed via D0-D15 pins. Also, with BS16# asserted, the 80386 automatically performs data transfers larger than 16 bits or misaligned 16 bits transfers in multiple cycles as needed. Note that 16-bit memory or I/O devices must be connected on D0-D15 pins.

Asserting BS16# only affects the 80386 when BE2# and/or BE3# are asserted during the cycle. Assertion of BS16# does not affect the 80386 if data transfer is only performed via D0-D15. On the other hand, the 80386 is affected by assertion of the BS16# pin, depending in which byte enable pins are asserted during the current bus cycle. For example, asserting BS16# during "upper half only" reads causes the 80386 to read data on the D0-D15 pins and ignores data on the D16-D31. Data that would have been read from D16-D31 (as indicated by BE2# and BE3#) will instead be read from D0-D15.

A 32-bit-wide memory can be interfaced to the 80386 by utilizing its BS16#, BE0#-BE3#, and A2-A31 pins. Each 32-bit memory word starts at a byte address that is a multiple of 4. BS16# is connected to HIGH (negated) for all bus cycles for 32-bit transfers. A2-A31 and BE0#-BE3# are used for addressing the memory.

For 16-bit memories, each 16-bit memory word starts at an address which is a multiple of 2. The address is decoded to assert BS16# only during bus cycles for 16-bit transfers.

A2-A31 can be used to address 16-bit memory also. A1 and two-byte enable signals are also required.

To obtain A1 and two-byte enables for 16-bit transfers, BE0#-BE3# should be decoded as in Table 4.11.

Figure 4.16 shows a block diagram interfacing 16- and 32-bit memories to 80386.

Finally, if an operand is not aligned such as a 32-bit doubleword operand beginning at an address not divisible by 4, then multiple bus cycles are required for data transfer.

4.2.11 80386 Read and Write Cycles

The 80386 performs data transfer during bus cycles (also called read or write cycles).

Two choices of address timing are dynamically selectable. These are nonpipelined and pipelined. One of these timing choices is selectable on a cycle-by-cycle basis with the Next Address (NA#) input.

After a bus idle state, the 80386 always uses nonpipelined address timing. However, the NA# may be asserted by an external device to select pipelined address timing, for the next cycle is made available before the present bus cycle is terminated by the 80386 by asserting READY#.

TABLE 4.11 Generating A1, BHE#, and BLE# for Addressing 16-Bit Devices

80386 signals				16-bit bus signals			
BE3#	BE2#	BE1#	BE0#	A1	BHE#	BLE# (A0)	Comments
H*	H*	H*	H*	x	x	x	x — no active bytes
H	H	H	L	L	H	L	
H	H	L	H	L	L	H	
H	H	L	L	L	L	L	
H	L	H	H	H	H	L	
H*	L*	H*	L*	x	x	x	x — not contiguous bytes
H	L	L	H	L	L	H	
H	L	L	L	L	L	L	
L	H	H	H	H	L	H	
L*	H*	H*	L*	x	x	x	x — not contiguous bytes
L*	H*	L*	H*	x	x	x	x — not contiguous bytes
L*	H*	L*	L*	x	x	x	x — not contiguous bytes
L	L	H	H	H	L	L	
L*	L*	H*	L*	x	x	x	x — not contiguous bytes
L	L	L	H	L	L	H	
L	L	L	L	L	L	L	

Note: BLE# asserted when D0-D7 of 16-bit bus is active; BHE# asserted when D8-D15 of 16-bit bus is active; A1 low for all even words; A1 high for all odd words.

Key: x = don't care
H = high voltage level
L = low voltage level
* = a nonoccurring pattern of Byte Enables; either none are asserted, or the pattern has Byte Enables asserted for noncontiguous bytes

In general, the 80386 samples NA# input during each bus cycle to select the desired address timing for the next bus cycle.

Physical data bus width (16- or 32-bit) is selected by the 80386 by sampling the BS16# (bus size 16) input pin near the end of the bus cycle. Assertion of BS16# indicates a 16-bit data bus, while negation of BS16# means a 32-bit data bus.

A read or write cycle is terminated by the 80386 on a low READY# (assertion) from the external device. Until the READY# is asserted, the 80386 inserts wait states to permit adjustment

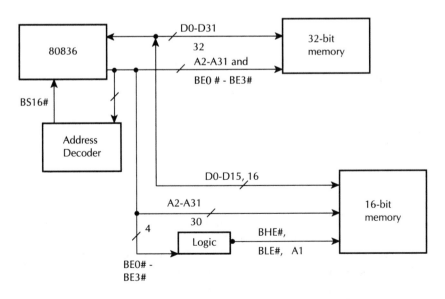

FIGURE 4.16 Interfacing 80386 16- and 32-bit memories.

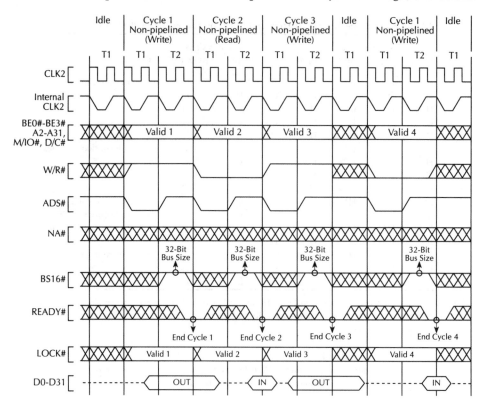

FIGURE 4.17   Bus cycles with nonpipelined address (zero wait states).

for the speed of any external device when a read cycle is terminated, and the 80386 latches the information present at its data pins. When a write cycle is acknowledged, the 80386 write data remain valid throughout phase one of the next bus state, to provide write data hold time.

To illustrate the concept of 80386 bus cycle timing, a mixture of read and write cycles with nonpipelined address timing is shown in Figure 4.17.

This diagram shows the fastest possible cycles with nonpipelined address timing having two bus states (T1 and T2) per bus cycle. In phase one T1, the address signals and bus control signals are valid and the 80386 activates ADS# low to indicate their availability.

During read cycle, the 80386 tristates its data signals to permit driving by the external device being addressed. During write cycle, the 80386 places data on the data bus starting in phase two of T1 until phase one of the bus state following cycle acknowledgment.

4.2.12 80386 Modes

The 80386 can be operated in real, protected, or virtual 8086 mode. These modes are described below.

4.2.12.a 80386 Real Mode

Upon reset or power-up, the 80386 operates in real mode. In real mode, the 80386 can access all the 8086 registers along with the 80386 32-bit registers. The memory addressing, memory size, and interrupts of 80386 in this mode are the same as those of the 80286 in real mode.

The 80386 can execute all the instructions in real mode. The main purpose of real mode is to initialize the 80386 for protected mode operation.

In real mode, the 80386 can directly address up to one megabyte of memory. The address lines A2-A19, BE0#-BE3# are used by the 80386 in this mode. Paging is not provided in real

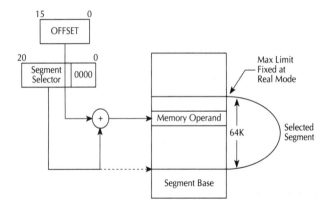

FIGURE 4.18   Real address mode addressing.

mode. Therefore, linear addresses are identical to physical addresses. The 20-bit physical address is formed by adding the shifted (four times to the left) segment registers to an offset as shown in Figure 4.18.

All segments in real mode are exactly 64K bytes wide. Segments can be overlapped in this mode. There are two memory areas which are reserved in real mode for system initialization and interrupt pointer table. Addresses 00000H thru 002FFH are reserved for the interrupt pointer table, while addresses FFFFFF0H thru FFFFFFFFH are reserved for system initialization. Many of the exceptions listed in Table 4.6 are not applicable to real mode. Exceptions 10, 11, 12, and 14 will never occur in this mode. Also, other exceptions have minor variations as follows:

Function	Interrupt number	Related instructions	Return address location
Interrupt table Limit too small	8	INT vector is not within table limit	Before instruction
Segment overrun exception	13	Word memory reference with offset = FFFFH or an attempt to execute an instruction past the end of a segment	Before instruction

4.2.12.b Protected Mode

The total 80386 capabilities are available when the 80386 operates in protected mode. This mode increases the linear space to four gigabytes (2^{32} bytes) and permits the execution of virtual memory programs of 64 tetra-bytes (2^{46} bytes). Also, in protected mode, the 80386 can run all existing 8086 and 80286 programs with on-chip memory management and protection features. The protected mode includes new instructions to support multitasking operating systems. The main difference between protected mode and real mode from a programmer's viewpoint is the increased memory space and a differing addressing mechanism. Similar to real mode, protected mode also includes two elements (16-bit selector for determining a segment's base address and a 32-bit offset or effective address) to obtain a 32-bit linear address. This 32-bit linear address is either used as the 32-bit physical address or, if paging is enabled, the paging mechanism translates this 32-bit linear address to a 32-bit physical address. Figure 4.19 shows the protected mode addressing mechanism. The selector is used to specify an index into a table defined by the operating system. The table includes the 32-bit base address of a given segment. The physical address is obtained by summing the base address obtained from the table with the offset.

With the paging mechanism enabled, the 80386 provides an additional memory management mechanism. The paging feature manages large 80386 segments.

The paging mechanism translates the protected linear addresses from the segmentation unit into physical addresses. Figure 4.20 shows this translation scheme.

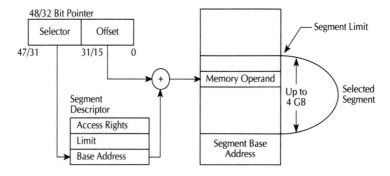

FIGURE 4.19   Protected mode addressing.

Let us now discuss 80386 segmentation, protection, and paging features.

Segmentation provides both memory management and protection. All information about the segments is stored in an 8-byte data structure called a descriptor. All the descriptors are stored in tables identified by the 80386 hardware. There are three types of tables holding 80386 descriptors: global descriptor table (GDT), local descriptor table (LDT), and interrupt descriptor table (IDT). These tables are memory arrays of variable lengths. Their sizes can vary from 8 bytes to 64K bytes. Each table can store up to 8192 8-byte descriptors. The upper 13 bits of a selector are used as an index into the descriptor table. The tables have associated registers which store a 32-bit linear base address and a 16-bit limit for each table. Each table has a set of registers, namely, GDTR (32-bit), LDTR (16-bit), and IDTR (32-bit), associated with it. The 80386 instructions LGDT, LLDT, and LIDT are used to load the base and 16-bit limit of the global, local, and interrupt descriptor tables into the appropriate registers. The SGDT, SLDT, and SIDT instructions store the base and limit values.

The GDT contains descriptors which are available to all the tasks in the system. In general, the GDT contains code and data segments used by the operating system, task state segments, and descriptors for LDTs in a system.

LDTs store descriptors for a given task. Each task has a separate LDT, while the GDT contains descriptors for segments which are common to all tasks.

The IDT contains the descriptors which point to the location of up to 256 interrupt service routines. Every interrupt used by a system must have an entry into the IDT. The IDT entries are referenced via INT instructions, external interrupt vectors, and exceptions.

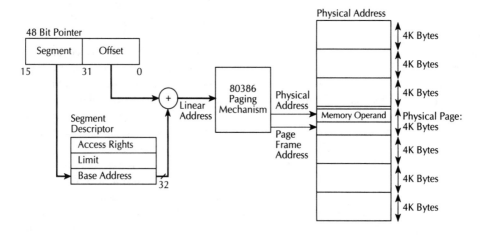

FIGURE 4.20   Paging and segmentation.

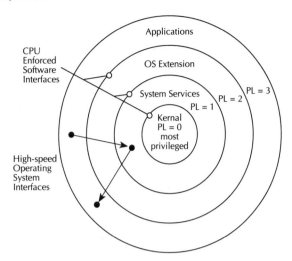

FIGURE 4.21 Four-level hierarchical protection.

The object to which the selector points is called a descriptor. Descriptors are eight bytes wide containing attributes about a given segment. These attributes contain the 32-bit base linear address of the segment, segment length, protection level, read/write/execute privileges, the default operand size (16 or 32 bits), and segment type.

In order to provide operating system compatibility between the 80286 and 80386, the 80386 supports all of the 80286 segment descriptors. The only differences between the 80286 and 80386 formats are that the values of the type fields and the limit and base address fields have been expanded for the 80386.

The 80286 system segment descriptors contain a 24-bit base address and 16-bit limit, while the 80386 system segment descriptors have a 32-bit base address, a 20-bit limit field, and a granularity bit. Note that the segment length is page granular if the granularity bit is one; otherwise, the segment length is byte granular.

By supporting 80286 segments the 80386 is able to execute 80286 application programs on an 80386 operating system. This is possible because the 80386 automatically can differentiate between the 80286-type and 80386-type descriptors. In particular, if the upper word of a descriptor is zero, then that descriptor is an 80286-type descriptor.

The only other differences between the 80286 and 80386 descriptors are the interpretation of the word count field of call gates and the B bit. The word count field specifies the number of 16-bit quantities to copy for 80286 call gates and 32-bit quantities for 80386 call gates. The B bit controls the size of pushes when using a call gate. If B = 0, then pushes are 16 bits, while pushes are 32 bits for B = 1.

The 80386 provides four protection levels for supporting a multitasking operating system to isolate and protect user programs from each other and the operating system. The privilege level controls the use of privileged instructions, I/O instructions, and access to segments and segment descriptors. The 80386 includes the protection as part of its memory management unit. The 80386 also provides an additional type of protection when paging is enabled.

The four-level hierarchical privilege system is shown in Figure 4.21. It is an extension of the user/ supervisor privilege mode used by minicomputers. Note that the user/supervisor mode is supported by the 80386 paging mechanism. The Privilege Levels (PL) are numbered 0 thru 3. Level 0 is the most privileged level.

The 80386 provides the following rules of privilege to control access to both data and procedures between levels of a task:

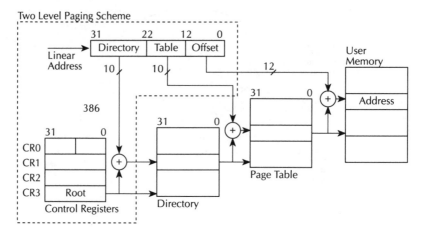

FIGURE 4.22   Paging mechanism.

- Data stored in a segment with a privilege level x can be accessed only by code executing at a privilege level at least as privileged as x.
- A code segment/procedure with privilege level x can only be called by a task executing at the same or a higher privilege level than x.

The 80386 supports task gates (protected indirect calls) to provide a secure method of privilege transfers within a task.

The 80386 also supports a rapid task switch operation via hardware. It saves the entire state of the machine (all of the registers, address space, and a link to the previous task), loads a new execution state, performs protection checks, and commences execution in the new task in approximately 17 microseconds.

Paging is another type of memory management for virtual memory multitasking operating systems. The main difference between paging and segmentation is that paging divides programs/data into several equal-sized pages, while segmentation divides programs/data into several variable-sized segments.

There are three elements associated with the 80386 paging mechanism. These are page directory, page tables, and the page itself (page frame). The paging mechanism does not have memory fragmentation since all pages have the same size of 4K bytes. Figure 4.22 shows the 80386 paging mechanism.

There are four 32-bit control registers (CR0-CR3) associated with the paging mechanism. CR2 is the page fault linear address register and contains the 32-bit linear address which caused the last page fault detected.

CR3 is the page directory physical base address register and contains the physical starting address of the page directory. The lower 12 bits of CR3 are always zero to ensure that the page directory is always page aligned. CR1 is reserved for future Intel processors. CR0 contains 6 defined bits for control and status purposes. The low-order 16 bits of CR0 are known as the machine status word and include special control bits such as the enable bit and the protection enable bit.

The page directory is 4K bytes long and permits up to 1024 page directory entries. Each page directory entry contains the address of the next level of tables, page tables, and information about the page table. The upper 10 bits of the linear address (A22-A31) are used as an index to select the correct page directory entry.

Each page table is 4K bytes long and holds up to 1024 page table entries. Page table entries contain the starting address of the page frame and statistical information about the page such as whether the page can be read or written in supervisor or user mode. Address bits A12-A21 are used as index to select one of the 1024 page table entries. The 20 upper-bit page frame address is concatenated with the lower 12 bits of the linear address to form the physical address. Page tables can be shared between tasks and swapped to disks.

The lower 12 bits of the page table entries and page directory entries contain statistical information about pages and page tables, respectively. As an example, the P (present) bit indicates whether a page directory or page table entry can be used in address translation. If P = 1, the entry can be used in address translation, and if P = 0, the entry cannot be used for translation and all other 31 bits are available for use by the software. These 31 bits can be used to indicate where on a disk the page is located.

The 80386 provides a set of protection attributes for paging systems. The paging mechanism provides two levels of protection: user and supervisor. The user level corresponds to level 3 of the segmentation-based protection and the supervisor level combines all of the other protection levels (0, 1, 2). Programs executing at level 0, 1, or 2 bypass the page protection, although segmentation-based protection is still enforced by hardware.

The 80386 takes care of the page address translation process, relieving the burden from an operating system in a demand-paged system. The operating system is responsible for setting up the initial page tables and the handling of any page faults. The operating system initializes the tables by loading CR3 with the address of the page directory and allocates space for the page directory and the page tables. The operating system also implements a swapping policy and handles all of the page faults.

4.2.12.c Virtual 8086 Mode

The virtual 8086 mode permits the execution of 8086 applications while taking full advantage of the 80386 protection mechanism. In particular, the 80386 permits concurrent execution of 8086 operating systems and applications, an 80386 operating system, and both 80286 and 80386 applications. For example, in a multiuser 80386-based microcomputer, one person can run an MD-DOS spreadsheet, another person can use MS-DOS, and a third person can run multiple UNIX utilities and applications.

One of the main differences between 80386 real and protected modes is how the segment selectors are interpreted. In virtual 8086 mode, the segment registers are used in the same way as the real mode. The contents of the segment register are shifted 4 times to the left and added to the offset to obtain the linear address.

The paging hardware permits the simultaneous execution of several virtual mode tasks and provides protection.

The paging hardware allows the 20-bit linear address produced by a virtual mode program to be divided up into 256 pages. Each one of the pages can be located anywhere within the maximum 4-gigabyte physical address space of the 80386.

The paging hardware also permits sharing of the 8086 operating system code by several 8086 applications. All virtual mode programs execute at privilege level 3. Therefore, virtual mode programs are subject to all of the protection checks defined in protected mode. This is different from real mode which executes programs in level 0.

4.3 80386 System Design

In this section, the 80386 is interfaced to typical memory and I/O chips. As mentioned in the last section the 80386 address and data lines are not multiplexed. There is a total of thirty address pins (A2-A31) on one chip. A0 and A1 are decoded internally to generate four byte enable outputs, BE0#, BE1#, BE2# and BE3#. In real mode, the 80386 utilizes 20-bit addresses and A2 through A19 address pins are active and the address pins A20 through A31 are used in real mode at reset, high for CS-bsed accesses, low for others, and always low after CS changes. In the protected mode, on the other hand, all address pins A2 through A31 are active.

In both modes, A0 and A1 are decoded internally. In all modes, the 80386 outputs on the byte enable pins to activate appropriate portions of the data bus to transfer byte (8-bit), word (16-bit), and double-word (32-bit) data as follows:

Byte Enable Pins	Data Bus
BE0#	D0-D7
BE1#	D8-D15
BE2#	D16-D23
BE3#	D24-D31

The 80386 supports dynamic bus sizing. This feature connects the 80386 with 32-bit or 16-bit data buses for memory or I/O. The 80386 32-bit data bus can be dynamically switched to a 16-bit bus by activating the BS16# input from high to low by a memory or I/O device. In this case, all data transfers are performed via D0-D15 pins. 32-bit transfers take place as two consecutive 16-bit transfers over data pins D0 through D15. On the other hand, the 32-bit memory or I/O device can activate the BS16# pin HIGH to transfer data over D0-D31 pins.

For reading a byte, the 80386 makes one of BE0#-BE3# active. For a word read (aligned:even address), the 80386 makes two byte enable outputs (BE0#-BE3#) active. On the other hand, for a 32-bit aligned read, the 80386 activates all byte enable outputs BE0#-BE3#.

The 80386 duplicates data on some 16-bit write operations in order to enhance performance of the data bus. This is illustrated in the table below:

Transfer Size	A1	A0	$\overline{BE3}$	$\overline{BE2}$	$\overline{BE1}$	$\overline{BE0}$	D_{31}-D_{34}	D_{23}-D_{16}	D_{15}-D_{8}	D_{7}-D_{0}
Byte	0	0	1	1	1	0				X
Byte	0	1	1	1	0	1			X	
Byte	1	0	1	0	1	1		X		DD
Byte	1	1	0	1	1	1	X		DD	
Word	0	0	1	1	0	0			X	X
Word	0	1	1	0	0	1		X	X	
Word	1	0	0	0	1	1	X	X	DD	DD
Dword	0	0	1	0	0	0		X	X	X
Dword	0	1	0	0	0	1	X	X	X	
Dword	0	0	0	0	0	0	X	X	X	X

Note: DD = Data Duplication.

In the above table, the 80386 duplicates data for three cases. For example, in case of write cycle for byte when data is written on D_{23}-D_{16} pins, the same data is duplicated on D_7-D_0 pins. On the other hand, writing a byte on D_{31}-D_{24} pins, the same data is written on D_8-D_{15} pins. Finally when 16-bit data is written on D_{31}-D_{16} pins, the same data is duplicated by the 80386 on D_{15}-D_0 pins.

The 80386 address pins A1 and A0 specify the four address of a four byte (32-bit) word consider the following:

D_{31}	D_{23}	D_{16},D_{15}	D_8,D_7	D_0	
					Data Pins

The contents of the memory addresses which include 0, 4, 8, … with A1A0 = 00_2 are transferred by D_0-D_7. Similarly, the contents of addresses which include 1, 5, 9, …, with A1A0 = 01_2 are transferred over D_{15}-D_8. On the other hand, the contents of memory addresses 2, 6, 10, … with A1A0 = 10_2 are transferred over D_{16}-D_{23} while contents of addresses 3, 7, 11, … with

A1A0 = 11_2 are transferred over D_{24}-D_{31}. Note that A1A0 is encoded from $\overline{BE3}$ - $\overline{BE0}$ pins. The following figure depicts this:

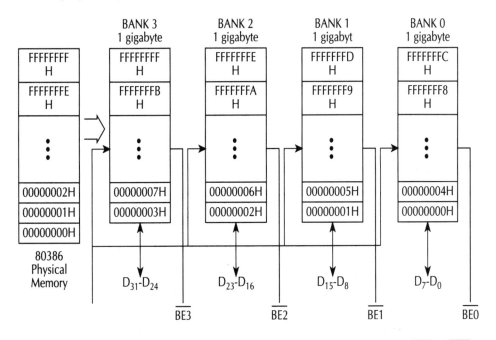

In each bank, a byte can be accessed by enabling one of the byte enables, $\overline{BE0}$ - $\overline{BE3}$. For example, in response to execution of a byte-MOVE instruction such as MOV (00000006H), BL, the 80386 outputs low on $\overline{BE2}$ and high on $\overline{BE0}$, $\overline{BE1}$ and $\overline{BE3}$ and the contents of address BL is written to address 00000006H. On the other hand, when the 80386 executes a MOVE instruction such as MOV (00000004H), AX, the 80386 drives $\overline{BE0}$ and $\overline{BE1}$ to low. The contents of 00000004H and 00000005H are transferred from AL and AH via D_0-D_7 and D_8-D_{15} respectively. For 32-bit transfer, the 80386 executing a MOVE instruction from an aligned address such as MOV (00000004H), EAX, the 80386 drives all bus enable pins ($\overline{BE0}$-$\overline{BE3}$) to low and written to the contents of four bytes (00000004H through 00000007H) from EAX. Byte (8-bit), aligned word (16-bit), and aligned double-word(32-bit) are transferred by the 80386 in a single bus cycle.

The 80386 performs misaligned transfers in two bus cycles. For example, the 80386 executing a misaligned word MOVE instruction such as MOV(00000003H), AX drives $\overline{BE3}$ to low in the first bus cycle and writes the contents of 00000003H from bank 3 into AL in the first bus cycle. The 80386 then drives $\overline{BE0}$ to low in the second bus cycle and writes the contents of 00000004H from bank 0 into AH in the second bus cycle. This transfer takes two bus cycles.

A 32-bit misaligned transfer such as MOV (00000002H, EAX, on the other hand, takes two bus cycles. In the first bus cycle, the 80386 enables $\overline{BE2}$ and $\overline{BE3}$, and writes the contents of address 00000002H and 00000003H from banks 2 and 3 respectively into low 16-bits of EAX. In the second cycle, the 80386 enables $\overline{BE0}$ and $\overline{BE1}$ to low and then writes the contents of address 00000004H and 00000005H into the upper 1 bits of EAX.

4.3.1 80386 Memory Interface

Figure 4.23 shows the basic memory interface block diagram.

The bus control logic provides the control signals for READY#, NA#, address latches, data buffers, and memory chips. The bus control logic activates READY# to LOW to signal the completion of the 80386 bus cycle and also outputs low on NA# (Next Address Request) to

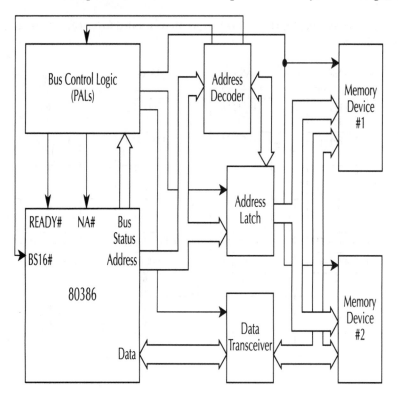

FIGURE 4.23   Basic memory interface block diagram.

activate address pipelining. The address decoder generates chip select signals and BS16# signal based on the address outputs of the 80386. These conventional memory interfacing concepts can be used to interface the 80386 to ROMs, EPROMs, and static RAMs.

The bus control logic decodes the 80386 status outputs (W/R#, M/IO#, and D/C#) and sends a command signal for the type of bus cycle according to Table 4.9 as per the following:

1. Memory read command (MRDC#) signal is generated from memory data read (M/IO#=1, D/C#=1, W/R#=0) or memory code read (M/IO#=1, D/C#=0, W/R#=0) cycle. MRDC commands the selected memory device to output data.
2. I/O read cycle (M/IO#0, D/C#=1, W/R#=0) generates the I/O read command (IORC#) output. IORC# commands the selected I/O device to output data.
3. Memory write command (MWTC#) is obtained from memory write cycle (M/IO#=1, D/C#=1, W/R#=1). MWTC# tells the selected memory device to receive data on the data bus.
4. I/O write command (IOWC#) is obtained from the IO write cycle (M/IO#=0, D/C#=1, W/R#=1). IOWC# tells the selected I/O device to receive data on the data bus.
5. INTA# is generated from Interrupt Acknowledge cycle (M/IO#=0, D/C#=0, W/R#=0). INTA# is sent back to the Interrupt controller such as the 80259A. A second INTA# cycle tells a device such as the 8257A to place the interrupt vector on the bus. PALs can be used to design bus control logic.

Address latches maintain the 80386 address for the duration of the bus cycle and are required to pipeline address since the address for the next bus cycle appears on the address lines before the end of the current bus cycle. Latches such as 74 × 373 can be used. Note that the 80386 can be run without address pipelining to eliminate the need for address latching but the system will run inefficiently.

Standard 8-bit transceivers (74×245) provide isolation and additional drive currents for the 80386 data bus. Transceivers are necessary to prevent the contention on the data bus that occurs if some devices are slow to remove data from the data bus after a read cycle. Transceivers can be omitted if the data float time of the device is short enough and the load on the 80386 data pins meets device specifications. Figure 4.24 shows memory interface of a typical 80386-based microcomputer.

Three 74AS373 octal latches are used to maintain the address for the duration of the bus cycle. The 74AS373 latch Enable (LE) input is controlled by the 80386 ALE output. The 74AS373 output Enable (OE#) is always active.

Two 74F139 decoders (2 to 4) are used. The top 74F139 decoder has A31 and $\overline{M/IO}$ as inputs. When the 80386 executes a memory-oriented instruction, it outputs HIGH on the M/IO pin and if A31 is one,. then the Y2 output of the decoder goes to LOW. The Y2 output of the decoder is connected to the chip select 1 wait state (CS1 WS) of the PAL 16R8B. The CS1 WS is also latched by the 74AS373 and is used as the chip select for the 27128 EPROMs. Note that the 80386 reset vector is included in high memory (FFFFFFF0H). Therefore, the memory maps for EPROMS must include these high memory addresses. Also, when M/IO#=1 and A31=0, the upper decoder generates chip select 0 wait state (CS0 WS). This signal is latched and used as chip select for 651628, 2KX8 static RAMs.

The address pins, byte enable signals (BE3 - BE0), chip select outputs of the decoder and W/R are latched by the 74AS373 at the falling edge of ALE . Note that 80386 address lines A1 through A14 are used to address each EPROM.

The PAL 1 and PAL2 devices are used to design the bus control logic. PAL1 follows the 80386 bus cycles and generates the overall bus cycle timing. PAL2 generates most of the control signals required for memory interface. The PAL equations are given in Intel 80386 hardware reference manual. These equations can be assembled by a PAL assembler program. The assembled code to the PAL is then applied by using a standard PROM programmer with a special PAL enhancement. The write control logic for the SRAMs are implemented by using 74F32 quad OR gates. The memory write command (MWTC) is applied to one input of each one of the OR gates. The other input to each one of the OR gates is the appropriate byte enable (BE3 - BE0) signal. For example, when both MWTC and BE3 are low, then WE input of the left most static RAM is enabled and data can then be written to that RAM.

Four 74AS245 transceivers are used to provide buffering of the data lines. Then DEN output generated by PAL2 is used to enable the transceivers and the latched 80386 W/R is used to control the direction of data transfer.

Two 27128's are used to provide $16K \times 16$ of EPROM. These chips are connected to the 80386 D0-D15 pins. The 27128 requires 14-bit address, A0–A13. MRDC (Memory Read Command) output generated by PAL2 is used to enable the 27128 OE Chip select signal (CS1 WS) from the input of PAL1 is latched and then converted to CS inputs of the 27128's. CS1 WS is also connected to the 80386 BS16 input to indicate to the 80386 that all read cycles are 16-bit.

The 27128 memory map can be determined as follows:

Memory Map for 27128-1

```
A31 A30 . . . . . . A15   A14 . . . . . . . A1 A0
1                                              1
```

don't cares	all zeros
assume 1's	to ones

= FFFF8001H, FFFF8003H, ... , FFFFFFFFH.

Memory Map for 27128-2

FFFF8000H, FFFF8002H, ... , FFFFFFFEH

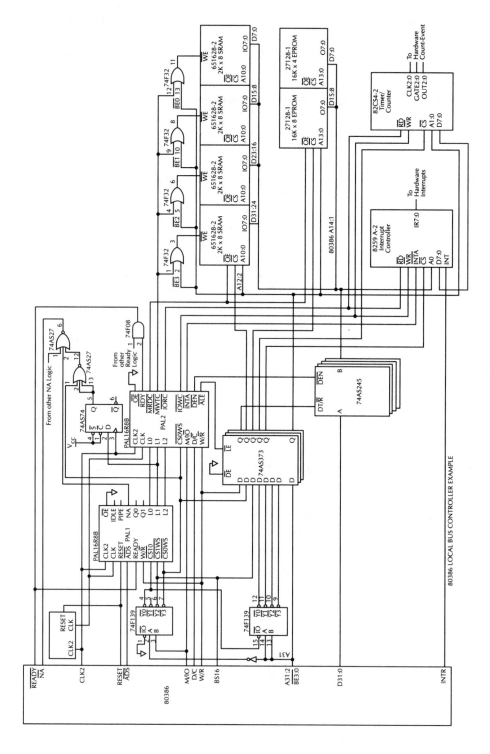

FIGURE 4.24   Basic memory interface for an 80386-based microcomputer.

Four 2K × 8 static RAM chips are used to provide a total of 8K bytes of RAM storage. Each RAM has 11-bit address input connected to the 80386 A12-A2 pins. When data is read from SRAMs, the MRDC signals activate all four RAMs and the 32-bit data from the selected location is placed on the 80386 D_{31}-D_0 pins. The 80386 then reads a byte, word, or double word from the bus.

During write to SRAMs, the 80386 may not output a 32-bit word. The 80386 in this case outputs the appropriate data size (byte, word or double word). It also sends appropriate BE0-BE3 signals to indicate which portion of the data bus carries the data. MWTC and WE signals ensure that data are written into the appropriate SRAM. Note that SRAMs are enabled by the CS0WS signal. This means that data read and write bus cycle do not require any wait states.

Let us determine the memory map for RAMs:

```
BANK 0   A31 A30 . . . A13   A12 .  . . A2 A1 A0
         0   _____/  _____/  0 0
             don't cares       all zeros to ones
             assume 1's
           = 7FFFE000H, 7FFFE004H, . . . . . ,
BANK 1     7FFFE001H, 7FFFE005H, . . . . .
BANK 2     7FFFE002H, 7FFFE006H, . . . . .
BANK 3     7FFFE003H, 7FFFE007H, . . . . .
```

When an 80386-based microcomputer system uses a large main memory of several megabytes, it is usually designed with high capacity, slow-speed dynamic RAMs and EPROMs. Although access times of DRAMs and EPROMs can be as fast as 60ns and 120ns respectively, the 80386 microcomputer system with these chips is expensive and too slow for running with zero wait states. The details of the 80386's interfacing to DRAMs can be found in Intel manuals and is beyond the scope of this chapter.

An 80386 microprocessor at a speed of 25MHz would require DRAMs at 40ns access time for zero wait-state. This is why, for large memory, wait states are introduced in all bus cycles while accessing memory. These degrade the overall performance of the 80386-based micro-computer.

A cache memory subsystem can be implemented in the 80386-based microcomputer to improve overall system performance while utilizing inexpensive, slow-speed DRAMs in main memory.

A cache memory subsystem includes a small amount of fast static RAM and a large amount of DRAM. The cache memory subsystem contains a fast SRAM between the 80386 and the slower main memory (DRAMs) along with a cache controller. The cache controller such as Intel 82385 includes the logic to implement the cache memory. Cache sizes are either 32K bytes or 64K bytes.

One of the two most widely used cache organizations, namely the direct-mapped cache or two-way set associative cache, is utilized in an 80386 system. Details of the 80386 cache implementation are provided in the 80386 hardware reference manual.

4.3.2 80386 I/O

The 80386 can use either a standard I/O or a memory-mapped I/O technique.

The address decoding required to generate chip selects for devices using standard I/O is often simpler than that required for memory-mapped devices. But, memory-mapped I/O offers more flexibility in protection than standard I/O does.

The 80386 can operate with 8-, 16-, and 32-bit peripherals.

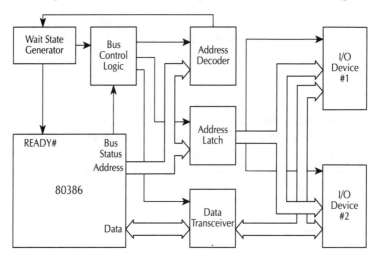

FIGURE 4.25　Basic I/O interface block diagram.

Eight-bit I/O devices can be connected to any of the four 8-bit sections of the data bus. Figure 4.25 shows 80386 I/O interface block diagram.

In the figure, the 80386's 32-bit data bus is multiplexed onto an 8-bit bus.

For efficient operation, 32-bit I/O devices should be assigned to addresses that are even multiples of four. If I/O devices are located on adjacent word boundaries, address decoding must generate the BS16# signal so that the 80386 performs a 16-bit bus cycle.

The block diagram is very similar to the 80386 memory interface block diagram. The purpose of various blocks in the figure has already been explained earlier in this section.

For standard I/O, the 80386 includes three types of I/O instructions. These are direct, indirect, and string I/O instructions which include the following:

Direct

For 8-bit:	IN AL, PORT
	OUT PORT, AL
For 16-bit:	IN AX, PORT
	OUT PORT, AX
For 32-bit:	IN EAX, PORT
	OUT PORT, EAX

Indirect

For 8-bit:	IN AL, DX
	OUT DX, AL
For 16-bit:	IN AX, DX
	OUT DX, AX
For 32-bit:	IN EAX, DX
	OUT DX, EAX

String

For 8-bit:	INSB,	(ES:DI) ← ((DX))
		DI ← DI $\pm$ 1
	OUTSB	((DX)) ← (ES:SI)
		SI ← SI $\pm$ 1
For 16-bit:	INSW,	(ES:DI) ← ((DX))
		(DI) ← DI $\pm$ 2
	OUTSW	(ES:SI) ← ((DX))

$$(SI) \leftarrow SI \pm 2$$

For 32-bit: INSD, $(ES:EDI) \leftarrow ((DX))$
$$EDI \leftarrow EDI \pm 4$$
OUTSD, $((DX)) \leftarrow (ES:ESI)$
$$ESI \leftarrow ESI \pm 4$$

The 82C55A Programmable Peripheral Interface (PPI) can be interfaced with the 80386 for obtaining parallel ports for either standard or memory-mapped I/O. The pin diagram and the features provided by the 82C55A are same as the 8255 which were described in Chapter 3. The 82C55A will be interfaced to the 80386 for simple I/O operation.

In summary, the 82C55A contains three 8-bit parallel ports namely Port A, Port B, and Port C. These ports can be configured as input or output ports by writing 1 or 0 respectively in the corresponding bits in the control register. The **control register** bits can be defined as follows:

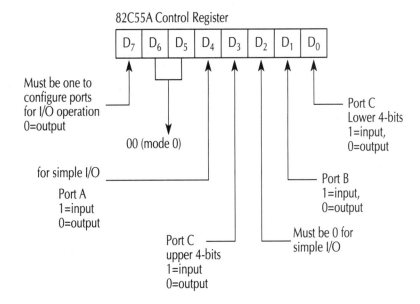

For example, outputting 89H to the control register will configure Ports A and B as output ports and Port C as an input port.

The 82C55A ports and control register are selected by the two-bit register select inputs (A0 and A1 inputs of the 82C55A) as follows:

A1	A0	
0	0	Port A
0	1	Port B
1	0	Port C
1	1	Control Register

Figure 4.26 shows 82C55A-80386 interface using standard I/O.
Now let us determine the I/O port addresses in the 82C55A-4.

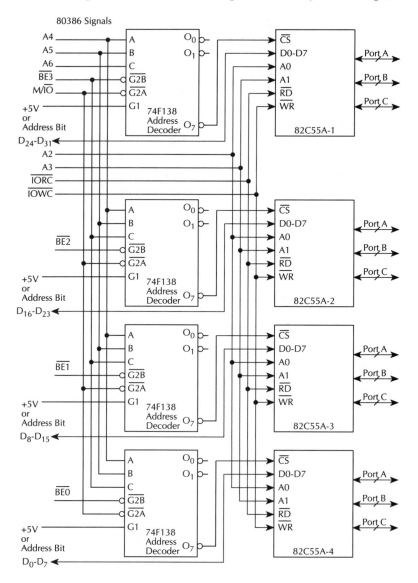

FIGURE 4.26   80386-82C55A interface using standard I/O.

A7	A6	A5	A4	A3	A2	A1	A0
X	1	1	1	0	0	0	0

↑

don't care   Port A

assume 1   =F0H

Note that, since $\overline{BE0}$ is used to select the 82C55A-4, 80386 A_1 and A_0 pins will be zeros. Therefore in the 82C55A-4 the port addresses can be obtained as follows:

8-bit Address	Port Name
F0H	Port A
F4H	Port B
F8H	Port C
FCH	Control Register

The seven other unused outputs of each address decoder can be used to enable three 82C55As in each category. Similarly, the port addresses for the other 82C55A's can be obtained as follows:

	8-bit Address	Port Name
	F1H	Port A
82C55A-3	F5H	Port B
	F9H	Port C
	FDH	Control Register
	F2H	Port A
82C55A-2	F6H	Port B
	FAH	Port C
	FEH	Control Register
	F3H	Port A
82C55A-1	F7H	Port B
	FBH	Port C
	FFH	Control Register

Note that $\overline{IORC}$ and $\overline{IOWC}$ are the outputs of PAL2 of the 80386 memory interface of Figure 4.24.

Using memory-mapped I/O, a read-mode 80386 can be interfaced to 82C55's in Figure 4.26 by connecting MRDC and MWTC outputs of PAL2 of Figure 4.24 to RD and WR of the 82C55A respectively. The G2A input line of the 74F138 can be connected through an inverter to unused 80386 address line such as A19, G1 can be connected to the 80386 M/IO pin so that when M/IO =1 and A19=1, I/O ports are selected. The unused outputs 0_0 through 0_6 of the 74F138 decodes can be connected to other peripherals. The schematic of Figure 4.26 can be changed to memory-mapped I/O by connecting each of the 82C55's as above.

Example 4.14

Write an 80386 assembly language program to input two 32-bit data via Ports 5274H and 5270H, logically OR them together and then output the 32-bit result to Port 5270H.

Solution

```
MOV   DX, 5274H   ;   Initialize DX
IN    EAX, DX     ;   Input first 32-bit data
MOV   EBX, EAX    ;   Save data
MOV   DX, 5270H   ;   Initialize DX
IN    EAX, DX     ;   Input second 32-bit data
```

```
OR    EAX, EBX    ;  Or the two data
OUT   DX, EAX     ;  Output to Port 5270H
HLT
```

Example 4.15

Prior execution of the 80386 INSD instruction, assume the following data:

```
EDI = 00000032H
Contents of address [ES:EDI] = 37124426H
DX = 0020H, DF = 0
Contnents of PORT 0020H = EEEF2752H
```

What are the contents of EDI, [ES:EDI], DX, DF, and PORT 0020H after execution of the INSD?

Solution

```
DF=0,  EDI=00000036H
Contents of [ES:EDI]=EEEF2752H
DX=0020H
Contents of PORT 0020H=EEEF2752H
```

Example 4.16

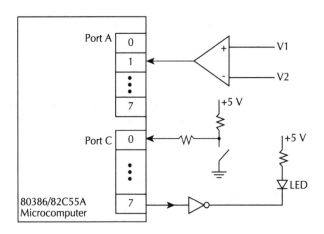

For the above, write an 80386 assembly language program such that if $V_1 > V_2$, input the switch via bit 0 of 82C55A Port C and if the switch is closed, turn the LED ON and if the switch is open, turn the LED OFF.

Solution

```
MOV   AL, 90H      ; Configure Port A as input, Port C
                   ; lower
```

```
        OUT    CNTRL, AL    ;  4-bit  as  input  and  upper  4-bit  as
                            ;  output
CHK:    IN     AL, PORTA    ;  Input Port A
        AND    AL, 02H      ;  check V₁ > V₂
        CMP    AL, 02H      ;  check if V₁ > V₂
        JNZ    CHK          ;  Loop if V₁ < V₂
        IN     AL, PORT C   ;  Input switch
        BTC    AL, 0        ;  complement bit 0 of Port C
        RCR  AL, 2
        OUT    PORT C, AL   ;  output to Port C
        HLT
```

Example 4.17

Repeat Example 4.16 using interrupt I/O. Assume that the comparator output is connected to 80386 INTR pin. Use INT255.

Solution

The block diagram for servicing interrupt is as follows:

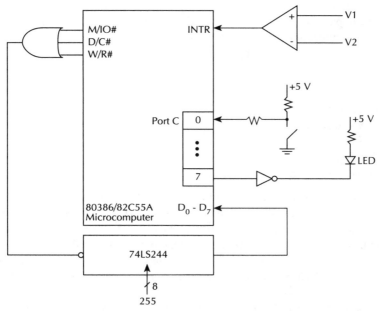

Note that M/IO#=0, D/C#=0, and W/R#=0 indicate Interrupt Acknowledge

Assume all segment register are initialized. Write service routine at IP=3000H and CS=0200H.

Main Program

```
        MOV    SP, VALUE    ;  Initialize SP
        STI                 ;  Enable Interrupts
        MOV    AL, 90H      ;  Configure
```

```
            OUT    CNTRL, AL    ;  Port C lower 4-bit as input,
                                ;  upper 4-bit as output
WAIT        JMP    WAIT         ;  wait for interrupt
            HLT
Service  Routine
            ORG    30000200H
BEGIN:      IN     AL, PORT C   ;  Input switch
            BTC    AL, 0        ;  complement switch
            RCR    AL, 2
            OUT    PORT C, AL   ;  output to LED
            IRET
            ORG    000003FCH
            DD     02003000H    ;  Initialize IP = 0200H and
                                ;  CS = 3000H
```

4.4 Coprocessor Interface

The performance of an 80386 system is enhanced by addition of a numeric coprocessor such as the 80287 or the 80387. The numeric coprocessor executes numeric instructions in parallel with the 80386. The 80386 automatically passes on these instructions to the coprocessor as it encounters them through I/O ports 800000F8H, 800000FAH, and 800000FCH.

The 80287 performs 16-bit transfers with a 16-bit data bus while the 80387 performs 32-bit transfers with a 32-bit data bus. The 80387 provides additional enhancements to the 80287 and includes full compatibility with the IEEE Floating-Point Standard draft 10.

The 80387 can utilize seven types of data using any 80386 addressing mode. These 80387 data types are listed in the following:

1. 32-bit Sort real
2. 64-bit Long real
3. 80-bit Temporary real
4. 16-bit Word Integer
5. 32-bit Short Integer
6. 64-bit Long Integer
7. 80-bit Packed decimal

The 32-bit short real (single precision) format contains the 23-bit significand (bits 0-22), 8-bit biased exponent (bits 23-30), and the sign bit (bit 31). The 64-bit Long real (double precision) includes the 52-bit significand (bits 0-51), the 11-bit biased exponent (bits 52-62), and the sign bit (bit 63). The 80-bit Temporary real (extended precision) contains the 64-bit significand (bits 0-63), the 15-bit biased exponent (bits 64-78), and the sign bit (bit 79).

The 80-bit packed decimal contains 18 decimal digits (bits 0-71) and the sign bit (bit 79). Bits 72 thru 78 are undefined.

The 16-bit word integer includes the 15-bit integer (bits 0-14) along with the sign bit (bit 15) while the 32-bit short integer contains the 31-bit integer (bits 0-30) and the sign bit (bit 31). Finally, the 64-bit Long integer contains the 63-bit integer (bits 0-62) along with the sign bit (bit 63).

4.4.1 Coprocessor Hardware Concepts

The 80386 samples its ERROR# input during initialization to determine which coprocessor is present. The 80287 and 80387 require different interfaces and therefore, slightly different protocols.

For coprocessor cycles, the address pin A_{31} is HIGH. I/O addresses 800000F8H and 800000FCH are used for transferring data to and from a coprocessor. The 80386 automatically generates these addresses for coprocessor instructions. Also, the 80386 provides chip-select signals for the coprocessor using $A_{31}=1$ and M/IO#=0.

The 80386 utilizes three input signals for controlling data transfers with coprocessors. These are BUSY#, Coprocessor Request (PEREQ), and ERROR#.

The BUSY# indicates that the coprocessor is presently executing an instruction and therefore cannot accept another instruction. A new instruction, therefore, cannot overrun the execution of the current coprocessor instruction.

PEREQ indicates that the coprocessor needs to perform data transfer to or from memory. Since the coprocessor is never a bus master, all I/O transfers are performed by the 80386.

If a coprocessor math instruction results in an error that is not masked by the coprocessor's control register, the ERROR# signal is asserted. The coprocessor data sheets describe these errors and explain how to mask them by writing programs.

The interfacing characteristics of the 80386 with the 80387 Numeric coprocessor is described in the following.

The 80387 runs at an internal clock frequency of up to 16MHz. It is designed to run either fully synchronously or pseudo synchronously with the 80386. In the pseudo synchronous mode, the interface logic of the 80387 runs with the 80386 clock signal while internal logic runs with a different clock signal.

Figure 4.27 shows a typical 80386-80387 interface schematic. The main interfaces are described below:

- The 80386 and 80387 BUSY#, ERROR#, and PEREQ signals are directly connected.
- The 82384 RESET output is connected to the 80386 RESET and 80387 RESET IN signals.
- The 80387 chip select inputs, NPS1# and NPS2 are respectively connected to the 80387 M/IO# and A31. With M/IO#=0 and A31=1, the 80386 selects the coprocessor.
- The command input (CMD#) of the 80387 distinguishes data from commands. The 80386 A2 is connected directly to the 80387 CMD#. The 80386 outputs address 800000FCH when reading or writing data while address 800000F8H is used when writing a command or reading status.
- The 80387 uses READY# and ADS# pins to track bus activity and determine when W/R#, NPS1#, NPS2 and status Enable (STEN) can be sampled. STEN is an 80387 chip select and is pulled HIGH. If multiple 80387's are used by one 80386, STEN can be used to activate one 80387 at a time.
- Ready out (READY0#) is an optional signal that can be used to generate the wait states required by a coprocessor. When the 80386 encounters a coprocessor instruction, it automatically generates one or more I/O cycles to addresses 800000F8H and 800000FCH. The 80386 performs all bus cycles to memory and transfers data to and from the 80387. All 80387 transfers are 32-bit wide. For 16-bit memories, the 80386 automatically performs the necessary conversion before transferring data with the 80387.

Read cycles (transfer from 80387 to the 80386) require at least one wait state while write cycles to the 80386 require no wait states. This requirement is automatically reflected in the state of the 80387 READY0# output which can be used to generate the required wait states.

Upon hardware reset, the 80386 before executing the first instruction, checks its ERROR# input to determine the type of coprocessor present. If the 80386 samples ERROR# low, it assumes that an 80387 is present. On the other hand, a high ERROR# input indicates either an 80287 is present or no coprocessor is used.

The coprocessor type can be determined via software by checking the ET bit in the machine status word. The 80387 is present if ET=1.

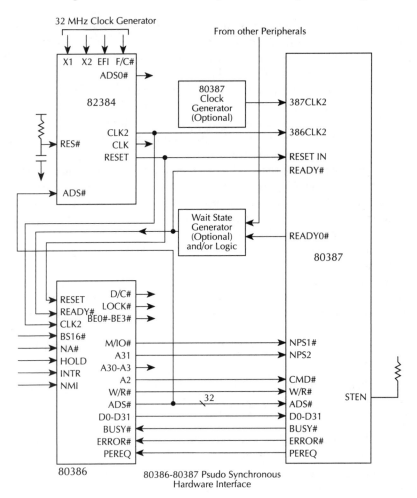

FIGURE 4.27   80386 system with 80387 coprocessor.

If the 80386 finds that an 80387 is present, an 80386 program must be written to execute the FNINT instruction to reset 80387 ERROR# output before any coprocessor transaction occurs.

When 80386 finds that either an 80287 is present or a coprocessor is not used, an 80386 program must be executed to determine the presence of an 80287 in order to set an internal status. An example of an 80386 instruction sequence to check the presence of an 80287 is given below:

```
;Initialization Routine to Detect An 80287 Numeric Processor
FND_287:    FNINIT              ;   INITIALIZE NUMERIC
                                ;   PROCESSOR
```

```
           FSTSW    AX           ;   RETRIEVE 80287 STATUS WORD
           OR       AL,AL        ;   TEST LOW-BYTE 80287
                                 ;   EXCEPTION FLAGS. IF ALL
                                 ;   ZERO, THEN 80287 PRESENT AND
                                 ;   PROPERLY INITIALIZED.
                                 ;   IF NOT ALL ZERO, THEN 80287
           JZ       GOT_287      ;   ABSENT. BRANCH IF
                                 ;   80287 PRESENT
           SMSW     AX           ;   NO NUMERIC PROCESSOR
           OR       AX,04H       ;   SET EM BIT IN MACHINE
           LMSW     AX           ;   STATUS WORD TO ENABLE
                                 ;   SOFTWARE EMULATION OF 80287
           JMP      CONTINUE
GOT_287:   SMSW     AX           ;   NUMERIC PROCESSOR PRESENT
           OR       AX,02H       ;   SET MP BIT IN MACHINE
           LMSW     AX           ;   STATUS WORD TO PERMIT
                                 ;   NORMAL 80287 OPERATION
CONTINUE:                        ;   AND OFF WE GO ....
```

In the above instruction sequence, the 80386 assumes that the 80287 is present. Therefore it executes an FNINT instruction. Next, the 80386 reads the 80287 status word. If an 80287 is present, the lower 8 bits of this word (the exception flags) are all zeros. If an 80287 is not present, these data lines are floating. If a pull-up resistor is connected to at least one of these lines, the absence of the 80287 is confirmed by at least one high bit in the lower eight bits of the status word. The routine then sets or resets the Emulate Coprocessor (EM) bit of the CR0 register of the 80386, depending on whether or not the 80287 is present.

4.4.2 Coprocessor Registers

Figure 4.28 shows the 80387 registers.

The 80387 contains three types of registers:

1. Eight 80-bit floating point stack registers
2. One 16-bit status word, one 16-bit control word and one 16-bit tag register
3. Four 32-bit error-pointer registers namely FIP, FCS, FOO, and FOS which store the instruction and memory operand causing an exception

The 80386 floating-point instructions consider the eight 80-bit registers as a stack of accumulators. The current stack top is called ST or ST(0). After a PUSH, ST(0) becomes ST(1) and all other ST(i)'s are incremented by one. After a POP, ST(1) before the POP becomes the new ST(0), and all ST(i)'s are decremented by one. A 3-bit field name TOP in the status word defines the register number ST(0) of the current stack top. If TOP=000_2, a push will decrement TOP to 111_2 and store a new value into ST(7). ST(7) is the present stack top. On the other hand, if TOP=111_2, a POP will read a value from ST(7) and then increment top to 000_2 so that ST(0) becomes the new top of stack

The 16-bit tag register includes eight 2-bit fields—one for each physical floating point register. This two-bit field indicates whether the corresponding physical floating-point register (0-7)

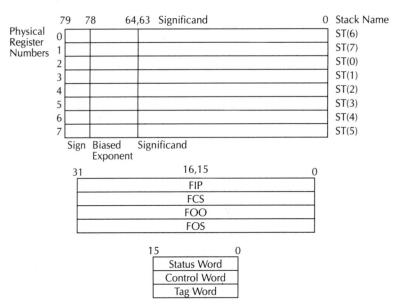

FIGURE 4.28 80387 registers.

rather than ST-relative names contains a valid, zero or special floating-point number(infinity, NAN) or is empty. The tag fields are used to indicate overflow or underflow of ST(i)'s. A stack overflow occurs if a PUSH decrements TOP to point to an ST(i) that is not empty. A stack underflow occurs if an empty ST(i) is popped. Stack underflow or overflow generates an invalid operation exception.

The control word register can be written by the program to control the 80387 operation. The control word contains exception mask bits for situations such as invalid operation, zero divide, overflow, and underflow. If the 80387 encounters an exception, the mask bit for that exception is checked to determine if the exception should be sent to a program error handler if the mask bit=0 or handled by an 80387 error handler if the mask bit=1.

The control word also contains information such as precision control and rounding control. The two-bit precision control field specifies that the results of arithmetic operations are rounded to 24-bit (short real), 53 bits (long real), or 64-bits (temporary real) precision before storing into the destination. All other operations use temporary real precision or a precision specified in the instruction.

On the other hand, the two bit rounding control field specifies that the results of floating-point operations are approximated or rounded to nearest toward minus infinity, plus infinity, or truncate toward zero.

The 80387 updates the bits of the status word register that can be checked by the program to check for special conditions. Bits 11, 12, and 13 specify the TOP field containing the three bit address of the physical register stack (0-7) corresponding to ST(0). The other bits in the status register include information such as indication of exception due to floating-point operations and floating-point condition code bits.

The four 32-bit 80387 Error Pointer register (FIP, FCS, FOO, and FOS) hold pointers to the last 80387 instruction executed along with its data. FCS stores the selector while FIP stores the offset. FOS, on the other hand, stores selector while FOO contains the offset.

4.4.3 80387 Instructions

The 80387 floating point instructions can be classified into six groups. These are data transfer, arithmetic, comparison, transcendental, loading, and control instructions. Some of these instructions are described in the following:

i) **Data Transfer Instructions**

-FLD mem or ST(n)	mem can be short, long or temporary real. This instruction loads the specified real number onto the new stack top. **Example:** Consider FLD mem32. If prior to execution of this FLD, mem32 = $4.1572 * 10^3$ with TOP = 4 then after FLD, mem32 = $4.1572 * 10^3$, ST = $4.1572 * 10^3$ with TOP = 3
-FXCH ST(n)	This instruction exchanges the contents of ST(n) with the stack top. **Example:** Consider FXCH ST(5). If prior to execution of this instruction, ST = $5.71252 * 10^{-87}$, ST(5) = $-2.78164 * 10^{82}$ with TOP = 5 then after this FXCH, ST = $-2.78164 * 10^{82}$, ST(5) = $5.71252 * 10^{-87}$ with TOP = 5.
-FST mem or ST(n)	The stack top is stored into specified memory or ST(n)

ii) **Arithmetic Instructions**

-FABS	This instruction converts the stack top to its absolute value. **Example:** If prior to execution of FABS, ST = $-5.372 * 10^{-300}$ then after FABS, ST = $5.372 * 10^{-300}$
-FADD FADD ST(n)	Performs real addition. • FADD; ST ← ST + ST(1) • FADD ST(n); ST ← ST + ST(n) **Example:** If prior to execution of FADD, ST=$3.17300 * 10^4$, ST(1) = $2.167 * 10^3$ then after FADD, ST=$3.38970 * 10^4$, ST(1) = $2.167 * 10^3$
-FDIV FDIV: FDIVP ST,ST(n): FDIV ST(n): FIDIV const:	Performs real division. ST ← ST(1)/ST ST ← ST(n)/ST. Stack is popped. ST ← ST(n)/ST ST ← ST/const. Integer divide **Example:** If prior execution of FDIV, ST = $4.240 * 10^{-4}$, ST(1) = $8.480 * 10^{+3}$ then after FDIV, ST = $5.000 * 10^{-8}$, ST(1) = $8.480 * 10^{+3}$

-FIMUL const:	Integer multiply. ST ← ST * const
-FMUL:	Performs multiplication of real numbers.
FMUL:	ST ← ST(1) * ST
FMUL ST(n):	ST ← ST * ST(n)
	Example:

If prior to execution of FMUL,
$ST = 2.500 * 10^{-10}$, $ST(1) = 2.500 * 10^{+7}$
then after FMUL,
$ST = 6.250 * 10^{-3}$, $ST(1) = 2.500 * 10^7$

-FSUB	Performs real subtraction
FSUB:	ST ← ST(1) − ST
FSUB ST(n):	ST ← ST − ST(n)
	Example:

If prior execution of FSUB,
$ST = 7.140 * 10^{10}$, $ST(1) = 8.217 * 10^{11}$
then after FSUB,
$ST = 7.503 * 10^{11}$

-FSQRT	ST ← sqrt(ST)
	Example

If prior execution of FSQRT,
$ST = 2.25 * 10^6$
then after FSQRT,
$ST = 1.5 * 10^3$

iii) Comparison Instructions

-FCOM ST(n)   Compares ST(n) numerically with top of stack. The condition codes C3, C2, and C0 (bits 14,10, and 8 in the status word) are set according to the following:

	C3	C2	C0
ST > ST(n)	0	0	0
ST < ST(n)	0	0	1
ST = ST(n)	1	0	0

iv) Transcendental Instructions

-F2XMI:	$ST \leftarrow 2^{ST} - 1$
	The range of the values of ST prior execution of F2XMI is −0.5 to +0.5.
-FCOS	This instruction computes the cosine of ST. ST is assumed to contain real numbers in radians. The result replaces the original ST.
-FSIN	This instruction computes the sine of ST. ST is assumed to contain real number in radians. The result replaces ST.
-FSINCOS	ST ← COS(ST)
	ST(1) ← SIN(ST)
-FYL2X	$ST = ST(1) * LOG_2(ST)$
	The stack is popped and the new stack top is replaced with the result.
-FYL2XP1	$ST = ST(1) * LOG_2(ST + 1.0)$
	The stack is popped and the new stack top is replace with the result of this computation.

v) Loading Instructions

-FLD n   The stack is pushed. The constant value, n is loaded into the new top of stack as for the following:

FLD1	Load 1.0
FLDL2E	Load Log_2e
FLDL2T	Load 2^{10}
FLDLG2	Load 10^2
FLDLN2	Load e^2
FLDPI	Load π
FLDZ	Load 0.0

vi) **Control Instructions**

-FLDCW mem16   Loads the control word with the contents of mem 16.

Example:

If prior execution of FLDCW mem16,

CW = 2547H, mem16 = 25F2H

then after FLDCW mem16,

CW = 25F2H, mem16 = 25F2H.

Example 4.18

Write an 80386 assembly language program using 80387 floating point instructions to compute $Y = \sqrt{X^2 - Z^2}$.

Solution

```
FLD     Z       ;   Load Z onto stack top
FLD     ST(1)   ;   Load a copy of Z to stack
FMUL            ;   Compute Z²
FLD     X       ;   Load X onto stack
FLD     ST(0)   ;   Load a copy of X to stack
FMUL            ;   compute X²
FSUB            ;   Compute X² - Z²
FSQRT           ;   Y=SQRT(X² - Z²)
HLT
```

Example 4.19

Write an 80386 assembly language program using 80387 floating point instructions to compute the volume of a sphere = $(4/3) * \pi * r^3$ where r is the radius of the sphere.

Solution

```
FLD     r           ;   Load r to stack top
FLD     ST(1)       ;   Make a copy of r on stack
FLD     ST(2)       ;   Make another copy of r
FMUL                ;   compute r²
FMUL                ;   compute r³
MOV     const, 4    ;   Load multiplier
FIMUL   const       ;   Compute 4 * r³
FMOV    const, 3    ;   Load divisor
FIDIV               ;   Compute (4/3) * r³
FLDPI               ;   Load π
FMUL                ;   Compute volume
HLT
```

QUESTIONS AND PROBLEMS

4.1 Write an 80186 assembly program to multiply a 16-bit signed number in BX by 00F3H. Assume that the result is 16 bits wide.

4.2 Identify the peripheral functional blocks integrated into the 80186.

4.3 What is the relationship between internal and external clocks of the 80186?

4.4 Identify the basic differences between 8086 and 80186.

4.5 Identify the main differences between the 80186 and 80286.

4.6 How much physical and virtual memory can the 80286 address?

4.7 What is the difference between the 80286 real address mode and PVAM? Explain how these two modes can be switched back and forth.

4.8 Explain how the 80286 determines where in memory the global descriptor table and the present local descriptor table are located.

4.9 Discuss briefly the 80286 protection mechanism.

4.10 Explain the meaning of 80286 call gates.

4.11 What is the purpose of 80286 CAP, $\overline{\text{COD}/\text{INTA}}$ pins?

4.12 Identify the 80286 pins used for interfacing it to a coprocessor.

4.13 Discuss the issues associated with isolating a user program from a supervisor program and then describe the 80286's protection features for protection. Assume that no task switching is involved. Also, assume that the supervisor program will perform all I/O operations and be present in the virtual memory space.

4.14 Compare the features of the 80386 with those of the 80286 from the following point of view: registers, clock rate, number of pins, number of instructions, modes of operation, memory management, and protection mechanism.

4.15 What are the basic differences between the 80386 real, protected, and virtual 8086 modes?

4.16 Assume the following register contents:

$$[\text{EBX}] = 0000\ 2000\text{H}$$
$$[\text{ECX}] = 0500\ 0000\text{H}$$
$$[\text{EDX}] = 5000\ 5000\text{H}$$

prior to execution of each of the 80386 instructions listed below. Determine the effective address after execution of each instruction and identify the addressing modes of both source and destination:

  i) **MOV [EBX * 2] [ECX], EDX**
  ii) **MOV [EBX * 4] [ECX + 20H], EDX**

4.17 Determine the effect of each of the following 80386 instructions:

    i) **MOVZX ECX, BX**
        assume [ECX] = F1250024H
                [BX] = F130H

prior to execution of the MOVZX instruction.

    ii) **SHLD CX, BX, 0**
        if [CX] = 0025H, [BX] = 0F27H

prior to execution of the SHLD instruction.

4.18 Given the following data prior to the execution of each of the following instructions, determine the contents of the specified registers and carry flag after execution:

    i) **Prior to execution: [EAX]=0527AF12H**
                            [EBX]=0071FFD1H
                            CF=1
      After execution of BTC EAX, EBX
    ii) **Prior execution: [BX]=F216H**
                      CF=1
      After execution of BTR BX, 20H
    iii) **Prior to execution: [EDX]=F214721FH**
                        CF=1
      After execution of: BTS EDX, 35H

4.19 Write an 80386 assembly language program to divide a signed 64-bit number in EBX:EAX by a 16-bit signed number in AX. Store the 32-bit quotient and remainder in memory locations.

4.20 Write an 80386 assembly program to compute Xi^2/N where N = 100 and Xi's are signed 32-bit numbers. Assume that Xi^2 can be stored as a 32-bit signed number without overflow.

4.21 Write an 80386 assembly program to input 100 32-bit string data via a port addressed by DX. The program will then store the data in memory locations addressed by [DS] and [ESI].

4.22 Find an 80386 compare instruction with the appropriate addressing mode to replace the following 8086 instruction sequence:

        MOV CL, 3
        SAL ESI, CL
        MOV EAX, [ESI]

4.23 Write an 80386 assembly language program to compute the following: X = Y + Z − 20EEH where X, Y, and Z are 64-bit variables. The upper 32 bits of Y and Z are stored at 3000H and 3008H each followed by the lower 32 bits. Store the upper 32-bit of the 64-bit result at location at 4000H followed by the lower 32 bits.

4.24 Assume that the 80386 registers BL, CX, and EDX contain a signed byte, a signed word, and a signed 32-bit word in two's complement form respectively. Write an 80386 assembly language program that will generate the signed result of operation BL + CX − EDX → BL.

4.25 Write an 80386 assembly language program to compute:

$$ECX=6*BX+EDX/EAX$$

where all numbers are signed numbers. Discard the remainder of EDX/EAX.

4.26   Write an 80386 assembly language program that checks all 32 bits of EAX from right to left (bit 0 to bit 31). The program will set the corresponding bit in EBX based on the location of the first one bit found in EAX. If the entire 32 bits of EAX are zero, clear EBX to all zeros.

4.27   Discuss how the following situation will be handled by the 80386: The 80386 executing an instruction causes both a general protection fault (interrupt 13) and coprocessor segment overrun (interrupt 9).

4.28   How does the 80386 generate the 32-bit physical address from A2-A31 and BE0#-BE3#?

4.29   What are the purposes of NA#, D/C#, BS16#, and ERROR# pins?

4.30   For 16- and 32-bit transfers, what is the logic level of the BS16# pin?

4.31   Discuss briefly the 80386 segmentation unit, paging unit, and protection.

4.32   How many bits are required for the address in real and protected modes?

4.33   Discuss the basic differences between the 80286 and 80386 descriptors.

4.34   What is the four-level hierarchical protection in protected mode?

4.35   Discuss briefly the 80386 virtual mode.

4.36   Consider the following pins and signals:

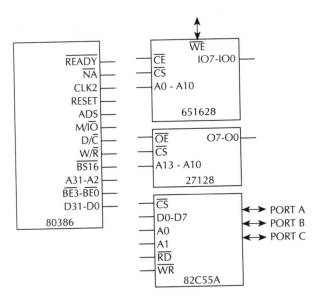

Draw a simplified diagram connecting the above chips to include the following:

-The 651628 to contain addresses
XXXXXXX1H, XXXXXXX5H, XXXXXXX9H . . . . .

. . . .

-The 27128 to include addresses
XXXXXXX0H, XXXXXXX4H, XXXXXXX8H, ...

. . . .

-The 82C55A to contain port addresses
Port A = X1H
Port B = X5H
Port C = X9H

Assume all don't cares to be zeros. Use PALs, latches, and other components as required. Determine memory and I/O maps.

4.37

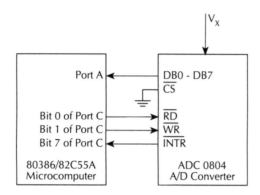

The ADC 0804 can be started by sending a HIGH to LOW transition at the $\overline{WR}$ pin. The conversion is completed when INTR is LOW. The 8-bit data at the DB0-DB7 pins can be read by the microcomputer by sending a LOW at the $\overline{RD}$ pin.

Write an 80386 assembly language program to start the 0804 by the microcomputer and input the converted data via Port A.

4.38 Repeat problem 4.37 by using:

  i) **NMI**
  ii) **INTR**

4.39 Assume 80386/80387 system. Write an 80386 assembly language program using floating point instructions to compute the mass of a sphere with radius r and density of $Log_e{}^2$. Mass is equal to density times volume.

4.40 Assume 80386/80387 system. Write an 80386 assembly language program to compute the logarithm with a base other than 2; or log base 'n' of X (Log_nX).

5

MOTOROLA MC68000

This chapter describes the details of the Motorola 68000 microprocessor. The basic architecture, addressing modes, instruction set, and interfacing features of the Motorola 68000 are included.

5.1 Introduction

The MC68000 is Motorola's first 16-bit microprocessor. All 68000 address and data registers are 32 bits wide and its ALU is 16 bits wide. The 68000 is designed using HMOS technology. The 68000 requires a single 5V supply. The processor can be operated from a maximum internal clock frequency of 25 MHz. The 68000 is available in several frequencies. These include 6 MHz, 8 MHz, 10 MHz, 12.5 MHz, 16.67 MHz, and 25 MHz. The 68000 does not have on-chip clock circuitry and, therefore, requires a crystal oscillator or external clock generator/driver circuit to generate the clock.

The 68000 has several different versions. These include 68008, 68010, and 68012. The 68000 and 68010 are packaged in a 64-pin DIP (Dual In-line Package) with all pins assigned or in a 68-pin quad pack or Pin Grid Array (PGA) with some unused pins. The 68000 is also packaged in 68-terminal chip carrier. The 68008 is packaged in a 48-pin dual in-line package while the 68012 is packaged in 84-pin grid array. The 68008 provides the basic 68000 capabilities with inexpensive packaging. It has an 8-bit data bus which facilitates interfacing of this chip to inexpensive 8-bit peripheral chips.

The 68010 provides hardware-based virtual memory support and efficient looping instructions. Like the 68000, it has a 16-bit data bus and a 24-bit address bus.

The 68012 includes all the 68010 features with a 31-bit address bus.

The clock frequencies of the 68008, 68010, and 68012 are the same as the 68000.

The following table summarizes the basic differences among the 68000 family members:

	68000	68008	68010	68012
Data size (bits)	16	8	16	16
Address bus size (bits)	24	20	24	31
Virtual memory	No	No	Yes	Yes
Directly addressable	16 Mbytes	1 Mbyte	16 Mybtes	2 gigabytes memory

In order to implement operating systems and protection features, the 68000 can be operated in two modes. These are supervisor and user modes. The supervisor mode is also called the operating system mode. In this mode, the 68000 can execute all instructions. The 68000

277

TABLE 5.1   68000 User and Supervisor Modes

	Supervisor mode	User mode
Enter mode by	Recognition of a trap, reset, or interrupt	Clearing status bit S
System stack pointer	Supervisor stack pointer	User stack pointer registers A0–A6
Other stack pointers	User stack pointer and registers A0–A6	

Instructions available	All including
	    STOP
	    RESET
	    MOVE to/from SR
	    ANDI to/from SR
	    ORI to/from SR
	    EORI to SR
	    MOVE USP to (An)
	    MOVE to USP
	    RTE
		All except those listed under supervisor
		    mode

Function code pin FC2	1	0

operates in one of these modes based on the S-bit of the Status register. When the S-bit is one, the 68000 operates in the supervisor mode. On the other hand, the 68000 operates in the user mode when S = 0.

Table 5.1 lists the basic differences between 68000 user and supervisor modes.

From Table 5.1 it can be seen that the 68000 executing a program in supervisor mode can enter the user mode by modifying the S-bit of the Status register to zero via an instruction. Instructions such as MOVE to SR, ORI to SR, EORI to SR can be used to accomplish this. On the other hand, the 68000 executing a program in user mode can enter the supervisor mode only via recognition of a trap, reset, or interrupt. Note that upon hardware reset, the 68000 operates in the supervisor mode and can execute all instructions. An attempt to execute privileged instructions (instructions that can only be executed in supervisor mode) in user mode will automatically generate an internal interrupt (trap) by the 68000.

The logical level in the 68000 Function Code pin (FC2) indicates to the external devices whether the 68000 is currently operating in user or supervisor mode. The 68000 has three function code pins (FC2, FC1, and FC0) which indicate to the external devices whether the 68000 is accessing supervisor program/data, user program/data, or performing an interrupt acknowledge cycle. These three pins are used for memory protection and for enabling an external chip such as the 74LS244 to provide an interrupt address vector.

The 68000 can operate on six different data types. These are bit, 4-bit BCD digit, 8-bit packed BCD, 8-bit byte, 16-bit word, and 32-bit long word.

The 68000 provides 56 basic instructions. With 14 addressing modes, 56 instructions, and 6 data types, the 68000 includes more than 1000 op codes. The fastest instruction is MOVE reg, reg and is executed in 500 ns at 8-MHz clock. The slowest instruction is 32-bit by 16-bit divide, which is executed in 21.25 μs at 8-MHz clock.

Like the 8-bit Motorola microprocessors such as Motorola 6800 and 6809, the 68000 supports memory-mapped I/O. Thus, the 68000 instruction set does not include any IN or OUT instructions. This allows the full instruction set to be used by I/O.

The 68000 is a general-purpose register-based microprocessor since any data register can be used as an accumulator or as a scratch pad register. Even though the 68000 program counter is 32 bits wide, only the low-order 24 bits are used for PC. With 24 bits as address, the 68000 can directly address 16 megabytes (2^{24}) of memory.

5.2 68000 Programming Model

The register architecture of the 68000 is shown in Figure 5.1.

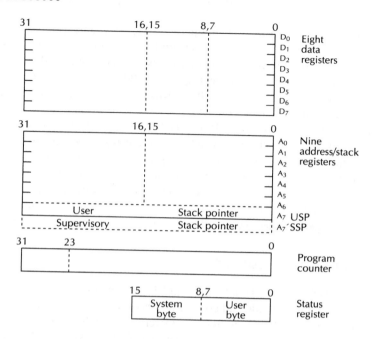

FIGURE 5.1   MC68000 programming model.

The 68000 chip contains eight 32-bit data registers (D0-D7) and nine 32-bit address registers (A0-A7, A7′). The 68000 uses A7 or A7′ as the user or supervisor stack pointer, depending on the mode of operation. Data items such as bytes (8 bits), words (16 bits), long words (32 bits), and BCD numbers (8 bits) are usually stored in the data registers. On the other hand, the address of the operand is usually stored in an address register. Since the address sizes used by 68000 instructions can be either 16 or 24 bits, the address registers can only be used as 16- or 32-bit registers. While using the 32-bit address registers as addresses, the 68000 discards the uppermost eight bits (bits 24 thru 31).

The 68000 status register consists of two bytes. These are a user byte and a system byte (Figure 5.2). The user byte includes the usual condition codes such as C, V, N, Z, and X. The meaning of C, V, N, and Z flags is obvious. However, the X-bit (extend bit) has a special meaning. The 68000 does not have any ADDC or SUBC instructions; rather, it has ADDX or SUBX instructions. For arithmetic operations, the carry flag C and the extend flag X are affected in an identical manner. This means that one can use ADDX or SUBX to include carries or borrows while adding or subtracting high-order long words (32 bits) in multiprecision additions or subtractions.

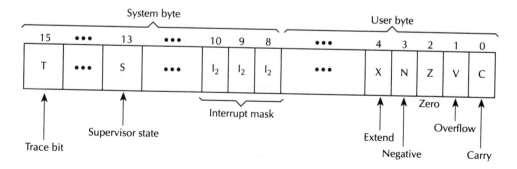

FIGURE 5.2   68000 status register.

The system byte contains a 3-bit interrupt mask (I2, I1, I0), a supervisor flag (S), and a trace flag (T). Interrupt mask is set according to interrupt level recognized. A3-A1 pins of the 68000 provide the current interrupt level being serviced. The 68000 interrupt pins IPL2 IPL1 IPL0 = 000 indicate nonmaskable interrupt while IPL2 IPL1 IPL0 = 111 means no interrupt. The other combinations of IPL2, IPL1, and IPL0 provide the 68000 maskable interrupt levels. It should be pointed out that signals on IPL2, IPL1, and IPL0 pins are inverted and then reflected on I2, I1, and I0, respectively. When the S-bit in SR is 1, the 68000 operates in the supervisor mode. As mentioned before, when S = 0, the 68000 assumes user mode of operation. When the TF (trace flag) is set to one, the 68000 generates an internal interrupt (trap) after execution of each instruction. A debugging routine can be written at the interrupt address vector to display registers and/or memory after execution of each instruction. This provides single-stepping facility. The 68000 can be placed in the single-step mode by setting the TF bit in SR to one by executing a logical privileged instruction such as ORI # $8000, SR in the supervisor mode.

5.3 68000 Addressing Structure

The 68000 supports 8-bit bytes, 16-bit words, and 32-bit long words as shown in Figure 5.3. Byte addressing contains both odd and even addresses (0, 1, 2, 3, 4, ...); and word and long-word addressing includes only even addresses in increments of 2 (0, 2, 4, 6, ...). As an example of the addressing structure, consider MOVE.L D0, $102050. If [D0] = $12345678, then after this MOVE, [$102050] = $12, [$102051] = $34, [$102052] = $56, and [$102053] = $78.

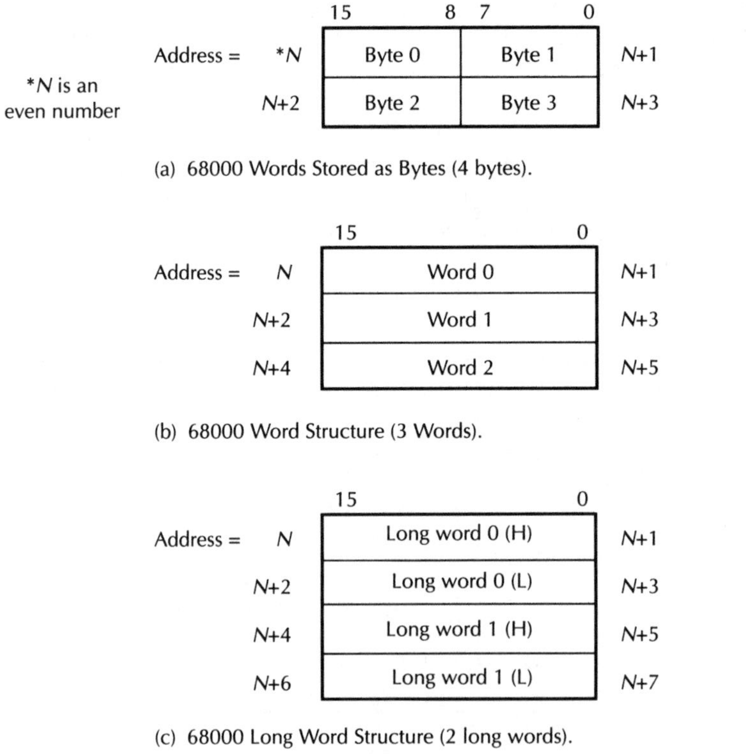

		15	8	7	0	
Address =	*N	Byte 0		Byte 1		N+1
*N is an even number	N+2	Byte 2		Byte 3		N+3

(a) 68000 Words Stored as Bytes (4 bytes).

		15	0	
Address =	N	Word 0		N+1
	N+2	Word 1		N+3
	N+4	Word 2		N+5

(b) 68000 Word Structure (3 Words).

		15	0	
Address =	N	Long word 0 (H)		N+1
	N+2	Long word 0 (L)		N+3
	N+4	Long word 1 (H)		N+5
	N+6	Long word 1 (L)		N+7

(c) 68000 Long Word Structure (2 long words).

For byte addressing (not shown in this figure), each byte can be uniquely addressed with bit 0 as the least significant bit and bit 7 as the most significant bit.

FIGURE 5.3   68000 addressing structure.

TABLE 5.2   68000 Addressing Modes

Mode	Generation	Assembler syntax
1. Register direct addressing		
Data register direct	EA = Dn	Dn
Address register direct	EA = An	An
2. Address register indirect addressing		
Register indirect	EA = (An)	(An)
Postincrement register indirect	EA = (An), An ← An + N	(An)+
Predecrement register indirect	An ← An − N, EA = (An)	−(An)
Register indirect with offset	EA = (An) + d16	d(An)
Indexed register indirect with offset	EA = (An) + (Ri) + d8	d(An, Ri)
3. Absolute data addressing		
Absolute short	EA = (Next word)	xxxx
Absolute long	EA = (Next two words)	xxxxxxxx
4. Program counter relative addressing		
Relative with offset	EA = (PC) + d16	d
Relative with index and offset	EA = (PC) + (Ri) + d8	d(Ri)
5. Immediate data addressing		
Immediate	DATA = Next word(s)	#xxxx
Quick immediate	Inherent data	#xx
6. Implied addressing		
Implied register	EA = SR, USP, SP, PC	

Notes: EA = effective address; An = address register; Dn = data register; Ri = address or data register used as index register; SR = status register; PC = program counter; USP = user stack pointer; SP = active system stack pointer (user or supervisor); d8 = 8-bit offset (displacement); d16 = 16-bit offset (displacement); N = 1 for byte, 2 for words, and 4 for long words; ( ) = contents of; and ← = replaces

5.4   68000 Addressing Modes

Table 5.2 lists the 14 addressing modes of the 68000. The addressing modes are divided into six basic groups. These are register direct, address register indirect, absolute, program counter relative, immediate, and implied.

The 68000 contains three types of instructions: zero-operand, single-operand, and two-operand. The zero-operand instructions have no operands in the operand field. A typical example is CLC (Clear carry) instruction. The single-operand instructions contain the effective address, EA, in the operand field. The EAs of these instructions are calculated by the 68000 using the addressing mode specified for this operand. For two-operand instructions, one of the operands usually contains the EA and the operand is usually a register or memory location. The EAs in these instructions are calculated by the 68000 based on the addressing modes used for the EAs. Some two-operand instructions have an EA in both operands. This means that the operands in these instructions can use two different addressing modes.

As mentioned before, the 68000 address registers do not support byte-sized operands. Therefore, when an address register is used as a source, either the low-order word or the entire long-word operand is used depending on the operation size. On the other hand, when an address register is used as the destination, the entire register is affected regardless of the operation size. If the operation size is a word, and the destination is an address register, then the 68000 performs the 16-bit operation, places the result in the low 16 bits of the address register, and then sign-extends the address register to 32 bits. An example is MOVEA.W #$8050,A5. In this case, the source operand is 16-bit immediate data 8050_{16} which is moved to the low 16 bits of A5, and the result is then sign-extended to 32 bits so that [A5] = $FFFF8050_{16}$. Data registers support data operands of byte, word, and long-word size.

5.4.1 Register Direct Addressing

In this mode, the eight data registers (D0-D7) or seven address registers (A0-A6) contain the data operand. For example, consider ADD $005000, D0. The destination of this instruction is in data register (direct mode).

Now, if [005000] = 0002_{16}, [D0] = 0003_{16}, then after execution of ADD $005000, D0 the contents of D0 = 0002 + 0003 = 0005. Note that in the above instruction, the $ symbol is used to represent hexadecimal numbers by Motorola. Also, note that instructions using address registers are not available for byte operations. In addition, in the 68000, the first operand of a two-operand instruction is the source and the second operand is the destination. Recall that in the 8086, the first operand is the destination while the second operand is the source.

5.4.2 Address Register Indirect Addressing

There are five different types of address register indirect mode. In the register indirect mode, an address register contains the effective address. For example, consider CLR (A1). If [A1] = $003000, then after execution of CLR (A1), the contents of memory location $003000 will be cleared to zero.

The postincrement address register indirect mode increments an address register by 1 for byte, 2 for word, and 4 for long word after it is used. For example, consider CLR.L (A0) + If [A0] = 005000_{16}, then after execution of CLR.L (A0) +, the contents of locations 005000_{16} through 005003_{16} are cleared to zero and [A0] = 005004_{16}. The postincrement mode is typically used with memory arrays stored from LOW to HIGH memory locations. For example, in order to clear 1000_{16} words starting at memory location 003000_{16}, the following instruction sequence can be used:

```
            MOVE.W #$1000,D0   ; Load length of data
                                 into D0
            MOVEA.L#$3000,A0   ; Load starting address
                                 into A0
REPEAT      CLR.W (A0)+        ; Clear a location
                                 pointed to by A0 and
                                 increment A0 by 2
            SUBQ#1,D0          ; Decrement D0 by 1
            BNE REPEAT         ; Branch to REPEAT if Z = 0, else
                               ; go to next instruction
            —
            —
            —
```

Note that in the above, CLR.W (A0)+ automatically points to the next location by incrementing A0 by 2 after clearing a memory location.

The predecrement address register indirect mode, on the other hand, decrements an address register by 1 for byte, 2 for word, and 4 for long word before using a register. For example, consider CLR.W – (A0). If [A0] = 002004_{16}, then during execution of CLR.W – (A0), the content of A0 is first decremented by 2; that is, [A0] = 002002_{16}, and the 16-bit contents of memory location 002002_{16} are then cleared to zero.

The predecrement mode is used with arrays stored from HIGH to LOW memory locations. For example, in order to clear 1000_{16} words starting at memory location 4000_{16} and below, the following instruction sequence can be used:

```
            MOVE.W #$1000,D0   ; Load length of data into D0
            MOVEA.L #$4002, A0 ; Load starting address plus 2
                               ; into A0
```

```
REPEAT  CLR.W - (A0)        ;  Decrement A0 by 2 and clear
                            ;  the memory location addressed
                            ;  by A0
        SUBQ#1, D0          ;  Decrement D0 by one
        BNE REPEAT          ;  If Z = 0, branch to REPEAT;
        -                   ;  otherwise, go to next
        -                   ;  instruction
        -
```

In the above, CLR.W – (A0) first decrements A0 by 2 and then clears the location. Since the starting address is 004000_{16}, A0 must initially be initialized with 004002_{16}.

It should be pointed out that the predecrement and postincrement modes can be combined in a single instruction. A typical example is MOVE.W(A5) +, –(A3).

The two other address register modes provide accessing of the tables by allowing offsets and indexes to be included with an indirect address pointer. The address register indirect with offset mode determines the effective address by adding a 16-bit signed integer to the contents of an address register. For example, consider MOVE.W $10 (A5), D3. If [A5] = 00002000_{16}, $[002010_{16}] = 0014_{16}$, then after execution of MOVE.W $10(A5), D3, register D3 will contain 0014_{16}. A5 is unchanged.

The indexed register indirect with offset determines the effective address by adding an 8-bit signed integer and the contents of a register (data or address register) to the contents of an address (base) register. This mode is usually used when the offset from the base address register needs to be varied during program execution. The size of the index register can be a 16-bit or a 32-bit value.

As an example, consider MOVE.W $10(A4, D3.W), D4. Note that in this instruction A4 is the base register and D3.W is the 16-bit index register (sign extended to 32 bits). This register can be specified as 32 bits by using D3.L in the instruction, and 10_{16} is the 8-bit offset which is sign-extended to 32 bits. If [A4] = 00003000_{16}, [D3] = 0200_{16}, $[003210_{16}] = 0024_{16}$, then the above MOVE instruction will load 0024_{16} into low 16 bits of register D4. A4 and D3 are unchanged.

The address register indirect with offset mode can be used to access one dimensional array such as a table where the offset (maximum 16 bits) can be the starting address of the table (fixed number) and the address register can hold the index number in the table to be accessed. Note that the starting address, plus the index number, provides the address of the element to be accessed in the table. For example, consider MOVE.W $3400 (A5), D1. If A5 contains 04, then this move instruction transfers the contents of 3404 (i.e., the fifth element, 0 being the first element) into low 16 bits of D1. The indexed register indirect with offset, on the other hand, can be used to access two dimensional arrays such as matrices.

5.4.3 Absolute Addressing

In this mode, the effective address is part of the instruction. The 68000 has two absolute addressing modes: absolute short addressing in which a 16-bit address is used (the address is sign-extended to 32 bits before use) and absolute long addressing in which a 24-bit address is used. For example, consider ADD $2000, D2 as an example of absolute short mode. If [$2000] = 0012_{16}, [D2] = 0010_{16}, then after execution of ADD $2000, D2, the address $2000 is sign-extended to 32 bits, whose low 24 bits are used as the address. Register D2 will then contain 0022_{16}. The absolute long addressing is used when the address size is more than 16 bits. For example, MOVE.W $240000, D5 loads the 16-bit contents of location $240000 into low 16 bits of D5. The absolute short mode includes an address ADDR in the range $0 \leq$ ADDR $\leq$ $7FFF or $FF8000 $\leq$ ADDR $\leq$ $FFFFFF. Note that a single instruction may use both short and long absolute modes, depending on whether the source or destination address is less than, equal to, or greater than the 16-bit address. A typical example is MOVE.W $500002, $1000.

5.4.4 Program Counter Relative Addressing

The 68000 has two program counter relative addressing modes: relative with offset, and relative with index and offset. In relative with offset, the effective address is obtained by adding the contents of the current PC with a sign-extended 16-bit displacement. This mode can be used when the displacement needs to be fixed during program execution. Typical branch instructions such as BEQ, BRA, and BLE use relative mode with offset. This mode can also be used by some other instructions. For example, consider ADD*+$30, D5 in which the source operand is relative to the offset mode. Note that typical assemblers use the symbol * to indicate offset. Now suppose that the current PC contents are 002000_{16}, the contents of 002030_{16} are 0005_{16}, and the low 16 bits of D5 contain 0010_{16}; after execution of this ADD instruction, D5 will contain 0015_{16}.

In relative with index and offset, the effective address is obtained by adding the contents of the current PC, a signed 8-bit displacement (sign-extended to 32 bits), and the contents of an index register (address or data register). The size of the index register can be 16 or 32 bits wide. For example, consider ADD.W $4 (PC, D0.W), D2. If $[D2] = 00000012_{16}$, $[PC] = 002000_{16}$, $[D0]_{low\ 16\ bits} = 0010_{16}$, and $[002014] = 0002_{16}$, then after this ADD, $[D2]_{low\ 16\ bits} = 0014_{16}$. This mode is used when the displacement needs to be changed during program execution.

5.4.5 Immediate Data Addressing Mode

There are two immediate modes available with the 68000. These are the immediate and quick immediate modes. In the immediate mode, the operand data are constant data, which is part of the instruction. For example, consider ADD #$0005, D0. If $[D0] = 0002_{16}$, then after this ADD instruction, $[D0] = 0002_{16} + 0005_{16} = 0007_{16}$. Note that the # symbol is used by Motorola to indicate the immediate mode.

The quick immediate mode allows one to increment or decrement a register by a number from 0 to 7. For example, ADDQ #1, D0 increments the contents of D0 by 1. Note that the data is inherent in the op code with the op code length of one word (16-bit). Data 0 to 7 are represented by three bits in the op code.

5.4.6 Implied Addressing

The instructions using this implicit mode do not require any operand, and registers such as PC, SP, or SR are implicitly referenced in these instructions. For example, RTE returns from an exception routine to the main program by using implicitly the PC and SR.

All 68000 addressing modes of Table 5.2 can further be divided into four functional categories as follows:

- Data Addressing Mode. An addressing mode is said to be a data addressing mode if it references data objects. For example, all 68000 addressing modes, except the address register direct mode, fall into this category.
- Memory Addressing Mode. An addressing mode that is capable of accessing a data item stored in the memory is classified as memory addressing mode. For example, the data and address register direct addressing modes cannot satisfy this definition.
- Control Addressing Mode. This refers to an addressing mode that has the ability to access a data item stored in the memory without the need to specify its size. For example, all 68000 addressing modes except the following are classified as control addressing modes:
  - Data register direct
  - Address register direct
  - Address register indirect with postincrement

Addressing Mode	Addressing Categories			
	Data	Memory	Control	Alterable
Data register direct	X	-	-	X
Address register direct	-	-	-	X
Address register indirect	X	X	X	X
Address register indirect with postincrement	X	X	-	X
Address register indirect with predecrement	X	X	-	X
Address register indirect with displacement	X	X	X	X
Address register indirect with index	X	X	X	X
Absolute short	X	X	X	X
Absolute long	X	X	X	X
Program counter with displacement	X	X	X	-
Program counter with index	X	X	X	-
Immediate	X	X	-	-

FIGURE 5.4   68000 addressing-functional categories.

- Address register indirect with predecrement
- Immediate
- Alterable Addressing Mode. If the effective address of an addressing mode is written into, then that mode is called alterable addressing mode. For example, the immediate and the program counter relative addressing modes will not satisfy this definition.

The addressing modes are classified into the four functional categories as shown in Figure 5.4.

5.5 68000 INSTRUCTION SET

The 68000 instruction set contains 56 basic instructions. Table 5.3 lists them in alphabetical order. Table 5.4 lists those affecting the condition codes. The repertoire is very versatile and offers an efficient means to handle high-level language data structures (such as arrays and linked lists). Note that in order to identify the operand size of an instruction, the following is placed after a 68000 mnemonic: .B for byte, .W or none for word, .L for long word. For example:

```
ADD.B           D0, D1  ;   [D1]8 ← [D0]8 + [D1]8
ADD.W or ADD    D0, D1  ;   [D1]16 ← [D0]16 + [D1]16
ADD.L           D0, D1  ;   [D1]32 ← [D0]32 + [D1]32
```

TABLE 5.3   68000 Instruction Set

Instruction	Size	Length (words)	Operation
ABCD – (Ay), –(Ax)	B	1	–[Ay]10 + – [Ax]10 + X → [Ax]
ABCD Dy, Dx	B	1	[Dy]10 + [Dx]10 + X → Dx
ADD (EA), (EA)	B, W, L	1	[EA] + [EA] → EA
ADDA (EA), An	W, L	1	[EA] + An → An
ADDI #data, (EA)	B, W, L	2 for B, W 3 for L	data + [EA] → EA
ADDQ #data, (EA)	B, W, L	1	data + [EA] → EA
ADDX – (Ay), –(Ax)	B, W, L	1	–[Ay] + –[Ax] + X → [Ax]
ADDX Dy, Dx	B, W, L	1	Dy + Dx + X → Dx
AND (EA), (EA)	B, W, L	1	[EA] ∧ [EA] → EA
ANDI #data, (EA)	B, W, L	2 for B, W 3 for L	data ∧ [EA] → EA
ANDI #data8, CCR	B	2	data8 ∧ [CCR] → CCR
ANDI #data16, SR	W	2	data16 ∧ [SR] → SR if s = 1; else trap
ASL Dx, Dy	B, W, L	1	C ← [Dy] ← 0, X ← number of shifts determined by [Dx]
ASL #data, Dy	B, W, L	1	C ← [Dy] ← 0, X ← number of shifts determined by # data
ASL (EA)	B, W, L	1	C ← [[EA]] ← 0, X ← shift once
ASR Dx, Dy	B, W, L	1	[Dy] → C, → X number of shifts determined by [Dx]
ASR#data, Dy	B, W, L	1	[Dy] → C, → X number of shifts determined by immediate data
ASR (EA)	B, W, L	1	[[EA]] → C, → X shift once
BCC d	B, W	1 for B 2 for W	Branch to PC + d if carry = 0; else next instruction
BCHG Dn, (EA)	B, L	1	[bit of [EA], specified by Dn]′ → Z [bit of [EA] specified by Dn]′ → bit of [EA]
BCHG #data, (EA)	B, L	2	Same as BCHG Dn, [EA] except bit number is specified by immediate data
BCLR Dn (EA)	B, L	1	[bit of [EA]]′ → Z 0 → bit of [EA] specified by Dn
BCLR #data, (EA)	B, L	2	Same as BCLR Dn, [EA] except the bit is specified by immediate data
BCS d	B, W	1 for B 2 for W	Branch to PC + d if carry = 1; else next instruction
BEQ d	B, W	1 for B 2 for W	Branch to PC + d if Z = 1; else next instruction
BGE d	B, W	1 for B 2 for W	Branch to PC + d if greater than or equal; else next instruction

TABLE 5.3 68000 Instruction Set (*continued*)

Instruction	Size	Length (words)	Operation
BGT d	B, W	1 for B 2 for W	Branch to PC + d if greater than; else next instruction
BHI d	B, W	1 for B 2 for W	Branch to PC + d if higher; else next instruction
BLE d	B, W	1 for B 2 for W	Branch to PC + d if less or equal; else next instruction
BLS d	B, W	1 for B 2 for W	Branch to PC + d if low or same; else next instruction
BLT d	B, W	1 for B 2 for W	Branch to PC + d if less than; else next instruction
BMI d	B, W	1 for B 2 for W	Branch to PC + d if $N = 1$; else next instruction
BNE d	B, W	1 for B 2 for W	Branch to PC + d if $Z = 0$; else next instruction
BPL d	B, W	1 for B 2 for W	Branch to PC + d if $N = 0$; else next instruction
BRA d	B, W	1 for B 2 for W	Branch always to PC + d
BSET Dn, (EA)	B, L	1	[bit of [EA]]' → Z 1 → bit of [EA] specified by Dn
BSET #data, (EA)	B, L	2	Same as BSET Dn, [EA] except the bit is specified by immediate data
BSR d	B, W	1 for B 2 for W	PC → −[SP] PC + d → PC
BTST Dn, (EA)	B, L	1	[bit of [EA] specified by Dn]' → Z
BTST #data, (EA)	B, L	2 2 for W	Same as BTST Dn, [EA] except the bit is specified by data
BVC d	B, W	1 for B 2 for W	Branch to PC + d if $V = 0$; else next instruction
BVS d	B, W,	1 for B 2 for W	Branch to PC + d if $V = 1$; else next instruction
CHK (EA), Dn	W	1	If Dn < 0 or Dn > [EA], then trap
CLR (EA)	B, W, L	1	0 → EA
CMP (EA), Dn	B, W, L	1	Dn − [EA] → Affect all condition codes except X
CMP (EA), An	W, L	1	An − [EA] → Affect all condition codes except X
CMPI #data, (EA)	B, W, L	2 for B, W 3 for L	[EA] − data → Affect all flags except X-bit
CMPM (Ay) +, (Ax)+	B, W, L	1	[Ax]+ − [Ay]+ → Affect all flags except X; update Ax and Ay
DBCC Dn, d	W	2	If condition false, i.e., $C = 1$, then Dn − 1 → Dn; if Dn ≠ −1, then PC + d → PC; else PC + 2 → PC
DBCS Dn, d	W	2	Same as DBCC except condition is $C = 1$
DBEQ Dn, d	W	2	Same as DBCC except condition is $Z = 1$
DBF Dn, d	W	2	Same as DBCC except condition is always false
DBGE Dn, d	W	2	Same as DBCC except condition is greater or equal
DBGT Gn, d	W	2	Same as DBCC except condition is greater than
DBHI Dn, d	W	2	Same as DBCC except condition is high
DBLE Dn, d	W	2	Same as DBCC except condition is less than or equal
DBLS Dn, d	W	2	Same as DBCC except condition is low or same
DBLT Dn, d	W	2	Same as DBCC except condition is less than
DBMI Dn, d	W	2	Same as DBCC except condition is $N = 1$
DBNE Dn, d	W	2	Same as DBCC except condition $Z = 0$
DBPL Dn, d	W	2	Same as DBCC except condition $N = 1$
DBT Dn, d	W	2	Same as DBCC except is always true
DBVC Dn, d	W	2	Same as DBCC except condition is $V = 0$
DBVS Dn, d	W	2	Same as DBCC except condition is $V = 1$

TABLE 5.3   68000 Instruction Set (*continued*)

Instruction	Size	Length (words)	Operation
DIVS (EA), Dn	W	1	Signed division $[Dn]32/[EA]16 \rightarrow$ $[Dn]0\text{-}7$ = quotient $[Dn]8\text{-}15$ = remainder
DIVU (EA), Dn	W	1	Same as DIVS except division is unsigned
EOR Dn, (EA)	B, W, L	1	Dn $\oplus$ [EA] $\rightarrow$ EA
EORI #data, (EA)	B, W, L	3 for L 2 for B, W	data $\oplus$ [EA] $\rightarrow$ EA
EORI #d8, CCR	B	2	d8 $\oplus$ CCR $\rightarrow$ CCR
EORI #d16, SR	W	2	d16 $\oplus$ SR $\rightarrow$ SR if S = 1; else trap
EXG Rx, Ry	L	1	Rx $\leftrightarrow$ Ry
EXT Dn	W, L	1	Extend sign bit of Dn from 8-bit to 16-bit or from 16-bit to 32-bit depending on whether the operand size is B or W
JMP (EA)	Unsized	1	[EA] $\rightarrow$ PC Unconditional jump using address in operand
JSR (EA)	Unsized	1	PC $\rightarrow$ –[SP]; [EA] $\rightarrow$ PC Jump to subroutine using address in operand
LEA (EA), An	L	1	[EA] $\rightarrow$ An
LINK An, # –d	Unsized	2	An $\leftarrow$ –[SP]; SP $\rightarrow$ An; SP – d $\rightarrow$ SP
LSL Dx, Dy	B, W, L	1	
LSL #data, Dy	B, W, L	1	Same as LSL Dx, Dy except immediate data specify the number of shifts from 0 to 7
LSL (EA)	B, W, L	1	Same as LSL Dx, Dy except left shift is performed only once
LSR Dx, Dy	B, W, L	1	
LSR #data, Dy	B, W, L	1	Same as LSR except immediate data specifies the number of shifts from 0 to 7
LSR (EA)	B, W, L	1	Same as LSR, Dx, Dy except the right shift is performed once only
MOVE (EA), (EA)	B, W, L	1	$[EA]_{source} \rightarrow [EA]_{destination}$
MOVE (EA), CCR	W	1	[EA] $\rightarrow$ CCR
MOVE CCR, (EA)	W	1	CCR $\rightarrow$ [EA]
MOVE (EA), SR	W	1	If S = 1, then [EA] $\rightarrow$ SR; else TRAP
MOVE SR, (EA)	W	1	If S = 1, then SR $\rightarrow$ [EA]; else TRAP
MOVE An, USP	L	1	If S = 1, then An $\rightarrow$ USP; else TRAP
MOVE USP, An	W, L	1	[USP] $\rightarrow$ An
MOVEM register list, (EA)	W, L	2	Register list $\rightarrow$ [EA]
MOVEM (EA), register list	W, L	2	[EA] $\rightarrow$ register list
MOVEP Dx, d (Ay)	W, L	2	Dx $\rightarrow$ d[Ay]
MOVEP d (Ay), Dx	W, L	2	d[Ay] $\rightarrow$ Dx
MOVEQ #d8, Dn	L	1	d8 sign extended to 32-bit $\rightarrow$ Dn
MULS (EA)16, (Dn)16	W	1	Signed 16 × 16 multiplication [EA]16 *[Dn]16 $\rightarrow$ [Dn]32
MULU (EA)16, (Dn)16	W	1	Unsigned 16 × 16 multiplication [EA]16 *[Dn]16 $\rightarrow$ [Dn]32
NBCD (EA)	B	1	0 – [EA]10 – X $\rightarrow$ EA
NEG (EA)	B, W, L	1	0 – [EA] $\rightarrow$ EA
NEGX (EA)	B, W, L	1	0 – [EA] – X $\rightarrow$ EA
NOP	Unsized	1	No operation
NOT (EA)	B, W, L	1	[EA]' $\rightarrow$ EA
OR (EA), (EA)	B, W, L	1	[EA]V[EA] $\rightarrow$ EA
ORI #data, (EA)	B, W, L	2 for B, W 3 for L	data V[EA] $\rightarrow$ EA

TABLE 5.3 68000 Instruction Set (*continued*)

Instruction	Size	Length (words)	Operation
ORI #d8, CCR	B	2	d8VCCR → CCR
ORI #d16, SR	W	2	If S = 1, then d16VSR → SR; else TRAP
PEA (EA)	L	1	[EA]16 sign extend to 32 bits → –[SP]
RESET	Unsized	1	If S = 1, then assert RESET line; else TRAP
ROL Dx, Dy	B, W, L	1	C ← [Dy] ← (rotate left diagram)
ROL #data, Dy	B, W, L	1	Same as ROL Dx, Dy except immediate data specifies number of times to be rotated from 0 to 7
ROL (EA)	B, W, L	1	Same as ROL Dx, Dy except [EA] is rotated once
ROR Dx,Dy	B, W, L	1	[Dy] → C (rotate right diagram)
ROR #data, Dy	B, W, L	1	Same as ROR Dx, Dy except the number of rotates is specified by immediate data from 0 to 7
ROR (EA)	B, W, L	1	Same as ROR Dx, Dy except [EA] is rotated once
ROXL Dx, Dy	B, W, L	1	X, C ← [Dy] ← (rotate left through extend diagram)
ROXL #data, Dy	B, W, L	1	Same as ROXL Dx, Dy except immediate data specifies number of rotates from 0 to 7
ROXL (EA)	B, W, L	1	Same as ROXL Dx, Dy except [EA] is rotated once
ROXR Dx, Dy	B, W, L	1	X, [Dy] → C (rotate right through extend diagram)
ROXR #data, Dy	B, W, L	1	Same as ROXR Dx, Dy except immediate data specifies number of rotates from 0 to 7
ROXR (EA)	B, W, L	1	Same as ROXR Dx, Dy except [EA] is rotated once
RTE	Unsized	1	If S = 1, then [SP]+ → SR; [SP]+ → PC, else TRAP
RTR	Unsized	1	[SP] +→ CC; [SP] +→ PC
RTS	Unsized	1	[SP] +→ PC
SBCD –(Ay), –(Ax)	B	1	–(Ax)10 — (Ay)10 – X → (Ax)
SBCD Dy, Dx	B	1	[Dx]10 – [Dy]10 – X → Dx
SCC (EA)	B	1	If C = 0, then 1s → [EA] else 0s → [EA]
SCS (EA)	B	1	Same as SCC except the condition is C = 1
SEQ (EA)	B	1	Same as SCC except if Z = 1
SF (EA)	B	1	Same as SCC except condition is always false
SGE (EA)	B	1	Same as SCC except if greater or equal
SGT (EA)	B	1	Same as SCC except if greater than
SHI (EA)	B	1	Same as SCC except if high
SLE (EA)	B	1	Same as SCC except if less or equal
SLS (EA)	B	1	Same as SCC except if low or same
SLT (EA)	B	1	Same as SCC except if less than
SMI (EA)	B	1	Same as SCC except if N = 1
SNE (EA)	B	1	Same as SCC except if Z = 0
SPL (EA)	B	1	Same as SCC except if N = 0
ST (EA)	B	1	Same as SCC except condition always true
STOP #data	Unsized	2	If S = 1, then data → SR and stop; TRAP if executed in user mode
SUB (EA), (EA)	B, W, L	1	[EA] – [EA] → EA
SUBA (EA), An	W, L	1	An – [EA] → An
SUBI #data, (EA)	B, W, L	2 for B,W 3 for L	[EA] – data → EA
SUBQ #data, (EA)	B, W, L	1	[EA] – data → EA

TABLE 5.3   68000 Instruction Set (*continued*)

Instruction	Size	Length (words)	Operation
SUBX – (Ay), – (Ax)	B, W, L	1	–[Ax] -- [Ay] – X → [Ax]
SUBX Dy, Dx	B, W, L	1	Dx – Dy – X → Dx
SVC (EA)	B	1	Same as SCC except if V = 0
SVS (EA)	B	1	Same as SCC except if V = 1
SWAP Dn	W	1	Dn [31:16] ↔ Dn [15:0]
TAS (EA)	B	1	[EA] tested; N and Z are affected accordingly; 1 → bit 7 of [EA]
TRAP #vector	Unsized	1	PC → –[SSP], SR → –[SSP], (vector) → PC; 16 TRAP vectors are available
TRAPV	Unsized	1	If V = 1, then TRAP; else next instruction
TST (EA)	B, W, L	1	[EA] – 0 → condition codes affected; no result provided
UNLK An	Unsized	1	An → SP; [SP]+ → An

All 68000 instructions may be classified into eight groups, as follows:

i) Data movement instructions
ii) Arithmetic instructions
iii) Logical instructions
iv) Shift and rotate instructions
v) Bit manipulation instructions
vi) Binary-coded decimal instructions

TABLE 5.4   68000 Instructions Affecting the Condition Codes

Instruction	X	N	Z	V	C
ABCD	+	U	+	U	—
ADD, ADDI, ADDQ, ADDX	+	+	+	+	+
AND, ANDI	—	+	+	0	0
ASL, ASR	+	+	+	+	+
BCHG, BCLR, BSET, BTST	—	—	+	—	—
CHK	—	+	U	U	U
CLR	—	0	1	0	0
CMP CMPA, CMPI, CMPM	—	+	+	+	+
DIVS, DIVU	—	+	+	+	0
EOR, EORI	—	+	+	0	0
EXT	—	+	+	0	0
LSL, LSR	+	+	+	0	+
MOVE (EA), (EA)	—	+	+	0	0
MOVE TO CC	+	+	+	+	+
MOVE TO SR	+	+	+	+	+
MOVEQ	—	+	+	0	0
MULS, MULU	—	+	+	0	+
NBCD	+	U	+	U	+
NEG, NEGX	+	+	+	+	+
NOT, OR, ORI	—	+	+	0	0
ROL, ROR	—	+	+	0	+
ROXL, ROXR	+	+	+	0	+
RTE, RTR	+	+	+	+	+
SBCD	+	U	+	U	+
STOP	+	+	+	+	+
SUB, SUBI, SUBQ, SUBX	+	+	+	+	+
SWAP	—	+	+	0	0
TAS	—	+	+	0	0
TST	—	+	+	0	0

Note: +, affected; — , not affected; U, undefined.

TABLE 5.5   68000 Data Movement Instructions

Instruction	Size	Comment
EXG Rx, Ry	L	Exchange the contents of two registers; Rx or Ry can be any address or data register; no flags are affected
LEA (EA), An	L	The effective address [EA] is calculated using the particular addressing mode used and then loaded into the address register; [EA] specifies the actual data to be loaded into An
LINK An, #-displacement	Unsized	The current contents of the specified address register are pushed onto the stack; after the push, the address register is loaded from the updated SP; finally, the 16-bit sign-extended displacement is added to the SP; a negative displacement is specified to allocate stack
MOVE (EA), (EA)	B, W, L	[EA]s are calculated by the 68000 using the specific addressing mode used; [EA]s can be register or memory location; therefore, data transfer can take place between registers, between a register and a memory location, and between different memory locations; flags are affected; for byte-size operation, address register direct is not allowed; An is not allowed in the destination [EA]; the source [EA] can be An for word or long-word transfers
MOVEM reg list, (EA) or (EA), reg list	W, L	Specified registers are transferred to or from consecutive memory locations starting at the location specified by the effective address
MOVEP Dn, d (Ay) or d (Ay), Dn MOVEP Dn, d (Ay) or d (Ay), Dn	W, L	Two [W] or four [L] bytes of data are transferred between a data register and alternate bytes of memory, starting at the location specified and incrementing by 2; the high-order byte of data is transferred first, and the low-order byte is transferred last. This instruction has the address register indirect with displacement-only mode
MOVEQ #data, Dn	L	This instruction moves the 8-bit inherent data into the specified data register; the data are then sign-extended to 32 bits
PEA (EA)	L	Computes an effective address and then pushes the 32-bit address onto the stack
SWAP Dn	W	Exchanges 16-bit halves of a data register
UNLK An	Unsized	An $\rightarrow$ SP; [SP] $+\rightarrow$ An

- [EA] in LEA [EA], An can use all addressing modes except Dn, An, [An] +, –[An], and immediate.
- Destination [EA] in MOVE [EA], [EA] can use all modes except An, relative, and immediate.
- Source [EA] in MOVE [EA], [EA] can use all modes.
- Destination [EA] in MOVEM reg list, [EA] can use all modes except Dn, An, [An]+, relative, and immediate. Source [EA] in MOVEM [EA], reg list can use all modes except Dn, An, –[An], and immediate.
- [EA] in PEA [EA]) can use all modes except Dn, An, [An]+, –[An], and immediate.

vii)   Program control instructions
viii) System control instructions

5.5.1 Data Movement Instructions

These instructions allow data transfers from register to register, register to memory, memory to register, and memory to memory. In addition, there are also special data movement instructions such as MOVEM (Move multiple registers). Typically, byte, word, or long-word data can be transferred. Table 5.5 lists the 68000 data movement instructions.

5.5.1.a MOVE Instructions

The format for the basic MOVE instruction is MOVE.S (EA), (EA), where S = .L, .W, or .B. (EA) can be a register or memory location depending on the addressing mode used. Consider MOVE.B D2, D0 which uses a data register direct mode for both the source and destination. Now if $[D2] = 03_{16}$, $[D0] = 01_{16}$, then after execution of this MOVE instruction $[D2] = 03_{16}$ and $[D0] = 03_{16}$.

There are several variations of the MOVE instruction. For example, MOVE.W CCR, (EA) moves the content of the low-order byte of the SR, i.e., CCR, to the low-order byte of the destination operand, and the upper byte of SR is considered as zero. The source operand is a word. Similarly, MOVE.W (EA), CCR moves an 8-bit immediate number or low-order 8-bit data from a memory location or register into the condition code register; the upper byte is ignored. The source operand is a word. Data can also be transferred between (EA) and SR or USP (in the supervisor mode when s = 1) using the following privileged instructions:

```
MOVE.W (EA), SR
MOVE.W SR, (EA)
MOVE.L USP, An
MOVE.L An, USP
```

MOVEA.W or .L (EA), An can be used to load an address into an address register. Word size source operands are sign-extended to 32 bits. Note that (EA) is obtained using an addressing mode. As an example, MOVEA.W #$8000, A0 moves the 16-bit word 8000_{16} into the low 16 bits of A0 and then sign-extends 8000_{16} to the 32-bit number $FFFF8000_{16}$.

Note that sign extension means extending bit 15 of 8000_{16} from bit 16 through bit 31. As mentioned, sign extension is required when an arithmetic operation between two signed binary numbers of different sizes is performed. (EA) in MOVEA can use all addressing modes. MOVEM instruction can be used to PUSH or POP multiple registers to or from the stack. For example, MOVEM.L D0-D7/A0-A6, – (SP) saves the contents of all of the 8 data registers and 7 address registers in the stack. This instruction stores address registers in the order A6-A0 first, followed by data registers in the order D7-D0, regardless of the order in the list. MOVEM.L (SP)+, D0-D7/A0-A6 restores the contents of the registers in the order D0-D7,A0-A6 regardless of the order in the list. MOVEM instruction can also be used to save a set of registers in memory. In addition to the above predecrement and postincrement modes for the effective address, MOVEM instruction allows all the control modes. If the effective address uses one of the control modes, such as absolute short, then the registers are transferred starting at the specified address and up through higher addresses. The order of transfer is from D0 to D7 and then from A0 to A7. For example, MOVEM.W A4/D1/D3/A0-A2, $8000 transfers the low 16-bit contents of D1, D3, A0, A1, A2, and A4 to locations $8000, $8002, $8004, $8006, $8008, and $800A, respectively.

The MOVEQ.L #d8, Dn moves the immediate 8-bit data into the low byte of Dn. The 8-bit data are then sign-extended to 32 bits. This is a one-word instruction. For example, MOVEQ.L #$FF, D0 moves $FFFFFFFF into D0.

In order to transfer data between the 68000 data registers and 6800 (8-bit) peripherals, the MOVEP instruction can be used. This instruction transfers two to four bytes of data between a data register and alternate byte locations in memory, starting at the location specified in increments of 2. Register indirect with displacement is the only addressing mode used with this instruction. If the address is even, all the transfers are made on the high-order half of the data bus. The high-order byte from the register is transferred first and the low-order byte is transferred last. For example, consider MOVEP.L $0050 (A0), D0. If [A0] = 00003000_{16}, $[003050] = 01_{16}$, $[003052] = 03_{16}$, $[003054] = 02_{16}$, $[003056] = 04_{16}$, then after the execution of the above MOVEP instruction, D0 will contain 01030204_{16}.

5.5.1.b EXG and SWAP Instructions

The EXG.L Rx, Ry instruction exchanges the 32-bit contents of Rx with that of the Ry. The exchange is between two data registers, two address registers, or between an address and a data register. The EXG instruction exchanges only 32-bit-long words. The data size (L) does not have to be specified after the EXG instruction since this instruction has only one data size. No flags are affected.

The SWAP.W Dn instruction, on the other hand, exchanges the low 16 bits of Dn with the high 16 bits of Dn. All condition codes are affected.

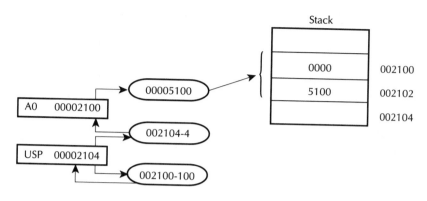

FIGURE 5.5   Execution of the link instruction.

5.5.1.c LEA and PEA Instructions

The LEA.L (EA), An moves an effective address (EA) into the specified address register. The (EA) can be calculated based on the addressing mode of the source. For example, LEA $00456074, A1 moves $00456074 to A1. This instruction is equivalent to MOVEA.L #$00456074, A1. Note that $00456074 is contained in the PC.

LEA instruction is very useful when address calculation is desired during program execution. (EA) in LEA specifies the actual data to be loaded into An, whereas (EA) in MOVEA specifies the address of actual data. For example, consider LEA $06 (A2, D5.W), A0. If [A2] = 00003000_{16}, [D5] = 0044_{16}, then the LEA instruction moves $0000304A_{16}$ into A0. On the other hand, MOVEA $06(A2 D5.W), A0 moves the contents of $0000304A_{16}$ into A0. Therefore, it is obvious that if address calculation is required, the instruction LEA is very useful.

The PEA (EA) computes an effective address and then pushes it onto the stack. This instruction can be used when the 16-bit address used in absolute short mode is required to be pushed onto the stack. For example, consider PEA $8000 in the user mode. If [USP] = $00005004, then $8000 is sign-extended to 32 bits and pushed to stack. The low-order 16 bits ($8000) are pushed at 005002_{16} and the high-order 16 bits ($FFFF) are pushed at 005000_{16}.

5.5.1.d LINK and UNLK Instructions

Before calling a subroutine, the main program quite often transfers values of certain parameters to the subroutine. It is convenient to save these variables onto the stack before calling the subroutine. These variables can then be read from the stack and used by the subroutine for computations. The 68000 LINK and UNLK instructions are used for this purpose. In addition, the 68000 LINK instruction allows one to reserve temporary storage for the local variables of a subroutine. This storage can be accessed as needed by the subroutine and be released using UNLK before returning to the main program. The LINK instruction is usually used at the beginning of a subroutine to allocate stack space for storing local variables and parameters for nested subroutine calls. The UNLK instruction is usually used at the end of subroutine before the RETURN instruction to release the local area and restore the stack pointer contents so that it points to the return address.

The LINK An, #-displacement causes the current contents of the specified An to be pushed onto the system (user or supervisor) stack. The updated SP contents are then loaded into An. Finally, a sign-extended 2's complement displacement value is added to the SP. No flags are affected. For example, consider LINK A0, #-$100. If [A0] = 00005100_{16} [USP] = 00002104_{16} then after execution of the LINK instruction the situation shown in Figure 5.5 occurs.

This means that after the LINK instruction, [A0] = 00005100_{16} is pushed onto the stack, the [updated USP] = 002100_{16} is loaded into A0. USP is then loaded with 002000_{16} and, therefore,

100 locations are allocated to subroutine at the beginning of which the above LINK instruction can be used. Note that A0 cannot be used in the subroutine.

The UNLK instruction at the end of this subroutine before the RETURN instruction releases the 100_{16} locations and restores the contents of A0 and USP to those prior to using the LINK instruction. For example, UNLK A0 will load [A0] = 00002100_{16} into USP, the two stack words 00005100_{16} into A0, and USP is then incremented by 4 to contain 00002104_{16}. Therefore, the contents of A0 and USP prior to using the LINK are restored. In the above example, after execution of the LINK, addresses $001FFF_{16}$ and below can be used as the stack. Sixty-four ($100) locations starting at 002000_{16} and above can be reserved for storing the local variables of the subroutine. These variables can then be accessed with an address register such as A0 as a base pointer using the address register indirect with displacement mode such as MOVE.W d(A0), D2 for read and MOVE.W D2, d(A0) for write.

5.5.2 Arithmetic Instructions

These instructions allow:

- 8-, 16-, or 32-bit additions and subtractions
- 16-bit by 16-bit multiplication (both signed and unsigned)
- Compare, clear, and negate instructions
- Extended arithmetic instructions for performing multiprecision arithmetic
- Test (TST) instruction for comparing the operand with zero
- Test and set (TAS) instruction which can be used for synchronization in multiprocessor system

The 68000 arithmetic instructions are summarized in Table 5.6.

5.5.2.a Addition and Subtraction Instructions

- Consider ADD.W $245000, D0. If $[245000_{16}] = 2014_{16}$ and $[D0] = 1004_{16}$, then after execution of this ADD instruction, the low 16 bits of D0 will contain 3018_{16}.
- ADDI instruction can be used to add immediate data to register or memory location. The immediate data follows the instruction word. For example, consider ADDI.W #$0062, $500000. If $[500000_{16}] = 1000_{16}$, then after execution of this ADDI instruction, memory location 500000_{16} will contain 1062_{16}. ADDQ, on the other hand, adds a number from 0 to 7 to the register or memory location in the destination operand. This instruction occupies 16 bits and the immediate data 0 to 7 is specified by 3 bits in the instruction word. For example, consider ADDQ.B#4, D0. If $[D0]_{low byte} = 60_{16}$, then after execution of this ADDQ, low byte of register D0 will contain 64_{16}.
- For ADD or SUB, if the destination is Dn, then the source (EA) can use all modes; if the destination is a memory location (all modes except Dn, An, relative, and immediate), then the source must be Dn.
- All subtraction instructions subtract source from destination. For example, consider SUB.W D0, $500000. If $[D0]_{low word} = 3003_{16}$ and $[500000_{16}] = 5005_{16}$, then after execution of this SUB instruction, memory location 500000_{16} will contain 2002_{16}.
- Consider SUBI.W# $0004, D0. If $[D0]_{low word} = 5004_{16}$, then after execution of this SUBI instruction, D0 will contain 5000_{16}. Note that the same result can be obtained by using a SUBQ.W# $4, D0. However, in this case the data 4 is inherent in the instruction word.

5.5.2.b Multiplication and Division Instructions

The 68000 instruction set includes both signed and unsigned multiplication of integer numbers.

TABLE 5.6 68000 Arithmetic Instructions

Instruction	Size	Operation
Addition and Subtraction Instructions		
ADD (EA), (EA)	B, W, L	[EA] + [EA] → EA
ADDI #data, (EA)	B, W, L	[EA] + data → EA
ADDQ #d8, (EA)	B, W, L	[EA] + d8 → EA
		d8 can be an integer from 0 to 7
ADDA (EA), An	W, L	An + [EA] → An
SUB (EA), (EA)	B, W, L	$[En]_{source} - [EA]_{dest} \to EA_{dest}$
SUBI # data, (EA)	B, W, L	[EA] − data → EA
SUBQ #d8, (EA)	B, W, L	[EA] − d8 → EA
		d8 can be an integer from 0 to 7
SUBA (EA), An	W, L	An − [EA] → An
Multiplication and Division Instructions		
MULS (EA), Dn	W	[Dn]16 * [EA]16 → [Dn]32 (signed multiplication)
MULU (EA), Dn	W	[Dn]16 * [EA]16 → [Dn]32 (unsigned multiplication)
DIVS (EA), Dn	W	[Dn]32 / [EA]16 → [Dn]32 (signed division, high word of Dn contains remainder and low word of Dn contains the quotient)
DIVU (EA), Dn	W	[Dn]32 / [EA]16 → [Dn]32 (unsigned division, remainder is in high word of Dn and quotient is in low word of Dn
Compare, Clear, and Negate Instructions		
CMP (EA), Dn	B, W, L	Dn − [EA] → No result; affects flags
CMPA (EA), An	W, L	An − [EA] → No result; affects flags
CMPI #data, (EA)	B, W, L	[EA] − data → No result; affects flags
CMPM (Ay) +, (Ax) +	B, W, L	[Ax]+ − [Ay]+ → No result; affects flags; Ax and Ay are incremented depending on operand size
CLR (EA)	B, W, L	0 → EA
NEG (EA)	B, W, L	0 − [EA] → EA
Extended Arithmetic Instructions		
ADDX Dy, Dx	B, W, L	Dx + Dy + X → Dx
ADDX − (Ay), − (Ax)	B, W, L	−[Ax] + −[Ay] + X → [Ax]
EXT Dn	W, L	If size is W, then sign-extended low byte of Dn to 16 bits; if size is L, then sign extend low 16 bits of Dn to 32 bits
NEGX (EA)	B, W, L	0 − [EA] − X → EA
SUBX Dy, Dx	B, W, L	Dx − Dy − X → Dx
SUBX − (Ay), − (Ax)	B, W, L	−[Ax] − −[Ay] − X → [Ax]
Test Instruction		
TST (EA)	B, W, L	[EA] − 0 → flags affected
Test and Set Instruction		
TAS (EA)	B	If [EA] = 0, then set Z = 1; else Z = 0, N = 1 and then always set bit 7 of [EA] to 1

Note: If [EA] in the ADD or SUB instruction is an address register, the operand length is WORD or LONG WORD. [EA] in any instruction is calculated using the addressing mode used. All instructions except ADDA and SUBA affect condition codes.

- Source [EA] in the above ADDA and SUBA can use all modes.
- Destination [EA] in ADDI and SUBI can use all modes except An relative, and immediate.
- Destination [EA] in ADDQ and SUBQ can use all modes except relative and immediate.
- [EA] in all multiplication and division instructions can use all modes except An.
- Source [EA] in CMP and CMPA instructions can use all modes.
- Destination [EA] in CMPI can use all modes except An, relative, and immediate.
- [EA] in CLR and NEG can use all modes except An, relative, and immediate.
- [EA] in NEGX can use all modes except An, relative, and immediate.
- [EA] in TST can use all modes except An, relative, and immediate
- [EA] in TAS can use all modes except An relative, and immediate.

- MULS (EA), Dn multiplies two 16-bit signed numbers and provides a 32-bit result. For example, consider MULS #-2, D0. If [D0] = 0004_{16}, then after this MULS, D0 will contain the 32-bit result $FFFFFFF8_{16}$ which is –8 in decimal. MULU (EA), Dn, on the other hand, performs unsigned multiplication. Consider MULU (A1), D2. If [A1] = 00002000_{16}, [002000] = 0400_{16}, and [D2] = 0300_{16}, then after this MULU, D2 will contain 32-bit result $000C0000_{16}$.
- Consider DIVS #4, D0. If [D0] = -9_{10} = $FFFFFFF7_{16}$, then after this DIVS, register D0 will contain

D0	FFFF	FFFE
	16-bit remainder = -1_{10}	16-bit quotient = -2_{10}

Note that in 68000, after DIVS, the sign of the remainder is always the same as the dividend unless the remainder is equal to zero. Therefore, in the example above, since the dividend is negative (-9_{10}), the remainder is negative (-1_{10}). Also, division by zero causes an internal interrupt automatically. A service routine can be written by the user to indicate an error. N = 1 if the quotient is negative, and V = 1 if there is an overflow. The DIVU instruction is the same as the DIVS instruction except that the division is unsigned. For example, consider DIVU #2, D0. If [D0] = 11_{10} = $0000000B_{16}$, then after execution of this DIVU, register D0 contains:

D	0001	0005
	16-bit remainder	16-bit quotient

As with the DIVS, division by zero using DIVU causes trap. Also, V = 1 if there is an overflow.

5.5.2.c Compare, Clear, and Negate Instructions

- The COMPARE instructions affect condition codes, but do not provide the subtraction result. Consider CMPM.W (A1)+, (A2)+. If [A1] = 00400000_{16}, [A2] = 00500000_{16}, [400000] = 0008_{16}, [500000] = 2006_{16}, then after this CMP instruction N = 0, C = 0, X = 0, V = 0, Z = 0, [A1] = 00400002_{16}, and [A2] = 00500002_{16}.
- CLR.L D3 clears all 32 bits of D3 to zero.
- Consider NEG.W (A1). If [A1] = 00500000_{16}, [500000_{16}] = 0007_{16}, then after this NEG instruction, the low 16 bits of location 500000_{16} will contain $FFF9_{16}$.

5.5.2.d Extended Arithmetic Instructions

- ADDX and SUBX instructions can be used in performing multiple precision arithmetic since there are no ADDC (add with carry) or SUBC (subtract with borrow) instructions. For example, in order to perform a 64-bit addition, the following two instructions can be used:

```
ADD.L  D2, D3  ; Add low 32 bits of data and store in D3
ADDX.L D4, D5 ; Add high 32 bits of data along with any
              ; carry from the low 32-bit addition and
              ; store result in D5
```

Note that in the above D5 D3 contains one 32-bit data and D4 D2 contains the other 32-bit data. The 32-bit result is stored in D5 D3.

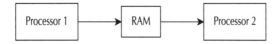

FIGURE 5.6 Two 68000s interfaced via shared RAM.

- Consider EXT.W D5. If [D5] low byte = $F8_{16}$, then after the EXT, [D5] = $FFF8_{16}$.

.5.2.e Test Instruction

- Consider TST.W (A1). If [A1] = 00700000_{16}, [700000] = $F000_{16}$, then after the TST.W (A1), the operation $F000_{16}$-0000_{16} is performed internally by the 68000, Z is cleared to zero, and N is set to 1. V and C flags are always cleared to zero.

.5.2.f Test and Set Instruction

A multiprocessor system includes a mechanism to safely access resources such as memory shared by two or more processors. This mechanism is called mutual exclusion.

Mutual exclusion allows a processor to lock out other processors from accessing a shared resource. This shared resource is also called a Critical Program Section, since once a processor starts executing the program sequence in this section, it must complete it before another processor accesses it.

A binary flag called semaphore stored in the shared memory is used to indicate whether the shared memory is free (semaphore = 0) to be accessed or busy (semaphore = 1; the shared memory being used by another processor). Testing and setting the semaphore is a critical operation and must be executed in an indivisible cycle by a processor so that the other processors will not access the semaphore simultaneously.

The test and set instruction, along with a hardware lock mechanism, can initialize a semaphore. The test and set instruction tests and sets a semaphore. The processor, when executing this instruction, generates a signal to provide the lock mechanism during the execution time of the instruction. This prevents other processors from altering the semaphore between the processor's testing and setting the semaphore.

The 68000 provides the TAS instruction to test and set a semaphore. During execution of the TAS, the 68000 generates LOW on the $\overline{AS}$ (address strobe) pin which can be used to lock out other processors from accessing the semaphore.

Let us explain the application of the TAS instruction. TAS(EA) is usually used to synchronize two processors in multiprocessor data transfers. For example, consider two 68000-based microcomputers with shared RAM shown in Figure 5.6.

Suppose that it is desired to transfer the low byte of D0 from processor 1 to the low byte of D2 in processor 2. A memory location, namely, TRDATA, can be used to accomplish this. First, processor 1 can execute the TAS instruction to test the byte in the shared RAM with address TEST for zero value. If it is zero, the processor 1 can be programmed to move the low byte of D0 into location TRDATA in the shared RAM. The processor 2 can then execute an instruction sequence to move the contents of TRDATA from the shared RAM into the low byte of D2. The following instruction sequence will accomplish this:

	Processor 1 Routine		Processor 2 Routine
Proc 1	TAS TEST	Proc 2	TAS TEST
	BNE Proc 1		BNE Proc 2
	MOVE.B D0, TRDATA		MOVE.B TRDATA, D2
	CLR.B TEST		CLR.B TEST
	—		—
	—		—
	—		—

TABLE 5.7   68000 Logical Instructions

Instruction	Size	Operation
AND (EA), (EA)	B, W, L	$[EA] \wedge [EA] \rightarrow EA$
ANDI #data, (EA)	B, W, L	$[EA] \wedge$ #data $\rightarrow$ EA; [EA] cannot be address register
ANDI #data8, CCR	B	CCR $\wedge$ #data $\rightarrow$ CCR
ANDI #data16, SR	W	SR $\wedge$ #data $\rightarrow$ SR
EOR Dn, (EA)	B, W, L	Dn $\oplus$ [EA] $\rightarrow$ EA; [EA] cannot be address register
EORI #data, (EA)	B, W, L	$[EA] \oplus$ #data $\rightarrow$ EA; [EA] cannot be address register
EORI #data16, SR	W	Data16 $\oplus$ SR $\rightarrow$ SR if s = 1; else trap
NOT (EA)	B, W, L	One's complement of [EA] $\rightarrow$ EA
OR (EA), (EA)	B, W, L	$[EA] \vee [EA] \rightarrow EA$
ORI #data, (EA)	B, W, L	$[EA] \vee$ #data $\rightarrow$ EA; [EA] cannot be address register
ORI #data8, CCR	B	CCR $\vee$ #data8 $\rightarrow$ CCR
ORI #data16, SR	W	SR $\vee$ #data $\rightarrow$ SR

- Source [EA] in AND and OR can use all modes except An.
- Destination [EA] in AND or OR or EOR can use all modes except Dn, An, relative, and immediate.
- Destination [EA] in ANDI, ORI, and EORI can use all modes except An, relative, and immediate.
- [EA] in NOT can use all modes except An, relative, and immediate.

Note that in the above, TAS TEST checks the byte addressed by TEST for zero. If [TEST] = 0, then Z is set to one; otherwise Z = 0 and N = 1. After this, bit 7 of [TEST] is set to 1. Note that a zero value of TEST indicates that the shared RAM is free for use and the Z bit indicates this after the TAS is executed. In each of the above instruction sequences, data is transferred using the MOVE instruction, TEST is cleared to zero so that the shared RAM is free for use by the other processor. Note that bit 7 of memory location TEST is called the semaphore.

In order to avoid testing of the TEST byte simultaneously by two processors, the TAS is executed in a Read-Modify-Write cycle. This means that once the operand is addressed by the 68000 executing the TAS, the system bus is not available to the other 68000 until the TAS is completed.

5.5.3 Logical Instructions

These instructions include logical OR, EOR, AND, and NOT as shown in Table 5.7.

- Consider AND.W D2, D6. If $[D2] = 0005_{16}$, $[D6] = 0FF1_{16}$, then after execution of this AND, the low 16 bits of D6 will contain 0001_{16}.
- Consider ANDI.B #$80, CCR. If $[CCR] = 0F_{16}$, then after this ANDI, register CCR will contain 00_{16}.
- Consider EOR.W D2, D6. If $[D2] = 000F_{16}$ and $[D6] = 000F_{16}$, then after execution of this EOR, register D6 will contain 0000_{16}, and D2 will remain unchanged at $000F_{16}$.
- Consider NOT.B D0. If $[D0] = 04_{16}$, then after execution of this NOT instruction, the low byte of D0 will contain FB_{16}.
- Consider ORI #$1008, SR. If $[SR] = A011_{16}$, then after execution of this ORI, register SR will contain $B019_{16}$. Note that this is a privileged instruction since the high byte of SR containing the control bits is changed and therefore can only be executed in the supervisor mode.

5.5.4 Shift and Rotate Instructions

The 68000 shift and rotate instructions are listed in Table 5.8.

TABLE 5.8 68000 Shift and Rotate Instructions

Instruction	Size	Operation
ASL, Dx, Dy	B, W, L	C ← [Dy] ← 0 ; X ←
		Shift [Dy] by the number of times to left specified in Dx; the low 6 bits of Dx specify the number of shifts from 0 to 63
ASL #data, Dn	B, W, L	Same as ASL Dx, Dy except that the number of shifts is specified by immediate data from 0 to 7
ASL (EA)	B, W, L	[EA] is shifted one bit to left; the most significant of [EA] goes to x and c, and zero moves into the least significant bit
ASR Dx, Dy	B, W, L	→ [Dy →] → C ; → X
		Arithmetically shift [Dy] to the right by retaining the sign bit; the low 6 bits of Dx specify the number of shifts from 0 to 63
ASR #data, Dn	B, W, L	Same as above except the number of shifts is from 0 to 7
ASR (EA)	B, W, L	Same as above except [EA] is shifted once to the right
LSL Dx, Dy	B, W, L	C ← [Dy] ← 0 ; X ←
		Low 6 bits of Dx specify the number of shifts from 0 to 63
LSL #data, Dn	B, W, L	Same as above except the number of shift is specified by immediate data from 0 to 7
LSL (EA)	B, W, L	[EA] is shifted one bit to the left
LSR Dx, Dy	B, W, L	0 → [Dy →] → C ; → X
		Same as LSL Dx, Dy except shift is to the right
LSR #data, Dn	B, W, L	Same as LSL #data, Dn, except shift is to the right by immediate data from 0 to 7
LSR (EA)	B, W, L	Same as LSL [EA] except shift is to the right
ROL Dx, Dy	B, W, L	C ← [Dy ←] ← (wrap)
		Low 6 bits of Dx specify the number of times [Dy] to be shifted
ROL #data, Dn	B, W, L	Same as above except that the immediate data specify that [Dn] to be shifted from 0 to 7
ROL (EA)	B, W, L	[EA] is rotated once to the left
ROR Dx, Dy	B, W, L	→ [Dy →] → C (wrap)
		Low 6 bits of Dx specify the number of shifts from 0 to 63
ROR #data, Dn	B, W, L	Same as ROL #data, Dn except the shift is to the right by immediate data from 0 to 7
ROR (EA)	B, W, L	[EA] is rotated once to the right
ROXL Dx, Dy	B, W, L	C ← [← Dy] ← X (wrap)
		Low 6 bits of Dx contain the number of rotates from 0 to 63
ROXL #data, Dy	B, W, L	Same as above except immediate data specify number of rotates from 0 to 7
ROXL (EA)	B, W, L	[EA] is rotated one bit to right

TABLE 5.8   68000 Shift and Rotate Instructions (*continued*)

Instruction	Size	Operation
ROXR Dx, Dy	B, W, L	
		Same as ROXL Dx, Dy except the rotate is to the right
ROXR #data, Dy	B, W, L	Same as ROXL #data, Dy, except rotate is to the right by immediate data from 0 to 7
ROXR (EA)	B, W, L	Same as ROXL [EA] except rotate is to the right

Note: [EA] in ASL, ASR, LSL, LSR, ROL, ROR, ROXL, and ROXR can use all modes except Dn, An, relative, and immediate.

- All the instructions in Table 5.8 affect N and Z flags according to the result. V is reset to zero except for ASL.
- Note that in the 68000 there is no true arithmetic shift left instruction. In true arithmetic shifts, the sign bit of the number being shifted is retained. In the 68000, the instruction ASL does not retain the sign bit, whereas the instruction ASR retains the sign bit after performing the arithmetic shift operation. Consider ASL.W D3, D0. If $[D3]_{low\ 16\ bits} = 0003_{16}$, $[D0]_{low\ 16\ bits} = 87FF_{16}$, then after this ASL instruction $[D0]_{low\ 16\ bits} = 3FF8_{16}$, $C = 0$, and $X = 0$. Note that the sign of the contents of D0 is changed from 1 to 0, and therefore the overflow is set. ASL sets the overflow bit to indicate sign change during the shift. In the example, the sign bit of D0 is changed after shifting [D0] three times. ASR, on the other hand, retains the sign bit. For example, consider ASR.W #2, D5. If $[D5] = 8FE2_{16}$, then after this ASR, the low 16 bits of $[D5] = E3F8_{16}$, $C = 1$, and $X = 1$. Note that the sign bit is retained. ASL (EA) or ASR (EA) shifts (EA) one bit to the left or right, respectively. For example, consider ASL.W (A2). If $[A2] = 00004000_{16}$ and $[004000] = F000_{16}$, then after execution of the ASL, $[004000] = E000_{16}$, $X = 1$, and $C = 1$. On the other hand, after ASR.W (A2), memory location 004000_{16} will contain 7800_{16}, $C = 0$, and $X = 0$. Note that only memory-alterable modes are allowed for (EA). Also, only 16-bit operands are allowed for (EA) when the destination is memory location.
- LSL and ASL instructions are the same in the 68000 except that with the ASL, V is set to 1 if there is a sign change of the number during the shift.
- Consider LSR.W# 0002, D0. If $[D0] = F000_{16}$, then after the LSR, $[D0] = 3C00_{16}$, $X = 0$, and $C = 0$.
- Consider ROL.B# 02, D2. If $[D2] = B1_{16}$ and $C = 1$, then after execution of the ROL, the low byte of $[D2] = C7_{16}$ and $C = 0$. On the other hand, with $[D2] = B1_{16}$ and $C = 1$, consider ROR.B # 02, D2. After execution of this ROR, register D2 will contain EC_{16} and $C = 0$.
- Consider ROXL.W D2, D1. If $[D2] = 0003_{16}$, $[D1] = F201$, $C = 0$, and $X = 1$, then the low 16 bits after execution of this ROXL are $[D1] = 900F_{16}$, $C = 1$, and $X = 1$.

5.5.5 Bit Manipulation Instructions

The 68000 has four bit manipulation instructions, and these are listed in Table 5.9.

- In all the above instructions, the 1's complement of the specified bit is reflected in the Z flag. The specified bit is then 1's complemented, cleared to zero, set to one, or unchanged by BCHG, BCLR, BSET, or BTST, respectively. In all the instructions in Table 5.10, if (EA) is Dn, then length of Dn is 32 bits; otherwise, the length of the destination is one byte.

TABLE 5.9 68000 Bit Manipulation Instructions

Instruction	Size	Operation
BCHG Dn, (EA) BCHG #data, (EA)	B,L	A bit in [EA] specified by Dn or immediate data is tested; the 1's complement of the bit is reflected in both the Z flag and the specified bit position
BCLR Dn, (EA) BCLR #data, (EA)	B,L	A bit in [EA] specified by Dn or immediate data is tested and the 1's complement of the bit is reflected in the Z flag; the specified bit is cleared to zero
BSET Dn, (EA) BSET #data, (EA)	B,L	A bit in [EA] specified by Dn or immediate data is tested and the 1's complement of the bit is reflected in the Z flag; the specified bit is then set to one
BTST Dn, (EA) BTST #data, (EA)	B,L	A bit in [EA] specified by Dn or immediate data is tested; the 1's complement of the specified bit is reflected in the Z flag

- [EA] in the above instructions can use all modes except An, relative, and immediate.
- If [EA] is memory location, then data size is byte; if [EA] is Dn then data size is long word.

- Consider BCHG.B #2, $003000. If $[003000] = 05_{16}$, then after execution of this BCHG instruction, $Z = 0$ and $[003000] = 01_{16}$.
- Consider BCLR.L # 3, D1. If $[D1] = F210E128_{16}$, then after execution of this BCLR, register D1 will contain $F210E120_{16}$ and $Z = 0$.
- Consider BSET.B #0, (A1). If $[A1] = 00003000_{16}$, $[003000] = 00_{16}$, then after execution of this BSET, memory location 003000 will contain 01_{16} and $Z = 1$.
- Consider BTST.B #2, $002000. If $[002000] = 02_{16}$, then after execution of this BTST, $Z = 0$ and $[002000] = 02_{16}$.

5.5.6 Binary-Coded Decimal Instructions

The 68000 instruction set contains three BCD instructions, namely, ABCD for adding, SBCD for subtracting, and NBCD for negating. These instructions operate on packed BCD operands and always include the extent (X) bit in the operation. The BCD instructions are listed in Table 5.10.

- Consider ABCD D1, D2. If $[D1] = 25_{16}$, $[D2] = 15_{16}$, $X = 0$, then after execution of the ABCD instruction, $[D2] = 40_{16}$, $X = 0$, and $Z = 0$.
- Consider SBCD – (A2), – (A3). If $[A2] = 00002004_{16}$, $[A3] = 00003003_{16}$, $X = 1$, $[002003] = 05_{16}$, $[003002] = 06_{16}$, then after execution of this SBCD, $[003002] = 00_{16}$, $X = 0$, and $Z = 1$.
- Consider NBCD (A1). If $[A1] = [00003000_{16}]$, $[003000] = 05$, $X = 1$, then after execution of the NBCD, $[003000] = FA_{16}$.

5.5.7 Program Control Instructions

These instructions include branches, jumps, and subroutine calls as listed in Table 5.11.

TABLE 5.10 68000 Binary-Coded Decimal Instructions

Instruction	Operand size	Operation
ABCD Dy, Dx	B	$[Dx]10 + [Dy]10 + X \rightarrow Dx$
ABCD – (Ay), – (Ax)	B	$-[Ax]10 + -[Ay]10 + X \rightarrow [Ax]$
SBCD Dy, Dx	B	$[Dx]10 - [Dy]10 - X \rightarrow Dx$
SBCD – (Ay), –(Ax)	B	$-[Ax]10 - -[Ay]10 - X \rightarrow [Ax]$
NBCD (EA)	B	$0 - [EA]10 - X \rightarrow EA$

Note: [EA] in NBCD can use all modes except An, relative, and immediate.

TABLE 5.11   68000 Program Control Instructions

Instruction	Size	Operation
Bcc d	B,W	If condition code cc is true, then PC + d →PC; the PC value is current instruction location plus 2; d can be 8- or 16-bit signed displacement; if 8-bit displacement is used, then the instruction size is 16 bits with the 8-bit displacement as the low byte of the instruction word; if 16-bit displacement is used, then the instruction size is two words with 8-bit displacement field (low byte) in the instruction word as zero and the second word following the instruction word as the 16-bit displacement There are 14 conditions such as BCC (Branch if Carry Clear), BEQ (Branch if result equal to zero, i.e., Z = 1), and BNE (Branch if not equal, i.e., Z = 0) Note that the PC contents will always be even since the instruction length is either one word or two words depending on the displacement widths
BRA d	B,W	Branch always to PC + d where PC value is current instruction location plus 2; as with Bcc, d can be signed 8 or 16 bits; this is an unconditional branching instruction with relative mode; note that the PC contents are even since the instruction is either one word or two words
BSR d	B,W	PC → –[SP] PC + d → PC The address of the next instruction following PC is pushed onto the stack; PC is then loaded with PC + d; as before, d can be 8 or 16 bits; this is a subroutine call instruction using relative mode
DBcc Dn, d	W	If cc is false, then Dn – 1 → Dn, and if Dn = –1, then PC + 2 → PC; If Dn ≠ –1, then PC + d → PC; if cc is true, PC + 2 → PC
JMP (EA)	Unsized	[EA] → PC This is an unconditional jump instruction which uses control addressing mode
JSR (EA)	Unsized	PC → –[SP] [EA] → PC This is a subroutine call instruction which uses control addressing mode
RTR	Unsized	[SP] + → CCR [SP]+ → PC Return and restore condition codes
RTS	Unsized	Return from subroutine [SP] + → PC
Scc (EA)	B	If cc is true, then the byte specified by [EA] is set to all ones, otherwise the byte is cleared to zero

- [EA] in JMP and JSR can use all modes except Dn, An, (An) +, –(An), and immediate.
- [EA] in Scc can use all modes except An, relative, and immediate.

- Consider Bcc d. There are 14 branch conditions. This means that cc in Bcc can be replaced by 14 conditions providing 14 instructions. These are BCC, BCS, BEQ, BGE, BGT, BHI, BLE, BLS, BLT, BMI, BNE, BPL, BVC, and BVS. It should be mentioned that some of these instructions are applicable to both signed and unsigned numbers, some can be used with only signed numbers, and some instructions are applicable to only unsigned numbers as shown below:

For both signed and unsigned numbers	For signed numbers	For unsigned numbers
BCC d (Branch if C = 0)	BGE d (Branch if greater or equal)	BHI d (Branch if high)
BCS d (Branch if C = 1)	BGT d (Branch if greater than)	BLS d (Branch if low or same)
BEQ d (Branch if Z = 1)	BLE d (Branch if less than or equal)	
BNE d (Branch if Z = 0)	BLT d (branch if less than)	
	BMI d (Branch if N = 1)	
	BPL d (Branch if N = 0)	
	BVC d (Branch if V = 0)	
	BVS d (Branch if V = 1)	

After signed arithmetic operations such as ADD or SUB the instructions such as BEQ, BNE, BVS, BVC, BMI, and BPL can be used. On the other hand, after unsigned arithmetic operations, the instructions such as BCC, BCS, BEQ, and BNE can be used.

If V = 0, BPL and BGE have the same meaning. Likewise, if V = 0, BMI and BLT perform the same function.

The conditional branch instructions can be used after typical arithmetic instructions such as subtraction to branch to a location if cc is true. For example, consider SUB.W D1, D2. Now if [D1] and [D2] are unsigned numbers, then

BCC d can be used if [D2] > [D1]
BCS d can be used if [D2] < [D1]
BEQ d can be used if [D2] = [D1]
BNE d can be used if [D2] ≠ [D1]
BHI d can be used if [D2] > [D1]
BLS d can be used if [D2] ≤ [D1]

On the other hand, if [D1] and [D2] are signed numbers, then after SUB.W D1, D2, the following branch instructions can be used:

BEQ d can be used if [D2] = [D1]
BNE d can be used if [D2] ≠ [D1]
BLT d can be used if [D2] < [D1]
BLE d can be used if [D2] ≤ [D1]
BGT d can be used if [D2] > [D1]
BGE d can be used if [D2] ≥ [D1]

Now, as an example consider BEQ* + \$20. If [PC] = 000200_{16}, then after execution of the BEQ instruction, program execution starts at 000220_{16} if Z = 1; if Z = 0, program execution continues at 000200. Note that * is used by some assemblers to indicate displacement.

- The instruction BRA and JMP are unconditional JUMP instructions. The BRA (Branch Always) instruction uses the relative addressing mode, whereas the JMP uses only control addressing modes. For example, consider BRA * + \$20. If [PC] = 000200_{16}, then after execution of the BRA, program execution starts at 000220_{16}.

  Now consider JMP (A1). If [A1] = 00000220_{16}, then after execution of the JMP, program execution starts at 000220_{16}.

- The instructions BSR and JSR are subroutine CALL instructions. BSR uses relative mode, whereas JSR uses absolute addressing mode. Consider the following program segment:

Main program			Subroutine
	—	PROC	MOVEM.L D0-D7/A0-A6, –(SP)
	—		—
	JSR PROC		— ⎫ main body of
START	—		— ⎬ the subroutine
	—		MOVEM.L (SP)+ , D0-D7/A0-A6
			RTS

In the above, JSR SUB instruction calls the subroutine called PROC. In response to JSR, the 68000 pushes the current PC contents called START onto the stack and loads the starting address PROC of the subroutine into PC. The first MOVEM in the PROC pushes all registers onto the stack and after the subroutine is executed, the second MOVEM instruction pops all the registers back. Finally, RTS pops the address START from the stack into PC, and the program control is returned to the main program. Note that BSR PROC could have been used instead of JSR PROC in the main program. In that case, the 68000 assembler would have considered the PROC with BSR as a displacement rather than as an address with the JSR instruction.

The use of stack to save temporary variables during subroutine execution can be facilitated by utilizing a stack frame. The stack frame is a set of locations in stack used for saving return addresses, local variables, and I/O parameters. Local variables such as loop counters used in the subroutine are used by the subroutine and are not returned to the main program. The stack frame can be utilized by the subroutine to access a new set of parameters each time the main program calls it.

In 68000, the stack frame can be created by the main program and the subroutine using the LINK and UNLK instructions. The subroutine can access the variables on the stack by using displacements from a base register called a frame pointer. The 68000 LINK instruction can be used to create a stack frame and define the frame pointer.

As an example of LINK/UNLK, the instruction LINK A5, #-100 creates a stack frame of 100_{10} with A5 as the frame pointer. The instruction such as MOVE.W d (A5), D1 and MOVE.W D1,d(A5) can then be used to access the stack frame.

A typical instruction sequence illustrating the use of LINK and UNLK is given below:

Main Program

```
            MOVE.L  Const,-(USP)    ; Push address of 32-bit
                                    ; constant to be passed
            PEA.W   ADDR            ; Push starting address of
                                    ; a table to be passed
            JSR     START           ; Jump to the subroutine
                                      START
              —
              —
              —
```

Subroutine

```
START       LINK    A3,#-12         ; Allocate 12 bytes
              —
              —
              —
            MOVE.L  DATA1,-4(A3),    ; Push local variable 1
            MOVE.L  DATA2,-8(A3),    ; Push local variable 2
            MOVE.L  DATA3,-12(A3)    ; Push local variable 3
Instructions   MOVEA.L 8(A3), A4    ; Obtain ADDR
arbitrarily    MOVE.L  (A4), D1     ; Read data from table
chosen         SUBQ.L  #5,-12(A3)   ; Subtract 5 from local
                                      variable 3
            MOVE.L  12(A3), D4      ; Read the 32-bit constant
              —
              —
              —
            UNLK    A3              ; Restore original values
            RTS                     ; RETURN
```

The following illustrates the stack contents at various points:

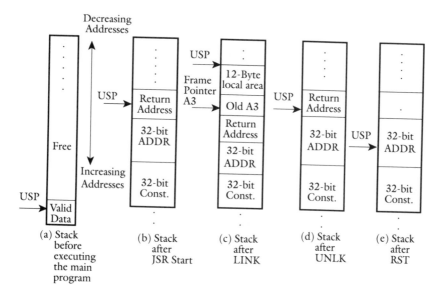

(a) Stack before executing the main program

(b) Stack after JSR Start

(c) Stack after LINK

(d) Stack after UNLK

(e) Stack after RST

In the main program, it is assumed that the 68000 is in user mode and, therefore, USP is used in the program. It is also assumed that the main program passes to the subroutine a 32-bit constant, and the starting address ADDR of a table to be accessed by the subroutine. The subroutine allocates 12 bytes to save three 32-bit local variables by using the LINK instruction. Several instructions are arbitrarily chosen (to illustrate the concept) to be executed by the subroutine to read data from the table, decrement the local variable 3 by 5, and read the 32-bit constant so that they can be used in the instruction sequence that follows (not shown in the subroutine). The UNLK instruction restores the original conditions and the RTS allows the program to return to the right place in the program.

- DBcc Dn, d tests both the condition codes and the value in a data register. DBcc first checks if cc (NE, EQ, GT, etc.) is satisfied. If satisfied, the next instruction is executed. If cc is not satisfied, the specified data register is decremented by 1. If [Dn] = -1, then next instruction is executed; if Dn $\neq -1$, then a branch is taken to PC + d. For example, consider DBNE D5, * -4 and [D5] = 00003002_{16}, [PC] = 002006_{16}. Now if Z = 1, then [D5] = 00003001_{16}; since [D5] $\neq -1$, program execution starts at 002002_{16}. There is a false condition in the DBcc instruction, and this instruction is the DBF (some assemblers use DBRA for this). In this case, the condition is always false. This means that after execution of this instruction, Dn is decremented by 1 and if [Dn] = -1, then the next instruction is executed; if [Dn] $\neq -1$, then branch to PC + d is taken.
- Consider SPL(A5). If [A5] = 00200020_{16} and N = 0, then after execution of the SPL, memory location 200020_{16} will contain 11111111_2.

5.5.8 System Control Instructions

The 68000 contains some system control instructions which include privileged instructions and trap instructions. Note that the privileged instructions can only be executed in the supervisor mode. The system control instructions are listed in Table 5.12.

- The RESET instruction, when executed in supervisor mode, outputs a signal on the RESET pin of the 68000 in order to initialize the external peripheral chips. The 68000 reset pin is bidirectional. The 68000 can be reset by asserting the reset pin using hardware, whereas the peripheral chips can be reset using the software RESET instruction.

TABLE 5.12   68000 System Control Instructions

Instruction	Size	Operation
Privileged Instructions		
RESET	Unsized	If supervisory state, then assert reset line; else TRAP
RTE	Unsized	If supervisory state, then restore SR and PC; else TRAP
STOP #data	Unsized	If supervisory state, then load immediate data to SR and then STOP; else TRAP
ORI to SR		
MOVE to/from USP		
ANDI to SR		These privileged instructions were discussed earlier
EORI to SR		
MOVE (EA) to/from SR		
Trap and Check Instructions		
TRAP # vector	Unsized	$PC \rightarrow - [SSP]$
		$SR \rightarrow - [SSP]$
		Vector address $\rightarrow PC$
TRAPV	Unsized	TRAP if $V = 1$
CHK (EA), Dn	W	If $Dn < 0$ or $Dn > [EA]$, then TRAP
Condition Code Register		
ANDI to CCR		
EORI to CCR		
MOVE (EA) to/from CCR		Already explained earlier
ORI to CCR		

Note: (EA) in CHK can use all modes except An.

- MOVE.L USP, (An) or MOVE.L (An), USP can be used to save, restore, or change the contents of USP in supervisor mode. The USP must be loaded in supervisor mode since MOVE USP is a privileged instruction.
- Consider TRAP # n. There are 16 TRAP instructions with n ranging from 0 to 15. The hexadecimal vector address is calculated using the following equation:

$$\text{Hexadecimal vector address} = 80 + 4^*n$$

The TRAP instruction first pushes the contents of PC and then the SR onto the system (user or supervisor) stack. The hexadecimal vector address is then loaded into PC. The TRAP is basically a software interrupt.

One of the 16 trap instructions can be executed in the user mode to execute a supervisor program located at the specified trap routine. Using the TRAP instruction, control can be transferred to the supervisor mode from the user mode.

There are other traps which occur due to certain arithmetic errors. For example, division by zero automatically traps to location 14_{16}. On the other hand, an overflow condition, i.e., if $V = 1$, will trap to address $1C_{16}$ if the instruction TRAPV is executed.

- The CHK (EA), Dn instruction compares [Dn] with (EA). If $[Dn]_{low\ 16\ bits} < 0$ or if $[D]_{low\ 16\ bits} > (EA)$, then a trap to location 0018_{16} is generated. Also, N is set to 1 if $[Dn]_{low\ 16\ bits} < 0$ and N is reset to zero if $[Dn]_{low\ 16\ bits} > (EA)$. (EA) is treated as a 16-bit 2's complement integer. Note that program execution continues if $[Dn]_{low\ 16\ bits}$ lies between 0 and (EA).

Consider CHK (A5), D2. If $[D2]_{low\ 16\ bits} = 0200_{16}$, $[A5] = 00003000_{16}$, $[003000_{16}] = 0100_{16}$, then after execution of the CHK, the 68000 will trap since $[D2] = 0200_{16}$ is greater than $[003000] = 0100_{16}$.

The purpose of the CHK instruction is to provide boundary checking by testing if the content of a data register is in the range from zero to an upper limit. The upper limit used in the instruction can be set equal to the length of the array. Then every time the array is accessed, the CHK instruction can be executed to make sure that the array bounds have not been violated.

The CHK instruction is usually placed following the computation of an index value to ensure that the limits of this index value are not violated. This allows checking whether or not the address of an array being accessed is within the array boundaries when the address register indirect with index mode is used to access an element in the array. For example, the instruction sequence shown below will allow accessing of an array with base address in A3 and array length of 125_{10} bytes:

```
        —

        —

        —

CHK  #124,  D5
MOVE.B  0  (A3,  D5.W),  D4

        —

        —

        —

        —
```

In the above, if low 16 bits of D5 is greater than 124, the 68000 will trap to location 0018_{16}.

In the above, it is assumed that D5 is computed prior to execution of the CHK instruction. Also, the 68000 assembler requires that a displacement value of 0 be specified as in the instruction MOVE.B 0(A3, D5.W), D4.

5.6 68000 Stacks

The 68000 supports stacks and queues with the address register indirect postincrement and predecrement addressing modes.

In addition to SPs, all seven address registers A0-A6 can be used as stack pointers by using appropriate addressing modes.

Subroutine calls, traps, and interrupts automatically use the system stack pointers: USP when S = 0 and SSP when S= 1. Subroutine calls push PC onto the stack, while RTS pops PC from the system stack. Traps and interrupt push both PC and SR onto the system, while RTE pops PC and SR from the system stack.

These stack operations fill data from high memory to low memory. This means that the system SP is predecremented by 2 (word) or 4 (long word) with push and post incremented by 2 (for word) or 4 (for long word) after pop. As an example, suppose that a 68000 call instruction (JSR or BSR) is executed when PC = $0031 F200; then after execution of the subroutine call the stack will push the PC as follows:

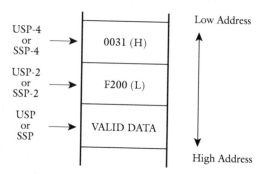

Note that the system stack pointer always points to valid data.

Stacks can be created by the user by using address register indirect with postincrement or predecrement modes. Using one of the seven (A0-A6) address registers, the user may create stacks which can be filled from either high memory to low memory or vice versa.

Stacks from high to low memory are implemented with predecrement mode for push and postincrement mode for pop. On the other hand, stacks from low to high memory is implemented with postincrement for push and predecrement for pop. For example, consider the following stack growing from high to low memory addresses in which A5 is used as the stack pointer:

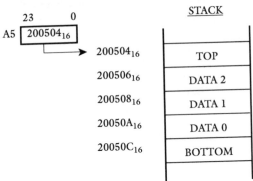

In order to push 16-bit content 0504_{16} of memory location 305016_{16}, the instruction MOVE.W \$305016, –(A5) can be used as follows:

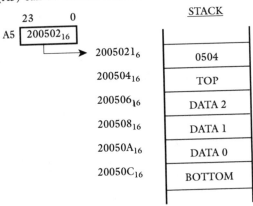

The 16-bit data 0504_{16} can be popped from the stack into low 16 bits of D0 by using MOVE.W (A5)+, D0. A5 will contain 200504_{16} after the POP. Note that in this case, the stack pointer A5 points to valid data.

Next, consider the stack growing from low to high memory addresses in which, say, A6 is used as the stack pointer:

STACK

305004_{16}	BOTTOM
305006_{16}	DATA 0
305008_{16}	DATA 1
$30500A_{16}$	TOP
$30500C_{16}$	FREE

23   0 → $30500C_{16}$

A6 $30500C_{16}$

Now, in order to PUSH 16-bit contents 2070_{16} of the low 16 bits of D5, the instruction MOVE.W D5, (A6)+ can be used as follows:

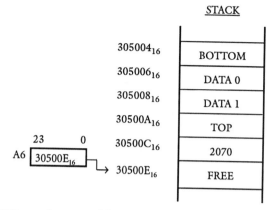

The 16-bit data 2070_{16} can be popped from the stack into 16-bit memory location 417024_{16} by using MOVE.W– (A6), $417024.

Note that in this case, the stack pointer A6 points to the free location above the valid data.

Example 5.1

Determine the effect of each one of the following 68000 instructions:

- **CLR D0**
- **MOVE.L D1, D0**
- **CLR.L (A0)+**
- **MOVE – (A0), D0**
- **MOVE 20(A0), D0**
- **MOVEQ.L #$D7, D0**
- **MOVE 21(A1, A0.L), D0**

Assume the following initial configuration before each instruction is executed. Also, assume all numbers in hex.

$$[D0] = 22224444, \quad [D1] = 55556666$$
$$[A0] = 00002224, \quad [A1] = 00003333$$
$$[002220] = 8888, \quad [002222] = 7777$$
$$[002224] = 6666, \quad [002226] = 5555$$
$$[002238] = AAAA, \quad [00556C] = FFFF$$

Solution

Instruction	Effective address	Net effect (HEX)
CLR D0	Destination EA = D0	[D0] = 22220000
MOVE.L D1, D0	Destination EA = D0	[D0] ← 55556666
	Source EA = D1	
CLR.L (A0)+	Destination EA = [A0]	[002224] = 0000
		[002226] = 0000
		[A0] ← 00002228
MOVE –(A0), D0	Source EA = [A0] – 2	[A0) ← 00002222
	Destination EA = D0	[D0] = 22227777
MOVE 20 (A0), D0	Source EA = [A0] + 20_{10} (or 14_{16})	[D0] = 2222AAAA
	= 002238	
	Destination EA = D0	
MOVEQ.L#$D7, D0	Source data = $D7_{16}$	[D0] ← FFFFFFD7
	Destination EA = D0	
MOVE 21(A0, A1.L), D0	Source EA = [A0] + [A1] + 21_{10}	[D0] = 2222FFFF
	= 556C	
	Destination EA = D0	
	. . .	

Example 5.2

Write a 68000 assembly program segment that implements each one of the following Pascal segments. Assume the following information about the variables involved in this problem.

Variable	Comments
X	Address of a 16-bit signed integer of array 1 of 10 elements
Y	Address of a 16-bit signed integer of array 2 of 10 elements
SUM	Address of the sum
	(a) If X $\geq$ Y, then X := X + 10;
	else Y :=Y – 12.
	. . .
	(b) Sum := 0; for i := 0 to 9, do
	sum := sum + A (i).
	. . .

Solution

(a)

```
        LEA X, A0        ;  Point A0 to X
        LEA Y, A1        ;  Point A1 to Y
        MOVE (A0), D0    ;  MOVE [X] into D0
        CMP (A1), D0     ;  COMPARE [X] WITH [Y]
        BGE THRPT
        SUBI #12, (A1)   ;  Execute else part
        BRA NEXT
THRPT   ADDI#10, (A0)    ;  Execute then part
NEXT    . . .
```

(b)

```
           LEA SUM, A1       ;  Point to SUM
           LEA Y, A0         ;  Point A0 to Y [0]
           CLR D0            ;  Clear the sum to zero
           MOVEQ.L #9,D1     ;  Initialize D1 with loop
                                  limit
     LOOP  ADD (A0) +, D0    ;  Perform the iterative
           DBF D1,LOOP       ;  Summation
           MOVE D0, (A1)     ;  Transfer the result
           . . .
```

Note that condition F in DBF is always false and thus we exit from the loop only when the content of the register D1 becomes –1. Therefore, we repeat the addition process for 10 times as desired.

. . .

Example 5.3

Write 68000 assembly program to clear 85_{16} consecutive bytes.

Solution

```
        ORG $2000
        MOVEA.L#$3000,A0   ;  LOAD A0 WITH $3000
        MOVE #$84, D0      ;  MOVE 84₁₆ INTO D0
LOOP    CLR.B(A0)+         ;  CLEAR [3000₁₆] AND
                           ;  POINT TO NEXT ADDRESS
```

```
              DBF  D0, LOOP        ;  DECREMENT AND BRANCH
FINISH   JMP FINISH                ;  HALT
```

Note that the 68000 has no HALT instruction in the user mode. The 68000 has the STOP instruction in supervisor mode. Therefore, the unconditional JUMP to some location such as FINISH JMP FINISH in the above program must be used. Since DBF is a word instruction and considers D0's low 16-bit word as the loop count, D0 must be initialized by a word MOVE rather than byte MOVE, even though 84_{16} can be accommodated in a byte. Also, one should be careful about using MOVEQ, since MOVEQ sign extends a byte to a long word.

Example 5.4

Write 68000 assembly program to compute

$$\sum_{i=1}^{N} X_i\, Y_i$$

where Xi and Yi are signed 16-bit numbers, N = 100. Assume no overflow and that the result is 32-bit wide.

Solution

```
              ORG  $1000
              MOVEQ.L#99,D0       ;  MOVE 99₁₀ INTO D0
              LEA  P,  A0         ;  LOAD ADDRESS P INTO A0
              LEA  Q,  A1         ;  LOAD ADDRESS Q INTO A1
              CLR.L D1            ;  Initialize D1 to zero
LOOP          MOVE (A0)+, D2      ;  MOVE [X] TO D2
              MULS (A1)+, D2      ;  D2 ← [X] * [Y]
              ADD.L D2, D1        ;  D1 ← Σ Xi Yi
              DBF  D0, LOOP       ;  DECREMENT AND BRANCH
FINISH   JMP FINISH               ;  HALT
```

Example 5.5

Write a 68000 subroutine to compute

$$Y = \sum_{i=1}^{N} \frac{X_i^{\,2}}{N}$$

Assume the Xi's are 16-bit signed integers and N = 100. The numbers are stored in consecutive locations. Assume A0 points to the Xi's and SP is already initialized. Also, assume no overflow.

Solution

```
SQRE     MOVEM.L D2/D3/A0,-(SP)   ;  Save registers
         CLR.L D1                 ;  Clear sum
         MOVEQ.L #99,D2           ;  Initialize loop count
LOOP     MOVE.W (A0)+, D3         ;  Move Xis into D3
         MULS D3,D3               ;  Computer Xi²
```

```
        ADD.L D3, D1
        DBF D2, LOOP            ;   Store sum of Xi² into D1
        DIVU #100, D1           ;   Compute ∑ Xi²/N
        MOVEM.L (SP)+, D2/D3/A0 ;   Restore registers
        RTS
```

Example 5.6

Write a 68000 assembly language program to move block of data length 100_{10} from the source block starting at location 002000_{16} to the destination block starting at location 003000_{16}.

Solution

```
        MOVEA.L #$2000, A4     ;   Load A4 with source address
        MOVEA.L #$3000, A5     ;   Load A5 with destination
                              ;   address
        MOVEQ.L #99, D0        ;   Load D0 with count -1 = 99₁₀
START   MOVE.W  (A4)+, (A5)+   ;   MOVE source data to destination
        DBF D0, START          ;   Branch if D0 ≠ -1
END     JMP END                ;   HALT
                                   . . .
```

Example 5.7

Write a 68000 assembly language program to add two words: each contains two ASCII digits. The first word is stored in two consecutive locations with the low byte pointed to by A0 at offset 0300_{16}, while the second word is stored in two consecutive locations with the low byte pointed to by A1 at 0700_{16}. Store the result in hex in memory pointed to by A1.

Solution

```
        MOVEQ.B #1,D2         ;   Initialize loop
        MOVEA.L #$0300,A0     ;   Initialize A0
        MOVEA.L #$0700,A1     ;   Initialize A1
        MOVEE.W #0,CCR        ;   Clear X-bit
START   MOVE.B  (A0)+,D0      ;   Move data
        ANDI.B  #$0F,D0       ;   Mask off upper nibble
        ANDI.B  #$0F,(A1)     ;   Mask off upper nibble
        ADDX.B  D0,(A1)+      ;   Add data
        DBF D2,START
END     JMP END              ;   Halt
```

Example 5.8

Write 6800 assembly language program to compare a source string of 50_{10} words pointed to by A0 with a destination string pointed to by A1. The program should be halted as soon as a match is found or the end of string is reached.

Solution

```
        MOVEQ.B #49,D0       ;   Initialize loop count
BEGIN   CMPM (A0)+,(A1)+     ;   Compare string
        DBEQ D0,BEGIN        ;   Compare until match or end of
                             ;   string
FINISH  JMP FINISH           ;   Halt
```

Example 5.9

Write a subroutine in 68000 assembly language which can be called by a main program. The subroutine will multiply a signed 16-bit number by a signed 8-bit number. The main program will call this subroutine, store the result in two memory words pointed to by A0 and stop. Assume A1 and A2 contain the signed 8-bit and 16-bit data respectively. Assume that the stack pointer is already initialized. Assume supervisor mode.

Solution

Main program

```
             MOVE.B  (A1),D0                 ;  Move 8-bit data
             MOVE.W  (A2),D1                 ;  Move 16-bit data
             CALL MULTI                      ;  Call subroutine
             MOVE.L  D1,(A0)                 ;  Store result
FINISH       JMP FINISH                      ;  Halt
```

Subroutine

```
MULTI        MOVEM.L  D0/D1/A0/A1/A2,-(A7)   ;  Save registers
             EXT.W D0                        ;  Sign extend D0.B
             MULS.W  D0,D1                   ;  Signed multiply
             MOVEM.L  (A7)+,D0/D1/A0/A1/A2   ;  POP registers
             RTS
```

Example 5.10

Write 68000 assembly language program to subtract two 64-bit numbers. Assume D3D4 and D1D0 contain the two 64-bit numbers. Store the result in D3D4.

Solution

```
             SUB.L  D0,D4        ;  Subtract
                                 ;  Low 32-bit
             SUBX.L  D1,D3       ;  Subtract
                                 ;  High 32-bit
FINISH       JMP FINISH          ;  HALT
```

Example 5.11

Write 68000 assembly language program that will perform the following operation:

$$5 * X - 6 * Y + (Y/8) \rightarrow D1.L$$

where X is unsigned 8-bit number stored in lowest byte of D0 and Y is 16-bit signed number stored in upper 16-bit number in D1 respectively. Discard remainder of Y/8.

Solution

```
             ANDI.W  #$00FF,D0   ;  Mask off high 16 bits
             MULU.W  #5,D0       ;  D0.L ← 5 * X
             SWAP.W  D1          ;  Data in low 16-bit
             MOVE.W  D1,D2       ;  Save D1
             MULS.W  #-6,D1      ;  D1.L ← -6 * Y
             ADD.L  D0,D1        ;  D1.L ← 5 * X - 6 * Y
             LSR.W  #3,D2        ;  D2.W ← Y/8
```

```
        ANDI.L #$0000FFFF,D2   ;   Mask off high 16 bits of D2
        ADD.L D2,D1            ;   Store result in D1.L
FINISH  JMP FINISH
```

Example 5.12

Write 6800 assembly program to convert a number from Fahrenheit degrees in bytes to Celsius degrees. The source byte is assumed to reside in memory pointed to by A0 and the result to be stored in use the equation C = [(F − 32)/9] * 5.

Solution

```
        MOVE.B (A0),D0   ;   Get degrees F.
        EXT.W D0         ;   sign extend
        SUBI.W #32,D0    ;   subtract 32
        MULS.W #5,D0     ;   D0 ← (F - 32) * 5
        DIVS #9,D0       ;   D0 ← D0/9
FINISH  JMP FINISH
```

5.7 68000 Pins and Signals

The 68000 is housed in one of the following packages:

- 64-pin dual in-line package (DIP)
- 68-Terminal Chip Carrier
- 68-pin Quad Pack
- 68-pin Grid Array

Figure 5.7 shows the pin diagrams of the 64-pin DIP package. Pin diagrams for the other three packages are shown in Appendix B.

The 68000 is provided with several Vcc and ground pins. Power is thus distributed in order to reduce noise problems at high frequencies.

In order to build a prototype to demonstrate that the paper design for the 68000-based microcomputer is correct, one must use either wire-wrap or solder for the actual construction. Prototype board must not be used. This is because at high frequencies above 4 MHz, there will be noise problems due to stray capacitances.

D0-D15 is the 16-bit data bus. Note that Dn indicates a data pin of the 68000 during hardware discussions while Dn refers to a data register of the 68000 during software discussions. All transfers to and from memory and I/O devices are conducted over the 16-bit bus. A1-A23 are the 23 address lines. A0 is obtained by encoding UDS (Upper Data Strobe) and LDS (Lower Data Strobe) lines. The 68000 operates on a single-phase TTL level clock at 4 MHz, 6 MHz, 8 MHz, 10 MHz, 12.5 MHz, 16.67 MHz, or 25 MHz. The 68000 also has a lower-power HCMOS version called the MC68HC000 which can run at 8 MHz, 10 MHz, 12.5 MHz, and 16.67 MHz.

There is no on-chip clock generator/driver circuitry and therefore the clock must be generated externally. This clock input is utilized internally by the 68000 to generate additional clock signals for synchronizing the 68000's internal operation.

Figure 5.8 shows the 68000 CLK waveform and clock timing specifications.

The clock is at TTL compatible voltage. The clock timing specification provides data for three different clock frequencies: 8 MHz, 10 MHz, and 12.5 MHz.

The 68000 CLK input can be provided by a crystal oscillator or by designing an external circuit. Figure 5.9 shows a simple oscillator to generate the 68000 CLK input. The clock circuit

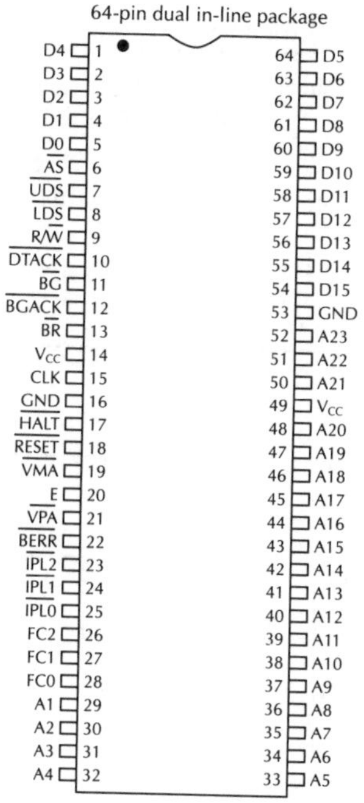

64-pin dual in-line package

```
           D4 ▯ 1   •            64 ▯ D5
           D3 ▯ 2               63 ▯ D6
           D2 ▯ 3               62 ▯ D7
           D1 ▯ 4               61 ▯ D8
           D0 ▯ 5               60 ▯ D9
           A̅S̅ ▯ 6               59 ▯ D10
          U̅D̅S̅ ▯ 7               58 ▯ D11
          L̅D̅S̅ ▯ 8               57 ▯ D12
         R̅/̅W̅ ▯ 9               56 ▯ D13
       D̅T̅A̅C̅K̅ ▯ 10              55 ▯ D14
           B̅G̅ ▯ 11              54 ▯ D15
        B̅G̅A̅C̅K̅ ▯ 12             53 ▯ GND
           B̅R̅ ▯ 13              52 ▯ A23
          Vcc ▯ 14              51 ▯ A22
          CLK ▯ 15              50 ▯ A21
          GND ▯ 16              49 ▯ Vcc
         H̅A̅L̅T̅ ▯ 17              48 ▯ A20
        R̅E̅S̅E̅T̅ ▯ 18             47 ▯ A19
          VMA ▯ 19              46 ▯ A18
            E ▯ 20              45 ▯ A17
          V̅P̅A̅ ▯ 21              44 ▯ A16
         B̅E̅R̅R̅ ▯ 22             43 ▯ A15
         IPL2 ▯ 23              42 ▯ A14
         IPL1 ▯ 24              41 ▯ A13
         IPL0 ▯ 25              40 ▯ A12
          FC2 ▯ 26              39 ▯ A11
          FC1 ▯ 27              38 ▯ A10
          FC0 ▯ 28              37 ▯ A9
           A1 ▯ 29              36 ▯ A8
           A2 ▯ 30              35 ▯ A7
           A3 ▯ 31              34 ▯ A6
           A4 ▯ 32              33 ▯ A5
```

FIGURE 5.7   64-pin grid array.

uses two inverters connected in series. Inverter 1 is biased in its transition region by the resistor R. Inverter 1 inputs the crystal output (sinusoidal) to produce logic pulse train at the output of inverter 1. Inverter 2 sharpens the wave and drives the crystal. For this circuit to work, HCMOS logic (74HC00, 74HC02, or 74HC04) must be used and a coupling capacitor should be connected across the supply terminals to reduce the ringing effect during high-frequency switching of the HCMOS devices. Additionally, the output of this oscillator is fed to the clock input of a D-flip-flop (74LS74) to further reduce the ringing. Hence, a clock signal of 50% duty cycle at a frequency of 1/2 the crystal frequency is generated.

The 68000 consumes about 1.5 watts of power.

The 68000 signals can be divided into five functional categories. These are

1. Synchronous and asynchronous control lines
2. System control lines
3. Interrupt control lines
4. DMA control lines
5. Status lines

5.7.1 Synchronous and Asynchronous Control Lines

The 68000 bus control is asynchronous. This means that once a bus cycle is initiated, the external device must send a signal back in order to complete it. The 68000 also contains three synchronous control lines that facilitate interfacing to synchronous peripheral devices such as Motorola's inexpensive MC6800 family. Note that synchronous operation means that bus control is synchronized or clocked using a common system clock signal. In 6800 family peripherals, this common clock is a phase 2 (Ø2) or an E clock signal depending on the particular chip used.

With synchronous control, all READ and WRITE operations must be synchronized with the common clock. However, this may create problems when interfacing slow peripheral devices. This problem does not arise with asynchronous bus control.

AC Electrical Specifications — Clock Timing

Characteristic	Symbol	8 MHz		10 MHz		12.5 MHz		Unit
		Min	Max	Min	Max	Min	Max	
Frequency of Operation	1	4.0	8.0	4.0	10.0	4.0	12.5	MHz
Cycle Time	t_{cyc}	125	250	100	250	80	250	ns
Clock Pulse Width	t_{CL}	55	125	45	125	35	125	ns
	t_{CH}	55	125	45	125	35	125	
Rise and Fall Times	t_{CR}	—	10	—	10	—	5	ns
	t_{Cl}	—	10	—	10	—	5	

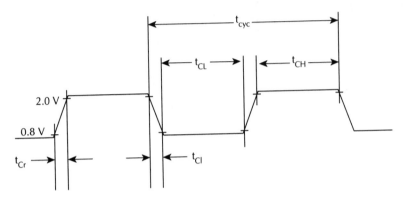

Note: Timing measurements are referenced to and from a low voltage of 0.8 volts and a high voltage of 2.0 volts, unless otherwise noted. The voltage swing through this range should start outside and pass through the range such that the rise or fall will be linear between 0.8 and 2.0 volts.

FIGURE 5.8   Clock input timing diagram.

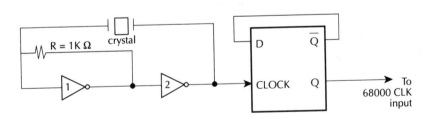

FIGURE 5.9   External clock circuitry.

The 68000 utilizes either synchronous or asynchronous bus cycles for data transfer. The purpose of using the synchronous bus cycles is to interface Motorola's inexpensive MC6800 peripheries to the 68000. On the other hand, the asynchronous bus protocol of the 68000 too is used to interface non-6800 type peripherals to the 68000. Once the 68000 starts a bus cycle, it expects one of three signals to tell it how to terminate the cycle. These signals are VPA (synchronous operation), DTACK (Asynchronous operation), or BERR (Bus Error; when no DTACK is received from the periperal device). Synchronous and Asynchronous operations are described in the following.

The 68000 has three control lines to transfer data over its bus in a synchronous manner. These are E (enable), VPA (valid peripheral address), and VMA (valid memory address).

The E clock corresponds to phase 2 clock of the 6800. The E clock is output at a frequency that is 1/10th of the 68000 input clock. The VPA is an input which tells the 68000 that a 6800 device is being addressed and therefore data transfer must be synchronized with the E clock. The VMA is processor's response to VPA. VPA is asserted when the memory address is valid. This also tells the external device that the next data transfer over the data bus will be synchronized with the E clock. VPA can be generated by decoding the address pins and address strobe (AS). Note that the 68000 asserts AS low when the address on the address bus is valid. VMA is typically used as part of the chip select of the 6800 peripheral. This ensures that the 6800 peripherals are selected and deselected at the correct time. The 6800 peripheral interfacing sequence is provided in the following:

1. The 68000 initiates a cycle by starting a normal read or write cycle.
2. The 6800 peripheral defines the 6800 cycle by asserting the 68000 VPA input. If the VPA is asserted as soon as possible after the assertion of AS, then VPA will be recognized as being asserted on the falling edge of S4 (third cycle). If the VPA is not asserted at the falling edge of S4 (third cycle), the 68000 inserts wait states until VPA is recognized by the 68000 as asserted. DTACK should not be asserted while VPA is asserted. The 6800 peripheral must remove VPA within one clock after AS is negated.
3. The 68000 monitors enable (E) until it is low. The 68000 then synchronizes all read and write operations with the E clock. VMA output pin is asserted low by the 68000.
4. The 68000 peripheral waits until E is active (HIGH) and then transfers the data.
5. The 68000 waits until E goes low (on a read cycle the data is latched as E goes low internally). The 68000 then negates VMA, AS, UDS, and LDS. The 68000 thus terminates the cycle and starts the next cycle.

Asynchronous operation is not dependent on a common signal. The 68000 utilizes the asynchronous control lines to transfer data between the 68000 and peripheral devices via handshaking. Using asynchronous operation, the 68000 can be interfaced to any peripheral chip regardless of its speed.

The 68000 provides five lines to control address and data transfers asynchronously. These are AS (Address Strobe), R/W (Read/Write), DTACK (Data Acknowledge), UDS (Upper Data Strobe), and LDS (Lower Data Strobe).

The 68000 outputs $\overline{AS}$ to notify the peripheral device when data are to be transferred. $\overline{AS}$ is active LOW when the 68000 provides a valid address on the address bus. The R/W is HIGH for read and LOW for write. The $\overline{DTACK}$ is used to tell the 68000 that a transfer is to be performed.

When the 68000 wants to transfer data asynchronously, it first generates the required address on the address lines in order to select the peripheral device and then activates the $\overline{AS}$ line.

Since the $\overline{AS}$ line tells the peripheral chip when to transfer data, the $\overline{AS}$ line should be part of the address decoding scheme. After enabling the $\overline{AS}$, the 68000 enters the wait state until it receives the $\overline{DTACK}$ from the selected peripheral device. On receipt of the $\overline{DTACK}$, the 68000 knows that the peripheral device is ready for data transfer. The 68000 then utilizes the R/W, $\overline{UDS}$, $\overline{LDS}$ and data lines to transfer data.

$\overline{UDS}$ and $\overline{LDS}$ are defined as follows:

$\overline{UDS}$	$\overline{LDS}$	Data Transfer occurs via	Address
1	0	D0-D7 pins for byte	Odd
0	1	D8-D15 pins for byte	Even
0	0	D0-D15 pins for word or Long word	Even

$\overline{UDS}$ and $\overline{LDS}$ are used to segment the memory into bytes instead of words. When the $\overline{UDS}$ is asserted, contents of even addresses are transferred on the high-order 8 lines of the data bus, D8-D15. The 68000 internally shifts these data to the low byte of the specified register. When $\overline{LDS}$ is asserted, contents of odd addresses are transferred on the low-order 8 lines of the data bus, D0-D7. During word and long word transfers, both $\overline{UDS}$ and $\overline{LDS}$ are asserted and information is transferred on all 16 data lines, D0-D15. However, an additional cycle is required for a long word 32-bit transfer. Note that during byte memory transfers, A0 corresponds to $\overline{UDS}$ for even addresses (A0 = 0) and to $\overline{LDS}$ for odd addresses (A0 = 1). The circuit in Figure 5.10 shows how even and odd memory chips are interfaced to the 68000.

5.7.2 System Control Lines

The 68000 has three control lines, namely, $\overline{BERR}$ (Bus Error), $\overline{HALT}$, and $\overline{RESET}$, that are used to control system-related functions.

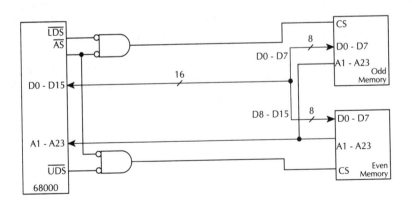

FIGURE 5.10   Interfacing of the 68000 to even and odd addresses.

The BERR is an input to the 68000 that is used to inform the processor that there is a problem with the instruction cycle currently being executed. With asynchronous operation, this problem may arise if the 68000 does not receive DTACK or VPA from a peripheral device. An external timer can be used to activate the BERR pin if the external device does not send DTACK within a certain period of time. On receipt of the BERR, the 68000 does one of the following:

- Automatically reruns the instruction cycle which caused the error
- Executes an error service routine

The troubled instruction cycle is rerun by the 68000 if it receives a HALT signal along with the BERR signal. On receipt of LOW on both HALT and BERR pins, the 68000 completes the current instruction cycle and then places its buses into the high impedance state. On removal of both HALT and BERR (that is, when both HALT and BERR are HIGH), the 68000 reruns the troubled instruction cycle. The cycle can be rerun repeatedly if both BERR and HALT are enabled/disabled continually.

On the other hand, an error service routine is executed only if the BERR is received without HALT. In this case, the 68000 will branch to a bus error vector address where the user can write a service routine. If two simultaneous bus errors are received due to rerun of the troubled instruction cycle via the BERR pin without HALT, the 68000 automatically goes into the HALT state until it is reset.

The HALT line can also be used by itself to perform single-stepping or to provide DMA. When HALT input is activated, the 68000 completes the current instruction and goes into a high impedance state until HALT is returned to HIGH. By enabling/disabling the HALT line continually, the single-stepping debugging can be accomplished. However, since most 68000 instructions consist of more than one instruction cycle, single-stepping using HALT is not normally used. Rather, the trace bit in the status register is used to single-step the complete instruction.

One can also use HALT to perform microprocessor-halt DMA. Since the 68000 has separate DMA control lines, DMA using the HALT line will not normally be used.

The HALT pin can also be used as an output signal. The 68000 will assert the HALT pin LOW when it goes into a HALT state as a result of a catastrophic failure. The double bus error (activation of BERR twice) is an example of this type of error. When this occurs, the 68000 places its bus into a high impedance state until it is reset. The HALT line informs the peripheral devices of the catastrophic failure.

The RESET line of the 68000 is also bidirectional. In order to reset the 68000, both the RESET and HALT pins must be asserted at the same time. The 68000 executes a reset service routine automatically for loading the PC with the starting address of the program. The 68000 RESET pin can also be used as an output line. A LOW can be sent to this output line by executing the RESET instruction in the supervisor mode in order to reset external devices connected to the 68000. The execution of the RESET instruction does not affect any data, address, or status register. Therefore, the RESET instruction can be placed anywhere in the program whenever the external devices need to be reset.

In order to reset the 68000, both the RESET and HALT pins must be asserted simultaneously by an external circuit. Figure 5.11 shows the timing diagram for the 68000 reset operation.

Upon hardware reset, the 68000 performs the following:

1. The 68000 reads four words from addresses $000000, $000002, $000004, and $000006. The 68000 loads the supervisor stack pointer high and low words with the contents of locations $000000 and $000002, respectively. Also, the program counter high and low words are loaded with the contents of locations $000004 and $000006, respectively.
2. The 68000 initializes the status register to an interrupt mask level of seven.
3. No other registers are affected by hardware reset.

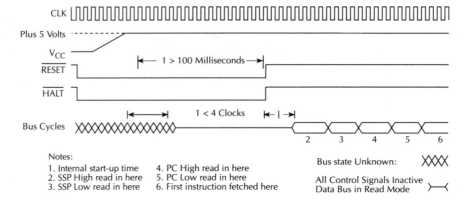

Notes:
1. Internal start-up time
2. SSP High read in here
3. SSP Low read in here
4. PC High read in here
5. PC Low read in here
6. First instruction fetched here

Bus state Unknown:
All Control Signals Inactive
Data Bus in Read Mode

FIGURE 5.11   68000 reset operation timing diagram.

When a RESET instruction is executed, the 68000 drives the $\overline{\text{RESET}}$ pin low for 124 clock periods. In this case, the processor is trying to reset the rest of the system. Therefore, there is no effect on the internal state of the 68000. All of the 68000 internal registers and the status register are unaffected by the execution of a reset instruction. All external devices connected to the $\overline{\text{RESET}}$ line will be reset at the completion of the reset instruction.

Asserting the $\overline{\text{RESET}}$ and $\overline{\text{HALT}}$ lines for 10 clock cycles will cause a processor reset, except when Vcc is initially applied to the 68000. In this case, an external reset must be applied for at least 100 milliseconds.

The reset circuit (used for 8085) depicted in Figure 5.12a satisfies the 68000 reset requirements mentioned above.

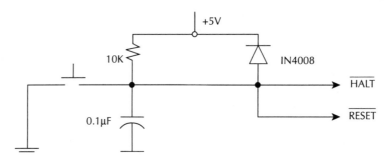

FIGURE 5.12a   68000 RESET circuit (simple).

The above circuit is similar to the 8085 reset circuit except that the output goes to both $\overline{\text{RESET}}$ and $\overline{\text{HALT}}$ lines of the 68000. A more accurate RESET circuit is shown in Figure 5.12b.

The Motorola MC1455 in Figure 5.12b is a timer chip that provides accurate time delays or oscillation. The timer is precisely controlled by external resistors and capacitors. The timer may be triggered by an external trigger input (falling waveform) and can reset by external reset input (falling waveform).

The reset circuit in Figure 5.12b will assert the 68000 $\overline{\text{RESET}}$ pin for at least 10 clock cycles. The internal block diagram of the MC1455 is shown in Figure 5.12c.

When the input voltage (Vcc) to the trigger comparator (COMPB) falls below 1/3 Vcc, the comparator output (COMPA) triggers the flip-flop so that its output becomes low. This turns the capacitor discharge transistor OFF and drives the digital output to the HIGH state. This condition permits the capacitor to charge at an exponential rate set by the RC time constant.

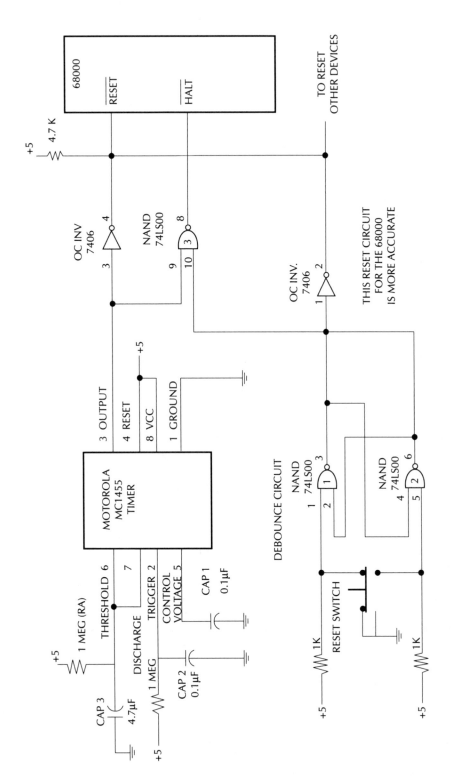

FIGURE 5.12b 68000 RESET circuit.

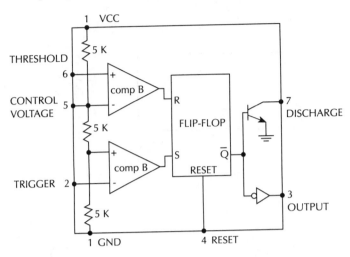

FIGURE 5.12c   MC1455 internal block diagram.

When the capacitor voltage reaches $^2/_3$ Vcc, the threshold comparator resets the flip-flop. The action discharges the timing capacitor and returns the digital output to the LOW state. The output will be HIGH for t = 1.1 R_AC seconds, where R_A = 1M and C = 4.7 μF.

The MC1455 can be connected so that it will trigger itself and the capacitor voltage will oscillate between $^1/_3$ Vcc and $^2/_3$ Vcc. Once the flip-flop has been triggered by an input signal, it cannot be retriggered until the present timing period has been completed. A reset pin is provided to discharge the capacitor, thus interrupting the timing cycle. The reset pin should be tied to Vcc when not in use. With proper trigger input as in the figure, the MC1455 output will stay HIGH for 1.1 R_AC = 5.17 s (1.1 * 10^6 * 4.7 * 10^{-6}). The 68000 requires the RESET and HALT lines to be low for at least 10 cycles. If the 68000 clock cycle is 0.125 μs (8-MHz clock), then the 68000 RESET and HALT pins must be LOW for at least 0.125 μs * 10 = 1.25 μs. Since the MC1455 output is connected to the 68000 RESET pin through an inverter, the RESET pin will be held LOW for 5.17 s (greater than 1.25 μs). Hence, the timing requirement for the 68000 RESET pin is satisfied. The HALT pin is activated by NANDing the MC1455 true output and the debouncing circuit output. The HALT pin is LOW when both inputs to NAND gate #3 are HIGH. The MC1455 output is HIGH for 5.17 s and the output of the debounce circuit is HIGH when the push button is activated. This will generate a LOW at the HALT pin for 5.17 s (greater than 1.25 μs). The timing requirements of the 68000 RESET and HALT pins will be satisfied by the reset circuit of Figure 5.12b.

Note that when the reset circuit is not activated, the bottom input of AND gate #2 is HIGH and the top input of AND gate #1 is LOW (grounded to LOW, see Figure 5.12b). Since a NAND gate always produces a HIGH output when one of the inputs is LOW, the output of NAND gate #1 will be HIGH. This will make the top input of NAND gate #2 HIGH and thus the output of NAND gate #2 will be LOW, which in turn will make the output of NAND gate #3 HIGH. Therefore, the 68000 will not be reset when the push button is not activated. Upon activation of the push button, the bottom input of NAND gate #2 is LOW (grounded to LOW); this will make the output of NAND gate #2 HIGH. Hence, both inputs of NAND gate #3 will be HIGH providing a LOW at the HALT pin for 5.17 s (greater than 1.25 μs).

TABLE 5.13 Function Code Lines

FC2	FC1	FC0	Operation
0	0	0	Unassigned
0	0	1	User data
0	1	0	User program
0	1	1	Unassigned
1	0	0	Unassigned
1	0	1	Supervisor data
1	1	0	Supervisor program
1	1	1	Interrupt acknowledge

5.7.3 Interrupt Control Lines

IPL0, IPL1, and IPL2 are interrupt control lines. These lines provide for seven interrupt priority levels (IPL2, IPL1, IPL0 = 111 means no interrupt and IPL2, IPL1, IPL0 = 000 means nonmaskable interrupt). IPL2, IPL1, IPL0 = 001 through 110 provides six maskable interrupts. The 68000 interrupts are discussed later in this chapter.

5.7.4 DMA Control Lines

BR (Bus Request), BG (Bus Grant), and BGACK (Bus Grant Acknowledge) lines are used for DMA purposes. The 68000 DMA will be discussed later in this chapter.

5.7.5 Status Lines

The 68000 has three output lines called the function code pins (FC2, FC1, and FC0). Table 5.14 shows how these lines tell external devices whether user data, user program, supervisor data, or supervisor program is being addressed. These lines can be decoded to provide user or supervisor programs and/or data, and interrupt acknowledge as shown in Table 5.13.

The FC2, FC1, and FC0 pins can be used to partition memory into four functional areas: user data memory, user program memory, supervisor data memory, and supervisor program memory. Each memory partition can directly access up to 16 megabytes, and thus the 68000 can be used to directly address up to 64 megabytes of memory. This is shown in Figure 5.13.

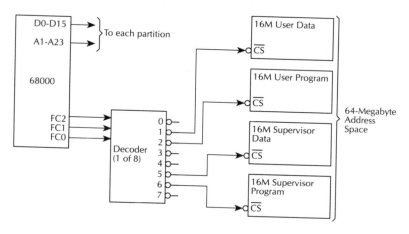

FIGURE 5.13 Partitioning 68000 address space using FC2, FC1, and FC0 pins.

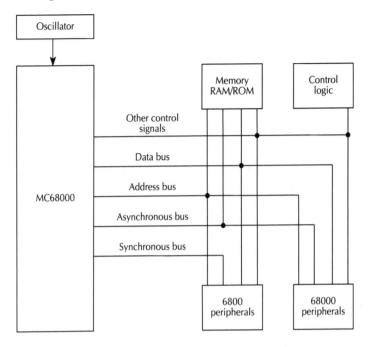

FIGURE 5.14   68000 basic system.

5.8 68000 System Diagram

Figure 5.14 shows a simplified version of the 68000 basic system diagram.

5.9 Timing Diagrams

The 68000 family of processors (68000, 68008, 68010, and 68012) uses a handshaking mechanism to transfer data between the processors and the peripheral devices. This means that all these processors can transfer data asynchronously to and from peripherals of varying speeds. Figure 5.15 shows 68000 read and write cycle timing diagrams.

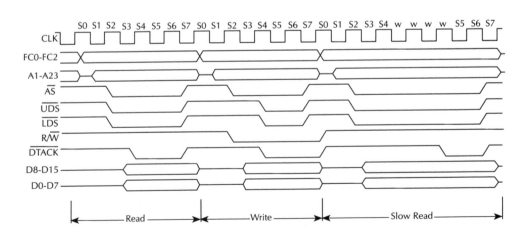

FIGURE 5.15   Read and write cycle timing diagrams.

During the read cycle, the 68000 obtains data from a memory location or an I/O port. If the instruction specifies a word such as MOVE.W$020504, D1 or a long word such as MOVE.L $030808, D0, the 68000 reads both upper and lower bytes at the same time by asserting UDS and LDS pins. When the instruction is for a byte operation the 68000 utilizes an internal bit to find which byte to read and then outputs the data strobe required for that byte. For byte operations, when the address is even (A0 = 0), the 68000 asserts UDS and reads data via D8-D15 pins into low byte of the specified data register. On the other hand, for reading a data byte from an odd address (A0 = 1), the 68000 outputs LOW on LDS and reads data byte via D0-D7 pins into low byte of the specified data register. For example, consider MOVE.B $507144, D5. The 68000 outputs LOW on UDS (since A0 = 0) and high on LDS. The memory chip's 8 data lines must be connected to the 68000 D8-D15 pins. The 68000 reads the data byte via D8-D15 pins into the low byte of D5. Note that for reading a data byte from an odd location by executing an instruction such as MOVE.B $507145, D5, the 8 data lines of the memory chip must be connected to the 68000 D0-D7 pins. The 68000, in this case, outputs low on LDS (A0 = 0) and high on UDS, then reads the data byte into low byte of D5.

Now, let us discuss the read timing diagram of Figure 5.15. Consider Figure 5.15 for word read timing. During S0, address and data signals are in the high impedance state. At the start of S1, the 68000 outputs the address on its address pins (A1-A23). During S0, the 68000 outputs FC2-FC0 signals. AS is asserted at the start of S2 to indicate valid address on bus. AS can be used at this point to latch the signals on the address pins. The 68000 asserts UDS and LDS pins to indicate a word transfer. The 68000 also outputs high on the R/W pin to indicate a read operation.

The 68000 now waits for the peripheral device to assert DTACK. Upon placing data on the data bus, the peripheral device asserts DTACK. The 68000 samples the DTACK signal at the end of S4. If DTACK is not asserted by the peripheral device, the processor automatically inserts wait states (W).

However, upon assertion of DTACK, the 68000 negates AS, UDS, and LDS signals and then latches the data from data bus into an internal register at the end of the next cycle. Once the selected peripheral device senses that the 68000 has obtained data from the data bus (by recognizing the negation of AS, UDS, or LDS), the peripheral device must negate DTACK immediately, so that it does not interfere with the start of the next cycle.

If DTACK is not asserted by a peripheral at the end of state 4 (Figure 5.15), the 68000 inserts wait states. The 68000 outputs valid addresses on the address pins and keeps asserting AS, UDS, and LDS until the peripheral asserts DTACK. The 68000 always inserts an even number of wait states if DTACK is not asserted by the peripheral, since all 68000 operations are performed using two clock states per clock cycle. Note that in Figure 5.15, the 68000 inserts 6 wait states or 3 cycles.

Consider Figure 5.15 for 68000 write word timing. The 68000 outputs the address of the location to be written into the address bus at the start of S1. During S0, the 68000 places the proper function code values at the FC2, FC1, and FC0 pins. If the 68000 used the data bus in the previous cycle, then it places all data pins in the high impedance state and then outputs LOW on AS and R/W pins. At the start of S3, the 68000 places data on D0-D15 pins. The 68000 then asserts UDS and LDS pins at the beginning of S4.

For the memory or I/O device, if DTACK is not asserted by memory or I/O device by the end of S4, the 68000 automatically inserts wait states into the write cycle.

The 68000 provides a special cycle called the read-modify-write cycle during execution of only the TAS instruction. This instruction reads a data byte, sets condition codes according to the byte value, sets bit 7 of the byte, and then writes the byte back into memory. The TAS instruction can be used in providing data transfer between two 68000 processors using shared RAM. The data byte mentioned above is held in the shared RAM. The read/modify/write cycle is indivisible. That is, it cannot be interrupted by any other bus request.

5.10 68000 Memory Interface

One of the advantages of the 68000 is that it can easily be interfaced to memory chips. This is because the 68000 goes into a wait state if DTACK is not asserted by the memory devices at the end of S4.

A simplified schematic showing an interface of a 68000 to two 2716s and two 6116s is shown in Figure 5.16. The 2716 is a 2 K×8 EPROM and the 6116 is a 2 K×8 static RAM. For a 4-MHz clock, each cycle is 250 ns. The 68000 samples DTACK at the falling edge of the S4 (third clock cycle) and latches data at the falling edge of S6 (fourth clock cycle). AS is used to assert DTACK. AS goes to LOW after 500 ns (two clock cycles). The time delay between AS going LOW and the falling edge of S6 is 500 ns.

Since the access times of the 2716 and 6116 are, respectively, 450 and 120 ns, delay circuits for DTACK are not required. As an example, the 68000-2716 timing parameters with various 68000 clock frequencies are as follows:

Case	68000 frequency	Clock cycle	Time before first DTACK is sampled	Comment
1	6 MHz	166.7 ns	3(166.7) = 500.1 ns	No timing problem
2	8 MHz	125 ns	3(125) = 375 ns	No timing problem since the 68000 latches data after 500 ns
3	10 MHz	100 ns	3(100) = 300 ns	Not enough time for the 2716 to place data on bus; needs delay circuit
4	12.5 MHz	80 ns	3(80) = 240 ns	Same as case 3
5	16.67 MHz	60 ns	3(60) = 180 ns	Same as case 3
6	25 MHz	40 ns	3(40) = 120 ns	Same as case 3

Note that LDS and UDS must be used as chip selects as in the figure. They must not be connected to A0 of the memory chips, since in that case half of the memory in each chip will be wasted.

Let us determine the memory map of Figure 5.16. Assume the don't care values of A23-A14 to be zeros.

Memory map for even 2716 (A12 must be zero to select even 2716 and A13 must be zero to deselect even 6116)

A23	A22	A21	A20	A19	A18	A17	A16	A15	A14	A13	A12	A11 ... A1	A0
0	0	0	0	0	0	0	0	0	0	0	0	can be from all zeros to all ones	0

The memory map includes the addresses $000000, $000002, $000004, ... $000FFE.

Memory map for odd 2716 (A12 must be 0 to select odd 2716 and A13 must be 0 to deselect odd 6116)

A23	A22	A21	A20	A19	A18	A17	A16	A15	A14	A13	A12	A11 ... A1	A0
0	0	0	0	0	0	0	0	0	0	0	0	can be from all zeros to all ones	1

The memory map includes the addresses $000001, $000003, $000005, ... $000FFF.

Memory map for even 6116 (A12 must be one to deselect even 2716 and A13 must be one to select even 6116)

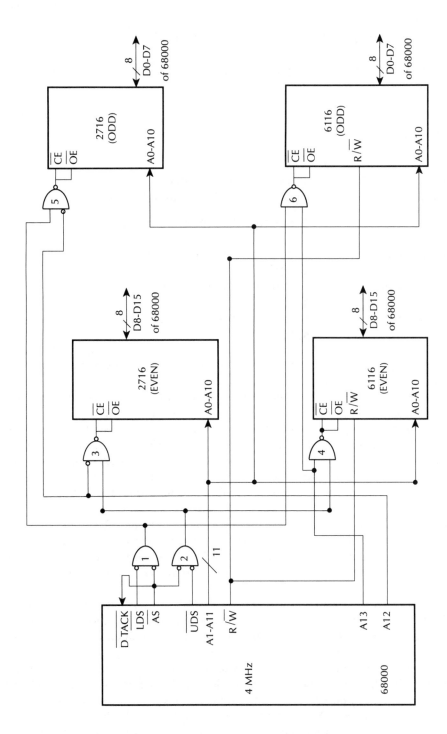

FIGURE 5.16 68000 interface to 2716/6116.

A23	A22	A21	A20	A19	A18	A17	A16	A15	A14	A13	A12	A11	...	A1	A0
0	0	0	0	0	0	0	0	0	0	1	1				0

can be from all zeros to all ones

The memory map includes the addresses $003000, $003002, . . . , $003FFE.

Memory map for odd 6116 (A12 must be one to deselect odd 2716 and A13 must be one to select to odd 6116)

A23	A22	A21	A20	A19	A18	A17	A16	A15	A14	A13	A12	A11	...	A1	A0
0	0	0	0	0	0	0	0	0	0	1	1				1

can be from all zeros to all ones

The memory map includes the addresses $003001, $003003, . . . , $003FFF.

In summary, the memory for the schematic in Figure 5.16 is shown in the following:

EVEN 2716 EPROM	$000000, $000002, $000004, ... $000FFE
ODD 2716 EPROM	$000001, $000003, $000005, ... $000FFF
EVEN 6116 RAM	$003000, $003002, $003004, ... $003FFE
ODD 6116 RAM	$003001, $003003, $003005, ... $003FFF

In the following, some examples will be considered to illustrate the use of memory map of the schematic of Figure 5.6 using the 68000 MOVE instruction.

Consider MOVE.B $000004, D1. Upon execution of this MOVE instruction, the 68000 reads the [$000004] from even 2716 to the low byte of D1. In order to execute the instruction, the processor places the upper 23 bits of $000004 on its A23-A1 pins, and asserts UDS and AS. Since A12 = 0, the output of AND gate 2 = 1, the output of AND gate 3 generates a LOW, making CE and OE of the even 2716 LOW. When the 68000 samples DTACK (asserted by AS in Figure 5.16), data placed on the 68000 D8-D15 lines by the even 2716 are read by the processor into low byte of D1. Similarly, a byte read operation from the odd 2716 can be explained.

Consider MOVE.W $000004, D1. The 68000 asserts both UDS and LDS in this case. The outputs of AND gates 3 and 5 are LOW. Both the even 2716 and the odd 2716 are selected. Data placed on D0-D7 pins and D8-D15 pins of the 68000 from locations $000004 and $000005 of the even 2716 and odd 2716 are read by the processor, respectively, into bits 8-15 and bits 0-7 of D0.

Consider MOVE.B D2, $003001. The 68000 asserts LDS, AS and outputs HIGH on A12 and A13. The output of AND gate 6 is LOW and thus selects odd 6116. Since R/W = 0, the odd 6116 writes the low byte of D2 from 68000 D0-D7 pins into location $003001. Similarly, other byte operations from the RAMs can be illustrated.

Consider MOVE.L D3, $003002. The 68000 asserts AS, and both UDS and LDS. It also outputs HIGH on A12 and A13. Both RAMs are selected. The high 16 bits of the 32-bit data placed on D0-D15 pins by the 68000 are written into locations #003002 (byte 0) and $003003 (byte 1), and the low 16 bits of the 32-bit data placed on D0-D15 pins by the 68000 are written into locations $003004 (byte 2) and $003005 (byte 3), respectively, of the even and odd 6116s. Similarly, the other long word operations can be illustrated. Note that a long-word write or read is done by the 68000 with 2 bus cycles.

5.11 68000 Programmed I/O

As mentioned before, the 68000 uses memory-mapped I/O. Programmed I/O can be achieved in the 68000 using one of the following ways:

1. By interfacing the 68000 asynchronously with its own family of peripheral devices such as the MC-68230, Parallel Interface/Timer chip.
2. By interfacing the 68000 synchronously with 6800 peripherals such as the MC6821 (note that synchronization means that every READ or WRITE operation is synchronized with the clock).

5.11.1 68000-68230 Interface

The MC68230 parallel interface/timer (PI/T) provides double buffered parallel interfaces and a timer for 68000 systems. Note that double buffering means that the ports have dual latches. Double buffering allows simultaneous reading of data from a port by the microprocessor and placing of data into the same port by an external device via handshaking. Double buffering is most useful in situations where a peripheral device and the processor are capable of transferring data at nearly the same speed. If there is a large difference in speed between the microprocessor and the peripheral, little or no benefit of double buffering is achieved. In these cases, however, there is no penalty for using double buffering. Double buffering permits the fetch operation of the data transmitter to be overlapped with the store operation of the data receiver.

The parallel interfaces provided by the 68230 can be 8 or 16 bits wide with unidirectional or bidirectional modes. In the unidirectional mode, a data direction register configures each port as an input or output. In the bidirectional mode, the data direction registers are ignored and the direction is determined by the state of four handshake pins.

The 68230 allows use of interrupts, and also provides a DMA request pin for connection to a DMA controller chip such as the MC68450. The timer contains a 24-bit-wide counter. This counter can be clocked by the output of a 5-bit (divide by 32) prescaler or by an external timer input pin (TIN).

Table 5.14 provides the signal summary and Figure 5.17 shows the 68230 pin diagram. The 68230 is a 48-pin device.

The purpose of D0-D7, R/W, and CS pins is obvious. RS1-RS5 are five register select input pins for selecting the 23 internal registers.

During read or interrupt acknowledge cycles, DTACK is asserted after data have been provided on the data bus and during write cycles it is asserted after data have been accepted at the data bus. A pullup resistor is required to maintain DTACK high between bus cycles.

Upon activation of the RESET input, all control and data direction registers are cleared and most internal operations are disabled by the assertion of RESET LOW.

The clock pin has the same specifications as the 68000.

PA0-PA7 and PB0-PB7 pins provide two 8-bit ports that may be concatenated to form a 16-bit port in certain modes. The ports may be controlled in conjunction with handshake pins H1-H4. A simple example of a handshake operation for input of data is for the I/O device to indicate to the port that new data are available at the port by activating H1 to HIGH. After input of data by the 68000, the H2 output of the 68230 is set to HIGH to indicate to the I/O device that data have been read and it may now provide another data byte to the port.

PC0-PC7 pins can be used as eight general purpose I/O pins or any combination of six special function pins and two general purpose I/O pins (PC0, PC1).

Port C can be configured as input or output by the port C data direction register.

The alternate function pins TIN, TOUT, and TIACK are timer I/O pins. For example, the PC2 pin can also be used as a timer input TIN. When the 68230 timer is used as an event

TABLE 5.14   Signal Summary of the MC68230

Signal name	Input/output	Active state	Edge/level sensitive	Output states
CLK	Input		Falling and rising edge	
$\overline{CS}$	Input	Low	Level	
D0-D7	Input/output	High = 1, low = 0	Level	High, low, high impedance
$\overline{DMAREQ}$	Output	Low		High, low
DTACK	Output	Low		High, low, high impedance[a]
H1(H3)[b]	Input	Low or high	Asserted edge	
H2(H4)[c]	Input or output	Low or high	Asserted edge	High, low, high impedance
PA0-PA7,[c]	Input/output	High = 1, low = 0	Level	High, low, high impedance
PB0-PB7,[c] PC0-PC7	Input or output input or output			
$\overline{PIACK}$	Input	Low	Level	
PIRQ	Output	Low		Low, high impedance[a]
RS1-RS5	Input	High = 1, low = 0	Level	
R/$\overline{W}$	Input	High read, low write	Level	
$\overline{RESET}$	Input	Low	Level	
TIACK	Input	Low	Level	
TIN (external clock)	Input		Rising edge	
TIN (run/halt)	Input	High	Level	
TOUT (square wave)	Output	Low		High, low
TOUT ($\overline{TIRQ}$)	Output	Low		Low, high impedance[a]

[a] Pullup resistors required.

[b] H1 is level sensitive for output buffer control in modes 2 and 3.

[c] Note these pins have internal pullup resistors.

counter, the counter value can be decremented upon application of pulses at TIN by external circuitry. This means that TIN is the timer clock input. TIN can also be configured as the timer run/halt input. In this case, a HIGH at TIN enables the 68230 internal timer clock. Therefore, the 68230 timer runs when TIN = 1. On the other hand, a low on TIN disables the 68230 internal clock and stops the timer. TOUT may provide an active low timer interrupt request

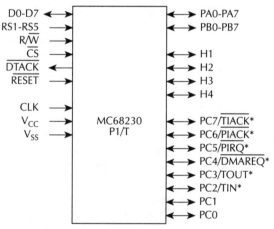

* Individual Programmable Dual-Function Pin

FIGURE 5.17   Logical pin assignment.

TABLE 5.15 68230 Register Addressing Assignments

Register		Register select bit 5	4	3	2	1	Accessible	Affected by Reset	Read cycle
Port General Control Register	(PGCR)	0	0	0	0	0	R W	Yes	No
Port Service Request Register	(PSRR)	0	0	0	0	1	R W	Yes	No
Port A Data Direction Register	(PADDR)	0	0	0	1	0	R W	Yes	No
Port B Data Direction Register	(PBDDR)	0	0	0	1	1	R W	Yes	No
Port C Data Direction Register	(PCDDR)	0	0	1	0	0	R W	Yes	No
Port Interrupt Vector Register	(PIVR)	0	0	1	0	1	R W	Yes	No
Port A Control Register	(PACR)	0	0	1	1	0	R W	Yes	No
Port B Control Register	(PBCR)	0	0	1	1	1	R W	Yes	No
Port A Data Register	(PADR)	0	1	0	0	0	R W	No	a
Port B Data Register	(PBDR)	0	1	0	0	1	R W	No	a
Port A Alternate Register	(PAAR)	0	1	0	1	0	R	No	No
Port B Alternate Register	(PBAR)	0	1	0	1	1	R	No	No
Port C Data Register	(PCDR)	0	1	1	0	0	R W	No	No
Port Status Register	(PSR)	0	1	1	0	1	R W[b]	Yes	No
Timer Control Register	(TCR)	1	0	0	0	0	R W	Yes	No
Timer Interrupt Vector Register	(TIVR)	1	0	0	0	1	R W	Yes	No
Counter Preload Register High	(CPRH)	1	0	0	1	1	R W	No	No
Counter Preload Register Middle	(CPRM)	1	0	1	0	0	R W	No	No
Counter Preload Register Low	(CPRL)	1	0	1	0	1	R W	No	No
Count Register High	(CNTRH)	1	0	1	1	1	R	No	No
Count Register Middle	(CNTRM)	1	1	0	0	0	R	No	No
Count Register Low	(CNTRL)	1	1	0	0	1	R	No	No
Timer Status Register	(TSR)	1	1	0	1	0	R W[b]	Yes	No

Note: R = read; W = write.

[a] Mode dependent.

[b] A write to this register may perform a special status resetting operation.

output or a general purpose square output, initially high. TIACK is an active low high-impedance input used for timer interrupt acknowledge.

Ports A and B have an independent pair of active low interrupt request (PIRQ) and interrupt acknowledge (PIACK) pins. PIRQ is an output pin and is used by the 68230 when it implements an interrupt-driven parallel I/O configuration. Therefore, PC2-PC7 pins may or may not be available for general purpose I/O. The 68230 PIRQ pin can be connected to one of the 68000 IPL pins and the 68230 PIACK pin can be connected to the NANDed output of 68000 FC2 FC1 FC0 pins. Since FC2 FC1 FC0 = 111_2 indicates interrupt acknowledge, PIACK is asserted when the 68000 is ready to service the interrupt. The DMAREQ pin provides an active low direct memory access controller request pulse for three clock cycles, compatible with MC68450 DMA controller chip.

Tables 5.15 provides the 68230 register addressing assignments.

The 68230 ports can be configured for various modes of operation. For example, consider ports A and B. Bits 6 and 7 of the port general control register, PGCR (R0) are used for configuring ports A and B in one of four modes as follows:

PGCR bits 7	6	
0	0	Mode 0 (unidirectional 8-bit mode)
0	1	Mode 1 (unidirectional 16-bit mode)
1	0	Mode 2 (bidirectional 8-bit mode)
1	1	Mode 3 (bidirectional 16-bit mode)

The other pins of PGCR are defined as follows:

PGCR

bit 5		H34 Enable
	0	Disabled
	1	Enabled
bit 4		H12 Enable
	0	Disabled
	1	Enabled
bit 3		
	0	H4 sense at high level when negated and low level when asserted
	1	H4 sense at low level when negated and high level when asserted

bit 2 (H3 sense),   same definition as H4 sense
bit 1 (H2 sense)
and bit 3 (H1 sense)

The modes 0 and 2 configure ports A and B as unidirectional or bidirectional 8-bit ports. Modes 1 and 3, on the other hand, combine ports A and B together to form a 16-bit unidirectional or bidirectional port. Ports configured as unidirectional must further be programmed as submodes of operation using bits 7 and 6 of PACR (R6) and PBCR (R7) as follows:

For unidirectional 8-bit mode (mode 0)

	Bit 7 of PACR or PBCR	Bit 6 of PACR or PBCR	
Submode 00	0	0	Pin-definable double-buffered input or single-buffered output
Submode 01	0	1	Pin-definable double-buffered output or nonlatched input
Submode 1X	1	X	Bit I/O (pin-definable single-buffered output or nonlatched input)

Note that in the above X means don't care. Nonlatched inputs are latched internally but the bit values are not available at the port.

The submodes define the ports as parallel input ports, parallel output modes, or bit-configurable I/O ports. In addition to these, the submodes further define the ports as latched input ports, interrupt-driven ports, DMA ports, and with various I/O handshake operations.

Figure 5.18 shows a simplified schematic for the 68000-68230 interface for simple I/O operation. A23 is chosen to be HIGH to select the 68230 chip so that the port addresses are different than the 68000 reset vector addresses 000000_{16} through 000006_{16}. The configuration in the figure will provide even port addresses, since $\overline{UDS}$ is used for enabling the 68230 CS.

From the figure, addresses are for registers PGCR (R0), PADDR (R2), PBDDR (R3), PACR (R6), PBCR (R7), PADR (R8), and PBDR (R9).

Consider PGCR.

A23	A22	A21	A20	....	A6	A5	A4	A3	A2	A1	A0	
1	0	0	0	....	0	0	0	0	0	0	0	= \$800000

RS5 — RS1 (under A5–A1), $\overline{UDS}$ (under A0)

Similarly,

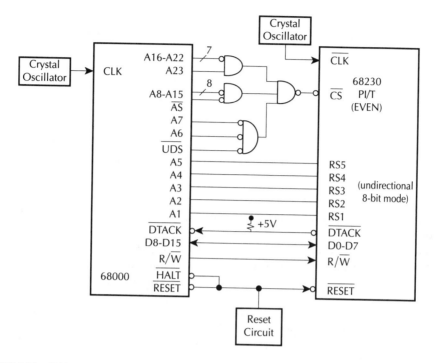

FIGURE 5.18   68000-68230 interface.

$$
\begin{aligned}
\text{address for PADDR} &= \$800004 \\
\text{address for PBDDR} &= \$800006 \\
\text{address for PACR} &= \$80000C \\
\text{address for PBCR} &= \$80000E \\
\text{address for PADR} &= \$800010 \\
\text{address for PBDR} &= \$800012
\end{aligned}
$$

As an example, the following instruction sequence will select mode 00, submode 1X and configure bits 0-3 of port A as outputs, bits 4-7 as inputs, and port B as an output port:

```
PGCR    EQU  $800000
PADDR   EQU  $800004
PBDDR   EQU  $800006
PACR    EQU  $80000C
PBCR    EQU  $80000E
        MOVE.B  #$00,  PGCR   ; Select mode 0
        MOVE.B  #$0FF, PACR   ; Port A bit I/O submode
        MOVE.B  #$0FF, PBCR   ; Port B bit I/O submode
        MOVE.B  #$0F,  PADDR  ; Configure Port A bits 0-3 as
                              ; outputs and bits 4-7 as inputs
        MOVE.B  #$0FF, PBDDR  ; Configure Port B as an output
                              ; port.
```

5.11.2 Motorola 68000-6821 Interface

The Motorola 6821 is a 40-pin peripheral interface adapter (PIA) chip. It is provided with an 8-bit bidirectional data bus (D0-D7), two register select lines (RS0, RS1), read/write line

TABLE 5.16   6821 Register Definition

| RS1 | RS0 | Control register bits 2 | | Register selected |
		CRA-2	CRB-2	
0	0	1	X	I/O port A
0	0	0	X	Data direction register A
0	1	X	X	Control register A
1	0	X	1	I/O port B
1	0	X	0	Data direction register B
1	1	X	X	Control register B

Note: X = Don't care.

(R/W), reset line (RESET), an enable line (E), two 8-bit I/O ports (PA0-PA7) and (PB0-PB7), and other pins.

There are six 6821 registers. These include two 8-bit ports (ports A and B), two data direction registers, and two control registers. Selection of these registers is controlled by the RS0 and RS1 inputs together with bit 2 of the control register. Table 5.16 shows how the registers are selected.

In Table 5.16 bit 2 in each control register (CRA-2 and CRB-2) determines selection of either an I/O port or the corresponding data direction register when the proper register select signals are applied to RS0 and RS1. A 1 in bit 2 allows access of I/O ports, while a 0 selects the data direction registers.

Each I/O port bit can be configured to act as an input or output. This is accomplished by setting a 1 in the corresponding data direction register bit for those bits which are to be output and a 0 for those bits which are to be inputs.

A RESET signal sets all PIA registers to 0. This has the effect of setting PA0-PA7 and PB0-PB7 as inputs.

There are three built-in signals in the 68000 which provide the interface with the 6821. These are the Enable (E), Valid Memory Access (VMA), and Valid Peripheral Access (VPA).

The Enable signal (E) is an output from the 68000. It corresponds to the E signal of the 6821. This signal is the clock used by the 6821 to synchronize data transfer. The frequency of the E signal is one tenth of the 68000 clock frequency. Therefore, this allows one to interface the 68000 (which operates much faster than the 6821) with the 6821. The Valid Memory Address (VMA) signal is output by the 68000 to indicate to the 6800 peripherals that there is a valid address on the address bus.

The Valid Peripheral Address (VPA) is an input to the 68000. This signal is used to indicate that the device addressed by the 68000 is a 6800 peripheral. This tells the 68000 to synchronize data transfer with the Enable signal (E).

Let us now discuss how the 68000 instructions can be used to configure the 6821 ports. As an example, bit 7 and bits 0-6 of port A can be configured, respectively, as input and outputs using the following instruction sequence:

```
BCLR.B #2, CRA       ;   ADDRESS DDRA
MOVE.B #$7F, DDRA     ;   CONFIGURE PORT A
BSET.B #2, CRA        ;   ADDRESS PORT A
```

Once the ports are configured to the designer's specification, 6821 can be used to transfer data from an input device to the 68000 or from the 68000 to an output device by using the MOVE.B instruction as follows:

```
MOVE.B (EA), Dn   Transfer 8-bit data from an input port to
                  the specified data register Dn.
```

MOVE.B Dn, (EA) **Transfer 8-bit data from the specified data register Dn to an output port.**

Figure 5.19 shows a block diagram of how two 6821s are interfaced to the 68000 in order to generate four 8-bit I/O ports.

In Figure 5.19, I/O port addresses can be obtained as follows. When A23 is HIGH and $\overline{AS}$ is LOW, the OR gate output will be LOW. This OR gate output is used to provide $\overline{VPA}$. The inverted OR gate output, in turn, makes CS1 HIGH on both the chips. Note that A23 is arbitrarily chosen. A23 is chosen to be HIGH to enable CS1 so that the addresses for the ports and the reset vector are not the same. Assuming the don't care address lines A22-A3 to be zeros, the addresses for the I/O ports, control registers, and the data direction registers for the even 6821 can be obtained as follows:

Port name	Memory address							
	A23	A22	...	A3	A2	A1	A0	
I/O port A/DDRA	1	0	...	0	0	0	0	$= 800000_{16}$
CRA	1	0	...	0	0	1	0	$= 800002_{16}$
I/O port B/DDRB	1	0	...	0	1	0	0	$= 800004_{16}$
CRB	1	0	...	0	1	1	0	$= 800006_{16}$

Note that in the above, A0 = 0 for even addressing. Also, from Table 5.16, for accessing DDRA, bit 2 of CRA with memory address 800002_{16} must be set to 1 and then port A can be configured with appropriate data in 800000_{16}. Note that port A and its data direction register, DDRA, have the same address, 800000_{16}. Similarly, port B and DDRB have the same address 800004_{16}. Bit 2 in CRA or CRB identifies whether address 800000_{16} or 800004_{16} is an I/O port or data direction register.

Similarly, the addresses for the ports, control registers, and the data direction register for the odd 6821 (A0 = 1) can be determined as follows: port A/DDRA (800001_{16}), CRA (800003_{16}), port B/DDRB (800005_{16}), and CRB (800007_{16}).

5.12 68000/2716/6116/6821-Based Microcomputer

Figure 5.20a shows the schematic of a 68000-based microcomputer with 4K EPROM, 4K Static RAM, and four 8-bit I/O ports.

Let us explain the various sections of the hardware schematic. Two 2716-1 and two 6116 chips are required to obtain the 4K EPROM and 4K RAM. The LDS and UDS pins are ORed with the memory select signal to enable the chip selects for the EPROMs and the RAMs.

Address decoding is accomplished by using a 3 × 8 decoder. The decoder enables the memory or I/O chips depending on the status of A12-A14 address lines and $\overline{AS}$ line of the 68000. $\overline{AS}$ is used to enable the decoder. I0 selects the EPROMs, I1 selects the RAMs, and I2 selects the I/O ports.

When addressing memory chips, $\overline{DTACK}$ input of the 68000 must be asserted for data acknowledge. The 68000 clock in the hardware schematic is 8 MHz. Therefore, each clock cycle is 125 nanoseconds. In Figure 5.20a, $\overline{AS}$ is used to enable the 3 × 8 decoder. The outputs of the decoder are gated to assert 68000 $\overline{DTACK}$. This means that $\overline{AS}$ is indirectly used to assert $\overline{DTACK}$. From the 68000 read timing diagram of Figure 5.15, $\overline{AS}$ goes to LOW after approximately two cycles (250 ns for 8-MHz clock) from the beginning of the bus cycle. With no wait states, the 68000 samples $\overline{DTACK}$ at the falling edge of S4 (375 ns) and, if recognized, the 68000 latches data at the falling edge of S6 (500 ns). If the $\overline{DTACK}$ is not recognized at the falling edge of S4, the 68000 inserts one cycle (125 ns in this case) wait state, samples $\overline{DTACK}$ at the end of SW (wait state), and, if recognized, latches data at the end of S6 (625 ns),

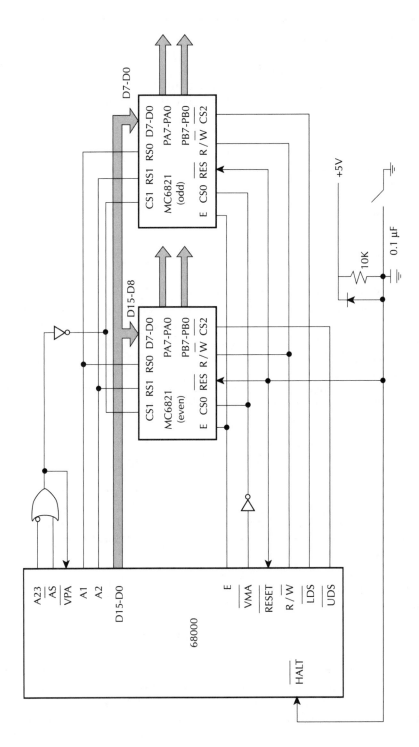

FIGURE 5.19 68000 I/O port block diagram.

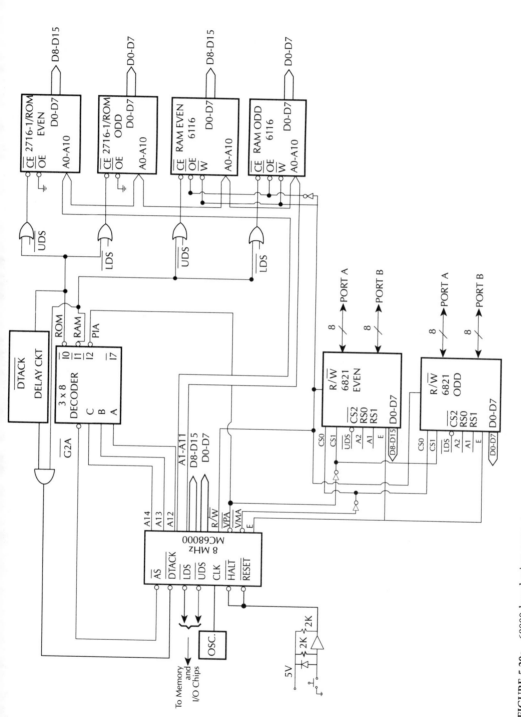

FIGURE 5.20a   68000-based microcomputer.

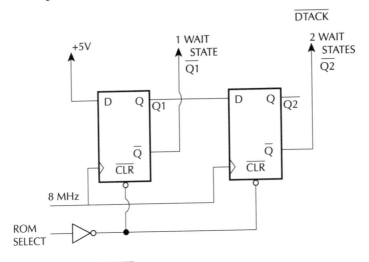

FIGURE 5.20b   Delay circuit for $\overline{\text{DTACK}}$.

and the process continues. Since the access time of 2716-1 is 350 ns, $\overline{\text{DTACK}}$ recognition by the 68000 at the falling edge of SW (wait state) (500 ns) and, hence, latching of data at the falling edge of S6 (625 ns) will satisfy the timing requirement. This means that the decoder output I0 for ROM select must go Low at the end of S6. Therefore, AS must be delayed by 250 ns, i.e., two cycles (S2 through S6).

A delay circuit (Figure 5.20b) is designed using a 74LS175-D-Flip-Flop. AS activates the delay circuit. The input is then shifted right two bits to obtain a two-cycle wait state to allow sufficient time for data transfer. $\overline{\text{DTACK}}$ assertion and recognition are delayed by two cycles during data transfer with the EPROMs. A timing diagram for the $\overline{\text{DTACK}}$ delay circuit is shown in Figure 5.20c.

When the ROM is not selected by the decoder, then clear pin is asserted (output of inverter). So, Q is forced LOW and $\overline{Q}$ is high. Therefore, $\overline{\text{DTACK}}$ is not asserted. When the processor is addressing the ROMs, then the output of the inverter is HIGH so the clear pin is not asserted.

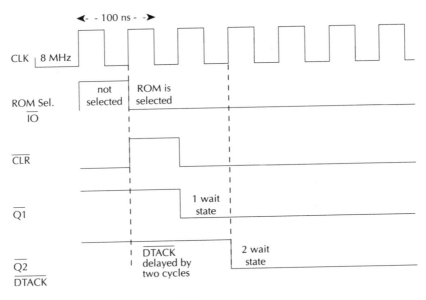

FIGURE 5.20c   Timing diagram for the $\overline{\text{DTACK}}$ delay circuit.

The D flip-flop will accept a high at the input, and Q2 will output high and $\overline{Q2}$ will output low. Now that $\overline{Q2}$ is low, it can assert $\overline{DTACK}$. Q1 will provide one wait state and Q2 will provide two wait states. Since the 2716-1 EPROM has a 350-ns access time and the microprocessor is operating at 8 MHz (125 ns clock cycle), two wait states are inserted before asserting $\overline{DTACK}$ (2*125 = 250 ns). Therefore, $\overline{Q2}$ can be connected to the $\overline{DTACK}$ pin through an AND function.

No wait state is required for RAMs since the access time for the RAMs is only 120 nanoseconds.

Four 8-bit I/O ports are obtained by using two 6821 chips. When the I/O ports are selected, the $\overline{VPA}$ pin is asserted instead of $\overline{DTACK}$. This will acknowledge to the 68000 that it is addressing a 6800-type peripheral. In response, the 68000 will synchronize all data transfer with the E clock.

The memory and I/O maps for the schematic are shown below:

Memory Mapping

A23—A15	A14	A13	A12	A11—A1	$\overline{LDS}$ or $\overline{UDS}$ A0	
0—0	0	0	0	0—0	0	ROM (EVEN) = 2K
0—0	0	0	0	1—1	0	$000000, $000002, $000004, . . ., $000FFE.
0—0	0	0	0	0—0	1	ROM (ODD) = 2K
0—0	0	0	0	1—1	1	$000001, $000003, $000005, . . ., $000FFF
0—0	0	0	1	0—0	0	RAM (even) = 2K
0—0	0	0	1	1—1	0	$001000, $001002, . . ., $001FFE
0—0	0	0	1	0—0	1	RAM (odd) = 2K
0—0	0	0	1	1—1	1	$001001, $001003, . . ., $001FFF

Memory Mapped I/O

A23—A15	A14	A13	A12	A11—A3	RS1 A2	RS0 A1	$\overline{LDS}$ or $\overline{UDS}$ A0	Register Selected (Address) EVEN
0—0	0	1	0	0—0	0	0	0	Port A or DDRA = $002000
0—0	0	1	0	0—0	0	1	0	CRA = $002002
0—0	0	1	0	0—0	1	0	0	Port B or DDRB = $002004
0—0	0	1	0	0—0	1	1	0	CRB = $002006
								ODD
0—0	0	1	0	0—0	0	0	1	Port A or DDRA = $002001
0—0	0	1	0	0—0	0	1	1	CRA = $002003
0—0	0	1	0	0—0	1	0	1	Port B or DDRB = $002005
0—0	0	1	0	0—0	1	1	1	CRB = $002007

Note that upon hardware reset, the 68000 loads the supervisor SP high and low words, respectively, from addresses $000000 and $000002 and the PC high and low words, respectively, from locations $000004 and $000006. The memory map of Figure 5.20a contains these reset vector addresses in the even and odd 2716s.

Example 5.13

Assume that in the configuration of Figure 5.18, port A has three inputs and an LED connected to bits 0-3. Port B has an LED connected to bit 3. Write 68000 assembly program to

1. Turn the port A LED ON and port B LED OFF if port A has an even number of high switch inputs.
2. Turn the port A LED OFF and port B LED ON if port A has an odd number of high switch inputs.
3. Turn both LEDs off if there are no high switch inputs.

Assume EPROM/RAM in the system.

Solution

```
PGCR    EQU $800000
PADDR   EQU $800004
PBDDR   EQU $800006
PACR    EQU $80000C
PBCR    EQU $80000E
PADR    EQU $800010
PBDR    EQU $800012
        MOVE.B #$0,   PGCR    ; Select mode 0
        MOVE.B #$0FF, PACR    ; Port A bit I/O submode
        MOVE.B #$FF,  PBCR    ; Port B bit I/O submode
        MOVE.B #$08,  PADDR ; Configure port A
        MOVE.B #$08,  PBDDR ; Configure port B
        MOVE.B PADR,  D1      ; Get port A switches
        ANDI.B #$07,  D1      ; Mask high five bits
                             ; Are all three inputs low
        BEQ LOW              ; If so, turn both LEDs off.
        CMPI.B #$03, D1      ; Are high switch inputs even
        BEQ EVEN             ; If so, turn port A LED ON
                             ;   and port B LED OFF.
        CMPI.B #$05, D1      ; Are high switch inputs even
        BEQ EVEN             ; If so, turn port A LED ON
                             ;   and port B LED OFF
        CMPI.B #$06, D1      ; Are high switch inputs even
        BEQ EVEN             ; If so, turn port A LED ON
                             ;   and port B LED OFF.
        CMPI.B #$07, D1      ; Are high switch inputs ODD
        BEQ ODD              ; If so, turn port A LED OFF
                             ;   and port B LED ON
        CMPI.B #$04, D1      ; Are high switch inputs even
        BEQ ODD              ; If so, turn port A LED ON
                             ;   and port B LED OFF
        CMPI.B $01, D1       ; Are high switch inputs ODD
        BEQ ODD              ; If so, turn port A LED OFF
                             ;   and port B LED ON
        CMPI.B $02, D1       ; Are high switch inputs ODD
        BEQ ODD              ; If so, turn port A LED OFF
                             ;   and port B LED ON
ODD     MOVE.B #$00, PADR    ; Turn port A LED OFF
        MOVE.B #$08, PBDR    ; Turn port B LED ON
```

```
                JMP  FINISH            ;  Halt
    EVEN        MOVE.B #$08, PADR      ;  Turn port A LED ON
                MOVE.B #$00, PBDR      ;  Turn port B LED OFF
                JMP  FINISH            ;  Halt
    LOW         MOVE.B #$00, PADR      ;  Turn port A LED OFF
                MOVE.B #$00, PBDR      ;  Turn port B LED OFF
    FINISH  JMP  FINISH                ;  Halt
```

5.13 68000 Interrupt I/O

The 68000 services interrupts in the supervisor mode. The 68000 interrupt I/O can be divided into two types: external interrupts and internal interrupts.

5.13.1 External Interrupts

The 68000 provides seven levels of external interrupts, 1 through 7. The external hardware provides an interrupt level using the pins IPL0, IPL1, IPL2. Like other processors, the 68000 checks for and accepts interrupts only between instructions. It compares the value of inverted IPL0- IPL2 with the current interrupt mask contained in the bits 10, 9, and 8 of the status register.

If the value of the inverted IPL0- IPL2 is greater than the value of the current interrupt mask, then the processor acknowledges the interrupt and initiates interrupt processing. Otherwise, the 68000 continues with the only current interrupt level if pending. Interrupt request level zero (IPL0- IPL2 all HIGH) indicates that no interrupt service is requested. An inverted IPL2, IPL1, IPL0 of 7 is always acknowledged and has the highest priority. Therefore, interrupt level 7 is "nonmaskable". Note that the interrupt level is indicated by the interrupt input pins (inverted IPL2, IPL1, IPL0).

To ensure that an interrupt will be recognized, the following interrupt rules should be considered:

1. The incoming interrupt request level must be at a higher priority level than the mask level set in the interrupt mask bits (except for level 7, which is always recognized).
2. The IPL0- IPL2 pins must be held at the interrupt request level until the 68000 acknowledges the interrupt by initiating an interrupt acknowledge (IACK) bus cycle.

Interrupt level 7 is edge-triggered. On the other hand, the interrupt levels 1 to 6 are level sensitive. But as soon as the status register is saved the processor updates its interrupt mask to the same level.

The 68000 does not have any EI (Enable Interrupt) or DI (Disable Interrupt) instructions. Instead, the level indicated by I2 I1 I0 in the SR disables all interrupts below or equal to this value and enables all interrupts above this. For example, in the supervisor mode, I2, I1, and I0 can be modified by using instructions such as AND with SR. If I2, I1, and I0 are modified to contain 100_2, then interrupt levels 1 to 4 are disabled and levels 5 to 7 are enabled. Note that I2 I1 I0 = 111 disables all interrupts except level 7.

Upon hardware reset, the 68000 operates in supervisor mode and sets I2 I1 I0 to 111_2 and disables the interrupt levels 1 through 6. Note that if I2 I1 I0 is modified to 110_2, the 68000 also disables levels 1 through 6 and level 7 is, of course, always enabled.

Once the 68000 has decided to acknowledge an interrupt request, it pushes PC and SR onto the stack, enters supervisor state by setting S-bit to 1, clears TF to inhibit tracing, and updates the priority mask bits and also the address lines A3-A1 with the interrupt level. The 68000 then asserts AS to inform the external devices that A3-A1 has the interrupt level. The processor sets FC2 FC1 FC0 to 111 to run an IACK cycle for 8-bit vector number acquisition. The 68000 multiplies the 8-bit vector by 4 to determine the pointer to locations containing the starting address of the service routine. The 68000 then branches to the service routine. The last instruction of the service routine should be RTE which pops PC and SR back from the stack. In order to explain how the 68000 interrupt priorities work, assume that I2 I1 I0 in SR has the value 011_2. This means that levels 1, 2, and 3 are disabled and levels 4, 5, 6, and 7 are enabled. Now, if the 68000 is interrupted with level 5, the 68000 pushes PC and SR, updates I2 I1 I0 in SR with 101_2, and loads PC with the starting address of the service routine.

Now, while in the service routine of level 5, if the 68000 is interrupted by level 6 interrupt, the 68000 pushes PC and SR onto the stack. The 68000 then completes execution of the level 6 interrupt. The RTE instruction at the end of the level 6 service routine pops old PC and old SR and returns control to the level 5 interrupt service routine at the right place and continues with the level 5 service routine.

External logic can respond to the interrupt acknowledge in one of the following ways: by requesting automatic vectoring or by placing a vector number on the data bus (nonautovector), or by indicating that no device is responding (Spurious Interrupt). If the hardware asserts $\overline{\text{VPA}}$ to terminate the IACK bus cycle, the 68000 directs itself automatically to the proper interrupt vector corresponding to the current interrupt level. No external hardware is required for providing interrupt address vector. This is known as autovectoring. The vectors for the seven autovector levels are given below:

	I2	I1	I0
Level 1 ← Interrupt vector $19 for	0	0	1
Level 2 ← Interrupt vector $1A for	0	1	0
Level 3 ← Interrupt vector $1B for	0	1	1
Level 4 ← Interrupt vector $1C for	1	0	0
Level 5 ← Interrupt vector $1D for	1	0	1
Level 6 ← Interrupt vector $1E for	1	1	0
Level 7 ← Interrupt vector $1F for	1	1	1

During autovectoring, the 68000 asserts $\overline{\text{VMA}}$ after assertion of $\overline{\text{VPA}}$ and then completes a normal 68000 read cycle.

In a nonautovector situation, the interrupting device uses external hardware to place a vector number on data lines D0-D7, and then performs a $\overline{\text{DTACK}}$ handshake to terminate the IACK bus cycle. The vector numbers allowed are $40 to $FF, but Motorola has not implemented a protection on the first 64 entries so that user-interrupt vectors may overlap at the discretion of the system designer. The 68000 multiplies this vector by 4 and determines the pointers to an interrupt address vector.

During the IACK cycle, the 68000 always checks the $\overline{\text{VPA}}$ line for LOW, and if $\overline{\text{VPA}}$ is asserted, the 68000 obtains the interrupt vector address using autovectoring. If $\overline{\text{VPA}}$ is not asserted, the 68000 checks $\overline{\text{DTACK}}$ for LOW. If $\overline{\text{DTACK}}$ is asserted, the 68000 obtains the interrupt address vector using nonautovectoring.

Another way to terminate an interrupt acknowledge bus cycle is with the $\overline{\text{BERR}}$ (Bus Error) signal. Even though the interrupt control pins are synchronized to enhance noise immunity, it is possible that external system interrupt circuitry may initiate an IACK bus cycle as a result of noise. Since no device is requesting interrupt service, neither $\overline{\text{DTACK}}$ nor $\overline{\text{VPA}}$ will be asserted to signal the end of the nonexisting IACK bus cycle. When there is no response to an IACK bus cycle after a specified period of time (monitored by the user by an external timer), the $\overline{\text{BERR}}$ can be asserted. This indicates to the processor that it has recognized a spurious interrupt. The 68000 provides 18H as the vector to fetch for the starting address of this exception handling routine.

The 68000 determines the interrupt address vector for each of the above cases as follows. After obtaining the 8-bit vector n, the 68000 reads the long word located at memory address

| Vector Number(s) | Address | | | Assignment |
	Dec	Hex	Space	
0	0	000	SP	Reset Initial SSP
–	4	004	SP	Reset Initial PC
2	8	008	SD	Bus Error
3	12	00C	SD	Address Error
4	16	010	SD	Illegal Instruction
5	20	014	SD	Zero Divide
6	24	018	SD	CHK Instruction
7	28	01C	SD	TRAPV Instruction
8	32	020	SD	Privilege Violation
9	36	024	SD	Trace
10	40	028	SD	Line 1010 Emulator
11	44	02C	SD	Line 1111 Emulator
12*	48	030	SD	(Unassigned, Reserved)
13*	52	034	SD	(Unassigned, Reserved)
14*	56	038	SD	(Unassigned, Reserved)
15	60	03C	SD	Uninitialized Interrupt Vector
16-23*	64	04C	SD	(Unassigned, Reserved)
	95	05F		–
24	96	060	SD	Spurious Interrupt
25	100	064	SD	Level 1 Interrupt Autovector
26	104	068	SD	Level 2 Interrupt Autovector
27	108	06C	SD	Level 3 Interrupt Autovector
28	112	070	SD	Level 4 Interrupt Autovector
29	116	074	SD	Level 5 Interrupt Autovector
30	120	078	SD	Level 6 Interrupt Autovector
31	124	07C	SD	Level 7 Interrupt Autovector
32-47	128	080	SD	TRAP Instruction Vectors
	191	08F		–
48-63*	192	0C0	SD	(Unassigned, Reserved)
	255	0FF		–
64-255	256	100	SD	User Interrupt Vectors
	1023	3FF		–

*Vector numbers 12, 13, 14, 16 through 23, and 48 through 63 are reserved for future enhancements by Motorola. No user peripheral devices should be assigned to these numbers.

FIGURE 5.21 68000 exception map. 'SP' means supervisor program space; 'SD' means supervisor data space.

4*n. This long word contains the address of the service routine. Therefore, the address is found using indirect addressing. Note that the spurious interrupt and bus error interrupt due to troubled instruction (when no DTACK is received by the 68000) have two different vectors. Spurious interrupt occurs when the BUS ERROR pin is asserted during interrupt processing.

During IACK cycle, FC2 FC1 FC0 = 111. A1-A3 has the interrupt level. The vector number is provided on the D0-D7 pins by external hardware. Note that during nonautovectoring, if VPA goes to LOW, the 68000 ignores it.

5.13.2 Internal Interrupts

The internal interrupt is a software interrupt. This interrupt is generated when the 68000 executes a software interrupt instruction called TRAP or by some undesirable event such as division by zero or execution of an illegal instruction.

5.13.3 68000 Exception Map

Figure 5.21 shows an interrupt map of the 68000. Vector addresses $00 through $2C include vector addresses for reset, bus error, trace, divide by 0, etc., and addresses $30 through $4C are unassigned. The RESET vector requires four words (addresses 0, 2, 4, and 6) and other vectors require only two words. As an example of how the 68000 determines the interrupt address vector, consider autovector 1. After VPA is asserted, if $IPL2-0 = 110$ (level = 001), the 68000 automatically obtains the 8-bit vector 25_{10} (19_{16}) and multiplies 19_{16} by 4 to obtain the 24-bit address 000064_{16}. The 68000 then loads the 16-bit contents of location 000064_{16} and the next location 000066_{16} into PC. For example, if the user wants to write the service routine for autovector 1 at address 271452_{16}, then $XX27_{16}$ and 1452_{16}, must, respectively, be stored at 000064_{16} and 000066_{16}. Note that XX in $XX27_{16}$ are two don't care nibbles (4 bits).

After hardware reset, the 68000 loads the supervisor SP high and low words, respectively, from addresses 000000_{16} and 000002_{16}, and the PC high and low words, respectively, from 000004_{16} and 000006_{16}. The assembler directive, Define Constant (DC) can be used to load PC and SSP. For example, the following instruction sequence loads SSP with \$004100 and PC with \$001000:

```
ORG    $000000
DC.L   $00004100
DC.L   $00001000
```

5.13.4 68000 Interrupt Address Vector

Suppose that the user decides to write a service routine starting at address \$123456 using autovector 1. Since the autovector 1 uses addresses \$000064 and \$000066, the numbers \$0012 and \$3456 must be stored in locations \$000064 and \$000066, respectively. The DC.L assembler directive can be used to load \$123456 into location \$000064 as follows:

```
ORG    $000064
DC.L   $00123456
```

5.13.5 An Example of Autovector and Nonautovector Interrupts

As an example to illustrate the concept of autovector and nonautovector interrupts, consider Figure 5.22. In this figure, I/O device 1 uses nonautovector and I/O device 2 uses autovector interrupts. The system is capable of handling interrupts from eight devices, since an 8-to-3 priority encoder such as the 74LS148 is used. Suppose that I/O device 2 drives the I/O2 LOW in order to activate line 3 of this encoder. This, in turn, interrupts the processor. When the 68000 decides to acknowledge the interrupt, it drives FC0-FC2 HIGH. The interrupt level is reflected on A1-A3 when AS is activated by the 68000. IACK3 and I/O2 signals are used to generate VPA. Once the VPA is asserted, the 68000 obtains the interrupt vector address using autovectoring.

In case of I/O1, line 5 of the priority encoder is activated to initiate the interrupt. By using appropriate logic, DTACK is asserted using IACK5 and I/O1. The vector number is placed on D0-D7 by enabling an octal buffer such as the 74LS244 using IACK5. The 68000 inputs this vector number and multiplies it by 4 to obtain the interrupt address vector.

5.14 68000 DMA

Three DMA control lines are provided with the 68000. These are BR (Bus Request), BG (Bus Grant), and BGACK (Bus Grant Acknowledge).

The BR line is an input to the 68000. The external device activates this line to tell the 68000 to release the system bus.

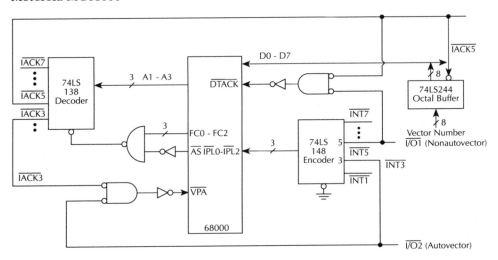

FIGURE 5.22   Autovector and nonautovector interrupts.

At least one clock period after receiving the $\overline{\text{BR}}$, the 68000 will enable its $\overline{\text{BG}}$ output line to acknowledge the DMA request. However, the 68000 will not relinquish the bus until it has completed the current instruction cycle. The external device must check AS (Address Strobe) line to determine the completion of instruction cycle by the 68000. When AS becomes HIGH, the 68000 will tristate its address and data lines and will give up the bus to the external device.

After taking over the bus, the external device must enable BGACK line. The BGACK line tells the 68000 and other devices connected to the bus that the bus is being used. The 68000 buses stay in a tristate condition until BGACK becomes HIGH.

5.15 68000 Exception Handling

A 16-bit microcomputer is usually capable of handling unusual or exceptional conditions. These conditions include situations such as execution of illegal instruction or division by zero and all interrupts. In this section, the exception handling capabilities of the MC68000 are described.

The 68000 exceptions can be divided into three groups, namely, groups 0, 1, and 2. Group 0 has the highest priority and group 2 has the lowest priority. Within the first two groups, there are additional priority levels. A list of 68000 exceptions along with individual priorities is shown below:

Group 0   Reset (highest level in this group), Address Error (next level), and Bus Error (lowest level)

Group 1   Trace (highest level), Interrupt (next level), Illegal op code (next level), and Privilege Violation (lowest level)

Group 2   TRAP, TRAPV, CHK, and ZERO DIVIDE (no individual priorities assigned in group 2)

Exceptions from group 0 always override an active exception from group 1 or group 2. Group 0 exception processing begins at the completion of the current bus cycle (two clock cycles). Note that the number of cycles required for a READ or WRITE operation is called a bus cycle. This means that during an instruction fetch if there is a group 0 interrupt, the 68000 will complete the instruction fetch and then service the interrupt.

Group 1 exception processing begins at the completion of the current instruction.

Group 2 exceptions are initiated through execution of an instruction. Therefore, there are no individual priority levels within group 2. Exception processing occurs when a group 2 interrupt is encountered, provided there are no group 0 or group 1 interrupts.

When an exception occurs, the 68000 saves the contents of the program counter and status register onto the stack and then executes a new program whose address is provided by the exception vectors. Once this program is executed, the 68000 returns to the main program using the stored values of program counter and status register.

Exceptions can be of two types: internal or external.

The internal exceptions are generated by situations such as division by zero, execution of illegal or unimplemented instructions, and address error. As mentioned before, internal interrupts are called traps.

The external exceptions are generated by bus error, reset, or interrupts. The basic concepts associated with interrupts, relating them to the 68000, have already been described. In this section we will discuss the other exceptions.

In response to an exception condition, the processor executes a user-written program. In some microcomputers, one common program is provided for all exceptions. The beginning section of the program determines the cause of the exception and then branches to the appropriate routine. The 68000 utilizes a more general approach. Each exception can be handled by a separate program.

As mentioned before, the 68000 has two modes of operation: user mode and supervisor mode. The operating system runs in supervisor mode and all other programs are executed in user mode. The supervisor mode is, therefore, privileged. Several privileged instructions such as MOVE to SR can only be executed in supervisor mode. Any attempt to execute them in user mode causes a trap.

We will now discuss how the 68000 handles exceptions which are caused by external reset, instructions causing traps, bus and address errors, tracing, execution of privileged instructions in user mode, and execution of illegal/unimplemented instructions.

The reset exception is generated externally. In response to this exception, the 68000 automatically loads the initial starting address into the processor.

The 68000 has a TRAP instruction which always causes an exception. The operand for this instruction varies from 0 to 15. This means that there are 16 TRAP instructions. Each TRAP instruction is normally used to call subroutines in an operating system. Note that this automatically places the 68000 in supervisor state. TRAPs can also be used for inserting breakpoints in a program. Two other 68000 instructions can traps if a particular condition is true. These are TRAPV and CHK. TRAPV generates an exception if the overflow flag is set. The TRAPV instruction can be inserted after every arithmetic operation in a program for causing a trap whenever there is the possibility of an overflow. A routine can be written at the vector address for the TRAPV to indicate to the user that an overflow has occurred. The CHK instruction is designed to ensure that access to an array in memory is within the range specified by the user. If there is a violation of this range, the 68000 generates an exception.

A bus error occurs when the 68000 tries to access an address which does not belong to the devices connected to the bus. This error can be detected by asserting the BERR pin on the 68000 chip by an external timer when no DTACK is received from the device after a certain period of time. In response to this, the 68000 executes a user-written routine located at an address obtained from the exception vectors. An address error, on the other hand, occurs when the 68000 tries to READ or WRITE a word (16-bit) or long word (32-bit) at an odd address. The address error has a different exception vector from the bus error.

The trace exception in the 68000 can be generated by setting the trace bit in the status register, in response to the trace exception after execution of every instruction. The user can write a routine at the exception vectors for the trace instruction to display registers and memory. The trace exception provides the 68000 with the single-stepping debugging feature.

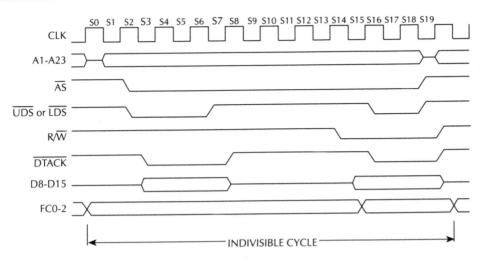

FIGURE 5.23   MC68000 read-modify-write cycle for TAS.

As mentioned before, the 68000 has some privileged instructions which must be executed in supervisor mode. An attempt to execute these instructions causes privilege violations. Finally, the 68000 causes an exception when it tries to execute an illegal or unimplemented instruction. It is common practice to use the illegal instruction $4AFA as a breakpoint.

5.16 Multiprocessing with 68000 Using the TAS Instruction and AS (Address Strobe) Signal

Earlier, the 68000 TAS instruction was discussed. The TAS instruction supports the software aspects of interfacing two or more 68000s via shared RAM. When TAS is executed, an indivisible read-modify-write cycle is performed. The timing diagram for this specialized cycle is shown in Figure 5.23. During both the read and the write portions of the cycle, the AS remains LOW, and the cycle starts as the normal read cycle.

However, in the normal read, the AS going inactive indicates the end of the read. During execution of the TAS, the AS stays LOW throughout the cycle, and therefore AS can be used in the design as a bus locking circuit.

Due to bus locking, only one processor at a time can perform a TAS operation in a multiprocessor system. The TAS instruction supports semaphore operations (globally shared resources) by checking a resource for availability and reserving or locking it for use by a single processor. The TAS instruction can, therefore, be used to allocate memory space reservations.

The TAS instruction execution flow for allocating memory is shown in Figure 5.24a and b. The shared RAM of Figure 5.24b is divided into M sections. The first byte of each section will be pointed to by (EA) of TAS (EA) instruction. In the flowcharts, (ea) first points to the first byte of section 1. The instruction TAS (ea) is then executed.

The TAS instruction checks the most significant bit (N bit) in (EA). N = 0 indicates that the section 1 is free; N = 1 means section 1 is busy. If N = 0, then section 1 will be allocated for use. On the other hand, if N = 1, section 1 is busy; a program will be written to subtract one section length from (EA) to check the next section for availability. Also, (EA) must be checked with the value TASLOCM. If (EA) < TASLOCM, then no space is available for allocation. However, if (EA) > TASLOCM, TAS is executed and the availability of that section is determined.

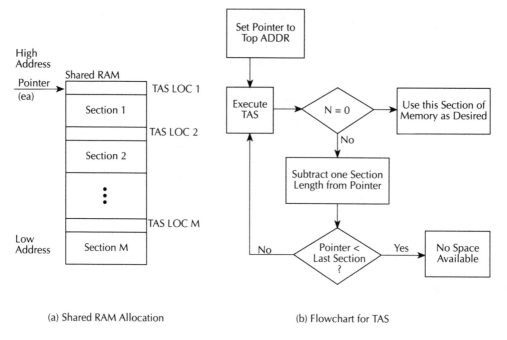

(a) Shared RAM Allocation   (b) Flowchart for TAS

FIGURE 5.24   Memory allocation using TAS.

In a multiprocessor environment, the TAS instruction provides software support for interfacing two or more 68000 via shared RAM. The AS signal can be used to provide the bus locking mechanism.

Examples of 68000 programmed and interrupt I/O are provided below.

Example 5.14

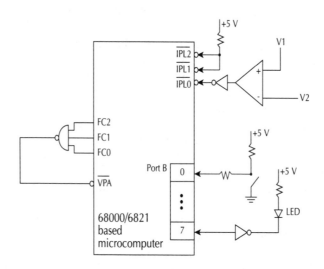

i)   In the above figure, the 68000/6821 based microcomputer is required to perform the following:

If $V_1 > V_2$, turn the LED ON if the switch is open or turn the LED OFF if the switch is closed. Write 68000 assembly language program to accomplish the above.

ii)   Repeat part I) using autovector level 1 and nonautovector (vector $40). Use port B for LED and switch as above. Assume supervisor mode.

Write the main program and service routine in 68000 assembly language program respectively at addresses $642F14 and $251F20.

Also, initialize supervisor stack pointer at $941760.

Solution

i)
```
        BCLR.B  #2,CRA      ; Address DDRA
        MOVE.B  #0,DDRA     ; Configure Port A as input
        BSET.B  #2,CRA      ; Address Port A
        BCLR.B  #2,CRB      ; Address DDRB
        MOVE.B  #$80,DDRB   ; Configure Port B
        BSET.B  #2,CRB      ; Address Port B
START   MOVE.B  PORTA,D0    ; Input Comparator
        ANDI.B  #$01,D0     ; Check if high
        BEQ START
        MOVE.B  PORTB,D1    ; Input switch
        ROXR.B  #2,D1       ; Align switch status
        MOVE.B  D1,PORTB    ; Output to LED
FINISH  JMP FINISH
```

ii)   Using Autovectoring Level 1

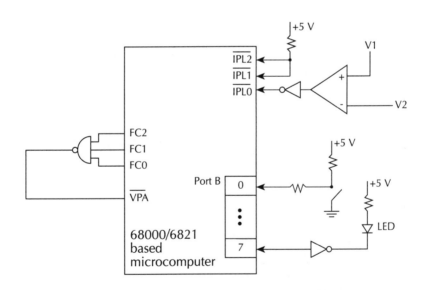

Main Program
```
        ORG $642F14
        BCLR.B  #2,CRB          ; Address DDRB
        MOVE.B  #$80,DDRB       ; Configure Port B
        BSET.B  #2,CRB          ; Address Port B
        ANDI.W  #$0F8FF,SR      ; Enable interrupts
WAIT    JMP WAIT                ; Wait for interrupts
FINISH  JMP FINISH              ; HALT
```

Service Routine

```
ORG $251F20
MOVE.B PORTB,D0    ; Input switch
ROXR.B #2,D0       ; Align switch
MOVE.B D0,PORTB    ; Output to LED
RTE
```

Reset Vector

```
ORG 0
DC.L $00941760
DC.L $00642F14
```

Service Routine Vector

```
ORG $000064
DC.L $00251F20
```

iii)   Using Nonautovectoring (Vector $40)

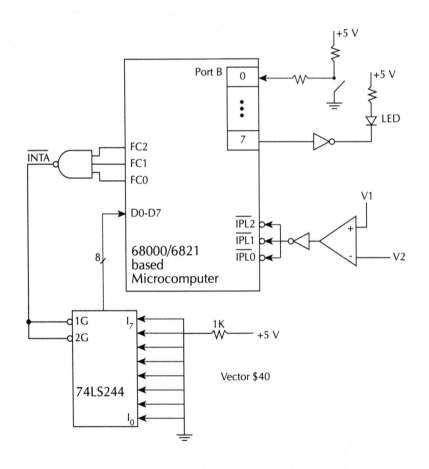

Main Program

```
ORG $642F14
BCLR.B #2,CRB      ; Address DDRB
MOVE.B #$80,DDRB   ; Configure Port B
```

```
        BSET.B  #2,CRB          ;   Address Port B
        ANDI.W  #$0F8FF,SR      ;   Enable interrupts
WAIT    JMP  WAIT               ;   Wait for interrupts
FINISH  JMP  FINISH             ;   Halt
```

Service Routine
```
        ORG  $251F20
        MOVE.B  PORTB,D0        ;   Input switch
        ROXR.B  #2,D0           ;   Align switch
        MOVE.B  D0,PORTB        ;   Output to LED
        RTE
```

Reset Vector
```
        ORG  0
        DC.L  $00941760
        DC.L  $00642F14
```

Service Routine Vector
```
        ORG  $000100
        DC.L  $00251F20
```

Questions and Problems

5.1 Assume that $[D0] = 25774411_{16}$. What will be the contents of D0 after execution of each of the following instructions:

  i) **CLR.B D0**

 ii) **CLR D0**

iii) **CLR.L D0**

5.2 Determine the contents of registers and the locations affected by each of the following instructions:

  i) **MOVE.L −(A2), (A3)+**

  Assume the following data prior to execution of the MOVE:

$$
\begin{array}{ll}
[A2] = \$300504, & [A3] = \$510718, \\
[\$300500] = \$01, & [\$300501] = \$F1, \\
[\$300502] = \$72, & [\$300503] = \$A1, \\
[\$510718] = \$53, & [\$510719] = \$20, \\
[\$51071A] = \$31, & [51071B] = \$27
\end{array}
$$

 ii) **MOVEA.W D1, A4**

  Assume the following data prior to execution of the MOVEA:

$$
\begin{array}{l}
[D1] = \$37158470 \\
[A4] = \$F1218234
\end{array}
$$

iii) **MOVEA.L A2, A3**

  Assume the following data prior to execution of the MOVEA:

$$
\begin{array}{l}
[A2] = \$1234F144 \\
[A3] = \$20718714
\end{array}
$$

5.3 Identify the following 68000 instructions as privileged or nonprivileged:

  i) **MOVE SR, (A2)**

 ii) **MOVE CCR, (A0)**

iii) `LEA.L  (A2),  A5`
iv) `MOVE.L A2,  USP`

5.4 What are the contents of register D1 after execution of the following two instructions? Assume [D1] = $34788480 prior to the execution of the instructions:

> `EXT.L  D1`
> `MOVEQ.L #$2F,  D1`

5.5 Find the contents of D1 after execution of the following DIVS instruction:

> `DIVS  (A1),  D1`

Assume [A1] = $205014, [$205014] = $FF, [$205015] = $FE, [D1] = $00000005 prior to execution of the instruction. Identify the quotient and remainder of the result in D1. Comment on the sign of the remainder.

5.6 Write a 68000 assembly program to divide an 8-bit signed number in low byte of D1 by an 8-bit signed number in low byte of D2. Store quotient and remainder in D1.

5.7 Write a 68000 assembly language program to add two 128-bit numbers. Assume that the first number is stored in consecutive memory locations starting at $605014. The second number is stored in consecutive memory locations starting at $708020. Store the result in memory locations beginning at $708020. Assume that all data storage to follow the conventional manner; that is highest byte to be stored as the lower address.

5.8 Write a 68000 assembly program to add top-two 32 bits of the stack. Store the 32-bit result onto the stack. Assume user mode.

5.9 Write a 68000 assembly program to multiply an 8-bit signed number in low byte of D1 by a 16-bit signed number in the high word of D5. Store the result in D3.

5.10 Write a 68000 assembly program to add twenty 32-bit numbers stored in consecutive memory locations starting at address $502040. Store the 32-bit result onto the stack. Assume that for each 32-bit number, the lowest address stores the highest byte of the number.

5.11 Write 68000 assembly language to find the minimum value of a string of ten signed 16-bit numbers using indexed addressing.

5.12 Write a 68000 assembly program to compare two strings of twenty ASCII characters. The first string is stored starting at $003000. The second string is stored starting at $004000. The ASCII character in location $003000 of string 1 will be compared with the ASCII character in location $004000 of string 2, [$003001] to be compared with [$004001], and so on. Each time there is a match, store $EEEE onto the stack; otherwise store $0000.

5.13 Write a 68000 assembly program to divide a 27-bit unsigned number in high 27 bits of D0 by 16_{10}. Do not use any divide instruction. Neglect the remainder. Store quotient in the low 27-bits of D0.

5.14 Write a subroutine in 68000 assembly language to compute

$$Z = \sum_{i=1}^{100} (Xi - Yi)$$

Assume that Xi's and Yi's are signed 16-bit and stored in consecutive locations starting at $020054 and $305116, respectively. Assume A0 and A1 point to Xi's and Yi's, respectively, and SP is already initialized. Also write the main program in 68000 assembly language to perform all initializations, call the subroutine, and then compute Z/100.

5.15 Write a subroutine in 68000 assembly language to subtract two unsigned eight-digit BCD numbers. The BCD number 1 is stored at location starting from $300000 thru $300003, with the least significant digit at $300003 and the most significant digit at $300000. Similarly, the BCD number 2 is stored at location starting from $400000 through $400003, with the least significant digit at $400003 and the most significant digit at $400000. The BCD number 2 is to be subtracted from BCD number 1. Store result in D1.

5.16 Write a subroutine in 68000 assembly to convert a 3-digit unsigned BCD number to binary. The most significant digit is stored in memory location starting at $003000, the next digit is stored at $003001, and so on. Store the binary result in D3.

Use the value of the 3-digit BCD number in $V = D2 \times 10^2 + D1 \times 10^1 + D0 = (D2 \times 10 + D1) \times 10 + D0$.

5.17 Write a recursive subroutine (A subroutine calling itself) in 68000 assembly language to find the factorial of an 8-bit number n by using $n! = n(n-1)(n-2) \ldots 1$. Store the result in D0.

5.18 Determine the status of $\overline{LDS}$, $\overline{UDS}$, $\overline{AS}$, FC2-FC0, and address lines immediately after execution of the following instruction sequence (before the 68000 tristates these lines to fetch the next instruction):

```
MOVE   #$2000,  SR
MOVE.B D2,  $030001
  -
  -
```

Assume the 68000 is in supervisor mode prior to execution of the above instructions.

5.19 Write a 68000 assembly program to output the contents of memory locations $003000 and $003001 to two seven-segment displays connected to two 8-bit ports A and B of a 68000/6821 system. Assume that displays are connected to ports A and B the same as shown in the figure of problem 5-39. Note that only the connections between Port A and one seven-segment display is shown in the figure.

5.20 Assume that in the configuration of Figure 5.19, port A and port B each has three switches and an LED connected to bits 0 through 3. Write 68000 assembly program to
  i) Turn the port A LED ON and port B LED OFF if port A has an even and port B has an odd number of high switch inputs.
  ii) Turn the port A LED OFF and port B LED ON if port A has an odd and port B has an even number of high switch inputs.
  iii) Turn both LEDs ON if both ports A and B have even number of high switch inputs.
  iv) Turn both LEDs OFF if both ports A and B have odd number of high switch inputs.

5.21 Interface a 68000 to 2716s, 6116s, and a 68230 to provide 4K EPROM, 4K RAM, and two 8-bit I/O ports. Draw a neat schematic and determine memory and I/O maps. Assume 16.67 MHz internal clock for the 68000.

5.22 If the $\overline{IPL2}\ \overline{IPL1}\ \overline{IPL0}$ pins are interrupted by an external device with the code 001_2 when the interrupt mask value I2I1I0 is 3_{10}, will the interrupt be serviced immediately or ignored by the 68000?

5.23   Discuss briefly the various 68000 exceptions.

5.24   Write a service routine for reset in 68000 assembly language that will initialize all data and address registers to zero, supervisor SP to $3F0728, user SP to $1F0524, and then jump to $002000.

5.25   Assume the following stack and register values before occurrence of an interrupt:

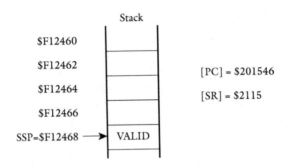

If an external device requests an interrupt by asserting the $\overline{IPL\,2}$ $\overline{IPL\,1}$ $\overline{IPL\,0}$ pins with the value 000, determine the contents of SSP and SR during interrupt and after execution of RTE at the end of the service routine of the interrupt. Draw the memory layouts showing where SSP points and the stack contents during and after the interrupt. Assume that the stack is not used by the service routine.

5.26   Suppose that two pumps (P1,P2) and two LEDs (L1,L2) are to be connected to a 68000-based microcomputer. Each pump has a 'pump running' output to indicate the ON/OFF status. The microcomputer runs the pumps via bits 2 and 3 of 8-bit port A. Two LEDs L1 (for P1) and L2 (for P2) are connected to bits 0 and 1 of 8-bit port B to indicate whether each pump is running. Assume that the pump can be turned ON by HIGH and turned OFF by LOW.

   i)   Using programmed I/O, draw a block diagram and write a 68000 assembly program to accomplish the above.
   ii)   Using interrupt I/O, draw a block diagram. Write the main program and service routine in 68000 assembly language to accomplish the above. The main program will perform all initializations and then start the pumps.

5.27   Compare the basic features of the 68000 with those of 68008, 68010, and 68012.

5.28   Write a 68000 assembly language program to add a 32-bit number stored in D0 (bits 0 through 15 containing the high-order 16 bits of the number and bits 16 through 31 containing the low-order 16 bits) with another 32-bit number stored in D1 (bits 0 through 15 containing the low-order 16 bits of the number and bits 16 through 31 containing the high-order 16 bits). Store the result in D0.

5.29   Write a subroutine in 68000 assembly language using the TAS instruction to find, reserve, and lock a memory segment for the main program. The memory is divided into four segments (0, 1, 2, 3) of 8 bytes each. The first byte of each segment includes a flagbyte to be used by the TAS instruction. In the subroutine, a maximum of four 8-byte memory segments must be checked for a free segment. Once a free segment is found, the TAS instruction is used to set the flagbyte. The starting address of the free segment must be stored in A5, and D5 must be cleared to zero to indicate a free segment. If no free block is found, a nonzero value of all ones must be stored in D5.

5.30 What are the remainder and quotient, and which registers contain them after execution of the following instruction sequence?

```
CLR  D0
MOVE #-5,D1
MOVE #2,D2
DIVS D2,D1
```

5.31 Write a 68000 assembly language program to divide $A5721624 by $F271. Store the remainder and quotient onto the user stack. Assume that the numbers are signed and stored in the stack as follows:

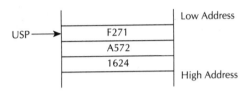

5.32 Write 68000 assembly language program to compute $X = Y + Z - \$30FE$ where X, Y, Z are 64-bit variables. The upper 32 bits of Y and Z are stored respectively in locations $005000 and $005008 followed by the lower 32 bits. Store the upper 32 bits of the 64-bit result at address $006000 followed by the lower 32 bits.

5.33 Assume that registers D0, D1, and D2 contain a signed byte, a signed word, and a signed 32-bit number respectively. Write 68000 assembly language program that will compute the signed 32-bit result: $D0.B + D1.W - D2.L \rightarrow D2.L$.

5.34 Write 68000 assembly language program to compute $X = 5 * Y + (Z/W)$ where address $005000, $005002, and $005004 store the 16-bit signed integers Y, Z and W. Store the 32-bit result in memory starting at address $005006. Discard the remainder of Z/W.

5.35 Write a 68000 assembly language program to add two 48-bit data values in memory as shown in figure below:

Store the result at the address pointed to by A2. The operation is given by:

$25	10	02	07	04	02
$02	04	07	1F	22	4A
$27	14	09	26	26	4C

Assume the data pointer and data are already initialized.

5.36 Assume the pins and signals for a 68000, 2716 (odd) and 6821 (even) shown in the figure below. Connect the chips and draw a neat schematic. Determine the memory and I/O maps. Assume a 16.67 MHz internal clock for the 68000.

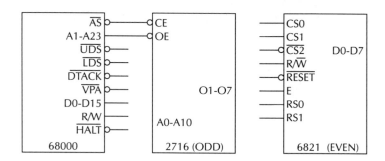

5.37 Assume the memory and I/O maps of Figure 5.20. Interface the following A/D to the 68000/2716-1/6116/6821 based microcomputer:

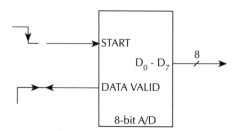

Write a 68000 assembly language program to input the A/D converter and turn ON an LED connected to bit 5 of Port A of even 6821 if the number read from A/D is ODD; otherwise turn the LED OFF. Assume that the LED is turned ON by a HIGH and turned OFF by a LOW.

5.38 Assume a 68000/6821 based system. Write a 68000 assembly program to input 16-bit data via Port A and Port B, and then divide this by the 8-bit data in the highest byte (bits 31-24) of D0. Assume all numbers to be signed.

5.39 A 68000/6821 based microcomputer is required to drive a common anode seven-segment display connected to Port A as follows:

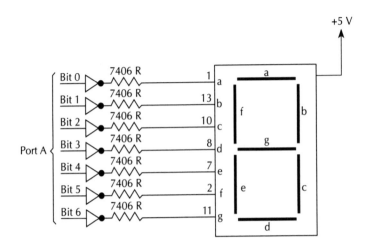

Write 68000 assembly language program to display a single hexadecimal digit (0 to F) from address $003000. Use a look up table.

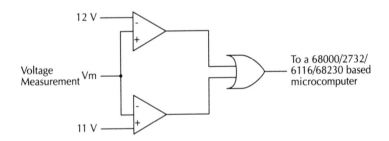

5.40

In the above figure, if Vm > 12V, turn an LED ON connected at bit 2 of Port A. On the other hand, if Vm < 11V, turn the LED OFF. Use registers and memory locations of your choice. Draw a block diagram showing the microcomputer and the connection of the figure to its ports. Also, write 68000 assembly language programs to accomplish the above using:

  a) Polled I/O
  b) Autovector level 7
  c) Non-autovector (vector $45)

6

MOTOROLA MC68020

This chapter first describes in detail the hardware, software, and interfacing features associated with the MC68020.

Topics include MC68020 architecture, addressing modes, instruction set, I/O, coprocessors, and system design.

6.1 Introduction

The MC 68020 is Motorola's first 32-bit microprocessor. The MC68020 is designed to execute all user object code written for previous members of the MC 68000 family.

The MC68020 is manufactured using HCMOS (combining HMOS and CMOS on the same device). The MC68020 consumes a maximum of 1.75 watts. It contains 200,000 transistors on a $^3/_8$" piece of silicon. The chip is packaged in a square ($1.345" \times 1.345"$) pin grid array (PGA) and contains 169 pins (114 pins used) arranged in a 13×13 matrix.

The processor speed of the MC68020 can be 12.5 MHz, 16.67 MHz, 20 MHz, 25 MHz, or 33 MHz. The chip must be operated from a minimum frequency of 8 MHz. Like the MC68000, it does not have any on-chip clock generation circuitry. The MC68020 contains 18 addressing modes and 101 instructions. All addressing modes and instructions of the MC68000 are included in the MC68020. The MC68020 supports coprocessors such as the MC68881/MC68882 floating-point and MC68851 memory management unit (MMU) coprocessors.

The following features of MC68020 are compared with those of MC68000:

Characteristic	68000	68020
Technology	HMOS	HCMOS
Size	$3-^1/_8" \times ^7/_8"$	1.345" × 1.345" (square size)
Number of pins	64, 68	169 (13 × 13 matrix; pins come out at the bottom of the chip; 114 pins currently used)
Control unit	Nanomemory (two-level control memory)	Nanomemory (two-level control memory)
Clock	6 MHz, 8 MHz, 10 MHz, 12.5 MHz, 16.67 MHz, 20 MHz, 25 MHz, or 33 MHz (no minimum requirements)	12.5 MHz, 16.67 MHz, 20 MHz, 25 MHz, or 33 MHz (must be 8 MHz minimum)
ALU	One 16-bit ALU	Three 32-bit ALUs
Address bus size	24 bits with A0 encoded from $\overline{\text{UDS}}$ and $\overline{\text{LDS}}$	32 bits; no encoding of A0 required
Data bus size	Uses D0-D7 for odd addresses and D8-D15 for even addresses during byte transfers; for word and long word uses D0-D15	8, 16, and 32 bits (byte, word, long word transfers occur, respectively, via D24-D31 lines, D16-D31 lines, and D0-D31 lines)
Instruction and data access	All word and long word accesses must be at even addresses for both instructions and data; for byte, instruction must be at even	Instructions must be accessed at even addresses; data accesses can be at any address for byte, word, and long word

(*continued*)

Characteristic	68000	68020
Instruction and data access (continued)	addresses, and data can be at either odd or even addresses	
Instruction cache	None	128-entry 16-bit word cache; at the start of an instruction fetch, the 68020 always outputs LOW on the ECS (external cycle start) pin and accesses the cache; if the instruction is found in the cache, the 68020 inhibits outputting LOW on the AS pin; otherwise the 68020 sends LOW on the AS pin and reads the instruction from the main memory
Directly addressable memory	16 megabytes	4 gigabytes (4, 294, 964, 296 bytes)
Registers	8 32-bit data registers 7 32-bit address registers 2 32-bit SPs 1 32-bit PC (24 bits used) 1 16-bit SR	8 32-bit data registers 7 32-bit address registers 3 32-bit SPs 1 32-bit PC (all bits used) 1 16-bit SR 1 32-bit VBR (vector base register) 2 3-bit function code registers (SFC and DFC) 1 32-bit CAAR (cache address register) 1 32-bit CACR (cache control register)
Addressing modes	14	18
Instruction set	56 instructions	101 instructions
Stack pointers	USP, SSP	USP, MSP (master SP), ISP (interrupt SP)
Status register	T, S, I0, I1, I2, X, N, Z, V, C	T0, T1, S, M, I0, I1, I2, X, N, Z, V, C

T1	T0	
0	0	No tracing
0	1	Trace on jumps
1	0	Trace on instruction execution
1	1	Undefined

S	M	
0	X	USP (X is don't care; can be 0 or 1)
1	0	ISP
1	1	MSP

Characteristic	68000	68020
Coprocessor interface	Emulated in software; that is, by writing subroutines, coprocessor functions such as floating-point arithmetic can be obtained	Can directly be interfaced to coprocessor chips. Coprocessor functions, such as floating-point arithmetic can be obtained via 68020 instructions
FC2, FC0, FC1 pins	FC2, FC0, FC1 = 111 means interrupt acknowledge	FC2, FC0, FC1 = 111 means CPU space cycle and then by decoding A16-A19, one can obtain breakpoints, coprocessor functions, and interrupt acknowledge

Some of the 68020 characteristics tabulated above will now be explained:

- The three independent ALUs are provided for data manipulation and address calculations.
- A 32-bit barrel shift register (occupies 7% of silicon) is included in the 68020 for very fast shift operations regardless of the shift count.
- The 68020 has three SPs. In the supervisor mode (when $S = 1$), two SPs can be accessed. These are MSP (when $M = 1$) and ISP (when $M = 0$). The ISP can be used to simplify and speed up task switching for operating systems. The 68020 user SP (USP) is used for the same purpose as the 68000 USP in the user mode.
- The vector base register (VBR) is used in interrupt vector computation. For example, in the 68000 the interrupt address vector is obtained by multiplying an 8-bit vector number by 4. In the 68020, on the other hand, the interrupt address vector is obtained by using VBR+4*8-bit vector number.

- The SFC (source function code) and DFC (destination function code) registers are 3 bits wide. These registers allow the supervisor to move data between address spaces. In supervisor mode, 3-bit addresses can be written into SFC or DFC using instructions such as MOVEC A1, SFC. The MOVES.W(A0), D0 can then be used to move a word from a location within the address space specified by SFC and (A0) to D0. The 68020 outputs [SFC] to the FC2, FC1, and FC0 pins. By decoding these pins via external decoder, the desired source memory location addressed by (A0) can be moved to D0. Now, if this data in D0 is to be moved to another space, then the following instructions will accomplish this:

```
MOVEC A3, DFC
MOVES.W D0, (A5)
```

Note that there is no MOVES mem, mem instruction. SFC and DFC allow one to move data from one space to another. Since in the above, MOVES.W D0, (A5) outputs [DFC] to FC2, FC1, and FC0 pins which can be used to enable the chip containing the memory location addressed by (A5). [D0] is then moved to this location.

The new addressing modes in the 68020 include scaled indexing, 32-bit displacements, and memory indirection. In order to illustrate the concept of scaling, consider moving the contents of memory location 50_{10} to A1. Using the 68000, the following instruction sequence will accomplish this:

```
MOVEA.W #10, A0         ; Lead starting address of a
                        ;  table to A0
MOVE.W #10, D0          ; Load index value to D0
ASL #2, D0             ; Scale index
MOVEA.L 0(A0, D0.W), A1 ; Access data
```

The scaled indexing can be used with the 68020 to perform the same as follows:

```
MOVEA.W #10, A0             ; Load starting address of
                           ;  a table to D0
MOVE.W #10, D0             ; Load index value to D0
MOVE.L (0, A0, D0.W*4), A1 ; Access data
```

Note that [D0] in the above is scaled by 4. Scaling 1, 2, 4, or 8 can be obtained.

- The new 68020 instructions include bit field instructions to better support compilers and certain hardware applications such as graphics, 32-bit multiply and divide instructions, pack and unpack instructions for BCD, and coprocessor instructions. Bit field instructions can be used to input data from A/D converters and eliminate wasting main memory space when the A/D converter is not 32-bits.

- FC2, FC1, FC0 = 111 means CPU space cycle. The 68020 makes CPU space access for breakpoints, coprocessor operations, or interrupt acknowledge cycles. The CPU space classification is generated by the 68020 based upon execution of breakpoint instructions, coprocessor instructions, or during the interrupt acknowledge cycle. The 68020 then decodes A19-A16 to determine the type of CPU space. For example, FC2, FC1, FC0 = 111 and A19, A18, A17, A16 = 0010 mean coprocessor instruction.

- For performing floating-point operations, the 68000 user must write subroutines using the 68000 instruction set. The floating-point capability in the 68020 can be obtained by connecting the Motorola 68881 floating-point coprocessor chip. The 68020's two coprocessor chips include the 68881 (floating-point) and 68851 (memory management). The MC68020 can have up to eight coprocessor chips. When a coprocessor is connected to the 68020, the coprocessor instructions are added to the 68020 instruction set automatically, and this is transparent to the user. For example, when the 68881 floating-point coprocessor is added to the 68020, instructions such as FADD (floating-point ADD) are available to the user. The programmer can then execute the instruction:

FADD FD0, FD1

Note that registers FD0 and FD1 are in the 68881. When the 68020 encounters the FADD instruction, it writes a command in the command register in the 68881, indicating that the 68881 has to perform this operation. The 68881 then responds to this by writing in the 68881 response register. Note that all coprocessor registers are memory-mapped. The 68020 thus can read the response register and obtain the result of the floating-point ADD from the appropriate location.

6.2 Programming Model

Figure 6.1 shows the MC68020 programming model. The user model has sixteen 32-bit general-purpose registers (D0-D7 and A0-A7), a 32-bit program counter (PC), and a condition code register (CCR) contained within the supervisor status register (SR). The supervisor model has two 32-bit supervisor stack pointers (1SP and MSP), a 16-bit status register (SR), a 32-bit vector base register (VBR), two 3-bit alternate function code registers (SFC and DFC), and two 32-bit cache handling (address and control) registers (CAAR and CACR). General-purpose registers D0-D7 are used as data registers for operation on all data types. General-purpose registers A0-A6, user stack pointer (USP) A7, interrupt stack pointer (1SP) A7', and master stack pointer (MSP) A7" are address registers that may be used as software stack pointers or base address registers.

The status register (Figure 6.2) consists of a user byte (condition code register CCR) and a system byte. The system byte contains control bits to indicate that the processor is in the trace mode (T1, T0), supervisor/user state (S), and master/interrupt state (M). The user byte consists of the following condition codes: carry (C), overflow (V), zero (Z), negative (N), and extend (X).

The bits in 68020 user byte are set at reset in the same way as the 68000 user byte. The bits I2, I1, I0, and S have the same meaning as the 68000. In the 68020, two trace bits (T1, T0) are included as opposed to one trace bit (T) in the 68000. These two bits allow the 68000 to trace on both normal instruction execution and jumps. The 68020 M-bit is not included in the 68000 status register.

The vector base register (VBR) is used to locate the exception processing vector table in memory.

The MC68020 distinguishes address spaces as supervisor/user and program/data. To support full access privileges in the supervisor mode, the alternate function code registers (SFC and DFC) allow the supervisor to access any address space by preloading the SFC/DFC registers appropriately.

The cache registers (CACR and CAAR) allow software manipulation of the instruction cache. The CACR provides control and status accesses to the instruction cache, while the CAAR holds the address for those cache control functions that require an address.

6.3 Data Types, Organization, and CPU Space Cycle

The MC68000 family supports data types of bits, byte integers (8 bits), word integers (16 bits), long word integers (32 bits), and binary coded decimal (BCD) digits. In addition to these, four

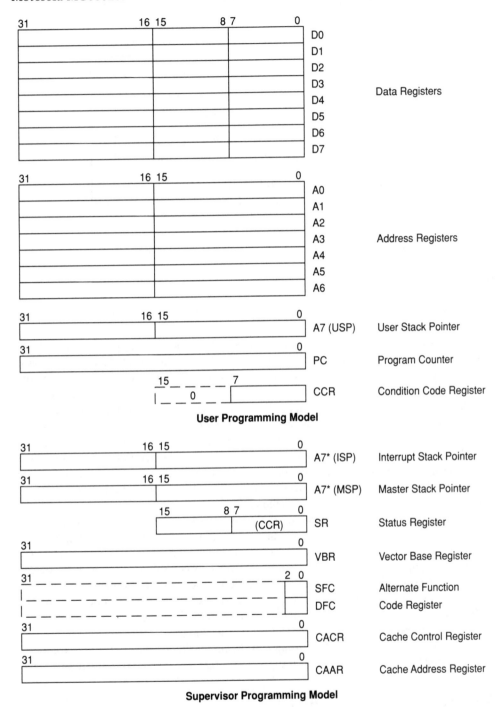

FIGURE 6.1 MC68020 programming model.

new data types are supported by the MC68020: variable-width bit field, packed BCD digits, quad words (64 bits), and variable-length operands.

Data stored in memory are organized on a byte-addressable basis, where the lower addresses correspond to higher-order bytes. The MC68020 does not require data to be aligned on even byte boundaries, but data that are not aligned are transferred less efficiently.

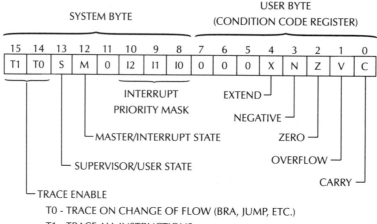

FIGURE 6.2   MC68020 status register.

Instruction words must be aligned on even byte boundaries. Figure 6.3 shows how data are organized in memory.

Table 6.1 shows decoding of the function code pins. The function code pins define the user/supervisor program and data spaces in the same way as the MC68000, except that FC2 FC1 FC0 = 111 for the MC68020 defines a new cycle called the CPU space cycle. Note that for the MC68000, FC2, FC1, FC0 = 111 provides the interrupt acknowledge cycle. CPU space is not intended for general instruction execution, but is reserved for processor functions. The CPU space has been subdivided into 16 types of access. The type of CPU access is indicated by address bits (A19-A16) in combination with the CPU space function code (FC2 FC1 FC0 = 111).

Table 6.2 defines the four different types of CPU accesses. The definition of regions in the CPU space makes it possible to acknowledge break points and interrupts and to communicate with coprocessors and other special devices (such as the MMU) without dictating memory organization for user- and supervisor-related activity.

The MC68020 has three stack pointers: the user stack pointer (USP) register A7, the interrupt stack pointer (ISP) register A7′, and the master stack pointer (MSP) register A7″. During normal operation most codes will be executed in user space and programs will use the A7 stack for temporary storage and parameter passing between software routines (modules). The ISP register is only used when an exception occurs, such as an external interrupt when control is passed to supervisor mode and the relevant exception process is performed. The MSP holds process-related information for the various tasks and allows for the separation of task-related and non-task-related exception process stacking.

When the master stack is enabled through bit (M) in the SR, all noninterrupting exceptions, such as divide by zero, software traps, and privilege violation, are placed in the master stack.

6.4 MC68020 Addressing Modes

Figure 6.4 lists the MC68020's 18 addressing modes. Table 6.3 compares the addressing modes of the MC68000 with those of the MC68020.

Since MC68000 addressing modes are covered in detail with examples in Chapter 5, the MC68020 modes which are not available in the MC68000 are covered in the following discussion.

6.4.1 Address Register Indirect (ARI) with Index and 8-Bit Displacement
Assembler syntax: (d8, An, Xn. size * SCALE)

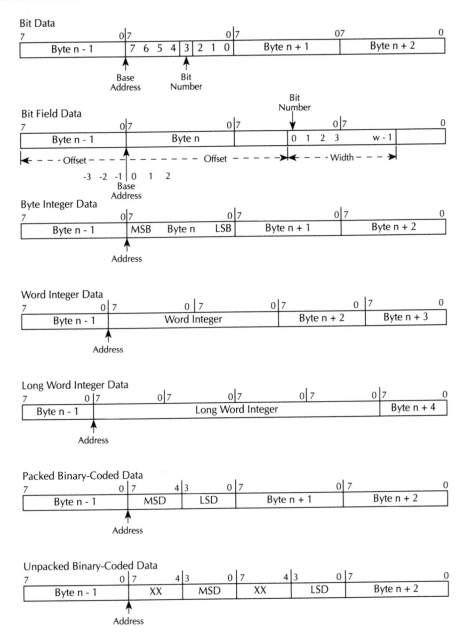

FIGURE 6.3 Memory data organization.

```
EA = (An) + (Xn.size * scale value) + d8
Xn can be W or L.
```

If index register (An or Dn) is 16 bits, then it is sign-extended to 32 bits and then multiplied by 1, 2, 4, or 8 prior to being used in EA calculation. d8 is also sign-extended to 32 bits prior to EA calculation. An example is

```
MOVE.W (0, A2, D2. W*2), D1
```

TABLE 6.1   Processing State Address Space

Function code			Address space
FC2	FC1	FC0	
0	0	0	(Undefined, reserved)[a]
0	0	1	User data space
0	1	0	User program space
0	1	1	(Undefined, reserved)[a]
1	0	0	(Undefined, reserved)[a]
1	0	1	Supervisor data space
1	1	0	Supervisor program space
1	1	1	CPU space

[a] Address space 3 is reserved for user definition, while 0 and 4 are reserved for future use by Motorola.

TABLE 6.2   CPU Space Address Access Encodings

CPU Space Access Types | Function Code | Address Bus

Breakpoint Acknowledge:
Function Code [1 1 1] — Address Bus, bits 31..23: 0 0 0 0 0 0 0 0 0 0 0 0 0 | 19..16: 0 0 0 0 | 0 0 0 0 0 0 0 0 0 0 0 0 | 4..2: BKPT# | 0 0

Access Level Control:
Function Code [1 1 1] — bits 31..: 0 0 0 0 0 0 0 0 0 0 0 0 0 | 0 0 0 1 | 0 0 0 0 0 0 0 0 | 6..0: MMU REG

Coprocessor Comm:
Function Code [1 1 1] — bits 31..: 0 0 0 0 0 0 0 0 0 0 0 0 0 | 0 0 1 0 | 15..13: CPID | 0 0 0 0 0 0 0 | 4..0: CIR Register

Interrupt Acknowledge:
Function Code [1 1 1] — bits 31..: 1 1 1 1 1 1 1 1 1 1 1 1 1 | 1 1 1 1 | 1 1 1 1 1 1 1 1 1 1 1 1 | 3..1: Level | 1 0: 1

CPU Space Type Field (bits 19–16)

TABLE 6.3   Addressing Modes, MC68000 Vs. MC68020

Addressing modes available		68000	68020
Data Register Direct	Dn	Yes	Yes
Address Register Direct	An	Yes	Yes
Address Register Indirect (ARI)	(An)	Yes	Yes
ARI with Postincrement	(An)+	Yes	Yes
ARI with Predecrement	−(An)	Yes	Yes
ARI with Displacement (16-bit displ)	(d, An)	Yes	Yes
ARI with Index (8-bit displ)	(d, An, Xn)	Yes[a]	Yes
ARI with Index (Base Displ: 0, 16, 32)	(bd, An, Xn)	No	Yes
Memory Indirect (Postindexed)	([bd, An], Xn, od)	No	Yes
Memory Indirect (Preindexed)	([bd, An, Xn], od)	No	Yes
PC Indirect with Displ. (16-Bit)	(d, PC)	Yes	Yes
PC Indirect with Index (8-Bit Displ)	(d, PC, Xn)	Yes[a]	Yes
PC Indirect with Index (Base Displ)	(bd, PC, Xn)	No	Yes
PC Memory Indirect (Postindexed)	([bd, PC], Xn, od)	No	Yes
PC Memory Indirect (Preindexed)	([bd, PC, Xn], od)	No	Yes
Absolute Short	(xxx).W	Yes	Yes
Absolute Long	(xxx).L	Yes	Yes
Immediate	#<data>	Yes	Yes

[a] 68000 has no scaling capability; 68020 can scale Xn by 1, 2, 4 or 8

Addressing Modes	Syntax

Register Direct
  Data Register Direct
  Address Register Direct
	Dn
	An

Register Indirect
  Address Register Indirect
  Address Register Indirect with Post Increment
  Address Register Indirect with Predecrement
  Address Register Indirect with Displacement
	(An)
	(An)+
	-(An)
	(d16, An)

Register Indirect with Index
  Address Register Indirect with Index (8-Bit Displacement)
  Address Register Indirect with Index (Base Displacement)
	(d8, An, Xn)
	(bd, An, Xn)

Memory Indirect
  Memore Indirect Post-Indexed
  Memory Indirect Pre-Indexed
	([bd, An], Xn, od)
	([bd, An, Xn], od)

Program Counter Indirect with Displacement	(d16, PC)

Program Counter Indirect with Index
  PC Indirect with Index (8-Bit Displacement)
  PC Indirect with Index (Base Displacement)
	(d8, PC, Xn)
	(bd, PC, Xn)

Program Counter Memory Indirect
  PC Memory Indirect Post-Indexed
  PC Memory Indirect Pre-Indexed
	([bd, PC,], Xn, od)
	([bd, PC, Xn], od)

Absolute
  Absolute Short
  Absolute Long
	xxx.W
	xxx.L

Immediate	#(data)

NOTES:

Dn = Data Register, D0-D7

An = Address Register, A0-A7

$d8, d16$ = A twos-complement, or sign-extended displacement; added as part of the effective address calculation; size is 8 (d8) or 16 (d16) bits; when omitted, assemblers use a value of zero.

Xn = Address or data register used as an index register; form is Xn.SIZE*SCALE, where SIZE is .W or .L (indicates index register size) and SCALE is 1, 2, 4, or 8 (index register is multiplied by SCALE); use of SIZE and/or SCALE is optional.

bd = A twos-complement base displacement; when present, size can be 16 or 32 bits.

od = Outer displacement, added as part of effective address calculation after any memory indirection; use is optional with a size of 16 or 32 bits.

PC = Program Counter

$(data)$ = Immediate value of 8, 16, or 32 bits.

$(\ )$ = Effective Address.

$[\ ]$ = Use as indirect address to long word address.

FIGURE 6.4 MC68020 addressing modes.

Suppose that [A2] = $5000 0000, [D2.W] = $1000, and [$5000 2000] = $1571; then after execution of the MOVE, [D1]$_{low\ 16\text{-}bit}$ = $1571, since EA = $5000 0000 + $1000 * 2 + 0 = $5000 2000.

6.4.2 ARI with Index (Base Displacement, bd: Value 0 or 16 Bits or 32 Bits)

```
Assembler syntax: (bd, An, Xn.size*SCALE)
EA = (An) + (Xn.size * SCALE) + bd
```

The figure below shows the use of ARI with index, Xn and base displacement, bd for accessing tables or arrays:

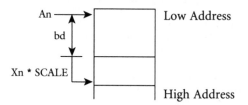

An example is **MOVE.W ($5000, A2, D1.W * 4), D5**. If [A2] = $3000 0000, [D1.W] = $0200, and [$3000 5800] = $0174, then after this **MOVE**, [D5]$_{low\ 16\ bits}$ = $0174, since EA = $5000 + $3000 0000 + $0200 * 4 = $3000 5800.

6.4.3 Memory Indirect

Memory indirect is distinguished from address register indirect by use of square brackets ([ ]) in the assembler notation.

The concept of memory indirect mode is depicted below:

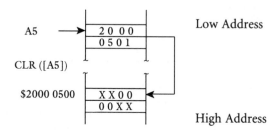

In the above, register A5 points to the effective address $2000 0501. Since CLR ([A5]) is a 16-bit clear instruction, two bytes in location $2000 0501 and $2000 0502 are cleared to zero.

Memory indirect mode can be indexed with scaling and displacements. There are two types of memory indirect with scaled index and displacements: postindexed memory indirect mode and preindexed memory indirect mode.

For postindexed memory indirect mode, an indirect memory address is first calculated using the base register (An) and base displacement (bd). This address is used for an indirect memory access of a long word followed by adding a scaled indexed operand and an optional outer displacement (od) to generate the effective address. bd and od can be zero, 16 bits, or 32 bits.

In this memory indirect mode, indexing occurs after memory indirection.

Assembler syntax:

```
([bd, An], Xn.size Size * Scale, od)
EA = ([bd + An]) + Xn.size * Scale + od)
```

An example is **MOVE.W ([$0004,A1], D1.W * 2, 2), D2.** If [A1] = $2000 0000, [$2000 0004] = $0000 3000, [D1.W] = $0002, [$0000 3006] = $1A40, then after execution of the above **MOVE,** intermediate pointer = (4 + $2000 0000) = $2000 0004, [$2000 0004], which is $0000 3000 used as a pointer. Therefore, EA = $0000 3000 + $0000 0004 + $2 = $00003006; hence, [D2]$_{low\ 16\ bits}$ = $1A40.

For memory indirect preindexed, the scaled indexed operand is added to the base register (An) and base displacement (bd). This result is then used as an indirect address into the data space. The 32-bit value at this address is fetched and an optional outer displacement (od) is added to generate the effective address. The indexing, therefore, occurs before indirection.

Assembler syntax:

```
([bd, An, Xn.size * Scale], od)
EA = (bd + An + Xn.size * Scale) + od
```

As an application of memory indirect preindexed mode, consider several I/O devices in a system. The addresses of these devices can be held in a table pointed to by An, bd, and Xn. The actual programs for the devices can be stored in memory pointed to by the respective device addresses and od.

As an example of memory indirect preindexed mode, consider **MOVE.W ([$0004, A2, D1.W * 4], 2), D5.** If [A2] = $3000 0000, [D1.W] = $0002, [$3000 000C] = $0024 1782, [$0024 1784] = $F270, then after execution of the above **MOVE,** intermediate pointer = $3000 0000 + 4 + $0002 * 4 = $3000 000C. Therefore, [$3000 000C] which is $0024 1782 is used as a pointer to memory. EA = $0024 1782 + 2 = $0024 1784. Hence, [D5]$_{low\ 16\ bits}$ = $F270. Note that in the above, bd, Xn, and od are sign-extended to 32 bits if one (or more) of them is 16 bits wide before the calculation.

6.4.4 Memory Indirect with PC

In this mode, PC (program counter) is used to form the address rather than an address register. The effective address calculation is similar to address register indirect.

6.4.4.a PC Indirect with Index (8-Bit Displacement)

The effective address is obtained by adding the PC contents, the sign-extended displacement, and the scaled indexed (sign-extended to 32 bits if it is 16 bits before calculation) register.

Assembler syntax:

```
(d8, PC, Xn.size * Scale)
EA = (PC) + (Xn.size * SCALE) + D8
```

For example, consider **MOVE.W D2, (2, PC, D1.W * 2).** If [PC] = $4000 0020, [D1.W] = $0020, [D2.W] = $20A2, then after this **MOVE,** EA = 2 + $4000 0020 + $0020 * 2 = $4000 0062. Hence, [$4000 0062] = $20A2.

6.4.4.b PC Indirect with Index (Base Displacement)

This address of the operand is obtained by adding the PC contents, the scaled index register contents, and the base displacement.

Assembler syntax:

```
(bd, PC, Xn.size * Scale)
EA = (PC) + (Xn.size * SCALE) + bd
```

Xn and bd are sign-extended to 32 bits if either or both are 16 bits.

As an example, consider **MOVE.W (4, PC, D1.W * 2), D2**. If [PC] = $2000 0004, [D1.W] = $0020, [2000 0048] = $2560, then after this **MOVE**, [D2.W] = $2560.

6.4.4.c PC Indirect (Postindexed)

An intermediate memory pointer in program space is calculated by adding PC (used as a base register) and bd. The 32-bit content of this address is used in the EA calculation. EA is obtained by adding the 32-bit contents with a scaled index register and od. Note that bd, od, and index register are sign-extended to 32 bits before using in calculation if one (or more) is 16 bits.

Assembler syntax:

```
([bd, PC], Xn.size * Scale, od)
EA = ([bd + PC] + Xn.size * Scale + od)
```

Consider another example: **MOVE.W ([2, PC], D1.W*4, 0), D1**. If [PC] = $3000 0000, [D1.W] = $0010, [$3000 0002] = $2040 0050, [$2040 0090] = $A240, then after this **MOVE**, [D1.W] = $A240.

6.4.4.d PC Indirect (Preindexed)

The scaled index register is added to the PC and bd. This sum is then used as an indirect address into the program space. The 32-bit value at this address is added to od to find EA.

Assembler syntax:

```
([bd, PC, Xn.size * Scale], od)
EA = (bd + PC + Xn.size * Scale) + od
```

od, bd, and the index register are sign-extended to 32 bits if one (or more) of them is 16 bits before the EA calculation.

As an example, consider **MOVE.W ([4, PC, D1.W * 2], 4), D5**. If [PC] = $5000 0000, [D1.W] = $0010, [$5000 0024] = $2050 7000, [$2050 7004] = $0708, then after this **MOVE**, [D5.W] = $0708.

A summary of PC modes is provided below:

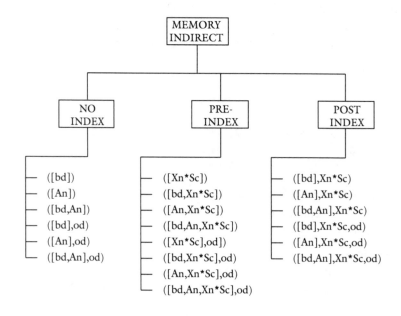

Example 6.1

Show the contents of registers A2, D4, A5, and the affected memory location(s), after execution of the following instruction:

$$\text{MOVEA.W} \quad (0, \quad \text{A2}, \quad \text{D4.L}), \quad \text{A5}$$

Assume prior to execution of the MOVEA instruction:

```
[A2]           =      $0571   6660
[D4]           =      $3072   8400
[A5]           =      $7271   5554
[$35E3 EA58]   =      $05
[$35E3 EA59]   =      $07
[$35E3 EA60]   =      $F7
[$35E3 EA61]   =      $F1
[$35E3 EA62]   =      $40
```

Solution

Effective address:

```
D4.L = $3072 8400
A2.L = $0571 6660
d    = $0000 0000
EA   = $35E3 EA 60
```

Therefore, after execution of the MOVEA instruction:

```
[A2] = $0571  6660
[D4] = $3072  8400
[A5] = $FFFFF7F1
```

Example 6.2

The following MC68000 instruction sequence:

```
MOVEA.L  6 (USP),  A1
MOVE.W  (A1),  D5
```

is used by a subroutine to access a parameter whose address has been passed into A1 and then moves the parameter to D5.

Find the equivalent MC68020 instruction.

Solution

```
MOVE.W  ([6,  USP]),  D5
```

Example 6.3

Find a MC68020 compare instruction with the appropriate addressing mode to replace the following MC68000 instruction sequence:

```
ASL.L    #3, D6
CMP.L    0 (A1, D6.L), D3
```

Solution

The equivalent MC 68020 instruction is

```
CMP.L (0, A1, D6.L * 8), D3
```

Example 6.4

Find the contents of register A1, D0, D5 and the affected memory locations after execution of:

```
MOVE.B (0, A1, D0.W * 2), D5
```

Assume the following data prior to execution of the MOVE instruction:

```
[A1]         =      $0000 2000
[D0]         =      $0000 0004
[D5]         =      $7124 8002
[$0000 2006] =          $51
[$0000 2007] =          $74
[$0000 2008] =          $82
[$0000 2009] =          $FO
```

Solution

Effective address:

```
= d8 + A1.L + D0.W *2
= 0 + $0000 2000 + $0000 0004 * 2
= $0000 2008
```

Therefore,

```
[A1] = $0000 2000
[D0] = $0000 0004
[D5] = $7124 8082
```

6.5 68020 Instructions

The MC 68020 instruction set includes all 68000 instructions, plus some new ones. Some of the 68000 instructions are enhanced.

The 68020 new instructions beyond those of the 68000 can be grouped as follows:

1. 68020 new privileged MOVE instructions
2. RTD instruction
3. CHK/CHK2 and CMP/CMP2 instructions
4. TRAPcc instructions
5. Bit field instructions
6. PACK and UNPK instructions
7. Multiplication and division instructions
8. 68000 enhanced instructions
9. 68020 Advanced Instructions (to be discussed in Section 6.6)

- BKPT instructions
- CALLM/RTM instructions
- CAS/CAS2 instructions
- Coprocessor instructions

TABLE 6.4   MC68020 Privileged Move Instructions

Instruction	Operand size	Operation	Notation
MOVEC	32	Rc → Rn	MOVEC.L Rc, Rn
		Rn → Rc	MOVEC.L Rn, Rc
MOVES	8, 16, 32	Rn → destination using DFC	MOVES.S Rn, (EA)
		Source using SFC → Rn	MOVES.S(EA), Rn

6.5.1 New Privileged Move Instruction

The new privileged move instructions are executed by the 68020 in the supervisor mode.

The MOVE instructions are summarized in Table 6.4. The MOVEC instruction was added to allow the new supervisor registers (Rc) to be accessed. Since these registers are used for system control, they are generally referred to as control registers and include the vector base register (VBR), the source function code and destination function code registers (SFC, DFC), the master, interrupt, and user stack pointers (MSP, ISP, USP), and the cache control and address registers (CACR, CAAR). Register (Rn) can be either an address register or data register.

The operand size indicates that these MOVEC operations are always long-word. Notice only register-to-register operations are allowed.

A control register (Rc) can be copied to an address or data register (Rn), or vice versa. When copying the 3-bit, SFC, or DFC register into Rn, all 32 bits of the register are overwritten and the upper 29 bits are "0".

The MOVE to alternate space instruction (MOVES) allows the operating system to access any addressed space defined by the function codes.

It is typically used when an operating system running in the supervisor mode must pass a pointer or value to a previously defined user program or data space.

The MOVES instruction allows register-to-memory or memory-to-register operations. When a memory-to-register occurs, this instruction causes the contents of the source function code register to be placed on the external function hardware pins.

For a register-to-memory move, the processor places the destination function code register on the external function code pins.

The MOVES instruction can be used to MOVE information from one space to another. For example, in order to move the 16-bit content of a memory location addressed by A0 in supervisor data space (FC2 FC1 FC0 = 101) to a memory location addressed by A1 in user data space (FC2 FC1 FC0 = 001), the following instruction on sequence can be used:

```
MOVEQ.L #5, D0      ;  Move source space 5 to D0
MOVEQ.L #1, D1      ;  Move dest space 1 to D1
MOVEC.L D0, SFC     ;  Initialize SFC
MOVEC.L D1, DFC     ;  Initialize DFC
MOVES.W (A0), D2    ;  Move memory location addressed by (A0)
                         and SFC to D2
MOVES.W D2, (A1)    ;  Move D2 to a memory location addressed
                         by (A1) and DFC
```

In the above, the first four instructions initialize SFC to 101_2 and DFC to 001_2. Since there is no MOVES mem, mem instruction, the register D2 is used as a buffer for the memory-to-memory transfer. MOVES.W (A0), D2 transfers [SFC] to FC2 FC1 FC0 and also reads the content of a memory location addressed by SFC and (A0) to D2. The 68020 FC2 FC1 FC0 pins can be decoded to enable the appropriate memory bank containing the memory location addressed by (A0). Next, MOVES.W D2, (A1) outputs [DFC] to FC2 FC1 FC0 and then moves [D2] to a memory location addressed by (A1) contained in a memory bank which can be enabled by decoding FC2, FC1, and FC0 pins.

Example 6.5

Find the content of memory location $5000 2000 after execution by MOVE.W SR, [A6]. Assume the following data prior to execution of the MOVE instruction:

$$[A6] = \$5000 \ 2000 \ [SR] = \$26A1$$
$$[\$5000 \ 2000] = \$02, \ [\$5000 \ 2001] = \$F1$$

Also, assume supervisor mode.

Solution

SR is moved to a memory location pointed to by $5000 2000. After execution:

$$[\$5000 \ 2000] = \$26$$
$$[\$5000 \ 2001] = \$A1$$

Example 6.6

Find the content of DFC after execution of MOVEC.L A5, DFC. Assume the following data prior to execution of the instruction:

$$[DFC] = 100 \,_2, \ [A5] = \$2000 \ 0105$$

Solution
After execution of the MOVEC, [DFC] = 101.

Example 6.7

Find the contents of D5 and the function code pins FC2, FC1, and FC0 after execution of MOVES.B D5, (A5). Assume the following data prior to execution of the MOVES:

$$[SFC] = 101 \,_2, \ [DFC] = 100 \,_2$$
$$[A5] = \$7000 \ 0023$$
$$[D5] = \$718F \ 2A05$$
$$[\$7000 \ 0020] = \$01, \ [\$7000 \ 0021] = \$F1$$
$$[\$7000 \ 0022] = A2$$
$$[\$7000 \ 0023] = \$2A$$

Solution
After execution of the above MOVES:

$$FC2 \ FC1 \ FC0 = 100 \,_2$$
$$[\$7000 \ 0023] = \$05$$

TABLE 6.5 RTD Instruction

Instruction	Operand size	Operation	Notation
RTD	Unsized	(SP) → PC, SP + 4 + d → SP	RTD # ⟨displacement⟩

6.5.2 Return and Delocate Instruction

Return and delocate instruction RTD is useful when a subroutine has the responsibility to remove parameters off the stack that were pushed onto the stack by the calling routine. Note that the calling routine's JSR (jump to subroutine) or BSR (branch to subroutine) instructions do not automatically push parameters onto the stack prior to the call, as do the CALLM instructions. Rather, the pushed parameters must be placed there using the MOVE instruction. Table 6.5 shows the format of the RTD instruction.

The RTD instruction operates as follows:

1. Read the long word from memory pointed to by the stack pointer.
2. Copy it into the program counter.
3. Increment the stack pointer by 4.
4. Sign-extend the 16-bit immediate data displacement to 32 bits.
5. Add it to the stack pointer.

Since parameters are pushed onto the stack to lower memory locations, only a positive displacement should be added to the SP when removing parameters from the stack. The displacement value (16 bits) is sign-extended to 32 bits.

Example 6.8

Write a 68020 instruction sequence to illustrate the use of RTD instruction by using BSR instruction and pushing three 32-bit parameters onto the stack.

Solution

```
MOVE.L PAR1, -(SP)      BSR SUBR          RTD#12
MOVE.L PAR2, -(SP)
MOVE.L PAR3, -(SP)
```

Calling routine pushes parameters on stack; this causes the stack pointer to be decremented by 12

Calling routine calls subroutine and the PC stacked; the subroutine then accesses the parameters to perform the task

Last instruction of the subroutine returns and delocates the parameters off stack by adding (12) to the stack pointer

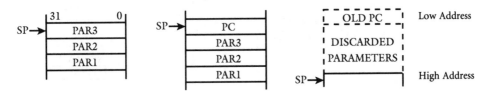

6.5.3 CHK/CHK2 and CMP/CMP2 Instructions

The enhanced MC68020 check instruction (CHK) now compares a 32-bit two's complement integer value residing in a data register (Dn) against a lower bound (L.B.) value of zero and against an upper bound (U.B.) value of the programmer's choice. These bounds are located at the effective address (EA) specified in the instruction format.

The CHK instruction has the following format:

CHK.S   (EA),Dn

where (.S) is the operand size designator which is either word (.W) or long word (.L).

If the data register value is less than zero (Dn <0) or if the data register value is greater than the upper bound (Dn >UB), then the processor traps through exception vector 6 (offset \$18) in the exception vector table. Of course, the operating system or the programmer must define a check service handler routine at this vector address. The condition codes after execution of the CHK are affected as follows:

If $D_n < 0$ then N = 1.
If $D_n >$ UB (Upper Bound) then N = 0.
If $0 \leq D_n \leq$ UB then N is undefined. X is unaffected and all other flags are undefined and
    program execution continues with the next instruction.

This instruction can be used for maintaining array subscripts since all subscripts can be checked against an upper bound (i.e., UB = array size minus one). If the compared subscript is within the array bounds ($0 \leq$ subscript value $\leq$ UB value), then the subscript is valid, and the program continues normal instruction execution. If the subscript value is out of array limits (i.e., 0 > subscript value, or the subscript value > UB value), then the processor traps through the CHK exception.

Example 6.9

Find the effects of execution of the MC68020 CHK instruction: CHK.L (A5), D3, where A5 contains a memory pointer to the array's upper bound value. Register D3 contains the subscript value to be checked against the array bounds. Assume the following data prior to execution of the CHK instruction:

**[D3] = \$0150 7126,   [A5] = \$00710004,
[\$0071 0004] = \$0150 0000**

Solution

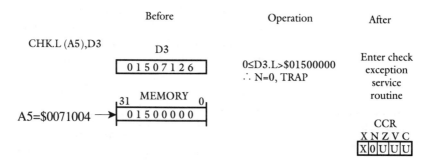

The long-word array subscript value \$01507126 contained in data register D3 is compared against the long-word upper-bound value \$01500000 pointed to by address register A5. Since the value \$01507126 contained in D3 exceeds the upper-bound value \$01500000 pointed to by A5, the N-bit is cleared. (The remaining CCR bits are either undefined or not affected.) This out-of-bound condition causes the program to trap to a check exception service routine. The operation of the CHK is summarized in Table 6.6.

TABLE 6.6 CHK Instruction Operation

Instruction	Operand size	Operation	Notation
CHK	16, 32	If Dn < 0 or Dn > source then TRAP	CHK (Ea), Dn

The 68020 CMP.S (EA), Dn (S = B or W or L) subtracts (EA) from Dn and affects the condition codes without any result being saved.

Both the CHK2 and the CMP2 instructions have similar formats (CHK2.S (EA),Rn) and (CMP2.S (EA),Rn). They compare a value contained in a data or address register designated by (Rn) against two (2) bounds chosen by the programmer. The size of the data to be compared (.S) may be specified as either byte (.B), word (.W), or long word (.L). As shown in the figure below, the lower-bound value (LB) must be located in memory at the effective address (EA) specified in the instruction, and the upper-bound value (UB) must follow immediately at the next higher memory address [i.e., UB addr. = LB. addr + SIZE where SIZE = B (+1), W (+2), or L = (+4)] as follows:

If the compared register is a data register (i.e., Rn = Dn) and the operand size (.S) is a byte or word, then only the appropriate low-order part of the data register is checked. If the compared register is an address register (i.e., Rn = An) and the operand size (.S) is a byte or word, then the bound operands are sign extended to 32 bits, and the extended operands are compared against the full 32 bits of the address register. After execution of CHK2 and CMP2, the condition code flags are affected as follows:

Carry = 1 if the contents of Dn are out of bounds
= 0 otherwise
Z = 1 if the contents of Dn are equal to either bound
= 0 otherwise.

In the case where an upper bound equals the lower bound, the valid range for comparison becomes a single value. The only difference between the CHK2 and CMP2 instructions is that for comparisons determined to be out of bounds, CHK2 causes exception processing utilizing the same exception vector as the CHK instructions, whereas the CMP2 instruction execution only affects the condition codes.

In both instructions, the compare is performed for either signed or unsigned bounds. The MC68020 automatically evaluates the relationship between the two bounds to determine which kind of comparison to employ. If the programmer wishes to have the bounds evaluated as signed values, the arithmetically smaller value should be the lower bound. If the bounds are to be evaluated as unsigned values, the programmer should make the logically smaller value the lower bound.

The following CHK2 and CMP2 instruction examples are identical in that they both utilize the same registers, comparison data, and bound values. The difference is how the upper and lower bounds are arranged.

Example 6.10

Determine the effects of CMP2.W (A2), D1. Assume the following data prior to execution of CMP2: [D1] = $5000 0200, [A2] = $0000 7000, [$0000 7000] = $B000, [$0000 7002] = $5000.

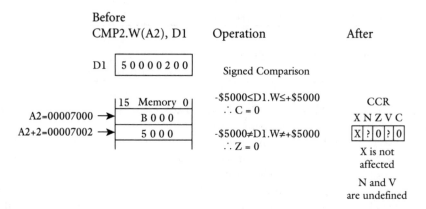

Before
CMP2.W(A2), D1   Operation   After

D1 | 5 0 0 0 0 2 0 0 |

Signed Comparison

|15 Memory 0|
A2=00007000 →| B 0 0 0 |
A2+2=00007002 →| 5 0 0 0 |

-$5000≤D1.W≤+$5000
∴ C = 0

-$5000≠D1.W≠+$5000
∴ Z = 0

CCR
X N Z V C
| X | ? | 0 | ? | 0 |
X is not
affected

N and V
are undefined

In this example, the word value $B000 contained in memory (as pointed to by address register A2) is the lower bound and the word value immediately following $5000 is the upper bound. Since the lower bound is the arithmetically smaller value, the programmer is indicating to the 68020 to interpret the bounds as signed numbers. The 2's complement value $B000 is equivalent to an actual value of –$5000. Therefore, the instruction evaluates the word contained in data register D1 ($0200) to determine if it is greater than or equal to the upper bound, +$5000, or less than or equal to the lower bound, –$5000. Since the compared value $0200 is within bounds, the carry bit (C) is cleared to zero. Also, since $0200 is not equal to either bound, the zero bit (Z) is cleared. The figure below shows the range of valid value that D1 could contain:

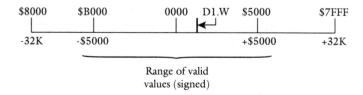

$8000 $B000   0000   D1.W $5000 $7FFF

-32K   -$5000   +$5000   +32K

Range of valid
values (signed)

A typical application for the CMP2 instruction would be to read in a number of user entries and verify that each entry is valid by comparing it against the valid range bounds. In the above CMP2 example, the user-entered value would be in register D1, and register A2 would point to a range for that value. The CMP2 instruction verifies if the entry is in range by clearing the CCR carry bit if it is in bounds and setting the carry bit if it is out of bounds.

Example 6.11

Find the effects of execution of CHK2.W (A2), D1. Assume the following data prior to execution of CHK2:

[D1] = $5000 0200, [A2] = $0000 7000, [$0000 7000] = $5000
[$0000 7002] = $B000

Solution
This time, the value $5000 is located in memory as the lower bound, and the value $B000 as the upper bound. Now, since the lower bound contains the logically smaller value, the programmer is indicating to the 68020 to interpret the bound values as unsigned number, representing only a magnitude. Therefore, the instruction evaluates the word value contained in register D1 ($0200) to determine if it is greater than or equal to lower bound $5000 or less

Instruction	Operand Size	Operation	Notation
CMP2	8, 16, 32	Compare Rn < source - lower bound or Rn > source - upper bound and set ccr	CMP2 (EA), Rn
CHK2	8, 16, 32	If Rn < source - lower bound or Rn > source - upper bound then TRAP	CHK2 (EA), Rn

FIGURE 6.4   Operation of CMP2 and CHK2.

than or equal to the upper bound $B000. Since the compared value $0200 is less than $5000, the carry bit is set to indicate an out-of-bounds condition and the program traps to the CHK/CHK2 exception vector service routine. Also, since $0200 is not equal to either bound, the zero bit (Z) is cleared. The range of valid values that D1 would contain is shown below:

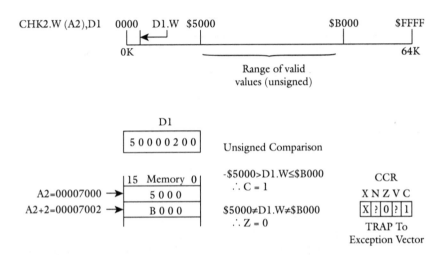

A typical application for the CHK2 instruction would be to cause a trap exception to occur if a certain subscript value is not within the bounds of some defined array. Using the CHK2 example format just given, if we define an array of 100 elements with subscripts ranging from 50_{10}–40_{10}, and if the two words located at (A2) and (A2 + 2) contain 40 and 49, respectively, and register D1 contains 100_{10}, then execution of the CHK2 instruction would cause a trap through the CHK/CHK2 exception vector. The operation of the CMP2 and the CHK2 instructions is summarized in Figure 6.4.

6.5.4 Trap On Condition Instructions

The new trap on condition (Trap cc) instruction has been added to allow a conditional trap exception on any of the following conditional conditions, as shown in Table 6.7.

These are the same conditions that are allowed for the set on condition (Scc) and the branch on condition (Bcc) instructions. The TRAPcc instruction evaluates the selected test condition based on the state of the condition code flags, and if the test is true, the MC68020 initiates exception processing by trapping through the same exception vector as the TRAPV instruction (vector 7, offset $1C, Vector address = VBR + offset). The trap on cc instruction format is TRAPcc (or) TRAPcc (.S) # <data>, where (.S) is the operand size designator, which is either word (.W) or long word (.L).

A summary of the TRAPcc instruction operation is shown in Figure 6.5.

TABLE 6.7   Conditions for TRAPcc

CC	Carry clear	$\overline{C}$
CS	Carry set	C
EQ	Equal	Z
F	Never true	0
GE	Greater or equal	$N \cdot V + \overline{N} \cdot \overline{V}$
GT	Greater than	$\underline{N} \cdot \underline{V} \cdot \overline{Z} + \overline{N} \cdot \overline{V} \cdot \overline{Z}$
HI	High	$\overline{C} \cdot \overline{Z}$
LE	Less or equal	$Z + N \cdot \overline{V} + \overline{N} \cdot V$
LS	Low or same	$C + \underline{Z}$
LT	Less than	$N \cdot \overline{V} + \overline{N} \cdot V$
MI	Minus	$\underline{N}$
NE	Not equal	$\overline{Z}$
PL	Plus	$\overline{N}$
T	Always true	$\underline{1}$
VC	Overflow clear	$\overline{V}$
VS	Overflow set	V

Instruction	Operand Size	Operation	Notation
TRAPcc	None	If cc then TRAP	TRAPcc
	16		TRAPcc.W #<data>
	32		TRAPcc.L #<data>

FIGURE 6.5   TRAPcc operation.

6.5.5 Bit Field Instructions

The bit field instructions allow an operation such as clear, set, one's complement, input, insert, and test one or more bits in a string of bits (bit field).

Table 6.8 lists all the bit field instructions. Note that the condition codes are affected according to the value in field before execution of the instruction. All bit field instructions affect the N and Z bits as shown for BFTST. C and V are always cleared. X is always unaffected.

For all instructions: Z = 1, if all bits in a field prior to execution of the instruction are zero; Z = 0 otherwise. N = 1 if the most significant bit of the field prior to execution of the instruction is one; N = 0 otherwise.

(EA)   address of the byte that contains bit 0 of the array
offset   #(0 to 31) or Dn(-2^{31} to $2^{31} - 1$)
width   #(1 to 32) or Dn (1 to 31, mod 32)

TABLE 6.8   Bit Field Instructions

Instruction	Operand size	Operation	Notation
BFTST	1-32	Field MSB → N, Z = 1 if all bits in field are zero; Z = 0 otherwise	BFTST (EA){offset:width}
BFCLR	1-32	0's → field	BFCLR (EA){offset:width}
BFSET	1-32	1's → field	BFSET (EA){offset:width}
BFCHG	1-32	Field → field	BFCHG (EA){offset:width}
BFEXTS	1-32	Field → Dn; sign extended	BFEXTS (EA){offset:width}, Dn
BFEXTU	1-32	Field → Dn; zero extended	BFEXTU(EA){offset:width}, Dn
BFINS	1-32	Dn → field	BFINS Dn, (EA) {offset:width}
BFFFO[a]	1-32	Scan for first bit set in field	BFFFO (EA) {offset:width}, Dn

[a] The offset of the first bit set in bit field is placed in Dn; if no set bit is found, Dn contains the offset plus field width.

where # and Dn respectively indicate immediate and data register modes.

Immediate offset is from 0 to 31, while offset in Dn can be specified from -2^{31} to $2^{31}-1$. All instructions are unsized. They are useful for memory conservation, graphics, and communications.

As an example, consider BFCLR $5002 {4:12}. Assume the following memory contents:

	7	6	5	4	3	2	1	0	← bit number
$5001	1	0	1	0	0	0	0	1	
$5002 (Base Address) →	1	0	0	1	1	1	0	0	
$5003	0	1	1	1	0	0	0	1	
$5004	0	0	0	1	0	0	1	0	

Bit 7 of the base address $5002 has the offset 0. Therefore, bit 3 of $5002 has offset value of 4. Bit 0 of location $5001 has offset value -1, bit 1 of $5001 has the offset value -2, and so on. The above BFCLR instruction clears 12 bits starting with bit 3 of $5002. Therefore, bits 0 to 3 of location $5002 and bits 0 to 7 of location $5003 are cleared to zero. Therefore, the above memory contents are as follows:

	7	6	5	4	3	2	1	0	
$5001	1	0	1	0	0	0	0	1	offset 4
$5002	1	0	0	1	0	0	0	0	width 12
$5003	0	0	0	0	0	0	0	0	offset 16
$5004	0	0	0	1	0	0	1	0	

The use of bit field instructions may result in memory savings. For example, assume that an input device such as a 12-bit A/D converter interfaced via a 16-bit port of an MC68020-based microcomputer. Now, suppose that one million pieces of data are to be collected from this port. Each 12 bits can be transferred to a 16-bit memory location or bit field instructions can be used.

Using 16-bit location for each 12-bit:
Memory bytes required:
= 2 * 1 million
= 2 million bytes
Using bit fields:
12 bits = 1.5 bytes
Memory requirements = 1.5 * 1 million
= 1.5 million bytes
Savings = 2 million bytes − 1.5 million bytes
= 500,000 bytes

Example 6.12

Find the effects of:

```
BFCHG    $5004      {D5 : D6}
BFEXTU   $5004      {2 : 4}, D5
BFINS    D4, (A0)   {D5 : D6}
BFFFO    $5004      {D6 : 4}, D5
```

Assume the following data prior to execution of each of the above instructions:

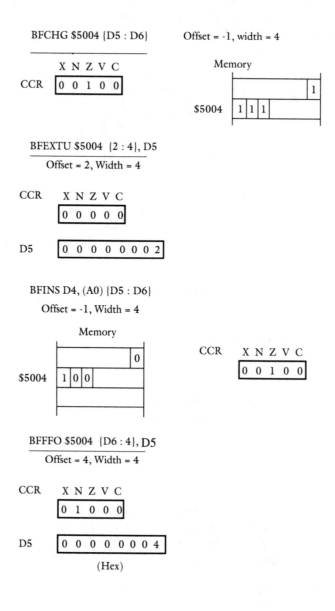

Register contents are given in hex, CCR and memory contents in binary, and offset to the left of memory in decimal.

Solutions

BFCHG $5004 {D5 : D6}   Offset = -1, width = 4

BFEXTU $5004 {2 : 4}, D5
Offset = 2, Width = 4

BFINS D4, (A0) {D5 : D6}
Offset = -1, Width = 4

BFFFO $5004 {D6 : 4}, D5
Offset = 4, Width = 4

(Hex)

TABLE 6.9 Pack and Unpack Instructions

Instruction	Operand size	Operation	Notation
PACK	16 → 8	Unpacked source + # data → packed destination	PACK –(An), –(An), #<data> PACK Dn, Dn, #<data>
UNPK	8 → 16	Packed source → unpacked source, unpacked source + # data → unpacked destination	UNPK –(An), –(An), #<data> UNPK Dn, Dn, #<data>

Note: Condition codes are not affected.

6.5.6 Pack and Unpack Instructions

Table 6.9 lists the details of PACK and UNPK instructions. Both instructions have three operands and are unsized. They do not affect the condition codes. The PACK instruction converts two unpacked BCD digits to two packed BCD digits. The UNPK instruction reverses the process and converts two packed BCD digits to unpacked BCD digits. Immediate data can be added to convert numbers from one code to another. That is, these instructions can be used to translate codes such as ASCII or EBCDIC to BCD and vice versa.

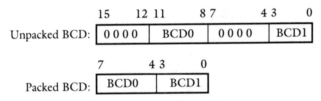

#data - Appropriate constants can be used to translate from ASCII
or EBCDIC to BCD or from BCD to ASCII or EBCDIC.

The PACK and UNPK instructions are useful when an I/O device such as an ASCII keyboard is interfaced to an MC68020-based microcomputer. Data can be entered into the microcomputer via the keyboard in ASCII codes. The PACK instruction can be used with appropriate adjustments to convert these ASCII codes into BCD. Arithmetic operations can be performed inside the microcomputer and the result will be in PACKED BCD. The UNPK instruction can similarly be used with appropriate adjustment to convert packed BCD to ASCII codes.

Example 6.13

Find the effects of execution of the following PACK and UNPK instructions:

1. **PACK D0, D5, # $0000**
2. **PACK – (A1), – (A4), # $0000**
3. **UNPK D4, D6, # $3030**
4. **UNPK – (A3), – (A2), # $3030**

Assume the following data:

MEMORY

```
        31                0
D0    [ X X X X  32  37 ]
        31                0
D5    [ X X X X X X  26 ]
        31                0
D4    [ X X X X X X  35 ]
        31                0
D6    [ X X X X X X  27 ]
        31                0
A2    [ 3 0 0 5 0 0   A3 ]
        31                0
A3    [ 5 0 7 1 2 4   B9 ]
        31                0
A1    [ 5 0 7 1 2 4   B3 ]
        31                0
A4    [ 3 0 0 5 0 0   A1 ]
```

	7 0
	:
	:
$507124B1	32
$507124B2	37
$507124B3	00
$507124B4	27
$507124B5	02
$507124B6	07
$507124B7	27
$507124B8	27

Solution

1.   PACK D0, D5 # $000

$$[D0] = 32\ 37$$
                              low
                              word
$$+\ 00\ 00$$
                         ─────────
$$32\ 37$$
                              ↓ ↓
$$[D5] = 27$$

Note that ASCII code for 2 is 32 and for 7 is 37. Hence the above PACK instruction converts ASCII code to packed BCD.

2.   PACK - (A1), - (A4), $0000

$[\$5071\ 24B2] = 37$ + 3237
$[\$5071\ 24B1] = 32$ + 0000
                              ──────
                               3237
                              ↙ ↙

Therefore, $[3005\ 00A0] = 27$ Packed BCD

Hence, the above instruction with the specified data converts two ASCII digits to their equivalent packed BCD.

3.   UNPK D4, D6 # $3030

$[D4] =$ XXXXXX 35
                              03   05
$+$ 30   30
                           ─────────
                              33   35

Therefore, after this UNPK

$[D6] =$ XXXX 33 35

$[D4] =$ XXXXXX 35

Therefore, this instruction with the assumed data converts from packed BCD 35 to ASCII 33 35.

4.   UNPK - (A3), - (A2), # $3030
                                                     [$5071 24B8] = 27

02	07
30	30
32	37

Hence,
[$300 500 A2] = 37
[$300 500 A1] = 32

This instruction with the assumed data converts two packed BCD digits to their equivalent ASCII digits.

6.5.7 Multiplication and Division Instructions

The MC68020 includes the following signed and unsigned multiplication instructions:

```
MULS.W  (EA),  Dn 16 × 16 → 32,  (EA)16 * Dn16 → Dn32
  or
MULU

MULS.L  (EA),  Dn 32 × 32 → 32,  (EA)  * Dn → Dn
  or                         Holds low 32 bits of the result
MULU                             after multiplication; upper 32 bits
                                 of the result are discarded

MULS.L  (EA),  Dn:Dm 32 × 32 → 64,  (EA) * Dn → Dm:Dn
  or                         Holds 32-bit multiplicand before
                                 multiplication and low 32 bits of
MULU                             the product after multiplication
Holds high 32 bits
  of the product after   Holds 32-bit multiplier before
multiplication           multiplication
```

(EA) in the above can be all modes except An. The condition codes N, Z, V are affected, C is always cleared to zero, and X is unaffected for both MULS and MULU. For signed multiplication, overflow (V = 1) can only occur for 32×32 multiplication producing a 32-bit result if the high-order 32 bits of the 64-bit product are not the sign extension of the low-order 32 bits. In the case of unsigned multiplication, overflow (V = 1) can occur for 32×32 multiplication producing a 32-bit result if the high-order 32 bits of the 64-bit product are not zero.

Both MULS and MULU have a word form and a long-word form. For the word form (16 × 16) the multiplier and multiplicand are both 16 bits and the result is 32 bits. The result is saved in the destination data register. For 32 bit × 32 bit, the multiplier and multiplicand are both 32 bits and the result is either 32 bits or 64 bits. When the result is 32 bits for a 32-bit × 32-bit operation, the low-order 32 bits of the 64-bit product is provided.

The signed and unsigned division instructions of the MC68020 include the following: Source is the divisor and destination is the dividend. The result (remainder and quotient) is stored in the destination.

```
DIVS.W  (ea),  Dn 32/16 → 16r:16q
  or
DIVU

DIVS.L  (ea),  Dn 32/32 → 32q (no remainder is
  or                               provided)
DIVU
```

```
DIVS.L (ea),  Dr:Dq  64/32  →  32r:32q
 or
DIVU
```

```
DIVSL.L (ea),  Dr:Dq  32/32  →  32r:32q
 or
DIVUL                  Contains the 32-bit dividend
```

The destination contains the dividend and (EA) contains the divisor. (EA) in the above instructions can use all modes except An.

The condition codes for either signed or unsigned division are affected as follows: N = 1 if the quotient is negative; N = 0 otherwise. N is undefined if overflow or divide by zero. Z = 1 if the quotient is zero; Z = 0 otherwise. Z is undefined for overflow or divide by zero. V = 1 for division overflow; V = 0 otherwise. X is unaffected, and C = 0 (always).

Division by zero causes a trap. If overflow is detected before completion of the instruction, V is set to one, but the operands are unaffected.

Both signed and unsigned division instructions have a word form and three long-word forms.

For the word form, the destination operand is 32 bits and the source operand is 16 bits. The 32-bit result in Dn contains the 16-bit quotient in the low word and the 16-bit remainder in the high word. The sign of the remainder is the same as the sign of the dividend.

```
For DIVS.L (EA), Dn
       or
     DIVU,
```

both destination and source operands are 32 bits. The result in Dq contains the 32-bit quotient and the remainder is discarded.

```
For DIVS.L (EA), Dr:Dq
       or
     DIVU,
```

the destination is 64 bits contained in any two data registers and the source is 32 bits. The 32-bit register Dr (D0-D7) contains the 32-bit remainder and the 32-bit Dq (D0-D7) contains the 32-bit quotient.

Example 6.14

Find the effects of the following multiplication and division instructions:

1. **MULU.L # $2, D5** if [D5] = $FFFFFFFF
2. **MULS.L # $2, D5** if [D5] = $FFFFFFFF
3. **MULU.L # $2, D5:D2**
   if [D5] = $2ABC 1800
   and [D2] = $FFFFFFFF
4. **DIVS.L # $2, D5** if [D5] = $FFFFFFFC
5. **DIVS.L # $2, D2:D0**
   if [D2] = $FFFFFFFF
   and [D0] = $FFFFFFFC
6. **DIVSL.L # $2, D6:D1** if [D1] = $0004 1234
   and [D1] = $FFFFFFFE

Solution

1. **MULU.L # $2, D5 if [D5] = $FFFFFFFF**

$$\begin{array}{r}
\text{\$FFFFFFFF} \\
* \quad \text{\$00000002} \\
\hline
00000001 \quad \text{FFFFFFFE}
\end{array}$$

↓	↓
V = 1	Low 32-bit
since	result in D5
this is	
nonzero	

Therefore, [D5] = $FFFFFFFE, N = 0 since the most significant bit of result is 0, Z = 0 since the result is not zero, V = 1 since the high 32 bits of the 64-bit product are not zero, C = 0 (always), and X = not affected.

2. **MULS.L # $2, D5 if [D5] = $FFFFFFFF**

$$\begin{array}{rr}
\text{\$FFFFFFFF} & (-1) \\
* \quad \text{\$00000002} & (+2) \\
\hline
\text{\$FFFFFFFF} \qquad \text{\$FFFFFFFE} & (-2)
\end{array}$$

↓

Result in D5

Therefore, [D5] = $FFFFFFFE, X = unaffected, C = 0, N = 1, V = 0, and Z = 0.

3. **MULU.L # $2, D5:D2 if [D5] = $2ABC 1800 and [D2] = $FFFFFFFF**

$$\begin{array}{r}
\text{\$FFFFFFFF} \\
* \quad \text{\$00000002} \\
\hline
00000001 \quad \text{FFFFFFFE} \\
\text{D5} \qquad\quad \text{D2}
\end{array}$$

N = 0, Z = 0, V = 1 since high 32 bits of the 64-bit product are not zero, C = 0, and X = not affected.

4. **DIVS.L # $2, D5 if [D5] = $FFFFFFFC**

$$\begin{array}{r}
-2 \\
\hline
\text{FFFF FFFE} \\
00000002 \, \overline{)\, \text{FFFF FFFC}} \\
+2 \qquad\qquad -4
\end{array}$$

[D5] = $FFFF FFFE, X = unaffected, N = 1, Z = 0, V = 0, and C = 0 (always).

5. **DIVS.L # $2, D2:D0 if [D2] = $FFFF FFFF and [D0] = $FFFF FFFC**

$$\begin{array}{r}
-2 \\
\hline
Q = \text{FFFF FFFE}, \; R = \text{0000 0000} \\
0000\ 0002 \, \overline{)\, \text{FFFF FFFF FFFF FFFC}} \\
2 \qquad\qquad\qquad\qquad -4
\end{array}$$

D2 = $0000 0000 = remainder, D0 = $FFFF FFFE = quotient, X = unaffected, Z = 0, N = 1, V = 0, and C = 0 (always).

TABLE 6.10   Enhanced Instructions

Instruction	Operand size	Operation
BRA label	8, 16, 32	PC + d → PC
Bcc label	8, 16, 32	If cc is true, then PC + d → PC; else next instruction
BSR label	8, 16, 32	PC → – (SP); PC + d → PC
CMPI.S # data, (EA)	8, 16, 32	Destination – # data → CCR is affected
TST.S (EA)	8, 16, 32	Destination – 0 → CCR is affected
LINK.S An, # – d	16, 32	An → – (SP); SP → An, SP + d → SP
EXTB.L Dn	32	Sign extend byte to long word

Note: S can be B, W, L. In addition to 8- and 16-bit signed displacements for BRA, Bcc, and BSR like the 68000, the 68020 also allows signed 32-bit displacements. Link is unsized in the 68000. (EA) in CMPI and TST support all MC68000 modes plus PC relative. Examples are CMPI.W#$2000, (START, PC). In addition to EXT.W Dn and EXT.L Dn as with the 68000, the 68020 also provides the EXTB.L instruction.

6. **DIVSL.L # $2, D6:D1** if [D1] = $0004 1224 and [D6] = $FFFF FFFD

$$\begin{array}{r} \overset{-1}{\overbrace{Q = FFFFFFFF}} \\ 0000\ 0002\ \big| \overline{\quad FFFFFFD \quad} \qquad R = \overset{-1}{\overbrace{FFFFFFFF}} \\ -3 \end{array}$$

[D6] = $FFFFFFFF = remainder, [D1] = $FFFFFFFF = quotient, X = unaffected, N = 1, Z = 0, V = 0, and C = 0 (always).

6.5.8   MC68000 Enhanced Instructions

The MC68020 includes the enhanced version of the 68000 instructions listed in Table 6.10.

Example 6.15

Write a program in MC68020 assembly language to find the first one in a bit field which is greater than or equal to 16 bits and less than or equal to 512 bits. Assume the number of bits to be checked is divisible by 16. If no ones are found, store $0000 0000 in D3; otherwise store the offset of the first set bit in D3 and stop. Assume A2 points to start of the array and D2 contains the number of bits in the array.

Solution

```
        CLR.L  D3                     ;  D3 is offset in bits
                                      ;  Contains the first bit
                                      ;    number set
        DIVU  #16, D2                 ;  [D2] = number of searches
        SUBQ.W #1, D2                 ;  Decrement D2 by 1 for use
                                      ;    in DBNE later
        MOVEQ.L #16, D5               ;  Load field width into D5
START   BFFFO  (A2) {D3:D5}, D3       ;  Search for one in 16 bits
        DBNE  D2, START               ;  Decrement branch if not
                                      ;    equal
```

```
        BNE  STOP             ;  If no ones found, stop
        CLR.L  D3             ;  Store 0 in D3
FINISH JMP FINISH             ;  HALT
```

Example 6.16

Write a program in MC68020 assembly language to convert 20 packed BCD digits to their ASCII equivalent and store the result to memory location $F1002004. The data bytes start at $5000.

Solution

```
        MOVEA.L # $5000, A5        ;  Load starting address of
                                   ;   the BCD array into A5
        MOVEA.L # $F1002004, A6    ;  Load starting address
                                   ;   of the ASCII array
                                   ;   into A6
        MOVEQ.L #19, D2            ;  Load data length to D2
START   MOVE.B  (A5) +, D3         ;  Load BCD value
        UNPK D3, D3, # $3030       ;  Convert to ASCII
        MOVE.W D3, (A6) +          ;  Store ASCII data at
                                   ;   address pointed to by
                                   ;   A6
        DBF D2, START             ;  Decrement and branch if
                                   ;   false
FINISH JMP FINISH                 ;  Otherwise STOP
```

Example 6.17

Write a program in MC68020 assembly language to divide a signed 32-bit number in D0 by a signed 8-bit number in D1 by storing the division result in the following manner:

1. Store 32-bit quotient in D0 and neglect remainder.
2. Store 32-bit remainder in D1 and 32-bit quotient in D0.

Assume dividend and divisor are already in D0 and D1, respectively.

Solution

```
    1.  EXTB.L D1            ;  Sign extend divisor to 32-bit
        DIVS.L D1, D0        ;  32-bit quotient in D0 and
FINISH    JMP FINISH         ;  remainder is discarded and
                             ;  halt
    2.  EXTB.L D1            ;  Sign extend divisor to 32 bits
        DIVSL.L D1, D1:D0    ;  32-bit remainder in D1 and
                             ;   32-bit quotient in D0
FINISH    JMP FINISH         ;  Halt
```

Example 6.18

Write a 68020 assembly language program to compute the following:

$$D2.L:D1.L = (INTEGER + D1.L)*(D1.L - D0.B)$$

Assume that the location INTEGER and register D0 contain signed 32-bit numbers while the low byte of D0 holds an unsigned number less than 127_{10}: Store the 64-bit result in D2.L:D1.L.

Solution

```
        MOVE.L    INTEGER, D2    ;
        ADD.L     D1,D2          ; compute sum
        EXTB.L    D0             ; compute sum
        SUB.L     D0,D1          ; compute (D1.L-D0.B)
        MULS.L    D2,D2:D1       ; perform multiplication
FINISH  JMP       FINISH
```

Example 6.19

Write a program in 68020 assembly language to multiply a signed byte by a 32-bit signed number to obtain a 64-bit result. The numbers are respectively pointed to by the addresses that are passed on to the user stack by a subroutine pointed to by (USP+4) and (USP+6). Store the 64-bit result in D2:D1.

Solution

```
        MOVE.B ([4,USP]),D0  ; Move first data
        EXTB.L DO            ; sign extend to 32 bits
        MOVE.L ([6,USP]),D1  ; Move second data
        MULS.L D0,D2:D1      ; Store result in D2:D1
FINISH  JMP FINISH           ; Halt
```

6.6 68020 ADVANCED INSTRUCTIONS

This section provides a detailed description of the 68020 advanced instructions including BKPT, CAS/CAS2, TAS, CALLM/RTM, and coprocessor instructions.

6.6.1 Breakpoint Instruction

A breakpoint is a debugging tool that allows the programmer to check or pass over an entire section of a program. Execution of a breakpoint usually results in exception processing. Hence, the programmer can use any of the TRAP vectors as breakpoints. Also, any of the interrupt levels can be used by external hardware to cause a breakpoint.

The BKPT instruction is included with the 68010, 68012 and 68020 microprocessors.

In order to place a breakpoint in a program loop located in RAM, the operating system can temporarily remove an instruction word from the program and insert the BKPT instruction in its place. In the case of the 68010/68012, the operating system services the breakpoint exception routine since trap to illegal instruction exception is taken upon execution of the BKPT instruction. Such interruption by the operating system due to BKPT instruction is undesirable since execution of the program being debugged is slowed down. This is why the 68020 BKPT instruction includes more real-time debug support via external hardware such as the Motorola MC68851 Paged Memory Management Unit chip in which the program loop count and the saved operation word are stored for up to eight BKPT instructions (BKPT#0 thru BKPT#7). The operating system exchanges the operation word with one of the eight breakpoint instruction as follows:

Breakpoint number	Instruction	op code
0	BKPT#0	$4848
1	BKPT#1	$4849
2	BKPT#2	$484A
3	BKPT#3	$484B
4	BKPT#4	$484C
5	BKPT#5	$484D
6	BKPT#6	$484E
7	BKPT#7	$484F

The operating system then stores the operation word in a register of external hardware and initializes that loop counter. The 68020 generates a special cycle called breakpoint cycle in the CPU space (FC2 FC1 FC0 = 111) each time it executes the BKPT instruction in the program being debugged. Figure 6.6 shows how CPU space 0 is encoded during execution of a BKPT instruction. 68020 then outputs the breakpoint number on the address bus. The external hardware, if present, decrements the corresponding loop counter and places the operation word on the data bus. The 68020 inputs this operation word and continues with the execution of the program. As soon as the 68020 executes the same BKPT instruction for the last count (last loop counter value), external breakpoint hardware asserts the BERR signal generating an illegal instruction trap. This is shown in the flowchart of Figure 6.7

Example 6.20

Explain the breakpoint operation shown in Figure 6.8.

Solution

In Step 1 the illegal instruction BKPT is inserted into the program flow as requested by a user. In this example, the 16-bit op code for ADD.L D2, D3 is replaced with the 16-bit op code for BKPT #4.

In Step 2, the op code ADD.L D2, D3 is stored by the MC68020 in the CPU space long-word memory location corresponding to breakpoint number four, address $10.

Step 3 shows how long-word memory locations $00 to $1C in CPU space 0 are used to temporarily store the op codes replaced by breakpoint instructions 0 through 7. The displaced op code is stored in the upper 16 bits, and an optional count value can be loaded into the lower 16 bits through external hardware control.

In Step 4, each time BKPT #4 is executed, the MC68020 accesses CPU space 0 location $10. If the count value (bits 0–15) is not zero, the replaced op code is placed on the data bus (bits 16–31), the counter is decremented by one, and the appropriate DSACKX lines are asserted by the external hardware. The MC68020 reads this op word and executes it. If the count value is zero, the 68020 BERR pin can be asserted by external hardware to cause exception processing.

The breakpoint function is summarized in Table 6.11. A hardware implementation to take advantage of the BKPT instruction's looping feature is shown in Figure 6.9. A debug monitor maintains a small amount of breakpoint memory. It is used to hold up to eight replacement

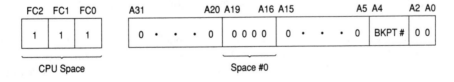

FIGURE 6.6 BKPT instruction format.

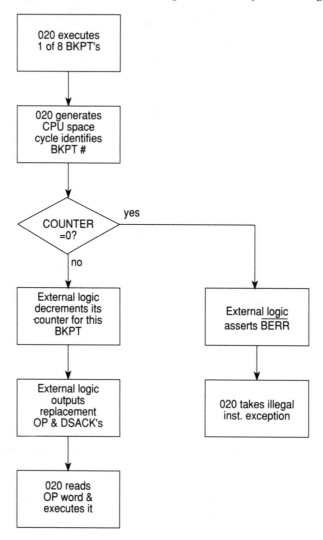

FIGURE 6.7   Breakpoint instruction flowchart.

op codes and eight counters. When a user wants a breakpoint to be encountered after say "10" passes of the op code, the debugger initiates a breakpoint counter to 10, replaces the user op code with a breakpoint instruction, saves the user op code in the breakpoint data memory, and executes the user's program. When the breakpoint op code is encountered, the MC68020 generates the breakpoint acknowledge bus cycle. The 74LS138 3-to-8 decoder shown on the right decodes the function codes as all ones (validated by the AS signal) and asserts its bottom output indicating a CPU bus space cycle. This output enables one of the inputs of the 74LS138 shown on the left. The top enable output is enabled when address lines A 16 through A 19 are all zero, indicating a type 0 CPU space cycle. This decoder enables one of eight breakpoint counters.

If the counter ≠ 0, then it is decremented by the external hardware and the replacement op code is returned on the data bus to the MC68020 with DSACK asserted. In this case, we started with a count of 10, so it is decremented to 9. When the breakpoint acknowledge cycle occurs on the 11th pass, the counter = 0 and this time bus error BERR signal is asserted by the external hardware. This forces illegal instruction processing and subsequent servicing of the breakpoint. With this type of hardware, operating system support is not required until the

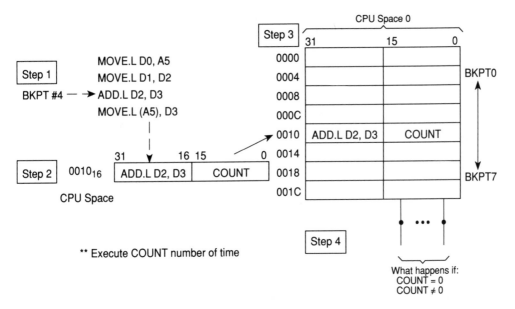

FIGURE 6.8   Executing a BKPT instruction.

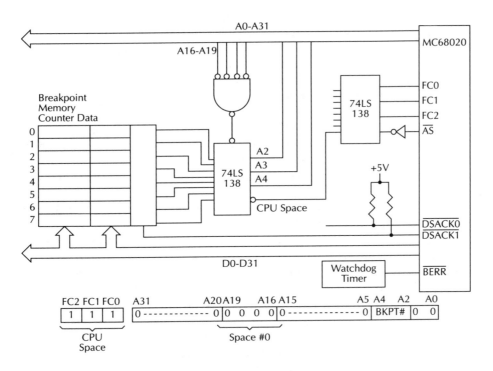

FIGURE 6.9   Breakpoint block diagram.

TABLE 6.11   Function of the Breakpoint

Instruction	Operand size	Operation	Notation
BKPT	None	If breakpoint cycle acknowledged then execute returned op word, else trap as illegal instruction	BKPT # data

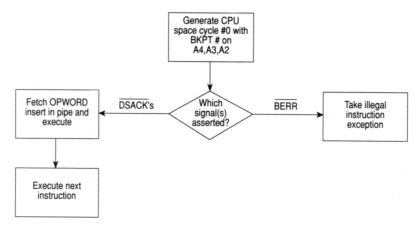

FIGURE 6.10   Breakpoint instruction flowchart.

count has been exhausted. This reduces the operating system overhead significantly. The MC68020 breakpoint instruction flow diagram is shown in Figure 6.10.

6.6.2 Call Module/Return from Module Instructions

For large programs, flowcharting does not provide an efficient software design tool. The flowcharts, however, can assist the programmer in dividing a large program into subprograms called modules (typically 20 to 50 lines). The task of dividing a large program into modules is called modular programming. Typical problems of modular programming include how to modularize a large program and then how to combine the individual modules.

One of the advantages of modular programming includes ease in writing, debugging, and testing a module rather than a large program.

By using modular programming, a program performing a specific function is divided into several smaller subfunctions. These subfunctions can be utilized once or several times during program execution. A main module is written and tested. The main module is used to call the subfunction modules in the sequence required to accomplish the overall function of the program.

Modular programming is supported by the MC68020 call module (CALLM) and the return from module (RTM) instructions.

The CALLM instruction creates a module stack frame on the stack (similar to an interrupt), stores the module state in that frame, and points to the address of a module descriptor (in the effective address) which contains control information for the entry into the called module.

The CALLM instruction loads the processor with the data provided by the module descriptor. Thus the CALLM instruction does not directly access the program module. Rather, it indirectly calls the routine via the module descriptor whose contents are maintained by the operating system. The module descriptor can be thought of as a gateway through which the calling program must gain access. The module being called can be thought of as a subroutine that is to act upon the arguments passed to it.

The CALLM instruction syntax is

CALLM # data,   (EA)

To use this instruction, an immediate data value (data 0–255) must be specified to indicate the number of bytes of argument parameters to be passed to the called module. The effective address (EA) that points to the external module descriptor must also be included. The program containing the CALLM instruction must define a RAM storage area for the module

TABLE 6.12 CALLM and RTM Instructions

Instruction	Operand size	Operation	Notation
CALLM	None	Save current module state on stack; load new module state from destination	CALLM data, (EA)
RTM	None	Reload saved module state in stack frame; place module data area pointer in Rn	RTM Rn

descriptor to reside in. The address of the module descriptor is known to the operating system at "run time". It is the operating system that loads the starting address of the library routine to be called into the module descriptor.

The return from module (RTM) instruction is the complement of the CALLM instruction. The RTM syntax is RTM Rn.

Register Rn (Rn can be data or address register) represents the module data area pointer. The RTM instruction recovers the previous module state from the stack frame and returns program execution in the calling module. The processor state (program counter, status word) comes from the previously stacked data. The operation of the module instructions is shown in Table 6.12.

Figure 6.11 shows an overview of the CALLM and RTM operation. When the CALLM instruction executes, it creates a module call stack frame and saves the current module state in that frame. It then loads the new module state from the module descriptor. The start address fetched from the module descriptor points to the module entry word of the library routine. The second word of the routine is its first op word. The last instruction of the routine is the RTM instruction.

The RTM instruction reloads the saved module state in the created stack frame, and user program execution continues with the next instruction.

6.6.3 CAS Instructions

The MC68020 compare and swap (CAS and CAS2) instructions provide support for multitasking and multiprocessing. The compare and swap instructions are used when several processors must communicate through a common block of memory, when globally shared data structures (such

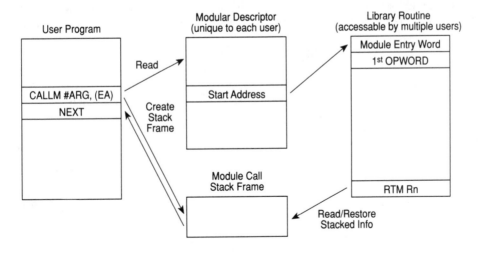

FIGURE 6.11 Overview of the CALLM and RTM operation.

as counters, stack pointers, or queue pointers) must be securely updated, or when multiple bus cycles may be required. A typical application is a counting semaphore (shared incrementer). Another application of the compare and swap instructions would be to manipulate pointers for system stacks and queues that use linked lists when a new item (element) is inserted or an existing element is deleted.

The CAS instruction uses two registers (Dc and Du) and the address (EA) of the globally shared operand variable (or pointer) to be protected. CAS operates on byte, word, and long word operands. The assembly level programming notation for the CAS instruction is

$$\texttt{CAS.S (.B or .W or .L)Dc, Du, (EA)}$$

The CAS instruction first compares the old "fetched" starting pointer in register Dc with the present starting pointer in the variable's original location (EA) to see if another task accessed the variable pointer and changed it while the first task was using it; if the two compared pointer values Dc and (EA) are equal (that is, pointer contents (EA) remain unchanged by interrupting tasks), then CAS passes the updated pointer value (located in Du) to the destination operand (EA) and sets the equal condition code flag in the CCR to $Z = 1$.

If the two compared values are not equal (the EA pointer has been changed by an interrupting task), then CAS copies the new changed pointer value contents of (EA) to register Dc and clears the equal condition code flag in the CCR to $Z = 0$.

A condensed version of the CAS operation is shown below:

```
CAS Dc, Du, (EA)
    1.   (EA) - Dc → cc
    2.   IF (EA) = Dc
         THEN Du → (EA)
         ELSE (EA) → Dc
```

Next, application of the CAS for queue insertion will be discussed.

Figure 6.12a shows how to insert a new entry in a queue using the MOVE instruction.

In a single user/single task environment the above can be accomplished by the following instruction sequence:

```
MOVE.L HEAD, (NEXT, A1)
MOVE.L A1, HEAD
```

But in a multiuser/multitask environment, the above instruction sequence may not accomplish the task. For example, more than one user may attempt to insert a new entry in an

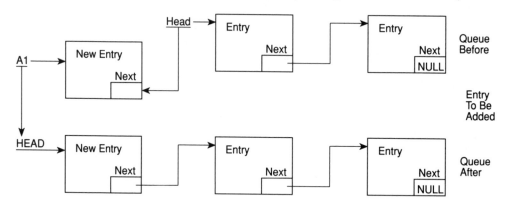

FIGURE 6.12a Inserting a new entry in queue (single user/single task).

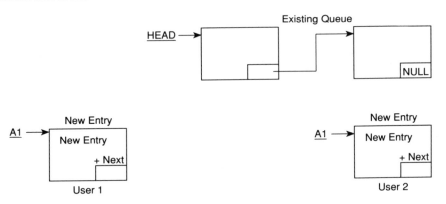

FIGURE 6.12b   Two users attempting to insert a new entry into an existing space.

existing queue. Suppose that after user 1 executes MOVE.L HEAD, (NEXT, A1), user 2 executes the instruction sequence

```
MOVE.L HEAD, (NEXT, A1)
MOVE.L A1, HEAD
```

and inserts a user 2 new entry in the existing queue before user 1 gets to MOVE.L A1, HEAD. This situation is depicted in Figures 6.12b and b.12c. This situation can be avoided by using the CAS instruction. The user 2 entry gets lost if the MOVE instruction is used for insertion. In Figure 6.12c, the HEAD (known to user 1 at the start) gets changed between the time user 1 established the forward link by the time user 1 updated HEAD.

The following instruction sequence uses CAS to insert a new entry in a queue in a multiuser/ multitask environment:

```
     MOVE.L HEAD, D0       ; Capture current HEAD
     MOVE.L A1, D1         ; Need value in Du for CAS
LOOP MOVE.L D0, (NEXT, A1) ; Establish forward link
```

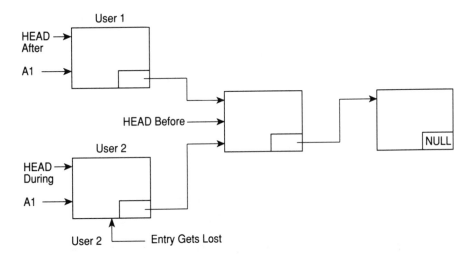

FIGURE 6.12c   Insertion of entries into an existing queue in a multiuser environment using MOVE (situation to avoid).

Case 1: No Intervention

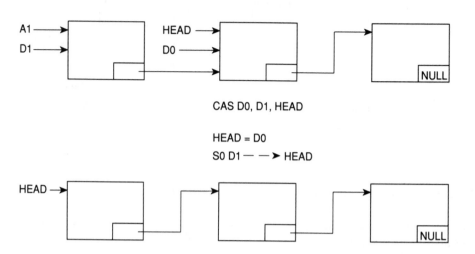

CAS D0, D1, HEAD

HEAD = D0
S0 D1 — —➤ HEAD

FIGURE 6.12d   Inserting an element in a queue using CAS without intervention.

```
CAS.L D0, D1, HEAD      ;  If HEAD unchanged, update
BNE LOOP                ;  You have new HEAD in D0, try
                           again
```

Figures 6.12d and 6.12e illustrate the use of CAS in the above instruction sequence, respectively, in a system with no intervention and simultaneous insertion by multiusers.

The CAS instruction cannot be interrupted. This ensures secure updates of variables in a multiprocessing system. Executing CAS causes the read-write-modify (RMC) pin to be asserted, which locks the bus. Other bus masters in the system must wait for RMC to be deasserted before they can take control of the bus. The CAS instructions read, compare, and

Case 2: "Simultaneous Insert"

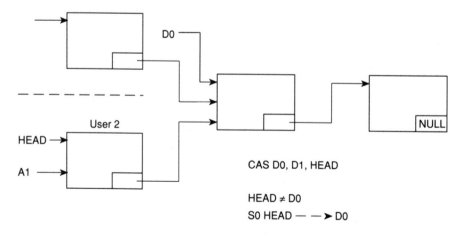

CAS D0, D1, HEAD

HEAD ≠ D0
S0 HEAD — —➤ D0

The program tries again, from loop. First instruction not necessary as CAS did it.

FIGURE 6.12e   Simultaneous insertion of elements.

store operations are performed while the bus is locked. This prevents other processors from interfering with the instruction while it performs its compare and swap operation.

Instruction CAS2 is identical to CAS except that it can be used to compare and update dual operands within the same indivisible cycle. The CAS2 instruction operates on word or long word operands. The notation for the CAS2 instruction is CAS2 .W Dc1:Dc2, Du1:Du2, (Rn1):(Rn2). With the CAS2 instruction, both comparisons must show a match for the contents of the update registers Du1 and Du2 to be stored at the operands' destination addresses in memory pointed to by the registers Rn1 and Rn2. If either comparison fails to match, both destination operands obtained from memory are copied to the compare registers Dc1 and Dc2. The CAS2 instruction memory references to operand destination addresses must be specified using register indirect addressing with either a data or address register used as a pointer to memory.

The condensed version of CAS2 Dc1:Dc2, Du1:Du2, (EA1):(EA2) operation is given below:

```
If (EA1) = Dc1 and (EA2) = Dc2, then
Du1 → (EA1) and Du2 → (EA2);
else (EA1) → Du1 and (EA2) → Du2
```

Example 6.21

Write a 68020 instruction sequence to delete an element from a linked list using the CAS2 instruction.

Solution

```
        LEA HEAD, A0          ; Load address of head pointer
                              ;   into A0
        MOVE.L (A0), D0       ; Move value of head pointer
                              ;   into D0
LOOP    TST.L D0              ; Check for null pointer
        BEQ EMPTY             ; If empty, no deletion required
        LEA (NEXT, D0.L), A1  ; Load address forward of link
                              ;   into A1
        MOVE.L (A1), D1       ; Put value of forward link
                              ;   in D1
        CAS2.L D0:D1, D1:D1,  ; If no change, update head
          (A0):(A1)          ;   and forward pointer
        BNE LOOP             ; D0 has new head, try again.
EMPTY   —     —
        —
        —
        —
```

After deleting an element:

TABLE 6.13   Summary of the TAS, CAS, and CAS2 Instructions

Instruction	Operand size	Operation	Notation
TAS	8	Destination –0 → CCR 1 → bit 7 of destination	TAS (EA)
CAS	8,16,32	Destination –Dc → CCR if Z, then Du → destination; else destination → Dc	CAS.S Dc, Du, (EA)
CAS2	16,32	dest1 – Dc1 → CCR; if Z, dest2 – Dc2 → CCR; if Z, Du1 → dest 1; Du2 → dest2; else Dest 1 → DC1, dest 2 → Dc2	CAS2.S Dc1:Dc2,Du1: Du2,(Rn1):(Rn2)

Note: S = B, W, or L.

Table 6.13 summarizes the operation of the CAS, CAS2, and TAS instructions.

6.6.4 Coprocessor Instructions

Table 6.14 lists the MC68020 coprocessor instructions. These instructions are available on the MC68020 system when a coprocessor such as the MC68881 (floating point) or MC68851 (paged memory management unit) is interfaced to the system. Note that cp in these instructions is replaced by F for floating point (MC68881) or P for paged (MC68851), depending on the coprocessor.

When MAIN obtains an F-line op word, it cooperates with a coprocessor to complete execution of the instruction

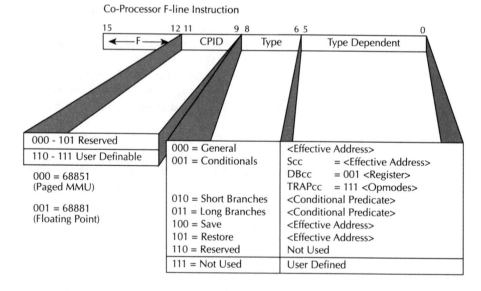

The F or P provides the 3-bit cp-id in the instruction. The GEN is replaced by the specific operation. For example, FMOVE is the MOVE instruction for the floating-point coprocessor. Upon execution of the cpGEN instruction, the MC68020 passes a command word to the coprocessor. The coprocessor then specifies the general data processing and movement instructions. The coprocessor finds the specific operation from the command word which follows the F-line instruction word. If the instruction requires an effective address for an operand to be fetched or stored, the third word follows the command word containing this effective address.

Note that F-line instruction word means that the high four bits of the instruction word are 1111_2. Normally, the coprocessor defines the specific instances of this instruction to provide

TABLE 6.14 Coprocessor Instructions

Instruction	Operand size	Operation	Notation
cpGEN	User defined	Pass command word to coprocessor and respond to coprocessor primitives	cpGEN (parameters defined by coprocessors)
cpBcc	16,32	If cpcc true, then PC + d → PC	cpBcc (label)
cpDBcc	16	If cpcc false, then (Dn − 1 → Dn; if Dn ≠ 1, then PC + d → PC)	cpDBccDn, (label)
cpScc	8	If cpcc true, then 1's → destination, else 0's → destination	cpScc (EA)
cpTRAPcc	None 16,32	If cpcc true, then TRAP	cpTRAPcc cpTRAPcc # (data)
cpSAVE[a]	None	Save internal state of coprocessor	cpSAVE (EA)
cpRESTORE[a]	None	Restore internal state of coprocessor	cpRESTORE (EA)

[a] Privileged instructions for operating system context switching (task switching) support.

• The 68020 (MAIN) and the coprocessor (CO) each do what they know how to do best

> MAIN: Tracks instruction stream
> Takes exceptions
> Takes branches
> CO: Does graphics manipulations
> Calculates transcendentals, floating point
> Does matrix manipulations

its instruction set. The condition codes may be modified by the coprocessors. An example of a cpGEN instruction for the MC68881 is

$$\text{FMOVE.L (A0) +, FP0}$$

In this instruction F replaces cp and MOVE replaces GEN in the cpGEN format. This instruction moves 4 bytes from the MC68020-based microcomputer memory, starting with a location addressed by the contents of A0 to low 32 bits of the 80-bit register (FP0) in the MC68881; A0 is then incremented by 4.

In the above for each type (bits 6–8), type dependent (bits 0–5) specifies effective address, conditional predicate, etc. For example, for general instructions (bits 8, 7, 6 = 000), type dependent (bits 0–5) specifies (ea).

The instruction FMOVE.L (A0) +, FP0 contains two words in memory as follows:

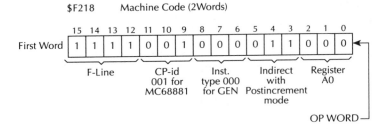

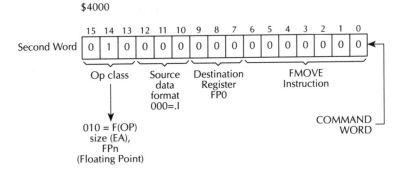

Therefore, the FMOVE.L (A0) +, FP0 contains two words in memory with the first word $F218 as the op word and the second word $4000 as the command word for the coprocessor.

The "cpBcc displacement" is the coprocessor conditional branch instruction. If the specified coprocessor condition is satisfied, program execution continues at location (PC) + displacement; otherwise the next instruction is executed. The displacement is a two's complement integer which may be either 16 or 32 bits. The coprocessor determines the specific condition from the condition field in the operation word. A typical example of this instruction is

FBEQ.L START

The cpDBcc (label) instruction works as follows. If cpcc is false, then Dn − 1 → Dn and if Dn ≠ − 1, then PC + d → PC, or if Dn = −1, then next instruction is executed. On the other hand, if cpcc is true, then the next instruction is executed. The coprocessor determines the specific condition from the condition word which follows the operation word. A typical example of this instruction is

FDBNE.W START

The operand size is 16 bits.

The cpScc.B (EA) instruction tests a specific condition. If the condition is true, the byte specified by (EA) is set to true (all ones); otherwise that byte is set to false (all zeros). The coprocessor determines the specific condition from the condition word which follows the operation word. (EA) in the instruction can use all modes except An, immediate, (d16, PC), (d8, PC, Xn), (bd, PC, Xn), and (bd, PC, Xn, od). An example of this instruction is

FSEQ.B $8000 1F20.

The cpTRAPcc or cpTRAPcc # data checks the specific condition on a coprocessor. If the selected coprocessor condition is true, the MC68020 initiates exception processing. The vector number is generated to reference the cpTRAPcc exception vector; the stacked PC is the address of the next instruction. If the selected condition is false, no operation is performed and the next instruction is executed. The coprocessor determines the specific condition from the word which follows the operation word. The user-defined optional immediate data (third word of the machine code for cpTRAPcc # data) is used by the trap handler routine. A typical example of this instruction is

FTRAPEQ

cpTRAPcc is unsized and cpTRAPcc # data has word and long word operands. cpSAVE and cpRESTORE are privileged instructions. Both instructions are unsized. cpSAVE (EA) saves the internal state of a coprocessor. (EA) can be predecrement on all alterable control addressing modes. This instruction is used by an operating system to save the context (internal state) of a coprocessor. The 68020 initiates a cpSAVE instruction by reading an internal register.

The cpRESTORE (EA) instruction restores the internal state of a coprocessor. (EA) can use postincrement or control addressing modes. This instruction is used by an operating system to restore the context of a coprocessor for both the user-visible and the user-invisible state.

6.7 MC68020 Cache/Pipelined Architecture and Operation

The MC68020 has a 256-byte direct-mapped instruction cache organized as 64 long word entries. Each cache entry consists of a tag field made up of the upper 24 address bits, the FC2 (user/supervisor) value, one valid bit (V), and 32 bits (two words) of instruction. Figure 6.13 shows the MC68020 on-chip cache organization. The 68020 cache only stores instructions; data or operands are fetched directly from main memory as needed.

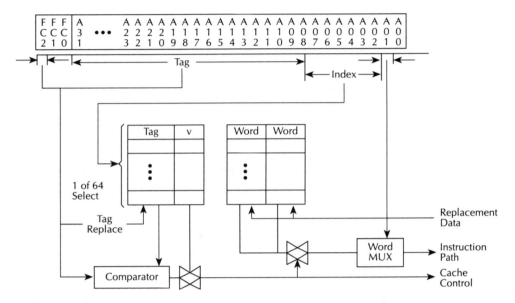

FIGURE 6.13   MC68020 on-chip cache organization.

When the 68020 fetches an instruction, it checks the cache first to determine if the word required is in the cache. First, one of 64 entries is selected by using the index field (A2–A7) of the access address as an index into the cache. Next, the 68020 compares address bits A31–A8 and function code bit FC2 with the 24-bit tag of the selected entry. If the function code and the address bit match, the 68020 sets the valid bit for the cache entry. This is called a cache hit. Finally, the 68020 uses address bit A1 to select the proper instruction word from the cache memory. If there is no match or valid bit is clear, a cache miss occurs and the instruction is fetched from external memory and put into the cache.

The MC68020 uses a 32-bit data bus and fetches instructions on long word address boundaries. Hence, each 32-bit fetch brings in two 16-bit instruction words which are then written into the on-chip cache. Subsequent prefetches will find the next 16-bit instruction word already present in the cache, and the related bus cycle is saved. Even when the cache is disabled, the subsequent prefetch will find the bus controller still holds the two instruction words and can satisfy the prefetch, again saving the related bus cycle. The bus controller provides an instruction hit rate of up to 50% even with the on-chip cache disabled.

Only the CPU uses the internal cache, so users have no direct access to the entries. However, several instructions allow the user to control the cache or dynamically disable it through the external hardware cache disable pin (CDIS). Typically, it is used by an emulator or bus state analyzer to force all bus cycles to be external cycles. The processor's two cache registers (CACR, CARR) can be programmed while in the supervisor mode by using the MOVEC instruction. Enabling, disabling, freezing, or clearing the cache is carried out by the cache control register (CACR). The CACR also allows the operating system to maintain and optimize the cache. The cache control (CACR) is shown in Figure 6.14.

FIGURE 6.14   Cache control register format. C = clear cache, CE = clear entry, F = freeze cache, and E = enable cache.

FIGURE 6.15   Cache address register format.

The clear cache (C) bit of the CACR is used to invalidate all entries in the cache. This is termed as "flushing the cache". Setting the (C) bit causes all valid bits (V) in the cache to be cleared, thus invalidating all entries.

The Clear Entry (CE) of the CACR is used in conjunction with the address specified in the cache address register (CAAR). When writing to and setting the (CE) bit, the processor uses the CAAR index field to locate the selected address in the CAAR and invalidate the associated entry by clearing the valid bit.

The freeze cache (F) bit of the CACR keeps the cache enabled, but cache misses are not allowed to update the cache entries. This bit can be used by emulators to freeze the cache during emulation execution. It could be used to lock a critical region of the code in the cache after it has been executed, providing the cache is enabled and the freeze bit is cleared. The enables cache (E) bit is used for system debug and emulation. This bit allows the designer to operate the processor with the cache disabled as long as the (E) bit remains cleared.

The cache address register (CAAR) format is shown in Figure 6.15.

The CAAR is used by the MC68020 to provide an address for the Clear Entry (CE) function as implemented in the CACR. The index portion of this register is used to specify which one of 64 entries to invalidate by clearing the associated cache entry valid bit (V).

Although the micromachine of the MC68020 is highly pipelined, the predominant pipeline mechanism is a three-stage instruction pipeline. Figure 6.16 shows the MC68020 pipeline organization. The pipeline is completely internal to the processor and is used as part of the

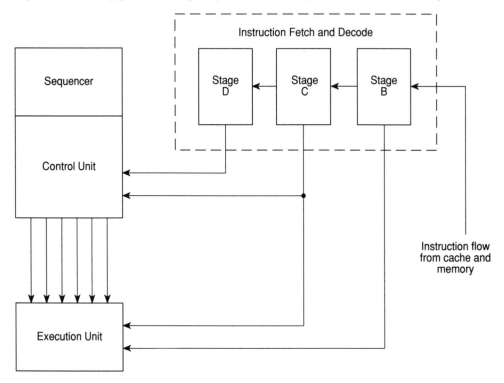

FIGURE 6.16   MC68020 pipeline.

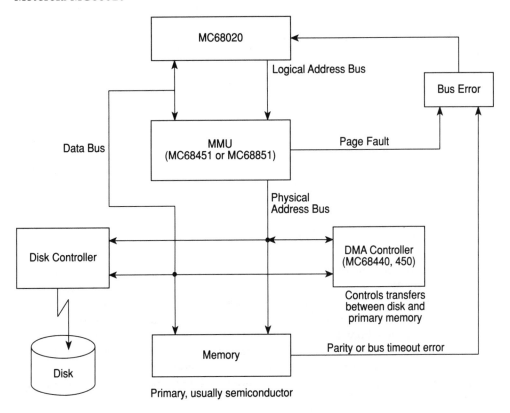

FIGURE 6.17 Typical virtual memory system, minimum configuration.

instruction fetching and decoding circuitry. Instructions from the on-chip cache, or from external memory (if the cache is disabled), go into the first stage of the pipeline and synchronously pass through the following two stages.

The pipeline output gives the 68020's control and execution unit a completely decoded instruction. The 68020 loads data and other operands into the pipeline so they are ready for immediate use. The pipeline speeds 68020 operation by making information available immediately. The benefit of the pipeline is to allow concurrent operations (parallelism) for up to three words of a single instruction or for up to three consecutive one-word instructions. Therefore, the performance benefits of a pipeline are maximized during the execution of in-line code.

6.8 MC68020 Virtual Memory

Virtual memory is a technique that allows all user programs executing on a processor to behave as if each had the entire 4-GB addressing range of the MC68020 at its disposal, regardless of the amount of physical memory actually present in the system. Virtual memory can be supported by providing a limited amount of high-speed physical memory that can be accessed directly by the processor while maintaining an image of a much larger "virtual" memory on a secondary storage device, such as high-capacity disk drives. Figure 6.17 shows a minimal system configuration for a typical virtual memory system. Also, any given instruction must be able to be aborted and restarted. When a processor attempts to access a location in the virtual memory map that is not resident in physical memory (this is called a "page fault"), the access to that location is temporarily suspended while data are fetched from secondary storage and placed into physical memory. Page faults force a trap to the bus error exception vector.

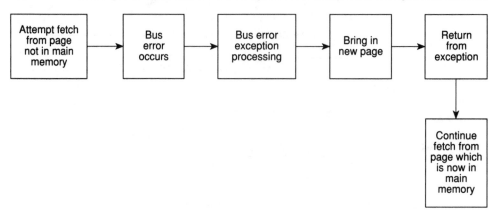

FIGURE 6.18   Bus error sequence.

The MC68020 processor has the abort capability via the bus error ($\overline{\text{BERR}}$) input to the processor. When $\overline{\text{BERR}}$ is asserted, exception processing causes the processor to save sufficient information to allow complete restoration of the faulted instruction. The faulted instruction will be recovered by the instruction continuation method, in which the faulted instruction is allowed to complete execution from the point of the fault. Instruction continuation is crucial for supporting virtual I/O devices in memory-mapped I/O systems. To handle instruction continuation properly, the processor must save certain internal processor information on the stack prior to running the exception code and return this information to the processor after executing the return-from-exception (RTE) instruction. Figure 6.18 shows the BUS error processing sequence that occurs when attempting to fetch an instruction from a page not in the main memory.

6.9 MC68020 Coprocessor Interface

The MC 68020's coprocessor interface is capable of extending the MC68020 instruction set and supporting new data types.

The interface between the main processor and a coprocessor is transparent to the user. The programmer does not have to be aware that a separate piece of hardware is executing some of the program code sequence. Hardware-implemented microcode within the MC68020 handles coprocessor interfacing so that a coprocessor can provide its capabilities to the programmer without appearing as external hardware, but rather as a natural extension to the main processor architecture.

When communicating with a coprocessor, the MC68020 executes bus cycles in CPU memory space to access a set of Coprocessor Interface Registers (CIRs). Table 6.15 shows the separate

TABLE 6.15   Coprocessor Interface Register

Register	Function	R / $\overline{\text{W}}$
Response	Requests action from CPU	$\underline{\text{R}}$
Control	CPU directed control	W
Save	Initiate save of internal state	R $\underline{}$
Restore	Initiate restore of internal state	$\underline{\text{R}}$ / W
Operation word	Current coprocessor instruction	$\underline{\text{W}}$
Command word	Coprocessor specific command	$\underline{\text{W}}$
Condition word	Condition to be evaluated	$\underline{\text{W}} \underline{}$
Operand	32-bit operand	R / W
Register select	Specifies CPU register or mask	R $\underline{}$
Instruction address	Pointer to coprocessor instruction	R / $\underline{\text{W}}$
Operand address	Pointer to coprocessor operand	R / $\underline{\text{W}}$

31		15		0
00	Response		Control	
04	Save		Restore	
08	Operation Word		Command	
0C	(Reserved)		Condition	
10	Operand			
14	Register Select		(Reserved)	
1B	Instruction Address			
1C	Operand Address			

FIGURE 6.19   Coprocessor interface register set map.

coprocessor interface registers along with their functions. As shown within this interface register set, the various registers are allocated to specific functions required for operating the coprocessor interface. There are registers specifically for passing information such as commands, operands, and EAs. Other registers are allocated for use during a context switch operation.

Figure 6.19 shows the coprocessor interface register set map and communication protocol between the main processor and the coprocessor necessary to execute a coprocessor instruction. The MC68020 implements the CIR communication protocol automatically, so the programmer is only concerned with the coprocessor's instruction and data type extensions to the MC68020 programmer's model.

Figure 6.20 shows the CIR set(s) (that is, more than one coprocessor) address map in CPU space.

The MC68020 indicates that it is accessing CPU memory space by encoding the function code lines high (FC0–FC2 = 111_2). Thus, the CIR set is mapped into CPU space the same way that a peripheral interface register set is generally mapped into data space. The address bus then selects the desired coprocessor chip.

Encoding of the address bus during coprocessor communication is shown in Figure 6.21. By using the cp-ID field on the address bus, up to eight separate coprocessors can be interfaced concurrently to the MC68020. Figure 6.21 also shows how this can be done. Interfacing to these separate coprocessors is a matter of decoding the relevant cp-ID field (address lines A13–A15) and the corresponding CPU space type field (address lines A16–A19) encoded within the coprocessor instruction, so that the MC68020 communicates with the relevant register set in CPU space.

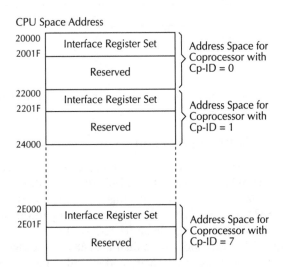

FIGURE 6.20   Coprocessor address map in CPU space.

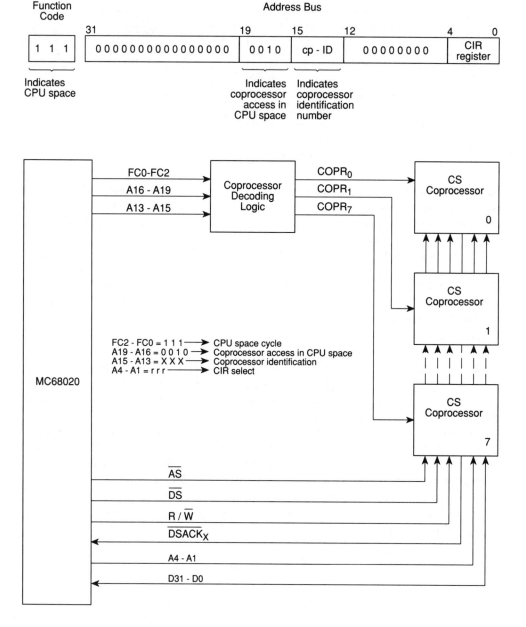

FIGURE 6.21   68020 interface to coprocessors. Coprocessor communication address bus encoding and multicoprocessor address decoding example.

6.9.1 MC68881 Floating-Point Coprocessor

The MC68881 HCMOS floating-point coprocessor implements IEEE standards for binary floating-point arithmetic. When interfaced to the MC68020, the MC68881 provides a logical extension to the MC68020 for performing floating-point operations. The MC68882 is an upgrade of the MCC68881 and provides in excess of 15 times the performance of the MC68881. It is a pin and software compatible upgrade of the MC68881.

The 68882 provides higher execution speed than the 68881. The 68882 contains a special on-chip hardware that converts between binary and extended floating-point numbers at a much higher speed than the 68881. Both chips are packaged in 68-pin PGA(Pin Grid Array).

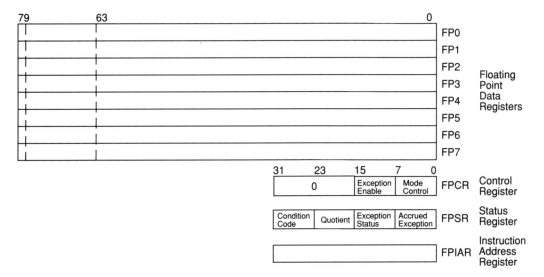

FIGURE 6.22a MC68881 programming model.

A summary of the MC68881 features is listed below:

- Eight general-purpose floating-point data registers, each supporting an 80-bit extended precision real data format (a 64-bit mantissa, plus a sign bit and a 15-bit signed exponent)
- A 67-bit arithmetic unit
- A 67-bit barrel shifter for fast shift operations
- instructions with 35 arithmetic operations
- Supports trigonometric and transcendental functions
- Supports seven data types: bytes, word, long word integers; single, double, and extended precision real numbers; and packed binary coded decimal string real numbers
- 22 constants including π, e, and powers of 10

The MC68881 is a non-DMA-type coprocessor which uses a subset of the general-purpose coprocessor interface supported by the MC68020. The MC68881 programming model is shown in Figure 6.22a.

The MC68881 programming model includes the following:

- Eight 80-bit floating-point registers (FP0–FP7). (These general-purpose registers are analogous to the MC68020 D0-D7 registers.)
- A 32-bit control register containing enable bits for each class of exception trap and mode bits, to set the user-selectable rounding and precision modes
- A 32-bit status register containing floating-point condition codes, quotient bits, and exception status information
- A 32-bit Floating-Point Instruction Address Register (FPIAR) containing the MC68020 memory address of the last floating-point instruction that was executed. (This address is used in exception handling to locate the instruction that caused the exception.)

The MC68881 can be interfaced as a coprocessor to the MC68020. Figure 6.22b provides the MC68020/68881 block diagram.

The MC68020 implements the coprocessor interface protocol in hardware and microcode. When the MC68020 encounters a typical MC68881 instruction, the MC68020 writes the

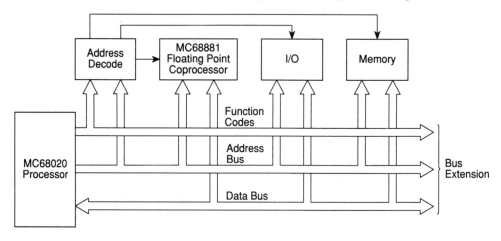

FIGURE 6.22b   Typical coprocessor configuration.

instruction to the memory-mapped command CIR and reads the response CIR. In this response, the BIU (Bus Interface Unit) translates any additional action required of the MC68020 by the MC68881. Upon satisfying the coprocessor requests, the MC68020 can fetch and execute subsequent instructions.

The MC68881 supports the following data formats:

> Byte Integer (B)
> Word Integer (W)
> Long Word Integer (L)
> Single Precision Real (S)
> Double Precision Real (D)
> Extended Precision Real (X)
> Packed Decimal String Real (P)

The capital letters included in parentheses denote the suffixes added to instructions in the assembly language source to specify the data format to be used

Figure 6.23 shows the MC68881 data formats.

These data formats are defined by IEEE standards. Integer data types do not include any fractional part of the number. The real formats contain an exponent part and a mantissa part. The single-real and double-real formats provide the sign of the mantissa, 8- and 11-bit exponents, and 23- and 52-bit fractional parts, respectively.

The extended real format is 96 bits wide with a 15-bit exponent, 64-bit mantissa, and a sign bit for the mantissa. The packed decimal real is 96 bits wide which provides sign for both mantissa and exponent parts. It includes a 3-digit (12 bits for BCD) base-10 exponent and a 17-digit (68 bits) base-10 mantissa. This format contains two bits to indicate infinity or not-a-number (NAN) representations. Note that NAN is a symbolic representation of a special number or situation in floating-point format. NANs include all numbers with nonzero fractions with the format's maximum exponent. The infinity data types include the zero fraction and maximum exponent.

Whenever an integer is used in a floating-point operation, the integer is automatically converted to an extended-precision floating-point number before being used. For example, the instruction FADD.W #2, FP0 converts the constant 2 to the floating-point number format in FP0, adds the two numbers, and then stores the result in FP0. This allows integer in floating-

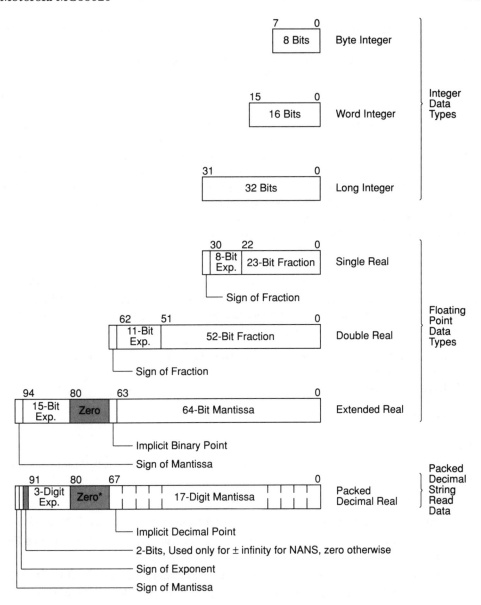

FIGURE 6.23 MC68881 data format summary.

point operations and also saves user memory, since an integer representation of a number is normally smaller than the equivalent floating-point representation.

The floating-point representation contains single-precision (32 bits), double-precision (64 bits), and extended-precision numbers (96 bits) as specified by the IEEE format. The single-precision and double-precision data types should be used in most calculations involving real numbers. The exponent is biased and the mantissa is in sign and magnitude form. Single and double precision require normalized numbers. Note that a normalized number has the most significant bit of the mantissa positioned such that the one lies to the left of the binary point.

Therefore, only the fractional part of the mantissa is stored in memory, which means that the most significant bit is implied and equal to one.

Extended-precision numbers are 96 bits wide but only 80 bits are used, and the unused 16 bits are for future expansion. Extended-precision numbers are for use as temporary variables, intermediate variables, or in situations where extra precision is required. Note that in extended precision, the 1 in 1 · f is explicit.

For example, a compiler might select extended-precision arithmetic for determining the value of the right side of an equation with mixed sized data, and then convert the answer to the data type on the left side of the equation. Extended-precision data should not be stored in large arrays due to the amount of memory required by each number. As with other data types, the packed BCD strings are automatically converted to extended-precision real values when they are input to the MC68881. This permits packed BCD number to be used as input to any operation such as FADD.P # − 2.012E + 18, FP1.

The MC68881 does not include any addressing modes. If the 68881 requests the 68020 to transfer an operand via the coprocessor interface, the 68020 provides the addressing mode calculations requested in the instruction.

Floating-point data registers FP0-FP7 always contain extended-precision values. Also, all data used in an operation are converted to extended precision by the MC68881 before the operation is performed. The MC68881 provides all results in extended precision. The MC68881 instructions can be grouped into six types:

1. MOVE instructions between the MC68881 and memory on the MC68020
2. Monadic operations
3. Dyadic operations
4. Branch, set, or trap conditionally
5. Miscellaneous

6.9.1.a 68881 Data Movement Instructions

The 68881 floating-point data movement instructions include FMOVE, FMOVEM, and FMOVECR.

Some examples of FMOVE instruction include:

- FMOVE.B LOC, FP3 transfers an 8-bit integer from memory address LOC to FP3.
- FMOVE.X FP4, FP0 transfers the extended-precision floating-point number from FP4 to FP0.
- FMOVE.S FP3, D1 moves a single precision floating-point number from FP3 to D1.
- FMOVE.P FP5, (A0) transfers a BCD floating-point number from FP5 into 12 consecutive bytes of memory starting at the address pointed to by A0.
- FMOVE.X FP0, -(USP) subtracts 12 from USP and transfers the extended precision floating-point number from FP6 to the user stack.

When data are moved from memory or from a 68020 data register to a floating-point register, the 68020 converts the data to an extended floating-point number.

Also, the FMOVE instruction from the 68881 floating-point register to memory or a 68020 data register convert data from the extended precision format to the form defined by the extension included with the instruction. Note that data are always represented in extended-precision form in a floating-point register.

The FMOVEM instruction transfers data between multiple 68881 floating-point registers and memory. Typical examples include:

```
FMOVEM (EA), FP1-FP4/FP6
FMOVEM FP0/FP1/FP5, (EA)
```

The registers FP0-FP7 are always moved as 96-bit extended data with no conversion.

Any combination of FP0-FP7 can be moved. (EA) can be control modes, predecrement or postincrement mode. For control or postincrement mode, the order of transfer is from FP7-FP0. For predecrement mode, the order of transfer is from FP0-FP7. Any combination of FPCR, FPSR, and FPIAR can also be moved by the MOVEM instruction. These registers are always moved in the order FPCR, FPSR, and FPIAR.

The 68881 FMOVECR #$mn, FPn instruction reads an extended precision constant from an ROM within the 68881/68882 into a floating-point register. The numeric values of some of the constants are selected from the following:

$mn	Constant
$00	π
$0B	$\text{Log}_{10}2$
$0C	e
$0D	Log_2E
$0E	Log_{10}E
$30	Log_e2
$31	Log_e10
$32	10^0
$33	10^1
$34	10^2
$35	10^4
$36	10^8

and so on.

For example the instruction FMOVEMCR .X #$0C, FP6 moves the extended precision value of e into FP6.

6.9.1.b Monadic

Monadic instructions have a single input operand. This operand may be in a floating-point data register, memory, or in an MC68020 data register. The result is always stored in a floating-point data register. Typical examples include:

```
FTAN  .  (fmt)  (EA),  FPn
  or
FTAN  .  X FPm,  FPn
  or
FTAN  .  X FPn
```

The FTAN instruction converts the source operand to extended precision (if necessary), computes the tangent of that number, and then stores the result in the destination floating-point data register.

Table 6.16a flowcharts the monadic function. Tables 6.16b and c list all MC68881 monadic instructions.

TABLE 6.16a   Monadic Functions

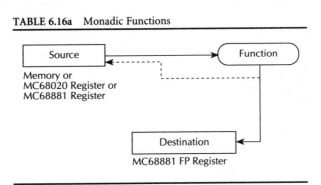

TABLE 6.16b   MC68881 — Monadic Instructions

	Transcendental functions
FACOS	ARC COSINE
FASIN	ARC SINE
FATAN	ARC TANGENT
FATANH	HYPERBOLIC ARC TANGENT
FCOS	COSINE
FCOSH	HYPERBOLIC COSINE
FETOX	E TO THE X POWER
FETOXM1	E TO THE $(X - 1)$ power
FLOG10	LOG TO THE BASE 10
FLOG2	LOG TO THE BASE 2
FLOGN	LOG BASE e OF X
FLOGNP1	LOG BASE e OF $(X + 1)$
FSIN	SINE
FSINCOS	SIMULTANEOUS SIN/COS
FSINH	HYPERBOLIC SINE
FTAN	TANGENT
FTANH	HYPERBOLIC TANGENT
FTENTOX	TEN TO THE X POWER
FTWOTOX	TWO TO THE X POWER

TABLE 6.16c   MC68881 — Monadic Instructions

	Nontranscendental functions
FABS	ABSOLUTE VALUE
FINT	INTEGER PART
FNEG	NEGATE
FSQRT	SQUARE ROOT
FNOP	NO OPERATION (SYNCHRONIZE)
FGETEXP	GET EXPONENT
FGETMAN	GET MANTISSA
FTST	TEST

6.9.1.c Dyadic Instructions

Dyadic instructions have two input operands. The first input operand comes from a floating-point data register, memory, or an MC68020 data register. The second input operand comes from a floating-point data register. The second input is also the destination floating-point data register. Typical examples include:

```
FCMP.L (fmt) (EA), FPn
   or
FCMP.X FPm, FPn
```

Table 6.17a flowcharts the dyadic function and Table 6.17b lists dyadic instructions.

TABLE 6.17a   MC68881 Dyadic Functions

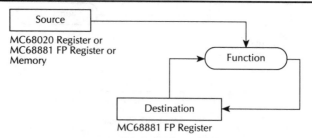

TABLE 6.17b   MC 68881 Dyadic Instructions

FADD	ADD
FCMP	COMPARE
FDIV	DIVIDE
FMOD	MOD
FMUL	MULTIPLY
FREM	IEEE REMAINDER
FSCALE	SCALE EXPONENT
FSUB	SUBTRACT

6.9.1.d BRANCH, Set, or Trap-On Condition

These instructions are similar to those of the MC68020 except that move conditions exist due to the special values in IEEE floating-point arithmetic. When the MC68020 encounters a floating-point conditional instruction, it passes the instruction to the MC68881 for performing the necessary condition checkup. The MC68881 then checks the condition and tells the MC68020 whether the condition is true or false. The MC68020 then takes the appropriate action.

The MC68881 conditional instructions are

```
FBcc.W      displ  Branch
  or .L

FDBcc.W     displ  Decrement and branch
  or .L

FScc.W      displ  Set byte according to condition
  or .L

FTRAPcc.W   displ  Trap-on condition
  or .L
```

All the above instructions can have 16- or 32-bit displacement.

cc is one of the 32 floating-point conditional test specifiers. Table 6.18 lists a few of them.

TABLE 6.18   Floating-Point Conditional Test Specifiers

Mnemonic	Definition
F	False
EQ	Equal
NE	Not equal
T	True
SF	Signaling false
SEQ	Signaling equal
GT	Greater than
GE	Greater than or equal
LT	Less than
LE	Less than or equal
GL	Greater or less than
GLE	Greater, less, or equal
NGLE	Not (greater, less, or equal)
NGL	Not (greater or less)
NLE	Not (less or equal)
NLT	Not (less than)
NGE	Not (greater or equal)
NGT	Not (greater than)
SNE	Signaling not equal
ST	Signaling true

Instruction: FADD.X FP0, FP1

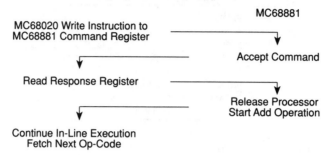

FIGURE 6.24a   MC68020/MC68881 concurrence example.

6.9.1.e Miscellaneous Instructions

These instructions include moves to and from the status, control, and instruction address registers. The virtual memory instructions FSAVE and FRESTORE that save and restore the internal state of the MC68881 also fall into this category. These instructions include:

```
FMOVE (EA), FPcr
FMOVE FPcr, (EA)
FRESTORE (EA)
```

FPcr means floating-point control register. The MC68881 does not perform addressing mode calculations. The MC68020 carries out this calculation as specified in the instruction.

Typical addressing modes include immediate, postincrement, predecrement, direct, and the indexed/indirect modes of the MC68020. Some addressing modes are restricted for some instructions. For example, PC relative mode is not permitted for a destination operand.

The MC68881 can execute an instruction concurrently or nonconcurrently with the MC68020 depending on the instruction being executed. Figures 6.24a and b show examples of concurrence and nonconcurrence.

Instruction: FMOVE.L FP1, (A0)+

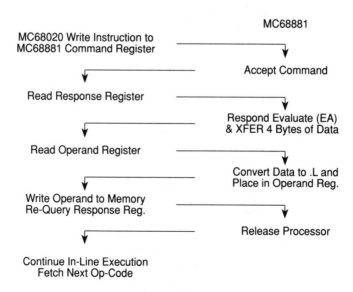

FIGURE 6.24b   MC68020/MC68881 nonconcurrence example.

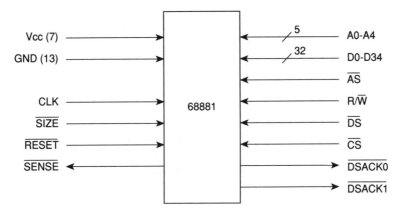

FIGURE 6.25   MC68881 pins and signals.

Figure 6.25 shows the MC68881 pins and signals.

The 68881 is included either in a 64-pin DIP or in a 68-PGA (Pin Grid Array) package. There are 7 Vcc (+5V) and 13 grounds for power distribution to reduce noise.

The five address lines A0-A4 are used by the MC68020 to select the coprocessor interface registers mapped in the MC68020 address space. These address pins select the registers as listed in Table 6.19.

When the MC68881 is configured to operate over an 8-bit data bus for processors such as MC68008, the A0 pin is used as an address signal for byte accesses of the MC68881 interface register. When the MC68881 is configured to operate over a 16-bit data bus (68000) or 32-bit data bus (68020), both A0 and SIZE pins are used, according to Table 6.20. The SIZE and A0 pins are used to configure the MC68881 for operation over 8-, 16-, or 32-bit data buses.

The address strobe AS goes LOW to indicate that there is a valid address on the address bus, and both CS and R/W are valid.

A low on DS indicates that there is valid data on the data bus during a write cycle.

If the bus cycle is a MC68020 read from MC68881, the MC68881 asserts DSACK1 and DSACK0 to indicate that the information on the data bus is valid. If the bus cycle is a MC68020 write to the MC68881, DSACK1 and DSACK0 are used to acknowledge acceptance of the data by the MC68881.

The MC68881 also uses DSACK0 and DSACK1 to dynamically indicate the device size on a cycle-by-cycle basis as discussed in Section 6.10.

TABLE 6.19   Coprocessor Interface Register Selection

A4-A0	Offset	Width	Type	Register
0000x	$00	16	Read	Response
0001x	$02	16	Write	Control
0010x	$04	16	Read	Save
0011x	$06	16	R/W	Restore
0100x	$08	16	—	(reserved)
0101x	$0A	16	Write	Command
0110x	$0C	16	—	(reserved)
0111x	$0E	16	Write	Condition
100xx	$10	32	R/W	Operand
1010x	$14	16	Read	Register select
1011x	$16	16	—	(reserved)
110xx	$18	32	Read	Instruction address
111xx	$1C	32	R/W	Operand address

TABLE 6.20   Data Bus Configuration

A0	Size	Data bus
—	Low	8-bit
Low	High	16-bit
High	High	32-bit

A low on MC68881 $\overline{\text{RESET}}$ pin clears the floating_point control, status, and instruction address registers. When performing power-up reset, external circuitry should keep the $\overline{\text{RESET}}$ pin asserted for a minimum of four clock cycles after Vcc is within tolerance. After Vcc is within tolerance for more than the initial power-up time, the $\overline{\text{RESET}}$ pin must be asserted for at least two clock cycles.

The MC68881 clock input is a TTL-compatible signal that is internally buffered for generation of the internal clock signals. The clock input should be a constant frequency square wave with no stretching or shaping techniques required. The MC68881 can be operated from a 12-, 16.67-, or 20-MHz clock.

The SENSE pin may be used as an additional ground pin for more noise immunity or as an indicator to external hardware that the MC68881 is present in the system. This signal is internally connected to the GND of the die, but it is not necessary to connect it to the external ground for correct device operation. If a pullup resistor (larger than 10 K ohm) is connected to this pin, external hardware may sense the presence of the MC68881 in a system.

Figure 6.26 shows the MC68020/MC68881 interface.

The A0 and SIZE pins are connected to Vcc for 32-bit operation. Note that A19, A18, A17, A16 = 0010_2 (indicating coprocessor function), FC2 FC1 FC0 = 111_2 (meaning CPU space cycle), and A15 A14 A13 = 001_2 (indicating 68881 floating-point coprocessor) are used to enable 68881 CS. The 68020 A19 and A18 pins are not used in the chip select decode since their values are 00_2.

The $\overline{\text{BERR}}$ pin is asserted for a low $\overline{\text{CS}}$ and a high $\overline{\text{SENSE}}$ signal. A trap routine can be executed to perform the coprocessor operations.

Example 6.22

Write 68020 assembly language program using 68881 floating-point instructions to compute the volume of a sphere = $4/3*\pi*r^3$ where r is the radius of the sphere stored in the 68020 register D0. Assume r is a single precision number.

Solution

```
        FMOVE.S  D0, FP0      ;  MOVE  r to FP0
        FMOVE.S  FP0, FP1     ;  Make another copy of r
        FSGLMUL.S FP1, FP1    ;  Compute r² by single precision
                             ;   multiply
        FSGLMUL.S FP0, FP1    ;  Compute r³ by single precision
                             ;   multiply
        FMOVE.S  #4E0, FP3    ;  Load constant 4
        FDIV.S   #3E0, FP2    ;  Compute 4/3
        FMOVECR.X #3E0, FP4   ;  Obtain π
        FSGLMUL.S FP2, FP4    ;  Compute 4/3π
        FSGLMUL.S FP1, FP4    ;  Compute volume
        FMOVE.S  FP4, D0      ;  Store volume in D0.
FINISH JMP FINISH
```

Example 6.23

For the following figure, write an MC68020 assembly program using floating-point coprocessor instructions; determine X and Θ.

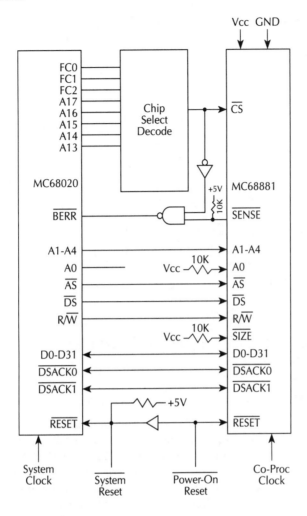

FIGURE 6.26 MC68020/MC68881 32-bit interface.

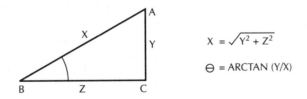

$$X = \sqrt{Y^2 + Z^2}$$

$$\Theta = \text{ARCTAN} (Y/X)$$

Assume X, Y, and Z are 32 bits wide.

Solution

```
MOVE.L Y, FP2      ;   MOVE Y TO FP2
MOVE.L Z, FP3      ;   MOVE Z TO FP3
FMOVE.L FP2, FP0   ;   MOVE Y TO FP0
FMOVE.L FP3, FP1   ;   MOVE Z TO FP1
FMUL.L FP0, FP0    ;   Y²
FMUL.L FP1, FP1    ;   Z²
FADD.L FP0, FP1    ;   Z² + Y²
FSQRT.L FP1        ;   X = Y² + Z²
FDIV.L FP3,FP2     ;   Θ is
```

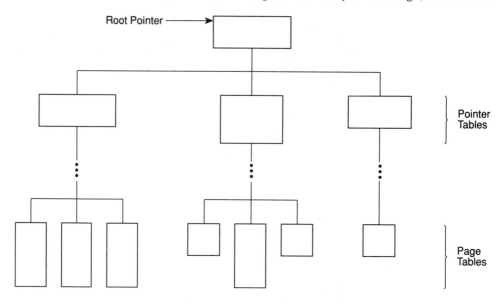

FIGURE 6.27   MC68851 translation table tree structure.

```
            FATAN FP2              ;   ARCTAN (Y/Z)
FINISH   JMP FINISH
```

6.9.2 MC68851 MMU

The MC68851 coprocessor is a demand Paged Memory Management Unit (PMMU) designed to support the MC68020-based virtual memory system. The 68851 is included in a 132-PGA package and can be operated at a frequency of either 12.5 or 16.67 MHz.

The main functions of the 68551 are to provide logical-to-physical address translation, protection mechanism, and to support the 68020 breakpoint operations.

The 68851 translates a logical address comprised of a 32-bit address and a 4-bit function code. The 68851 has four function code pins (discussed later), FC0-FC3, issued by the 68020 into a corresponding 32-bit physical address in main memory. The 68851 initiates address translation by searching for the page descriptor corresponding to the logical-to-physical mapping in the on-chip 64-entry full-associative Address Translation Cache (ATC). This cache stores recently used page descriptors. If the descriptor is not found in the ATC, then the 68851 aborts the logical bus cycle, signals the 68020 to retry the operation, and requests mastership of the logical bus. Upon receiving indication that the logical bus is free, the 68851 takes over the logical bus and executes bus cycles to search the translation tables in physical (main) memory pointed to by the relevant root pointer to locate the required translation descriptor that defines the page accessed by this logical address.

After obtaining the needed translation descriptor, the 68851 returns control of the logical bus to the 68020 to retry the previous bus cycle which can now be verified for access rights and be properly translated by the 68851.

The 68851 automatically searches the translation tables in case of descriptor misses in the ATC using hardware without any assistance from the operating system. The 68851 translation tables have a tree structure, as depicted in Figure 6.27.

As shown in the figure, the root of a translation table is pointed to by one of the three 64-bit root pointer registers: CPU root pointer, supervisor root pointer, or DMA root pointer.

The CPU root pointer points at the translation table tree for the currently executing task; the supervisor root pointer points to the operating systems translation table; and the DMA root pointer points to a DMA controller's (if present in the system) translation table.

All addresses contained in the translation tables are physical addresses. In Figure 6.29, table entries at the higher levels of the tree (pointer tables) contain pointers to other tables. Entries at the leaf level (page tables) contain page descriptors. The pointer table lookup normally uses the function codes as the index, but they may be suppressed. The 68851 includes the 4-bit bidirectional function code pins, FC0-FC3, which indicate the address space of the current bus cycle. The 4-bit function code consists of the three function code outputs (FC0-FC2) of the 68020 and a fourth bit that indicates that a DMA access is in progress. When the 68851 is bus master, it drives the function code pins as outputs with a constant value of FC3-FC0 = $5, indicating the supervisor data space.

The 68851 hierarchical protection mechanism monitors and enforces the protection/privilege mechanism. These may be up to 8 levels with privilege hierarchy, and the upper 3 bits of the incoming logical address define these levels, with level 0 as the highest privilege in the hierarchy and level 7 as the lowest level. Privilege levels of 0, 2, or 4 can also be implemented with the 68851, in which case the access level encoding is included in the upper zero, one, or two logical address lines, respectively. The 68851 access level mechanism, when enabled, compares the access level of the memory logical address with the current access level as defined in the Current Access Level (CAL) register. The current access level defines the highest privilege level that a task may assume at that time.

If the privilege level provided by the bus cycle is more privileged than allowed, then the 68851 terminates this access as a fault.

In the 68851 protection mechanism, the privilege level of a task is defined by its access level. Smaller values for access levels specify higher privilege levels. In order to access program and/ or data requiring a higher privilege level than the level of the current task, the 68851 provides CALLM (call module) and RTM (return from module) instructions. These instructions allow a program to call a module operating at the same or higher privilege level and to return from that module after completing the module function.

The 68851 provides a breakpoint acknowledge facility to support the 68020. When the 68020 executes a breakpoint instruction, it executes a breakpoint acknowledge cycle and reads a predefined address in the CPU space cycle. The 68851 decodes this address and responds by either providing a replacement op code for the breakpoint op code and asserting DSACKx inputs or by asserting 68020 BERR input to execute illegal instruction exception processing. The 68851 can be programmed to provide (1) the replacement op code n times ($1 \leq n \leq 255$) in a loop and then assert BERR or (2) assert BERR on every breakpoint.

The 68851 instructions provide an extension to the 68020 instruction set via the coprocessor interface. These instructions provide:

1. Loading and storing of MMU registers
2. MMU control functions
3. Access rights and conditionals checking

For example, the PMOVE instruction moves data to or from the 68851 registers using all 68020 addressing modes. PVALID compares the access level bits of an incoming logical address with those of the Valid-Access Level (VAL) register and traps if the address bits are less. PTEST searches the ATC and translation tables for an entry corresponding to a specific address and function code. The results of the test are placed in the 68851 status register which can be tested by various conditional branch and set instructions.

Optionally, the PTEST instruction can obtain the address of the page descriptor. A companion instruction, PLOAD, takes a logical address and function code, searches the translation table, and loads the ATC with an entry to translate the logical address.

PLOAD can be used to load the ATC before starting the memory transfer. This can speed up a DMA operation. PFLUSH and its variations clear the ATC of either all entries, entries with a specified function code, or those limited to a particular function code and logical address.

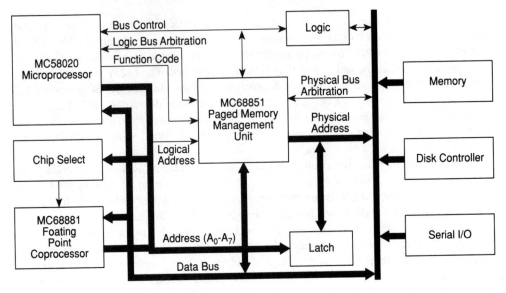

FIGURE 6.28   A 68020-based microcomputer with coprocessors.

PSAVE saves the contents of any register that reflects the current task's state and the internal state of the 68851 dealing with coprocessor and module call operations. PRESTORE restores the information saved by PSAVE. PSAVE and PRESTORE permit the context of the 68851 to be switched.

The PBcc, PDBcc, PScc, and PTRAPcc instructions have the same meaning as those of the 68020 except that the conditions are based on the 68851 condition codes.

Figure 6.28 shows a 68020-based microcomputer system which interfaces 68881 and 68851 chips to the 68020. The 68851 MMU is placed between the logical and physical address buses. The 68851 allows interfacing memory, disk controller, and serial I/O devices to the 68020.

6.10 MC68020 Pins and Signals

Figure 6.29 shows the MC68020 functional signal groups. Tables 6.21 lists these signals along with a description of each.

There are 10 VCC (+5V) and 13 ground pins to distribute the power in order to reduce noise.

Both the 32-bit address (A0-A31) and data (D0-D31) buses are nonmultiplexed. Like the MC68000, the three function code signals FC2, FC1, and FC0 identify the processor state (supervisor or user) and the address space of the bus cycle currently being executed as follows:

FC2	FC1	FC0	Cycle type
0	0	0	(Undefined, reserved)[a]
0	0	1	User data spare
0	1	0	User program space
0	1	1	(Undefined, reserved)[a]
1	0	0	(Undefined, reserved)[a]
1	0	1	Supervisor data space
1	1	0	Supervisor program space
1	1	1	CPU space

[a] Address space 3 is reserved for user definition, while 0 and 4 are reserved for future use by Motorola.

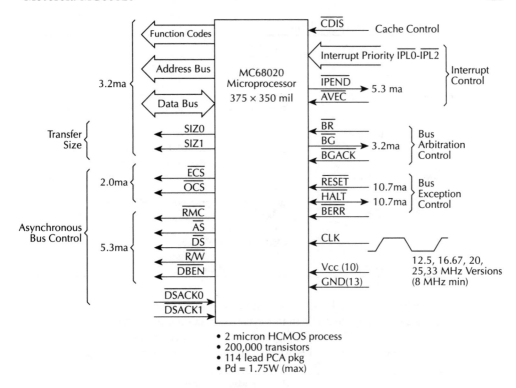

FIGURE 6.29 MC68020 functional signal groups.

TABLE 6.21 Hardware Signal Index

Signal name	Mnemonic	Function
Address bus	A0-A31	32-bit address bus used to address any of 4,294,967,296 bytes
Data bus	D0-D31	32-bit data bus used to transfer 8, 16, 24, or 32 bits of data per bus cycle
Function codes	FC0-FC2	3-bit function code used to identify the address space of each bus cycle
Size	SIZ0/SIZ1	Indicates the number of bytes remaining to be transferred for this cycle; these signals, together with A0 and A1, define the active sections of the data bus
Read-modify-write cycle	$\overline{\text{RMC}}$	Provides an indicator that the current bus cycle is part of an indivisible read-modify-write operation
External cycle start	$\overline{\text{ECS}}$	Provides an indication that a bus cycle is beginning
Operand cycle start	OCS	Identical operation to that of ECS except that OCS is asserted only during the first bus cycle of an operand transfer
Address strobe	$\overline{\text{AS}}$	Indicates that a valid address is on the bus
Data strobe	DS	Indicates that valid data is to be placed on the data bus by an external device or has been placed on the data bus by the MC68020
Read/write	$\text{R}/\overline{\text{W}}$	Defines the bus transfer as an MPU read or write
Data buffer enable	$\overline{\text{DBEN}}$	Provides an enable signal for external data buffers
Data transfer and size acknowledge	$\overline{\text{DSACK0}}/\overline{\text{DSACK1}}$	Bus response signals that indicate the requested data transfer operation are completed; in addition, these two lines indicate the size of the external bus port on a cycle-by-cycle basis

TABLE 6.21   Hardware Signal Index (*continued*)

Signal name	Mnemonic	Function
Cache disable	CDIS	Dynamically disables the on-chip cache to assist emulator support
Interrupt priority level	$\overline{\text{IPL0-IPL2}}$	Provides an encoded interrupt level to the processor
Autovector	$\overline{\text{AVEC}}$	Requests an autovector during an interrupt acknowledge cycle
Interrupt pending	$\overline{\text{IPEND}}$	Indicates than an interrupt is pending
Bus request	$\overline{\text{BR}}$	Indicates than an external device requires bus mastership
Bus grant	$\overline{\text{BG}}$	Indicates than an external device may assume bus mastership
Bus grant acknowledge	$\overline{\text{BGACK}}$	Indicates than an external device has assumed bus control
Reset	$\overline{\text{RESET}}$	System reset
Halt	$\overline{\text{HALT}}$	Indicates that the processor should suspend bus activity
Bus error	$\overline{\text{BERR}}$	Indicates an invalid or illegal bus operation is being attempted
Clock	CLK	Clock input to the processor
Power supply	VCC	+5 volt ± 5% power supply
Ground	GND	Ground connection

Note that in MC68000, FC2, FC1, FC0 = 111 indicates interrupt acknowledge cycle. In the MC68020, this means CPU space cycle. In this cycle, by decoding the address lines A19-A16, the MC68020 can perform various types of functions such as coprocessor communication, breakpoint acknowledge, interrupt acknowledge, and module operations as follows:

A19	A18	A17	A16	Function performed
0	0	0	0	Breakpoint acknowledge
0	0	0	1	Module operations
0	0	1	0	Coprocessor communication
1	1	1	1	Interrupt acknowledge

Note that A19, A18, A17, A16 = 0011_2 to 1110_2 is reserved by Motorola. In the coprocessor communication CPU space cycle, the MC68020 determines the coprocessor type by decoding A15-A13 as follows:

A15	A14	A13	Coprocessor type
0	0	0	MC68851 paged memory management unit
0	0	1	MC68881 floating-point coprocessor

The MC68020 offers a feature called dynamic bus sizing which enables designers to use 8- and 16-bit memory and I/O devices without sacrificing system performance. The key elements used to implement dynamic bus sizing are the internal data multiplexer, the SIZE output (SIZ0 and SIZ1 pins), and the DSACKX inputs (DSACK0 and DSACK1). The MC68020 uses these signals dynamically to interface to the various-sized devices (8-, 16-, or 32-bit) on a cycle-by-cycle basis. For example, if the MC68020 executes an instruction that reads a long-word operand, it will attempt to read all 32 bits during the first bus cycle. The MC68020 always assumes the memory or I/O size to be 32 bits when starting the bus cycle. Hence, it always transfers the maximum amount of data on all bus cycles. If the device responds that it is 32 bits, the processor latches all 32 bits of data and continues to the next operand. If the device responds that it is 16 bits wide, the MC68020 generates two bus cycles, obtaining 16 bits of data each time; an 8-bit transfer is handled similarly, but four bus cycles are required, obtaining 8 bits of data each time. Each device (8-, 16-, or 32-bit) assignment is fixed to particular sections of the data bus to minimize the number of bus cycles needed to transfer devices. For example,

TABLE 6.22   Dynamic Sizing Control Signals

	DSACK Codes and Results	
DSACK1	DSACK0	Result
H	H	Insert wait states in current bus cycle
H	L	Complete cycle — port size is 8 bits
L	H	Complete cycle — port size is 16 bits
L	L	Complete cycle — port size is 32 bits

	SIZE Output Encodings	
SIZ1	SIZ0	Size
0	1	Byte
1	0	Word
1	1	3 bytes
0	0	Long word

Note: To adjust the size of the physical bus interface, external circuits must issue strobe signals to gate data bus buffers. By using data strobe control, external logic can enable the proper section of the data bus.

the 8-bit devices transfer data via D31-D24 pins, the 16-bit devices via D31-D16 pins, and the 32-bit devices via D31-D0 pins.

A routing and duplication multiplexer takes the four byte of 32 bits and routes them to their required positions; depending on its bus size, the positioning of bytes is determined by SIZ1, SIZ0, and A1 and A0 address output pins.

The four signals added to support dynamic bus sizing are $\overline{DSACK0}$, $\overline{DSACK1}$, SIZ0, and SIZ1. Data transfer and device size acknowledge signals ($\overline{DSACK0}$ and $\overline{DSACK1}$) are used to terminate the bus cycle and to indicate the external size of the data bus (see Table 6.22). As the MC68020 steps through memory during the data operand transfer process, the two size line outputs (SIZ0 and SIZ1) indicate how many bytes are still to be transferred during a given bus cycle (Table 6.22).

The multiplexer routes and/or duplicates one or more bytes in the 32-bit data to permit any combination of aligned or misaligned transfers. The remaining number of bytes to be transferred during the second and subsequent cycles, if required, is defined by the SIZ0 and SIZ1 outputs. The address lines A0 and A1 define the byte position in Figure 6.30. For example, $A1A0 = 00_2$, $A1A0 = 01_2$, $A1A0 = 10_2$, and $A1A0 = 11_2$ indicate byte 0 (OP0), byte 1 (OP1), byte 2 (OP2), and byte 3 (OP3), respectively, of the 32-bit operand. A2-A31 indicate the long-word base address of that portion of the operand to be accessed.

Table 6.23 defines the data pattern along with SIZ1, SIZ0, A1, and A0 of the MC68020's internal multiplexor to the external data bus D31-D0.

In each cycle, the MC68020 outputs to the SIZ1 SIZ0 pins to indicate to the external device the number of bytes remaining to be transferred. The $\overline{DSACK1}$ and $\overline{DSACK0}$ inputs to the MC68020 from the device terminate the bus cycle and for the subsequent cycles (if required) indicate the device size. The A1 and A0 output address pins to the device from the MC68020 indicate which data pins are to be used in the data transfer. For example, an 8-bit device always transfers data to MC68020 via D31-D24 pins for all combinations of A1A0 (00_2, 01_2, 10_2, 11_2). On the other hand, the 16-bit device always transfers data via D31-D16 pins. However, in the first cycle, if the address is even ($A1A0 = 00_2$ or 10_2), 16-bit data transfer takes place using D31-D16 pins with the data byte addressed by the odd address via D23-D16 pins. On the other hand, if the starting address is odd ($A1A0 = 01_2$ or 11_2) for a 16-bit device, only a byte is transferred in the first cycle via D23-D16 pins. For a 32-bit device, in the first cycle, if $A1A0 = 00_2$, all 32-bit data are transferred via D0-D31 pins; if $A1A0 = 01_2$, three bytes are transferred via D23-D0 pins; if $A1A0 = 10_2$, two bytes are transferred via D15-D0 pins; if $A1A0 = 11_2$, only one byte is transferred via D7-D0 pins. Note that the MC68020's

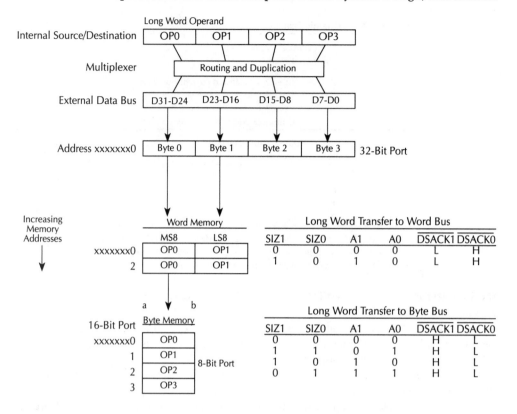

FIGURE 6.30   Dynamic bus sizing interface to port sizes.

TABLE 6.23   Internal to External Data Bus Multiplexor

MC68020 Register			
Byte 0	Byte 0	Byte 0	Byte 0

Multiplexor

External Data Pins

Transfer size	SIZ1	SIZ0	A1	A0	D31-D24	D23-D16	D15-D8	D7-D0
					Byte 3	Byte 3	Byte 3	Byte 3
					Byte 2	Byte 3	Byte 2	Byte 3
Byte	0	1	X	X	Byte 2	Byte 2	Byte 3	Byte 2
Word	1	0	X	0	Byte 1	Byte 2	Byte 3	Byte 0*
	1	0	X	1	Byte 1	Byte 1	Byte 2	Byte 3
3 bytes	1	1	0	0	Byte 1	Byte 2	Byte 1	Byte 2
	1	1	0	1	Byte 1	Byte 1	Byte 2*	Byte 1
	1	1	1	0	Byte 0	Byte 1	Byte 2	Byte 3
	1	1	1	1	Byte 0	Byte 0	Byte 1	Byte 2
Long word	0	0	0	0	Byte 0	Byte 1	Byte 0	Byte 1
	0	0	0	1	Byte 0	Byte 0	Byte 1*	Byte 0
	0	0	1	0				
	0	0	1	1				

Note: X = don't care; * = byte ignored on read, this byte output on write.

data transfers with even and odd starting addresses are known as aligned and misaligned transfers, respectively.

The MC68020 always starts transferring data with the most significant byte first. As an example, consider MOVE.L D3, $50005170. Since the address is even, this is an aligned transfer from the 32-bit data register D3 to an even memory address. In the first bus cycle, the MC68020 does not know the size of the external device and hence outputs all the data on D31-D0 pins, taking into consideration that the device size may be byte, word, or long word. The MC68020 outputs OP0, OP1, OP2, and OP3, respectively, on D31-D24, D23-D16, D15-D8, and D7-D8 pins. If the device is 8-bit, it will take the data OP0 from the D31-D24 pins and write to locations $50005170 in the first cycle. However, by the second cycle, the device asserts $\overline{DSACK1}$ and $\overline{DSACK0}$ as 10_2 indicating an 8-bit device; the MC68020 then transfers the remaining 24 bits via D31-D24 in three consecutive cycles. If the device is 32-bit, it obtains data bytes OP0-OP3 in one cycle. Now, let us consider a 16-bit device. During the first cycle, the MC68020 outputs A1A0 = 00_2 indicating an aligned transfer, and SIZ1 SIZ0 = 00_2 indicating 32-bit transfer. Therefore, in the first cycle, the device obtains OP0 and OP1 from D31-D24 and D23-D16, respectively. The device then asserts $\overline{DSACK1}$ and $\overline{DSACK0}$ as 01_2 to terminate the cycle and to indicate to the MC68020 that it is a 16-bit device. In the second cycle, the MC68020 outputs the A1A0 as 10_2, indicating that 16-bit data to be obtained by the device via D31-D16 pins and SIZ1 and SIZ0 as 10_2, indicating that two more bytes remain to be transferred. The MC68020 places the low two bytes (OP2 and OP3) from register D3 via the multiplexer on D31-D16 pins. The device takes these data and places them into locations $5000 5172 and $5000 5173, respectively, by activating $\overline{DSACK1}$ and $\overline{DSACK0}$ as 01_2, indicating completion of the cycle.

If [D3] = $03F1 2517 (OP0 = $03, OP1 = $F1, OP2 = $25, and OP3 = $17), then data transfer for MOVE.L D3, $5000 5170 takes place as shown in the following:

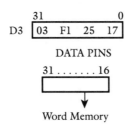

		MC68020				16-bit Memory	
D31	D16	SIZ1	SIZ0	A1	A0	$\overline{DSACK}$	$\overline{DSACK0}$
First cycle	03	F1	0	0	0	0	0	1
Second cycle	25	17	1	0	1	0	0	1

Now let us consider a misaligned transfer to a 16-bit device. For example, consider MOVE.L D4, $6017 2421. Assume [D4] = $7126E214, that is, OP0 = $71, OP1 = 26, OP2 = $E2, OP3 = $14. Now, suppose that the device is 16-bit. In the first cycle, the MC68020 outputs $717126E2 via the multiplexor; the multiplexor places these data on D31-D0 pins considering that the device may be 8-, 16-, or 32-bit as follows:

D31 : D24	D23 : D16	D15 : D8	D7 : D0
71	71	26	E2

This is because the device accepts $71 if it is 8-bit via D31-D24 pins, $71 via D24-D16 pins if it is 16-bit, and 24-bit data $7126E2 via D23-D0 pins if it is 32-bit.

For 8-bit and 32-bit devices, four and two cycles are required, respectively, to complete the long-word transfer. Now, let us consider the 16-bit device for this example in detail.

In the first cycle, the MC68020 outputs SIZ1, SIZ0 as 00_2, indicating a 32-bit transfer, and A1A0 as 01_2, indicating that a byte transfer is to take place via D23-D16 pins in the first cycle. The memory device obtains $71 from D23-D16 and writes these data to $60172421 and then activates DSACK1 DSACK0 as 01_2, indicating to the MC68020 that it is a 16-bit device. In the second cycle, the MC68020 outputs SIZ1 SIZ0 as 11_2, indicating that three more bytes remain to be transferred, and A1A0 as 10_2, indicating a 16-bit transfer is to take place via D31-D16.

The memory device activates DSACK1 DSACK0 as 01_2 to write $26E2 to locations $60172422 and $60172423. The MC68020 then terminates the cycle.

In the third cycle, the MC68020 outputs SIZ1 SIZ0 as 01_2, indicating a byte remaining to be transferred, and A1A0 as 00_2, indicating that the remaining byte transfer is to take place via D31-D24. The MC68020 then outputs $14 to D31-D24 pins. The device activates DSACK1 DSACK0 as 01_2 and writes $14 to location $60172424. The MC68020 then terminates the cycle.

Note that the MC68020 outputs the 32-bit address via its A31-A0 pins and the device uses this address to write data to the selected memory location. A1 and A0 are used by the device to determine which data lines are to be used for data transfer. For example, the 16-bit device transfers data via D16-D23 for an odd memory address and it transfers data via D31-D24 for an even memory location. Therefore, SIZ1, SIZ0, A1, and A0 must be used as inputs to the address decoding logic. This is discussed later.

The data transfer for the above example takes place as follows:

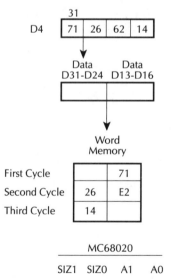

MC68020				16-Bit Memory	
SIZ1	SIZ0	A1	A0	DSACK1	DSACK0
0	0	0	1	0	1
1	1	1	0	0	1
0	1	0	0	0	1

Figure 6.31 shows a functional block diagram for the MC68020 interfaces to 8-, 16-, and 32-bit memory or I/O devices.

Aligned long-word transfers to 8-, 16-, and 32-bit devices are shown in Figure 6.32.

MC68020 byte addressing is summarized in Figure 6.33.

Figure 6.34 shows misaligned long-word transfers to 8-, 16-, and 32-bit devices.

Now, let us explain the other MC68020 pins.

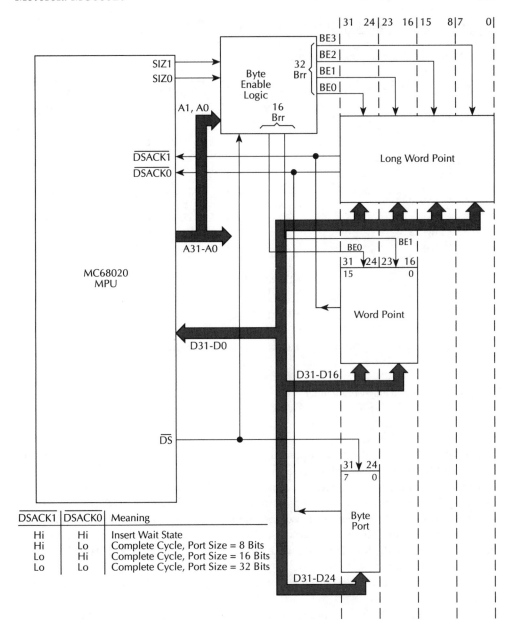

FIGURE 6.31   MC68020 dynamic bus sizing block diagram.

The ECS (external cycle start) pin is a MC68020 output pin. The MC68020 asserts this pin during the first one half clock of every bus cycle to provide the earliest indication of the start of a bus cycle. The use of ECS must be validated later with AS, since the MC68020 may start an instruction fetch cycle and then abort it if the instruction is found in the cache. In the case of a cache hit, the MC68020 does not assert AS, but provides A31-A0, SIZ1, SIZ0, and FC2-FC0 outputs.

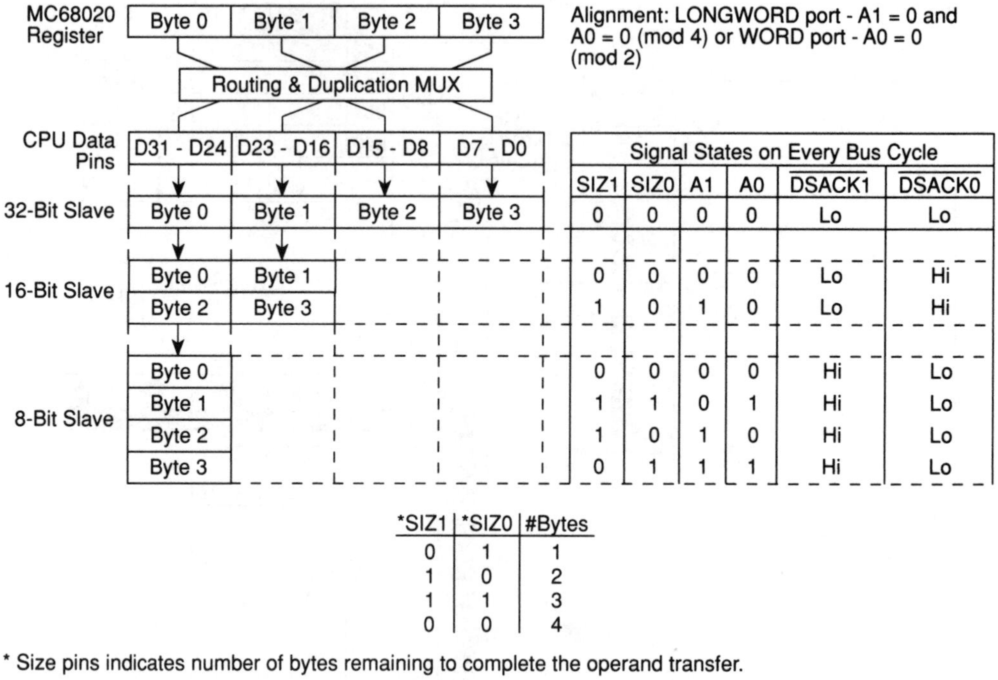

MC68020 Register: | Byte 0 | Byte 1 | Byte 2 | Byte 3 |

Alignment: LONGWORD port - A1 = 0 and A0 = 0 (mod 4) or WORD port - A0 = 0 (mod 2)

Routing & Duplication MUX

CPU Data Pins	D31 - D24	D23 - D16	D15 - D8	D7 - D0

	D31-D24	D23-D16	D15-D8	D7-D0	SIZ1	SIZ0	A1	A0	DSACK1	DSACK0
32-Bit Slave	Byte 0	Byte 1	Byte 2	Byte 3	0	0	0	0	Lo	Lo
16-Bit Slave	Byte 0	Byte 1			0	0	0	0	Lo	Hi
	Byte 2	Byte 3			1	0	1	0	Lo	Hi
8-Bit Slave	Byte 0				0	0	0	0	Hi	Lo
	Byte 1				1	1	0	1	Hi	Lo
	Byte 2				1	0	1	0	Hi	Lo
	Byte 3				0	1	1	1	Hi	Lo

(Signal States on Every Bus Cycle)

*SIZ1	*SIZ0	#Bytes
0	1	1
1	0	2
1	1	3
0	0	4

* Size pins indicates number of bytes remaining to complete the operand transfer.

FIGURE 6.32   Aligned long-word transfer.

The MC68020 asserts the OCS (operand cycle start) pin only during the first bus cycle of an operand transfer or instruction prefetch.

The MC68020 asserts the RMC (read-modify-write) pin to indicate that the current bus operation is an indivisible read-modify-write cycle. RMC should be used as a bus lock to ensure integrity of instructions which use read-modify-write operations such as TAS and CAS.

In a read cycle, the MC68020 asserts the DS (data strobe) pin to indicate that the slave device should drive the bus. During a write cycle, it indicates that the MC68020 has placed valid data on the data bus.

DEN (data buffer enable) is output by the MC68020 which may be used to enable external data buffers.

The CDIS (cache disable) input pin to the MC68020 dynamically disables the cache when asserted.

The interrupt pending (IPEND) input pin indicates that the value of the inverted IPL 2 - IPL0 pins is higher than the current I2I1I0 in SR or that a nonmaskable interrupt has been recognized.

The MC68020 AVEC input is activated by an external device to service an autovector interrupt. The AVEC has the same function as the MC68000 VPA.

The functions of the other signals such as AS, R/ W, IPL 2 - IPL0, BR, BG, and BGACK are similar to those of the MC68000.

The MC68020 system control pins are functionally similar to those of the MC68000. However, there are some minor differences. For example, for hardware reset, RESET and HALT pins need not be asserted simultaneously. Therefore, unlike the MC68000, RESET and HALT pins are not tied together in the MC68020 system.

RESET and HALT pins are bi-directional, open-drain (external pull-up resistances are required), and their functions are independent.

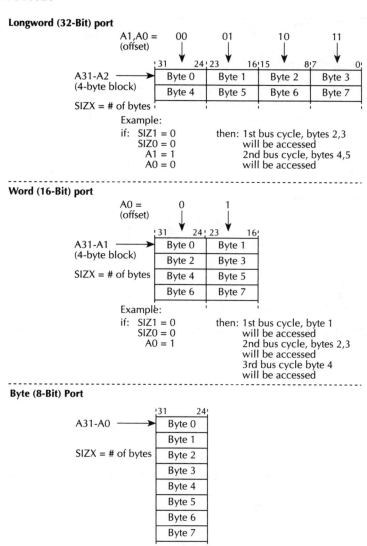

FIGURE 6.33   MC68020 byte addressing.

When $\overline{\text{HALT}}$ input is asserted by an external device, the following activities take place:

- All control signals become inactive.
- Address lines, R/W line, and function code lines remain driven with last bus cycle information.
- All bus activities stop after current bus cycle completion.

Assertion of $\overline{\text{HALT}}$ stops only external bus activities and the processor execution continues. That is, the MC68020 can continue with instruction execution internally if cache hits occur and if the external bus is not required.

The MC68020 asserts the $\overline{\text{HALT}}$ output for double bus fault. The $\overline{\text{BERR}}$ input pin, when asserted by an external device, causes the bus cycle to be aborted and strobes negated. If the $\overline{\text{BERR}}$ is asserted during operand read or write (not prefetch), exception processing occurs immediately. The $\overline{\text{BERR}}$ pin can typically be used to indicate a nonresponding device (no DSACKX received from the device), vector acquisition failure, illegal access determined by

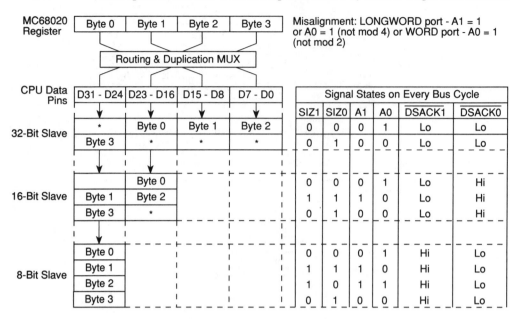

FIGURE 6.34   Misaligned long-word transfer.

memory management unit hardware such as access fault (protected memory scheme), and page fault (virtual memory system).

Figure 6.35 shows the MC68020 reset characteristics.

The RESET signal is a bi-directional signal. The RESET pin, when asserted by an external circuit for a minimum of 520 clock periods, resets the entire system including the MC68020. Upon hardware reset, the MC68020 completes any active bus cycle in an orderly manner and then performs the following:

- Reads the 32-bit contents of address $00000000 and loads it into the ISP (contents of $00000000 are loaded to the most significant byte of the ISP and so on)
- Reads the 32-bit contents of address $00000004 into the PC (contents of $00000004 to most significant byte of the PC and so on)
- Sets I2I1I0 bits of the SR to 111, sets S-bit in the SR to 1, and clears T1, T0, M bits in the SR
- Clears the VBR to $00000000
- Clears the cache enable bit in the CACR
- All other registers are unaffected by hardware reset

When the RESET instruction is executed, the MC68020 asserts the RESET pin for 512 clock cycles and the processor resets all the external devices connected to the RESET pin. Software reset does not affect any internal register.

Figure 6.36 shows a MC68020 reset circuit.

The Motorola MC3456 contains a dual timing circuit. The MC3456 uses an external resistor-capacitor network as its timing elements. Like the MC1455 timer used in the 68000 reset circuit, the MC3456 includes comparators and an R-S flip-flop. From the MC3456 data sheet, the RC values connected at the TRG input of the MC3456 will make output OP HIGH for $T = 1.1\, R_a C_a$ seconds, where R_a = resistor connected at the TSH = 1 Mohm, and C_a =

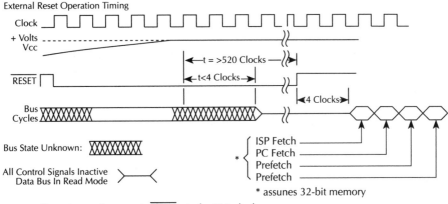

External Reset Operation Timing

Reset instruction asserts $\overline{\text{RESET}}$ pin for 512 clocks.

If reset instruction will not be executed,
Then $\overline{\text{RESET}}$ pin should be asserted for $\geq$ 10 clocks
Else $\overline{\text{RESET}}$ pin should be asserted for $\geq$ 520 clocks

Reset Flow Diagram

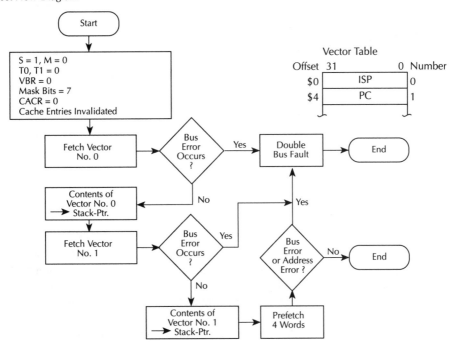

FIGURE 6.35 MC68020 reset characteristics (higher level exception).

capacitor connected at $\overline{\text{DIS}}$ = 0.47 µF. This means that OP will be HIGH for 517 msec (1.1 M *0.47 µF); the OP will then go back to LOW state. Therefore, the output of the inverter (connected at OP) will be LOW for 517 ms.

The push button connected to the input of the debouncing circuit, when not activated, will generate HIGH at the bottom input of AND gate #2 and LOW at the top of input of AND gate #1.

Since a NAND gate generates a HIGH with one of the inputs as LOW, the output of NAND gate #1 will be HIGH. This means that both inputs of NAND gate #2 will be HIGH. Therefore, the output of NAND gate #2 will be LOW. This LOW level is inverted and connected to a WIRED.OR circuit along with the inverted OP of the MC3456 at the MC68020 RESET pin.

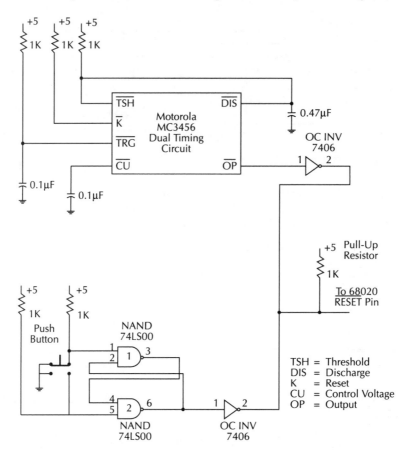

FIGURE 6.36   MC68020 reset circuit.

The output of the debouncing circuit will be HIGH when the push button is activated. For example, activation of the push button will generate a HIGH at the top input of NAND gate #1 and a LOW at the bottom input of NAND gate #2. The AND gate #2 will generate a HIGH at its output. Therefore, the inverted output will be LOW which is presented at the WIRED.-OR circuit of the 68020 RESET pin. A LOW will be provided at the 68020 RESET pin only when both inputs are LOW, that is, when the push button is activated and inverted OP of MC3456 is LOW.

The MC3456 timer will keep this $\overline{\text{RESET}}$ pin signal LOW for 517 msec. As mentioned before, the 68020 requires the RESET pin to stay LOW for at least 520 clock cycles. For a 60-ns (16-MHz) clock, the 68020 must then be LOW for at least 31.2 µs (520 * 60 ns). Since the reset circuit of Figure 6.39 outputs a LOW for 517 msec (>31.2 µs) upon activation of the push button, the 68020 RESET pin will be asserted properly.

Example 6.24

Determine the number of bus cycles, the bytes written to memory (in Hex), and signal levels of A1, A0, SIZ1, and SIZ0 pins that would occur when the instruction MOVE.L D1, (A0) with [D1] = $5012 6124 and [A0] = $2000 2053 is executed by the MC68020.

Assume   1.   32-bit memory
          2.   16-bit memory
          3.   8-bit memory

Indicate the bus cycles in which $\overline{\text{OCS}}$ is asserted.

Solution

1. 32-bit memory; misaligned transfer since starting address is odd

	A1	A0	SIZ1	SIZ0	MC68020 D31-D0 pins D31-D24	D21-D16	D15-D8	D7-D0
First bus cycle	1	1	0	0				50
Second bus cycle	0	0	1	1	12	61	24	

$\overline{\text{OCS}}$ is asserted in the first bus cycle.

2. 16-bit memory

	A1	A0	SIZ1	SIZ0	MC68020 D31-D16 pins D31-D24	D23-D16
First bus cycle	1	1	0	0		50
Second bus cycle	0	0	1	1	12	61
Third bus cycle	1	0	0	1	24	

$\overline{\text{OCS}}$ is asserted in the first bus cycle.

3. 8-bit memory

	A1	A0	SIZ1	SIZ0	MC68020 D31-D24 PINS
First bus cycle	1	1	0	0	50
Second bus cycle	0	0	1	1	12
Third bus cycle	0	1	1	0	61
Fourth bus cycle	1	0	0	1	24

$\overline{\text{OCS}}$ is asserted in the first bus cycle.

Example 6.25

Determine the contents of the PC, SR, MSP, and the ISP after a MC68020 hardware reset. Assume a 32-bit memory with the following data prior to the reset:

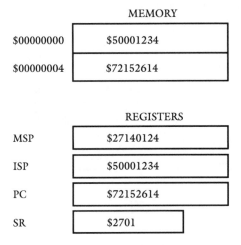

MEMORY

$00000000	$50001234
$00000004	$72152614

REGISTERS

MSP	$27140124
ISP	$61711420
PC	$35261271
SR	$0301

Solution

After hardware reset, the following are the memory and register contents:

MEMORY

$00000000	$50001234
$00000004	$72152614

REGISTERS

MSP	$27140124
ISP	$50001234
PC	$72152614
SR	$2701

```
                 15 14 13 12 11 10 9  8  7  6  5  4  3  2  1  0
Note that [SR] = $2701 = | 0| 0| 1| 0| 0| 1| 1| 1| 0| 0| 0| 0| 0| 0| 0| 1|
                          T1 T0 S  M     I2 I1 I0             X  N  Z  V  C
```

Therefore, T1T0 = 00, S = 1, M = 0, and I2I1IO = 111. Other bits are unaffected.

6.11 MC68020 Timing Diagrams

The MC68020 always activates all data lines. The MC68020 can perform either synchronous or asynchronous operation. Synchronous operation permits interfacing the devices which use the MC68020 clock to generate DSACKX and other asynchronous inputs. The asynchronous input setup and hold times must be satisfied for the assertion or negation of these inputs. The MC68020 then guarantees recognition of these signals on the current falling edge of the clock.

On the other hand, asynchronous operation provides clock frequency independence for generating DSACKX and other asynchronous inputs. This operation requires utilization of only the bus handshake signals (AS, DS, DSACKX, BERR, and HALT). In asynchronous operation, AS indicates the beginning of a bus cycle and DS validates data in a write cycle.

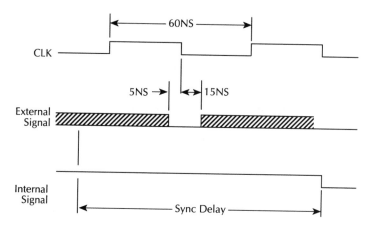

FIGURE 6.37a Synchronizing asynchronous inputs.

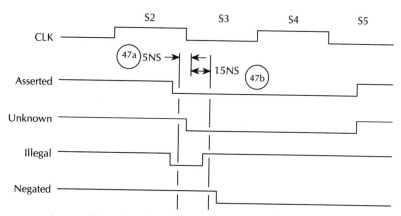

In gereral, signal level must be maintained from S2 into S5 or results will be unpredictable. An exception to this rule is bus error.

FIGURE 6.37b Asynchronous input recognition.

The SIZ1, SIZ0, A1, and A0 signals are decoded to generate strobe signals. These strobes indicate which data bytes are to be used in the transfer. The memory or I/O devices then place data on the right portion of the data bus for a read cycle or latch data in a write cycle. The selected device finally activates the DSACKX lines according to the device size to terminate the cycle. If no DSACKX is received by the MC68020, or the access is invalid, the external device can assert BERR to abort or BERR and HALT to retry the bus cycle. There is no limit on the time from assertion of AS to the assertion of DSACKX, since the MC68020 keeps inserting wait states in increments of one cycle until DSACKX is recognized by the processor.

For synchronization, the MC68020 uses a time delay to sample an external asynchronous input for high or low and then synchronizes this input to the clock.

Figures 6.37a and b show an example of synchronization and recognition of asynchronous inputs.

Note that for all inputs, there is a sample window of 20 ns during which the MC68020 latches the input level. In order to guarantee the recognition of a certain level on a particular falling edge of the clock, the input level must be held stable throughout this sample window of 20 ns. If an input changes during the sample window, the level recognized by the MC68020 is unknown or illegal. One exception to this rule is the delayed assertion of BERR where the signal must be stable through the window or the MC68020 may exhibit erratic behavior.

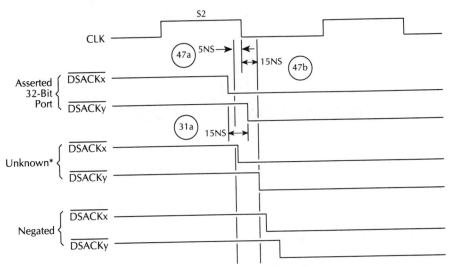

* But if asserted and if DSACKy asserted within 15NS, then 32-bit port

FIGURE 6.38a $\overline{\text{DSACKX}}$ input recognition.

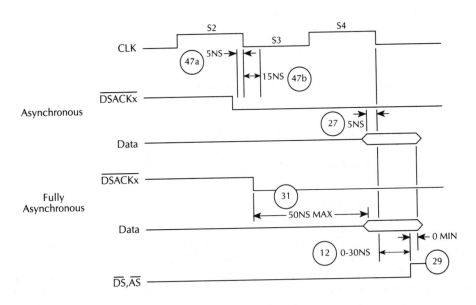

FIGURE 6.38b   Protocol for reading data.

Note that if the $\overline{\text{BERR}}$ is asserted during an instruction prefetch, the MC68020 delays bus error exception processing until the faulted data are required for execution. Bus error processing will take place for faulty access if change in program flow such as branching occurs, since the faulty data are not required. Also, after satisfying the setup and hold times, all input signals must meet certain protocols. For example, when $\overline{\text{DSACKX}}$ is asserted it must remain asserted until AS is negated. Figures 6.38a and b show timing of $\overline{\text{DSACKX}}$ input recognition and the MC68020's reading of data satisfying the required protocol.

In the timing diagrams of Motorola's 68020 manuals, parameter #47 (47a and 47b) provides the asynchronous input setup time of 20 ns. All numbers circled in the timing diagrams are the timing parameters provided in Motorola manuals.

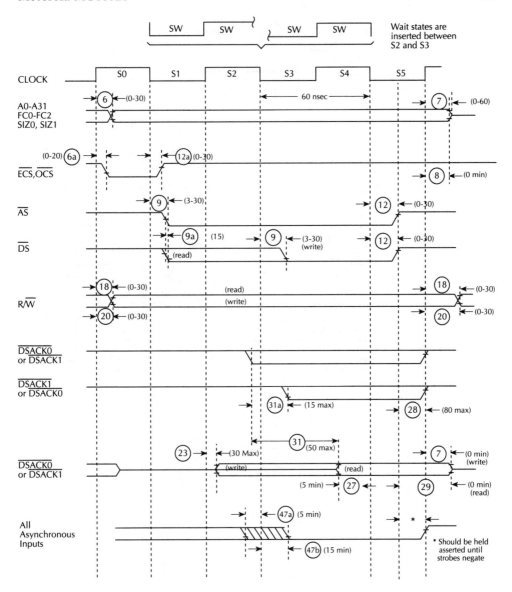

FIGURE 6.39a Asynchronous bus cycle timing. All time is in nanoseconds.

In Figure 6.38b (synchronous operation), assertion of DSACKX is recognized on the falling edge of S2; the MC68020 latches valid data on the falling edge of S4. For asynchronous operation, data are latched 50 ns (parameter 31) after assertion of DSACKX. If DSACKX or BERR is not asserted by the external device during the 20 ns window of the falling edge of S2, the 68020 inserts wait states until one of these input signals is asserted. A minimum of three clock cycles is required for a read operation. DSACKX remains asserted until AS negation is satisfied in Figure 6.39a.

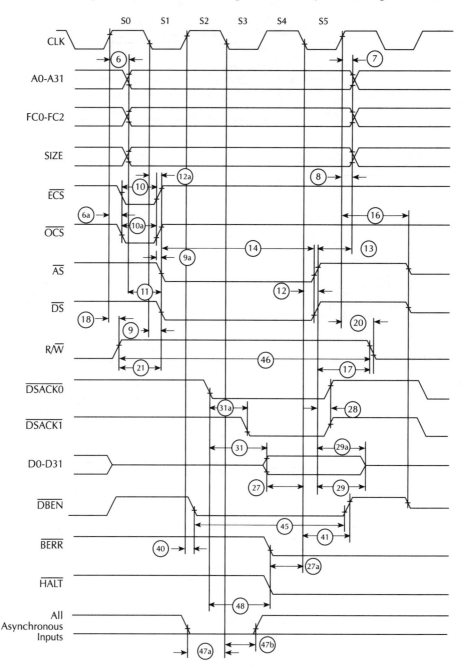

FIGURE 6.39b Read cycle timing diagram. Timing measurements are referenced to and from a low voltage of 0.8 volts and high voltage of 2.0 volts, unless otherwise noted. The voltage swing through this range should start outside and pass through the range such that the rise or fall will be linear between 0.8 and 2.0 volts.

Figure 6.39a shows asynchronous bus cycle timing along with various parameters. Figures 6.39b, c, and d show typical MC68020 read and write timing diagrams (general form) along with their AC specifications. Note that in Figures 6.39b and c signals such as SIZ0, SIZ1, DSACKX, D0-D31, A1, and A0, which precisely distinguish data transfers between 8-, 16-, and 32-bit devices, are kept in general form.

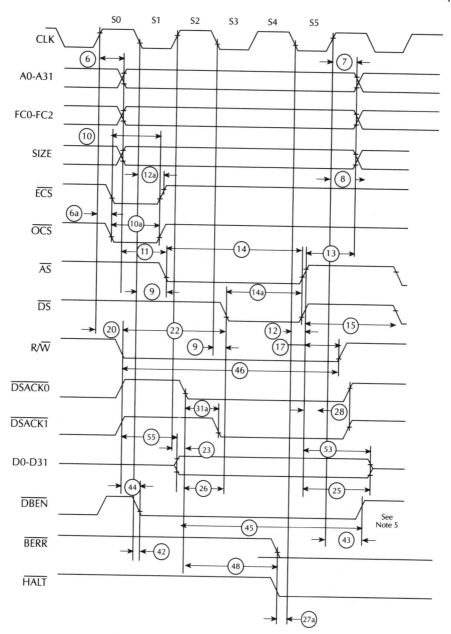

FIGURE 6.39c Write cycle timing diagram. Timing measurements are referenced to and from a low voltage of 0.8 volts and a high voltage of 2.0 volts, unless otherwise noted. The voltage swing through this range should start outside and pass through the range such that the rise or fall will be linear between 0.8 and 2.0 volts. "Note 5" refers to Figure 6.39d.

6.12 Exception Processing

The MC68020 exceptions are functionally similar to those of the MC68000 with some minor variations. The MC68020 exceptions can be generated by external or internal causes. Externally generated exceptions include interrupts, bus errors, reset, and coprocessor-detected

Num.	Characteristic	12.5 MHz		16.67 MHz		20 MHz		25 MHz		Unit
		Min	Max	Min	Max	Min	Max	Min	Max	
6	Clock High to Address/FC/Size/$\overline{RMC}$ Valid	0	40	0	30	0	25	0	25	ns
6A	Clock High to $\overline{ECS}$, $\overline{OCS}$ asserted	0	30	0	20	0	15	0	15	ns
7	Clock High to Address/Data/FC/$\overline{RMC}$/Size High Impedance	0	80	0	60	0	50	0	40	ns
8	Clock High to Address/FC/Size/$\overline{RMC}$ Invalid	0	–	0	–	0	–	0	–	ns
9	Clock Low to $\overline{AS}$, $\overline{DS}$ Asserted	3	40	3	30	3	25	3	20	ns
9A[1]	$\overline{AS}$ to $\overline{DS}$ Assertion (Read) (Skew)	-20	20	-15	15	-10	10	-10	10	ns
10	$\overline{ECS}$ Width Asserted	25	–	20	–	15	–	10	–	ns
10A	$\overline{OCS}$ Width Asserted	25	–	20	–	15	–	10	–	ns
10B[7]	$\overline{ECS}$, $\overline{OCS}$ Width Negated	20	–	15	–	10	–	5	–	ns
11[6]	Address/FC/Size/$\overline{RMC}$ Valid to $\overline{AS}$ Asserted (and $\overline{DS}$ Asserted, Read)	20	–	15	–	10	–	5	–	ns
12	Clock Low to $\overline{AS}$, $\overline{DS}$ Negated	0	40	0	30	0	25	0	20	ns
12A	Clock Low to $\overline{ECS}$, $\overline{OCS}$ Negated	0	40	0	30	0	25	0	20	ns
13	AS, DS Negated to Addess/FC/Size/$\overline{RMC}$ Invalid	20	–	15	–	10	–	5	–	ns
14	$\overline{AS}$ (and $\overline{DS}$ Read) Width Asserted	120	–	100	–	85	–	65	–	ns
14A	$\overline{DS}$ Width Asserted, Write	50	–	40	–	38	–	30	–	ns
15	$\overline{AS}$, $\overline{DS}$ Width Negated	50	–	40	–	38	–	30	–	ns
15A[8]	$\overline{DS}$ Negated to $\overline{AS}$ Assertedd	45	–	35	–	30	–	25	–	ns
16	Clock High to $\overline{AS}$/$\overline{DS}$/R/W/$\overline{DBEN}$ High Impedance	–	80	–	60	–	50	–	40	ns
17[6]	$\overline{AS}$, $\overline{DS}$ Negated to R/$\overline{W}$ High	20	–	15	–	10	–	5	–	ns
18	Clock High to R/$\overline{W}$ High	0	40	0	30	0	25	0	20	ns
20	Clock High to R/$\overline{W}$ Low	0	40	0	30	0	25	0	20	ns
21[6]	R/$\overline{W}$ High to $\overline{AS}$ Asserted	20	–		–		–		–	ns
22[6]	R/$\overline{W}$ Low to $\overline{DS}$ Asserted (Write)	90	–		–		–		–	ns
23	Clock High to Data Out Valid	–		–		–		–		ns
25[6]	$\overline{AS}$, $\overline{DS}$ Negated to Data Out Invalid	20	–	15	–	10	–	5	–	ns
25A[9]	$\overline{DS}$ Negated to $\overline{DBEN}$ Negated (Write)	20	–	15	–	10	–	5	–	ns
26[6]	Date out Valid to $\overline{DS}$ Asserted (Write)	20	–	15	–	10	–	5	–	ns
27	Data-In Invalid to Clock Low (Data Setup)	10	–	5	–	5	–	5	–	ns
27A	Late $\overline{BERR}$/$\overline{HALD}$ Asserted to Clock Low Setup Time	25	–	20	–	15	–	10	–	ns
28	$\overline{AS}$, $\overline{DS}$ Negated to $\overline{DSACKx}$/$\overline{BERR}$/$\overline{HALT}$/$\overline{AVEC}$ Negated	0	110	0	80	0	65	0	50	ns
29	$\overline{DS}$ Negated to Data-In Invalid (Date-In Hold Time)	0	–	0	–	0	–	0	–	ns
29A	$\overline{DS}$ Negated to Data-In (High Impedance)	–	80	–	60	–	50	–	40	ns
31[2]	$\overline{DSACKx}$ Asserted to Data-In Invalid	–	60	–	50	–	43	–	32	ns
31A[3]	$\overline{DSACKx}$ Asserted to $\overline{DSACKx}$ Valid ($\overline{DSACK}$ Asserted Skew)	–	20	–	15	–	10	–	10	ns

FIGURE 6.39d   Read and write cycle specifications.

errors. Internally generated exceptions are caused by certain instructions, address errors, tracing, and breakpoints. Instructions that may cause internal exceptions as part of their instruction execution are CHK, CHK2, CALLM, RTM, RTE, DIV, and all variations of the TRAP instruction. In addition, illegal instructions, privilege violations, and coprocessor violations cause exceptions. Table 6.24 lists the priority and characteristics of all MC68020 exceptions.

Num.	Characteristic		12.5 MHz		16.67 MHz		20 MHz		25 MHz		Unit
			Min	Max	Min	Max	Min	Max	Min	Max	
32	$\overline{\text{RESET}}$ Input Transition Time		–	1.5	–	1.5	–	1.5	–	1.5	Chks
33	Clock Low to $\overline{\text{BG}}$ Asserted		0	40	0	30	0	25	0	20	ns
34	Clock Low to $\overline{\text{BG}}$ Negated		0	40	0	30	0	25	0	20	ns
35	$\overline{\text{BR}}$ Asserted to $\overline{\text{BG}}$ Asserted ($\overline{\text{RMC}}$ Not Asserted)		1.5	3.5	1.5	3.5	1.5	3.5	1.5	3.5	Chks
37	$\overline{\text{BGACK}}$ Asserted to $\overline{\text{BG}}$ Negated		1.5	3.5	1.5	3.5	1.5	3.5	1.5	3.5	Chks
37A	$\overline{\text{BGACK}}$ Asserted to $\overline{\text{BR}}$ Negated		0	1.5	0	1.5	0	1.5	0	1.5	Chks
39	$\overline{\text{BG}}$ Width Negated		120	–	90	–	75	–	60	–	ns
39A	$\overline{\text{BG}}$ Width Asserted		120	–	90	–	75	–	60	–	ns
40	Clock High to $\overline{\text{DBEN}}$ Asserted (Read)		0	40	0	30	0	25	0	20	ns
41	Clock High to $\overline{\text{DBEN}}$ Negated (Read)		0	40	0	30	0	25	0	20	ns
42	Clock Low to $\overline{\text{DBEN}}$ Asserted (Write)		0	40	0	30	0	25	0	20	ns
43	Clock Low to $\overline{\text{DBEN}}$ Netgated (Write)		0	40	0	30	0	25	0	20	ns
44[6]	R/$\overline{\text{W}}$ Low to $\overline{\text{DBEN}}$ Asserted (Write)		20	–	15	–	10	–	5	–	ns
45[6]	$\overline{\text{DBEN}}$ Width Asserted	Read	80	–	60	–	50	–	40	–	ns
		Write	160		120		100		80		
46	R/$\overline{\text{W}}$ Width Asserted (Write and Read)			–		–		–		–	ns
47A	Asynchronous Input Setup Time		180	–	150	–	125	–	100	–	ns
47B	Asynchronous Input Hold Time		10	–	5	–	5	–	5	–	ns
48[4]	$\overline{\text{DSACKs}}$ Asserted to $\overline{\text{BERR}}$/$\overline{\text{HALT}}$ Asserted		–	35	–	30	–	25	–	20	ns
53	Date Out Hold from Clock High		0	–	0	–	0	–	0	–	ns
55	R/$\overline{\text{W}}$ Asserted to Data Bus Inpedance Changes		40	–	30	–	25	–	20	–	ns
56	$\overline{\text{RESET}}$ Pulse Width (Reset Instruction)		512	–	512	–	512	–	512	–	Chks
57	$\overline{\text{BERR}}$ Negated to $\overline{\text{HALT}}$ Negated (Rerun)		0	–	0	–	0	–	0	–	ns
58[10]	$\overline{\text{BGACK}}$ Negated to Bus Driven		1	–	1	–	1	–	1	–	Chks
59[10]	$\overline{\text{BG}}$ Negated to Bus Driven		1	–	1	–	1	–	1	–	Chks

Notes:

1. This number can be reduced to 5 nanoseconds if strobes have equal loads.
2. If the asynchronous setup time (#47) requirements are satisfied, the $\overline{\text{DSACKx}}$ low to data setup time (#31) and $\overline{\text{DSACKx}}$ low to $\overline{\text{BERR}}$ low setup time (#48) can be ignored. The data must only satisfy the data-in to clock low setup time (#27) for the following clock cycle. $\overline{\text{BERR}}$ must only satisfy the late $\overline{\text{BERR}}$ low to clock low setup time (#27A) for the following clock cycle.
3. This parameter specifies the maximum allowable skew between DSACK0 to DSACK1 asserted or DSACK1 to DSACK0 asserted, specification #47 must be met by DSACK0 to DSACK1.
4. In absence of $\overline{\text{DSACKx}}$, $\overline{\text{BERR}}$ is an asynchronous input using the asynchronous input setup time (#47).
5. DBEN may stay asserted on consecutive write cycles.
6. Actual value depends on the clock input waveform.
7. This is a new specification that indicates the minimum high time for $\overline{\text{ECS}}$ and $\overline{\text{OCS}}$ in the event of an internal cache hit followed immediately by a cache miss or operand cycle.
8. This is a new specification that guarantees operation with the MC68881, which specifies a minimum time for $\overline{\text{DS}}$ negated to $\overline{\text{AS}}$ asserted (specification #13A). Without this specification, incorrect interpretation of specifications #9A and #15 would indicated that the MC68020 does not meet the MC68881 requirements.
9. This is a new specification that allows a system designer to guarantee data hold times on the output side of data buffers that have output enable signals generated with $\overline{\text{DBEN}}$.
10. These are new specifications that allow system designers to guarantee that an alternate bus master has stopped driving the bus when the MC68020 regains control of the bus after an arbitration sequence.

FIGURE 6.39d (*continued*)

MC68020 exception processing is similar in concept to the MC68000 with some minor variations. In the MC68020 exception processing occurs in four steps and varies according to the cause of the exception. The four steps are summarized below:

1. During the first step, an internal copy is made of the SR, and the SR is set for exception processing. This means that the status register enters the supervisor state and tracing is disabled.

TABLE 6.24   Exception Priorities and Recognition Times

Exception priorities			Time of recognition
Group 0	.0	Reset	End of clock cycle
Group 1	.0	Address error	
	.1	Bus error	
Group 2	.0	BKPT #N, CALLM, CHK, CHK2, cp TRAPcc	Within an instruction cycle
		cp mid-instruction	
		cp protocol violation, divide-by-zero, RTE,	
		RTM, TRAP #N, TRAPV	
Group 3	.0	Illegal instruction, unimplemented LINE F,	Before instruction cycle begins
		LINE A, privilege violation, cp preinstruction	
Group 4	.0	cp post-instruction	End of instruction cycle
	.1	Trace	
	.2	Interrupt	

Note: Halt and bus arbitration are recognized at end of bus cycle. 0.0 is highest priority; 4.2 is lowest.

2. In the second step, the vector number of the exception vector is determined from either the exception requesting peripheral (nonautovector) or internally upon assertion of the AVEC (autovector) input. Note that in the MC68000, VPA is asserted for autovectoring. The vector base register points to the base of the 1-KB exception vector table which contains 256 exception vectors. The processor uses exception vectors as memory pointers to fetch the address of routines that handle the various exceptions.

3. In the third step, the processor saves PC and SR on the supervisor stack. For coprocessor exceptions, additional internal state information is saved on the stack as well.

4. The fourth step of the execution process is the same for all exceptions. The exception vector is determined by multiplying the vector number by four and adding it to the contents of the vector base register (VBR) to determine the memory address of the exception vector. The PC (and ISP for the reset exception) is loaded with the exception vector. The instruction located at the address given in the exception vector is fetched and the exception handling routine is thus executed.

Exception processing saves certain information on the top of the supervisor stack. This information is called the exception stack frame.

The MC68020 provides six different stack frames. The sizes of these frames vary from four-words to forty six words depending on the exception. For example, the normal four word stack frame is generated by interrupts, privilege violations etc. A six-word stack frame is generated by instruction-related exceptions such as CHK/CHK2 and zero divide.

The MC68020 utilizes the concept of two supervisor stacks pointed to by MSP and ISP. The M-bit (when S = 1) determines the active supervisor stack pointer. The MC68020 accesses MSP when S = 1, M = 0. The MSP can be used for program traps and other exceptions, while the ISP can be used for interrupts. The use of two supervisor stacks allows isolation of user processes or tasks and asynchronous supervisor I/O tasks.

IPL2, IPL1, IPL0, AVEC, and IPEND pins are used as the MC68020 hardware interrupt control signals (Figure 6.40). The MC68020 supports seven levels of prioritized interrupts encoded by using IPL2, IPL1, IPL0 pins like the 68000.

In Figure 6.40, when an interrupting priority level 1 through 6 is requested, the MC68020 compares the interrupt level to the interrupt mask to determine whether the interrupt should be processed. An interrupt recognized as valid does not force immediate exception processing; a valid interrupt causes IPEND to be asserted, signaling to external devices that the MC68020 has an interrupt pending. Exception processing for a pending interrupt that begins at the next instruction boundary of a higher priority exception is also not currently valid. The DESKEW logic in Figure 6.40 continuously samples the IPL2-IPL0 pins on every falling edge of the

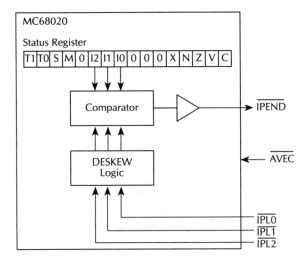

FIGURE 6.40   Interrupt control signals.

clock, but deskews or latches an interrupt request when it remains at the same level for two consecutive falling edges of the input clock. Figure 6.41 gives an example of the MC68020 interrupt deskewing logic.

Whenever the processor reaches an instruction execution boundary, it checks for a pending interrupt. If it finds one, the MC68020 begins an exception processing and executes an interrupt acknowledge cycle (IACK) with FC2 FC1 FC0 = 111_2 and A19 A18 A17 A16 = 1111_2. The MC68020 basic hardware interrupt sequence is shown in Figure 6.42a. Figure 6.42b shows the interrupt acknowledge flowchart. Before the interrupt acknowledge cycle is completed, the MC68020 must receive either AVEC, DSACKX, or BERR; otherwise it will execute wait states until one of these input pins is activated externally.

If AVEC is asserted, the MC68020 automatically obtains the vector address internally (autovectored). If the MC68020 DSACKX pins are asserted, the MC68020 takes an 8-bit vector from the appropriate data lines (D0-D7 for 32-bit device, D16-D23 for 16-bit device, and D24-D31 for 8-bit device). This is called nonautovectored interrupts and the MC68020 obtains the interrupt vector address by adding VBR with 4 * (8-bit vector).

Figure 6.43 shows an example of autovectored and nonautovectored interrupt logic. Finally, if the BERR pin is asserted, the interrupt is considered to be spurious and the MC68020 assigns the appropriate vector number for handling this.

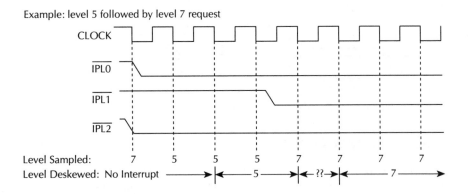

FIGURE 6.41   MC68020 interrupt deskewing logic.

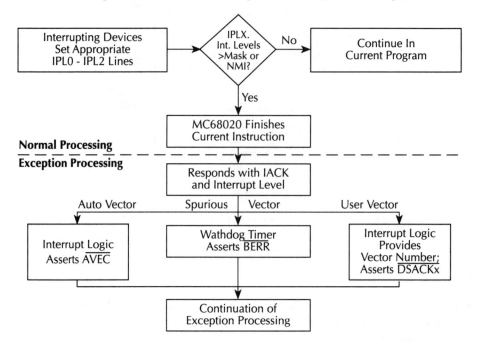

FIGURE 6.42a   MC68020 basic hardware interrupt sequence.

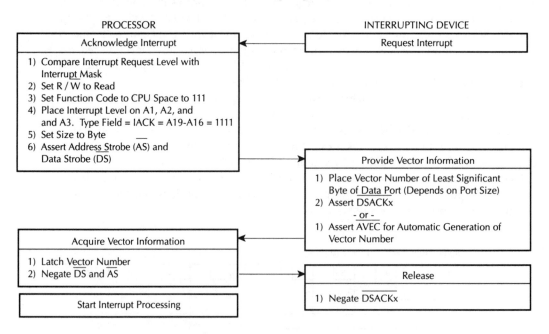

FIGURE 6.42b   Interrupt acknowledge sequence flowchart.

6.13 MC68020 System Design

The following MC68020 system design will use a 128-KB, 32-bit wide memory and 8-bit parallel I/O port. The memory system is partitioned into four unique address space encodings: user data, user program, supervisor data, and supervisor program. This design uses RAMs for memory accesses and EPROMs for program memory accesses. Each address space has 32 KB

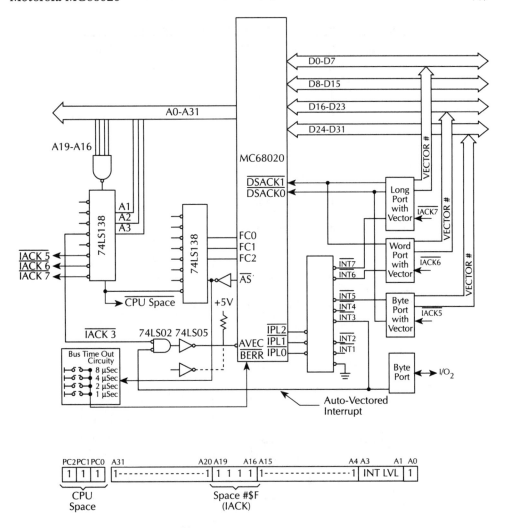

FIGURE 6.43 Autovectored and nonautovectored interrupt logic.

of memory available for use. Data I/O port space is appended to both user and supervisor data spaces (see MEMORY MAP); this is done by decoding the user/supervisor data space and address line A15.

The 32-bit-wide system memory consists of 4-byte-wide memories, each connected to its associated portion of the system data bus (D24-D31, D16-D23, D8-D15, and D0-D7). To manipulate this memory configuration, 32-bit data bus control byte enable logic is incorporated to generate byte strobes (DBBEE44, DBBEE33, DBBEE22, and DBBEE11) (Table 6.25).

The control byte enable state table shows the necessary individual byte strobe states as dictated by the MC68020's size (SIZ1, SIZ0) and address offset (A1, A0) encodings.

Karnaugh Maps (Table 6.26) for each data strobe signal have been created to identify the logic required to implement its state table requirement. The logic created for each data strobe is then combined into a complete 32-bit control logic schematic and connected to the memory structure as shown in the system hardware schematic diagram (Figure 6.44).

The system hardware design also identifies the required interconnections between the MC68020 MPU, the 74LS138 address space decoder, the EPSON SRM 20256LC (32KX8

TABLE 6.25   Control Byte Enable State Table for 32-Bit Device

SIZ1	SIZ0	A1	A0	DBBE11	DBBE22	DBBE33	DBBE44
0	1	0	0	1	0	0	0
		0	1	0	1	0	0
		1	0	0	0	1	0
		1	1	0	0	0	1
1	0	0	0	1	1	0	0
		0	1	0	1	1	0
		1	0	0	0	1	1
		1	1	0	0	0	1
1	1	0	0	1	1	1	0
		0	1	0	1	1	1
		1	0	0	0	1	1
		1	1	0	0	0	1
0	0	0	0	1	1	1	1
		0	1	0	1	1	1
		1	0	0	0	1	1
		1	1	0	0	0	1

Hardware Design
68020 System with 128K × 32-Bit Memory and 8-Bit I/O Port

MEMORY
—
MAP

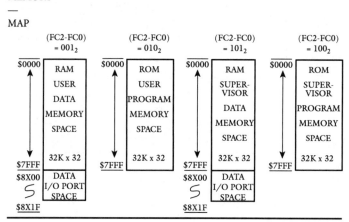

TABLE 6.26   K Maps for Strobe Signals for 32-Bit Devices

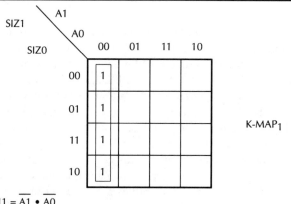

DBBE11 = $\overline{A1} \cdot \overline{A0}$

TABLE 6.26 K Maps for Strobe Signals for 32-Bit Devices (*continued*)

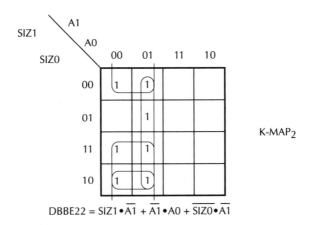

$$DBBE22 = SIZ1 \bullet \overline{A1} + \overline{A1} \bullet A0 + \overline{SIZ0} \bullet \overline{A1}$$

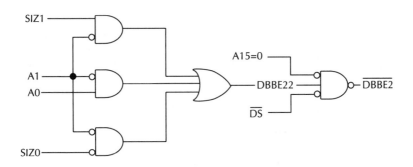

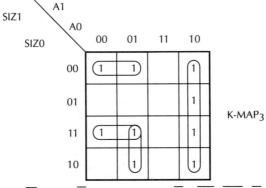

$$DBBE33 = A1 \bullet \overline{A0} + SIZ1 \bullet \overline{A1} \bullet A0 + SIZ1 \bullet SIZ0 \bullet \overline{A1} + \overline{SIZ1} \bullet \overline{SIZ0} \bullet \overline{A1}$$

TABLE 6.26   K Maps for Strobe Signals for 32-Bit Devices (*continued*)

$$DBBE44 = \overline{SIZ1} \cdot \overline{SIZ0} + A1 \cdot A0 + SIZ1 \cdot A1 + SIZ1 \cdot SIZ0 \cdot A0$$

CMOS SRAM with 120 ns access time) user/supervisor data RAMs, the 27C256 (32KX8 CHMOS EPROM with 120 ns access time user/supervisor program EPROMs, the 32-bit port control logic, and the MC68230 parallel I/O interface.

Since each memory is 32 KB × 8, only MPU address lines A2-A16 are connected. The 74LS138 selects memory banks to enable, as dictated by the decoder FC2-FC0 signals. Control logic-generated data strobes (DBBE4 - DBBE1) select which byte-wide portion of the data bus to activate. The 8-bit parallel I/O interface (MC68230) provides three bidirectional 8-bit ports as well as asynchronous handshake signals necessary for communication protocols.

The control byte enable logic diagram for generating DBBE1-DBBE4 is shown in Figure 6.45.

The enable logic uses A0, A1, SIZ0, and SIZ1 to decode $\overline{DBBE1}$ — $\overline{DBBE4}$. The memory is separated into four address space spaces: user data, user program, supervisor data, and supervisor program. The 68020 address pin A15 is used to select memory or I/O (A15 = 0 for memory, A15 =1 for I/O). The 68020 DSACK1 and DSACK0 pins for 32-bit memory (DSACK1 =0, DSACK0 =0) are asserted by $\overline{DBBE1}$ — $\overline{DBBE4}$ via appropriate logic shown in Figure 6.44. For the 68230 8-bit I/O chip, the 68020 DSACK1 and DSACK0 pins must be asserted as DSACK1 = 0 and DSACK1 = 1 for I/O transfer. Note that DSACK1 = 1 and DSACK0 insert wait states.

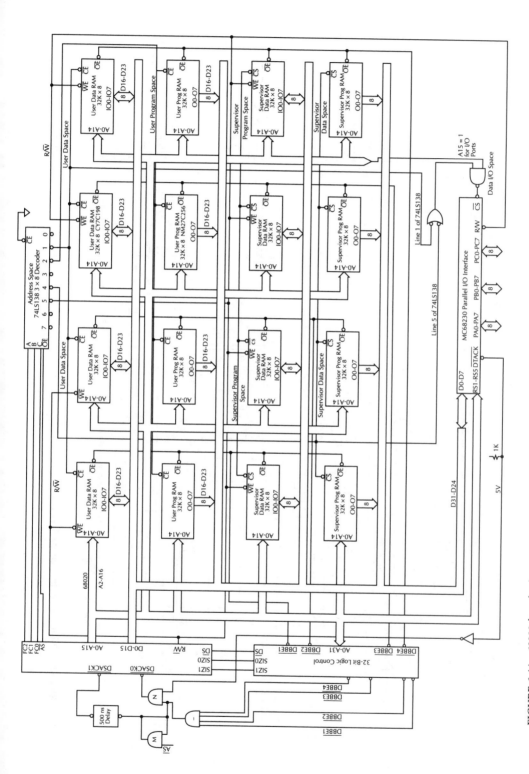

FIGURE 6.44 68020-based microcomputer.

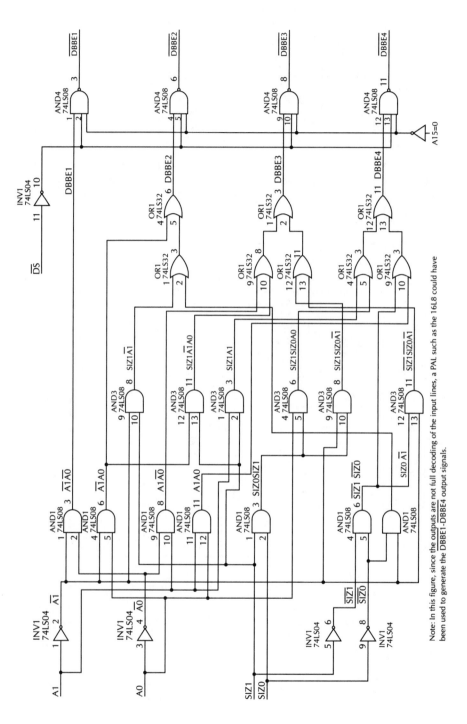

Note: In this figure, since the outputs are not full decoding of the input lines, a PAL such as the 16L8 could have been used to generate the DBBE1–DBBE4 output signals.

FIGURE 6.45   32-bit control byte enable logic.

The data enable signals ($\overline{\text{DBBE1}}$ — $\overline{\text{DBBE4}}$) are connected to the $\overline{\text{OE}}$ pin of each memory chip to ensure selection of correct memory segment. A sample window of at least 20 ns sample window is needed for DSACK1 and DSACK0; otherwise these signals will be unknown or illegal.

Each cycle of the 68000 with an 8-MHz clock is 125 ns. The 68020 samples $\overline{\text{DSACKX}}$ at the end of two clock cycles (250 ns for 8 MHz). If the 68020 DSACKX pins are asserted with at least 20 ns window, the 68020 will latch data at the end of three cycles (375 ns at the falling edge of S4).

Let us discuss the timing requirements of Figure 6.44.

The 27C256 EPROM and SRM 20256LC SRAM have access times of 120 ns each. The byte enable signals ($\overline{\text{DBBE1}}$ — DBBE4) are derived from the 68020 DS and some other signals as shown in Figure 6.45. Among these signals, DS takes the longest to go to a LOW (approximately 1.2 cycles for a read and 2 cycles for a write). This means that each memory chip is enabled by the appropriate DBBE1 — DBBE4 signals after 150 ns for read (EPROM and SRAM) and 250 ns for write (SRAM) for an 8 MHz 68020. With 120 ns access time, the write operation for the SRAMs will take the longest (375 ns). Since the 68020 latches data at the end of three cycles (375 ns for 8 MHz) without any wait states, appropriate delays for assertion of DSACKX along with at least 20 ns window must be provided. A delay of 500 ns with associated logic is included in Figure 6.44 to accomplish this. A ring counter can be used for the delay circuit.

The 68020-based microcomputer in Figure 6.44 includes 32-bit memory and 8-bit I/O. Note that DSACK1 and DSACK0 signals are asserted as 00 for 32-bit, 01 for 8-bit, and 11 to insert wait states.

Consider timing requirements for 32-bit memory in Figure 6.44. When a memory chip is selected for read or write, one or more of the enable signals ($\overline{\text{DBBE1}}$ — DBBE4) will be LOW. This will provide a LOW at the output of AND gate 1. This, in turn, will drive the DSACK0 pin of the 68020 to a LOW. The 68020 AS pin along with the output of AND gate 1 at inputs of AND gate 3 will enable the 500 ns delay circuit. This will drive the 68020 DSACK1 pin to a LOW after 500 ns providing at least 20 ns window for the 68020 DSACKX pins and appropriate timing for both EPROMs and SRAMs.

Next, consider timing requirements for the 8-bit I/O. After I/O is selected, the 68230 DSACK pin goes to a LOW and the memory enable lines ($\overline{\text{DBBE1}}$ — DBBE4) are all HIGH. The output of AND gate 1 and inverted DSACK will drive DSACK0 to a LOW. Also, the output of AND gate 1 along with the 68020 AS pin will provide a LOW at the 68020 DSACK1 pin after 500 ns indicating an 8-bit device. Thus, timing requirements for 8-bit I/O are satisfied.

QUESTIONS AND PROBLEMS

6.1 Find the contents of 68020 registers that are affected and the condition codes after execution of

    i)   **ADD.B D2, D3**
    ii) **ADD.W D5, D6**
    iii) **ADDA.L A2, A4**

Assume the following data prior to execution of each of the above instructions:

```
[D2] = $01F462F1
[D3] = $01001110
[D5] = $00008210
[D6] = $00001010
[A2] = $71240010
[A4] = $21040100
```

6.2 i) How many ALUs does the 68020 have? Comment on the purpose of each.
       ii) What is the purpose of the 68020 32-bit barrel shifter?

6.3 a) Summarize the basic differences between the 68000 and 68020.
       b) Discuss the differences between 68000 and 68020 debug capabilities implemented in their status registers. Will the instructions listed below cause trace or change of flow:

   i) **MOVE SR, D5**
   ii) **TRAPEQ START**

       when $Z = 1$?

6.4 Determine the number of bus cycles, the bytes written to memory (in Hex), and signal levels of A1, A0, SIZ1, and SIZ0 pins that would occur when the following 68020 instruction

```
MOVE.W D2, (A5) with
[D2] = $20161462 and
[A5] = $10057012
```

is executed. Assume:

   i) 16-bit memory
   ii) 8-bit memory

6.5 Show the contents of the affected 68020 registers and memory locations after execution of the instruction:

MOVE ($100, A5, D3.W *4), D1

Assume the following register and memory contents prior to execution of the above instruction:

```
[A5] = $0000F210
[D3] = $00001002
[D1] = $F125012A
[$00013318] = $4567
[$0001331A] = $2345
```

6.6 Show the contents of the affected 68020 registers and memory locations after execution of the instruction:

MOVE·B ([$102, A2, D0.W * 2],$206), D1

Assume the following register and memory contents prior to execution:

```
[A2] = $0000 0100
[D0] = $0000 0020
[D1] = $0000 0300
```

	Memory
$0000 0240	1567 0200
	0300 1500

$0200 0500	1756 1020
	2050 1F21
	1072 F217

6.7 A subroutine in the supervisor mode is required to read a parameter from the stack configuration given below:

```
                          15                0
A7' = $00002000  →  ┌─────────────────┐
     $000020002     │  ─    PC    ─    │
                    ├─────────────────┤
     $00002004      │       SR         │
                    ├─────────────────┤
     $00002006      │    PARAMETER     │
                    ├─────────────────┤
                    │       ═          │
                    │       ═          │
                    └─────────────────┘
```

Write a 68000 instruction sequence to read the parameter into D5 and then write the equivalent 68020 instruction. Assume that the (A7') and the offset of the parameter (6) are known.

6.8 The 68000 instruction sequence below searches a table of 10_{10} 32-bit elements for a match. The address register A5 points to element 0, D3 contains the length (10_{10}) to be searched, and D2 contains the number to be matched. Find the 68020 single instruction which can replace lines 3, 4, and 5.

```
Line
1                  MOVE.B  #10,  D3
2                  SUBQ.B  #1,  D3
3       START      MOVE D3, D5
4                  ASL.L  #2,  D5
5                  CMP.L  0(A5,  D5.L),  D2
6                  DBEQ D3,  START
                     -
                     -
                     -
```

6.9 Find the contents of D1, D2, A4, CCR and the memory locations after execution of the following 68020 instructions:

  i) **BFEXTS $5000 {8:16}, D4**
 ii) **BFINS D2, (A4) {D1:4}**
iii) **BFSET $5000 {D1:10}**

Assume the following data prior to execution of each of the above instructions:

$$[D1] = \$0000\ 0004 \qquad [D4] = \$0000\ 3000$$
$$[D2] = \$1234\ 5678 \qquad [A4] = \$0000\ 5000$$

	7							0
-16	0	1	1	0	1	1	1	1
-8	1	1	1	0	1	1	1	1
\$5000 →	0	0	1	0	1	0	0	1
+8	0	1	0	1	1	1	0	0
+16	1	0	1	0	1	0	1	1

6.10 Find the 68020 condition codes after execution of CMP2.W (A2), D7. Determine the range of valid values. Indicate whether the comparison is signed or unsigned. Also, indicate the register values along with upper and lower bounds on the following:

|_____|

lower bound   upper bound

Assume the following data prior to execution:

$$[D7] = \$F271\ 1020$$

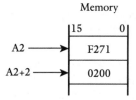

6.11 Find the 68020 condition codes and also determine if an exception occurs due to execution of

CHK2.W $5002, A1

Assume the following data prior to execution:

$$[A1] = \$0000\ F200$$

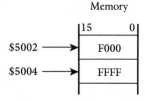

6.12 Fill in the missing hex values for the following 68020 instructions:

Address	Instruction	Label	Mnemonic
$0200 0200	$60 — —	START 0	BRA.B START2
.	.	.	
.	.	.	
.	.	.	
$0200 0206	$60 — —	START 1	BRA.W START0
.	— —	.	
.	.	.	
.	.	—	
$0200 020F	.	START 2	.
.			.
.			.

6.13 Identify the following 68020 instructions as valid or invalid. Comment if an instruction is invalid?

   i) **BFSET (A0) {-2:5}, D7**
  ii) **DIVS D5,D5**
 iii) **CHK.B D2, (A1)**

6.14 Determine the values of Z and C flags after execution of each of the following 68020 instructions:

   i) **CHK2.W (A5), D3**
  ii) **CMP2.L $2001, A5**

Assume the following data prior to execution of each of the instructions:

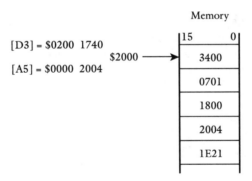

6.15 Write a 68020 assembly language program to compute

$$Y = \sum_{i=1}^{20} Xi^2 / N$$

Assume Xi's to be unsigned 32-bit numbers and the array starts at $0000 2000.

6.16 Write a 68020 assembly language program to translate each of 10 packed BCD digits to its ASCII equivalent. Assume that the BCD digits are stored at an address starting at $5000 and above. Store the ASCII bytes starting at $6000.

6.17   What are the minimum times for a read bus cycle and a write bus cycle for a 16.67-MHz 68020?

6.18   What are the functions of the 68020 VBR, CACR, and CAAR?

6.19   Determine the values of FC2, FC1, FC0, SR, MSP, ISP, PC, A31-A0, D31-D0, SIZ1, SIZ0, and R/W of the 68020 upon hardware reset for the first 3 bus cycles. Assume the following memory contents prior to reset:

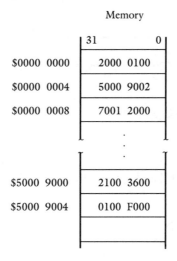

Memory

	31 0
$0000 0000	2000 0100
$0000 0004	5000 9002
$0000 0008	7001 2000
$5000 9000	2100 3600
$5000 9004	0100 F000

6.20   For a 25-MHz 68020

  i)   Determine the length of time an address is valid during assertion of $\overline{AS}$ for a read bus cycle. Assume no wait states.
  ii) Determine the length of time the data is valid when $\overline{DS}$ is asserted during a write bus cycle. Assume no states.

6.21 i)   What happens to the 68020 when $\overline{BERR}$ and $\overline{HALT}$ pins are asserted together?
  ii) Identify which of the following 68020 instructions cause the RMC signal to be asserted: CAS2, TAS, CHK2, CALLM.
  iii) Which signals cause RMC to be asserted?
  iv) What happens when 68020 IPEND is asserted?

6.22 i)   Which exceptions of the 68020 are not available on the 68000?
  ii) What is a throwaway stack frame? When is it created?
  iii) List two 68020 instructions which may cause a format error exception.
  iv) How would the 68020 get out of a double bus fault?

6.23   Draw a neat schematic to interface a 4K EPROM, 4K RAM, and 2 I/O ports. Use 2716, 6116, 68230, and a 68020. Determine memory and I/O maps.

6.24   Assume a 68020/68881 system. Write a program in 68020 assembly language to find the area of a circle, $A = \pi r^2$, where r is the 32-bit radius of the circle, in single precision.

6.25   What are the purposes of the 68020 CALLM and RTM instructions?

6.26 Determine the contents of the 68020 registers, memory locations and condition code register after execution of CAS.B D2, D4, (A0). Assume the following data prior to execution of the instruction:

$$[D0] = \$0000\ 0000$$
$$[D5] = \$AAAA\ AAAA$$
$$[D6] = \$0000\ 3AAB$$
$$[D1] = \$0000\ 0001$$

6.27 Determine the contents of the 68020 registers and memory locations after execution of CAS.W D5;D6, D0:D0, (A5); (A6). Assume following data prior to execution:

$$[D0] = \$0000\ 0000$$
$$[D5] = \$AAAA\ AAAA$$
$$[D6] = \$0000\ 3AAB$$
$$[D1] = \$0000\ 0001$$

6.28 i) Name two exception vectors for the MC68851 and the MC68881.
ii) What is the size of the 68020 on-chip cache?
iii) What is the 68020 cache access time?

6.29 What are the values of SIZ1, SIZ0, FC2, FC1, FC0, $R/\overline{W}$, and A31-A0 pins after execution of the 68020 BKPT #3 instruction?

6.30 Determine the values of FC2, FC1, FC0, and A31-A0 pins for a 68020 CPU space cycle with a floating-point coprocessor command register being accessed.

6.31 Assume a 68020/68881 system. Write a program in 68020 assembly language to find

$$\sqrt{X^2 + Y^2}$$

Assume X and Y are 32-bit integers.

6.32 Determine the effects of executing CAS.B D3,D5,(A0). Assume the following data prior to execution:

$$[D3] = \$2512\ 2551$$
$$[D5] = \$7015\ 2652$$
$$[A1] = \$5000\ 0004$$
$$[\$5000\ 0004] = \$5312\ \text{and}$$
$$[CCR] = \$0000\ 0010_2.$$

6.33 Write a 68020 assembly language instruction sequence using CAS for counting a semaphore. That is, the instruction sequence will increment a count in a shared location.

6.34 Determine the effects of executing CAS2.LD4:D5,D6:D4,(A5):(A6). Assume the following data prior to execution:

$$[D4] = \$0000\ 7000$$
$$[A5] = \$1234\ 5000$$
$$[D5] = \$0000\ 8000$$
$$[A6] = \$5000\ 0200$$
$$[D6] = \$0000\ 9000$$

6.35 Write a 68020 instruction sequence using CAS2 to insert an element in a double-linked list.

6.36 i)   What are the main functions of the 68851 MMU?
    ii) Summarize the address translation and protection mechanism of the 68851.
    iii) How does the 68851 support the 68020 breakpoint function?

6.37 For a 16-bit device, use K-maps to express the memory byte enable lines, $\overline{DBBE1}$ and $\overline{DBBE2}$ in terms of 68020 A1, A0, SIZ1, SIZ0 and $\overline{DS}$ signals in minimized form. Note that for each expression, all 68020 signals mentioned above may not be necessary.

7

MOTOROLA MC68030/MC68040, INTEL 80486 AND PENTIUM MICROPROCESSORS

This chapter describes the basic features of the Motorola 68030/68040, the Intel 80486, and the Pentium microprocessors.

7.1 Motorola MC68030

The MC68030 is a virtual memory microprocessor based on an MC68020 core with additional features. The MC68030 is designed by using HCMOS technology and can be operated at 16.67-, 20-, and 33-MHz clocks.

The MC68030 contains all features of the MC68020, plus some additional features. The basic differences between the MC68020 and MC68030 are listed below:

Characteristics	MC68020	MC68030
On-chip cache	256-byte instruction cache	256-byte instruction cache and 256-byte data cache
On-chip Memory Management Unit (MMU)	None	Paged data memory management (demand page of the MC68851)
Instruction set	101	103 (four new instructions are for the chip MMU; CALLM and RTM instructions are not support)

Like the MC68020, the MC68030 also supports 7 data types and 18 addressing modes. The MC68030 I/O is identical to the MC68020. The enhancements to the MC68020 such as instruction cache and MMU, along with the basic MC68030 features, are described in the following section.

7.1.1 MC68030 Block Diagram

Figure 7.1 shows the MC68030 block diagram and includes the major sections of the processor.

The bus controller includes all the logic for performing bus control functions and also contains the multiplexer for dynamic bus sizing. It controls data transfer between the MC68030 and memory or I/O devices at the physical address.

The control section contains the execution unit and associated logic. Programmable Logic Arrays (PLAs) are utilized for instruction decode and sequencing.

461

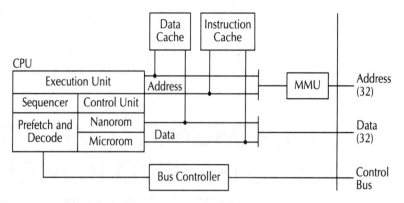

FIGURE 7.1   MC68030 block diagram.

The instruction and data cache units operate independently. They obtain information from the bus controller for future use. Each cache has its own address and data buses and thus permits simultaneous access. Both the caches are organized as 64 long word entries (256 bytes) with a block size of four long words. The data cache uses a write-through policy with no write allocation on cache misses.

The memory management unit maps address for page sizes from 256 bytes to 32K bytes. Mapping information stored in descriptors resides in translation tables in memory that are automatically searched by the MC68030 on demand. Most recently used descriptors are maintained in a 22-entry fully associative cache called the Address Translation Cache (ATC) in the MMU, permitting address translation and other MC68020 functions to occur simultaneously. The MMU utilizes the ATC to translate the logical address generated by the MC68030 into a physical address.

7.1.2 MC68030 Programming Model

Figure 7.2 shows the MC68030 programming model. The additional registers implemented in the 68030 beyond those of the 68020 are for supporting the MMU features. All common registers in the 68030 are the same as the 68020, except the cache control register which has additional control bits for the data cache and other new cache functions.

The 68030 additional registers are

- 32-bit translation control register (TC)
- 64-bit CPU root pointer (CRP)
- 64-bit supervisor root pointer (SRP)
- 32-bit transparent translation registers TT0 and TT1
- 16-bit MMU status register, MMUSR

The TC includes several fields that control address translation. These fields enable and disable address translation, enable and disable the use of SRP for the supervisor address space, and select or ignore the function codes in translating addresses. Other TC fields specify memory page sizes, the number of address bits used in translation, and the translation table structure.

The CRP holds a pointer to the root of the translation tree for the currently executing 68030 task. This tree includes the mapping information for the task's address space.

The SRP holds a pointer to the root of the translation tree for the supervisor's address space when the 68030 is configured to provide a separate address space for the supervisor programs.

Registers TT0 and TT1 can each specify separate memory blocks as directly addressable without address translation. Logical addresses in these areas are the same as physical addresses

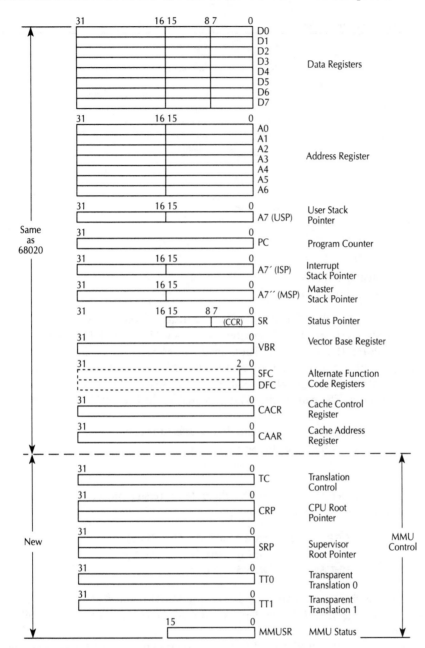

FIGURE 7.2 MC68030 programming model.

for memory access. Therefore, registers TT0 and TT1 provide fast movement of larger blocks of data in memory or I/O space, since delays associated with the translation scheme are not encountered. This feature is useful in graphics and real-time applications.

The MMUSR register includes memory management status information resulting from a search of the address translation cache or the translation tree for a particular logical address.

7.1.3 MC68030 Data Types, Addressing Modes, and Instructions

Like the MC68020, seven basic data types are supported on the MC68030:

TABLE 7.1 MC68030 New Instructions

Instruction	Operand syntax	Operand size	Operation
PFLUSH	(FC),#mask [(EA)]	None	Invalidate ATC entries at effective address
PLOAD	(FC),(EA),{R/ W }	None	Create ATC entry for effective address
PMOVE	Rn,(EA)	16, 32	Register n → destination
	(EA),Rn	16, 32	Source → register n
PTEST	(FC), (EA),#level, {R/ W }An	None	Information about logical address → PMMU status

- Bits
- Bit fields
- BCD digits
- Byte integers (8 bits)
- Word integers (16 bits)
- Long word integers (32 bits)
- Quad word integers (64 bits)

The 18 addressing modes of the MC68020 are also supported by the MC68030.

The MC68030 includes all MC68020 instructions (except CALLM and RTM), plus a subset of the MC68851 (PMMU) instructions. These instructions (Table 7.1) include PMOVE, PTEST, PLOAD, and PFLUSH, and they are compatible with the corresponding instructions in the MC68851 PMMU. The MC68020 requires the MC68851 coprocessor interface to execute these instructions. These instructions are executed by the MC68030 just like all other instructions.

All the MMU instructions are privileged and do not affect the condition codes. These new instructions are explained in the following sections.

7.1.3.a PMOVE Rn, (EA) or (EA), Rn

Rn can be any MMU register. (EA) uses control alterable addressing mode. The operand size depends on the MMU registers used as follows:

```
CRP, SRP   Quad word (64-bit)
TC, TT     Long word (32-bit)
MMUSR      Word (16-bit)
```

The PMOVE instruction moves data to and from the MMU registers. This instruction is normally used to initialize the MMU registers and to read the contents of the MMUSR for determining a fault.

As an example, consider PMOVE (A5), SRP. This instruction moves a 64-bit word pointed to by A5 to the supervisor root pointer.

7.1.3.b PTEST

The PTEST has four forms:

```
PTESTR (function code), (EA), #level
PTESTR (function code), (EA), #level,An
PTESTW (function code), (EA), #level
PTESTW (function code), (EA), #level,An
```

The PTEST instruction interrogates the MMU about a logical address and is normally used to determine the cause of a fault. The PTEST instruction executes a table search of the ATC or the translation tables to a specified level for the translation descriptor corresponding to the (EA) and indicates the results of the search in the MMU status register.

The PTESTR or PTESTW means that the search is to be done as if the bus cycle is a read or a write. The details of the operand are given below:

- The function code is specified in one of the following ways
  - Immediate (three bits in the command word)
  - Data register (three least-significant bits of the data register specified in the instruction)
  - Source function code register
  - Destination function code register
- The (EA) operand provides the address to be tested.
- The level operand defines the maximum depth of table or number of descriptors to be searched. Level 0 means searching ATC only while levels 1 to 7 indicate searching the translation tables only.

When the address register, An, is specified for a translation table search, the physical address of the last table entry (last descriptor) fetched is loaded into the address register.

The MMUSR includes the results of the search.

As an example of PTEST, consider

$$\textbf{PTESTR SFC, (A3), \#4, A5}$$

The function code for the page is in the Source Function Code register, SFC. The content of A3 is the logical address and the search is to be extended to 4 levels. Search is to be done as if for a read bus cycle and the physical address of the last entry checked is to go to A5.

The PTEST instruction is unsized and the condition codes are unaffected, but the MMUSR contents are changed.

.1.3.c PLOAD

The PLOAD instruction loads an entry into the ATC. This is normally used in demand paging systems to load the descriptors into the ATC before returning to execute the instruction that caused the page fault.

The two forms for the instruction are

$$\textbf{PLOADR (function code), (EA)}$$
$$\textbf{PLOADW (function code), (EA)}$$

The function code is specified in one of the following ways:

- Immediate
- Data register
- Source function code register
- Destination function code register

(EA) can be control alterable addressing modes only. The PLOAD instruction searches the ATC for (EA) and also searches the translation table for the descriptor with respect to (EA). It is used to load a descriptor from the translation tables to the address translation cache.

The condition codes and MMUSR contents are unaffected.

As an example, consider PLOADR SFC, (A5). The function code for the desired page is in SFC. The logical address for which the descriptor is desired is in A5 and the descriptor is to be loaded as for a read bus cycle.

7.1.3.d PFLUSH

The PFLUSH instruction invalidates cache entries. The forms of PFLUSH are

```
PFLUSHA
PFLUSH (function code), mask
PFLUSH (function code), mask, (EA)
```

The operands can be specified as follows:

- (Function code) can be immediate (3 bits), data register (least significant 3 bits), SFC, or DFC.
- mask when set to one uses corresponding bit in function code for matching.
- (EA) can be any one of the control alterable addressing modes.

The instruction is unsized and condition codes and MMUSR contents are unaffected.

The PFLUSHA instruction invalidates all ATC entries. When (function code) and mask are specified in the PFLUSH instruction, the instruction invalidates all entries for the specified function code or codes. When the PFLUSH instruction also specifies (EA), the instruction invalidates the page descriptor for that effective address entry in each specified function code.

The mask operand includes three bits corresponding to the function code bits. Each bit in the mask that is set to one means that the corresponding bit of the function code operand applies to the operation. Each bit in the mask with zero value means that a bit of function code operand and the function code that is ignored. As an example, consider PFLUSH #1,4. The mask operand of 100_2 causes the instruction to consider only the most significant bit of the function code operand. Since the function code is 001_2, function codes 000, 001, 010, and 011 are selected.

7.1.4 MC68030 Cache

The instruction cache in the 68030 is a 256-byte direct-mapped cache with 16 blocks. Each block contains four long words. Each long word can be accessed independently and thus 64 entries are possible with A1 selecting the correct word during an access. Figure 7.3 shows the 68030 instruction cache.

The index or a block (or line) is addressed by address lines A4 through A7 and each long word entry is selected by A2 and A3. The tag includes address lines A8 through A31 along with FC2.

A0 is not used since instructions must be at even addresses (A0 is always zero).

An entry means that a line hit has occurred and the valid bit (four valid bits, one for each long word) for the selected entry is set. If an entry miss occurs, the cache entry can be updated with the instructions read from memory. If the cache is disabled, no cache hits or updates will take place, and if the cache is enabled but frozen, hits can occur without updates.

CACR can be used to clear cache entries and enable or disable the caches. The system hardware can disable the on-chip caches at any time by asserting the CDIS input pin.

Figure 7.4 shows the organization of the 68030 data cache. The data cache is organized in the same way as the instruction cache except that all three function code bits are used to determine a line hit. The data cache can be updated for both read and write. If the data cache is disabled, no hits or updates will occur. However, when the data cache is enabled and frozen, hits can occur and updates for write but not read can take place. The data cache can be updated for both read and write if enabled and not frozen.

The data cache utilizes a write-through policy with no write allocation of data writes. This means that if a cache hit takes place on a write cycle, both the data cache and the external device

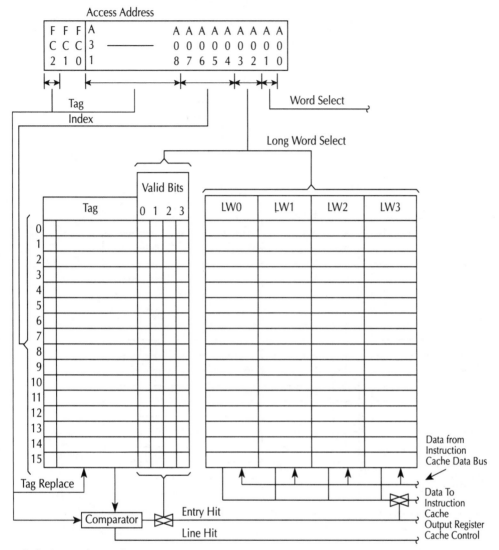

- Cache size = 64 longwords
- Line size = 4 longwords
- Set size = 1 (direct mapped)
- For an entry hit to occur
  - The access address tag field (A8-A31 and FC2) must match the tag field selected by the index field (A4-A7)
  - The selected longword entry (A2-A3) must be valid
  - The cache must be enabled in the Cache Control Register

FIGURE 7.3   MC68030 instruction cache.

are updated with the new data. If a write cycle genrates a miss in the data cache, only the external device is updated and no data cache entry is replaced or allocated for that address.

Let us now consider some examples of the 68030 cache. Consider Figure 7.5 showing an instruction cache entry hit. The figure shows the CLR.W(A1) instruction with op code 4251_{16} at PC = $11CCA29E_{16}$ to be accessed in user space (FC2 = 0). The most significant column in the tag field is the FC2 bit which is 1 for supervisor space and 0 for user space. The 68030 outputs 0 on FC2 and $11CCA29E_{16}$ on its address pins. Assume that the instruction cache is enabled. Since A3-A0 is 1110_2, the address bits A3-A2 are 11_2. This means LW3 in cache is

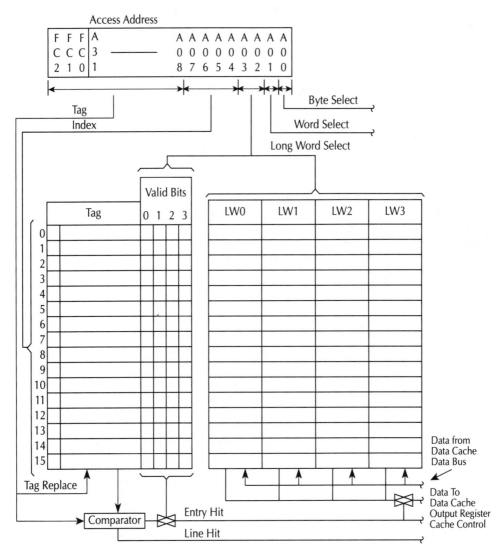

FIGURE 7.4   MC68030 data cache.

accessed. A31-A8 is $11CCA2_{16}$ which is used as the tag field with A7 through A4 (9_{10}) as the index field. A line hit occurs since the tag at index 9 in the cache matches A31 through A8 and FC2 = 0. Therefore, the valid bit 3 is set to one. The op code 4251_{16} is thus read into the cache output register.

Figure 7.6 illustrates an instruction cache miss. In this case, the op code 4251_{16} for CLR.W(A1) stored at the PC value of $276F1A64_{16}$ in supervisor space (FC2 = 1) is to be accessed. The 68030 outputs one on the FC2 pin and $276F1A64_{16}$ on the A31-A0 pins. A miss occurs since the tag in line 6 ($057CD2_{16}$) does not match the address bits A31-A8 ($276F1A_{16}$) and FC2 = 1 does not match the function code bit (0), the most significant bit in the tag field. Assume the cache is enabled and not frozen. The 68030 reads 4251_{16} into LW1 from external memory and updates

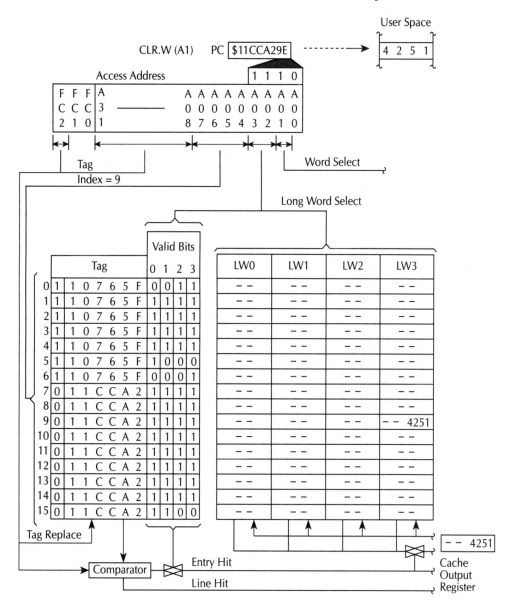

FIGURE 7.5 Instruction cache hit entry example.

the FC2 field and tag field with 1 and $276F1A_{16}$, respectively. The valid bits are updated as 0100_2.

Figure 7.7 shows an example of a data cache hit. The 68030 executes MOVE.W(A1), D0 to read data $C28F_{16}$ at address $75B4A176_{16}$ pointed to by A1 in user data space FC2 FC1 FC0 = 001. Note that the most significant column shows the value on the function code pins. Since A3A2 = 01, LW1 is accessed with tag value of $75B4A1_{16}$ and index value of 7. Assume that the cache is enabled. Index = 7 since the tag value in the cache matches A31-A8 and the function code values (FC2 FC1 FC0 = 001) match the function code field value of 1 in the cache, a cache hit occurs. The valid bit 1 is set to one and datum $C28F_{16}$ is placed in the cache output register. Figure 7.8 shows an example of data cache miss. The 68030 executes the instruction MOVE.W(A1), D0 to read the contents of $01F376B8_{16}$ pointed to by A1 into D0 in the supervisor data space (FC2 FC1 FC0 = 101_2). Since A3A2 = 10_2, LW2 in the cache is accessed. Assume the cache is enabled but

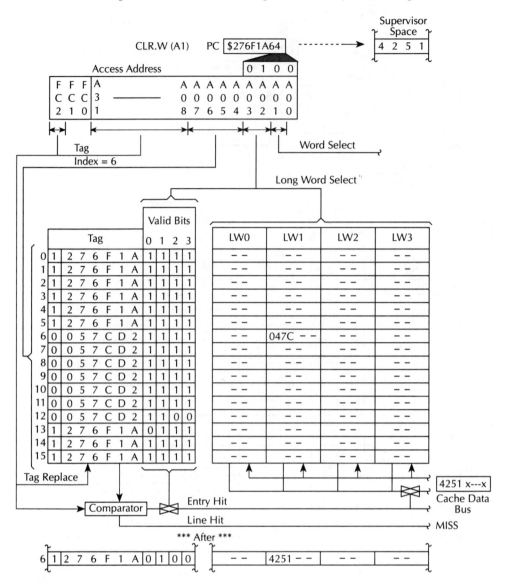

FIGURE 7.6   Instruction cache miss.

not frozen. Since the function code pins in cache do not match the 68030 function codes, cache miss occurs. Valid bit 2 is set to one. The 68030 obtains datum 1576_{16} from external memory into low word of D0 and then updates the function code and tag fields of the cache. The 68030 then invalidates LW0, LW1, LW3 and validates LW2 by writing 0010_2 in the valid bits.

7.1.5 68030 Pins and Signals

The 68030 is housed in a 13×13 PGA package for 16.67 MHz and RC Suffix Package for 20 MHz.

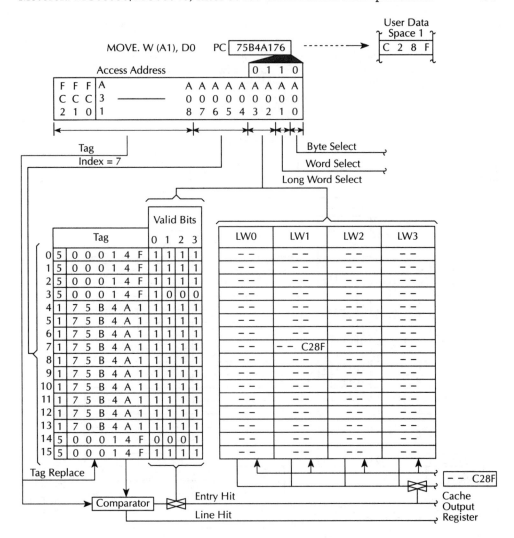

FIGURE 7.7   Data cache entry hit example.

Figure 7.9 shows the 68030 pin diagram. Figure 7.10 shows the 68030 functional signal groups. Table 7.2 summarizes the signal descriptions.

7.1.6 MC68030 Read and Write Timing Diagrams

The MC68030 provides three ways of data transfer between itself and peripherals. These are

- Asynchronous transfer
- Synchronous transfer
- BURST mode transfer

In asynchronous operation, the external devices connected to the system bus can operate at clock frequencies different from the MC68030 clock. Asynchronous operation requires only

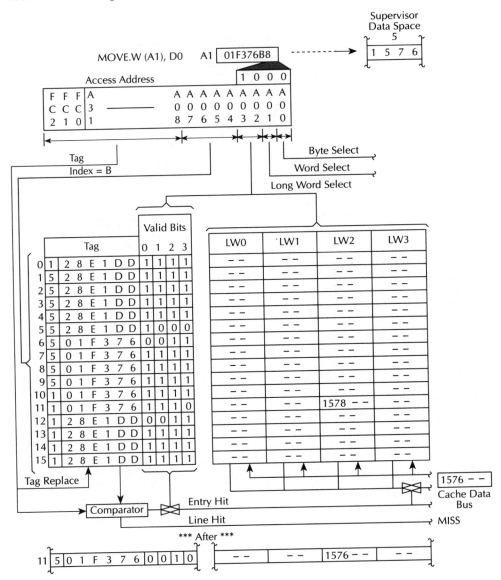

FIGURE 7.8   Data cache miss.

the handshake lines $\overline{AS}$, $\overline{DS}$, $\overline{DSACKx}$, $\overline{BERR}$ and $\overline{HALT}$. The asynchronous bus cycles of the MC68030 are similar to those of the MC68020. The MC68030 can transfer data in a minimum of three clock cycles. The dynamic bus sizing using the $\overline{DSACKx}$ signals can determine the amount of data transferred on a cycle-by-cycle basis.

Synchronous bus cycles are terminated with the $\overline{STERM}$ (synchronous termination) signal instead of $\overline{DSACKx}$ and always transfer 32-bit data in a minimum of two clock cycles instead of three clock cycles.

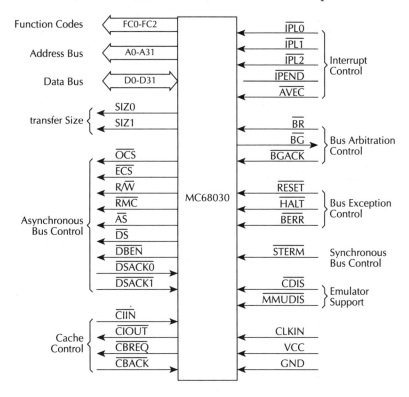

FIGURE 7.9 MC68030 functional signal groups.

The synchronous cycles terminated with STERM are for 32-bit devices only, while the synchronous cycles terminated by DSACKx can be for 8-, 16-, or 32-bit ports. The main difference between the use of STERM and DSACKx is that STERM can be asserted and data can be transferred earlier than for a synchronous cycle terminated with DSACKx. Wait cycles can be inserted by delaying the assertion of STERM if required.

BURST mode transfer is the most efficient technique of transferring data between the 68030 and external devices. This can be used to fill blocks of the instruction and data caches when the MC68030 asserts CBREQ (cache burst request).

BURST mode transfer takes place in synchronous operation and requires assertion of STERM to terminate each of its cycles. BURST mode is enabled by bits in the cache control register (CACR).

As an illustration of 68030 read/time timing diagrams, 68030 longword read and write cycles for asynchronous operation will be considered.

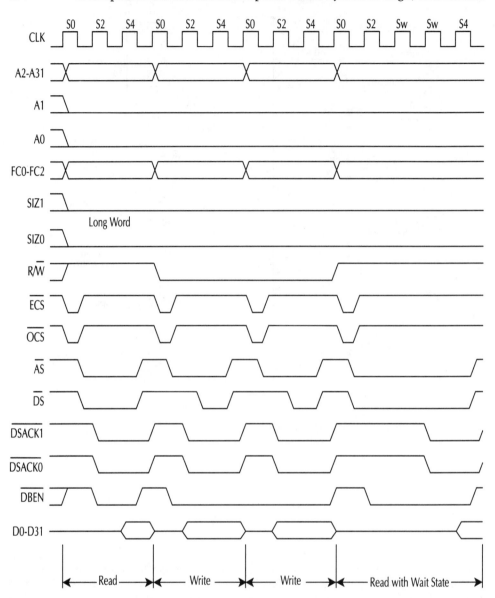

FIGURE 7.10   Asynchronous read-write cycles — 32-bit device.

Figure 7.10 depicts two write cycles (between two read cycles, with no idle time in-between) for a 32-bit device. The timing diagram in Figure 7.10 for read is explained as follows:

1. The read cycle for an instruction such as MOVE.L (A0), D1 starts in state 0 (S0). The MC68030 drives the external cycle start (ECS) pin LOW indicating the start of an external cycle. When the cycle is the first external cycle of a read operand, the MC68030 outputs low on the operand cycle start (OCS) pin. The MC68030 then places a valid address (content of register A0 in this case) on A0-A31 pins and valid function codes on FC0-FC2 pins. The MC68030 drives R/ W HIGH for read and drives data buffer enable (DBEN) inactive to disable data buffers. SIZ0 and SIZ1 signals become valid, indicating the number of bytes to be transferred. Cache inhibit out (CIOUT) becomes valid, indicating the state of the MMUCI bit in the address translation descriptor or in the selected TTx register.

TABLE 7.2 Signal Description

Signal name	Mnemonic	Function
Function codes	FC0-FC2	3-bit function code used to identify the address pace of each bus cycle
Address bus	A0-A31	32-bit address bus used to address any of 4,294,967,296 bytes
Data bus	D0-D31	32-bit data bus used to transfer 8, 16, 24, or 32 bits of data per bus cycle
Size	SIZ0/SIZ1	Indicates the number of bytes remaining to be transferred for this cycle; these signals, together with A0 and A1, define the active sections of the data bus
Operand cycle start	$\overline{\text{OCS}}$	Identical operation to that of $\overline{\text{ECS}}$ except that $\overline{\text{OCS}}$ is asserted only during the first bus cycle of an operand transfer
External cycle start	$\overline{\text{ECS}}$	Provides an indication that a bus cycle is beginning
Read/write	$\overline{\text{R}/\text{W}}$	Defines the bus transfer as an MPU read or write
Read-modify-write cycle	$\overline{\text{RMC}}$	Provides an indicator that the current bus cycle is part of an indivisible read-modify-write operation
Address strobe	$\overline{\text{AS}}$	Indicates that a valid address is on the bus
Data strobe	$\overline{\text{DS}}$	Indicates that valid data is to be placed on the data bus by an external device or has been placed on the data bus by the MC68020
Data buffer enable	$\overline{\text{DBEN}}$	Provides an enable signal for external data buffers
Data transfer and size acknowledge	$\overline{\text{DSACK0}}/\overline{\text{DSACK1}}$	Bus response signals that indicate the requested data transfer operation is completed; in addition, these two lines indicate the size of the external bus port on a cycle-by-cycle basis and are used for asynchronous transfers
Cache inhibit in	$\overline{\text{CIIN}}$	Prevents data from being loaded into the MC68030 instruction and data caches
Cache inhibit out	$\overline{\text{CIOUT}}$	Reflects the CI bit in ATC entries; indicates that external caches should ignore these accesses
Cache burst request	$\overline{\text{CBREQ}}$	Indicates a burst request for the instruction or data cache
Cache burst acknowledge	$\overline{\text{CBACK}}$	Indicates that accessed device can operate in burst mode
Interrupt priority level	IPL0 - IPL2	Provides an encoded interrupt level to the processor
Interrupt pending	$\overline{\text{IPEND}}$	Indicates that an interrupt is pending
Autovector	$\overline{\text{AVEC}}$	Requests an autovector during an interrupt acknowledge cycle
Bus request	$\overline{\text{BR}}$	Indicates that an external device requires bus mastership
Bus grant	$\overline{\text{BG}}$	Indicates that an external device may assume bus mastership
Bus grant acknowledge	$\overline{\text{BGACK}}$	Indicates that an external device has assumed bus mastership
Reset	$\overline{\text{RESET}}$	System reset; same as 68020
Halt	$\overline{\text{HALT}}$	Indicates that the processor should suspend bus activity
Bus error	$\overline{\text{BERR}}$	Indicates an invalid or illegal bus operation is being attempted
Synchronous termination	$\overline{\text{STERM}}$	Bus response signal that indicates a port size of 32 bits and that data may be latched on the next falling clock edge
Cache disable	$\overline{\text{CDIS}}$	Dynamically disables the on-chip cache to assist emulator support
MMU disable	$\overline{\text{MMUDIS}}$	Dynamically disables the translation mechanism of the MMU
Clock	CLK	Clock input to the processor
Power supply	Vcc	+5 volt ± 5% power supply
Ground	GND	Ground connection

2. During S1, the MC68030 activates $\overline{\text{AS}}$ LOW, indicating a valid address on A0-A31. The MC68030 then activates $\overline{\text{DS}}$ LOW and negates $\overline{\text{ECS}}$ and $\overline{\text{OCS}}$ (if asserted).

3. During S2, the MC68030 activates the $\overline{\text{DBEN}}$ pin to enable external data buffers. The selected device such as a memory chip utilizes $\overline{\text{DS}}$, SIZ0, SIZ1, $\overline{\text{R}/\text{W}}$, and $\overline{\text{CIOUT}}$ to place data on the data bus. The MC68030 outputs LOW on $\overline{\text{CIIN}}$ if appropriate. Any or all bytes on D0-D31 are selected by SIZ0, SIZ1, A0, and A1. At the same time, the selected device asserts the $\overline{\text{DSACKx}}$ signals. The MC68030 samples $\overline{\text{DSACKx}}$ at the falling edge of the S2 and if $\overline{\text{DSACKx}}$ is recognized, the MC68030 inserts no wait states.

As long as at least one of the $\overline{DSACKx}$ inputs is asserted by the end of S2 (satisfying the asynchronous setup time requirement), the MC68030 latches data on the next falling edge of the clock and ends the cycle.

If DSACKx is not recognized by the MC68030 by the beginning of S3, the MC68030 inserts wait states and DSACKx must remain HIGH throughout the asynchronous input setup and hold times around the end of the S2. During the MC68030's wait states, the MC68030 continually samples DSACKx on the falling edge of each of the subsequent cycles until one DSACKx is asserted.

4. With no wait states, at the end of S4, the MC68030 latches data.
5. During S5, the MC68030 negates AS, DS, and DBEN. The MC68030 keeps the address valid during S5 to provide address hold time for memory systems. R/W, SIZ0, and SIZ1, and FC0-FC2 also remain valid during S5.

The timing diagram in Figure 7.10 for write is explained as follows:

1. The MC68030 outputs LOW on both $\overline{ECS}$ and $\overline{OCS}$ and places a valid address and function codes on A0-A31 and FC2-FC0, respectively. The MC68030 also places LOW on R/W and validates SIZ0, SIZ1, and $\overline{CIOUT}$.
2. In S1, the MC68030 asserts AS and DBEN and negates $\overline{ECS}$ and $\overline{OCS}$ (if asserted).
3. During S2, the MC68030 outputs data to be written on D0-D31 and samples DSACKx at the end of S2.
4. During S3, the MC68030 outputs LOW on $\overline{DS}$, indicating that the data are stable on the data bus. As long as at least one of the $\overline{DSACKx}$ inputs is recognized by the end of S2, the cycle terminates after one cycle. If DSACKx is not recognized by the start of S3, the MC68030 inserts wait states. If wait states are inserted, the MC68030 continues to sample the DSACKx signals on the subsequent falling edges of the clock, until one of the DSACKx inputs is recognized. The selected device such as memory utilizes R/W, DS, SIZ0, SIZ1, and A0 and A1 to latch data from the appropriate portion of D0-D31 pins. The MC68030 generates no new control signals during S4.
5. During S5, the MC68030 negates AS and DS. The MC68030 holds the address and data valid to provide address hold time for memory systems. The processor also keeps R/W, SIZ0, SIZ1, FC0-FC2, and DBEN valid during S5.

7.1.7 MC68030 On-Chip Memory Management Unit

7.1.7.a MMU Basics

A Memory Management Unit (MMU) translates addresses from the microprocessor (logical) to physical addresses. Logical addresses are assigned to the task when it is linked, while physical addresses are assigned at the time the task is loaded into memory based on free physical addresses. The MMU keeps a task in its own address space. If a task attempts to go out of its own address space, the MMU asserts a bus error. Besides protection, this feature is valuable in demand paging systems.

An operating system must know the free physical memory addresses available so that it can load the next task to these spaces. One way of accomplishing this is by dividing the memory into contiguous blocks of equal size. These are called pages on the logical side of the MMU and page frames on the physical side. The system memory will contain page descriptions which point to the page frames and status of the page frames.

The page size determines the number of address bits to be translated by the MMU and the number which address memory directly. The lower bits of the logical address related to the page size are not translated by the MMU and go directly to the physical address bus as shown in Figure 7.11. Note that A8-A3 bits provide the page number and A0-A7 bits define the 256-byte page offset in this case.

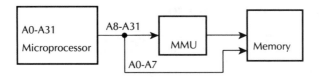

FIGURE 7.11 Logical to physical translation.

A translation table can be used to translate pages to page frames. Figure 7.12 shows an example of table translation of pages. The task control block contains a pointer to the starting address of the translation table. A logical address is translated by accessing the location in the translation table determined by starting address of translation + (page number * entry size). The entry accessed contains the page frame number. Note that entry size is 4 for 32 bits (4 bytes).

Each task has a translation table. Single or multiple translation tables can be used. Figure 7.13 shows an example of a single pointer level translation. The TCB contents $00010000 contain the starting address of the translation table. The 12-bit page offset ($C5E for P2) replaces the low 12 bits of the physical address directly. The logical address for P2 is translated by accessing the translation table at the location

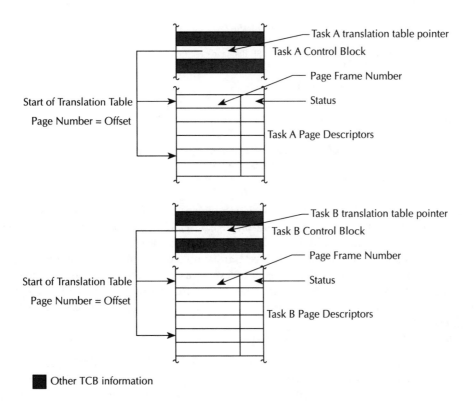

FIGURE 7.12 Table translation of pages.

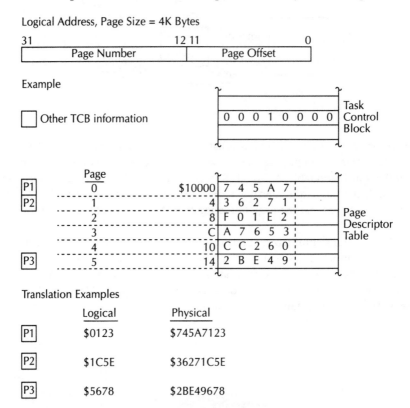

FIGURE 7.13   Single-pointer level translation.

```
= $10000 + (Page number * entry size)
= $10000 + ($00001 * 4)
= $100004
```

Therefore, location $100004 in the translation table is accessed and its content $36271 is obtained as the page frame number. The 12-bit page offset $C5E is concatenated with the page frame number to obtain the physical address in page P2 as $36271C5E.

One of the disadvantages of the single-level translation is that for most systems, the required table sizes would be too large. Therefore, multilevel translation tables are used. For example, consider the double pointer level translation table of Figure 7.14. In this case, the TCB contains a pointer to the starting address of the level 1 translation table. Level 1 descriptors contain pointers to level 2 page descriptors.

A logical address is translated in two steps:

1. A pointer is obtained from level 1 translation table as follows: pointer from TCB = starting address of level 1 translation table + (level 1 * entry size of level 1 table).
2. This pointer is used to access a location in level 2 table by adding the pointer obtained from level 1 table with (level 2 * entry size of level 2 table).

The location in table 2 thus obtained contains the page frame number. Consider the logical address:

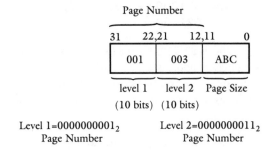

If TCB contents ($10000) point to the starting address of table 1, the level 1 location

```
= $10000 + level 1 * entry size of level 1
= $10000 + 1 * 4
= $10004
```

Therefore, if the content of $10004 is $21000 pointing to the starting address of the level 2 table, then the accessed location in table 2

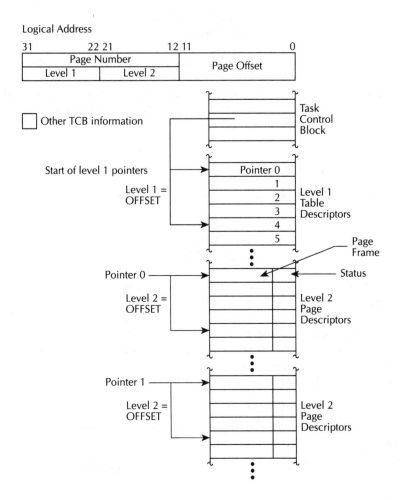

FIGURE 7.14 Double pointer level translation.

$$= \$21000 + (3_{10} * 4)$$
$$= \$2100C$$

Assume the content of $2100C is $25719. Then $25719 is concatenated with the page offset $ABC to obtain the 32-bit physical address $25719ABC.

An example of the double-pointer level translation is shown at the top of the next page. Assume a page size of 4K bytes. Let us translate the following logical addresses to physical addresses for the task shown above:

$0000072A and $00401A59

Consider the logical address $0000072A. This address has level 1 and level 2 page numbers of 0. This means that level 1 location

$$= \$00780000 + 0 * 4$$
$$= \$00780000$$

Therefore, [$00780000] points to the starting address of the level 2 table; then the accessed location in table 2

$$= \$00800000 + 0 * 4$$
$$= \$00800000$$

The content of location $00800000 ($176A5) is concatenated with the page offset $72A so that the physical address is $176A572A.

Next, consider the logical address $00401A59. The uppermost 10 bits ($0000\ 000001_2$) define the level 1 page number as one. The next 10 bits ($0000\ 000001_2$) define the level 2 page number as one. The page size is $A59. Level 1 location

$$= \$7800\ 0000 + (1 * 4)$$
$$= \$7800\ 0004$$

The content of location $7800 0004 ($00801000) points to the starting address of level 2 table. Level 2 location

$$= \$00801000 + 1 * 4$$
$$= \$0080\ 1004$$

The content of location $0080 1004 ($91024) is concatenated with the page offset $A59 to obtain the physical address $91024 A59.

The main advantage of the multiple translation table is that it reduces the required translation table size significantly. However, multiple memory accesses (two in the example of two-level translation) are required to obtain a descriptor. However, an address translation cache can be used to speed up memory access.

An MMU provides three basic functions:

- Translates page number to page frame number
- Restricts access, i.e., a task is restricted to its own address space.
- Provides write protection by not allowing write access to write-protected pages

Status information is included in the low bits of the translation table entry data to provide protection information.

7.1.7.b 68030 On-chip MMU

Figure 7.15 provides a block diagram of of the MC68030 and identifies the on-chip MMU.

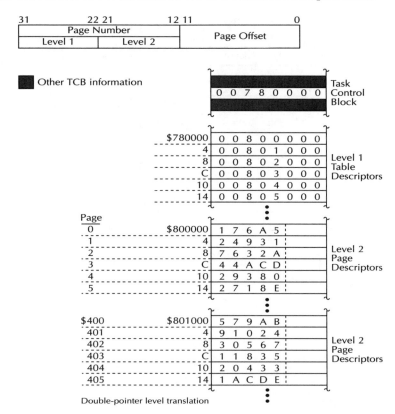

The 68030 on-chip MMU translates logical addresses to physical addresses. If the desired information is in the address translation cache, no time delay occurs due to MMU address translation. The page descriptor is obtained by the MMU by searching the translation tables if needed.

The pins used by the 68030 on-chip MMU are A31-A0, $\overline{\text{CIOUT}}$, and $\overline{\text{MMUDIS}}$.

The MMU outputs the translated physical addresses (from logical addresses) or the addresses for <u>fetching</u> translation data (when a table is searched) on the A31-A0 pins. The MMU asserts the $\overline{\text{CIOUT}}$ pin for pages which are defined as noncacheable. The $\overline{\text{MMUDIS}}$ pin, when asserted by an external emulator, disables the MMU.

The 68030 executes a normal bus cycle when the address translation information is in the ATC. However, if the address translation information is not in the ATC, the 68030 executes additional bus cycles to obtain the desired address translation information.

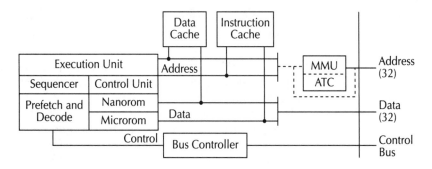

FIGURE 7.15   MC68030 on-chip MMU block diagram.

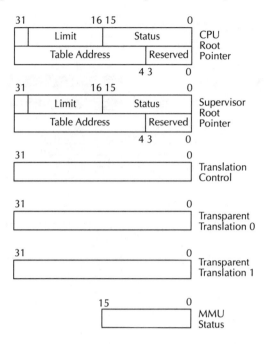

FIGURE 7.16   MC68030 MMU registers.

The 68030 includes three main elements in its MMU: a set of registers, an ATC, and table search logic.

Figure 7.16 shows the MMU registers.

The CRP or SRP points to the beginning of the task translation table and supervisor translation table (if enabled), respectively.

The translation control register controls the MMU functions such as MMU enable.

The transparent translation registers output specified logical addresses on the A31-A0 pins.

The MMUSR provides the results of execution of the MMU instructions PMOVE (EA), MMUSR, and PTEST. The MMUSR stores memory management status information resulting from a search of the address translation cache, or the translation tree, for a particular logical address.

As mentioned before, four MMU instructions — PMOVE, PTEST, PLOAD, and PFLUSH — are included in the 68030 instruction set. The MMU provides up to 5 levels of address translation tables. Address translation starts with the contents of the root pointers CRP or SRP.

Figure 7.17 provides a typical 5-level table translation scheme.

Figure 7.18 provides the ATC block diagram along with a simplified flowchart for the physical address translation.

The ATC is a content-addressable, fully associative cache with up to 22 descriptors. The ATC stores recently used descriptors so that a table search is not required if future accesses are required.

There are six types of descriptors used in MMU operation. These are ATC page descriptors, table descriptors, early termination page descriptors, invalid descriptors, and indirect descriptors. Some descriptors have a long and a short form.

Figure 7.19 shows the page descriptor summary.

Each ATC entry consists of a logical address and information from a corresponding page descriptor. The 28-bit logical or tag portion of each entry consists of three fields. These are the

Up to 5 level translation tables

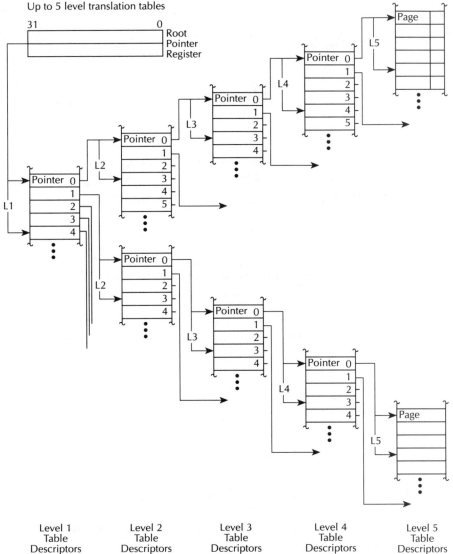

FIGURE 7.17 MC68030 MMU five-level translation scheme.

valid bit field (bit 27), function code field (bits 26-24), and 24-bit logical address field (bits 0-23). The V-bit indicates that the entry is valid if V = 1. The function code field includes the function code bits (FC0-FC2) corresponding to the logical address in this entry. The 24-bit logical address includes the most significant logical address bits for this entry. All 24 bits are used in comparing this entry to an incoming logical address when the page size is 256 bytes. For larger page sizes, some least significant bits of this field are ignored.

Table descriptors have short (32-bit) and long (64-bit) forms. In the 32-bit short form, the table descriptor includes four status bits (bits 0-3) and a 28-bit table address. The status bit provides information such as write protection and descriptor type.

The table address contains the 28-bit physical base address of a table of descriptors. The long table descriptor includes a 28-bit table address, a 16-bit status, and a 16-bit limit field. The status bits in the long form include additional information such as whether the table is a

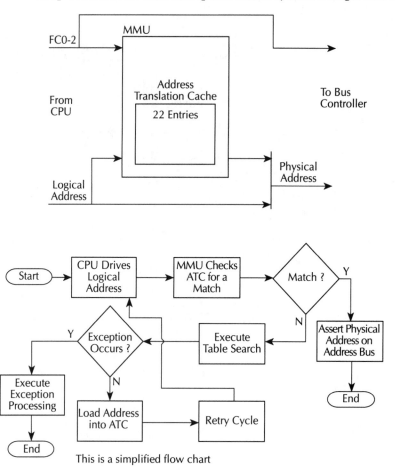

FIGURE 7.18 Address translation cache (ATC) and flowchart for translation.

supervisor-only table. The limit field includes a limit to which the index portion of an address is compared to detect an out-of-bounds index.

The page descriptors also have short and long formats. The short page format is identical to the short table format except that for page tables, page table and status information related to pages are used. The long page descriptor contains a 24-bit page address and a 16-bit status. The status bits provide information such as write protect, descriptor type, and identification of a modified page.

The early termination page descriptor (both short and long form) includes the page descriptor identification bits in the descriptor-type field of the status bits, but the descriptor resides in a pointer table. This means that the table in which an early termination page descriptor is located is not at the bottom level of the address translation tree. The invalid descriptors can also be short and long forms. These descriptors only contain the two descriptor-type status bits and are used with long or short page and table descriptors.

The indirect descriptors also include short and long forms. They contain the physical address of a page descriptor and two descriptor-type bits which identify an indirect descriptor that points to a short or long format page descriptor.

Table 7.3 summarizes differences between 68030 on-chip MMU and 68851.

Now let us discuss the details of the 68030 MMU registers. The CRP points to the first level translation table. It is normally loaded during a context (task) switch. The ATC is automatically flushed when the CRP is loaded. Figure 7.20 shows the details of the CRP.

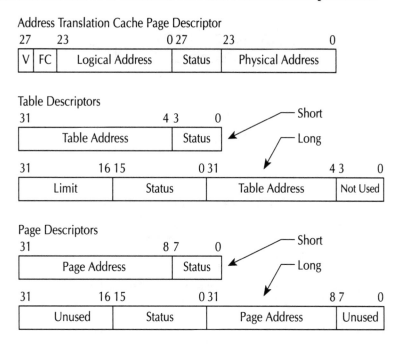

Address Translation Cache Page Descriptor

Table Descriptors

Page Descriptors

• Early Termination Page Descriptors
• Invalid Descriptors
• Indirect Descriptors

FIGURE 7.19   Page descriptor summary.

The limit field in the CRP limits the size of the next table. That is, it limits the size of the table index field. If the limit is exceeded during a table search, an ATC entry is created with bus error set. Two fields are assigned to the limit. These are L / U and limit. L / U = 1 means a lower range limit, while L / U = 0 indicates an upper range limit. If L / U = 0 and limit = $7FFF or L / U = 1 and limit = 0, the limit function is suppressed. The translation control register (TC) permits the user to control the translation of the access. TC is also used to enable and disable the main function of the MMU. Figure 7.21 shows the details of TC.

The SRP is similar to the CRP except that it is used only for supervisor access. This register must be enabled in the TC before use. The format of SRP is shown in Figure 7.22.

TABLE 7.3   68030 On-Chip MMU vs. 68851

Features of the 68851 not included on the 68030 MMU:
    No access level (no CALLM or RTM instructions)
    No breakpoint registers
    No DMA root pointer
    No task aliasing
    22 entry ATC instead of 64
    No lockable content-addressable memory (CAM) entries
    No shared globally entries
    Instructions supported
        PFLUSHA, PFLUSH, PMOVE, PTEST, PLOAD
    Instructions not supported (F-line trap)
        PVALID, PFLUSHR, PFLUSHES, PBcc, PDBcc, PScc,
        PTRAPcc, PSAVE, PRESTORE
    Control alterable effective addresses only for the
        MMU instructions

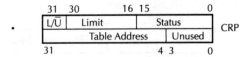

• Status Field

• DT Values

00   Not allowed. If loaded with this value, an MMU Configuration Error exception occurs.

01   Table address field is a page descriptor.

10   Table address field points to descriptors which are 4 bytes long.

11   Table address field points to descriptors which are 8 bytes long.

FIGURE 7.20   CRP details.

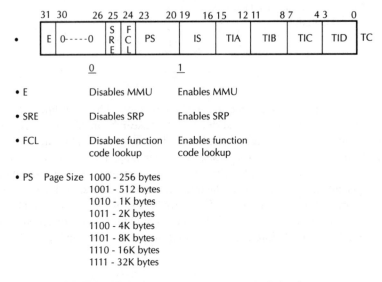

FIGURE 7.21   TC details.

FIGURE 7.22   SRP form.

| 31 | 24 | 23 | | 16 | 15 | 14 | 11 | 10 | 9 | 8 | 7 | 6 | 4 | 3 | 2 | 0 |

• | LOGICAL ADDRESS BASE | LOGICAL ADDRESS MASK | E | 0---0 | CI | R/W̄ | RWM | 0 | FC BASE | 0 | FC MASK |

	0	1
• E	Register disabled	Register enabled
• R/W̄	Match writes	Match reads
• RWM	Force matching of the R/W̄ bit	Don't - care the R/W̄ bit
• CI	Cache the data	Don't cache the data; assert CIOUT
• FC BASE	Function code value to be matched	
• FC MASK	0 - force caring of this bit	
	1 - don't care this bit	

• LOGICAL ADDRESS BASE
Base field of logical addresses to be transparently translated.
A24-A31 must be matched.

• LOGICAL ADDRESS MASK
Allows don't-caring of bits in the Logical Address Base
0 - force caring of this bit
1 - don't - care this bit

FIGURE 7.23   TT0 and TT1 details.

If a task is required to access physical memory directly, such as in graphics applications TT0 or TT1 can be used to specify the physical areas of memory to be accessed.

Figure 7.23 provides formats for TT0 and TT1.

The MMUSR includes the results of execution of the PTEST instruction for level = 0 or level ≠ 0.

Figure 7.24 provides the MMUSR details.

The ATC entries include two (logical and physical) 28-bit fields. The logical field includes a valid bit, function codes, and page number. The physical field contains four control/status bits and page frame number. Figure 7.25 includes the ATC entry structure.

The translation tables supported by the 68030 contain a tree structure. The root of a translation table tree is pointed to by one or two root pointer registers.

Table entries at higher levels of the tree contain pointers to other tables, while entries at the leaf level (page tables) contain page descriptors. The technique used for table searches utilizes portions of the logical address as an index for each level of the lookup. All addresses contained in the translation table entries are physical addresses.

Figure 7.26 shows the 68030 MMU translation table tree structure.

The function codes are usually used as an index in the first level of lookup in the table. However, this may be suppressed. In table searching, up to 15 of the logical address lines can be ignored. The number of levels in the table indexed by the logical address can be set from one to four and up to 15 logical address bits can be used as an index at each level. One main advantage of this tree structure is to deallocate large portions of the logical address space with a single entry at the higher levels of the tree.

The entries in the translation tables include status information with respect to the pointer for the next level of lookup or the pages themselves. These bits can be used to designate certain pages or blocks of pages as supervisor-only, write-protected, or noncacheable. The 68030 MMU exceptions include the following:

• The information in the MMU status register is the result of execution of the PTEST instruction.

- Level = 0　Search the ATC only
- Level ≠ 0　Search translation tables only

15	14	13	12	11	10	9	8	7	6	5	4	3			0
B	L	S	0	W	I	M	0	0	T	0	0	0		N	

Bit	Meaning	Level = 0	Level ≠ 0
B	Bus error	Bus error is set in Matching ATC entry	Bus error occured during table walk
L	Limit bit	Always cleared	An index exceeded the limit
S	Supervisor violation	Always cleared	Supervisor bit is set in descriptor
W	Write protect	Address is write protected	Address is write protected
I	Invalid bit	No entry in ATC or B is set	No translation in table, or B or L is set
M	Modified bit	ATC entry has M bit set	Page descriptor has M Bit set
T	Transparent	Address is within range of TT0 or TT1	Always cleared
N	Number of tables	0	Number of tables used in translation

FIGURE 7.24　MMUSR details.

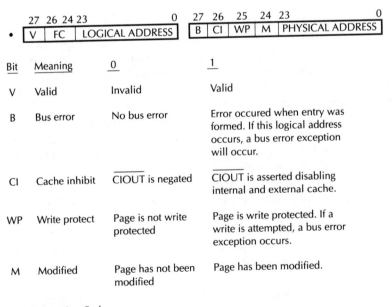

27	26	24	23		0		27	26	25	24	23		0
V	FC		LOGICAL ADDRESS				B	CI	WP	M	PHYSICAL ADDRESS		

Bit	Meaning	0	1
V	Valid	Invalid	Valid
B	Bus error	No bus error	Error occured when entry was formed. If this logical address occurs, a bus error exception will occur.
CI	Cache inhibit	$\overline{CIOUT}$ is negated	$\overline{CIOUT}$ is asserted disabling internal and external cache.
WP	Write protect	Page is not write protected	Page is write protected. If a write is attempted, a bus error exception occurs.
M	Modified	Page has not been modified	Page has been modified.

• FC Function Code

FIGURE 7.25　ATC entry structure.

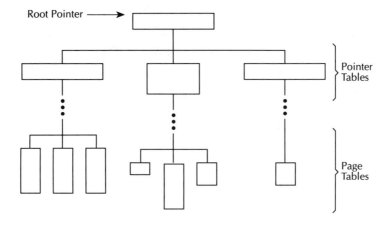

FIGURE 7.26   MC68030 MMU translation table tree structure.

- Bus error
- F-line
- Privilege violation
- Configuration error

Example 7.1

Assume user program space. The MC68030 executes CLR.W(A1) with PC contents and instruction cache contents as follows.

Given the op code for CLR.W(A1) is 4251_{16}, [PC] = \$02513080.

Instruction Cache Contents

Function Codes = 010_2

Index			valid bits 0 1 2 3	LWO	LW1	LW2	LW3
7	2	025130	1 1 1 1	4251xxxx	—	—	—
8	2	025130	0 1 1 1	4251xxxx	—	—	—
9	2	025130	1 1 1 1	———	—	—	—

When the instruction CLR.W(A1) is fetched by the 68030, will a cache hit or miss occur? Why?

Solution
A miss will occur since the valid bit is zero.

Example 7.2

Write an instruction sequence to freeze the data cache and disable the instruction cache.

Solution

```
MOVEC  CACR,D1   ;  Get current
                 ;  CACR
BCLR.L #0,D1     ;  Disable
                 ;  instruction
                 ;  cache
BSET.L #9,D1     ;  Freeze data cache
MOVEC D1, CACR   ;  Write to
                 ;  CACR
```

Example 7.3

Assume a page size of 4K bytes. Determine the physical address from the logical addresses of the task shown below:

Page			
0	$002000	12345	Page
1	$002004	7AB12	Descriptor
2	$002008	1BCA7	Table

Logical addresses to be translated are $000001A5, $00002370.

Solution

From logical addresses $000001A5, the page number is upper 20 bits, i.e., page number is zero. This is because the page size is given as 4K bytes and hence the lower 12 bits ($2^{12} = 4K$) are used as the page size. Since TCB is $002000, $12345 is concatenated with the lower 12 bits ($1A5) of the logical address $000001A5 to obtain the physical address $123451A5.

Similarly, the physical address for the logical address $00002370 is $1BCA7370.

Example 7.4

Determine the contents of SRP when enabled in TC to describe a page descriptor table located at $05721050 and limited to logical address 0 – $7000.

Solution

The format for SRP is

31 30		16, 15	2, 1 0
L / $\overline{\text{U}}$	Limit	0----	DT
Table Address		Unused	
31		3	0

L / $\overline{\text{U}}$ is 0 for upper limit range. LIMIT is $7000. DT must be 01 for page descriptor. Table address is $05721050. Hence, SRP is

31			0
0	7000	0001	
	05721050		

Example 7.5

What happens upon execution of the PTEST instruction with level 0 (i.e., search translation tables only)? Assume that the MMUSR contains $4005.

Solution

From Figure 7.24, level 0 means search translation tables only. Bit 14 defines the limit bit and bits 0–2 define the number of tables used in translation. [MMSUR] = $4005 indicates limit bit = 1 and number of tables used in translation is 5. Therefore, the index limit is exceeded when searching the five translation tables.

7.2 MC68040

This section presents an overview of the Motorola 68040 32-bit microprocessor. Emphasis is given to the coverage of its on-chip floating-point hardware.

7.2.1 Introduction

The MC68040 is Motorola's enhanced 68030, 32-bit microprocessor. The MC68040 is implemented in Motorola's latest HCMOS technology. Providing an ideal balance between speed, power, and physical device size, the MC68040 integrates an MC68030-compatible integer unit, an MC68881/MC68882-compatible floating-point unit (FPU), dual independent demand-paged memory management units (MMUs) for instruction and data stream accesses, and independent 4K-byte instruction and data cache. A high degree of instruction execution parallelism is achieved through the use of multiple independent execution pipelines, multiple internal buses, and separate physical caches for both instruction and data accesses.

The main on-chip features of the MC68040 include:

- MC68030-Compatible Integer Execution Unit
- MC68881/MC68882-Compatible Floating-Point Execution Unit
- Independent Instruction and Data Memory Management Units (MMUs)
- 4K-Byte Physical Instruction Cache and 4K-Byte Physical Data Cache Accessible Simultaneously
- Concurrent integer unit, FPU, MMU, and Bus Controller Operation which maximize throughput
- 32-Bit, nonmultiplexed external address and data buses with synchronous interface

The Intel equivalent of the 68040 is the 80486. Table 7.4 depicts a comparison of the main features and capabilities of the two processors. The most significant difference between the two devices is throughput. The 68040 executes more integer and floating point operations per second (at a clock rate of 25Mhz) than the 486. The primary reasons for the 040's outstanding performance are its pipelined integer and floating point units, as well as its advanced memory management implementation. Both these features are discussed in detail later.

7.2.2 Register Architecture/Addressing Modes

Figure 7.27 shows the MC68040 programming model. The registers are partitioned into two levels of privilege: user and supervisor. The user model has sixteen 32-bit general-purpose registers (D0-D7 and A0-A7), a 32-bit program counter (PC), and a condition code register (CCR) contained within the supervisor status register (SR). The MC68040 user programming model also incorporates the MC68882 programming model consisting of eight 80-bit floating-point data registers, a floating-point control register, a floating-point status register, and a

TABLE 7.4   Main features of Motorola MC68040 vs. Intel 80486

Chip	ALU/Data Bus	Clock Rate	Cache	FPU	Integer Throughput
68040	32/32 Bit	25-50 MHz	On-Chip	On-Chip	22 MIPS
80486	32/32 Bit	25-66 MHz	On-Chip	On-chip	15 MIPS

floating-point instruction address register. The supervisor model has two 32-bit stack pointers (ISP and MSP), a 16-bit status register (SR), a 32-bit vector base register (VBR), two 3-bit alternate function code registers (SFC and DFC), a 32-bit cache control register, a supervisor root pointer (SPR) and user root pointer (URP), a translation control register (TCR), four transparent translation registers: two for instruction accesses (ITT1-ITT0), and two for data access (DTT1-DTT0), and an MMU status register (MMUSR).

The key feature distinguishing the 68040 from its predecessors is its on-chip Floating Point Unit (FPU). The 68040 has 11 on-chip registers dedicated to support floating point operations: eight 80 bit data registers, and one each 32 bit floating point control (FPCR), status (FPSR), and instruction address (FPIAR) register.

The eight floating point registers always contain extended precision numbers. All external operands are converted to extended precision prior to use in any calculation, or storage in any FP register. The FPCR consists of one exception enable byte for traps during exception operations, and one mode byte for setting user-defined rounding modes. The FPCR may be modified and read by the user and is cleared by a hardware reset or a restore null state operation. The remaining 16-bits in the FPCR are reserved by Motorola for future definition.

The FPSR contains several status byte indicating status for the last calculation. The FPSR Condition Code byte indicates the results of an operation were negative, zero, infinity, or not a number (NAN). The Exception Status byte indicates floating point operational exceptions such as Branch/Set on Unordered (arithmetic operation attempted using NAN), Signaling NAN, and Operand error. The FPSR Accrued Exception byte contains a history of exceptions

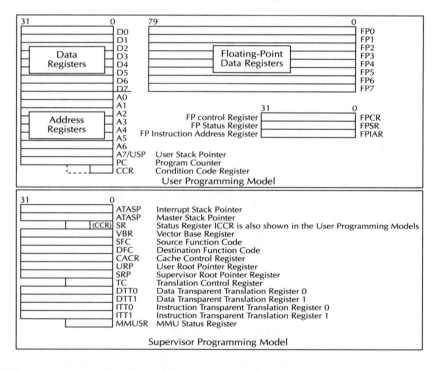

FIGURE 7.27   68040 programming model.

generated by previous operations. The FPSR Quotient byte is included for compatibility with the MC68881/2 floating point units, and contains the least significant 7 bits and the sign bit of the quotient from the previous operation.

The 68040 FPU operates simultaneously with the Integer Unit. In addition to this, the pipelined FPU can concurrently execute two instructions. This pipelined superscalar architecture increases the complexity of discovering which of several instructions executing concurrently may have generated an exception. The FPIAR holds the logical address of the instruction prior to its execution, to allow a floating point exception handler to locate the possible cause of an exception.

The FPU supports the same data as the 68881/68882. The MC68040 supports the same addressing mode as the 68020/68030.

7.2.3 Instruction Set/Data Types

The instructions provided by the MC68040 are listed in Table 7.5.

TABLE 7.5 68040 Instruction Set Summary

Mnemonic	Description
ABCD	Add Decimal with Extend
ADD	Add
ADDA	Add Address
ADDI	Add Immediate
ADDQ	Add Quick
ADDX	Add with Extend
AND	Logical AND
ANDI	Logical AND Immediate
ASL, ASR	Arithmetic Shift Left and Right
Bcc	Branch Conditionally
BCHG	Test Bit and Change
BCLR	Test Bit and Clear
BFCHG	Test Bit Field and Change
BFCLR	Test Bit Field and Clear
BFEXTS	Signed Bit Field Extract
BFEXTU	Unsigned Bit Field Extract
BFFFO	Bit Field Fine First One
BFINS	Bit Field Insert
BFSET	Test Bit Field and Set
BFTST	Test Bit Field
BRA	Branch
BSET	Test Bit and Set
BSR	Branch to Subroutine
BTST	Test Bit
CAS	Compare and Swap Operands
CAS2	Compare and Swap Dual Operands
CHK	Check Register Against Bounds
CHK2	Check Register Against Upper and Lower Bounds
CINV*	Invalidate Cache Entries
CLR	Clear
CMP	Compare
CMPA	Compare Address
CMPI	Compare Immediate
CMPM	Compare Memory to Memory
CMP2	Compare Register Against Upper and Lower Bounds
CPUSH*	Push then Invalidate Cache Entries
DBcc	Test Condition, Decrement and Branch
DIVS, DIVSL	Signed Divide

TABLE 7.5   68040 Instruction Set Summary (*continued*)

Mnemonic	Description
DIVU, DIVUL	Unsigned Divide
EOR	Logical Exclusive OR
EORI	Logical Exclusive OR Immediate
EXG	Exchange Registers
EXT, EXTB	Sign Extend
FABS*	Floating-Point Absolute Value
FADD*	Floating-Point Add
FBcc	Floating-Point Branch
FCMP	Floating-Point Compare
FDBcc	Floating-Point Decrement and Branch
FDIV*	Floating-Point Divide
FMOVE*	Move Floating-Point Register
FMOVEM	Move Multiple Floating-Point Registers
FMUL*	Floating-Point Multiply
FNEG*	Floating-Point Negate
FRESTORE	Restore Floating-Point Internal State
FSAVE	Save Floating-Point Internal State
FScc	Floating-Point Set According to Condition
FSORT*	Floating-Point Square Root
FSUB*	Floating-Point Subtract
FTRAPcc	Floating-Point Trap-On Condition
FTST	Floating-Point Test
ILLEGAL	Take Illegal Instruction Trap
JMP	Jump
JSR	Jump to Subroutine
LEA	Load Effective Address
LINK	Link and Allocate
LSL, LSR	Logical Shift Left and Right
MOVE	Move
MOVE16*	16-Byte Block Move
MOVEA	Move Address
MOVE CCR	Move Condition Code Register
MOVE SR	Move Status Register
MOVE USP	Move User Stack Pointer
MOVEC*	Move Control Register
MOVEM	Move Multiple Registers
MOVEP	Move Peripheral
MOVEQ	Move Quick
MOVES*	Move Alternate Address Space
MULS	Signed Multiply
MULU	Unsigned Multiply
NBCD	Negate Decimal with Extend
NEG	Negate
NEGX	Negate with Extend
NOP	No Operation
NOT	Logical Complement
OR	Logical Inclusive OR
ORI	Logical Inclusive OR Immediate
PACK	Pack BCD
PEA	Push Effective Address
PFLUSH*	Flush Entry(ies) in the ATCs
PTEST*	Test a Logical Address
RESET	Reset External Devices
ROL, ROR	Rotate Left and Right
ROXL, RORX	Rotate with Extend Left and Right
RTD	Return and Deallocate
RTE	Return from Exception
RTR	Return and Restore Codes

TABLE 7.5   68040 Instruction Set Summary (*continued*)

Mnemonic	Description
RTS	Return from Subroutine
SBCD	Subtract Decimal with Extend
Scc	Set Conditionally
STOP	Stop
SUB	Subtract
SUBA	Subtract Address
SUBI	Subtract Immediate
SUBQ	Subtract Quick
SUBX	Subtract with Extend
SWAP	Swap Register Words
TAS	Test Operand and Set
TRAP	Trap
TRAPcc	Trap Conditionally
TRAPV	Tap on Overflow
TST	Test Operand
UNLK	Unlink
UNPK	Unpack BCD

* MC68040 additions or alterations to the MC68030 and MC68881/
MC68882 instruction set.

Instructions in Table 7.5 which have an asterisk are unique to the 68040. The ptest and pflush MMU control instructions have new formats, but are similar to those of the 68030. The movel6 instruction allows for faster block transfers. Since this instruction operates only between 16 byte memory locations, and most compiler data types are 8 bytes or less in size, this instruction is not useful for most operations. Larger data structures such as arrays may benefit from use of the movel6 instruction, but only if they are aligned on a 16-byte boundary. The 68040 Translation Control Register is accessed with a movec instruction as opposed to pmove instruction on the 68030. On the 68040, the bsr and bra instructions are both one clock cycle faster than jsr and jmp respectively. Since each pair is usually interchangeable, the branch instructions are preferred.

Though the 68040 possesses an on-chip FPU, only a subset of the 68882 floating point coprocessor instructions are directly supported. The remaining functions such as transcendental (trigonometric and exponential) are emulated in software via a trap using Motorola's Floating-Point software package (FPSP) for the 68040. The FPSP emulates the floating-point instructions of the 68881/68882 which are not provided by the 68040. This is an illegal instruction mechanism used to indicate instructions not recognized by the hardware. The software emulator is also invoked by the underflow exception and a new unsupported-data type exception that handles denormalized and packed-decimal data representations in 68040. Motorola claims even these instructions will execute 25–100% faster on a 25 MHz 68040 than on a 33MHz 68882 FPU. One way to avoid using emulated floating point instructions is to call a library routine linked to the user program to do the FP operations. This approach is faster than emulation since the trap and decode times associated with emulation (which are comparable to the actual computing time) are eliminated. Also, emulation routines carry calculations to extended (80 bit) precision. This always adds extra calculation iterations, and thus extra time, but is not always needed by the application. Table 7.6 depicts the indirectly supported FP instructions. Note that loading of constant π is not directly supported by the 68040. Constant π can be indirectly loaded by using the MOVECR instruction.

The MC68040, with its integer unit and floating-point unit (FPU), support the operand data types shown in Table 7.7. The operand types supported by the integer unit include the data types supported by the MC68030 plus a new data type (16-byte block) for the MOVE16 instruction. The user instruction MOVE16 has been added to the instruction set to support 16 byte memory to memory data transfers. Integer unit operands can reside in registers, in

TABLE 7.6   Indirectly Supported Floating-Point
Instructions by the 68040

Mnemonic	Description
FACOS	Floating-Point Arc Cosine
FASIN	Floating-Point Arc Sine
FATAN	Floating-Point Arc Tangent
FATANH	Floating-Point Hyperbolic Arc Tangent
FCOS	Floating-Point Cosine
FCOSH	Floating-Point Hyperbolic Cosine
FETOX	Floating-Point e^x
FETOXL	Floating-Point $e^x - 1$
FGETEXP	Floating-Point Get Exponent
FGETMAN	Floating-Point Get Mantissa
FINT	Floating-Point Integer Part
FINTRZ	Floating-Point Integer Part, Round-to-Zero
FLOG10	Floating-Point Log_{10}
FLOG10	Floating-Point Log_2
FLOGN	Floating-Point Log_e
FLOGNP1	Floating-Point $\text{Log}_e (x + 1)$
FSQRT	Floating-Point Square Root
FMOD	Floating-Point Modulo Remainder
FMOVECR	Floating-Point Move Constant ROM
FREM	Floating-Point IEEE Remainder
FSCALE	Floating-Point Scale Exponent
FSGLDIV	Floating-Point Single Precision Divide
FSFLMUL	Floating-Point Single Precision Multiply
FSIN	Floating-Point Sine
FSINCOS	Floating-Point Simultaneous Sine and Cosine
FSINH	Floating-Point Hyperbolic Sine
FTAN	Floating-Point Tanget
FTANH	Floating-Point Hyperbolic Tangent
FTENTOX	Floating-Point 10^x
FTWOTOX	Floating Point 2^x

memory, or within the instructions themselves, and may be a single bit, a bit field, a byte, a word, a long word, a quad word, or a 16-byte block. Each integer data register is 32 bits wide. Byte operands occupy the lower 8 bits, word operands the lower order 16 bits, and long-word operands the entire 32 bits. The LSB of a long-word integer is addressed as bit zero and the MSB is addressed as bit 31. For bit fields, the MSB is addressed as bit zero, and the LSB is addressed as the width of the field minus one.

TABLE 7.7   68040 Data Types

Operand Data Type	Size	Supported By:	Notes
Bit	1 Bit	IU	—
Bit Field	1-32 Bits	IU	Field of Consecutive Bit
BCD	8 Bits	IU	Packed: 2 Digits Byte Unpacked: 1 Digit Byte
Byte Integer	8 Bits	IU, FPU	—
Word Integer	16 Bits	IU, FPU	—
Long-Word Integer	32 Bits	IU, FPU	—
Quad-Word Integer	64 Bits	IU	Any Two Data Registers
16-Byte	128 Bits	IU	Memory-Only, Aligned to 16-Byte Boundary
Single-Precision Real	32 Bits	FPU	1-Bit Sign, 8-Bit Exponent, 23-Bit Mantissa
Double-Precision Real	64 bits	FPU	1-Bit Sign, 11-Bit Exponent, 52-Bit Mantissa
Extended-Precision Real	80 Bits	FPU	1-Bit Sign, 15-Bit Exponent, 64-Bit Mantissa

Example 7.6

Write a 68040 assembly language program to compute the volume of a sphere to extended precision by using $V = 4/3 * \pi r^3$ where r is the radius of the sphere.

Solution

```
        ORG  $1000      ; initialize program at location hex
                          1000
        FMOVE (A7),fp0  ; move radius value from user stack
                          to fp register 0
        FMOVE (A7),fp1  ; and into fp register 1
        FMUL fp1,fp1    ; obtain r squared, store in fp 1
        FMUL fp0,fp1    ; obtain 4 cubed, store in fp 1
        FMOVE #4,fp2    ; load decimal constant 4 into fp
                          register 2
        FDIV #3,fp2     ; obtain 4/3, store in fp2
        FMOVE #22,fp4   ; load decimal constant 22 into fp
                          register 4
        FDIV #7,fp4     ; obtain π store in fp4
        FMUL fp2,fp4    ; obtain (4 * π)/3, store in fp4
        FMUL fp4,fp1    ; obtain ((4 * π)/3) * r cubed, store
                          in fp1
FINISH  JMP FINISH      ; halt
```

This program will leave the extended precision results of the routine in floating point register #1.

Example 7.7

Write a 68040 assembly language program to compute the hypotenuse of a right angle triangle by using

$$Z = \sqrt{x^2 + y^2}$$

Solution

The following program requires software emulation of the indirectly supported fsqrt floating point instruction to perform a Pythagorean theorem calculation of the length of the hypotenuse of a right angle triangle:

```
        ORG  $1000      ; initialize at hex location 1000
        FMOVE (A7)+,fp0 ; load adjacent side length into fp
                          reg 0
        FMUL fp0,fp0    ; square side 1, store in fp0
        FMUL fp2,fp2    ; square side 2, store in fp2
        FADD fp0,fp2    ; sum the squared side lengths and
                          store in fp reg 2
        FSQRT fp2       ; calculate length of hypotenuse
                          and store in fp 2
FINISH  JMP FINISH      ; halt
```

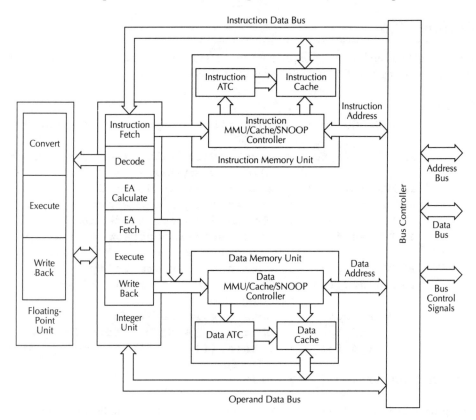

FIGURE 7.28   MC68040 block diagram.

7.2.4 68040 Processor Block Diagram

Figure 7.28 shows a block diagram of the MC68040. Instruction execution is pipelined in both the Integer Unit (IU) and FPU, which interface to fully independent data and instruction memory units. In the MC68040, high-usage instructions (such as branch instructions) execute in a minimum number of clock cycles. In the development of the IU, Motorola ran thousands of real world code traces to determine which instructions were used most often. The integer pipeline consists of six stages: Instruction prefetch, Instruction decode, Effective address calculate, Effective address fetch, Execute and Writeback. The six-stage IU pipeline executes instructions at an average of 1.3 instructions/clock cycle: about 3 times faster than the 68030 IU. This approaches the one instruction/clock cycle execution rate of a RISC architecture; an outstanding accomplishment for a CISC architecture processor. Figure 7.29 depicts the instruction pipeline of the Integer Unit.

The Floating Point Unit (FPU) pipeline consists of three stages; Operand Conversion, Execute, and Result Normalization. A floating point instruction will flow-through the IU until it reaches the execute stage. Here, the instruction and its operands are transferred to the operand conversion stage of the FPU. If the instruction requires data to be transferred back to the integer unit, then the IU execute stage must wait for the data. This may stall the IU pipeline by 10's of clock cycles. If not, the IU continues instruction execution in parallel with the FPU. The compiler may minimize these pipeline stalls by reordering integer and floating point instructions to minimize register and memory location dependencies between concurrently executing instructions. While the FPU pipeline is slower than the IU pipe, the on-chip FPU operations in the 68040 take about one half the number of clock cycles of the 68882 coprocessor which is required for floating point operations with the 680x0 series prior to the 68040.

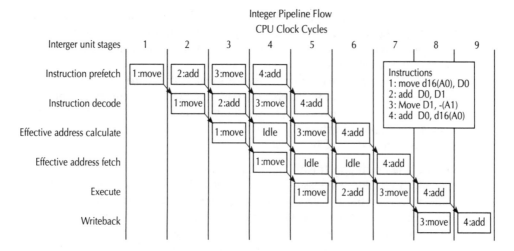

FIGURE 7.29 Integer unit instruction pipeline.

The instruction and data memory units operate independently. They consist of an MMU which utilizes the ATC to translate the logic address generated by the MC68040 into a physical address, while the bus snooper circuit ensures cache coherency in multimaster applications. This split cache or "harvard architecture" uses a subset of the 68030's MMU instruction set.

7.2.5 68040 Memory Management

The memory management function performed by the MMU is called demand paged memory management. Since a task specifies the areas of memory it requires as it executes, memory allocation is supported on a demand basis. If a requested area of memory is not currently mapped by the system, then the access causes a demand for the operating system to load to allocate the required memory image. The technique used by the MC68040 is paged memory management because physical memory is managed in blocks of a specified number of bytes, called page frames. The logical address space is divided into fixed-size pages that contain the same number of bytes as the page frames. The memory management software assigns a physical base address to a logical page. The system software then transfers data between secondary storage and memory, one or more pages at a time. Because of the phenomenon of locality of reference, instruction and data that are used in a program have a high probability of being reused within a short time. Additionally, instruction and data operands residing near the instruction and data currently in use also have a high probability of being utilized within a short period.

Memory management in the MC68040 has been improved by including separate independent paged MMUs for instruction and data accesses. The data and instruction MMUs support virtual memory systems by translating logical addresses to physical addresses using translation tables stored in memory. Each MMU contains an address translation cache (ATC) in which recently used logical-to-physical address translations are stored. For normal accesses, the translation process proceeds as follows for the accessed instruction or data memory unit:

1. Compare the logical address and privilege mode to the parameters in the transparent translation registers and use the logical address as a physical address for the access if one of the transparent translation registers match. Each transparent translation register can define a block of logical addresses that are used as physical addresses without translation.

2. Compare the logical address and privilege mode to the tag portions of the entries in the ATC. When the access address and privilege mode matches a tag in the ATC and no access violation is detected, the ATC outputs the corresponding physical address to the cache controller, which access the data within the cache.
3. When no transparent translation register nor valid ATC entry matches, the MMU aborts the current access and searches the translation tables in memory for the correct translation. If the table search completes without errors, the MMU stores the translation in the ATC and provides the physical address for the access, allowing the memory unit to retry the original access by repeating step 2.

Separate 4K-byte on chip instruction and data cache operate independently, and are accessed in parallel with address translation. Each cache and corresponding MMU reside on a separate internal address bus and data bus, allowing simultaneous access to both. Both four-way set associative caches have 64 sets of four, 16-byte lines. Each cache line contains an address tag, status information, and four long words of data (D0-D3). The address tag contains the upper 22-bits of the physical address. The processor fills the cache lines using burst mode access which transfers the entire line as four long words. This mode of operation not only fills the cache efficiently, but also captures adjacent instruction or data items that are likely to be required in the near future due to locality characteristics of the executing task.

Each memory unit access by the integer unit is translated from a logical address to a physical address to access the data in the cache. To minimize latency of the requested data, the lower untranslated bits of the logical address (which map directly to physical address bits) are used to access a set of cache lines in parallel with the translation of the upper logical address bits. Bits PA9-PA4 are used to index into the cache and select one of the 64 sets of four lines. The four tags from the selected cache set are compared to the translated physical address bits PA31-PA12 from the MMU and bits PA11 and PA10 of the untranslated page offset. The cache has a hit, if any one of the four tags match.

The caches improve the overall performance of the system by reducing the number of bus transfers required by the processor to fetch information from memory and by increasing the bus bandwidth available for other bus masters in the system. To further improve system performance the data cache in the MC68040 supports both write back and write through caching modes for storing write accesses.

Write back — The information is written only to the block in the cache. The modified cache block is written to main memory only when it is replaced.
Write through — The information is written to both the block in the cache and to the block in the lower-level memory.

The MC68040 implements a bus snooper that maintains coherency of the caches by monitoring the accesses by an alternate bus master and performing cache maintenance operations as requested by the alternate master. Matching cache lines can be invalidated during the alternate master access to memory, or memory can be inhibited to allow the MC68040 to respond to the access as a slave. By intervening in the access, the processor can update its internal caches for an external write (destination data) or supply cache data to the alternate bus master for an external read (source data). In this manner, the caches in the MC68040 are prevented from accumulating old or invalid copies of data (stale data), and external masters are allowed access to locally modified data within the caches that is no longer consistent with external memory (dirty data). Cache coherency is also supported by allowing memory pages to be specified as write through instead of write back; processor writes to write through pages always update external memory via an external bus access after updating the cache, keeping memory and cache data consistent.

7.2.6 Discussion and Conclusion

The 68040 represents a significant advance in the Motorola 680x0 processor family. The most common commercial applications for these devices are high-end Macintosh computers such as the Quadra series and aftermarket Macintosh accelerator boards such as the Radius Rocket. The FPU pipeline and interface overhead reduction have resulted in an on-chip floating-point unit which outperforms the 68030/68882 floating point capabilities, even when floating point instructions are emulated in software, the 68040 is clocked at a slower speed than the 68030. Memory management technologies such as the Address Translation Cache, write-back main memory block updates, and split instruction/data caches have reduced clock cycles/instruction to a rate (average = 1.3 clock cycles/instr) which rivals RISC architecture processors. This is very unusual, since a primary benefit of reduced instruction set computers is increased throughput due to a smaller number of instruction being fetched from memory. The combination of multistage pipelining, instruction sequencing, and memory management have produced a processor which outperforms its primary competition in both fixed and floating point throughput.

7.3 Intel 80486 Microprocessor

Intel 80486 is an enhanced 32-bit microprocessor with an on-chip floating-point hardware.

7.3.1 Intel 80486/80386 Comparison

Table 7.8 compares the basic features of the 80486 with those of 80386.

TABLE 7.8   80386 vs. 80486

Characteristic	80386	80486
Introduced in	1985; 386SX in 1988	1989
Principle features	Adds paging 32-bit extension, on-chip address translation, and greater speed to 80286 functions. 32-bit microprocessor	Adds on-chip cache, floating-point unit, and greater speed to 386 functions. 32-bit microprocessor
Data bus size accommodated	16-, 32-bit	8-, 16-, 32-bit
Parity	None	Generation and checking
Address bus size	32-bit	32-bit
Bus utilization with zero wait states	70%	50%
On-chip transistors	275,000	1.2 million
Directly addressable memory	4 Gigabytes	4 Gigabytes
Virtual memory size	64 Tetrabytes	64 Tetrabytes
Internal clock	Divides the external clock input by 2	Same as external clock input
Clock	25 MHz to 50 MHz	25 MHz to 66 MHz
Pins	100 for 80386SX and 168 for other 80386's	168
Address and data buses	non-multiplexed	non-multiplexed
Instruction Execution speed	4.5 clocks	1.8 clocks
Registers	8 32-bit general purpose registers 32-bit EIP and Flag register 6 16-bit segment registers 6 64-bit segment descriptor registers 4 32-bit system control registers (CR0-CR3)	All registers listed under the 80386 plus the following registers: 8 80-bit 8 2-bit 8 16-bit 3 16-bit 2 48-bit

TABLE 7.8   80386 vs. 80486 (*continued*)

Characteristic	80386	80486
Tested debug registers for built-in self test	2 Test registers 8 Debug registers for testing control ROMs, TLB, etc.	5 Test registers 8 Debug registers for testing control ROMs, TLB, cache, etc
Start of a bus cycle	Defined by activation of ADS output pin by the 80386	Defined by activation of ADS output pin by the 80486
Bus cycle definition	Bus cycle is defined by the encoding of M/IO#, D/C#, and W/R# pins	Bus cycle is defined in the same way except that the encoding of M/IO#, D/C#, and W/R# are different on the 80386 and 80486
Address	Defined by A2-A31 BE0#-BE3#	Same as the 80386
End of bus cycle	Indicated by the RDY# pin RDY# input indicates that an external device has presented valid data on the data bus or that the external device has accepted data	RDY# and BRDY# pins The RDY# pin has the same meaning as the 80386 except that the BRDY# input is used during a burst transfer. A maximum of 16 bytes can be transferred during the burst. The 80486 asserts the BLAST# output pin to end the burst
Direct Memory Access (DMA)	Two pins are used: HOLD input pin HLDA output pin	Three pins are used: HOLD input pin HLDA output pin BREQ output
Bus lock	LOCK# output pin allows the processor to complete multiple bus cycles without interruption via the HOLD pin	Two pins are used: LOCK# and PLOCK#. The 80486 LOCK# output pin has the same meaning as the 80386. The PLOCK# (bus pseudo-lock) output pin allows the processor to complete multiple bus cycles of aligned 64-bit data including floating-point
Bus backoff	Not available	The BOFF# input pin indicates that another bus master needs to complete a bus cycle in order for the 80486's current cycle to complete
Address HOLD	Not available	The AHOLD input pin causes the 80486 to float its address bus in the next clock cycle. This allows an external device to drive an address into the 80486 for internal cache line invalidation
On-chip memory management hardware	Yes	Yes
Operating modes: Real, Protected, and Virtual 8086 modes	Yes. Does not support maximum or minimum mode like the 8086	Same as the 80386
Addressing modes	11	11
On-chip floating-point hardware	No	Yes
Instructions	Listed in Chapter 4 including the floating-point instructions where the 80386 is interfaced to the 80386	All 80386 instructions including the floating-point instructions for the on-chip floating-point hardware plus six new instructions

7.3.2 Special Features of the 80486

Intel 80486 is a 32-bit microprocessor like the Intel 80386. It executes the complete instruction set of the 80386 and the 80387DX floating-point coprocessor. Unlike the 80386, the 80486 on-chip floating-point hardware eliminates the need for an external floating-point coprocessor chip and the on-chip cache minimizes the need for external cache and associated control logic.

The 80486 is object-code-compatible with the 8086, 8088, 80186, 80286, and 80386 processor. It can perform a complete set of arithmetic and logical operations on 8-, 16-, and 32-bit data types using a full-width ALU and eight general-purpose registers. Four-gigabytes of physical memory can be addressed directly via its separate 32-bit addresses and data paths. An on-chip memory management unit is added which maintains the integrity of memory in the multitasking and virtual-memory environments. Both memory segmentation and paging are supported.

As mentioned above, the 80486 has an internal 8K byte cache memory, which provides fast access to recently used instructions and data. The internal write through cache can hold 8K bytes of data or instructions. The internal cache can be invalidated or flushed, so that an external cache controller can maintain cache consistency in multiprocessor environments. Write-back and flush controls over an external cache are also provided. The internal cache and instruction prefetch buffer can be filled very rapidly from memory each clock cycle.

The on-chip floating-point unit performs floating-point operations on the 32-, 64-, and 80-bit arithmetic formats specified in IEEE standard, and is object-code-compatible with the 8087, 80287, 80386 coprocessors.

An instruction restart feature allows programs to continue execution following an interrupt generated by an unsuccessful memory access attempt; an important feature for supporting demand-paged virtual memory.

The fetching, decoding, execution, and address translation of instructions is overlapped within the 80486 processor using instruction pipe-lining. This allows a continuous execution rate of one clock cycle per instruction for most instructions.

The 80486 processor can operate in three modes (set in software): real, protected, and virtual 8086 mode.

After reset or power up, the 80486 is initialized in Real Mode. This mode has the same base architecture as the 8086, but allows access to the 32-bit register set of the 80486 processor. Nearly all of the 80486 processor instructions are available, but default operand-size is 16 bits. The main purpose of Real Mode is to set up the processor for Protected Mode.

The Protected Mode or the Protected Virtual Address Mode is where the complete capabilities of the 80486 become available. Segmentation and paging can both be used in Protected Mode. All exiting 8086, 80286, and 386 processor software can be run under the 80486 processor's hardware-assisted protection mechanism.

The Virtual 8086 Mode is a sub-mode for the Protect Mode. It allows 8086 programs to be run but adds the segmentation and paging protection mechanisms of Protected Mode. It is more flexible to run 8086 in this mode than in the Real Mode since it can simultaneously execute the 80486 operating system and both 80286 and 80486 processor applications.

The 80486 is provided with a bus backoff feature. Using this, the 80486 will float its bus signals if another bus master needs control of the bus during a 80486 bus cycle and then restarts its cycle when the bus again becomes available.

The 80486 includes dynamic bus sizing. Using this feature, external controllers can dynamically alter the effective width of the data bus with 8-, 16-, or 32-bit bus widths.

In terms of programming models, the Intel 386 DX or SX has very few differences with the 80486 processor. The 80486 processor defines new bits in the EFLAGS, CR0, and CR3 registers, and in entries in the first and second-level page tables. In the 386 processor, these bits were reserved, so the new architectural features should not be a compatibility issue.

The AC (alignment check) flag (bit number 18) is a new flag added to the 80486 EFLAGS register. All other flags in the 80486 EFLAGS are same as the 80386. The 80486 AC flag can be used in conjunction with an AM (Alignment Mask) bit in the 80486 CR0 control register to control alignment checking. The 80486 performs alignment checking when both the AC flag and AM bit are set to one. This checking is disabled when AM bit is cleared to zero. An alignment check exception is generated when reference is made to an unaligned operand such as a word at an odd byte address or a doubleword at an address which is not an integral multiple of four. This is the 80486 new exception vector 17.

On the 80386, loading a segment descriptor would always cause a locked read and write to set the accessed bit of the descriptor. On the 80486, the locked read and write occur only if the bits are not already set.

The following control register bits in CR0 and CR3 have been added beyond those of the 80386:

1. Five new bits have been defined in the CR0 register. These are NE (Numeric error), WP (Write Protect), AM (Alignment Mask), NW (Not Write-through) and CD (Cache disable). The NE bit enables the standard mechanism for reporting floating-point numeric errors when set. The WP bit, when set to one, write-protects user-level pages against supervisor-mode access. The purpose of the AM bit is already explained. The NW bit enables write-through and cache invalidation cycle when cleared to zero.

2. Two new bits have been defined in the 80486 CR3 control register. These are the PCD (Page-level Cache Disable) and the PWT (Page-level Writes Transparent) bit. The state of the PCD bit is driven on the PCD pin during bus cycles which are not paged, such as interrupt acknowledge cycles, when paging is enabled. The state of the PWT bit, on the other hand, is driven on the PWT pin during bus cycles which are not paged such as interrupt acknowledge cycles when paging is enabled.

7.3.3 80486 New Instructions Beyond Those of the 80386

There are six instructions added to the 80486 instruction set beyond those of the 80386 instruction set as follows:

1. Three New Application Instructions
   - BSWAP instruction
   - XADD instruction
   - CMPXCHG Instruction
2. Three New System Instructions
   - INVD Instruction
   - WBINVD Instruction
   - INVLPG Instruction

The 80386 can execute all its floating-point instructions when the 80387 is present in the system. The 80486, on the other hand, can directly execute all its floating-point instructions (same as the 80386 floating-point instructions) since it has the on-chip floating-point hardware.

The 80486's six new instructions are described in the following. Since the 80386 floating-point instructions are discussed in Chapter 4 and are the same as the 80486, they are not covered in this chapter.

The three new application instructions included with the 80486 are listed below:

1. BSWAP reg_{32} instruction reverses the byte order of a 32-bit register, converting a value in little/big endian form to big/little endian form. That is, the BSWAP instruction exchanges bits 7...0 with bits 32...24 and bits 15...8 with bits 23...16 of a 32-bit register.

Executing this instruction twice in a row leaves the register with the original value. When BSWAP is used with a 16-bit operand size, the result left in the destination operand is undefined. For example, if [EAX] = 12345678H, then after BSWAP EAX, the contents of EAX are 78563412H.

Note that little endian is a byte-oriented method where the bytes are ordered (left to right) as 3,2,1, and 0 with byte 3 being the most significant byte. Big endian on the other hand, is also a byte-oriented method where the bytes are ordered (left to right) as 0,1,2, and 3 with byte 3 being the most significant byte.

The BSWAP instruction speeds up execution of decimal arithmetic by operating on four digits at a time.

2. XADD dest, source

```
reg8/mem8,   reg8
    or
reg16/mem16,  reg16
reg32/mem32,  reg32
```

The XADD dest, source loads the destination into the source and then loads the sum of the destination and the original value of the source into the destination. For example, if [AX] = 0123H, [BX] = 9876H, then after XADD AX, BX, the contents of AX and BX are respectively 9999H and 0123H.

3. CMPXCHG dest, source

```
reg8/mem8,   reg8
reg16/mem16,  reg 16
    or
reg32/mem32,  reg32
```

The CMPXCHG instruction compares the accumulator (AL, AX or EAX register) with the destination. If they are equal, the source is loaded into the destination; Otherwise, the destination is loaded into the accumulator. For example, if [DX] = 4324H, [AX] = 4532H, and [BX] = 4532H, then after CMPXCHG BX, DX, the ZF flag is set to one and [BX] = 4324H.

There are three new system instructions used for managing the cache and TLB. These are described below:

1. INVD instruction flushes the internal cache and a special function bus cycle is issued which indicates that any external caches should also be flushed. Data held in write-back external caches is discarded.
2. WBINVD instruction flushes the internal cache and a special function bus cycle is used which indicates that any external cache should write-back its contents to main memory. Another special function bus cycle follows, directing the external cache to flush itself.
3. INVLPG instruction is used to invalidate a single entry in the TLB.

The form of the 80486 MOV instruction used to access the test register has changed beyond that of the 80386. Also, new test registers have been defined for the cache in the 80486 and the model of the 80486 TLB (Translation Lookaside buffer) accessed through the test register has changed. Note that the TLB in 80486 is a cache for page table entries.

7.4 Intel Pentium Microprocessor

Table 7.9 summarizes the basic differences between the basic features of 486 and Pentium processor families:

TABLE 7.9   Basic Differences Between 80486 and Pentium Processor Families

Feature	486 Processors	Pentium Processor
Clock	25 to 66 MHz	60 to 100 MHz
Address and data buses	32-bit data bus	32-bit address bus
	32-bit address bus	64 bit data bus
Pipeline model	Single	Dual
Internal cache	8K for both data and instruction	8K data
		8K instruction
Cache type	Write-through	Write back
Number of transistors	1.2 million	3.2 million
Performance at 66 MHz in MIPS (Millions of Instructions Per Second)	54 MIPS	112 MIPS
Number of pins	168	273

Microprocessors have served largely separate markets and purposes: business PCs and engineering workstations. The PCs have used Microsoft's DOS and Windows operating systems while the workstations have used various features of UNIX. The PCs have not been utilized in the workstation market because of their relatively modest performance, especially with regard to complicated graphics display and their floating-point performance. Workstations have been kept out of the PC market partially because of their high prices and hard-to-use system software.

The Pentium will bring the PCs up to workstation-class computational performance with sophisticated graphics. The Intel Pentium is a 32-bit microprocessor with a 64-bit data bus. The Intel Pentium, like its predecessor, the Intel 80486 is 100% object code compatible with other X86 systems.

The Pentium processor has three modes of operation. The mode determines which instructions and architecture features are accessible. These modes are: real-address mode (also called "real mode"), protected mode, and system management mode.

In the real-address mode, the Pentium processor runs programs written for 8086/8088, 80186/80188, or for the real-address mode of an 80286, 80386, or 80486. The architecture of the Pentium processor in this mode is identical to that of the 8086/8088, 80186, and 80188 processors.

In the protected mode, all instruction and architectural features of the Pentium are available to the programmer. Some of the architectural features of the Pentium processor include memory management, protection, multitasking, and multiprocessing. While in protected mode, the virtual-8086 mode can be enabled for any task. For the v86 mode, the Pentium can directly execute 'real address mode' 8086 software in a protected, multitasking environment.

The Pentium processor is provided with a System Management Mode (SMM) similar to the one used in the 80486SL line, which allows engineers to design for low power usage. SMM is entered through activation of an external interrupt pin (System Management Interrupt, SMI#). All registers are saved for later restoration. The SMM interrupt service routine is then executed from a separate address space. SMM is exited by executing a new instruction (RSM) executable from SMM. The Pentium then returns to the interrupted program.

In December 1994, Intel detected a flaw in the Pentium chip while performing certain division calculations. The Pentium is not the first chip that Intel had problems with. The first version of the Intel 80386 had a math flaw which Intel quickly fixed before any complaints. Some experts feel that Intel should have acknowledged the math problem in the Pentium when it was first discovered and then offered to replace the chips. In that case, the problem with the Pentium most likely would have been ignored by the users. However, Intel was heavily criticized by computer magazines when the division flaw was first detected in the Pentium.

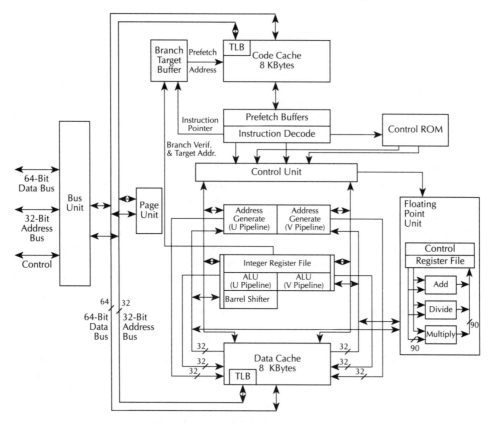

FIGURE 7.30 Pentium™ processor block diagram.

The flaw in the division algorithm in the Pentium was caused by a problem with a look-up table used in the division. Errors occur in the fourth through fifteenth significant decimal digits. This means that in a result such as 5.758346, the last three digits could be incorrect. For example, the correct answer for the operation 4,195,835 - (4,195,835 ÷ 3,145,727) × (3,145,727) is zero. The Pentium provided a wrong answer of 256. IBM claimed this problem can occur once every 24 days.

It is the author's opinion that the circuitry inside the 32-bit microprocessors are so complex that the math flaw in the Pentium is not unusual. Intel already regained its reputation by fixing the division flaw in the Pentium and has already been shipping replacement chips.

Intel shipped approximately 4 million original Pentium chips by the end of 1994. The manufacturing cost to Intel for replacement chip amounts to approximately $150 per chip and results in $100 per chip for shipping and upgrade services. Based on these numbers, the cost to Intel for a total recall of Pentium chips will be $1 billion which is almost 50% of Intel's 1994 earnings.

The author strongly believes in Intel products. Intel lived up to the expectations of the author and many other users of Intel products by correcting the division flaw in the Pentium chip.

7.4.1 Pentium Processor Block Diagram

Figure 7.30 shows a block diagram of the Pentium processor.

The Pentium microprocessor is based on a superscalar design. This means that the processor includes dual pipelining and executes more than one instruction per clock cycle. Note that scalar microprocessors such as the 80486 family have only one pipeline and execute one instruction per clock cycle and superscalar processors allow more than one instruction to be executed per clock cycle.

The Pentium microprocessor contains the complete 80486 instruction set along with some new ones which are discussed later. Pentium's on-chip memory management unit is completely compatible with that of the 80486.

The Pentium includes faster floating-point on-chip hardware than the 80486. Pentium's on-chip floating-point has been completely redesigned over the 80486. Faster algorithms provide up to ten times speed-up for common operations such as add, multiply, and load. The two instruction pipelines and on-chip floating-point unit are capable of independent operations. Each pipeline issues frequently-used instructions in a single clock cycle. The dual pipelines can jointly issue two integer instructions in one clock cycle or one floating-point instruction (under certain circumstances, two floating-point instructions) in one clock cycle.

Branch prediction is implemented in the Pentium by using two prefetch buffers: one to prefetch code in a linear fashion and one to prefetch code according to the contents of the Branch Target Buffer (BTB) so the required code is almost always prefetched before it is needed for execution. Note that the branch addresses are stored in the Branch Target Buffer (BTB).

The Pentium includes 8Kbytes of on-chip code cache and 8Kbytes of data cache. Each cache is two-way set associative. Also, each cache has a dedicated Translation Lookaside Buffer (TLB) to translate linear addresses to physical address used by each cache.

The block diagram shows the two instruction pipeline, the "U" pipe and the "V" pipe which are not equivalent and interchangeable. The U-pipe can execute all integer and floating-point instructions, while the V-pipe can only execute simple integer instructions and the floating-point exchange register contents (FXCH) instructions.

The instruction decode unit decodes the prefetched instructions so that the Pentium can execute them. The control ROM includes the microcode for the Penitum processor and has direct control over both pipelines.

A barrel shifter is included in the chip for fast shift operations.

7.4.2 Pentium Registers

The Pentium processor includes the same registers as the 80486. Three new system flags are added to the 32-bit EFLAGS register. These are the ID (ID flag, bit 21 of EFLAGS), the VIP (Virtual Interrupt Pending, bit 20 of EFLAGS) and the VIF (Virtual Interrupt Flag, bit 19 of EFLAAGS).

The ability of a program to set or clear the ID flag indicates that the processor supports the CPUID instruction. The CPUID instruction provides information about the family member and model of the microprocessor on which it is executing.

7.4.3 Pentium Addressing Modes and Instructions

The Pentium includes the same addressing modes as the 80386/80486.

The Penitum microprocessor includes three new application instructions and four new system instructions beyond those of the 80486.

The new application instructions are CMPXCHG8B, CPUID and RDTSC. The new system instructions are

   i) **MOV CR4,r32** and **MOV r32,CR4**
  ii) **RDMSR**
 iii) **WRMSR**
 iv) **RSM**

The details of RDTSC instruction is considered proprietary and confidential by Intel and is not provided in Intel manuals. Detailed information on this may be obtained from Intel upon signing a non-disclosure agreement.

CMPXCHG8B reg64 compares the 64-bit value in EDX:EAX with the 64-bit contents

or
mem64

of reg 64 or mem64. If they are equal, the 64-bit value in ECX:EBX is stored in reg64 or mem64; otherwise the content of reg64 or mem64 is loaded into EDX:EAX.

Certain features of the Pentium processor described in the 'Pentium Processor Data Book' are considered proprietary by Intel and may be obtained upon signing a non-disclosure agreement with Intel. These features are specific to the Pentium processor and may not be continued in the same way in future processors. Examples include functions for testability, performance monitoring, and machine check errors. These features are accessed through Model Specific Registers. The new instructions RDMSR and WRMSR are used to read from and write into theses registers. In order to use such model Specific features, software should check "Family" member by using the CPUID instruction. Software which uses these registers and functions may not be compatible with Intel's future processors. The RSM instruction resumes operation of a program interrupted by a System Management Mode interrupt.

Pentium floating-point instructions execute much faster than those of the 80486 instructions. For example, a 66MHz Pentium microprocessor provides about three times the floating-point performance of a 66MHz Intel 80486 DX2 microprocessor.

7.4.4 Pentium Vs. 80486 Basic Differences in Registers, Paging, Stack Operations, and Exceptions

7.4.4.a Registers of the Pentium processor vs. those of the 80486

This section discusses the basic differences between the Pentium and 80486 control, debug, and test registers. Two bits, namely CD (Cache Disable) and NW (Not Write-through) in the control register, CR0 are redefined in the Pentium. The values of zero in both CD and NW in CR0 implement a write-back strategy for the data cache in the Pentium. On the 80486, these values implement a write-through strategy.

One new control register, CR4 is included in the Pentium. CR4 contains bits that enable certain extensions to the 80486 provided in the Pentium processor. These extensions include functions such as debugging extensions that support I/O breakpoints and machine check exceptions for handling certain hardware error conditions.

The Pentium processor defines the type of breakpoint access by two bits in DR7 to perform breakpoint functions such as breakpoint on instruction execution only, break on data writes only and break on data reads or writes but not instruction fetches.

The implementation of test registers on the 80486 used for testing the cache and TLB has been redesigned using model specific registers in the Pentium processor.

7.4.4.b Paging

The Pentium processor provides an extension to the memory management/paging functions of the 80486 to support larger page sizes.

7.4.4.c Stack Operations

The Pentium microprocessor, 80486 and 80386 push a different value on the stack for a PUSH SP instruction than the 8086. The 32-bit processors push the value of the SP before it is decremented while the 8086 pushes the value of the SP after it is decremented.

7.4.4.d Exceptions

The Pentium processor implements new exceptions beyond those of the 80486. For example, a machine check exception is newly defined for reporting parity errors and other hardware errors.

External hardware interrupts on the Pentium may be recognized on different instruction boundaries due to the pipelined execution of the Pentium processor and possibly, an extra instruction passing through the v-pipe concurrently with an instruction in the u-pipe. When the two instructions complete execution, the interrupt is then serviced. Therefore, the EIP pushed onto the stack when servicing the interrupt on the Pentium processor may be different than that for the 80486 (i.e. it is serviced later).

The priority of exceptions is the same on both the Pentium and 80486.

7.4.5 Input/Output

The Pentium processor handles I/O in the same way as the 80486. The Pentium can use either standard I/O or memory mapped I/O. Standard I/O is accomplished by using IN/OUT instructions and a hardware protection mechanism.

When memory-mapped I/O is used, memory-reference instructions are used for input/output and the protection mechanism is provided via segmentation or paging.

The Pentium can transfer 8-, 16- or 32-bits to a device. Like memory-mapped I/O, 16-bit ports using standard I/O should be aligned to even addresses so that all 16 bits can be transferred in a single bus cycle. Like doublewords in memory-mapped I/O, 32-bit ports in Standard I/O should be aligned to address which are multiples of four. The Pentium supports I/O transfer to misaligned ports, but there is a performance penalty because an extra bus cycle must be used.

The INS and OUTS instructions move blocks of data between I/O ports and memory. The INS and OUTS instructions, when used with repeat prefixes, perform block input or output operations. The string I/O instructions can operate on byte (8-bit) strings, word (16-bit) strings, or double word (32-bit) strings.

When the Pentium is running in Protected mode, I/O operates as in real address mode with additional protection features as follows:

1. References to memory-mapped I/O ports, like any other memory reference, are subject to access protection and control by both segmentation and paging.
2. Using standard I/O, execution of an I/O instruction is subject to two protection mechanisms:
   a. The IOPL field in the EFLAGS register controls access to the I/O instructions.
   b. The I/O permission bit map of a TSS (Task State Segment) controls access to individual port in the I/O address space.

7.4.6 Applications With the Pentium

The performance of the Pentium's Floating-Point Unit (FPU) makes it appropriate for wide areas of numeric applications:

1. Business Data Processing
   Pentium's FPU can accept decimal operands and produce extra decimal results of up to 18 digits. This greatly simplifies accounting programming. Financial calculations that use power functions can take advantage of exponential and logarithmic functions.

2. Many mini and mainframe large simulation problems can be executed by the Pentium. These applications include complex electronic circuit simulations using SPICE and simulation of mechanical systems using finite element analysis.
3. Since the FPU of the Pentium can execute integer instructions concurrently, the Pentium can be used in graphics applications such as computer-aided design (CAD).
4. The Pentium's FPU can move and position machine control heads with accuracy in real time. Axis positioning can efficiently be performed by the hardware trigonometric support provided by the FPU. The Pentium can, therefore, be used for computer numerical control (CNC) machines.
5. The pipelined instruction feature of the Pentium processor makes it an ideal candidate for DSP (Digital Signal Processing) and related applications for computing matrix multiplications and convolutions.

Other possible application areas for the Pentium includes robotics, navigation, data acquisition, and Process Control.

QUESTIONS AND PROBLEMS

7.1   i)   Identify the the 68030 registers that are not included in the 68020.
    ii)   Name a 68030 register which is included in the 68020 but formatted in a different way from the 68020. Why is it structured differently?

7.2   What are the functions of the 68030 REFILL, STATUS, and STERM pins?

7.3   i)   What is the difference between the 68030 DSACK and STERM?
    ii)   What are the minimum bus access times in clock cycles for 68030 synchronous and asynchronous operations?

7.4   What conditions must be satisfied before a cache hit occurs for either instruction or data cache in the 68030?

7.5   Assume user data space. The 68030 executes the instruction CLR.W(A1) with [A1] = $20507002 and [$20507002] = $1234 with the following information in the data cache:

		Tag	Valid bits 0 1 2 3	LW0	LW1	LW2	LW3
0	1	205070	1 1 1 1	12345124	00121234	70001111	01112222
1	1	205070	0 1 1 1	77777777	12121212	FF0011EE	AAAA1111
2	1	5F1236	1 0 1 0	72101234	25252020	FEEEEEEE	00110011
3	1	021472	0 0 0 1	11223344	BBBBAAAA	71171625	00200000

Assume all numbers in hex.
Will a cache hit occur? Why or why not?

7.6   What are the functions of 68030 EBREQ and CBACK?

7.7   Write a 68030 instruction sequence to clear the instruction cache entry for address $00005040.

7.8   For a page size of 4K bytes, translate the 68030 logical address $00000240 to a physical address. Assume the following data:

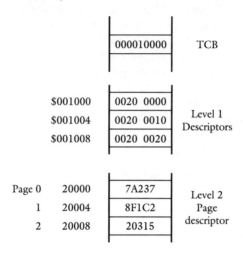

Assume all numbers in hex.

7.9   Identify the inputs and outputs that affect the 68030 MMU.

7.10   i)   How many levels of translation tables does the 68030 MMU provide?

ii)   What is the maximum number of page descriptor entries in the ATC?

7.11   Assume the following data contents in the 68030 CRP:

31						0
0	0 1 0 0				0 0 0 3	
	2 A 6 7				1 1 2	4
31					4, 3	0

Assume all numbers in hex.

i)   Is this a page or table descriptor?

ii)   What is the address of the next table?

iii)   What is the allowed range of logical addresses for the next table?

7.12   Determine the contents of the 68030 TC which will enable MMU, SRP, disable function code look-up, and will permit 2 levels of equal size look-up with a page size of 1K byte for a 30-bit address.

7.13   Show the contents of the 68030 TT0 for translating addresses $20000000–$21FFFFFF in supervisor data space. These addresses should be read-only and cacheable.

7.14   Find the following 68030 MMU instructions:

i)   for accessing an MMU register

ii)   for placing descriptors in the ATC

iii)   for invalidating one or more entries in the cache

iv)   for testing a descriptor either in the memory or in the cache

7.15   Write a 68030 instruction sequence for initializing the TC, CRP, and SRP registers from memory pointed to by A2. Can you initialize all MMU registers by this instruction? If not, list the register or registers.

7.16   Compare 68030 on-chip MMU features with those of the 68851 MMU.

7.17   What is meant by configuration error exception in the 68030?

7.18   In the 68030 MMU (demand paging system), what happens when a task attempts to access a page not in memory? What action should be taken by the demand paging operating system?

7.19   Discuss the main features of the 68040 as compared with those of the 68030.

7.20   What is the basic difference between the 68881/68882 floating-point coprocessor and the 68040 on-chip floating-point hardware?

7.21   Write a 68040 assembly language program to compute the area of a circle, $A = \pi r^2$ where r is the 32-bit radius of the circle in D0.

7.22   Write a 68040 assembly language program to compute:

$$20 \mathrm{Log}_{10}(V_1/V_2)$$

where V_1 and V_2 are 32-bit single precision numbers stored in D0 and D1 respectively. Store the result in D2.

7.23   Write a 68040 assembly language program to compute the quadratic formula:

$$x = \frac{-b \pm \sqrt{b^2 - 4ac}}{2a}$$

where a, b, and c are 32-bit integers.

7.24   Compare the on-chip hardware features of the 80486 and Pentium microprocessors.

7.25   What are the sizes of the address and data buses of the 80486 and Pentium?

7.26   Identify the main differences between the 80486 and Pentium.

7.27   What are the clock speed, pipeline model, number of on-chip transistors and number of pins on the 80486 and Pentium processors?

7.28   Discuss the application areas in which the Pentium-based PC's will be used.

7.29   Identify the main differences between the Intel 80386 and 80486.

7.30   Identify the pins on the 80386 and 80486 by which the bus cycle is defined.

7.31   What are the functions of the 80486 i) RDY# and BRDY# pins? ii) LOCK# and PLOCK#?

7.32   What is meant by the 80486 BUS BACKOFF feature?

7.33   What is meant by the 8086 real, protected, and virtual 8086 modes?

7.34   What is the meaning of 80486 alignment checking? How is it handled in the 80486?

7.35   How many new control bits have been added to control registers CR0 and CR3 by the 80486 beyond those of the 80386?

7.37   How many new instructions are added to the 80486 beyond those of the 80386? List them.

7.38   Given the following registers contents,

$$[EBX] = 7F27108AH$$
$$[BCX] = 2A157241H,$$

What is the content of ECX after execution of the following 80486 instruction sequence:

```
MOV  EBX,ECX
BSWAP  ECX
BSWAP  ECX
BSWAP  ECX
BSWAP  ECX
```

7.39   If [EBX] = 0123A212H, and [EDX] = 46B12310H, then what are the contents of EBX and EDX after execution of the 80486 instruction XADD EBX, EDX?

7.40   If [BX] = 271AH, [AX] = 712EH and [CX] = 1234H, what are the contents of CX after execution of the 80486 instruction CMPXCHG CX, BX?

7.41   What is the purpose of each of the 80486 INVD, WBINVD and INVLPG instructions?

7.42   What are three modes of the Pentium Processor? Discuss them briefly.

7.43   What is meant by the statement "The Pentium processor is based on a superscalar design."?

7.44   What are the purposes of the U-pipe and V-pipe of the Pentium processor?

7.45   Discuss how the branch prediction feature is implemented in the Pentium processor.

7.46   What are the sizes of the data and instruction caches in the Pentium?

7.47   How many new bits are added in the Pentium's EFLAGS register beyond those of the 80486? Discuss them.

7.48   How many new application and system instructions are added to the Pentium beyond those of the 80486? List them.

7.49   Discuss the main differences between the Pentium and 80486 control, debug, and test registers.

7.50   Discuss the new exceptions implemented in the Pentium beyond those of the 80486.

7.51   How is the I/O protection mechanism implemented in the standard I/O and Memory-mapped I/O of the Pentium? Discuss each.

7.52   Discuss two applications of the Pentium processor. Why is the Pentium an ideal candidate for these applications?

8

RISC MICROPROCESSORS: INTEL 80960, MOTOROLA MC88100 AND POWERPC

This chapter provides an overview of the hardware, software, and interfacing features associated with three popular RISC microprocessors namely, the Intel 80960 SA/SB, the Motorola MC88100, and the PowerPC. Finally, the basic features of typical 64-bit RISC mircoprocessors are discussed.

8.1 Basics of RISC

RISC is an acronym for Reduced Instruction Set Computer. This type of microprocessor emphasizes simplicity and efficiency. RISC designs start with a necessary and sufficient instruction set. The purpose of using RISC architecture is to maximize speed by reducing clock cycles per instruction. Almost all computations can be obtained from a few simple operations. The goal of RISC architecture is to maximize the effective speed of a design by performing infrequent operations in software and frequent functions in hardware, thus obtaining a net performance gain.

The following summarizes the typical features of a RISC microprocessor:

1. The microprocessor is designed using hardwired control with little or no microcode. Note that variable length instruction formats generally require microcode design. All RISC instructions have fixed formats, and therefore microcode design is not necessary.
2. A RISC microprocessor executes most instructions in a single cycle.
3. The instruction set of a RISC microprocessor typically includes only register-to-register, load, and store. All instructions involving arithmetic operations use registers, while load and store operations are utilized to access memory.
4. The instructions have simple fixed format with few addressing modes.
5. A RISC microprocessor has several general-purpose registers and large cache memories.
6. A RISC microprocessor processes several instructions simultaneously and thus includes pipelining.
7. Software can take advantage of more concurrency. For example, Jumps occur after execution of the instruction that follows. This allows fetching of the next instruction during execution of the current instruction.

RISC microprocessors are suitable for embedded applications. An embedded application is one in which the processor monitors and analyzes signals from one segment of the system and produces output required by another segment of the system; thus, it behaves as a controller that bridges various parts of the entire system. It performs all its functions without any user input.

RISC microprocessors are well suited for applications such as image processing, robotics, graphics, and instrumentation. The key features of the RISC microprocessors that make them ideal for these applications are their relatively low level of integration in the chip and instruction pipeline architecture. These characteristics result in low power consumption, fast instruction execution, and fast recognition of interrupts.

The state-of-the-art 64-bit RISC microprocessors include Digital Equipment Corporation's Alpha 21164, Motorola/IBM/Apple PowerPC 620 and Sun Microsystems Ultrasparc. Among these processors, the Alpha 21164 is the fastest with a maximum clock frequency of 300 MHz, four-way superscalar design, and 128-bit data bus. These processors are compared later in this chapter.

8.2 Intel 80960

The Intel 80960 family includes two types of 16-bit RISC microprocessors. These are the 80960SA and 80960SB processors. The 80960SA is designed as an Intra-agent communication (IAC) microprocessor. IAC messages can be sent for execution into the bus interface of a 80960SA processor from software executing on another processor.

The 80960SB, on the other hand, is designed as a floating-point RISC microprocessor and includes on-chip floating-point hardware.

The 80960SA contains 32 32-bit registers while the 80960SB includes an additional four floating-point registers with a total of 36 32-bit registers.

The 80960SA/SB comes in two speeds: 10MHz and 16MHz. The clock input is divided by 2 internally to generate the internal processor clock.

8.2.1 Introduction

This section covers the basic architecture of the chip, its instruction set, typical 80960 based system design utilizing a burst controller with burst and non-burst memories.

8.2.2 Key Performance Features

The following summarizes the main features of the 80960SA/SB:

8.2.2.a Load and Store Model

Most operations are performed on operands in CPU registers rather than in memory. All of the arithmetic, comparison, branching, and bit operations are performed with registers and literals (5-bit and floating-point). Only LOAD & STORE are memory reference instructions.

8.2.2.b Large Internal Register Sets

Large internal register sets featuring 32 32-bit general purpose and specific function registers are divided into two types: global and local. Both of these types can be used for general storage of operands. The only difference between global and local registers is the global registers retain their contents across procedure boundaries, whereas the processor allocates a new set of local registers each time a procedure is called.

8.2.2.c On-Chip Code and Data Caching

To reduce memory accesses, two features are added: an instruction cache and multiple sets of local registers. The former allows pre-fetching of blocks of instructions from the main memory while the latter allows the processor to perform most procedure calls without having to write the local registers out to the stack in memory.

.2.2.d Overlapped Instruction Execution

This is accomplished through a register scoreboarding scheme which enhances program execution speed. Register scoreboarding permits instruction execution to continue while data are being fetched from memory. When a load instruction is executed, the processor sets one or more scoreboard bits to indicate the target registers to be loaded. After the target registers are loaded, the scoreboard bits are cleared. While the target registers are being loaded, the processor is allowed to execute other instructions, called independent instructions, that do not use these registers. The net result of using this technique is that code can often be optimized in such a way as to allow some instructions to be executed in parallel.

.2.2.e Single Clock Instructions

Most of the commonly used instructions are executed in a minimum number of clock cycles (usually one clock).

For example, instructions, either 32 or 64-bits long, are aligned on 32-bit boundaries allowing instructions to be decoded in one clock cycle. This eliminates the need for an instruction-alignment stage in the pipeline resulting in over 50 instructions that can be executed in a single clock cycle.

.2.2.f Interrupt Model

To handle interrupts, the processor maintains an interrupt table of 248 interrupt vectors, of which 240 are available for general use. When an interrupt is generated, the processor uses a pointer from the interrupt table to perform an implicit call to an interrupt handler procedure. The processor automatically saves the state of the processor prior to receiving the interrupt, performs the interrupt routine, and then restores its previous state. A separate interrupt stack is also provided to segregate interrupt handling from application programs. Interrupt handling facilities feature prioritizing pending interrupts.

.2.2.g Procedure Call Mechanism

Each time a call instruction is issued, the processor automatically saves the current set of local registers and allocates a new set of local registers for the called procedure. Likewise, on the return from a procedure, the current set of local registers is deallocated and the local registers for the procedure being returned to are restored. Thus, on a procedure call, the program never has to explicitly save and restore those local variables.

.2.2.h Instruction Set and Addressing

The processor offers a full set of load, store, move, arithmetic, comparison, and branch instructions, with operations on both integer and ordinal data types. It also provides a complete set of Boolean and bit-field instructions, to simplify operations on bits and bit strings. The addressing modes are efficient and straightforward, while at the same time providing the necessary indexing and scaling modes required to address complex array and records.

The 32 address lines provide 4-gigabytes of address space for programs and data.

Table 8.1 lists the 80960SA/SB instruction set. The 80960SA does not include the floating-point instructions.

.2.2.i Floating Point Unit (Available with 80960SB only)

The on-chip floating point unit includes a full set of floating point operations including add, subtract, multiply, divide, trigonometric functions, and logarithmic functions. These operations are performed on single precision (32-bit), double precision (64-bit), and extended precision (80-bit) data. Four 80-bit floating-point registers are provided to hold extended precision values.

8.2.3 80960 SA/SB Registers

Figure 8.1 shows the 80960SB registers. The processor provides three types of data registers: global, floating-point, and local. As their names imply, the global registers constitute a set of general-purpose registers whose contents are retained across procedure boundaries. The

TABLE 8.1   80960SA/SB Instruction Set

Data Movement	Arithmetic	Logical	Bit and Bit Field
Load, Store, Move, Load Address	Add, Subtract, Multiply, Divide, Shift, Remainder, Modulo, Extended Multiply, Extended Divide	And, Not And, And Not, Or, Xor, Not Or, Or Not, Nor, Exclusive Nor, Not, Nand, Rotate	Set Bit, Clear Bit, Not Bit, Check Bit, Alter Bit, Scan for Bit, Scan Over Bit, Extract, Modify

Comparison	Branch	Call/Return	Fault
Compare, Conditional Compare, Compare and Increment, Compare and Decrement	Unconditional Branch, Conditional Branch, Compare and Branch	Call, Call extended, Call system, Return, Branch and Link	Conditional Fault, Synchronous Fault

Debug	Miscellaneous	Decimal	Conversion
Modify Trace Controls, Mark, Force Mark	Atomic Add, Atomic Modify, Flush Local Register, Modify Arithmetic Controls, Modify Process Control, Scan Byte For Equal, Test Condition Code	Move, Add with Carry, Subtract with Carry	Convert Real to Integer, Convert Integer to Real

Floating-Point	Synchronous
Move Real, Add, Subtract, Multiply, Divide, Remainder, Scale, Round, Square Root, Sine, Cosine, Tangent, Arctangent, Log, Log Binary, Log Natural, Exponent, Classify, Copy Real Extended, Compare	Synchronous Move, Synchronous Load

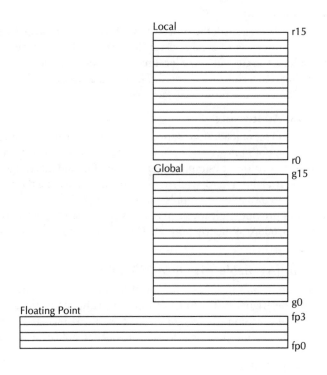

FIGURE 8.1   Local and global registers sets.

4 floating point registers are for extended precision floating point operations and are available only with the 80960SB. Their contents are also preserved across procedure calls. The 16 local registers are to hold local variables. For each procedure that is called, the processor allocates a separate set of local registers.

It should be noted that the global register g15 is reserved to hold the current frame pointer FP, while the others are available for general use. The local register r0 is used to hold the previous frame pointer (PFP), r1 is the stack pointer, r2 is used as a Return Instruction Pointer which is saved on the stack later and r3-r15 are available for general use.

Some special features of the 80960SB registers are provided in the following:

2.3.a Register Scoreboarding

The main purpose is to permit instructions to be executed concurrently provided that they are independent instructions.

2.3.b Instruction Pointer

The 32-bit r1 holds the address of the instruction currently being executed. Since the instructions are required to be aligned on a word boundary, the least significant 2 bits of IP are always 0. IP can not be read directly. However, IP can be used as an offset into address space. This addressing mode can be used with the load address (lda) instruction to read the current value of IP. When a break in instruction stream occurs due to an interrupt or procedure call, the IP contents will be stored in r2, and later saved on the stack.

2.3.c Process Control Register

The processor's process control register is made up of a set of 32 bits, as shown below:

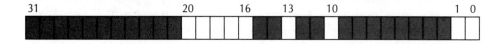

```
Bit 0       :   Trace Enable
Bit 1       :   Execution Mode, 0 = User, 1 = Supervisor
Bit 10      :   Trace Fault Pending
Bit 13      :   State Flag, 0 = Execution Mode, 1 = Interrupted
                Mode
Bit 20-16   :   Priority
```

2.3.d Arithmetic Control

The arithmetic control bits include the condition code, arithmetic status, integer overflow flag and mask, floating point overflow, underflow, zero divide, invalid-op, inexact flags, masks, and faults. The processor sets or clears these bits to show the results of certain operations. For example, the processor modifies the condition code flags after each fault. These bits are set by the currently running program to tell the processor how to respond to certain fault conditions.

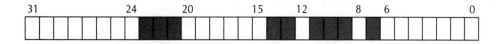

```
Bit 0-2     :   Condition Code
Bit 3-6     :   Arithmetic Status
```

```
Bit  8     :   Integer Overflow Flag
Bit  12    :   Integer Overflow Mask
Bit  15-20 :   Floating Point Condition Flags
Bit  24-29 :   Floating Point Condition Masks
Bit  30-31 :   Floating Point Normalizing and Rounding Mode
```

8.2.4 Data Types and Addresses

In order to be consistent with the data types included in the Intel 80960 manual, new terminologies such as ordinal and literals are introduced in this section.

8.2.4.a Data Types

The processor defines and operates on the following data types:

- Integer (8, 16, 32, and 64 bits)   signed whole numbers
- Ordinal (8, 16, 32, and 64 bits)   general-purpose, unsigned whole numbers
- Real (32, 64, and 80 bits)   conforms to IEEE single (32-bit), double (64-bits),
  (floating-point)   and extended precision (80-bit) floating point
                                        representations
- Bit/Bit Field   span of 1 or more bits within register boundary
- Decimal (ASCII digits)   decimal values in ASCII format.
- Triple-Word (96 bits)   consecutive bytes
- Quad-Word (128 bits)   consecutive bytes

8.2.4.b Literals

The processor recognizes two types of literals: ordinal and floating-point, which can be used as operands in some instructions. An ordinal literal can range from 0 to 31 (5 bits). When an ordinal literal is used as an operand, the processor expands it to 32 bits by adding leading zeros. If an ordinal literal is used in an instruction that requires integer operands, the processor treats the literal as a positive integer value.

For floating-point, the processor recognizes two literals: +0.0 and +1.0. These floating point literals can only be used with floating point instructions. Ordinal literals can also be used in converting integer to real to get more values.

8.2.4.c Register Addressing

A register may be used as an operand in an instruction by giving the register number (e.g., g0, f5, fp3). Both floating-point and non-floating-point instructions can reference global and local registers in this way. However, floating-point registers can only be referenced in conjunction with a floating-point instruction.

If the instruction requires more than one word, the reference is to the lowest number, which must be even when 2 words are required, must be multiples of four when 3 or 4 words are required. This is called "Register Alignment."

8.2.4.d Memory Addressing Modes

Table 8.2 lists the 80960 memory addressing modes.

8.2.4.d.i Absolute Offset.   Absolute offset is used to reference a memory location directly. An example is st g2, START which stores the word from register g2 into memory location START.

8.2.4.d.ii Register Indirect/Register Indirect with Offset.   This mode permits an address to be specified with an ordinal value (32 bits) in a register or a displacement added to a value in a register. The register value is called the address base (abase). An example of register indirect is the ldob (r1), r2 which loads an ordinal byte from memory location addressed by r1 into r2. An example of

TABLE 8.2 80960 Memory Addressing Modes

Mode	Description	Assembler Syntax
Absolute offset	offset	exp
Register indirect	abase	(reg)
Register indirect with offset	abase + offset	exp(reg)
Register indirect with index	abase + (index * scale)	(reg)[reg*scale]
Register indirect with index and displacement	abase + (index*scale) + displacement	exp(reg)[reg*scale]
Index with displacement	(index*scale) + displacement	exp[reg*scale]
IP (instruction pointer) with displacement	IP + displacement + 8	exp(IP)

register indirect with offset is stl g4, BEGIN(g2) which stores double word from g4, g5 stored at memory location addressed by BEGIN+(g2).

8.2.4.d.iii Register Indirect with Index/Register. This mode allows a scaled index value in another register to be added to the value in a register. The scale factor can be 1, 2, 4, 8, or 16. A displacement may also be added to the abase value and scaled index. An example is ldq (r1) [r2*4], r4 which loads a quad-word starting at the memory location addressed by (r1)+(r2 scaled by 4) into register r4 through r7.

An example of register indirect with index and displacement is st g1, VALUE(g3)[g4*4] which loads word in g1 into memory location addressed by (g3) +VALUE+(g4*4).

8.2.4.d.iv Index with Displacement. This mode allows a scaled index to be used with a displacement. The index is contained in a register and is multiplied by a scaling constant before the displacement is added to it. An example is ldis VALUE [r8*2], r10 which loads short integer at memory location addressed by VALUE+(r8*2) into r10.

8.2.4.d.v IP with Displacement. This mode is often used with load and store instructions to make them IP relative. With this mode the displacement plus a constant of 8 is added to the IP of the instruction. An example is st r1, VALUE (IP) which stores words in r1 at memory location addressed by 8+IP+VALUE.

8.2.5 80960SA/SB Instruction Set

The 80960 includes 182 instructions. An assembly-language statement consists of an instruction mnemonic, followed by from 0 to 3 operands, separated by commas. The following example illustrates the assembly-language statement for the addo instruction:

```
addo g1, g3, g5
```

adds the ordinal operands in global register g1 and g3 and stores the result in g5.

The instructions can be classified into four categories:

1. Data Movement
2. Conversion
3. Arithmetic and Logic Operations
4. Comparison and Control

The following provides a list of operands used in the instructions:

reg — global (g0, g1, . . . , g15) or local (r0, r1, . . . , r15) registers
freg — global (g0, g1, . . . , g15) or local (r0, r1, . . . , r15) registers or floating-point (fp0 thru fp3) registers
lit — integer or ordinal literal of the range 0. . .31
flit — floating-point literal of value 1.0 or 0.0

disp   — signed displacement of range -2^{22} to $+2^{22}-1$
mem   — address defined with the full range of addressing modes
addr — address
efa   — effective address

8.2.5.a Data Movement

The data movement instructions move data between the global and local registers, and between these registers and memory.

8.2.5.a.i *Load Instructions.*

$$
\left.\begin{array}{l}
\text{Load integer byte, ldib} \\
\text{(8-bit)} \\
\text{Load integer short, ldis} \\
\text{(16-bit)} \\
\text{Load ordinal byte, ldob} \\
\text{(8-bit)} \\
\text{Load ordinal short, ldos} \\
\text{(16-bit)}
\end{array}\right\} \text{mem, reg}
$$

The above instructions load a byte or half word (2 bytes) and convert it into a full 32-bit word. Integers are sign-extended, ordinals and zero-extended automatically. For example, ldib (r1), r0 loads the 8-bit integer in memory addressed by r1 into register r0.

```
Load   - ld  ⎫
Load long  - ldl ⎬ mem, reg
Load triple - ldt ⎪
Load quad  - ldq ⎭
```

The ld, ldl, ldt and ldq instructions copy 4, 8, 12, and 16 bytes, respectively, from memory into successive registers.

ldl mem, reg must specify an even numbered register (eg. g0, g2, . . ., g16).

ldt mem, reg and ldq mem, reg must specify a register that is a multiple of four (eg. g0, g4, g8, . . ., r0, r4, r8,. . .) For example, consider ldq 1254(r1), r4 loads the contents of memory location starting at address r1+1254 into register r4 thru r7.

8.2.5.a.ii *Store Instructions.*   Each load instruction has a corresponding store instruction which stores bytes or words from registers to memory.

The store instructions are listed below:

```
st   ⎫
stob  ⎪
stos  ⎪
stib  ⎬ reg, mem
stis  ⎪
stl   ⎪
stt   ⎪
stq  ⎭
```

The stob and stib, and stos and stis instructions store a byte and half-word (16-bit), respectively, from the low order bytes of the specified source register. The st, stl, stt, and stq instructions store 4, 8, 12, and 16 bytes, respectively, from successive registers to memory.

For the stl instruction, the specified source register must specify an even numbered register (e.g. g0, g2, . . ., g16). For the stt and stq instructions, the specified source register number must be a multiple of four such as g0, g4, g8. . ., g16.

As an example, the instruction st g4, 2478(g8) stores the word in register g4 into memory location starting at offset 2478+(g8).

2.5.a.iii MOVE. The move instructions copy data from a register or group of registers to another register or group of registers. These are listed below:

```
move word   - mov   ⎫
move longword - movl ⎪
                     ⎬  reg/lit, reg
move triple word - movt ⎪
move quad word - movq ⎭
```

The movl, movt, and movq instructions specify the source and destination registers as the first (lowest numbered) register of several successive registers. For the movl, these registers must be even numbered such as g0, g2, .., r0, r2, ... while for the movt and movq instructions, these registers must be an integral multiple of four such as g0, g4. .., r0, r4, ...

As an example, the instruction movt r4, g8 moves a triple word (three 32-bit) from registers r4, r5, r6 into g8, g9, g10.

2.5.a.iv Load Address. lda mem, reg computes an effective address specified with mem or efa and stores it in the destination, reg. Note that efa represents an effective address based on an addressing mode. This instruction loads a constant longer than 5 bits into a register. To load a register with a constant of 5 bits or less, the move instruction (mov) can be used with a literal as the source operand.

As an example, the instruction, lda 40(g7), g0 computes the effective address specified with 40+(g7) and stores it in g0.

lda 0x845, r4 loads the constant 845H into r4. Note that 0x indicates data in hexadecimal.

2.5.a.v Floating-Point Move (Available with 80960SB Only). The following move-real instructions (movr, movrl and movre) are provided for moving real number values between the global and local registers and the floating-point registers:

```
move real - movr       ⎫
move longreal - movrl  ⎬  freg/flit, freg
move extended real - movre ⎭
```

As an example, the following instruction sequence converts a real value in g1 to a long real value, which is stored in g8, g9.

```
movr   g1, fp0
movrl  fp0, g8
```

The two instructions cpysre and cpyrsre for real extended numbers are explained in the following:

```
cpysre   src1, src2, dst
   or
cpyrsre  freg/flit freg/flit freg
```

copies the absolute value of src1 into dst based on the sign of src2.

For cpysre: If src2 is positive then dst ← abs(src1); else dst ← abs(src1).
For cpyrsre: If src2 is negative then dst ← abs(src1); else dst ← abs(src1)

If the src1, src2, or dst operand specifies one of g0 thru g15 or r0 thru r15, this register (lowest) is the first of three successive registers. Also, this register number must be a multiple of 4 (e.g. g0, g4, g8, ..).

As an example, the instruction, cpysre g0, fp1, r4 means that the absolute value from g0g1g2 is copied to r4r5r6; the sign from fp1 is copied to r4r5r6.

8.2.5.b Conversion (Available With 80960SB Only)

As mentioned before, data can be converted from one length to another by means of the load and store instructions. For example, the ldis instruction loads a short integer from memory to a register and automatically converts the integer from a half word to a full word.

The 80960SB extended instruction set provides instructions to perform conversions between integer and real data types. These instructions are listed below:

```
Convert integer to real , cvtir      ⎫
Convert long integer to real, cvtilr  ⎬  reg/lit, freg
Convert real to integer, cvtri        ⎪
Convert real to long integer, cvtril  ⎭
Convert truncated real to integer, cvtzri       ⎫ freq/flit,reg
Convert truncated real to long integer, cvtzril ⎭
```

For the cvtilr instruction, the source operand specifies the first (lowest numbered) of two successive registers. This register must be even numbered (e.g. g0, g2, g4, . . .).

Converting an integer to long real format requires two instructions as follows:

1. cvtir or cvtilr can be used to convert the integer to extended real.
2. movrl can then be used to move the value from freg to two global or local registers.

For example the instruction sequence:

```
        cvtir g2, fp0
        movrl fp0, g4
```

converts an integer in g2g3 to real and stores it in fp0; movrl then converts the real value in fp0 to a long real value and stores the result in g4g5.

The cvtril and cvtzril instructions specify the destination operand as the first (lowest numbered) of two successive registers. This register must be even numbered. Also, the nontruncated version of cvtzri and cvtzril instructions round according to the current rounding mode in the Arithmetic Control register. The truncated version always rounds towards zero.

As an example, the following instruction sequence converts a long real value in g8g9 to a long integer in g2g3:

```
movrl g8, fp0   ;  long real source in g8g9 is converted to
                ;  extended-real format in fp0
cvtril fp0, g2  ;  extended real value in fp0 is converted
                   to long
                ;  integer in g2g3.
```

Synchronous Load and Move

Both the 80960SA and 80960SB include these instructions.

The 80960SA/SB executes the store instructions asynchronously with the memory controller. Once the processor outputs data for storing in main memory, it continues with execution of the next instruction in the program, and assumes that its bus control logic hardware will complete the operation. The 80960SA/SB includes four special instructions for performing memory operations that perform store and move operations synchronously with memory.

The synchronous load instruction, synld reg/addr, reg copies a word from the source into a register. When this instruction is performed, the processor waits until a condition code bit is set in the arithmetic control register indicating that the operation has been completed, before it begins executing the next instruction. The synld instruction is primarily used to read the contents of the interrupt-control register.

The following instructions

$$\left.\begin{array}{l} \text{Synmov} \\ \text{Synmovl} \\ \text{Synmovq} \end{array}\right\} \quad \text{reg/addr, reg/addr}$$

copies one (synmov), two (synmovl) or four (synmovq) words from memory location(s) specified by the source to the destination and waits for completion, including those operations initialed prior to this instruction. The primary function of these instructions is for sending IAC (Inter-agent communication) messages. The primary function of an IAC mechanism is to provide alternative to the external interrupt mechanism to communicate with the processor. Also, certain processor functions such as purging the instruction cache and setting breakpoint registers can only be done with the IAC mechanism. IAC messages are defined in such a way that processors can send them amongst themselves on the bus in a multiprocessor system. For example, a program on processor A can send a message to processor B telling it to flush its instruction cache. Without this facility, processor A would need to generate an interrupt to processor B to tell a program in processor B to flush the cache.

Since IAC messages carry out specific control functions that are not included in instructions, they are useful in single-processor systems. The 80960SA/SB can send an IAC message by writing the message to a special memory-mapped location. The memory mapping only occurs if one of the synchronous load/move instructions is used. A memory write to its specific memory-mapped location using one of these instructions does not cause a bus operation to occur; instead the data are interpreted by the processor as an IAC message and the message causes the same function to be performed by the processor. The function is performed synchronously (i.e. immediately after the synchronous load/move) instruction.

.2.5.c Arithmetic and Logic Operations

.2.5.c.i Table 8.3 lists 80960SA/SB add, subtract, multiply, divide, and shift instructions.

TABLE 8.3   80960SA/SB Add/Subtract/Multiply/Divide/Shift

Operations	Instructions and data types			
	Integer	Ordinal	Real	Long Real
add* src1, src2, dst reg/lit reg/lit reg dst ← src2 + src1 * = i or o or r or rl	addi	addo	addr	addrl
sub* src1, src2, dst reg/lit reg/lit reg dst ← src2 - src1 * = i or o or r or rl	subi	subo	subr	subrl
mul*src1, src2, dst freg/flit freg/flit freg dst ← src2 * src1 * = i or o or r or rl	muli	mulo	mulr	mulrl

TABLE 8.3   80960SA/SB Add/Subtract/Multiply/Divide/Shift (*continued*)

Operations	Instructions and data types			
	Integer	Ordinal	Real	Long Real
div* src1, src2, dst dst ← src2 / src1 reg/lit reg/lit reg * = i or o or r or rl No remainder is provided after div*, dst contain quotient.	divi	divo	divr	divrl

rem*src1, src2, dst
   freg/flit freg/flit freg
Performs src2 / src1 and stores the remainder in dst. The sign of
  the result (if nonzero) is the same as the sign of src2. Calculation
  of the remainder is done by repeated subtraction.
* = i or o or r or rl
	remi	remo	remr	remrl

signed integer modulo:
  modi src1, src2, dst
     reg/lit reg/lit reg
  dst = src2 − (src2 ÷ src1)*src1
	modi	—	—	—

Shift left
shl* Len, src, dst
   reg/lit reg/lit reg
Shifts src left by the number of bits specified in the Len operand
  and stores result in dst. For values greater than 32, the processor
  interprets the value as 32.
* = i or o
	shli	shlo	—	—

Shift right
shr* len, src, dst
   reg/lit reg/lit reg
shifts src right by the number of bits indicated with the len
  operand and stores the result in dst. For values of len greater
  than 32, the processor interprets the value as 32.
* = i or di or o
	shri
	shrdi
		shro	—	—

Details of Table 8.3

Note the instructions addr/addrl, subr/subrl, roundr/roundrl and sqrtr/sqrtrl are only available with the 80960SB.

For addrl, subrl, mulrl, divrl and remrl instructions, if src1, src2, or dst operand specifies one of the registers from g0 thru g15 or r0 thru r15, the register is the first (lowest numbered) of two successive registers. Also, this register must be even numbered (eg. go, g2, g4, . . .).

The binary results from subi and subo are identical except that subi can signal an integer overflow.

For the divi instruction, an integer overflow can be signaled.

The shlo instruction shifts zeros into the least-significant bit and the shro instruction shifts zeros into the most-significant bit.

The shli instruction shifts zeros into the least-significant bit; if the bits shifted out are not the same as the sign bit, an overflow is generated. If overflow occurs, the sign of the result is the same as the sign of the src operand.

The shri instruction performs an arithmetic shift operation by shifting the sign bit in from the most-significant bit.

The shrdi instruction is provided for dividing an integer by a power of 2. With shrdi, one is added to the result if the bits shifted out are non-zero and the operand is negative, which produces the correct result for negative operands.

Remi and modi differ when there is a negative operand: the result of remi has the same sign as the dividend; that of modi has the same sign as the divisor. For example, if r3 = 3, r4 = (−7):

"remi r3, r4, r5" stores (−1) to r5, (−7) = −2 * 3 + (−1)
"modi r3, r4, r5" stores 1 to r5, (−7) = −2 * 3 − 1

shrdi adds 1 to the result if bits shifted out are non-zero and operand is negative, which produces the correct result for negative operands (if division is desired).

2.5.c.ii Rotate Instruction. The operation of the rotate instruction is provided below:

Instruction	Operation
rotate len, src, dst reg/lit reg/lit reg	dst ← rotate (len mod 32 (src)) copies src to dst and rotates the bits in the dst as follows:

The len operand specifies the number that the dst operand is rotated. The len operand can be from 0 to 31.
The instruction can also be used to rotate bits to right.

2.5.c.iii Extended Arithmetic. There are four instructions for double precision integer arithmetic. These are described below:

1. Add ordinal with carry,
   addc src1, src2, dst
      reg/lit reg/lit reg
   Operation: dst ← src2 + src1 + carry
   Flags affected: carry (c) and overflow (v)
2. Subtract ordinal with carry,
   subc src1, src2, dst
      reg/lit reg/lit reg
   Operation: dst ← src2 − src1 + carry
   Flags affected: carry, overflow.
3. Extended multiply,
   emul src1, src2, dst
      reg/lit reg/lit reg
   Operation: dst + 1, dst ← src1 * src2
   The result is 64 bits and is stored in two adjacent registers. The dst operand specifies the lower numbered register, which receives the least significant bits of the result. The dst operand must be an even numbered register (ro, r2, r4, . . ., or g0, g2, . . .).
4. Extended Divide
   ediv src1, src2, dst
      reg/lit reg/lit reg
   Operation: dst ← Remainder of src2/src1
   dst + 1 ← Quotient of src2/src1
   Scr2 is a long ordinal (64 bits) which is contained in two adjacent registers. Src2 specifies the lower numbered register which contains the least significant bits of the operand. Src2 operand must be an even numbered register. Src1 value is a normal ordinal 32 bits. dst operand must be an even numbered register.

2.5.c.iv Floating-Point Arithmetic Instructions (Available with 80960SB Only). In addition to floating-point add (addr/addrl), subtract (subr/subrl), multiply (mulr/mulrl) and divide (divr/divrl) which were already explained, additional floating-point instructions are listed in Table 8.4.
Note that in Table 8.4,

For roundrl,
sqrtrl,
sinrl,
cosrl,   } → If the src or dst operand references a global or local register, this
tanrl   register is the first (lowest numbered) of two successive registers.
logbnrl,   This register must be even numbered (g0, g2,. . ., r0, r2. . .).
exprl.

For atanrl,
logrl,   } → If the src1, src2, or dst references a global or local register, this
logeprl,   register is the first (lowest numbered) of two successive registers.
scalerl,   This register must be even numbered (g0, g2, . . ., r0, r2, . . .).

TABLE 8.4 80960SB Floating-Point Arithmetic Instructions Beyond Add/Subtract/Multiply/Divide

Operation		Instructions and Data Types	
		Real	Long Real

Basic:	round* src, dst
	  freg/lit freg
	rounds src to the nearest integral
	  value depending on the rounding
	  mode and stores the result in dst.
	  * = r or rl
		roundr	roundrl

	sqrt* src, dst
	  freg/flit freg
	calculates the square root of src and
	  stores it in dst
	  * = r or rl
		sqrtr	sqrtrl

Trigonometric Operation			

Calculate the specified trigonometric function
of src and stores the result in dst. The src
value is in radians. The resulting dst value is
in the rang −1 to +1 inclusive for sine and
cosine
	Sin* src, dst
	  freg/flit freg
	  * = r or rl
		sinr	sinrl

	cos* src, dst
	  freg/flit freg
	  * = r or rl
		cosr	cosrl

For tangent, the source value is a finite real
number between −∞ to +∞
	tan* src, dst
	  freg/lit freg
	  * = r or rl
		tanr	tanrl

atan calculates arctangent of src2/src1 and
stores result in dst. The result is in radians
and lies between −π to π inclusive. The sign
of the result is same as the sign of src2.
	atan* src1, src2, dst
	  freg/lit freg/lit freg
	  * = r or rl
		atanr	atanrl

Operation (Logarithmic, Exponential, and Scale)			

Logbn calculates the $\log_2(src)$ and stores the
integral part of this value as real number in
dst.
	logbn* src, dst
	  freg/flit freg
	  * = r or rl
		logbnr	logbnrl

Log* calculates src2*$\log_2$(src1) and stores result
in dst. Compute y*$\log_2$(x).
	log*src1, src2, dst
	  freg/flit freg/flit freg
	  * = r or rl
		logr	logrl

logep* calculates rc2*$\log_2$(src1 + 1) and stores
result in dst. Compute y*$\log_2$(x + 1).
	logep*src1, src2, dst
	  freg/flit freg/flit freg
	  * = r or rl
		logepr	logeprl

exp* performs dst ← (2^{src} − 1). The source value must be within −0.5 to +0.5 inclusive. Compute 2^x − 1.	exp* src, dst freg/flit freg	expr	exprl

TABLE 8.4   80960SB Floating-Point Arithmetic Instructions Beyond Add/Subtract/Multiply/Divide (*continued*)

Operation (Logarithmic, Exponential and Scale)		Instructions and Data Types	
		Real	Long Real
scale* performs dst ← src2*(2^{src1}). src1 is integer, src2 and dst are reals. Multiply a floating-point value by a power of 2.	scale*src1, src2, dst reg/lit freg/flit freg	scaler	scalerl

8.2.5.c.v Logical, Bit/Bit Field Operations.   Table 8.5 lists these instructions:

TABLE 8.5   Logical Instructions

Instruction	Operation Performed
and src1, src2, dst reg/lit reg/lit reg	dst ← src2 ∧ src1
andnot src1, src2, dst reg/lit reg/lit reg	dst ← src2 ∧ (src)′
nand src1, src2, dst reg/lit reg/lit reg	dst ← (src2 ∧ src1)′
notand src1, src2, dst reg/lit reg/lit reg	dst ← (src2) ′ ∧ src1
or src1, src2, dst reg/lit reg/lit reg	dst ← src2 ∨ src1
ornot src1, src2, dst reg/lit reg/lit reg	dst ← src2 ∨ (src1)′
nor src1, src2, dst reg/lit reg/lit reg	dst ← (src2 ∨ src1)′
notor src1, src2, dst reg/lit reg/lit reg	dst ← (src2) ′ ∨ src1
xor src1, src2, dst reg/lit reg/lit reg	dst ← src2 ⊕ src1
xnor src1, src2, dst reg/lit reg/lit reg	dst ← (sr2 ⊕ src1)′

Note that in the above, ∧ = and, ∨ = or, ⊕ = exclusive or, ′ = NOT

Table 8.6 lists bit/bit field instructions.

TABLE 8.6   Bit/Bit Field Instructions

Instruction	Operation
alterbit bitpos, src, dst reg/lit reg/lit reg	copies the src to dst with one bit altered. The bitpos specifies the bit to be changed and the condition code determines the value the bit is to be changed to. If the condition code is 010_2, the selected bit is set to one; if the condition code is 000_2, the bit is cleared to zero.
chkbit bitpos, src reg/lit reg/lit	checks the bit in src specified by bitpos and sets the condition code according to the value found. If the bit is one, the condition code is set to 010_2; if the bit is zero, the condition code is cleared to 000_2.
clrbit bitpos, src, dst reg/lit reg/lit reg	copies src to dst with the bit specified by bitpos cleared to zero.
notbit bitpos, src, dst reg/lit reg/lit reg	copies src to dst with the bit specified by bitpos ones complemented.
scanbit src, dst reg/lit reg	searches src for most-significant set-bit. If the set-bit is found, its bit number is stored in dst and the condition code is set to 010_2. If src is zero, all ones are stored in dst and the condition code is cleared to 000_2.
setbit bitpos, src, dst reg/lit reg/lit reg	copies src to dst with the bit specified by bitpos set to one.

TABLE 8.6   Bit/Bit Field Instructions (*continued*)

Instruction	Operation
sparobit src, dst reg/lit reg	searches src for the most-significant clear-bit. If the clear-bit is found, its number is stored in dst and the condition code is set to 010_2. If the src value is all ones then all ones are stored in dst and the condition code is cleared to 000_2.
extract bitpos, len, src/dst reg/lit reg/lit reg	shifts a specified bit field in src/dst right and fills the bits to the left of the shifted bit field with zeros. The bitpos value specifies the least significant bit of the bit field to be shifted and the len value specifies the length of the bit field.
modify mask, src, src/dst reg/lit reg/lit reg	modifies selected bits in src/dst with bits from src. The mask operand selects the bits to be modified: Only the bits set in the mask operand are modified in src/dst. src/dst ← (src ∧ mask) ∨ (src/dst ∧ (mask)′)

8.2.5.c.vi Byte Operations.   The scanbyte instruction performs a byte-by-byte comparison of two ordinals to determine if two corresponding bytes are equal.

The format of the scanbyte is as follows:

scanbyte src1, src2
reg/lit reg/lit

The scanbyte performs a byte-by-byte comparison of src1 and src2 and sets the condition code to 010_2 if any two corresponding bytes are equal. If no corresponding bytes are equal, the condition code is cleared to 000_2.

The scanbyte operation is detailed below:

$$\text{If } (src1 \wedge 000000FF_{16}) = (src2 \wedge 000000FF_{16})$$
$$\text{or}$$
$$(src1 \wedge 0000FF00_{16}) = (src2 \wedge 0000FF00_{16})$$
$$\text{or}$$
$$(src1 \wedge 00FF0000_{16}) = (src2 \wedge 00FF0000_{16})$$
$$\text{or}$$
$$(src1 \wedge FF000000_{16}) = (src2 \wedge FF000000_{16})$$

then condition code = 010_2; else condition code = 000_2.

8.2.5.c.vii Decimal Arithmetic (Available with 80960SB Only).   These instructions operate on 32-bit decimal operands that contain an 8-bit ASCII-coded decimal in the least-significant byte.

dmovt src, dst reg reg	copies src to dst. The least significant byte of src is tested to find whether or not it is a valid ASCII digit (30_{16} thru 39_{16}). If the value is a valid ASCII decimal, the condition code is cleared to 000_2; otherwise, it is set to 010_2. This instruction is normally used iteratively to validate decimal strings.
daddc src1, src2, dst reg reg reg	adds bits 0 thru 3 of src2 and src1 (with bit 1 of condition code used here as carry bit). The result is stored in bits 0 thru 3 of dst. If there is a carry after addition, bit 1 of condition code is set to one. Bits 4 thru 31 of scr2 are copied to dst unchanged. The instruction assumes that the least significant 4 bits of src1 and src2 are valid BCD digits.

The daddc is intended to be used iteratively to add BCD values in which the least significant four bits of the operands represent valid BCD numbers from 0 to 9.

dsubc src1, src2, dst reg reg reg	subtracts bits 0 thru 3 of src1 and src2 as follows: dst ← src2 − src1 − 1 + C.

Bit 1 of condition code is used as C (carry bit). The other characteristics of dsubc are same as the daddc instruction.

The dsubc is intended to be used iteratively to subtract BCD values in which the least significant four bits of the operands represent valid BCD numbers from 0 to 9.

8.2.5.c.viii Atomic Instructions. In multiprocessor systems, a mechanism is required to allow programs to manipulate shared data in an indivisible manner so that when such an operation is underway, another processor cannot perform the same operation. The 80960 includes two instructions called atomic instructions to implement higher-level synchronization mechanisms, such as locks and semaphores.

The atmod src, mask, src/dst
reg reg/lit reg
addr

instruction copies the src/dst value into the memory location specified in src. The src is a register containing the address and thus the name reg addr in the instruction. The bits set in the mask operand select the bits to be modified in memory. The initial value from memory is stored in src/dst.

For example, atmod g1, g3, g6 performs the following:

g1 ← g1 ANDed by g3 where g1 contains the address of a word in memory.
g6 ← Initial value stored at address g1 in memory.

The read and write of memory are done atomically (i.e. other processors are prevented from accessing the word of memory specified with the src/dst operand until the operation has been completed).

The memory location in src is the address of the first byte (least significant byte) of the word to be modified.

The atadd src/dst, src, dst
reg reg/lit reg
addr

adds the src value (full word) to the value in memory specified by src/dst. The initial value from memory is stored in dst.

The read and write of memory are done atomically. The memory location in src/dst is the address of the first byte (least significant byte) of the word.

The atadd instruction, therefore, adds a value of a word in memory and returns the original value of the word. For example, atadd r2, r4, r9 performs the following:

r2 ← r4 + (r2) where r2 specifies the address of a word in memory;
r9 ← initial value stored at address r2 in memory.

The atomic read operation waits until the LOCK line on the external bus is not asserted and then asserts the LOCK line and performs the read. The atomic write operation performs a write operation and deasserts the LOCK line. This ensures that another processor cannot perform an atomic read operation between read and write to the word in memory specified with the src/dst operand until the operation has been completed.

8.2.5.d Comparison and Control

Though this 80960SA/SB RISC processor has a condition code register, it is not affected by most arithmetic and movement instructions. An explicit comparison instruction is needed for conditional branches. This feature has its advantage. Between the instruction that sets condition code and the instruction that performs conditional branching, many independent arithmetic operations can be inserted. That will increase the pipeline efficiency. Arithmetic instructions that change condition codes are: addc, subc, dmovt, daddc, dsubc.

8.2.5.d.i *Comparison.*	These instructions compare integer (signed numbers) and ordinals (unsigned numbers):

> **Compare Integer (cmpi)/Ordinal (cmpo)**
> cmpi	src1, src2
> or	reg/lit reg/lit
> cmpo

compares src2 and src1 values and sets the condition code according to the following:

Condition Code	Comparison
100	src1 < src2
010	src1 = src2
001	src1 > src2

> **Compare and Increment Integer(cmpinci)/Ordinal (cmpinco)**
> cmpinci	src1, src2, dst
> or	reg/lit reg/lit reg
> cmpinco

compares src2 and src1 values and sets the condition code according to the results of the comparison. Src2 is then incremented by one and the result is stored in dst.

The condition codes are affected by the comparison result in exactly the same way as the cmpi/cmpo.

> **Conditional Compare Integer (concmpi)/Ordinal (concmpo)**
> concmpi	src1, src2
> or	reg/lit reg/lit
> concmpo

compares src2 and src1 value if bit 2 of the condition code is not set. If the comparison is performed, the condition code is set according to the comparison results in the same way as cmpi/cmpo.

> **Compare and Decrement Integer (cmpdeci)/Ordinal (cmpdeco)**
> cmpdeci	rc1, src2, dst
> or	reg/lit reg/lit reg
> cmpdeco

compares src2 and src1 values and sets the condition code according to the comparison results in the same way as cmpi/cmpo. The src2 is then decremented by one and the result is stored in dst.

The following instructions are for real and long real floating-point numbers (available with only 80960SB microprocessor):

> **Compare real (cmpr/longreal (cmprl))**
> compr	src1, src
> or	freg/flit freg/flit
> comprl

compares src2 with src1 and sets the condition code according to the result as follows:

Condition Code	Comparison
100	src1 < src2
010	src1 = src2
001	src1 > src2
000	if either src1 or scr2 is a NaN

cmpr/cmprl clears the condition code flags to 000_2 for the unordered condition. Note that the unordered relationship is true when at least one of the two values compared is a NaN.

Compare Ordered Real (cmpor)/Ordered Long Real (cmporl)
cmpor   src1, src2
  or   freg/flit freg/flit
cmporl

compares src2 and src1 and sets the condition code in the same way as cmpr/cmprl.

Compor/comporl clears the condition code to 000_2 and an invalid-operation exception is signaled for the unordered condition. Note that the unordered condition is true when at least one of the two values being compared is a NaN.

Classify Real (cassr)/Long Real (classrl)
classr   src
  or   freg/flit
classrl

checks classification of real number in src and stores the class in arithmetic-status bits (3 through 6) of the arithmetic controls as follows:

A Status	Classification
S000	Zero
S001	Denormalized number
S010	Normal finite number
S011	Infinity
S100	Quiet NaN
S101	Signaling NaN
S110	Reserved operand

The S bit is set to the sign of the src operand.

For cmprl and cmporl and classrl instructions, if src1 or src2 for cmprl/cmporl or src for calssrl specifies a global or local register, this register is the first (lowest numbered) of two successive registers. Also, this register must be even numbered.

8.2.5.d.ii *Control Instructions.* The 80960SA/SB include the following unconditional branch instructions:

Branch (b)/Branch Extended (bx)
b targ   or   bx targ
disp   mem

branches to the instruction specified with the targ operand.

For the b instruction the range of targ operand is from -2^{23} to $(2^{23} - 4)$ bytes from the current IP. For bx, the targ can be farther than -2^{23} to $(2^{23} - 4)$ bytes for the current IP. Also, since the targ operand for bx is a memory type, full range of addressing words including register indirect mode can be used.

Branch and Link (bal)/Link Extended (balx)
bal targ   or   balx targ, dst
disp   mem reg

stores the address of the next instruction (next IP value) in a register and branches to the instruction specified with the targ operand. These instructions are intended for calling leaf procedures (procedures that do not call other procedures). Using the b or bx instruction, the leaf procedure can branch to the IP saved by bal or balx.

For bal, the address of the next instruction is saved in g14. The range of targ is from -2^{23} to $2^{23} - 4$.

The balx performs the same operation as the bal except that the address of the next instruction is stored in dst, allowing it to be stored in any available register. The range of targ can be farther -2^{23} to $(2^{23} - 4)$ bytes from the current IP.

Compare and Branch

These instructions compare two operands, then branch (or not) according to the result:

	Compare	
Branch If	Integer (Signed)	Ordinal (Unsigned)
Equal	cmpibe src1, src2, targ	cmpobe src1, src2, targ
Not Equal	cmpibne src1, src2, targ	cmpobne src1, src2, targ
Less	cmpibl src1, src2, targ	cmpobl src1, src2, targ
Less or Equal	cmpible src1, src2, targ	cmpoble src1, src2, targ
Greater	cmpibg src1, src2, targ	cmpobg src1, src2, targ
Greater or Equal	cmpibge src1, src2, targ	cmpobge src1, src2, targ
Ordered	cmpibo src1, src2, targ	cmpobo src1, src2, targ
Unordered	cmpibno src1, src2, targ	cmpobno src1, src2, targ

In the above instructions, src1 = reg/lit, src2 = reg, and targ = disp. These instructions compare src1 and src2, and set the condition code based on the result. If the AND of the condition code and the mask part of the instruction is not zero, the processor branches to targ; otherwise, the processor goes to the next instruction. Note that the condition code 000_2 indicates no condition and is the unordered condition while condition code = 111_2 is the 'ordered' condition. The terms 'ordered' and 'unordered' are used when comparing two floating-point numbers. If, when comparing two floating-point values, one of the values is a NaN (Not a number), the relationship is said to be 'unordered'; otherwise, the releationship is 'ordered'.

Bit Instructions:
Check Bit and Branch if SET, bbs bitpos, src, targ
Check Bit and Branch if Clear, bbc reg/lit reg disp

bbs and bbc instructions check the bit in src specified by bitpos and set the condition code according to the value. The processor then branches to targ according to the condition.

Test Condition Codes

These instructions cause a TRUE (1) to be stored in a destination register if the condition code matches. Otherwise, a FALSE (0) is stored.

teste dst Test if Equal   testne dst Test if Not Equal
testl dst Test if Less   testg   dst Test if Greater

testle dst Test if Less or Equal   **testge** dst Test if Greater or Equal
testo dst Test Ordered   **testno** dst Test if Unordered

In the above, dst = reg.

Conditional Fault

These instructions permit a fault to be generated explicitly according to the state of the condition-code bits:

faulte	Fault if Equal	**faultne**	Fault if Not Equal
faultl	Fault if Less	**faultg**	Fault if Greater
faultle	Fault if Less or Equal	**faultge**	Fault if Greater or Equal
faulto	Fault Ordered	**faultno**	Fault if Unordered

Call and Return

The processor offers an on-chip call return mechanism for making procedure calls to local procedures and kernel procedures. These instructions support that mechanism:

call	targ	Calls where targ = disp
callx	targ	Calls Extended where targ = mem
calls	targ	Calls System where targ = reg/lit
ret		Return

The **call** and **callx** instructions call local procedures. They differ only in addressing mode. The processor will allocate a new set of local registers and a new stack frame for the called procedure. The **calls** instruction operates similarly, except that it gets its target procedure address from the system procedure table. Depending on the type of entry being pointed to in the procedure table, the calls instruction can cause a supervisor procedure call to be executed.

The **ret** instruction performs a return from a called procedure to a calling procedure. The same instruction is used to return from local and supervisor calls and from implicit calls to interrupt and fault handlers. The processor takes care of all the details.

Debug

The processor supports debugging and program tracing. These are the debugging tools:

modte Modify Trace Control
mark Mark — generates a breakpoint trace event if breakpoint trace mode flag is enabled.
fmark Force Mark — generates a breakpoint trace event regardless of the breakpoint trace mode flag.

Processor Management

The processor provides several instructions for use in controlling processor-related functions.

modpc src, mask, src/dst stores the contents of src/dst in the process control register, with
    reg/lit reg/lit reg the bits set in the mask modified. The src/dst then contains the initial value of the process control register. The src/dst is a dummy operand and must be set equal to the mask operand. The processor must be in the supervisor mode for executing this instruction.

flushreg copies each local register set except the current set to its associated stack-frame in memory and marks them as invalid, meaning that they will be reloaded from memory if and when they become the current local register set.

modac mask, src, dst   places the contents of src in the Arithmetic control register with the bits set in the mask modified register. The dst then contains
     reg/lit reg/lit reg   the initial value of the Arithmetic controls.

Conditional Branches:

be	Branch if Equal	**bne**	Branch if Not Equal
bl	Branch if Less	**bg**	Branch if Greater
ble	Branch if Less or Equal	**bge**	Branch if Greater or Equal
bo	Branch if Ordered	**bno**	Branch if Unordered

These instructions are single-operand with the operand "targ" or "disp" defined in the same way as bal.

Value of π

The 80960SA/SB uses the value $41490FDA_{16}$ for π. The details of this computation are given in Intel i960SA/SB Microprocessor SA/SB reference manual. As an example, π can be located into a register such as r4 by using lda 0X41490fda, r4 where ox is used to represent hexadecimal number by the 80960SA/SB assembler.

80960 Assembler

The 80960 assembler uses the first operand of a two operand instruction as the source operand and the second operand as the destination. The assembler directive # is used before a comment. 0X before an immediate number is used to represent a hex number.

Example 8.1

Identify the addressing modes for the following 80960SA/SB instructions:
     i) ldl 4816(r3),g4
     ii) st r3,34(r8)[r4*4]

Solution
     i)   source   destination
        register indirect   register
     ii) source   destination
        register   register indirect with
                         scaled index and displacement.

Example 8.2

Determine whether the following 80960SA/SB instructions are valid or invalid. Comment.

     i) ldq (g8) [g9], r2
     ii) stl 46, 52(r5)

Solution

     i) Not valid since for ldq instruction, the destination must specify a register number that is multiple of 4 such as r0, r4, r8, . . . , g0, g4, g8, ... Since register r2 is not a multiple of 4, the instruction is invalid.
     ii) Valid since for stl, the source must be an even numbered register which is r6 in this case.

Example 8.3

Write an 80960SA/SB instruction sequence to read 32-bit elements 5, 6, and 7 from a table stored in memory into register r1, r2, and r3 respectively. Assume that register r5 points to the starting address containing element 0 (32-bit data) of the table.

Solution

```
        ldt 5(r5), r8    #r8 ← ((r5+5))
                         #r9 ← ((r5+6))
                         #r10 ← ((r5+7))
        mov r8, r1       #r1 ← r8
        mov r9, r2       #r2 ← r9
        mov r10, r3      #r3 ← r10.
```

Example 8.4

Write an assembly language program in 80960SA/SB assembly language to add two 64-bit numbers. Assume that the two 64-bit numbers are stored in r1, r0 and r3, r2 respectively. Store the 64-bit result in r0, r1.

Solution

```
        cmpo  1,  0      #  clears bit 1(carry bit)
                        #  of the condition code register
        addc  r0,r2,r0  #  r0 ← r2+r0+carry bit
        addc  r1,r3,r1  #  r1 ← r3+r1+carry bit
finish  b finish        #  halt
```

Example 8.5

Write an 80960 assembly language program to perform the following operation:

$$(A/B) + C * D$$

where A, B, C, D are stored in r0, r1, r2, r3 as 32-bit integers. Assume C*D generates 32-bit product. Discard remainder of A/B. Store the 32-bit result in r4.

Solution

```
        divi  r1,  r0,  r4   #  r4 ← r0/r1
        muli  r2,  r3,  r5   #  r5 ← r2*r3
        addi  r5,  r4        #  r4 ← r4 + r5
finish  b finish            #  stop
```

Example 8.6

Write a program in 80960SA/SB assembly language that copies bits 3-6 of register r1 into bits 31-28 of register r2.

Solution

```
        extract 3,  4,  r1   #  r1 = 000. . . 000aaaa
        shlo 28, r1, r1      #  r1 = aaaa000. . . 000
```

```
        shlo 28, 15, r3     # r3 = 1111000. . . 000
        modify r3, r1, r2   # r2 = aaaabbb. . . bbb
finish  b finish            # halt
```

Example 8.7

Write a program in 80960SA/SB assembly language to branch to a label 'start' if the 32-bit operand in register g2 is not a finite number.

Solution

```
        classr g2
        modac 0, 0, g1        # place condition code in g1
                              # arithmetic status in bits
                              #  3-6 of g1
        shro 3, g1, g1        # move arith status in bits
                              #  0-3 of g1
        and 7, g1, g1         # g1 = bits 0-2 of arith status
        cmpobge g1, 3, start  # branch if status not equal
                              # to s000, s001, or s010.
finish  b finish              # halt
```

Example 8.8

Write an 80960SA/SB instruction to add four 32-bit words of additional space to the stack.

Solution

$$\text{addo sp, 16, sp} \quad \text{\# sp} \leftarrow \text{sp + 16}$$

Note that the 80960SA/SB Intel assembler uses 'sp' to represent the stack pointer, r1. Also, sp in the above instruction is incremented in one-byte increments by the addo instruction so that sp must be incremented by 16.

Example 8.9

Write a program in 80960 SA/SB assembly language to convert a long-real value in r0 to long-integer value in r8.

Solution

```
        movrl ro, fp0      # long-real value in
                           # ro is converted to
                           # extended-real in fp0
        cvtril fp0, r8     # extended real-value
                           # in fpo is converted to
                           # long integer
finish  b finish           # halt
```

Example 8.10

Write a program in 80960SA/SB assembly language to compute the area of a circle by using $A = \pi r^2$ where 'A' is the area in 32-bit real to be stored in register r1 and 'r' is the radius of the circle stored in r0 as 32-bit real.

Solution

```
        mulr ro, ro, g10    #   calculate r²
        lda 0x41490fda,r1   #   Load π
        mulr g10, r1, r1    #   r1 contains the area
finish  b finish            #   halt
```

Example 8.11

Write an 80960SA/SB assembly language program to convert from polar coordinates to rectangular coordinates as follows:

$$x = r\cos\theta, \qquad y = r\sin\theta$$

where r0, r1 contain r and θ (in radians) of the polar coordinates and r2, r3 contain x, y of the rectangular coordinates respectively.

Solution

```
        cosr r1, g4         #   g4 = cos(r1)
        sinr r1, g5         #   g5 = sin(r1)
        mulr r0, g4, r2     #   r2 = r*cos(r1)
        mulr r0, g5, r3     #   r3 = r*sin(r1)
finish  b finish            #   halt
```

8.2.6 80960SA/SB Pins and Signals

Figure 8.2 shows the 80960SA/SB pins and signals.

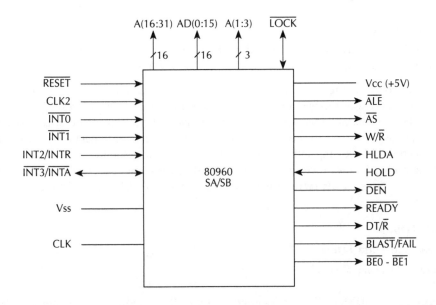

FIGURE 8.2 80960SA/SB pins and signals.

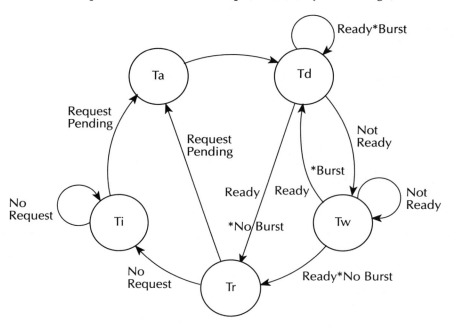

FIGURE 8.3   Basic L-bus states.

Some of the main features of the 80960SA/SB bus include the following:

- 32-bit addressing
- 16-bit multiplexed low 16-bit address/data bus.
- Two byte enables and an eight-word burst capability that allows transfers from 1 to 16 bytes in length.
- Basic bus states.

8.2.6.a Basic Bus States

There are five basic bus states: idle Ti, address Ta, data Td, recover Tr, and wait Tw as shown in Figure 8.3 assuming only one bus master resides on the bus.

- **Ti:** the processor enters this state when no address or data transfer is in progress.
- **Ta:** when processor receives a new request and starts transmitting address.
- **Td:** following Ta, the processor transmits or receives data if $\overline{\text{READY}}$ input is asserted. If not the processor enters wait state Tw and remains there until data is ready. Tw may be repeated allowing sufficient time for I/O devices to respond.
- **Tr:** following Td, the processor enters recovery state and comes back to Ti. In case of burst transactions, it exits Tw or Tr to transfer next data word. When done, it enters recovery state.

8.2.6.b Signals Groups

Address and Data Lines

The address/data signal consists of 35 lines.

A16–A31:   ADDRESS BUS carries upper 16-bit of 32-bit addresses to memory. No latch is required.

AD(0–15):   16-bit LOW ADDRESS/DATA BUS represents addresses in Ta and data in Td.

A(1-3):   ADDRESS BUS carries the burst addresses to memory. They are incremented during burst cycle indicating the next byte address of burst mode. They are duplicated with AD(1-3) during the address cycle.

Control Lines

Consists of 12 signals that permit the transfer of data.

$\overline{\text{ALE}}$:   ADDRESS LATCH ENABLE, active high, is asserted during Ta and deasserted before the beginning of the Td state. It floats when processor is not a bus master.

$\overline{\text{AS}}$:   ADDRESS STATUS indicates an address state is asserted every address state and deasserted during the following Td and driven HIGH during RESET.

DT / $\overline{\text{R}}$:   DATA TRANSMIT/RECEIVE indicates the flow of data. During READ operation at TA, Td, and Tr, it remains LOW. It is HIGH during Write operations.

$\overline{\text{DEN}}$:   DATA ENABLE enables data transceivers and is asserted during Td and Tw.

READY: INPUT from other devices indicates data on the bus ready to be read or written. If not asserted during a Td cycle, the Td cycle is extended for the next cycle by inserting a wait state, Tw.

$\overline{\text{LOCK}}$:   BUS LOCK, prevents other devices from gaining the control of the bus. Asserted when processor performs READ MODIFIED WRITE or INTERRUPT ACKNOWLEDGE. WAITS if LOCK is asserted by other devices.

$\overline{\text{BE1}}$ – $\overline{\text{BE0}}$:   BYE ENABLE indicates which data bytes (up to two) on the bus take part in the current bus cycle. BE1 corresponds to AD15-A8 and BE0 corresponds to AD7-AD0.

W / $\overline{\text{R}}$:   Instructs memory or I/O devices to write or read data on the bus. It is asserted during Ta and remains valid during subsequent Td cycles.

HOLD, HLDA: Used for DMA.

$\overline{\text{BLAST}}$ / $\overline{\text{FAIL}}$:   Indicates that an error occurred during the initialization. The failure state is indicated by a combination of $\overline{\text{BLAST}}$ asserted and both $\overline{\text{BE}}$ signals not asserted. $\overline{\text{FAIL}}$ is asserted while the processor performs a self-test. If the self test is successful, the $\overline{\text{FAIL}}$ is deasserted.

CLK2/CLK: The 80960SA/SB uses two clock signals (CLK2 and CLK). CLK2 provides the input clock to the 80960 and is double the specified processor frequency. CLK is the clock input signal for the peripheral devices and is the operating frequency of the processor.

The four interrupt pins of the 80960SA/SB are $\overline{\text{INT0}}$, INT1, INT2/INTR, and $\overline{\text{INT3}}$/ $\overline{\text{INTA}}$. The on-chip control register determines how these interrupts are used by the processor. The 80960SA/SB can be interrupted using any of the two methods as follows:

1. Receipt of a signal on any or all of the four direct interrupts ($\overline{\text{INT0}}$, INT1, INT2, and $\overline{\text{INT3}}$).
2. Receipt of a signal on the interrupt request (INTR) utilizes $\overline{\text{INTA}}$ to obtain an interrupt vector from an external device such as the 8259. The setting of the on-chip Interrupt Control Register selects one of the methods.

The $\overline{\text{RESET}}$ pin must be asserted for at least 41 CLK2 cycles. Upon hardware reset, the 80960 performs a self test if $\overline{\text{INT0}}$ INT1 $\overline{\text{INT3}}$ $\overline{\text{LOCK}}$ = 1x11. If the self test fails, the 80960SA/SB enters the stopped state. Otherwise, the 80960 performs a checksum test of 16 words fetched from memory at physical address 00000000_{16}. After a successful checksum test, the 80960SA/SB uses some of the previously fetched words as addresses to initial data structures.

8.2.7 Basic READ and WRITE

READ:

1. During Ta state:
   - The processor places address on the address and address/data lines.
   - It asserts $\overline{ALE}$ used to latch address.
   - It asserts $\overline{AS}$.
   - $W / \overline{R}$ is low indicating read operation.
   - $DT / \overline{R}$ is low and used as direction input to data transceivers.
2. During Td state:
   - The processor reads data on the AD(0-15) pins.
   - It asserts $\overline{DEN}$ which is used to enable data transceivers.
   - The processor asserts $\overline{BE1} - \overline{BE0}$ to specify which bytes the processor uses when reading the data word.
   - $\overline{READY}$ is asserted by external logic to indicate data is ready to be read. If not asserted, Tw is generated and repeated until $\overline{READY}$ is asserted.
3. The recovery states follow the data state allowing adequate time for external devices to remove their data from the bus before the 80960SA/SB generates the next address on the bus. $W / \overline{R}$, $DT / \overline{R}$, and $\overline{DEN}$ become inactive.

WRITE:

1. During Ta state:
   - The processor places address on address and address/data lines.
   - It asserts $\overline{ALE}$ used to latch address.
   - It then asserts $\overline{AS}$.
   - $W / \overline{R}$ is HIGH indicating WRITE operation.
   - $DT / \overline{R}$ is HIGH and used as direction input to data transceivers.
2. During Td state:
   - The processor places data on the AD(0-15) pins.
   - The processor asserts $\overline{BE1} - \overline{BE0}$ to specify which bytes the processor is writing in the word.
   - It asserts $\overline{DEN}$ used to enable data transceivers.
   - $\overline{READY}$ is asserted by external logic to indicate data written. If not asserted, Tw is generated and repeated until $\overline{READY}$ is asserted. Data is held on the bus.
3. During Tw $\overline{READY}$ remains asserted and data is written into memory or storage device.
4. The recovery states follow the data state. $W / \overline{R}$, $DT / \overline{R}$, and $\overline{DEN}$ become inactive.

Burst READ and WRITE

This is an enhancement feature of the 80960SA/SB processor. It supports burst transactions that read or write up to eight 16-bit words at a maximum rate of one word per processor cycle.

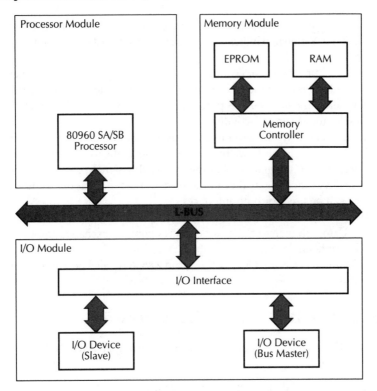

FIGURE 8.4 Basic80960SA/SB system configuration.

8.2.8 80960SA/SB-Based Microcomputer

Figure 8.4 depicts a typical 80960SA/SB system block diagram. The various components are described in the following.

The 80960SA/SB processor performs bus operations using multiplexed address and data signals and provides all the necessary control signals. For example standard control signals, such as Address Latch Enable ($\overline{\text{ALE}}$), Address Status ($\overline{\text{AS}}$), Write/Read command ($\text{W}/\overline{\text{R}}$), Data Transmit/Receive ($\text{DT}/\overline{\text{R}}$), and Data Enable ($\overline{\text{DEN}}$) are provided by the 80960SA/SB processor. The 80960SA/SB processor also generates byte enable signals that specify which bytes on the data lines are valid for the transfer.

To transfer control of the bus to an external bus master, the 80960SA/SB processor provides two arbitration signals: hold request (HOLD) and hold acknowledge (HLDA). After receiving HOLD, the processor grants control of the bus to an external bus master by asserting HLDA.

A memory module can consist of the memory controller, Erasable Programmable Read Only Memory (EPROM), and static or dynamic Random Access Memory (RAM). The memory controller first conditions the bus signals for memory operation. It demultiplexes the address and data lines, generates the chip select signals from the address and $\overline{\text{BE}}$ signals, detects the start of the cycle for burst mode operation, and latches the byte enable signals.

The memory controller generates the control signals for EPROM, SRAM, and DRAM. In particular, it provides the control signals, multiplexed row/column address, and refresh control for dynamic RAMs. The controller can be designed to accommodate the burst transaction of the 80960SA/SB processor. In addition to supplying the operation signals, the controller generates the $\overline{\text{READY}}$ signal to indicate that data has been transferred to or from the 80960SA/SB processor.

The 80960SA/SB processor directly addresses up to 4G bytes of physical memory.

The I/O module consists of the I/O components and the interface circuit. I/O components can be used to allow the 80960SA/SB processor to use most of its clock cycles for computational and system management activities.

The interface circuit performs several functions. It demultiplexes the address and data lines, generates the chip select signals from the address, produces the I/O read or I/O write command from the processor's W/$\overline{\text{R}}$ signal, latches the byte enable signals, and generates the $\overline{\text{READY}}$ signal. Because these functions are the same as some of the functions of the memory controller, the same logic can be used for both interfaces.

The 80960SA/SB processor uses memory-mapped addresses to access I/O devices. This allows the CPU to use many of the same instructions to exchange information for both memory and peripheral devices.

Typical I/O chips such as Intel 82C64 timer and Z8536 parallel port/timer can be used with the 80960SA/SB

8.3 Motorola MC88100 RISC Microprocessor

MC88100 is a 32-bit RISC microprocessor designed using HCMOS technology. The 88100 includes the following:

- Hardwired control design with no microcodes
- 20- or 25-Mhz internal clock frequency
- Packaged in 17×17 (180 pins used) PGA (Pin Grid Array) with a maximum size of 1.78″ × 1.78″
- Includes 51 instructions
- Contains four fully parallel on-chip execution units (pipelined)
- Unlike the 80960SA/SB, the 88100 does not support 80-bit extended floating-point format
- Unlike the 80960SA/SB, the 88100 does not include instructions for computing trigonometric and logarithmic functions
- User and supervisor modes
- 32-bit on-chip combinational multiplier
- Separate data and instruction buses that include 32-bit data bus, 32-bit instruction address bus, 32-bit data address bus, and 32-bit instruction bus (fixed instruction length of 32 bits)
- Directly interfaces to memory or to 88200 cache/memory management unit
- 4 gigabytes of directly addressable memory

The 88100 performs register-to-register operations for all data manipulation instructions. Source operands are contained in source registers or are included as an immediate value inherent in the instruction. A separate destination register stores the results of an instruction. This means that source operand registers can be reused in the subsequent instructions. Register contents can be read from or written to memory only with ld (load) and st (store) instructions. A xmem (memory exchange) instruction is included for semaphore testing and multiprocessor application.

The 88100 contains 51 instructions. All instructions are executed in one cycle. The instructions requiring more than one cycle are executed in effectively one cycle via pipelining. All instructions are decoded by hardware and no microcode is used.

The 88100 includes all data manipulation instructions as register-to-register or register plus immediate value instructions. This eliminates memory access delays in data manipulation. Only 10 memory addressing modes are provided: three modes for data memory, four modes for instruction memory, and three modes for registers.

All 88100 instructions are 32 bits wide. This fixed instruction format minimizes instruction decode time and eliminates the need for alignment. All instructions are fetched in a single memory access. The 88100 implements delayed branching to minimize pipeline delay. For pipelined architecture, branching instructions can slow down execution speed due to the time required to flush and refill the pipeline. The 88100 delayed branching feature allows fetching of the next instruction before the branch instruction is executed.

The 88100 provides two modes: supervisor and user. The supervisor mode is used by the operating system, while the application programs are executed in user mode.

The 88100 includes four execution units which operate independently and concurrently. The 88100 can perform up to five operations in parallel.

Scoreboard bits are associated with each of the general-purpose registers. When an instruction is executed or dispatched, the scoreboard bit of the destination register is set, reserving that register for that instruction. Other instructions are executed or dispatched as long as their source and destination operands have clear scoreboard bits. When an instruction completes execution, the scoreboard bit of the destination is cleared, thus freeing that register to be used by other instructions.

The 88100 memory devices can interface directly to memory. Most 88100 designs implement at least two 88200 CMMUs (one for data memory and one for instruction memory). The P-bus provides the interface to the 88200/memory system. The 88200 is an optional external chip that provides paged virtual memory support and data/instruction cache memory.

Conditional test results are provided to any specified, general-purpose register instead of a dedicated condition code register. Conditions are computed at the explicit request of the programmer using compare instructions. This eliminates contention between concurrent execution units accessing a dedicated condition code register.

8.3.1 88100/88200 Interface

Figure 8.5 shows typical 88100 interfaces to several 88200s. The PBUS (processor bus) contains logical addresses, while MBUS (memory bus) contains all physical addresses. Up to 4 88200s can reside on each PBUS. Note that in the figure, the MC88000 includes the entire RISC microprocessor family, with 88100 being the first microprocessor.

Figure 8.6 shows the 88100/88200 block diagram. Each unit in the 88100 can operate independently and simultaneously. Each unit may be pipelined.

The integer unit performs 32-bit arithmetic, logic, bit field, and address operations. All operations are performed in one clock cycle. The integer unit includes 21 control registers. The floating-point unit supports IEEE 754-1985 floating-point arithmetic, integer multiply, and divide. This unit contains 11 control registers with a five-stage add pipeline and a six-stage multiply pipeline. Six optional SFUs (special function units) are reserved in the architecture. The SFUs can be added to or removed from a given system with no impact on the architecture.

The data unit performs address calculation and data access and includes a three-stage pipeline.

The instruction unit fetches instruction codes and contains a two-stage pipeline.

The register file includes 32 32-bit general-purpose registers.

The sequencer uses a scoreboard to control register reads/writes. It dispatches instructions and recognizes exceptions.

Figure 8.7 shows the 88200 internal block diagram.

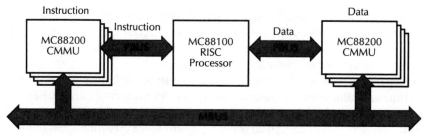

MC88100 32-Bit RISC Microprocessor
• 1.5 Micron HCMOS, 180 pins
• Highly pipelined
• Separate instruction and data buses (Harvard architecture)

MC88200 32-Bit Cache/Memory Management Unit
• 1.5 Micron HCMOS, 180 pins
• 56 entry Page Address Traslation Cache (PATC)
• 10 entry Block Address Translation Cache (BATC)
• 16 Kbyte code/data cache

M88000 Processor BUS
• Synchronous, non-multiplexed, pipelined
• 33-bit logical addresses, 32-bit data path
• 1 word each clock cycle, maximum transfer rate

M88000 Memory BUS
• Synchronous, multiplexed
• 32-bit physical addresses, 32-bit data path
• N words each N + 1 clock cycle maximum transfer rate

FIGURE 8.5   Typical 88100 interface to 88200s. Note that MC88000 represents the 88000 family which includes 88100, 88200, and all future products.

The block address translation cache (BATC) contains 10 entries and is fully associative with software replacement.

The page address translation cache (PATC) includes 56 entries and is fully associative with hardware replacement. The SRAM (static RAM) array contains 16K bytes of static RAM and is set associative.

8.3.2 88100 Registers

Figure 8.8 shows the 88100 registers. All registers are 32 bits wide.

Three types of registers are included:

- 32 32-bit registers, r0-r31, containing program data (source operand and instruction results). All of these registers except r0 (constant 0) have read/write access.
- Internal registers control instruction execution and data transfer
- Control registers in the various execution units containing status, execution control, and exception processing information

The internal registers cannot be directly accessible in software, while most control registers can be accessed in supervisor mode. The internal registers can only be modified and used indirectly.

The control registers include shadow registers and exception time registers, integer-unit control registers, and floating-point unit control registers.

The shadow registers are associated with several internal registers. Shadowing is utilized by the 88100 to keep track of the internal pipeline registers at each stage of the instruction execution.

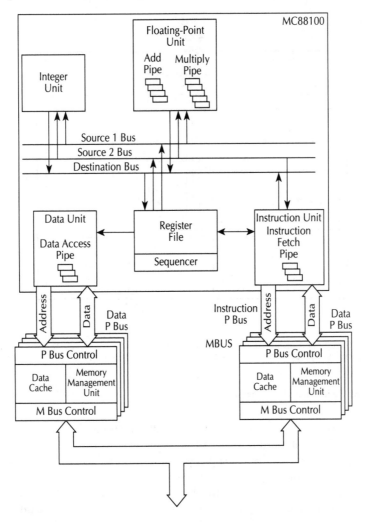

FIGURE 8.6   MC88100/MC88200 block diagram.

There are 21 32-bit control registers (cr0 through cr20) in the integer unit. Fourteen of these registers provide exception information for integer unit or data unit exceptions. The other seven registers include status information, the base address of the exception vector table, and general-purpose storage.

The floating point includes 11 control registers (fcr0-fcr8, fcr62, and fcr63). fcr0 through fcr8 contain exception information such as the exception type, source operands and results, and the instruction in progress. These registers can only be accessed in supervisor mode. Registers fcr62 and fcr63 are not privileged. These two registers can be used to enable user-supplied exception handler software and to report exception causes in user mode.

The supervisor programmer's model contains all general-purpose and control registers. The general-purpose registers provide data and address information, while the control registers provide exception recovery and status information for the integer unit.

In user mode, all general-purpose registers can be accessed. Two control registers (floating-point control and status) can be accessed in the user mode.

Among the 32 general-purpose registers, r0 always contains the value 0, r1 is loaded with the subroutine return address, and r2 through r31 are general-purpose.

Figure 8.9 shows the 88000 register data formats. The 88100 supports two types of data formats, namely, integer (signed or unsigned) and floating-point real numbers. The integers

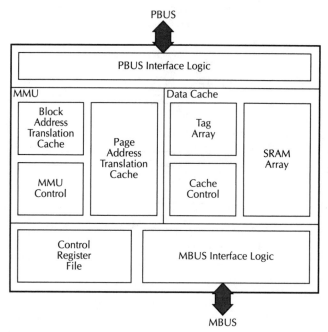

Cache Accesses in parallel with Address Translation.
Two Level Page Address Translation Tables for Supervisor and User Programs

FIGURE 8.7   88200 internal block diagram.

can be byte, half-word (16-bit), and word (32-bit). All operations affect all 32 bits of a general-purpose register. The half-word or byte pads the sign bit.

The floating-point data can be single precision (32-bit) and double precision (64-bit).

Figure 8.10 shows formats for Fer62 and Fer63.

The reserved bits in fcr62 and fcr03 are always read as zero. The FPcr defines the desired rounding mode and which exceptions are handled by user exception handles. The FPSR indicates which floating-point exceptions have occurred but were not processed by a user exception handler.

The 88100 general-purpose register convention is shown in Figure 8.11. r31 addresses the top of the stack. r30 contains the address of the current data frame in the stack.

Figure 8.12 shows the 88100 stack operation. The SP must always be 32-bit aligned. The stack grows from high memory to low memory addresses.

Next, consider the supervisor programmer model:

- The VBR contains the base address of the exception vector table.

- The 88100 does not automatically use SR0-SR3. They are reserved for operating system use.

Figure 8.13 shows the 88100 processor status register format. In Figure 8.13, Big Endian means the most significant byte at the highest byte address. Serial instruction is to complete before the next one begins. Note that not all adds/subtracts affect C. The SFDI bit enables or disables the floating-point unit. When SFDI = 1, attempted execution of any floating-point or integer multiply/divide instructions causes a floating-point precise exception.

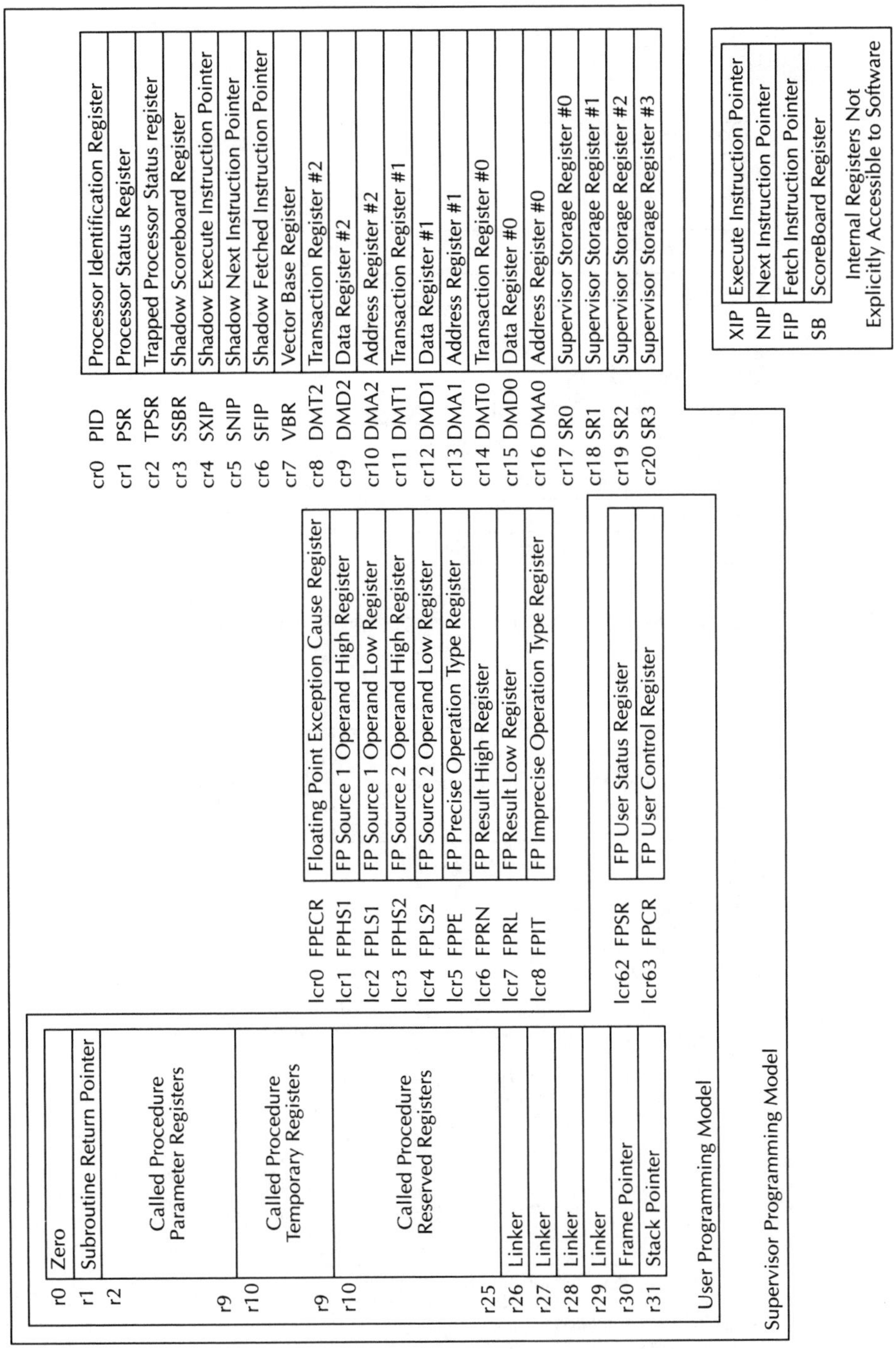

cr0	PID	Processor Identification Register
cr1	PSR	Processor Status Register
cr2	TPSR	Trapped Processor Status register
cr3	SSBR	Shadow Scoreboard Register
cr4	SXIP	Shadow Execute Instruction Pointer
cr5	SNIP	Shadow Next Instruction Pointer
cr6	SFIP	Shadow Fetched Instruction Pointer
cr7	VBR	Vector Base Register
cr8	DMT2	Transaction Register #2
cr9	DMD2	Data Register #2
cr10	DMA2	Address Register #2
cr11	DMT1	Transaction Register #1
cr12	DMD1	Data Register #1
cr13	DMA1	Address Register #1
cr14	DMT0	Transaction Register #0
cr15	DMD0	Data Register #0
cr16	DMA0	Address Register #0
cr17	SR0	Supervisor Storage Register #0
cr18	SR1	Supervisor Storage Register #1
cr19	SR2	Supervisor Storage Register #2
cr20	SR3	Supervisor Storage Register #3

lcr0	FPECR	Floating Point Exception Cause Register
lcr1	FPHS1	FP Source 1 Operand High Register
lcr2	FPLS1	FP Source 1 Operand Low Register
lcr3	FPHS2	FP Source 2 Operand High Register
lcr4	FPLS2	FP Source 2 Operand Low Register
lcr5	FPPE	FP Precise Operation Type Register
lcr6	FPRN	FP Result High Register
lcr7	FPRL	FP Result Low Register
lcr8	FPIT	FP Imprecise Operation Type Register

lcr62	FPSR	FP User Status Register
lcr63	FPCR	FP User Control Register

r0	Zero
r1	Subroutine Return Pointer
r2	Called Procedure Parameter Registers
r9	Called Procedure Temporary Registers
r10	
r9	Called Procedure Reserved Registers
r10	
r25	
r26	Linker
r27	Linker
r28	Linker
r29	Linker
r30	Frame Pointer
r31	Stack Pointer

User Programming Model

Supervisor Programming Model

XIP	Execute Instruction Pointer
NIP	Next Instruction Pointer
FIP	Fetch Instruction Pointer
SB	ScoreBoard Register

Internal Registers Not
Explicitly Accessible to Software

FIGURE 8.8 MC88100 register organization.

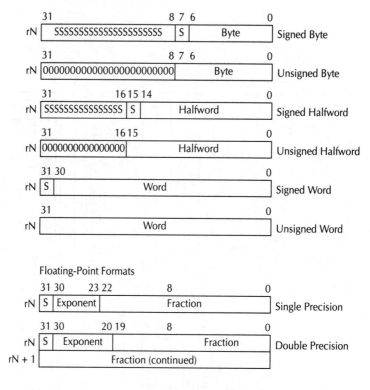

FIGURE 8.9   88100 register data formats. An "S" in the diagram above indicates a sign bit.

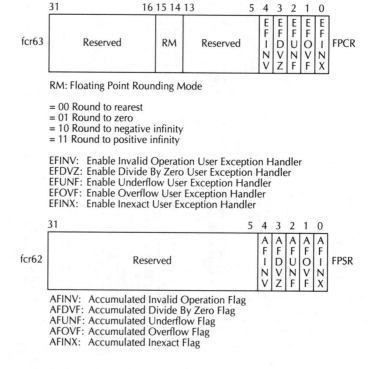

FIGURE 8.10   88100 fcr 62 and fcr 63 formats.

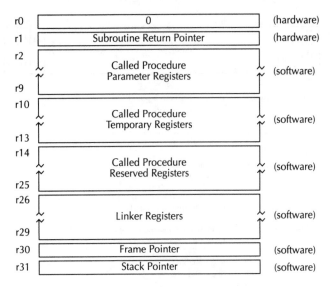

FIGURE 8.11   88100 register convention.

8.3.3 88100 Data Types, Addressing Modes, and Instructions

Tables 8.7a and 8.7b list the data types and addressing modes supported by the 88100. Table 8.8 summarizes the 88100 instructions.

The 51 instructions listed in Table 8.8 of the MC88100 can be divided into 6 classes: integer arithmetic, floating-point arithmetic, logical, bit field, load/store/exchange, and flow control.

These simple instructions must be used to obtain complex operations. Shift and rotate operations are special cases of bit field instructions. Only ld, st, and xmem can access memory.

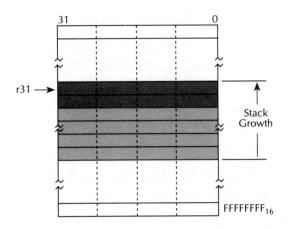

FIGURE 8.12   88100 stack operation.

31	30	29	28	27	26	10	9	4	3	2	1	0

```
M
O   B   S       D
D   O   E   C   E       Reserved           Reserved    S   M   I   S
E       R       X                                      F   X   N   F
                C                                      D   M   D   R
                                                       I           Z
```

MODE = 0	Processor is in user mode.	
= 1	Processor is in supervisor mode.	
BO = 0	Big Endian byte order in memory.	
= 1	Little Endian byte order in memory.	
SER = 0	Concurrent instruction execution.	
= 1	Serial instruction execution.	
C = 0	No carry / borrow generated.	
= 1	Carry / borrow generated.	
DEXC = 0	Data memory exception not pending.	
= 1	Data memory exception pending.	
SFD1 = 0	SFU1 enabled.	
= 1	SFU1 disabled.	
* MXM = 0	Misaligned memory accesses generate exceptions.	
= 1	Misaligned memory access truncate.	
	Misaligned address to next lower aligned address.	
IND = 0	Interrupt enabled.	
= 1	Interrupt disabled.	
SFRZ = 0	Shadow registers enabled.	
= 1	Shadow registers frozen.	

* 88100 can address each byte. For halfword, aligned address include all
  addresses in multiples of 2, for word in multiples of 4 and for
  doubleword in multiple of 8.

FIGURE 8.13   88100 processor status register format.

TABLE 8.7a   Data Types

Data type	Represented as
Bit fields	Signed and unsigned bit fields from 1 to 32 bits
Integer	Signed and unsigned byte (8 bits)
	Signed and unsigned half-word (16 bits)
	Signed and unsigned word (32 bits)
Floating-point	IEEE P754 single precision (32 bits)
	IEEE P754 double precision (64 bits)

TABLE 8.7b   Addressing Modes

Data addressing mode	Syntax
Register indirect with unsigned immediate	rD,rS1,imm16
Register indirect with index	rD,rS1,rS2
Register indirect with scaled index	rD,rS1[rS2]

Instruction addressing mode	Syntax
Register with 9-bit vector number	m5,rS1,vec9
Register with 16-bit signed displacement	m5,rS1,d16
Instruction pointer relative	d26
(26-bit signed displacement)	
Register direct	rS2

TABLE 8.8   Instruction Set Summary

Mnemonic	Description
Integer Arithmetic Instructions	
add	Add
addu	Add unsigned
cmp	Compare
div	Divide
divu	Divide unsigned
mul	Multiply
sub	Subtract
subu	Subtract unsigned
Floating-Point Arithmetic Instructions	
fadd	Floating-point add
fcmp	Floating-point compare
fdiv	Floating-point divide
fldcr	Load from floating-point control register
flt	Convert integer to floating point
fmul	Floating-point multiply
fstcr	Store to floating-point control register
fsub	Floating-point subtract
fxcr	Exchange floating-point control register
int	Round floating point to integer
nint	Floating-point round to nearest integer
trnc	Truncate floating point to integer
Logical Instructions	
and	AND
mask	Logical mask immediate
or	OR
xor	Exclusive OR
Bit-Field Instructions	
cir	Clear bit field
ext	Extract signed bit field
extu	Extract unsigned bit field
ff0	Find first bit clear
ff1	Find first bit set
mak	Make bit field
rot	Rotate register
set	Set bit field
Load/Store/Exchange Instructions	
ld	Load register from memory
lda	Load address
ldcr	Load from control register
st	Store register to memory
stcr	Store to control register
xcr	Exchange control register
xmem	Exchange register with memory
Flow Control Instructions	
bb0	Branch on bit clear
bb1	Branch on bit set
bcnd	Conditional branch
br	Unconditional branch
bsr	Branch to subroutine
jmp	Unconditional jump
jsr	Jump to subroutine
rte	Return from exception
tb0	Trap on bit clear
tb1	Trap on bit set
tbnd	Trap on bounds check
tcnd	Conditional trap

Also, only compare instructions affect condition codes. Most MC88100 instructions can have one of the three formats:

1. Triadic register instructions (three operands). The general format is

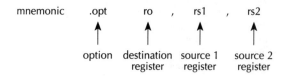

An example is add.ci r2, r7, r4. Note that Motorola's assembler expects the 88100 instructions in lower case. In the example, the mnemonic is add and the option is ci, meaning use 'carry in' in the operation. The source registers r7 and r4 remain unchanged unless one of them is used as destination. This add instruction adds [r7] and [r4] with carry and stores the result in r2.

2. Triadic register instructions with 16-bit field instruction. The general format is mnemonic.opt rD, rs1, imm16. Consider

   add.Co r2, r4, 0XA125

   0X before data A125 means that A125 is in hex. This notation is used by the Motorola assembler.

3. Dyadic register instructions (two operands). The general format is mnemonic . opt rd, rs2. An example is flt.sw r5, r6 — which converts the integer source word in r6 into floating-point in r5.

Table 8.9 lists the 88100 load, store, and exchange instructions.

TABLE 8.9   MC88100 LOAD, STORE, Exchange Instructions

Instructions		Exceptions
ld[.\<opts>]	rD,rS1,\<imm16>	
ld[.\<opts>][.\<space>]	rD,rS1,rS2	
ld[.\<opts>][.\<space>]	rD,RS1 [rS2]	
st[.\<size>]	rS,rS1,\<imm16>	Data access
st[.\<size>][.\<space>]	rS,rS1,rS2	Misaligned access
st[.\<size>][.\<space>]	rS,rS1[rS2]	Privilege violation
xmem[.bu]	rS,rS1,\<imm16>	
xmem[.bu][.\<space>]	rS,rS1,rS2	
xmem[.bu][.\<space>]	rS,rS1[rS2]	
ldcr	rD,crCRS	
stcr	rS,crCRD	
xcr	rD,rS1,crCRS/D	
		Privilege violation
fldcr	rD,crFCRS	
fstcr	rS,crFCRD	
fxcr	rD,rS1,crFCRS/D	

\<opts>	For	ld	\<size>	For	st
.b	—	Signed byte	.b	—	Byte
.bu	—	Unsigned byte	.h	—	Halfword
.h	—	Signed halfword	none	—	Word
.hu	—	Unsigned halfword	.d	—	Doubleword
none	—	Word			
.d	—	Doubleword			

TABLE 8.9   MC88100 LOAD, STORE, Exchange Instructions
(*continued*)

\<space\>	For	ld,st,xmem
.usr	—	Access user space regardless of mode bit in PSR
none	—	Access space indicated by PSR MODE bit

- rs = Source register.
- crCRS = Source control register.
- crCRD = Destination control register.
- crCRS/D = Source/destination control register.
- crFCRS = Source floating-point control register.
- crFCRD = Destination floating-point control register.
- crFCRS/D = Source/destination floating-point control register.
- Memory accesses for **xmem** are indivisible.

- ld loads a general-purpose register from data memory. There are three operands with this instruction. Two source operands are used to calculate the address. Three forms of ld use the three addressing modes available. The (.opt) for ld specifies the size of data read from memory.
- The st instructions are similar to ld instructions, except they are used to store source data.
- The exchange instructions (xmem, xcr, fxcr) swap the contents of a general-purpose register with data memory or with a control register.
- Consider ld r47, r31, 0X4. The mode used here is register indirect with unsigned immediate. If $[r31] = 00005000_{16}$, the effective address is computed by adding rsl (r31) with unsigned 16-bit immediate data. Therefore, the 88100 loads the register r47 with 32-bit data from a memory location addressed by 00005004_{16}; r31 is the SP. Therefore, the access occurs within the stack. Since the immediate data are unsigned, the accessed address cannot be less than r31. This means that the stack grows toward the lower address.
- Consider st.b rs, rs1, rs2. This instruction has register indirect with index mode.

The access address is (rs1) + (rs2) where rs1 is the base register and rs2 is the index register. For example, consider st.b r1, r0, r5. If $[r5] = 00010200_{16}$; then since r0 is always 0, the low 8-bit content of r1 is stored at address 00010200_{16}. The 88100 ignores any carry generated during address calculation. Note that in the above r0 is the base address and r5 is the index register.

Finally, consider st.hu rs, rs1 [rs2]. The mode is register indirect with index. The access address is (rs1) + (rs2)* (operand size). The scaling is specified by surrounding the index register rs2 by square brackets. Operand size is 1 for byte, 2 for halfword, 4 for word, and 8 for double word.

As an example, consider st. r5, r31 [r1]. If $[r1] = 00000003_{16}$, $[r31] = 00005000_{16}$, then the effective address is

$$00005000 + 4 * 00000003 = 0000500C_{16}$$

scaled by 4 for word since the instruction without any option specified means 32-bit word.

Therefore, the above store instruction stores the 32-bit contents of r5 into a memory location addressed by $0000500C_{16}$.

Table 8.10 shows the integer arithmetic instructions.

TABLE 8.10   MC88100 Integer Arithmetic
Instructions

Instructions		Exceptions
add[.<opt>]	rD,rS1,rS2	Integer overflow
add	rD,rS1,<imm16>	
addu[.<opt>]	rD,rS1,rS2	None
addu	rD,rS1,<imm16>	
sub[.<opt>]	rD,rS1,rS2	Integer overflow
sub	rD,rS1,<imm16>	
subu[.<opt>]	rD,rS1,rS2	None
subu	rD,rS1,<imm16>	
mul	rD,rS1,rS2	None
mul	rD,rS1,<imm16>	
div	rD,rS1,rS2	
div	rD,rS1,<imm16>	Integer divide
divu	rD,rS1,rS2	
divu	rD,rS1,<imm16>	
lda [.<size>]	rD,rS1,<imm16>	
lda [.<size>]	rD,rS1,rS2	None
lda [.<size>]	rD,rS1,[rS2]	

	<opt> FOR add/addu/sub/subu	
none	No carry	
.ci	Use carry in	
.co	Propagate carry out	
.cio	Use carry in and propagate carry out	

	<size> FOR lda	
.b	Scale rS2 by 1	
.h	Scale rS2 by 2	
none	Scale rS2 by 4	
.d	Scale rS2 by 8	

- **mul** yields correct signed and unsigned results.
- Division by zero signals the integer divide exception.
- An integer divide exception occurs when either source operand is negative for **div**.
- Unscaled **lda** is functionally equivalent to **addu**.

- Consider add [.<opt>] rd, rs1, rs2. Three options can be used with this instruction as follows:
  - add.Ci rD, rs1, rs2 adds the 32-bit contents of rs1 with rs2 and the C-bit in processor status register, and stores the 32-bit result in rD without providing any carry-out, and thus the C-bit in PSR is unchanged.
  - add.Co rD, rs1, rs2 adds the 32-bit contents of rs1 with rs2 without any carry-in and stores the result in rD and reflects the carry-out in the C-bit of PSR.
  - add.Cio rD, rs1, rs2 adds the 32-bit contents of rs1 with rs2 and the C-bit from the PSR and stores the result in rD and reflects any carry-out in the C-bit in the PSR.
- Consider add.Cio r2, r1, r5. If the C-bit in the PSR is 0, [r1] = 8000 F102$_{16}$, [r5] = F1101100$_{16}$, then after this add [r2] = 71110202$_{16}$ and the C-bit in the PSR is set to one.
- If no option is specified in an instruction such as add or addu, the carry-in is not included in addition and also no carry-out from the addition is provided. For example, consider addu r1, r5, 0XF112. The 16-bit immediate data F112$_{16}$ is converted to an unsigned 32-bit number 0000-F112$_{16}$ and is added with the 32-bit contents of r5. The

32-bit result is stored in r1. The carry-out is not provided. The add instruction performs signed arithmetic. If the result cannot be accommodated in a 32-bit integer, an integer overflow exception occurs. The immediate 16-bit data in an add instruction are sign-extended to 32 bits before addition.

- The sub instructions are similar to the add instructions. The content of rs2 is subtracted from the content of rs1 with the C-bit in the PSR as borrow if .Ci is used for <opt>.

- mul instructions multiply a 32-bit number (signed or unsigned) in rs2 by a 32-bit number (signed or unsigned) in rs1 and store the low 32-bit result in rD and discard the upper 32 bits of the result.

- div instructions perform signed division while divu carries out unsigned division. divu (unsigned division) instructions divide the 32-bit content of rs1 by the 32-bit content of rs2 or a 16-bit immediate value. The 32-bit quotient is stored in rD and the remainder is discarded. If the divisor is zero, an integer divide exception is taken and rD is unaffected. div (signed division) operates similarly except that the integer divide exception is taken if either the dividend (rs1) or the divisor (rs2) has a negative value. The exception handler must convert the negative value to positive, perform the signed integer divide, and convert the sign of the result.

- lda is the load address instruction. lda calculates the access address using one of three indirect addressing modes. lda loads rD with the access address. Unscaled lda is functionally equivalent to addu with the same operands.

Table 8.11 lists the 88100 floating-point arithmetic instructions.

- The MC88100 permits a mixture of single and double precision source and destination operands.
- **trnc** performs "round to zero" rounding.
- **nint** performs "round to nearest" rounding.
- **int** performs rounding specified by the RM field of the FPCR.
- fadd adds the contents of rs1 with rs2 and stores the result in rD.
- fsub subtracts the contents of rs2 from rs1 and stores the result in rD.
- fmul multiplies the contents of rs1 by rs2 and stores the result in rD.
- fdiv divides the contents of rs1 by rs2 and stores the quotient in rD.

TABLE 8.11   MC88100 Floating-Point Arithmetic Instructions

Instructions		Exceptions
fadd.<sizes>	rD,rS1,rS2	Floating point reserved operand
fsub.<sizes>	rD,rS1,rS2	Floating point overflow
		Floating point underflow
fmul.<sizes>	rD,rS1,rS2	Floating point inexact
fdiv.<sizes>	rD,rS1,rS2	Floating point divide by zero
trnc.<sizes>	rD,rS2	
nint.<sizes>	rD,rS2	Floating point integer conversion overflow
int.<sizes>	rD,rS2	Floating point reserved operand
flt.<sizes>	rD,rS2	Floating point inexact

<sizes> **FOR fadd/fsub/fmul/fdiv**
sss, ssd, sds, sdd, dss, dsd, dds, ddd
<sizes> **FOR trnc/nint/int**
ss, sd
<sizes> **FOR flt**
ss, ds

TABLE 8.12   MC88100 Logical Instructions

Instructions		Exceptions
and [.c]	rD,rS1,rS2	
and [.u]	rD,rS1,<imm16>	
mask [.u]	rD,rS1,<imm16>	
or [.c]	rD,rS1,rS2	None
or [.u]	rD,rS1,<imm16>	
xor [.c]	rD,rS1,rS2	
xor [.u]	rD,rS1,<imm16>	

• Option .c ones-complements the contents of rs2 before performing the operation.
• Option .u performs the specified logical operation between 16-bit immediate data <imm16> and the high 16 bits of rs1.

- The 88100 utilizes hardware to perform these IEEE floating-point computations.
- The 88100 allows single and double precision operands. <sizes> specify the operand sizes of rD, rs1, and rs2 as single or double precision. For example, .ssd means that rD and rs1 are single precision, while rs2 is double precision.
- trnc, nint, int, and flt provide conversions between integer and floating-point values. These instructions have two operands with one operand having a floating point value and the other having an integer value. trnc, nint, and int convert a floating-point format to equivalent format. The difference between them is the type of rounding performed. trnc rounds toward zero, and nint rounds to the nearest value, int rounds as specified by the RM field in FPCR.
- Two exceptions are provided for floating-point instructions. An integer conversion overflow exception occurs when the operand value cannot be expressed as a high word. A reserved operand exception occurs with certain floating-point values.
- flt converts a signed 32-bit number into a floating-point format. The integer operand size is always specified by <sizes> indicating signed word size. A 'd' for double or 's' for single defines the floating-point operand's precision.

Table 8.12 lists the MC88100 logical instructions.

If the option .u is omitted, the specified logical operation is performed between the 16-bit immediate data <imm16> and the low-order 16 bits of rs1.

- The mask always affects all 32 bits of rD. The mask logically ANDS the 16-bit immediate value with the low 16 bits or highest 16 bits of rs1 and clears the other 16 bits to zero. If the .u option is omitted, the AND is performed with the low-order 16 bits of rs1 and if .u is included, the AND is performed with the high-order 16 bits of rs1.
- When both operands are registers (rs1 and rs2) for AND, OR, and XOR, the 32-bit logical operation is performed. When the .c option is used, the MC88100 ones-complements the contents of rs2 before performing the operation. The MC88100 only performs a 16-bit operation when the second source operand is a 16-bit immediate. The .u option, when present in AND, OR, and XOR, performs the logical operation between the high 16-bit value of rs1 and the 16-bit immediate data and then stores the result in the high 16 bits of rD. The low 16 bits of rs1 are copied into the low 16 bits of rD. If the .u option is not present, only the low 16 bits of rs1 are used in the operation.

Table 8.13 lists the 88100 bit field instructions.

- W5 is five bit width and <05> is a five bit offset.
- The number of bits in a bit field is called width. Width can be from 1 to 32. The least significant bit in bit field is called the offset. When the sum of width and offset is greater than 32, the bit field may be imagined to extend beyond the most significant bit of the register.

TABLE 8.13 MC88100 Bit Field Instructions

Instructions	Exceptions	
clr	rD,rS1,w5 <05>	
clr	rD,rS1,rS2	
set	rD,rS1,w5 <05>	
set	rD,rS1,rS2	
ext	rD,rS1,w5 <05>	
ext	rD,rS1,rS2	
extu	rD,rS1,w5 <05>	None
extu	rD,rS1,rS2	
mak	rD,rS1w5 <05>	
mak	rD,rS1,rS2	
rot	rD,rS1, <05>	
rot	rD,rS1,rS2	
ff1	rD,rS2	
ff0	rD,rS2	

• ext and extu affect all 32 bits of rD.
• ext and extu perform shift right operations when the width equals 32.
• mak clears rD before inserting the bit field.
• mak performs shift left operations when the width equals 32.
• rot rotates the bit field to the right.

An isolated bit field ends in the least significant bit of a register with an implicit offset of zero. Bit fields may contain signed and unsigned values. For unsigned isolated bit fields, the high-order bits in a register are all zero and for signed isolated bit field, they are the 2's complement sign bit. In either case, the entire register contains the word value equivalent to the isolated bit field value. The 88100 includes instructions to isolate embedded bit field for arithmetic manipulation. The 88100 instruction moves an isolated bit field back into an embedded bit field. This is illustrated in the following:

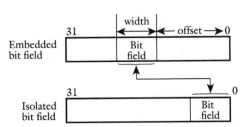

Two bit field formats are used. These are literal width with offset and register width with offset. The literal width with offset uses immediate width and offset values. An example is set r5, r1, 3 <7>. The destination register is r5; the source bit field is 3 bits wide with an offset of 7 in r1. With the Motorola assembler, the offset must be included in angle brackets < >.

The register width with offset uses three operands. An example is clr r1, r3, r4. The source bit field is in r3 with offset and width determined from r4. The MC88100 obtains the offset from bits 0-4 of r4 and the width from bits 5-9 of r4. The upper 22 bits of r4 are don't cares. Both formats use an offset of 0 to 31 and width of 1 to 32 with 32 encoded as 0. The destination register (r1 in this case) stores the final result. The content of the source register (r3) does not change after the operation.

• ext (signed) and extu (unsigned) instructions extract the register value from rs1 and convert it to an isolated bit field in rD. For a bit field width of 32 (encoded as 0), ext and extu perform a shift right operation. The content of rs1 is shifted to the right by the

TABLE 8.14   MC88100 Integer Compare

Instructions	Exceptions
cmp	rD,rS1,<imm16>
cmp	rD,rS1,rS2
cmp	

cmp Predicate Bit String

31 30 29 28 27 26 25 24 23 22 21 20 19 18 17 16 15 14 13 12	11	10	9	8	7	6	5	4	3	2	1	0
rD 0 0 0 0 0 0 0 0 0 0 0 0 0 0 0 0 0 0 0 0	hs	lo	ls	hi	ge	lt	le	gt	ne	eq	0	0

Predicate Bits	Bit Set If And Only If:	
eq	rS1 == rS2	
ne	rS1 ≠ rS2	
gt	rS1 > rS2	
le	rS1 <= rS2	
lt	rS1 < rS2	Signed Evaluation
ge	rS1 >= rS2	
hi	rS1 > rS2	
ls	rS1 <= rS2	
lo	rS1 < rS2	Unsigned Evaluation
hs	rS1 >= rS2	

number of bits specified in the offset and the result is stored in rD. extu performs a logical shift, while ext performs an arithmetic shift.

- mak creates an imbedded bit field in rD with an offset specified by an immediate value or by the content of rs2. The MC88100 stores the least significant bits of rs1 in the imbedded bit field. The bits outside the imbedded bit field in rD are cleared to zero. The mak is the inverse operation of ext and extu.
- The shift left operation may be performed by a bit field width of 32. The offset specifies the number of positions to be shifted.
- rot reads rs1 and rotates it to the right by the number of bits specified in <05> or in bits 0-4 of rs2. The result is stored in rD.
- ff1 finds the most significant set bit in rs2 and stores the bit number in rD.
- If all bits are cleared, the 88100 loads 32 into rD. ff0 operates similarly but finds the most significant clear bit.

Table 8.14 summarizes the 88100 Integer Compare instructions.

- cmp provides integer data comparison. The 88100 compares the rs1 contents with either an unsigned 16-bit immediate number or the content of rs2. The 16-bit immediate data are converted to a 32-bit value with zeros in the high 16 bits before comparison. The result of the comparison is stored in rD.

Table 8.15 shows the 88100 floating-point compare instructions. Table 8.16 lists the 88100 conditional branch instructions.

An instruction using the .n option must not be followed by another flow control instruction. (Error undetected by the MC88100.)

<cond> for bcnd/tcnd
eq0
ne0
gt0
lt0
ge0
le0

TABLE 8.15   MC88100 Floating-Point Compare

Instructions	Exceptions
fcmp.sss rD,rS1,rS2 fcmp.ssd rD,rS1,rS2 fcmp.sds rD,rS1,rS2 fcmp.sdd rD,rS1,rS2	Floating point reserved operand

fcmp Predicate Bit String

31 30 29 28 27 26 25 24 23 22 21 20 19 18 17 16 15 14 13 12 11 10 9 8 7 6 5 4 3 2 1 0

rD | 0 0 0 0 0 0 0 0 0 0 0 0 0 0 0 0 0 0 0 0 | ob in ib ou ge lt le gt ne eq cp nc

Predicate Bits	Bit Set If And Only If:
nc	operands are not comparable
cp	operands are comparable
eq	rS1 == rS2
ne	rS1 ≠ rS2
gt	rS1 > rS2
le	rS1 <= rS2
lt	rS1 < rS2
ge	rS1 >= rS2
ou	(rS1 > rS2 OR rS1 < 0) AND rS2 > 0
ib	rS1 <= rS2 AND rS1 >= 0 and rS2 > 0
in	rS1 < rS2 AND rS1 > 0 AND rS2 > 0
ob	(rS1 >= rS2 OR rS1 <= 0) AND rS2 > 0

- <d16> ‖ signed 16-bit displacement.
- <vec9> ‖ 9-bit vector number (0-511).
- <b5> ‖ 5-bit bit number (0-31).
- The .n option indicates "execute next". The next instruction executes whether or not the branch takes effect.
- tbnd traps if the value in rS1 is greater than the value in rS2 or <imm16>, or if rS1 is negative.
- bcnd tests the content of rs1 for = 0, ≠ 0, >0, <0, ≤0, and ≥0 and branches with 16-bit signed displacement if the condition is true. The 16-bit signed displacement is sign-extended to 32 bits, shifted twice to the left, and adds to address of bcnd to branch with a displacement of (2^{-18}) to $(2^{18} - 4)$ bytes. .n indicates 'execute next'. If .n is present and the condition is true, the bcnd executes the next instruction before taking the branch. The 'execute next' allows the 88100 to branch without flushing the execution pipeline and thus provides faster execution.

TABLE 8.16   MC88100 Conditional Flow Control Instructions

	Instructions	Exceptions
bcnd [.n]	<cond>,rS1,<d16>	None Trap vec9 Privilege violation
bb1 [.n]	<b5>,rS1,<d16>	None
bb0 [.n]	<b5>,rS1,<d16>	
tb1	<b5>,rS1,<vec9>	Trap vec9
tb0	<b5>,rS1,<vec9>	Privilege violation
tbnd	rS1,rS2	
tbnd	rS1,<imm16>	

TABLE 8.17   MC88100 Unconditional Flow Control Instructions

Instructions		Exceptions
br [.n]	<d26>	
bsr [.n]	<d26>	None
jmp [.n]	rS2	
jsr [.n]	rS2	
rte		Privilege violation

- <d26> ‖ signed 26-bit displacement.
- .n ‖ "execute next".
- **bsr** and **jsr** save the return address in r1.
- "**jmp rl**" performs return from subroutine.
- The last instruction of typical exception handlers is **rte**.

- tcnd also tests the content of rs1 for the condition but traps if the condition is true. The 88100 includes 512 vectors in the vector table. <vec9> specifies a 9-bit vector number from 0 to 511. .bb1 (branch on bit set) tests the content of rs1 for a set bit. If it is set, the 88100 takes a branch. <b5> specifies a bit number from 0 to 31. bb1 usually follows cmp and fcmp instructions. tbb1 is similar to bb1 except a trap is taken if the bit is set. bb0 (branch on bit clear) is similar to bb1 except a branch is taken if the specified bit is 0. tb0 is similar to tb1 except a trap is taken if the specified bit is 0.
- tbnd (trap on bound check) generates bound check violation if rs1 contents are out of bounds; 0 is the implicit lower bound. The upper bound is either unsigned 16 bits or is contained in rs2. The value of the upper bound is treated as an unsigned number.

Table 8.17 lists the unconditional jump instructions.

- br always unconditionally branches with signed 26-bit displacement with a range of (2^{-26}) to $(2^{+26} - 4)$ bytes.
- bsr is an unconditional subroutine call and saves the return address in r1. When [.n] is specified, the return address is the address of bsr plus 8.
- jmp branches to the address specified by the contents of rs2. The 88100 rounds the least 2 bits of rs2 to 00 before branching for alignment. However, the contents of rs2 are unchanged by the instruction. .jsr is similar to jmp except it is a subroutine jump to an address specified by the contents of rs2 and also saves the return address in r1.
- The 88100 does not provide any return from subroutine instruction. jump r1 fetches the next instruction from the return address saved by bsr or jsr.
- rte provides an orderly termination of an exception handler. It uses the shadow registers to restore the state that existed before the exception. rte can only clear the mode bit in the PSR and ensures that the instruction is executed in user mode.

Example 8.12

Show the contents of registers and memory after the 88100 executes the following instructions:

- i) st.h r2, r3, r5
- ii) xmem r2, r2[r3]

  Assume the following data:

Memory

2000	FF	25	71	02
2004	00	01	02	03
2008	A2	71	36	25

8000	01	02	A2	71
8004	B1	11	26	05

$[r2] = 00000004_{16}$

$[r3] = 00002000_{16}$

$[r5] = 00000002_{16}$

Solution

i) $[r3] + [r5] = 2002_{16}$. Consider st.h r2, r3, r5 where .h stands for half-word (16 bits). The low 16 bits of r2 is stored in 2002_{16} and 2003_{16}. Therefore

2000	FF	25	00	04

ii) xmem r2, r2[r3]. This is a 32-bit word operation. Hence, the scale factor is 4. The effective address

$$= [r2] + 4 * [r3]$$
$$= 0000\ 0004_{16} + 4 * 00002000_{16}$$
$$= 0000\ 8004_{16}$$

Therefore, after the xmem, $[r2] = B111\ 2605_{16}$ and $[00008004] = 0000\ 0004_{16}$.

Example 8.13

Write an instruction to logically shift right by 3 bits the value of r2 into r5.

Solution

Since r2 is 32 bits wide, extu performs logical right shift and the width 32 is encoded as 0. extu r5, r2, 0<3>.

Example 8.14

Write an MC88100 instruction sequence to logically AND the 32-bit contents of r7 with $F2710562_{16}$ and store the result in r5.

Solution

The instruction "and" logically ANDs 16-bit operands.

```
and r5, r7, 0x0562     ;  and logically ANDs 0562₁₆
                       ;  with low 16 bits of
                       ;  r7 and stores result in
                       ;  low 16 bits of r5
and.u r5, r7, 0xF271   ;  and.u logically ANDs
                       ;  F271₁₆ with high 16 bits
                       ;  of r7 and stores result
                       ;  in high 16 bits of r5
```

Example 8.15

Find an MC88100 instruction to load register r5 with the constant value 9.

Solution
  add r5, r0, 9.

Example 8.16

Find an MC88100 instruction to branch to the instruction with label START if the value in r5 is equal to zero.

Solution
  bcnd eq0, r5, START.

Example 8.17

Write an 88100 assembly language program to add two 64-bit numbers in registers r2 r3 and r4 r5. Store result in r4 r5.

Solution

```
        add r5,  r3,  r5      ;  add low 32-bits
        add.cio r4,  r2,  r4  ;  add next 32 bits with carry
finish  br finish
```

Example 8.18

Write an 88100 assembly language program to perform the following operation:

$$(A/B) + C^* D$$

where A, B, C, D are stored in r2, r3, r4, r5 as signed 32-bit integers. Assume C*D generates a 32-bit product. Discard the remainder of A/B. Store the 32-bit result in r6.

Solution

```
        div r8,  r2,  r3     ;  r8 ← r2/r3
        mul r7,  r4,  r5     ;  r7 ← r4*r5
        add.co r6,  r7,  r8  ;  store result in r6
finish  br finish            ;  halt
```

Example 8.19

Write a program in 88100 assembly language to compute the area of a circle by using $A = \pi r^2$ where A is the area in 32-bit single-precision to be stored in register r4 and r is the radius of the circle stored in r2 as a 32-bit floating-point number.

Solution

```
        fmul.sss r4,  r2,  r2 ;  calculate r²
        add r5,  r0,  7       ;  move 7 into r5
```

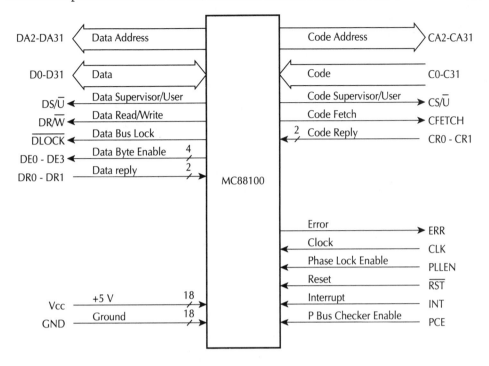

FIGURE 8.14 MC88100 signal functional diagram.

```
         flt.ss  r6, r5         ; convert 22 to floating-point
         fdiv.sss r8, r4, r5    ; compute r²/7
         add r7, r0, 22         ; move 22 into r7
         flt.ss r9, r7          ; convert 22 into floating-point
         fmul.sss r4, r8, r9    ; compute πr²
finish   br finish              ; halt
```

8.3.4 88100 Pins and Signals

Figure 8.14 shows the 88100 pins by functional group. Table 8.18 provides a brief description of the functions of these pins.

TABLE 8.18 Signal Index

Signal name	Mnemonic	Function
Data address bus	DA2-DA31	Provides the 30-bit word address to the data memory space; an entire data word (32 bits) is always addressed; individual bytes or half words are selected using the data byte strobe signals
Data bus	D0-D31	32-bit bidirectional data bus interfacing the MC88100 to the data memory space
Data supervisor/user	DS/$\overline{\text{U}}$	This signal selects between the supervisor select data address space and the user data address space; DS/$\overline{\text{U}}$ is determined by the value of the MODE bit in the processor status register, or by the **.usr** option of the **ld** and **st** instructions
Data read/write	DR/$\overline{\text{W}}$	Indicates whether the memory transaction is a read (DT0 = 1) or a write (DT0 = 0)
Data bus lock	$\overline{\text{DLOCK}}$	The memory lock pin is used by the **xmem** instruction in conjunction with the CMMU; when asserted, the CMMU maintains control of the memory bus during the two **xmem** accesses; data are guaranteed to be unaccessed between the read and write accesses of the **xmem** instruction

TABLE 8.18   Signal Index (*continued*)

Signal name	Mnemonic	Function
Data byte enable	DBE0-DBE3	Used during memory accesses, these signals indicate which bytes are accessed at the addressed location; DEB0-DEB3 are always valid during memory write cycles; a memory read is always 4 bytes wide, and the processor uses the enables to extract the valid data; that is, during an **ld** instruction, the memory system should drive all 32 data signals, regardless of whether 1, 2, or 4 bytes enables are asserted; when DEB0-DEB3 are negated, the transaction is a null; otherwise, the transaction is a valid load or store operation
Data reply	DR0-DR1	Indicates the status of the data memory transaction
Code address bus	CA2-CA31	Provides the 30-bit word address to the instruction memory space; all instructions are 32 bits wide and are aligned on 4-byte boundaries; therefore, the lower two bits of the address space are not required and are implied to be zero
Code bus	C0-C31	This read-only, 32-bit data bus interfaces the MC88100 to the instruction memory space; instructions are always 32 bits wide
Code supervisor/user select	CS/$\overline{\text{U}}$	Selects between the user and supervisor instruction memory spaces; when asserted, selects supervisor memory and when negated, user memory; this signal is determined by the value of the MODE bit in the processor status register
Code fetch	CFETCH	When asserted, signals that an instruction fetch is in progress; when negated, the transaction is a null transaction (code P bus idle)
Code reply	CR0-CR1	Signals the status of the instruction memory transaction
Error	ERR	Asserted when a bus comparator error occurs, ERR indicates that the desired signal level was not driven on the output pin; ERR is used in systems implementing a master/checker configuration of MC88100s
Clock	CLK	Internal clock normally phase locked to minimize skew between the external and internal signals; since CLK is applied to peripherals (such as CMMU devices), exact timing of internal signals is required to properly synchronize the device to the P bus
Phase lock enable	PLLEN	Asserted during reset to select phase locking, PLLEN controls the internal phase lock circuit that synchronizes the internal clocks to CLK
Reset	$\overline{\text{RST}}$	Used to perform an orderly restart of the processor; when asserted, the instruction pipeline is cleared and certain internal registers are cleared or initialized; when negated, the reset vector is fetched from memory, with execution beginning in supervisor mode
Interrupt	INT	Indicates that an interrupt request is in progress; when asserted, the processor saves the execution context and begins execution at the interrupt exception vector; software is responsible for handling all recognized interrupts (those between instructions when no higher priority exception occurs)
P bus checker enable	PCE	Used in systems incorporating two or more MC88100s redundantly; when negated, the processor operates normally and when asserted, the processor monitors (but does not drive) all of its outputs except ERR as inputs
Power supply	Vcc	+ 5-volt power supply
Ground	GND	Ground connections

The 88100 uses memory-mapped I/O. The 88100 can readily be interfaced to memory and I/O chips using its bus control signals.

8.3.5 88100 Exception Processing

The 88100 includes the following exceptions:

- Reset the hardware interrupts which are activated externally via the respective input pins
- Externally activated errors such as a memory access fault
- Internally generated errors such as divide by zero
- Trap instructions

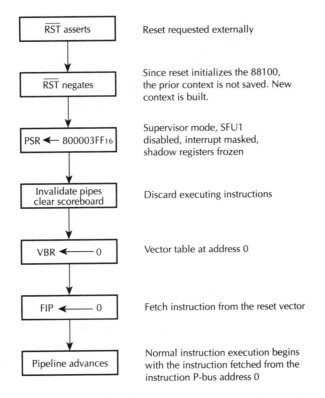

FIGURE 8.15 88100 reset exception flowchart.

Exceptions are processed by the 88100 after completion of the current instructions. When an exception is acknowledged, the 88100 freezes the contents of shadow and exception time control registers, disables interrupts and all SFUs, and enters the supervisor mode.

Figure 8.15 shows the 88100 reset flowchart. When the 88100 RST pin is asserted, all outputs go into high impedance state except ERR which indicates no error. Upon hardware reset, the 88100 initializes the PSR with appropriate data (800003FF$_{16}$ for supervisor mode operation, SFU1 disabled, interrupt masked, and shadow registers frozen), VBR with zero value, and fetches two 32-bit instructions from the reset vector 0.

VBR contains the address of the 88100 vector table. The vector table contains 512 vectors. Each vector corresponds to an exception. Each vector address contains the first two instructions of its exception routine. The instruction stored in the first vector is usually a branch instruction such as ' br.n START' (delayed branch) where START is the starting address of the exception routine. The second vector is normally used to save the current SP, and therefore the instruction such as stcr r31, cr17 is stored at the second vector. The first instruction of the exception routine should load the new SP to be used in the exception routine by using an instruction such as ldcr r31, cr18. The last instruction of the exception handling routine should be rte which returns control to the interrupting routine.

8.4 IBM/Motorola/Apple PowerPC 601

This section provides an overview of the basic features of PowerPC microprocessor. The PowerPC 601 is jointly developed by Apple, IBM, and Motorola. It is available from IBM as PP 601 and from Motorola as MPC 601.

The PowerPC 601 is the first implementation of the PowerPC family of Reduced Instruction Set Computer (RISC) microprocessors. There are two types of PowerPC implementations:

32-bit and 64-bit. The PowerPC 601 implements the 32-bit portion of the IBM PowerPC architectures and Motorola 88100 bus control logic. It includes 32-bit effective (logical) addresses, integer data types of 8, 16, and 32 bits, and floating-point data types of 32 and 64 bits. For 64-bit PowerPC implementations, the PowerPC architecture provides 64-bit integer data types, 64-bit addressing, and other features necessary to complete the 64-bit architecture.

The 601 is a pipelined superscalar processor and is capable of executing three instructions per clock cycle. A pipelined processor is one in which the processing of an instruction is broken down into discrete stages, such as decode, execute, and writeback (result of the operation is written back in the register file).

Because the tasks required to process an instruction are broken into a series of tasks, an instruction does not require the entire resources of an execution unit. For example, after an instruction completes the decode stage, it can pass on to the next stage, while the subsequent instruction can advance into the decode stage. This improves the throughput of the instruction flow. For example, it may take three cycles for an integer instruction to complete, but it there are no stalls in the integer pipeline, a series of integer instructions can have a throughput of one instruction per cycle. Each unit is kept busy in each cycle.

A superscalar processor is one in which multiple pipelines are provided to allow instruction to execute in parallel. The PowerPC 601 includes three execution units. These are a 32-bit integer unit (IU), a branch processing unit (BPU), and a pipelined floating-point unit (FPU).

The PowerPC 601 contains an on-chip, 32-Kbyte unified cache (combined instruction and data cache) and an on-chip memory management unit (MMU). It has a 64-bit data bus and 32-bit address bus. The 601 supports single-beat and four-beat burst data transfer for memory accesses. Note that a single-beat transaction indicates data transfer of up to 64 bits. The PowerPC 601 uses memory-mapped I/O. Input/Output devices can also be interfaced to the PowerPC 601 by using I/O controller. The 601 is designed by using an advanced, CMOS process technology and maintains full compatibility with TTL devices.

The main features of the PowerPC 601 are compared with a similar pipelined superscaler RISC microprocessor manufactured by Digital Equipment Corporation, the Alpha 21064. Finally, typical 64-bit RISC microprocessors are discussed.

8.4.1 PowerPC 601 Block Diagram

Figure 8.16 shows the functional block diagram of the PowerPC 601.

The 601 contains the following on-chip hardware:

  a. RTC (Real Time Clock)
  b. Instruction Unit
  c. Execution Unit
  d. Memory Management Unit (MMU)
  e. Cache Unit
  f. Memory Unit
  g. System Interface

8.4.1.a RTC (Real Time Clock)

The RTC has normally been an I/O device completely outside the CPU in the most earlier microcomputers. While the RTC appearing inside the microcomputer chip is common in single chip microcomputers, this is the first time the RTC is implemented inside a top-of-the line microprocessor such as the PowerPC. The implication is that modern multi-tasking operating systems require time keeping for task switching as well as keeping the calendar date.

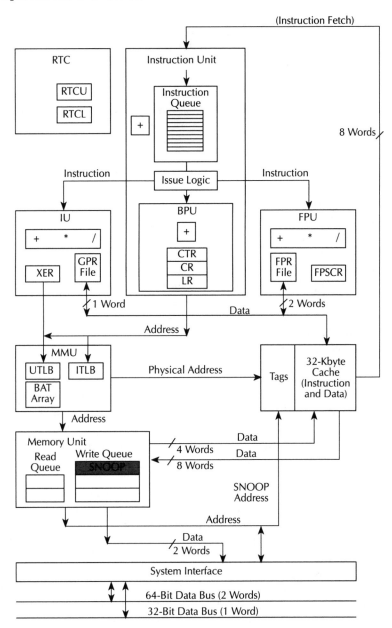

FIGURE 8.16 PowerPC 601 microprocessor block diagram.

The 601 real-time clock (RTC) on-chip hardware provides a measure of real-time in terms of time of day and data with a calendar range of 136.19 years. The RTC contains two registers. These are the RTC upper (RTCU) reigster and the RTC lower (RTCL) register. The RTCU register maintains the number of seconds from a point in time specified by software. The RTCL register counts nanoseconds. The contents of these registers can be copied to any 601 general purpose reigster.

8.4.1.b Instruction Unit

The 601 instruction unit computes the address of the next instruction to be fetched. The instruction unit includes an instruction queue and a Branch Processing Unit (BPU).

The instruction queue holds up to eight instructions and can be filled from the cache during a single cycle.

The BPU searches through the instruction queue for a conditional branch instruction and tries to resolve it early in order to achieve zero-cycle branch in many instances.

8.4.1.c Execution Unit

The 601 execution unit includes three on-chip hardware components. These are floating-point unit (FPU), integer unit (IU), and branch processing unit (BPU). These units operate independently and in parallel.

The FPU includes a single-precision multiply-add (combinatorial) array, the floating-point status and control register (FPSCR), and thirty-two 64-bit floating-point registers. The PowerPC FPU is a cominatorial unit that provides a complete product followed by a sum (with previously accumulated products) in a single clock cycle. Note that this is the heart of a DSP (Digital Signal Processor) chip. The implication is that the power PC chip can replace a DSP. IBM has intentions along these lines for multimedia work in PowerPC based PC's. It is expected that while the machine 'may' be fast enough, the software and timing complexity will be so great that distributing the tasks to other processors (like DSPs) can be more likely to result in an error free product. The PowerPC is pipelined so that most floating-point instructions can be issued back to back. The FPU has no feed-forwarding capabilities. In other words, as a floating point operation completes, another floating-point instruction that may be waiting for those results must wait for the data to be written into the register file before decode can begin. The multiply-add array allows the 601 to implement floating-point operations such as multiply, add, divide, and multiply-add. The 601 FPU supports all IEEE754 floating-point data types (normalized, denormalized, NaN, Zero, and infinity) in hardware.

The Integer Unit (IU) executes all integer instructions (computational and logical instructions). Most integer instructions are single-cycle instructions. The IU interfaces with the cache and MMU for all instructions that access memory. The IU executes one integer instruction at a time by utilizing its arithmetic logic unit (ALU), multiplier, divider, integer exception register (XER), and the general purpose register (GPR) file.

The branch processing unit (BPU) is used for prediction of 601 conditional march instructions early in order to achieve zero-cycle branch. The BPU contains an adder to compute branch target addresses and three special-purpose, user-control registers namely, the link register (LR), the count register (CTR), and the condition register (CR). The LR is used to save the return pointer (computed by the BPU) for subroutine calls. The LR also contains the branch target address for certain types of branch instructions such as branch conditional to link register instruction (bclrx). The CTR, on the other hand, contains the branch target address for some other instructions such as branch conditional to count register (bcctrx) instruction. The CR reflects the result of certain operations and provides a mechanism for testing and branching.

8.4.1.d Memory Management Unit (MMU)

The memory management unit (MMU) of the PowerPC 601 supports up to 4 peta bytes ($2^{**}52$) of virtual memory and 4 Gigabytes of physical memory.

The main functions of the 601 on-chip MMU hardware are to:

- Translate Logical (effective) addresses to physical addresses for memory accesses.
- Translate I/O accesses (most I/O accesses are assumed to be memory mapped).
- Translate I/O controller interface accesses.

- Provides access protection on blocks and pages of memory by the operating system in relation to the supervisor/user privilege level of the access and in relation to whether the access is load or store.

The 601 generates three types of accesses that require address translation. These are instuction accesses, data accesses to memory generated by load and store instructions, and I/O controller interface accesses generated by load and store instructions.

The 601 MMU supports demand-paged virtual memory. Virtual memory management allows execution of programs larger than the size of the physical memory. Demand-paged means that individual pages are loaded into physical memory from system memory only when they are first accessed by an executing program.

To accomplish the above functions, the 601 MMU hardware contains three components. These are UTLB (Unified Translation Lookaside Buffer), ITLB (Instruction Translation Lookaside Buffer) and BAT (Block Address Translation) array.

For instruction accesses, the 601 MMU first performs a lookup in the four entries of the ITLB for both block- and page-based physical address translations. Instruction accesses that miss in the ITLB and all data accesses cause a lookup in the UTLB and BAT array for the physical address translation. In most cases, the physical address translation resides in one of the TLBs and the physical address bits are readily available to the on-chip cache. However, if the physical address translation misses in the TLBs, the 601 automatically performs a search of the translation tables in memory using the information in the table search descriptor register (SDR1) and the corresponding segment register.

8.4.1.e Cache Unit

The PowerPC 601 includes a 32-Kbyte, eight-way set associative, unified (instruction and data) cache. The cache line size is 64 bytes, divided into two eight-word sectors, each of which can be snooped, loaded, cached out, or invalidated independently. Note that snooping means monitoring addresses driven by a bus master to detect the need for coherency actions. The 601 controls cacheability, write policy, and memory coherency. The cache uses a least recently used (LRU) replacement policy.

The instruction unit provides the cache with the address of the next instruction to be fetched. In the case of a cache hit, the cache returns the instruction and as many of the instructions following it as can be placed in the eight-word instruction queue up to the cache sector boundary.

The cache tag directory has one address port dedicated to the instruction fetch and load/ store accesses and one port dedicated to snooping transactions on the system interface.

8.4.1.f Memory Unit

The 601's on-chip memory unit consists of read and write queues that buffer operations between the external interface and the cache. These operations are comprised of operations resulting from load and store instructions that are cache misses, read and write operations required to maintain cache coherency, and table search operations. The read queue contains two elements and write queue contains three elements. The read queue receives requests from the cache unit for arbitration onto the 601 bus interface. Each element of the write queue can contain up to eight words (one sector) of data. One element of the write queue marked snoop is dedicated to writing cache sectors to system memory after a modified sector is hit by a snoop from another processor or snooping device on the system bus. The other two elements in the write queue are used for storing operations and writing back modified sectors that have been deallocated by updating the queue. That is, when a cache location is full, the least-recently used cache sector is deallocated by first being copied into the write queue and from there to system memory.

8.4.1.g System Interface

The 601 system interface includes a 32-bit address bus, a 64-bit data bus, and 52 control and information signals.

The 601 control and information signals allow for functions such as address arbitration, address start, address termination, address transfer, data arbitration, data start, and data termination.

In a multiprocessor system, the system interface supports bus pipelining. The 601 supports split bus transactions for systems with potential multiple bus masters. Allowing multiple bus transactions to occur simultaneously increases the available bus bandwidth for other activity and as a result, improves performance.

8.4.2. Byte and Bit Ordering

The PowerPC 601 supports both big- and little-endian byte ordering. The default byte and bit ordering is big-endian is shown below:

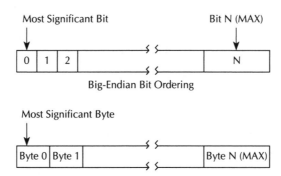

For example, to specify the ordering of four bytes (ABCD) within 32 bits, the 601 can use either the ABCD (big-endian) or DCBA (little-endian) ordering. The 601 big- or little-endian modes can be selected by setting the LM bit (bit 28) in the HID0 register.

Note that big-endian ordering (ABCD) assigns the lowest address to the highest-order eight bits of the multibyte data. On the other hand, little-endian byte ordering (DCBA) assigns the lowest address to the lowest order (rightmost) 8 bits of the multibyte data.

Note that Motorola 68XXX supports big-endian byte ordering while Intel 80XXX supports little-endian byte ordering.

8.4.3 PowerPC 601 Registers and Programming Model

Figure 8.17 shows the register of the PowerPC 601 32-bit implementation. For 64-bit implementation, most registers are 64-bit wide.

These registers can be accessed depending on the program's access privilege level (supervisor or user mode). The privilege level is determined by the privilege level (PR) bit in the machine status register (MSR). The supervisor mode of operation is typically used by the operating system while the user mode is used by the application software.

The PowerPC 601 programming model contains user- and supervisor-level registers as follows:

8.4.3.a User-Level Registers

The user-level register can be accessed by all software with either user or supervisor privileges.

User Programming Model				User-Level SPRs	

User Programming Model

FPR0
FPR1
⋮
FPR31
0   31

User-Level SPRs

SPR0	MQ Register [1]
SPR1	XER – Integer Exception Register
SPR4	RTCU – RTC Upper Register (for reading only)[1,3]
SPR5	RTCL – RTC Lower Register (for reading only)[1,3]
SPR8	LR – Link Register
SPR9	CTR – Count Register

0   31

GPR0
GPR1
⋮
GPR31
0   31

Condition Register
CR
0   31

Floating Point Status and Control Register
FPSCR
0   31

Supervisor Programming Model

Machine State Register[2]
MSR
0   31

Segment Registers
SR0
SR1
⋮
SR15
0   31

Supervisor-Level SPRs

SPR18	DSISR – DAE/Source Instruction Service Register
SPR19	DAR – Data Address Register
SPR20	RTCU – RTC Upper Register (for writing only)[1,3]
SPR21	RTCL – RTC Lower Register (for writing only)[1,3]
SPR22	DEC – Decrementer Register[4]
SPR25	SDR1 – Table Search Description Register 1
SPR26	SRR0 – Save and Restore Register 0
SPR27	SRR1 – Save and Restore Register 1
SPR272	SPRG0 – SPR General 0
SPR273	SPRG1 – SPR General 1
SPR274	SPRG2 – SPR General 2
SPR275	SPRG3 – SPR General 3
SPR282	EAR – External Access Register
SPR287	PVR – Processor Version Register
SPR528	IBAT0U – BAT 0 Upper [2]
SPR529	IBAT0L – BAT 0 Lower [2]
SPR530	IBAT1U – BAT 1 Upper [2]
SPR531	IBAT1L – BAT 1 Lower [2]
SPR532	IBAT2U – BAT 2 Upper [2]
SPR533	IBAT2L – BAT 2 Lower [2]
SPR534	IBAT3U – BAT 3 Upper [2]
SPR535	IBAT3L – BAT 3 Lower [2]
SPR1008	HID0[1]
SPR1009	HID1[1]
SPR1010	HID2 (IABR)[1]
SPR1013	HID5 (DABR)[1]
SPR1023	HID15 (PIR)[1]

[1] 601-only registers. These registers are not necessarily supported by other PowerPC processors.
[2] These registers may be implemented differently on other PowerPC processors. The PowerPC architecture defines two sets of BAT registers–eight IBATs and eight DBATs. The 601 implements the IBATs and treats them as unified BATs.
[3] RTCU and RTCL registers can be written only in supervisor mode, in which case different SPR numbers are used.
[4] DEC register can be read by user programs by specifying SPR6 in the mfspr instruction (for POWER compatibility).

FIGURE 8.17   PowerPC 601 microprocessor programming model-registers.

The 32 32-bit GPRs (General Purpose registers, GPR0-GPR31) can be used as the data source or destination for all integer instructions. They can also provide data for generating addresses.

The 32 32-bit FPRs (Floating-Point registers, FPR0 through FPR31) can be used as data source and destination for all floating point instructions.

The Floating-point status and control register (FPCSR) is a user-control register in the Floating-Point Unit (FPU). It contains floating-point status and control bits such as floating-point exception signal bits, exception summary bits, and exception enable bits.

The condition register (CR) is a 32-bit register, divided into eight 4-bit fields, CR0-CR7. These fields reflect the results of certain arithmetic operations and provide mechanisms for testing and branching.

The remaining user-level registers are 32-bit special purpose registers, SPR0, SPR1, SPR4, SPR5, SPR8, and SPR9.

SPR0 is known as the MQ register and is used as a register extension to hold the product for the multiplication instructions and the dividend for the divide instructions. The MQ register is also used as an operand of long shift and rotate instructions.

SPR1 is called the Integer exception register (XER). The XER is a 32-bit register that indicates carries and overflow bits for integer operations. It also contains two fields for load string and compare byte indexed instruction.

SPR4 and SPR5 respectively represent two 32-bit read-only register and hold the upper (RTCU) and lower (RTCL) portions of the Real Time Clock (RTC). The RTCU register maintain the number of second from a time specified by software. The RTCL register maintains the fraction of the current second in nanoseconds.

SPR8 is the 32-bit link register (LR). The link register can be used to provide the branch target address and to hold the return address after branch and link instructions.

SPR9 represents the 32-bit count register (CTR). The CTR can be used to hold a loop count that can be decremented during execution of certain branch instructions. The CTR can also be used to hold the target address for the branch conditional to count register instruction.

8.4.3.b Supervisor-Level Registers

The supervision-level registers can only be accessed by the programs executed with supervisor privileges. These include the following:

8.4.3.b.i Machine State Register (MSR).   The MSR is a 32-bit register that defines the state of the processor. When an exception occurs, the bits in MSR are changed according to the exception. The bits in the MSR indicate processor information such as privilege level and single step trace enable.

The privilege level (PR bit, bit 17 of MSR) indicates whether the 601 can execute both user and supervisor level instructions (PR = 0) or can only execute user level instructions (PR = 1).

The single-step trace enable (SE bit, bit 21 of MSR) indicates whether the processor executes instructions normally (SE = 0) or executes instructions in single step mode (SE = 1).

8.4.3.b.ii Segment Registers.   The 601 includes sixteen 32-bit registers (SR0-SR15). The bits in the segment register are interpreted differently depending on the value of the T-bit (bit 0). For example, if T= 0 in the selected segment register, the effective address is a reference to an ordinary memory segment. On the other hand, if T=1, in the selected segment register, the effective address is a reference to an I/O controller interface segment.

8.4.3.b.iii Supervisor-Lever SPRs.   Many of the SPRs can be accessed only by supervisor-level instructions. Any attempt to access these SPRs with user-level instructions will result in a "privilege exception".
These registers consist of the following:

- The 32-bit data access exception (DAE)/source instruction service register (DSISR) defines the cause of data access and alignment exceptions.
- Real-Time clock (RTC) register includes two 32-bit registers namely, RTC upper (RTCU) and RTC lower (RTCL). The registers can be read from by user-lever software, but can be written to only by supervisor-level software.
- Decrement register (DEC) is a 32-bit decrementing counter that provides a mechanism for causing a decrementer exception after a programmable delay. The 601 implements a separate clock input rather than the processor clock that serves both the DEC and the RTC facilities.
- The 32-bit Table search description register 1 (SDR1) specifies the page table format used in logical-to-physical address translation for pages.

- The 32-bit machine status save/restore register 0 (SPR0) is used by the 601 for saving the address of the instruction that caused the exception, and the address to return to when a return from Interrupt (rfi) instruction is executed.
- General SPRs, SPRG0-SPRG3, are 32-bit registers provided for operating system use.
- The 32-bit external access register (EAR) control facility through the External Control Input Word Indexed (eciwx) and External Control Output Word Indexed (ecowx) instructions.
- The 32-bit processor version register (PVR) is a read-only register that identifies the version (model) and revision level of the PowerPC processor.
- The eight 32-bit block address translation (BAT) register are grouped into four pairs of BATs (BAT0U-BAT3U and BAT0L-BAT3L).

  The block address translation mechanism in the 601 is implemented as a software controlled BAT array. The BAT array maintains the address translation information for four blocks of memory. The BAT array in the 601 is maintained by the system software and is implemented as a set of eight special-purpose registers (SPRs). Each block is defined by a pair of SPRs called upper and lower BAT registers.

- Five 32-bit hardware implementation registers (HID0-HID2, HID5, and HID15) are provided primarily for implementing debugging features such as break point and single stepping. HID15 holds the four-bit processor identification tag (PID) that is useful for differentiating processor in multiprocessor system designs.

8.4.4 PowerPC 601 Memory Addressing: Effective Address (EA) Calculation

The effective address (EA) is the 32-bit address computed by the processor when executing a memory access or branch instruction or when fetching the next sequential instruction.

Effective address computations for both data and instruction accesses use 32-bit unsigned binary arithmetic. If the sum of the effective address and the operand length exceeds the maximum effective address, a carry from bit is ignored. Arithmetic and logic instructions do not read or modify memory. To use the contents of a memory location in a computation, and then modify the same or another memory location, the memory contents must be loaded into a register, modified, and then written back to the target location using load or store instructions. This is a consequence of a RISC architecture. Note that the RISC ideology caused the lack of capability for instructions such as ADD instruction to directly modify memory. ADDs are register/register operations in a RISC while ADDs are memory operations in a CISC. The concept of alignment is also applied more generally to data in memory. For example, 12 bytes of data are said to be word-aligned if its address is a multiple of 4. The operand of a single-register memory access instruction has a natural boundary equal to the operand length. That is, the "natural" address of an operand is an integral multiple of the operand length.

A memory operand is said to be aligned if it is aligned at its natural boundary, otherwise it is misaligned. The PowerPC can transfer both aligned and misaligned data between the processor and memory. However, the placement (location and alignment) of operands in memory affects the relative performance of memory accesses. Best performance is guaranteed if memory operands are aligned on natural boundaries

Operands for single-register memory access instructions have the characteristics shown below:

Memory Operand (if aligned)	Length	Address (28–31)
Byte	8 bits	XXXX
Half word	2 bytes	XXX0
Word	4 bytes	XX00
Double word	8 bytes	X000

In the above, an "X" indicates that the bit can be 0 or 1 independence of the state of the bits in the address.

Load and store operations have two types of effective address generation :

8.4.4.a Register Indirect with Immediate Index Mode

Instructions using this mode contain a signed 16-bit index (d operand in the 32-bit instruction) which is sign extended to 32 bits, and added to the contents of a general purpose register specified by five bits in the 32-bit instruction (rA operand) to generate the effective address. A zero in the rA operand causes a zero to be added to the immediate index (d operand). The option to specify rA or 0 is shown in the instruction descriptions of the 601 user's manual as the notation (rA|0).

An example is lbz rD, d (rA) where rA specifies a general purpose register (GPR) containing an address, d is the 16-bit immediate index and rD specifies a general purpose register as destination. Consider lbz r1, 20 (r3). The effective address (EA) is the sum r(3|0) + 20. The byte in memory addressed by the EA is loaded into bits 31 through 24 of register r1. The remaining bits in r1 are cleared to zero. Note that registers r1 and r3 represent GPR1 and GPR3 respectively.

8.4.4.b Register Indirect with Index Mode

Instructions using this addressing mode add the contents of two general purpose registers (one GPR holds an address and the other GPR holds the index). An example is lbzx rD, rA, rB where rD specifies a GPR as destination, rA specifies a GPR as the index, and rB specifies a GPR holding an address. Consider lbzx r1, r4, r6. The effective address (EA) is the sum (r4|0) + (r6). The byte in memory addressed by the EA is loaded into register r1 (24-31). The remaining bits in register rD are cleared to 0.

PowerPC 601 conditional and unconditional Branch instructions compute the effective address (EA) or the next instruction address using various addressing modes. A few of them are described below:

8.4.4.b.i Branch Relative.   Branch instructions (32-bit wide) using the relative mode generate the address of the next instruction by adding an offset and the current program counter contents. An example of this mode is an instruction "**be start**" unconditionally jumps to the address PC + start.

8.4.4.b.ii Branch Absolute.   Branch instructions using this mode include the address of the next instruction to be executed. For example, the instruction **ba begin** unconditionally branches to the absolute address "begin" specified in the instruction.

8.4.4.b.iii Branch to Link Register.   Branch instructions using this mode branch to the address computed as the sum of the immediate offset and the address of the current instruction. The instruction address following the instruction is placed into the link register. For example, the instruction **bl start** unconditionally jumps to the address computed from current PC contents plus start. This return address is also placed in the link register.

8.4.4.b.iv Branch to Count Register.   Instructions using this mode branch to the address contained in the current register. Consider **bcctr BO,BI** means branch conditional to count register. This instruction branches conditionally to the address specified in the count register.

The BI operand specifies the bit in the condition register to be used as the condition of the branch. The BO operand specifies how the branch is affected by or affects condition or count registers. Numerical values specifying BI and BO can be obtained from the 601 manual.

Note that some instructions combine the link register and count register modes. An example is bcctrl BO, BI. This instruction first performs the same operation as the bcctr and then places the instruction address following the instruction into the link register. This instruction is a form of "conditional call" as the return address is saved in the link register.

8.4.5 PowerPC 601 Typical Instructions

The 601 instructions are divided into the following categories:

- a. Integer Instructions
- b. Floating-point Instructions
- c. Load/store Instructions
- d. Flow control Instructions
- e. Processor control Instructions

Integer instructions operate on byte (8-bit), half-word (16-bit), and word (32-bit) operands. Floating-point instructions operate on single-precision and double precision floating-point operands.

Since the PowerPC is based on the RISC as opposed to the CISC architecture, arithmetic and Logical instructions do not read or modify memory.

8.4.5.a Integer Instructions

The integer instructions include integer arithmetic, integer compare, integer rotate and shift, and integer logical instructions.

The integer arithmetic instructions always set integer exception register bit, CA, to reflect the carry out of bit 7. Integer instructions with the overflow enable (OE) bit set will cause the XER bits SO (summary overflow — overflow bit set due to exception) and OV (overflow bit set due to instruction execution) to be set to reflect overflow of the 32-bit result.

Some examples of integer instructions are provided in the following.

Note that rS, rD, rA, and rB in the following examples are 32-bit general purpose registers (GPR) of the 601 and SIMM is 16-bit signed immediate define:

- add rD, rA, SIMM

  Performs the following immediate operation: rD ← (rA|0) + SIMM; (rA|0) can be either (rA) or 0.
- add rD, rA, rB performs rD ← rA + rB
- add.rD, rA, rB adds with CR update as follows: rD ← rA + rB

  The dot suffix enables the update of the condition register.
- Subf rD, rA, rB performs rD ← rB – rA
- Subf.rD, rA, rB performs the same operation as Subf but updates the condition code register.
- addme rD, rA performs the (add to minus one extended)

  Operation: rD ← (rA) + FFFF FFFFH + CA bit in XER
- Subfme rD, rA performs the (subtract from minus one extended)

  Operation: rD ← (rA)′ + FFFF FFFFH + CA bit in XER where (rA)′ represents the ones complement of the contents of rA.
- mulhwu rD, rA, rB performs an unsigned multiplication of two 32-bit numbers in rA and rB. The high-order 32 bits of the 64-bit product are placed in rD.
- mulhw rD, rA, rB performs the same operation as the mulhwu except that the multiplication is for signed numbers.

- mullw rD, rA, rB places the low-order 32 bits of the 64-bit product (rA) * (rB) into rD. The low-order 32-bit products are independent whether the operands are treated as signed or unsigned integers.
- mulli rD, rA, SIMM places the low-order 32 bits of the 48-bit product (rA) * SIMM16 into rD. The low-order bits of the 32-bit product are independent of whether the operands are treated as signed or unsigned integers.
- divw rD, rA, rB divides the 32-bit signed dividend in rA by the 32-bit signed divisor in rB. The 32-bit quotient is placed in rD and the remainder is discarded.
- divwu rD, rA, rB is same as the divw instruction except that the division is for unsigned numbers.
- cmpi crfD, L, rA, SIMM compares 32 bits in rA with immediate value SIMM treating operands as signed integer. The result of comparison is placed in crfd field (0 for CR0, 1 for CR1 and so on) of the condition register. L = 0 indicates 32-bit operands while L=1 represents 64-bit operands. For example, cmpi 0, 0, rA, 200 compares 32-bits in register rA with immediate value 200 and CR0 is affected according to the comparison.
- xor, rA, rS, rB performs exclusive-or operation between the contents of rS and rB. The result is placed into register rA.
- extsb rA, rS places bits 21–31 of rS into bits 21–31 of rA. Bit 24 of rS is the sign extended through bits 0–23 of rA.
- slw rA, rS, rB shifts the contents or rS left the shift count specified by rB [27–31]. Bits shifted out of position 0 are lost. Zeros are placed in the vacated positions on the right. The 32-bit result is placed into rA.
- Srw rA, rS, rB is similar to slw rA, rS, rB except that the operation is for right shift.

8.4.5.b Floating-Point Instructions

Some of the 601 floating-point instructions are provided below:

- fadd frD, frA, frB adds the contents of the floating-point register, frA to the contents of the floating-point register frB. If the most significant bit of the resultant significand is not a one, the result is normalized. The result is rounded to the specified precision under control of the FPSCR register. The result is then placed in frD.

Note that this 'fadd' instruction requires one cycle in execute stage, assuming normal operations; however, there is an execute stage delay of three cycles if next instruction is dependent.

The 601 floating-point addition is based on "exponent comparison and add by one" for each bit shifted, until the two exponents are equal. The two significants are then added algebraically to form an intermediate sum. If a carry occur, the sum's significand is shifted right on bit position and the exponent is increased by one.

- fsub frD, frA, frB performs frA – frB, normalization and rounding of the result are performed in the same way as the fadd.
- fmul frD, frA, frC performs frD ← frA * frC.

Normalization and rounding of the result are performed in the same way as the fadd. Floating-point multiplication is based on exponent addition and multiplication of the significands.

- fdiv frD, frA, frB performs the floating-point division frD ← frA/frB. No remainder is provided. Normalization and rounding of the result are performed in the same way as the fadd instruction.
- fmsub frD, frA, frC, frB performs frD ← frA * frC – frB. Normalization and rounding of the result are performed in the same way as the fadd instruction.

8.4.5.c Load/Store Instructions

Some examples of the 601 load and store instructions are listed below:

- lhzx rD, rA, rB loads the half word (16- bit) in memory addressed by the sum (rA|0) + (rB0) into bits 16 through 31 of rD. The remaining bits of rD are cleared to zero.
- sthux rS, rA, rB stores the 16-bit half word from bits 16-31 of register rS in memory addressed by the sum (rA|0) + (rB). The value (rA|0) + rB is placed into register rA.
- lmw rD, d(rA) loads n (where n = 32-D and D = 0 through 31) consecutive words starting at memory location addressed by the sum (rA|0) + d into the general purpose register specified by rD through r31.
- stmu rS, d(rA) is similar to lmw except that the stmw stores n consecutive words.

8.4.5.d Flow Control Instructions

Flow control instructions include conditional and unconditional branch instructions. An example of one of these instructions is provided below:

- bc (branch conditional) BO, BI, target branch with offset target if the condition bit in CR specified by bit number BI is true (The condition 'true' is specified by a value in BO).

For example, bc 12, 0, target means that branch with offset target if the condition specified by bit 0 in CR (BI = 0 indicated result is negative) is true (specified by the value BO = 12 according to Motorola PowerPC 601 manual).

8.4.5.e. Processor Control Instructions

Processor control instructions are used to read from and write to the machine state register (MSR), condition register (CR), and special status register (SPRs). Some examples of these instructions are provided below:

- mfcr rD places the contents of the condition register into rD.
- mtmsr rS places the contents of rS into the MSR. This is a supervisor-level instruction.
- mfmsr rD places the contents of MSR into rD. This is a supervisor-level instruction.

8.4.6 PowerPC 601 Exception Model

All 601 exceptions can be described as either precise or imprecise and either synchronous or asynchronous. Asynchronous exceptions are caused by events external to the processor's execution. Synchronous exceptions, on the other hand, are handled precisely by the 601 and are caused by instructions; precise exception means that the machine state at the time the exception occurs is known and can be completely restored. That is, the instructions that invoke trap and system call exceptions complete execution before the exception is taken. When exception processing completes, execution resumes at the address of the next instruction.

An example of a maskable asynchronous, precise exception is the external interrupt. When an asynchronous, precise exception such as the external interrupt occurs, the 601 postpones its handling until all instructions and any exceptions associated those instructions complete execution.

System reset and machine check exceptions are two nonmaskable exceptions that are asynchronous and imprecise. These exceptions may not be recoverable or may provide a limited degree of recoverability for diagnostic purpose.

Asynchronous, imprecise exceptions have the highest priority with the synchronous, precise exceptions the next priority and the asynchronous, precise exceptions have the lowest priority.

The 601 exception mechanism allows the processor to change automatically to supervisor state as a result of exceptions. When exceptions occur, information about the state of the

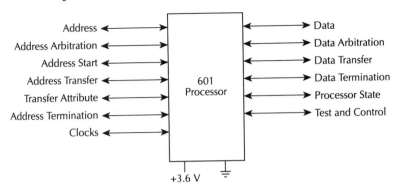

FIGURE 8.18   System interface.

processor is saved to certain registers rather than in memory as is usually done with other processors in order to achieve high speed. The processor then begins execution at an address (exception vector) predetermined for each exception. The exception handler at the specified vector is then processed with processor in supervisor mode.

8.4.7 601 System Interface

Figure 8.18 shows the system interface signals of the PowerPC 601.

The pins and signals of the PowerPC 601 include a 32-bit address bus, a 64-bit data bus, and 52 control and information signals.

The 601 control and information signals include the address arbitration, address start, address transfer, transfer attribute, address termination, data arbitration, data transfer, and data signals. Test and control signals provide diagnostics for selected internal circuitry.

The processor supports multiple master through a bus arbitration scheme that allows various devices to compete for the shared resource. The arbitration logic can implement priority protocols and can park masters to avoid arbitration overhead.

The following sections describe the 601 bus support for memory and I/O controller interface operations.

8.4.7.a Memory Accesses

Memory accesses allow transfer sizes of 8, 16, 24, 32, 40, 48, 56, or 64 bits in one bus clock cycle. Data transfers occur in either single-beat transactions or four-beat burst transactions. A single-beat transaction transfers as much as 64 bits. Single-beat transactions are caused by non cached accesses that access memory directly. An example is reading and writing when the cache is disabled. Burst transactions, which always transfer an entire cache sector (32 bytes), are initiated when a sector in the cache is read from or written to memory.

8.4.7.b I/O Controller Interface Operations

Both memory and I/O accesses can use the same bus transfer protocols. The 601 also has the ability to define memory areas as I/O controller interface areas.

The 601 uses $\overline{\text{TS}}$ pin for memory-mapped accesses and XATS pin for I/O controller interface accesses.

8.4.7.c 601 Signals

The 601 signals are grouped as follows:

1. Address arbitration signals — these signals provide arbitration for address bus mastership.
2. Address transfer start signals — these signals indicate that a bus master has begun a transaction on the address bus.
3. Address transfer signals — these signals consisting of the address bus, address parity, and address parity signals, are used to transfer the integrity of the transfer.
4. Transfer attribute signals — these signals provide information about the type of transfer, such as the transfer size, and whether the transaction is bursted or cache-inhibited.
5. Address transfer termination signals — these signals are used to acknowledge the end of the address phase of transaction. They also indicate whether a condition exists that requires the address phase to be replaced.
6. Data arbitration signals — these signals are used to arbitrate for data bus mastership.
7. Data transfer signals — these signals consisting of the data bus, data parity, and data parity error signals, are used to transfer the data and to ensure the integrity of the transfer.
8. Data transfer termination signals — In a single-beat transaction, these signals indicate the end of the tenure, while in burst access, the data termination signals apply to individual beats and indicate the end of the tenure only after the final data beat.
9. System status signals — these signals include the interrupt signal and reset signals.
10. Clock signals — these signals determine the system clock frequency. These signals can also be used to synchronize multiprocessor systems.
11. Processor state signals — these signals are used to set the reservation coherency bit. The 601 uses this bit for atomic memory updating by using its atomic instructions. Note that in multiprocessor systems, a mechanism is required to allow programs to manipulate shared data in an indivisible manner so that when such an operation is underway, another processor cannot perform the same operation. In order to implement higher-level synchronization mechanisms, such as locks and semaphores, atomic instructions are included in the 601. Memory can be updated atomically by setting a reservation on the load instructions (lwarx) and checking that the reservation is still valid (stwcx) is executed.

The reservation (RSRV) output signal is always driven by bus clock cycle and reflects the status of the reservation coherency bit in the reservation address register.

8.4.8 PowerPC 601 Vs. Alpha 21064

Both Motorola/IBM/Apple PowerPC 601 and Digital Equipment Corporation's Alpha 21064 are RISC-based superscalar microprocessors. That is, they can execute two or more instructions per cycle.

The PowerPC 601 contains powerful instructions while the Alpha 21064 includes a simplified instruction set with a very fast clock.

Both the PowerPC 601 and the Alpha are based on load/store architectures. This means that all instructions that access memory are either loads or stores, and all operate instructions are from register to register. They both have 32-bit fixed-length instructions along with 32 integer and 32 floating-point registers.

The PowerPC 601 includes two primary addressing modes. These are register plus displacement and register plus register. In addition, the 601 load and store instructions perform the load or store operation and also modify the index register by placing the just-computed effective address in it. The Alpha 21064, on the other hand, has only one primary addressing mode called register plus displacement. In Alpha, load and store instructions do not update the index register.

There are significant differences in the way the two microprocessors handle branching. In both architectures, branch target addresses are normally determined by using program counter

TABLE 8.19   Summary of Implementation Characteristics

Characteristic	PowerPC 601	Alpha 21064
Technology	0.6-micron CMOS	0.75-micron CMOS
Die size	1.09 cm square	2.33 cm square
Transistor count	2.8 million	1.68 million
Total cache (instruction + data)	32 Kbyte	16 Kbyte
Package	304-in QFP	431-pin PGA
Clock frequency	50 and 66 Mhz	150 to 200 Mhz

relative mode. That is, the branch target address is determined by adding a displacement to the program counter. However, as mentioned before, conditional branches in the 601 may test fields in the condition code register and the contents of a special register called the count register (CTR). A single 601 branch instruction can implement a loop-closing branch by decrementing the CTR, testing its value, and branching if it is nonzero.

In the Alpha 21064, on the other hand, conditional branches test a general-purpose register relative to zero or by low-order bit (a 1 or 0 in low order bit respectively mean odd or even) odd/even register contents. Thus, results of most instructions can be used directly by conditional branch instructions, as long as they are tested against zero or odd/even.

There are also differences in the way the return address is saved by certain control transfer instructions such as the subroutine call. For example, special jump instructions are used to save the return address in a general-purpose register. The 601, on the other hand, does this in any branch by setting the link (LK) bit to one. The return address is saved in the link register.

Next, the implantation characteristics of the 601 and 21064 are considered. Table 8.19 summarizes these differences.

Both the PowerPC 601 and Alpha 21064 utilize sophisticated pipelines. The 601 uses relatively short independent pipelines with more buffering while the 21064 includes longer pipelines with less internal buffering. The 601 does a lot of computation in each pipe stage. Furthermore, the clock of the Alpha is approximately three times faster than the 601.

The two microprocessors utilize different cache memory designs. For example, the 601 has a unified (combined) 32-Kbyte cache. That is, instructions and data reside in the same cache in the 601. The 21064, on the other hand, has separate data and instruction caches of 8 Kbytes each. Therefore, the 601 is expected to have a higher hit rate than the 21064.

Finally, the 601 offers high performance by utilizing sophisticated design tricks. The 21064 gains performance by design simplicity. For example, the 601 includes powerful instructions such as floating-point multiply-add and update load/store that perform more tasks with fewer instructions. The 21064 architecture's simplicity, on the other hand, lends itself better to very high clock rate implementations.

8.5 64-Bit RISC Microprocessors

Typical 64-bit RISC microprocessors include the Alpha 21164 (Digital Equipment Corporation), the PowerPC 620 (Motorola/IBM/Apple) and the Ultrasparc (Sun Microsystems). These 64-bit processors are ideal candidates for data crunching machines and high-performance desktop systems/workstations.

The number of instructions issued per cycle has been increasing steadily. For example, the PowerPC 601 can issue instructions to the integer, floating-point, and branch-processing units in one and the same cycle. The 64-bit RISC microprocessors can typical issue four instructions to six independent units or more.

These 64-bit processors include multiple integer units which allow multiple integer operations in each cycle. For example, the PowerPC 620 contains three integer units — two for single cycle and one for multiple cycle operations. Some 64-bit RISC processors such as the Ultrasparc include multiple floating-point units.

The clock frequencies of the 64-bit RISC microprocessors vary from 133 MHz to 300 MHz. These processors can issue a maximum of four instructions per cycle. In order to keep the data and instructions flowing, many 64-bit RISC processors such as the Alpha 21164 are provided with 128-bit data bus. The Alpha 21164 is the fastest microprocessor available today with a maximum of 300 MHz clock. The Alpha 21164 is a four-way superscalar processor.

Table 8.20 compares the various features of typical 64-bit RISC microprocessors.

TABLE 8.20   Comparison of Various Features of Typical 64-bit RISC Microprocessors

Features	Digital Equipment Corp. Alpha 21164	Motorola/IBM/Apple PowerPC 620	Sun Microsystems Ultrasparc
Clock speed	300 MHz	133 MHz	167 MHz
Millions of transistors	9.3	7	3.8
On-chip data/instruction cache, K Byte	8/8 Primary 96 Unified secondary	32/32	16/16
Power, W	50	30	30
Data bus size	128-bit	128-bit	128-bit
Address bus size	40-bit	40-bit	41-bit
Maximum number of instructions per cycle	4	4	4
Number of independent units (integer, floating-point, etc.)	4	6	9

Questions and Problems

8.1 Summarize the basic features of RISC microprocessors. Identify how some of these features are implemented in the 80960SA/SB and 88100.

8.2 What operations are controlled by the 80960SA/SB and 88100 register scoreboard?

8.3 Compare the main on-chip features of the 80960SA/SB with those of the 88100. Comment on the floating-point and real data types.

8.4 Identify the 80960SA/SB and 88100 stack pointers.

8.5 Compare 80960SA/SB cache with the 88200 cache.

8.6 Assume a 80960SA/SB with the condition code 010. Write an instruction to set bit 20 to one in register g8 and store the result in g8.

8.7 What happens after execution of the following instructions:
   i)   cmpo ox10, r7
   ii) cmpinco r12, g4, g7

8.8 Find the contents of r8 with (r2) > (r4) after the following 80960SA/SB instruction sequence is executed:

        compo r2, r4
        testg r8

8.9   Find 80960SA/SB single instructions which are equivalent to the following instruction sequences:
   i)   cmpi 0,g0
        ble begin
   ii) chkbit 1, g8
        be start

8.10   For the following 80960SA/SB instruction, what will be the size of the result: addr ro, r1, fp1.

8.11   Assume a 80960SA/SB. Find the operation performed along with the register in which the result is stored after execution of each of the following instructions:
   i)   logbnrl r8, fp0
   ii) logepr r0, r4, fp0
   iii) modi r1, r2, r3
   iv) notand g3, g4, g6
   v)   sqrtrl r2, fp1

8.12   What functions are performed by the following 80960SA/SB pins: BE_0-BE_1, BLAST/FAIL, A(1:3).

8.13   Discuss the 80960SA/SB interrupts.

8.14   Describe briefly the functional blocks included in the 88200. What does the 88200 provide to an 88100 system?

8.15   What is the maximum number of 88200s that can be present in one 88000 processing mode?

8.16   How many pipelines are in the 88100?

8.17   Since there is no return from subroutine instruction, how does the 88100 return from subroutine?

8.18   What 88100 floating-point control registers can be accessed by user mode programs?

8.19   Show the contents of registers and memory locations after the 88100 executes the following instructions:
   i)   st.h r1, r2, r0
   ii) ld.h r1, r2, 0X0A
        Assume [r1] = 0000 0020$_{16}$
               [r2] = 0000 0300$_{16}$
All numbers are in hexadecimal.

8.20   Find the contents of r5 after execution of the following 88100 instructions:
   i)   mask.u r5, r2, 0XFFFF
   ii) mask r5, r2, r6
        Assume [r5] = AAAA 0100$_{16}$
               [r2] = 0020 05FF$_{16}$
               [r6] = 7777 7777$_{16}$

        prior to execution of the above instructions.

All numbers are in hexadecimal.

8.21   What is the effect of the 88100 tb1 r0, r1, 200 instruction?

8.22   What are the functions of the 88100 CS/$\overline{U}$, BE0-BE3, C0, and C2 pins?

8.23   What 88100 registers are affected by hardware reset?

8.24   Discuss briefly the 88100 exceptions.

8.25   Write an 80960SA/SB or 88100 assembly language instruction sequence to logically shift the content of r2 into r1 to the right by 8 bits.

8.26   Write an assembly language program in 80960SA/SB or 88100 assembly language to subtract a 64-bit number in r4 r5 from another 64-bit number in r0 r1. Store result in r0 r1.

8.27   Write a program in 80960SA/SB or 88100 assembly language to compute the volume of a sphere V = 4/3 πr^3 where r is the 32-bit radius stored in register r2.

8.28   Write a program in 80960SA/SB assembly language to perform the following: A + (B/C) where A is an 80-bit floating-point number contained in a floating-point register. B and C are respectively 64-bit floating-point numbers stored in r2 r3 and r4 r5 respectively. Store 80-bit result in a floating-point register. Discard the remainder of B/C.

8.29   Repeat 8.28 for the 88100 except that A, B, and C are 64-bit floating-point numbers. Assume that the number A is in fp1. Store the result in fp2.

8.30   Write program in 80960SA/SB or 88100 assembly language to compute the roots of the quadratic equation ax² + bx + c = 0 by using

$$x = \frac{-b \pm \sqrt{b^2 - 4ac}}{2a}$$

Assume that register r4, r5, r6 respectively contain a, b, and c. Store the roots in registers of your choice.

8.31   Discuss the types of PowerPC architectures.

8.32   How many execution units are included in the PowerPC 601? Comment.

8.33   How does the PowerPC 601 achieve zero-cycle branch?

8.34   What do you mean by the unified cache of the 601? What is its size?

8.35   What is meant by the snoop controller of the 601? What is its purpose?

8.36   List the user-level and supervisor-level registers of the 601?

8.37   How does the 601 MSR indicate the following:
   i)   The 601 executes both the user- and supervisor-level instructions.
   ii)   The 601 executes only the user-level instructions.

8.38 Explain the operation performed by each of the following instruction:
- i) lbz r2,30(r4)
- ii) lbzx r1,r2
- iii) add.r1,r2,r3
- iv) divwu r2,r3,r4
- v) extsb r1,r2
- vi) fsub fr2,fr3,fr4

8.39 Repeat Examples 8.13 through 8.19 using the 601 assembly language instructions of the PowerPC.

8.40 Discuss briefly the exceptions included in the PowerPC 601.

8.41 What is the purpose of the reservation coherency bit of the 601?

8.42 Compare the basic features of the 601 with the Alpha 21064.

8.43 Compare the basic features of the Alpha 21164, the PowerPC620, and the Ultrasparc.

PERIPHERAL INTERFACING

This chapter describes interfacing characteristics of a microcomputer with typical peripheral devices such as a hexadecimal keyboard, display, DMA controller, printer, CRT (Cathode Ray tube) terminal, and coprocessor.

9.1 Keyboard Interface

9.1.1 Basics of Keyboard and Display Interface to a Microprocessor

A common method of entering programs into a microcomputer is via a keyboard. A popular way of displaying results by the microcomputer is by using seven segment displays. The main functions to be performed for interfacing a keyboard are

1. Sense a key actuation.
2. Debounce the key.
3. Decode the key.

Let us now elaborate on the keyboard interfacing concepts. A keyboard is arranged in rows and columns. Figure 9.1 shows a 2 × 2 keyboard interfaced to a typical microcomputer. In Figure 9.1, the columns are normally at a HIGH level. A key actuation is sensed by sending a LOW to each row one at a time via PA0 and PA1 of port A. The two columns can then be input via PB2 and PB3 of port B to see whether any of the normally HIGH columns are pulled LOW by a key actuation. If so, the rows can be checked individually to determine the row in which the key is down. The row and column code in which the key is pressed can thus be found.

The next step is to debounce the key. Key bounce occurs when a key is pressed or released — it bounces for a short time before making the contact. When this bounce occurs, it may appear to the microcomputer that the same key has been actuated several times instead of just once. This problem can be eliminated by reading the keyboard after 20 ms and then verifying to see if it is still down. If it is, then the key actuation is valid.

The next step is to translate the row and column code into a more popular code such as hexadecimal or ASCII. This can easily be accomplished by a program.

There are certain characteristics associated with keyboard actuations which must be considered while interfacing to a microcomputer. Typically, these are two-key lockout and N-key rollover. The two-key lockout takes into account only one key pressed. An additional key

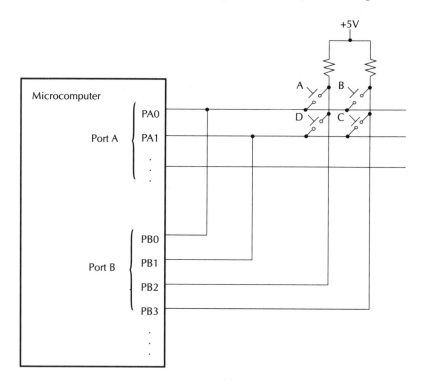

FIGURE 9.1   A 2 × 2 keyboard interfaced to a microcomputer.

pressed and released does not generate any codes. The system is simple to implement and most often used. However, it might slow down the typing since each key must be fully released before the next one is pressed down. On the other hand, the N-key rollover will ignore all keys pressed until only one remains down.

Now let us elaborate on the interfacing characteristics of typical displays. The following functions are to be typically performed for displays:

1. Output the appropriate display code.
2. Output the code via right entry or left entry into the displays if there is more than one display.

The above functions can easily be realized by a microcomputer program. If there is more than one display, they are typically arranged in rows. A row of four displays is shown in Figure 9.2. In the figure, one has the option of outputting the display code via right entry or left entry. If it is entered via right entry, then the code for the most significant digit of the four-digit display should be output first, then the next digit code, and so on. Note that the first digit will be shifted three times, the next digit twice, the next digit once, and the last digit (least significant digit in this case) does not need to be shifted. The shifting operations are so fast that visually all four digits will appear on the display simultaneously. If the displays are entered via left entry, then the least significant digit must be output first and the rest of the sequence is similar to the right entry.

FIGURE 9.2   A row of four displays.

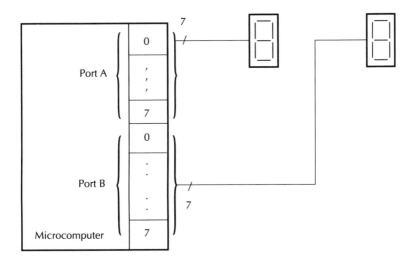

FIGURE 9.3   Nonmultiplexed hexadecimal displays.

Two techniques are typically used to interface a hexadecimal display to the microcomputer. These are nonmultiplexed and multiplexed. In nonmultiplexed methods, each hexadecimal display digit is interfaced to the microcomputer via an I/O port. Figure 9.3 illustrates this method.

BCD to seven-segment conversion is done in software. The microcomputer can be programmed to output to the two display digits in sequence. However, the microcomputer executes the display instruction sequence so fast that the displays appear to the human eye at the same time.

Figure 9.4 illustrates the multiplexing method of interfacing the two hexadecimal displays to the microcomputer.

In the multiplexing scheme, seven-segment code is sent to all displays simultaneously. However, the segment to be illuminated is grounded.

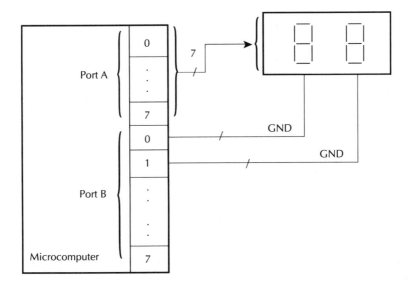

FIGURE 9.4   Multiplexed displays.

The keyboard and display interfacing concepts described here can be realized by either software or hardware. In order to relieve the microprocessor of these functions, microprocessor manufacturers have developed a number of keyboard/display controller chips such as the Intel 8279. These chips are typically initialized by the microprocessor. The keyboard/display functions are then performed by the chip independent of the microprocessor.

The amount of keyboard/display functions performed by the controller chip varies from one manufacturer to another. However, these functions are usually shared between the controller chip and the microprocessor.

9.1.2 8086 Keyboard Interface

In this section, an 8086-based microprocomputer is designed to display 4 hexidecimal digits entered via a keypad (16 keys). Figure 9.5 shows the hardware schematic.

9.1.2.a Hardware

The 8086/8255 microcomputer is designed using standard I/O. For simplicity, only seven address lines are used to directly access the system memory. Therefore, only 128 bytes of system memory can be accessed by the microcomputer. Furthermore, RAM is not available in the system, although RAM should have been used in the design for interrupts and subroutines. However, a small system like this will work without any RAM. Finally, only 8-bit even addressed I/O ports are available in the system. The ports are configured as follows:

1. Port A is configured as an input port to receive the row-column code
2. Port B is configured as an output port to display the key(s) pressed
3. Port C is configured as an output port to control the row-column code

Table 9.1 shows memory and I/O maps.

The system is designed to run at 2 Mhz. Debouncing is provided to avoid unwanted oscillation caused by the opening and closing of the key contacts. In order to ensure stability of the input signal, a delay of 20 msec is used for debouncing the input.

The following diagram shows the internal layout of the keypad used:

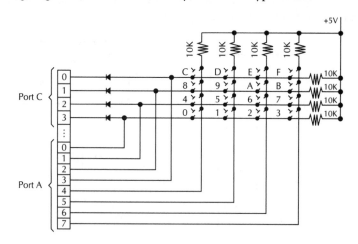

9.1.2.b Software

The program begins by performing all necessary initializations. Next, it makes sure that all the keys are opened (not pressed). A delay loop of 20 msec is included for debouncing. The initial loop counter values is calculated as follows:

```
mov    reg/imm      (4 cycles)
loop   label        (19/5 cycles)
```

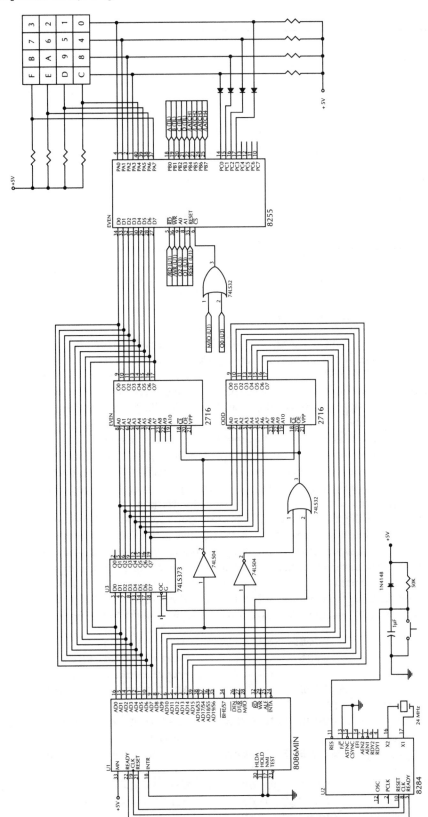

FIGURE 9.5 8086 Keyboard interface.

TABLE 9.1   Memory and I/O Maps

Memory Map:

A_{19}	A_{18}	A_{17}	A_{16}	A_{15}	A_{14}	A_{13}	A_{12}	A_{11}	A_{10}	A_9	A_8	A_7	A_6	A_5	A_4	A_3	A_2	A_1	A_0
1	1	1	1	1	1	1	1	1	1	1	1	0	0	0	0	0	0	0	0
1	1	1	1	1	1	1	1	1	1	1	1	0	0	0	0	0	0	0	1
1	1	1	1	1	1	1	1	1	1	1	1	0	0	0	0	0	0	1	0
—	—	—	—	—	—	—	—	—	—	—	—	—	—	—	—	—	—	—	—
1	1	1	1	1	1	1	1	1	1	1	1	1	1	1	1	1	1	1	0
1	1	1	1	1	1	1	1	1	1	1	1	1	1	1	1	1	1	1	1

I/O Map:

A_7	A_6	A_5	A_4	A_3	A_2	A_1	A_0	Port Selected	Hex
1	1	1	1	1	0	0	0	Port A	F8
1	1	1	1	1	0	1	0	Port B	FA
1	1	1	1	1	1	0	0	Port C	FC
1	1	1	1	1	1	1	0	CSR	FE

$$20 \text{ msec} * (2{,}000{,}000 \text{ cycle/msec}) = 40{,}000 \text{ cycles}$$
$$4 + 19 * (\text{count} - 1) + 5 = 40{,}000$$

$$\rightarrow \text{count} = 38DZ_{16}$$

The next three lines detect for a key closure. If a key closure is detected, it is first debounced. It is necessary to determine exactly which key is pressed. To do this, a sequence of row-control codes (0FH, 0EH, 0DH, 07H) are output via Port C. The row-column code is input via Port A to determe if the column code changes corresponding to each different row code. If the column code is not 0FH (changed), the input key is identified. The program then indexes through a look-up table to determine the row-column code. If the code is found, the corresponding index value which equals to the input key's face value (a single hexidecimal digit) is buffered. Since the microcomputer does not have access to any RAM, the input key's face value along with three previously entered values are saved into register **bx**. The upper 4 bits of **bh** correspond to the most significant digit (least current input), while the lower 4 bits correspond to the next significant digit. The upper 4 bits of **bl** correspond to the second to the least significant digit. Finally, the lower 4 bits of **bl** correspond to the least significant digit (most current input). All four digits are output via Port B (one digit at a time). In other words, the displays (four TIL311's) are refreshed for every key input. The program is written such that it will continuously scan for input key and update the display for each new input. Note that lower-case letters are used to represent the 8086 registers in the program. For example, al, ah, and ax in the program represent the 8086 AL, AH, and AX registers, respectively.

A listing of the 8086 assembly language program is given in the following:

```
                          .MODEL  SMALL
                          .8086           ; restrict to 8086
                                            instructions only!!!
0000                      .CODE           ; ASSUME CS:CODE, DS:DATA
0000            main      PROC
= 00F8                    PORTA EQU 0F8h  ; hex keyboard input (row/
                                            coln)
= 00FA                    PORTB EQU 0FAh  ; LED displays/controls
                                            (c3-c0)
= 00FC                    PORTC EQU 0FCh  ; hex keyboard row controls
= 00FE                    CSR EQU 0FEh    ; control status register
= 00F0                    OPEN EQU 0F0h   ; row/coln codes if all
                                            keys are opened
```

```
0000              start:                    ;
0000  B0 90               mov al, 90h       ; config PortA,B,C as i/o/o
0002  E6 FE               out CSR, al       ;
0004  2A C0               sub al, al        ; clear al
0006  E6 FA               out PORTB, al     ; enable/initialize displays
0008  2B DB               sub bx, bx        ; clear bx (content for
                                            ;   displays)
000A              scan-key:
000A  2A C0               sub al, al        ; clear al
000C  E6 FC               out PORTC, al     ; set row controls to zero
000E              key-close:
000E  E4 F8               in al, PORTA      ; read PORTA
0010  3C F0               cmp al, OPEN      ; Are all keys opened?
0012  75 FA               jnz key-close     ; repeat if closed
0014  B9 38D2             mov cx, 38d2h     ; delay of 20 msec
0017  E2 FE     delay1:   loop delay 1      ; debounce key opened
0019              key-open:
0019  E4 F8               in al, PORTA      ; read PORTA
001B  3C F0               cmp al, OPEN      ; Are all keys closed?
001D  74 FA               jz key-open       ; repeat if opened
001F  B9 38D2             mov cx, 38d2h     ; delay of 20 msec
0022  E2 FE     delay2:   loop delay2       ; debounce key closed
0024  B0 FF               mov al, 0FFh      ; set al to all one's
0026  F8                  clc               ; clear carry
0027              next-row:
0027  D0 D0               rcl al, 1         ; set up row mask
0029  8A c8               mov cl, al        ; save row mask in al
002B  E6 FC               out PORTC, al     ; set a row to zero
002D  E4 F8               in al, PORTA      ; read PORTA
002F  8A D0               mov dl, al        ; save row/coln codes in
cl
0031  24 0F               and al, 0Fh       ; mask row code
0033  3C 0F               cmp al, 0Fh       ; Is coln code affected?
0035  75 05               jnz decode        ; if yes, decode coln code
0037  8A C1               mov al, cl        ; restore row mask to al
0039  F9                  stc               ; if no, set carry
003A  EB EB               jmp next-row      ; check next row
003C              decode:
003C  BE FFFF             mov si, -1        ; initialize index register
003F  B1 0F               mov cl, 0Fh       ; set up counter
0041              search:
0041  46                  inc si            ; increment index
0042  2E: 3A 94           cmp dl,           ; index thru table of
  0075 R                  [TABLE+si]        ;   codes
0047  E0 F8               loopne search     ; loop if not found
0049  B1 04               mov cl, 04h       ; amount to be shifted (1
                                            ;   digit)
004B  D3 E3               shl bx, cl        ; advanced [bx] by 1 digit
004D  03 DE               add bx, si        ; append current digit
004F  8A c7               mov al, bh        ; extract 1st/2nd digits
0051  24 F0               and al, 0F0h      ; mask 2nd digit
0053  D2 E8               shr al, cl        ; move digit to a3-a0
```

```
0055  0C 70              or al, 70h       ;  enable /L3 (set low)
0057  E6 FA              out PORTB, al    ;  display 1st digit (MSD)
0059  8A C7              mov al, bh       ;  extract 1st/2nd digits
005B  24 0F              and al, 0Fh      ;  mask 1st digit
005D  0C B0              or al, 0B0h      ;  enable /L2 (set low)
005F  E6 FA              out PORTB, al    ;  display 2nd digit
0061  8A C3              mov al, bl       ;  extract 3rd/4th digits
0063  24 F0              and al, 0F0h     ;  mask 4th digit
0065  D2 E8              shr al, cl       ;  move digit to a3-a0
0067  0C D0              or al, 0D0h      ;  enable /L1 (set low)
0069  E6 FA              out PORTB, al    ;  display 3rd digit
006B  8A C3              mov al, bl       ;  extract 3rd/4th digits
006D  24 0F              and al, 0Fh      ;  mask 3rd digit
006F  0C E0              or al, 0E0h      ;  enable /L0 (set low)
0071  E6 FA              out PORTB, al    ;  display 4th digit (LSD)
0073  EB 95              jmp scan-key     ;  return to scan another
                                          ;    key input
0075  E7       TABLE     DB 0E7h          ;  code for 0
0076  EB                 DB 0EBh          ;  code for 1
0077  ED                 DB 0EDh          ;  code for 2
0078  EE                 DB 0EEh          ;  code for 3
0079  D7                 DB 0D7h          ;  code for 4
007A  DB                 DB 0DBh          ;  code for 5
007B  DD                 DB 0DDh          ;  code for 6
007C  DE                 DB 0DEh          ;  code for 7
007D  B7                 DB 0B7h          ;  code for 8
007E  BB                 DB 0BBh          ;  code for 9
007F  BD                 DB 0BDh          ;  code for A
0080  BE                 DB 0BEh          ;  code for B
0081  77                 DB 77h           ;  code for C
0082  7B                 DB 7Bh           ;  code for D
0083  7D                 DB 7Dh           ;  code for E
0084  7E                 DB 7Eh           ;  code for F
0085  B8 4C00            mov ax, 4C00h    ;  these two lines are
                                          ;    required
0088  CD 15              int 21           ;  to exit DOS
008A           main      ENDP             ;  end of procedure
                         END main         ;  end of program
```

9.2 DMA Controllers

As mentioned before, direct memory access (DMA) is a type of data transfer between the microcomputer's main memory and an external device such as disk without involving the microprocessor. The DMA controller is an LSI (Large-Scale Integration) chip in a microcomputer system which supports DMA-type data transfers. The DMA controller can control the memory in the same way as the microprocessor, and, therefore, the DMA controller can be considered as a second microprocessor in the system, except that its function is to perform I/O transfers. DMA controllers perform data transfers at a very high rate. This is because several functions for accomplishing the transfer are implemented in hardware. The DMA controller is provided with a number of I/O ports. A typical microcomputer system with a DMA controller is shown in Figure 9.6.

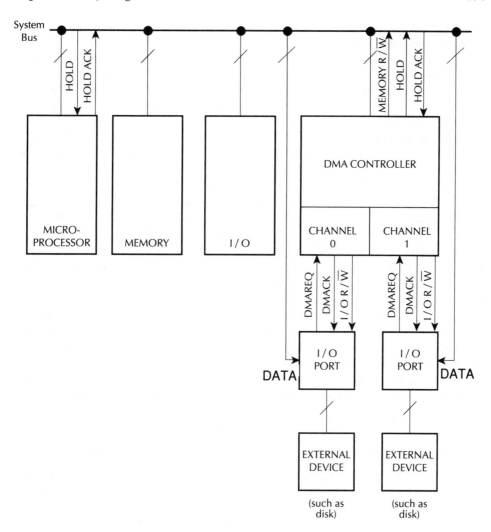

FIGURE 9.6 A microcomputer system with a DMA controller.

The DMA controller in the figure connects one or more ports directly to memory so that data can be transferred between these ports and memory without going through the microprocessor. Therefore, the microprocessor is not involved in the data transfer.

The DMA controller in the figure has two channels (Channel 0 and 1). Each channel contains an address register, a control register, and a counter for block length. The purpose of the DMA controller is to move a string of data between the memory and an external device. In order to accomplish this, the microprocessor writes the starting address of memory where transfer is to take place in the address register, and controls information such as the direction of transfer in the control register and the length of data to be transferred in the counter.

The DMA controller then completes the transfer independent of the microprocessor. However, in order to carry out the transfer, the DMA controller must not start the transfer until the microprocessor relinquishes the system bus and the external device is ready.

The interface between an I/O port and each channel has typically a number of control signals which include DMAREQ, DMACK, and I/O read/write signals. When the I/O port is ready with an available buffer to receive data or has data ready to write into memory location, it activates the DMAREQ line of the DMA controller. In order to accomplish the transfer, the

DMA controller sends the DMACK to the port, telling the port that it can receive data from memory or send data to memory.

DMACK is similar to a chip select. This is because when the DMACK signal on the port is activated by the DMA controller, the port is selected to transfer data between the I/O device and memory. The main difference between a normal and DMA transfer is that the read or write operations have opposite meanings — that is, if the DMA controller activates the read line of the port, then data are read from a memory location to the port. However, this is a write operation as far as the port is concerned. This means that a read from a memory location is a write to the port. Similarly, a write to a memory location is equivalent to a read from the port. The figure shows two types of $R/\overline{W}$ signals. These are the usual memory $R/\overline{W}$ signal and the I/O $R/\overline{W}$ for external devices. The DMA controller activates both of these lines at the same time in opposite directions. That is, for reading data from memory and writing into a port, the DMA controller activates the memory $R/\overline{W}$ HIGH and I/O $R/\overline{W}$ LOW. The I/O ports are available with two modes of operation: non-DMA and DMA.

For non-DMA (microprocessor-controlled transfers), the ports operate in a normal mode. For DMA mode, the microprocessor first configures the port in the DMA mode and then signals the DMA controller to perform the transfer. The $R/\overline{W}$ line is complemented for providing proper direction of the data transfer during DMA transfer.

The DMA controller has a HOLD output signal and a HOLD ACK input signal. The port, when ready, generates the DMAREQ's signal for the DMA controller. The DMA controller then activates the HOLD input signal of the microprocessor, requesting the microprocessor to relinquish the bus, and waits for a HOLD ACK back from the microprocessor.

After a few cycles, the microprocessor activates the HOLD acknowledge and tristates the output drivers to the system bus. The DMA controller then takes over the bus. The DMA controller:

1. Outputs the starting address in the system bus
2. Sends DMACK to the I/O port requesting DMA
3. Outputs normal $R/\overline{W}$ to memory and complemented $R/\overline{W}$ to the I/O port

The I/O port and memory then complete the transfer. After the transfer, the DMA controller disables all the signals including the HOLD on the system bus and tristates all its bus drivers. The microprocessor then takes over the bus and continues with its normal operation.

For efficient operation, the DMA controller is usually provided with a burst mode in which it has control over the bus until the entire block of data is transferred.

In addition to the usual address, control, and counter registers, some DMA controllers are also provided with data-chain registers which contain an address register, a control register, a counter, and a channel identification. These data-chain registers store the information for a specific channel for the next transfer. When the specified channel completes a DMA transfer, its registers are reloaded from the data-chain registers and the next transfer continues without any interruption from the microprocessor. In order to reload the data-chain registers for another transfer, the microprocessor can check the status register of the DMA controller to determine whether the DMA controller has already used the contents of the data-chain registers. In case it has, the microprocessor reinitializes the data-chain registers with appropriate information for the next block transfer and the process continues.

In order to illustrate the functions of a typical DMA controller just described, Motorola's MC68440 dual channel DMA controller is described.

The MC68440 is designed for the MC68000 family microprocessors to move blocks of data between memory and peripherals using DMA.

The MC68440 includes two independent DMA channels with built-in priorities that are programmable. The MC68440 can perform two types of DMA: cycle stealing and burst. In addition, it can provide noncontinuous block transfer (continue mode) and block transfer restart operation (reload mode).

Figure 9.7 shows a typical block diagram of the MC68000/68440/68230 interface to a disk.

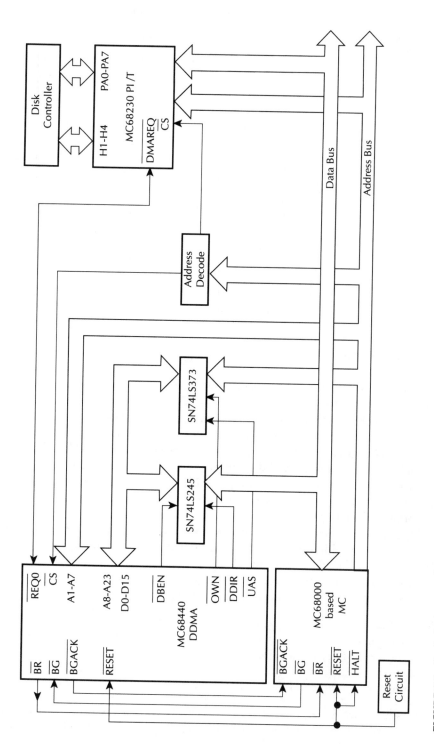

FIGURE 9.7 Typical system configuration.

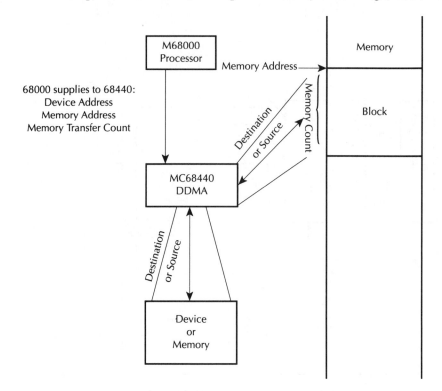

FIGURE 9.8   Data block format.

Data transfer between the disk and the memory takes place via port A of the MC68230, using handshaking signals H1-H4.

The A8/D0 through A23/D15 lines are multiplexed. The MC68440 multiplex control signals $\overline{OWN}$, $\overline{UAS}$ (upper address strobe), $\overline{DBEN}$ (data buffer enable), and $\overline{DDIR}$ (data direction) are used to control external demultiplexing devices such as 74LS245 bi-directional buffer and 74LS373 latch to separate address and data information on the A8/D0-A23/D15 lines. The MC68440 has 17 registers plus a general control register for each of the two channels and is selected by the lower address lines (A1-A7) in the MPU mode. A1-A7 also provides the lower 7 address outputs in the DMA mode.

A1-A7 lines can select 128 (2^7) registers; however, with A1-A7 lines, only seventeen registers with addresses are defined in the range from 00_{16} through FF_{16} and some addresses are not used. As an example, the addresses of the channel status register and the channel priority register are, respectively, 00_{16} and $2D_{16}$.

The MC68440 registers contain information about the data transfer such as:

1. Source and destination addresses along with function codes
2. Transfer count
3. Operand size and device port size
4. Channel priority
5. Status and error information on channel activity

The processor service request register (PSRR) of the MC68230 defines how the $\overline{DMAREQ}$ pin should be used and how the DMA transfer should take place, whether via handshaking or ports.

A data block contains a sequence of bytes or words starting at a particular address with the block length defined by the transfer count register. Figure 9.8 shows the data block format.

There are three phases of a DMA transfer. These are channel initialization, data transfer, and block termination. During channel initialization, the MC68000 loads the MC68440 registers with control information, address pointers, and transfer counts, and then starts the channel.

During the transfer phase, the MC68440 acknowledges data transfer requests and performs addressing and bus controls for the transfer. Finally, the block termination phase takes place when the transfer is complete. During this phase, the 68440 informs the 68000 of the completion of data transfer via a status register. During the three phases of a data transfer operation, the MC68440 will be in one of the three modes of operation. These are idle, MPU, and DMA. The MC68440 goes into the idle mode when it is reset by an external device and waits for initialization by the MC68000 or an operand transfer request from a peripheral.

The MPU mode is assumed by the MC68440 when its $\overline{CS}$ (chip select) is enabled by the MC68000. In this mode, the MC68440 internal registers can be read or written for controlling channel operation and for checking the status of a block transfer.

The MC68440 assumes the DMA mode when it takes over the bus to perform an operand transfer.

In Figure 9.7, upon reset, the MC68440 goes into idle mode. In order to initialize the MC68440 registers, the MC68000 outputs appropriate register addresses on the bus. This will enable the MC68440 $\overline{CS}$ line and places the MC68440 in the MPU mode. The MC68000 initializes the MC68440 registers in this mode. The MC68000 then executes the RESET instruction to place the MC68440 back to the idle mode.

The MC68000 now waits for a transfer request from the 68230. When the 68000 desires a DMA transfer between the disk and memory, it enables the $\overline{CS}$ line of the 68230. The 68230, when ready, activates the $\overline{DMAREQ}$ line low, which in turn drives the $\overline{REQ0}$ line of the MC68440 to low. The MC68440 then outputs low on its $\overline{BR}$ line requesting the MC68000 to relinquish the bus. The MC68000, when ready, sends a low on its $\overline{BG}$ pin. This tells the MC68440 to take over the bus. The MC68440 then enters the DMA mode and sends a low on its $\overline{BGACK}$ pin to inform the MC68000 of its taking over the bus. The MC68440 transfers data between the disk and memory via the MC68230. Each time a byte is transferred, the MC68440 decrements the transfer counter register and increments the address register. When the transfer is completed, the MC68440 updates a bit in the status register to indicate this. It also asserts the $\overline{DTC}$ (data transfer complete) to indicate completion of the transfer.

The MC68440 $\overline{DTC}$ pin can be connected to the MC68230 PIRQ pin. The MC68230 then outputs high to the MC68440 $\overline{REQ0}$ pin, which in turn places a HIGH on the MC68000 $\overline{BR}$, and the MC68000 takes over the bus and goes back to normal operation.

9.3 Printer Interface

Microprocessors are typically interfaced to two types of printers: serial and parallel.

Serial printers print one character at a time, while parallel printers print a number of characters on a single line so fast that they appear to be printed simultaneously. Depending on the character generation technique used, printers can be classified as impact or nonimpact. In impact printers, the print head strikes the printing medium, such as paper, directly, in order to print a character. In nonimpact printers, thermal or electrostatic methods are used to print a character.

Printers can also be classified based on the character formation technique used. For example, character printers use completely formed characters for character generation, while matrix printers use dots or lines to create characters.

The inexpensive serial dot matrix impact printer is very popular with microcomputers. An example of such a printer is the LRC7040 manufactured by LRC, Inc. of Riverton, Wyoming. The LRC7040 can print up to 40 columns of alphanumeric characters. The printer includes four major parts. These are the frame, the printhead, the main drive, and the paper handling

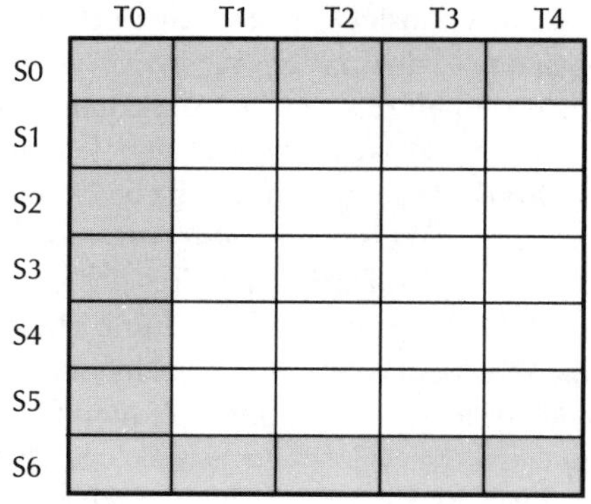

FIGURE 9.9 5 × 7 Dot matrix pattern for generating the character 'c'.

components. The LRC7040 provides 8 inputs in the basic configuration. One input turns the main drive motor ON or OFF, while the other seven inputs control the print solenoids for the printhead, using TTL drivers.

The LRC7040 utilizes a 5 × 7 matrix of dots to generate characters. The columns are labeled T0 through T4 and rows are labeled S0 through S6. Each row corresponds to one of the solenoids. The entire printhead assembly is moved from left to right across the paper so that at some time the printhead is over the column T0, then it's over column T1, and so on.

A character is generated by energizing the proper solenoids at each one of the columns T0 through T4. Figure 9.9 shows how the character C is formed.

At T0, solenoids S0 through S6 are ON and at T1 through T4 solenoids S0 and S6 are active to form the character C. A number of characters can be formed by the microcomputer by sending appropriate data to the printhead to generate the correct pattern of active solenoids for each of the five instants of time. The code for the character C consists of 5 bytes of data in the sequence $7F_{16}$, 41_{16}, 41_{16}, 41_{16}, 41_{16} as follows:

	S6	S5	S4	S3	S2	S1	S0	
Column T0	1	1	1	1	1	1	1	$= 7F_{16}$
Column T1	1	0	0	0	0	0	1	$= 41_{16}$
Column T2	1	0	0	0	0	0	1	$= 41_{16}$
Column T3	1	0	0	0	0	0	1	$= 41_{16}$
Column T4	1	0	0	0	0	0	1	$= 41_{16}$

Note that in the above, it is assumed that a 1 will turn a solenoid ON and a 0 will turn it OFF. Also it is assumed that S7 is zero.

The interface signals to the printer include a pair of wires for each solenoid, a pair of wires for each motor (main drive motor and line feed motor), a pair of wires indicating the state of the HOME microswitch, and a pair of wires indicating the state of the LINEFEED microswitch.

Paper feed is accomplished by activating the line feed motor. The LINEFEED microswitch is activated by the print logic when the actual paper feed takes place. The control logic can use the trailing edge of the signal generated by the LINEFEED microswitch to turn the line feed motor OFF. The LRC7040 also has an automatic line feed version.

The HOME microswitch is activated HIGH when the printhead is at the left-hand edge of the paper. When the printhead is over the print area and moves from left to right, the HOME microswitch is deactivated to zero.

The solenoids must be driven from 40 ± 4 volts with a peak current of 3.6 A. An interface circuit is required at the microcomputer's output to provide this drive capability.

There are two ways of interfacing the printer to a microcomputer. These are

1. Direct microcomputer control
2. Indirect microcomputer control using a special chip called the Printer Controller

The direct microcomputer control interfaces the printer via its I/O ports and utilizes mostly software. The microcomputer performs all the functions required for printing the alphanumeric characters.

Indirect microcomputer control, on the other hand, utilizes a printer control chip such as the Intel 8295 Dot Matrix Printer Controller. The benefits of each technique depend on the specific application.

The direct microcomputer approach provides an inexpensive interface and can be appropriate when the microcomputer has a light load. The indirect microcomputer approach, on the other hand, may be useful when the microcomputer has a heavy load and cost is not a major concern.

9.3.1 LRC7040 Printer Interface Using Direct Microcomputer Control

The steps involved in starting a printing sequence by the microcomputer are provided below:

1. The microcomputer must turn the Main Drive motor (MDM) ON by sending a HIGH output to the MDM.
2. The microcomputer is required to detect a HIGH at the HOME microswitch. This will ensure that the printhead is at the left-hand margin of the print area.
3. The microcomputer is then required to send five bytes of data for an alphanumeric character in sequence to energize the solenoids. Each solenoid requires a pulse of about 400 ms to generate a dot on the paper. A pause of about 900 ms is required between these pulses to provide a space between dots.

Figure 9.10 shows a block diagram interfacing the LRC7040 printer to an MC68000/6821/ 6116/2716-based microcomputer.

Using the hardware of Figure 9.10, an MC68000 assembly program can be written to send the start pulse for the main drive motor, detect the HOME microswitch, and then, by utilizing the timing requirements of 400 μs and 900 μs of the printer, a hexadecimal digit (0 to F) stored in a RAM location can be printed. An MC68000 assembly language program for printing the character C is shown in Figure 9.11 assuming the 68000 user mode so that USP can be initialized.

The program assumes a look-up table which stores the 5-byte code for the character C starting at $003000. Furthermore, the program assumes that the delay routines DELAY400 for 400 μs and DELAY900 for 900 μs are available. The program prints only one character C and then stops. The program is provided for illustrating the direct microcomputer control technique for printing.

9.3.2 LRC7040 Printer Interface to a Microcomputer Using the 8295 Printer Controller Chip

With direct microcomputer control, the microcomputer spends time in a "wait" loop for polling the status of the HOME signal from the LRC7040 printer. In order to unload the microcomputer of polling the printer status and other functions, typical LSI printer controller chips such as the Intel 8295 can be used.

The 8295 is a dot matrix printer controller. It provides an interface for microprocessors such as the 8085 and 8086 to dot matrix impact printers such as the LRC7040. The 8295 is packaged in a 40-pin DIP and can operate in a serial or parallel communication mode with the 8085 or 8086. In parallel mode, command and data transfers to the printer by the processor occur via

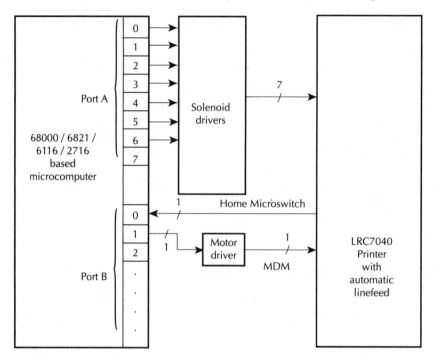

FIGURE 9.10   MC68000-based microcomputer interface to the printer.

polling, interrupts, or DMA using commands. The processor specifies the format of the printed character and controls all printer functions such as linefeed and carriage return.

The 8295 includes a 40-character buffer. When the buffer is full or a carriage return is received, a line is printed automatically. The 8295 has the buffering capability of up to 40 characters and contains a 7×7 matrix character generator, which includes 64 ASCII characters. The mode selection (serial or parallel) is not software programmable and is inherent in system hardware. For example, by connecting the 8295 $\overline{\text{IRQ}}/\text{SER}$ pin to ground, the serial mode is enabled; otherwise the parallel mode is enabled. The two modes cannot be mixed in a single application. Note that the $\overline{\text{IRQ}}/\text{SER}$ pin is also the 8295 interrupt request to the processor in the parallel mode.

9.3.2.a 8295 Parallel Interface

Two 8295 registers (one for input and the other for output) can be accessed by the processor in the parallel mode. The registers are selected as follows:

$\overline{\text{RD}}$	$\overline{\text{WR}}$	$\overline{\text{CS}}$	Register Selected
1	0	0	Input data register
0	1	0	Output status register

Two types of data can be written in the input data register by the processor:

1. A command to be executed. The command can be 0XH or 1XH. For example, the command 08H will enable DMA mode. On the other hand, the command 11_{16} will enable normal left-to-right printing for printers whose printhead home is on the right.
2. A character data (defined in the 8295 data sheet) such as 37H for '7' or 41H for 'A' to be stored in the character buffer for printing.

The 8295 status is available in the output status register at all times. Typical status bits indicate whether the input buffer is full or DMA is enabled. For example, the IBF (Input buffer

```
                    ORG $500000              ;   Address DDRA
                    BCLR.B#$2, CRA           ;   Configure Bits 0-6
                    MOVE.B # $7F, DDRA       ;   As outputs and access Port A
                    BSET.B # $2, CRA         ;   Address DDRB
                    BCLR.B # $2, CRB         ;   Configure bit 0 and
                    MOVE.B # $02, DDRB       ;   Bit 1 of Port B as input
                    BSET.B # $2, CRB         ;   and output respectively
                                             ;   and access Port B
                    MOVEA.L # $040000, A1    ;   Initialize
                    MOVE.L A1, USP           ;   USP
                    MOVE.B # $02, PORT B     ;   Turn MDM on
        HOME        MOVE.B PORT B, D1        ;   Input HOME switch
                    ANDI.B # $01, D1         ;   Wait for HOME
                    BEQ HOME                 ;   Switch to be on
                    MOVEA.L # $3000, A2      ;   Move starting address
                                             ;     of character
                    MOVE.B $05, D3           ;   Initialize
                                             ;   Character
                                             ;   Counter
    CHARACTER       MOVE.B # $00, PORT A     ;   Generage
                    MOVE. B (A2)+, PORT A    ;   Solenoid pulses
                    CALL DELAY 400           ;   Generate 400µs
                    MOVE.B # $00, PORT A     ;   Pulses
                    CALL DELAY900            ;   Delay 900µs
                    SUBQ.B # $01, D3         ;   Subtract character counter
                    BNE CHARACTER            ;   Loop to output all five bytes
                    MOVE.B # $00, PORT B     ;   Turn MDM OFF
        STOP        JMP STOP                 ;   Halt
                    ORG $003000              ;
                    DB $7F, $41, $41, $41, $41  ;  5-byte
                    END                      ;   Code for C
```

FIGURE 9.11   Assembly language program for printing the character C.

full; bit 1 of the status register) is set to one whenever data are written to the input data register. When IBF = 1, no data should be written to the 8295. The DE bit (DMA Enabled; bit 4 of the status register) is set to one whenever the 8295 is in DMA mode. Upon completion of the DMA transfer, the DE is cleared to zero.

The 8295 IRQ/$\overline{\text{SER}}$ pin is used for interrupt driven systems. This output is activated HIGH when the 8295 is ready to receive data. Using polling in parallel mode, the 8295 IRQ/$\overline{\text{SER}}$ pin can be input via the processor I/O port and data can be sent to the 8295 input data register.

Using interrupt in parallel mode, the 8295 IRQ/$\overline{\text{SER}}$ pin can be connected to a processor's interrupt pin to provide an interrupt-driven system.

Using polled or interrupt techniques in parallel mode, the processor typically communicates with the 8295 by performing the following sequence of operations in the main program (polled) or service routine (interrupt):

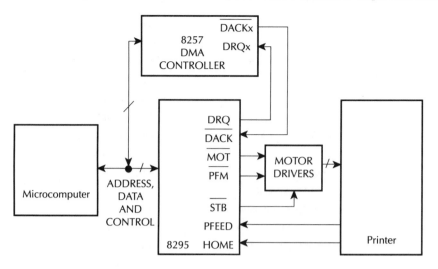

FIGURE 9.12   8295 DMA transfer.

- The processor reads the 8295 status register and checks the IBF flag for HIGH.
- If IBF = 1, the processor waits in a wait loop until IBF = 0. The processor writes data to be printed to the 8295 input data register. The IBF flag is then set to one indicating no data should be written.

Data can also be transferred from the main memory to the 8295 via the DMA method using a DMA controller such as 8257.

The processor initializes the 8257 by sending a starting address and a block length. The processor also enables the 8295 DMA channel by sending it the "ENABLE DMA" command (08H) followed by two bytes specifying the block length to be transferred (low byte first). The 8295 will then activate the DMA request line of the 8257 without any further involvement by the main processor. The DMA enable (DE) flag in the status register will be HIGH until the data transfer is completed. As soon as the data transfer is completed, the DE flag is cleared to zero and $\overline{\text{IRQ}/\text{SER}}$ is set to HIGH. The 8295 then goes back to the non-DMA mode of operation.

Figure 9.12 shows a block diagram of the 8295 DMA transfer.

Typical control signals between the 8257 and the processor include HOLD, HLDA, $\overline{\text{RD}}$, and $\overline{\text{WR}}$. The 8295 control signals for the processor include $\overline{\text{CS}}$, $\overline{\text{RD}}$, $\overline{\text{WR}}$, RESET, and $\overline{\text{IRQ}/\text{SER}}$. $\overline{\text{CS}}$, $\overline{\text{RD}}$, and $\overline{\text{WR}}$ pins are used to select either 8295 input or output registers. The 8295 control signals for the printer include $\overline{\text{MOT}}$, $\overline{\text{PFM}}$, $\overline{\text{STB}}$, PFEED, and HOME.

The 8295 $\overline{\text{MOT}}$ output pin, when LOW, drives the motor. The MOT output is automatically in LOW on power-up. This will make the 8295 HOME input pin HIGH, indicating that the printhead of the printer is in HOME position.

The $\overline{\text{PFM}}$ signal, when LOW, drives the paper feed motor, and this is LOW on power-up. The PFEED is an 8295 input and indicates status of paper feed. A LOW on the $\overline{\text{PFM}}$ indicates that the paper feed mechanism is 'disabled' and a one indicates that the S1 through S7 signals, when LOW, drive the seven solenoids of the printer. Each character datum, when written into the 8295 input data register, is automatically converted to the five-byte code by the 8295 and provides the proper ON/OFF sequence for the solenoids. The $\overline{\text{STB}}$ output is used to determine the duration of solenoid activation and is automatically provided by the 8295.

9.3.2.b 8295 Serial Mode

The 8295 serial mode is enabled by connecting the $\overline{\text{IRQ}/\text{SER}}$ pin to LOW. The serial mode is enabled immediately upon power-up. The serial baud rate is programmed by the D2, D1,

D0 data lines. For example, D2 D1 D0 = 001 means 150 baud (bits/sec) and is used to set the serial transfer data rate. In this mode, $\overline{RD}$ must be tied to high and $\overline{CS}$ and $\overline{WR}$ must be tied to ground. The processor needs a UART (Universal asynchronous receiver transmitter) such as the 8251. The 8295 $\overline{DACK}$ / SIN signal (data input for serial mode) must be connected to the 8251 TXD output (8251 transmit data output bit). Also, the 8295 $\overline{DRQ}$ / CTS (clear to send in serial mode) must be connected to the 8251 $\overline{CTS}$ output. Note that a UART chip converts parallel to serial data and vice versa. The processor must wait for 8295 $\overline{CTS}$ to go LOW before sending data via TXD.

9.4 CRT (Cathode Ray Tube) Controller and Graphics Controller Chips

The CRT terminal is extensively used in microcomputer systems as an efficient man-machine interface. The users communicate with the microcomputer system via the CRT terminal. It basically consists of a typewriter keyboard and a CRT display. In order to relieve the microprocessor from performing the tedious tasks of CRT control, manufacturers have designed an LSI chip called the CRT Controller. This chip simplifies and minimizes the cost of interfacing the CRT terminal to a microcomputer.

The CRT controller supports all the functions required for interfacing a CRT terminal to a microprocessor. The microprocessor and the CRT controller usually communicate via a shared RAM. The microprocessor writes the characters to be displayed in this RAM; the CRT controller reads this memory using DMA and then generates the characters on the video display. The CRT controller provides functions such as clocking and timing, cursor placement, and scrolling. The CRT controller chip includes several registers that can be programmed to generate timing signals and video interface signals required by a terminal. The display functions are driven by clock pulses generated from a master clock. The CRT controller chip normally produces a special symbol such as a blinking signal or an underline on the CRT. This signal is commonly called the 'cursor'. It can be moved on the screen to a specific location where data need to be modified. The scrolling function implemented in the CRT controller moves currently displayed data to the top of the screen as new data are entered at the bottom.

In this section, fundamentals of CRT, character generation techniques, and graphics controllers are discussed.

A typical CRT controller such as the Intel 8275 is then considered to illustrate its basic functions. Finally, the graphics functions provided by Intel 82786 are covered.

9.4.1 CRT Fundamentals

A CRT consists of an evacuated glass tube, a screen with an inner fluorescent coating, and an electron gun for producing electron beams. When the electrons generated by the gun are focused on the fluorescent inner coating of the screen, an illuminated phosphor dot is produced.

The position of the dot can be controlled by deflecting the electron beam by using an electromagnetic deflection technique. A complete display is produced by moving the beam horizontally and vertically across the entire surface of the screen and at the same time by changing its intensity.

Most modern CRT terminals generate the display by using horizontal and vertical scans. In the horizontal scan, the beam moves from the upper left-hand corner to the extreme right-hand of the line and thus travels across the screen. The beam then goes off and starts at the left of the next lower line for another scan. After several horizontal scans, the beam reaches the bottom of the screen to complete one vertical scan. The beam then disappears from the screen and begins another vertical scan from the top. This type of scan is also called 'raster' scan. This

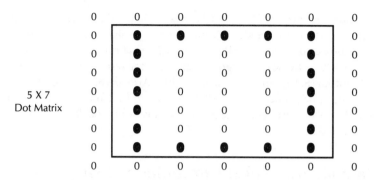

FIGURE 9.13 Generation of '0' using 5 × 7 dot matrix.

is because the display is produced on the screen by continuously scanning the beam across the screen for obtaining a regular pattern of closely spaced horizontal lines, or raster covering the entire screen. One of the most common examples of a raster display is the home TV set. The typical bandwidth used in these TV sets is 4.5 MHz. The raster displays used with microcomputers include a wider bandwidth from 10 MHz to 20 MHz for displaying detailed information. In most modern CRT terminals, each sweep field contains the entire picture or text to be displayed.

In order to display characters, the screen is divided by horizontal and vertical lines into a dot matrix. A matrix of 5 × 7 or 7 × 9 dots is popular for representing a character. For example, a 5 × 7 dot matrix can be used to represent the number '0' as shown in Figure 9.13.

To provide space around the character, one top, one left, one right, and one bottom line are left blank. Each character is generated using 5 × 7 dot matrix. Therefore, each character requires 35 dots, which can be turned ON or OFF depending on the dot pattern required by the character. The pattern of dots is usually stored in ROM. A ROM pattern for '0' is shown in Figure 9.14.

One character requires a 35-bit word. Each row is addressed by three bits. After reading each row data, it is transferred to a parallel to serial shift register. These data are then shifted serially by a clock to the CRT. For a standard 64-character set with each character represented by a 5 × 7 dot matrix, a total of 2240-bit (64 × 7 × 5) ROM is required. Each character in the 64 (2^6)-

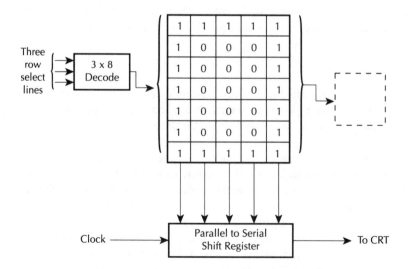

FIGURE 9.14 ROM pattern for '0'.

character set can be addressed using 6 address lines and three row select or scan lines (row select counter typically used) are required to identify the dot row of the character. The ROM, addressing logic, and parallel to serial shift register are referred to as a character generator. Also, a memory known as display memory is required in the CRT to store the character data to be displayed. When a character is entered via the CRT keyboard, it is stored in the display memory.

Graphics can display any figure on the CRT screen. An example of such a VLSI chip is the Intel 82786. In this chip, linked lists are used to update the display and can thus generate displays at a high speed.

Most modern graphics use the bit mapping technique rather than character generation. In order to understand bit mapping, consider a CRT screen as divided into 512 by 128 dots. Each dot is called a Pixel, or picture element, which can be illuminated by an electron beam. Each dot is a single bit in a 64K ($512 \times 128 = 65,536$) by 1 RAM and is called a bit plane. If a '1' is stored in a specific bit location, the associated Pixel is turned ON (white). On the other hand, if a '0' is stored, the corresponding Pixel is turned OFF (black). The video refresh circuitry implemented in the VLSI chip converts the ones and zeros in the bit plane to whites and blacks on the CRT screen.

Resolution is an important factor to be considered in graphics. In order to provide various colors and intensity, more than one bit is utilized in representing a Pixel. For example, Apple's 68000-based LISA microcomputer uses four bits per Pixel on a 364×720 Pixel screen. Therefore, a high speed RAM of over 1 megabit ($364 \times 720 \times 4 = 1,048,320$) is required to support such a resolution.

Therefore, graphics generation requires a bit-mapped RAM array and the LSI video interface chip. The software involves determining the information written to the bit plane array to generate the desired graphic display. Most graphics systems generate figures by combinations of straight-line segments. The software is required to generate a straight line by identifying each Pixel and writing information to its corresponding bit-map positions.

The concepts associated with CRT controllers and graphics described above are illustrated by using the Intel 8275 and Intel 82786 in the following.

9.4.2 Intel 8275 CRT Controller

The INTEL 8275 is a single chip (40-pin) CRT controller. It provides the functions required to interface CRT raster scan displays with Intel microcomputer systems using the 8051, 8085, 8086, and 8088. It refreshes the display by storing (buffering) the information to be displayed from memory and controls the display position on the screen. The 8275 provides raster timing, display row buffering, visual attribute decoding, cursor timing, and light pen detection. The 8275 can be interfaced with the Intel 8257 DMA controller and character generator ROM for dot matrix decoding.

Figure 9.15 shows the 8275's interface to a microcomputer system and the 8257.

The 8275 obtains display characters from memory and displays them on a row-by-row basis. There are two row buffers in the 8275. It uses one row for display, and at the same time fills the other row with the next row of characters to be displayed. The number of display characters per row and the number of character rows are software programmable.

The 8275 utilizes the 8257 DMA controller to fill the row buffer that is not being used for display. It displays character rows one line at a time.

The 8275 controller provides visual attribute codes such as graphics symbols, without the use of the character generator, and blinking, highlighting, and underlining of characters. The raster timing is controlled by the 8275. This is done by generating the horizontal retrace and vertical retrace signals on the HRTC and the VRTC pins.

The 8275 provides the light pen input and associated registers. The light pen input is used to read the registers. A command can be used to read the light pen registers. The light pen consists

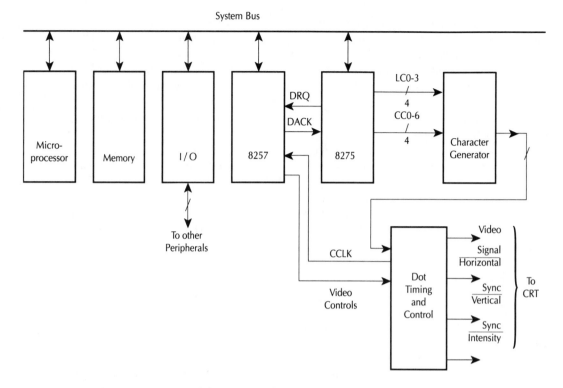

FIGURE 9.15   Microcomputer/8275/8257 interface.

of a microswitch and a tiny light sensor. When the light pen is pressed against the CRT screen, the microswitch enables the light sensor. When the raster sweep reaches the light sensor, it enables the light pen output. If the output of the light pen is presented to the 8275 LPEN pin, the row and character position coordinates are stored in a pair of registers. These registers can be read by a command. A bit in the status register in the 8275 is set, indicating detection of the light pen signal. The 8275 can generate a cursor. The cursor location is determined by a cursor row register and a character position register which are added by a command to the controller.

The cursor can be programmed to appear on the display in many forms such as a blinking underline and a nonblinking underline. The 8275 does not provide a scrolling function.

The 8275 outputs the line count (LC0-LC3) and character code (CC0-CC6) signals for the character generation. The LC0-LC3 signals are contents of the 8275 line counter which are used to address the character generator for the line positions on the screen. The CC0-CC6 outputs of the 8275 are the contents of the row buffers used for character selection in the character generator.

The 8275 video control signals typically include line attribute codes, highlight, and video suppression. The two line attribute codes (LA0 and LA1 pin outputs) must be decoded by the dot timing logic to produce the horizontal and vertical line combinations for the graphic displays defined by the character attribute codes. The video suppress (VSP pins) output signal is used to blank the video signal to the CRT. The highlighted (HLGT) output signal is used to intensify the display at a specific position on the screen, as defined by the attribute codes.

The dot timing and interface logic must provide the character clock (CCLK pin) input of the 8275 for proper timing.

9.4.3 Intel 82786 Graphics Controller

The Intel 82786 is a single VLSI chip providing bitmapped graphics. It is designed for microcomputer graphics applications, including personal computers, engineering worksta-

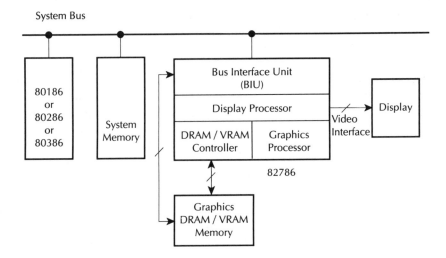

FIGURE 9.16 80186/80286/80386 interface to 80786.

tions, terminals, and laser printers. The 82786 is designed using Intel's CHMOS III process. It is capable of both drawing and refreshing raster displays. It supports high resolution displays with a 25-MHz Pixel clock and can display up to 256 colors simultaneously. It can be interfaced to all Intel microcomputers such as 80186, 80286, and 80386. Figure 9.16 shows a block diagram of the 80186/80286/80386 interfaces to the 82786.

The 82786 includes three basic components. These are a display processor (DP), a graphics processor (GP), and a bus interface unit (BIU), with a DRAM/VRAM controller.

The display processor controls the CRT timing and provides the serial video data stream for the display. It can assemble several windows (portions of bitmaps) on the screen from different bitmaps scattered across the memory accessible to it.

The graphics processor executes commands from a graphics command block placed in memory by the 80186/80286/80386 and updates the bitmap memory for the display processor. The graphics processor has high level video display interface-like commands and can draw graphical objects and text at high speeds.

The BIU controls all communication between the 82786, 80186/80286/80386, and memory. The BIU contains a DRAM/VRAM controller that can perform block transfers. The display processor and graphics processor use the BIU to access the bitmaps in memory.

The system bus connects the 80186/80286/80386 and system memory to the 82786. The video interface connects the 82786 to the CRT or other display. The video interface is controlled directly by the display processor. The 82786 can be programmed to generate all the CRT signals for up to 8 bits/Pixel (256 colors) displays. The other interfaces are controlled by the BIU. The BIU interfaces the graphics and display processors to the 80186/80286/80386 and system memory as well as the graphics memory via the internal DRAM/VRAM controller.

The dedicated graphics DRAM/VRAM memory provides the 82786 with fast access to memory without contention with the microprocessor and system memory.

Usually, the bitmaps to be drawn and displayed, the characters, and commands for the 82786 are all stored in this memory. The 82786 DRAM/VRAM controller interfaces directly with a number of dynamic RAMS without external logic.

Figure 9.15 shows the most common configuration. The microprocessor can access the system memory, while the 82786 accesses its dedicated graphics memory simultaneously. However, when the microprocessor accesses the graphics memory, the 82786 cannot access the system memory. Also, when the 82786 accesses the system memory, the microprocessor cannot access the graphics memory.

If DMA capability is provided, the 82786 can operate in either slave or master mode. In the slave mode, the microprocessor or DMA controller can access the 82786 internal registers or dedicated graphics memory through the 82786. In the master mode, the 82786 can access the system memory.

The microprocessor software can access both system and graphics memory in the same way. When the microprocessor accesses the 82786, the 82786 runs in slave mode.

In slave mode, the 82786 appears like an intelligent DRAM/VRAM controller to the microprocessor. The microprocessor can chip-select the 82786 and the 82786 will acknowledge when the cycle is completed by asserting a READY signal for the microprocessor.

The 82786 graphics and display processor accesses both system memory and graphics memory in the same way. When the 82786 accesses system memory, the 82786 must run in master mode.

In the master mode, the 82786 acts as a second microprocessor controlling the system bus. The 82786 activates the HOLD line to take control of the system bus. When the microprocessor asserts HLDA line, the 82786 takes over the bus. When the 82786 is finished with the bus, it will disable the HOLD line and the microprocessor can remove HLDA and take over the bus.

The 82786 provides two different video interfaces when using standard DRAMS. The 82786 reads the video data from memory and internally serializes the video data to generate the serial video data stream. When using VRAMS, the 82786 loads the VRAM shift register. Periodically, the shift register and external logic then generate the serial video data stream.

9.5 Coprocessors

In Chapters 1 and 7, the basics of coprocessors, along with the functions provided by Motorola coprocessors such as the MC68851 and MC68881, are covered. In this section, a brief overview of the Intel coprocessors will be provided.

Intel offers a number of coprocessors which include numeric coprocessors such as 8087/80287/80387, DMA coprocessors such as the 82258, and graphic coprocessors such as the 82786. In the following, a brief overview of Intel's numeric coprocessors, which include the 8087, 80287, and 80387, is given.

9.5.1 Intel 8087

Intel 8087 numeric data coprocessor is designed using HMOS III technology and is packaged in a 40-pin DIP. When an 8087 is present in a microcomputer system, it adds 68 numeric processing instructions and eight 80-bit registers to the microprocessor's register set. The 8087 can be interfaced to Intel microprocessors such as the 8086/8088 and 80186/80188.

The 8087 supports seven data types which include 16-, 32-, and 64-bit integers, 32-, 64-, and 80-bit floating point, and 18-digit BCD operands. The 8087 is compatible with the IEEE floating-point format. It includes several arithmetic, trigonometric, exponential, and logarithmic instructions.

The 8087 is treated as an extension to the microprocessor, providing additional register data types, instructions, and control at the hardware level. At the programmer's level, the microprocessor and the 8087 are viewed as a single processor. For the 8086/8088, the microprocessor's status (S0-S2) and queue status lines (QS0-QS1) enable the 8087 to monitor and decode instructions in synchronization. For 80186/80188 systems, the queue status signals of the 80186/80188 are synchronized to the 8087 by the 8288 bus controller. The 8087 can operate in parallel with and independent of the microprocessor. For resynchronization, the 8087's BUSY pin tells the microprocessor that the 8087 is executing an instruction and the microprocessor's WAIT instruction tests this signal to ensure that the 8087 is ready to execute subsequent instructions. The 8087 can interrupt the microprocessor when it detects an error or exception. The 8087's interrupt register line is typically connected to the microprocessor through an 8259 programmable interrupt controller for 8086/8088 systems and INT for 80186/80188 systems.

The 8087 uses the request/grant lines of the microprocessor to gain control of the system bus for data transfer.

The microprocessor controls overall program execution while the 8087 utilizes the coprocessor interface to recognize and perform numeric operations.

9.5.2 Intel 80287

Intel 80287 is an enhanced 8087 that extends the 80286 microprocessor. The 80287 adds over 50 instructions to the 80286 instruction set. The 80287 is designed using HMOS technology and is housed in a 40-pin special package called 'CERDIP'.

The 80287 supports the IEEE floating-point format. The 80287 expands the 80286 data types to include 32-bit, 64-bit, and 80-bit floating point, 32-bit and 64-bit integers, and 18-digit BCD operands. It extends the 80286 instruction set to trigonometric, logarithmic, exponential, and arithmetic instructions for all data types.

The 80287 executes instructions in parallel with an 80286. The 80287 has two operating modes like the 80286. Upon reset, the 80287 operates in real address mode. It can be placed in the protected virtual address mode by executing an instruction on the 80286. The 80287 cannot be placed back to the real address mode unless reset. Once in protected mode, all memory references for numeric data or status information follow the 80286 memory management and protection rules and thus the 80287 extends the 80286 protected mode.

The 80287 receives instructions and data via the data channel control signals (PEREQ — Processor Extension Data Channel Operand Transfer Request); PEACK — Processor Extension Data Channel Operand Transfer Acknowledge; BUSY; NPRD — Numeric Processor RD; NPWR — Numeric Processor WR). When in protected mode, all information received by the 80287 is validated by the 80286 memory management and protection unit. When the 80287 detects an exception, it will indicate this to the 80286 by asserting the ERROR signal.

The 80286/80287 is programmed as a single processor. All memory addressing modes, physical memory, and virtual memory of the 80286 are available in the 80287.

9.5.3 Intel 80387

Intel 80387 is a numeric coprocessor that extends the 80386 processor with floating-point, extended integer, and BCD data types. It is compatible with IEEE floating point. The 80387 includes 32-, 64-, and 80-bit floating point, 32- and 64-bit integers, and 18-digit BCD operands. It extends the 80386 instruction set to include trigonometric, logarithmic, exponential, and arithmetic instructions of all data types. The 80387 can operate in the real, protected, or virtual 8086 modes of the 80386. It is designed using CHMOS III technology and is packaged in a 68-pin PGA (Pin Grid Array).

The 80387 operates in the same manner whether the 80386 is executing in real address mode, protected mode, or virtual 86 mode. All memory access is handled by the 80386; the 80387 operates on instructions and values passed to it by the 80386. Therefore, the 80387 is independent of the 80386 mode.

The 80387 includes three functional units that can operate in parallel. The 80386 can be transferring commands and data to the 80387 bus control logic for the next instruction while the 80387 floating-point unit is performing the current numeric instruction. This parallelism improves system performance.

The 80387 adds to an 80386 system additional data types, registers, instructions, and interrupts. All communication between the 80386 and 80387 is transparent to application software. Thus, the 80387 greatly enhances the 80386 capabilities.

Questions and Problems

9.1   Interface a hexadecimal matrix keyboard and four LED displays to an 8086/8255-based microcomputer.
- i)   Draw a hardware schematic of the design. Show only the pertinent signals.
- ii)   Write an 8086 assembly language program to display the hex digit on the display from 0-F each time a digit is pressed on the keyboard.

9.2   Describe the basic functions of a DMA Controller. How does it control the I/O R/$\overline{W}$ and memory R/$\overline{W}$ signals? Why is the DMA Controller faster than the microprocessor for data transfer?

9.3   Describe briefly the main features of Motorola's MC68440 DMA controller.

9.4   Draw a functional block diagram showing the pertinent signals of the MC68020/68230/68440 interface.

9.5   Define the MC68440 modes of operation.

9.6   Which mode and which address lines are required by the MC68440 to decode the register addresses? Why does the MC68440 require more address lines than it requires for register address decoding?

9.7   Draw a functional block diagram of the MC68440/68008 interface.

9.8   What is the difference between the following?
- i)   Serial and Parallel printers
- ii)   Impact and Nonimpact printers
- iii)   Character and Matrix printers

9.9   Assume an LRC7040 printer. Draw a functional block diagram of the LRC7040 printer to an 8086-based microcomputer. Write an 8086 assembly language program to print the hexadecimal digit '0' on the printer.

9.10   Draw a functional block diagram of the 8295 printer controller interface to an 8085-based microcomputer.

9.11   How are the 8295 input data register and output status registers accessed? What are the functions of these registers?

9.12   How are the 8295 serial and parallel modes of operation selected?

9.13   In the 8295 parallel mode, describe briefly how printers are interfaced for polled, interrupt, or DMA operation.

9.14   Summarize the basics of a CRT. What is the main difference between character generation displays and graphics displays?

9.15   What are the typical functions of a CRT controller? Relate these typical functions to the Intel 8275.

9.16 Draw a functional block diagram showing an 8086-based microcomputer interface to an 8275. Show only pertinent signals.

9.17 What is meant by bitmapping? How does it apply to graphics?

9.18 Describe briefly the functions provided by the Intel 82786 graphics controller.

9.19 Draw a functional block diagram showing an 80386/82786 interface. Show only pertinent signals.

9.20 Summarize the basic differences between the Intel 8087, the 80287, and the 80387 numeric coprocessors. Why are these three separate chips for the same coprocessor family provided by Intel?

10

DESIGN PROBLEMS

This chapter includes a number of design problems that utilize external hardware. The systems are based on typical microprocessors such as the 8085, 8086, and 68000.

The concepts presented can be extended to other microprocessors.

10.1 Design Problem No. 1

0.1.1 Problem Statement

An 8085-based digital voltmeter is designed which will measure a maximum of 5 V DC via an A/D converter and then display the voltage on two BCD displays. The upper display is the integer part (0 to 5 V DC) and the lower display is the fractional part (0.0 to 0.9 V DC).

0.1.2 Objective

A digital voltmeter capable of measuring DC voltage up to and including 5 V will be built and tested. The voltmeter is to be implemented using the Intel 8085 microprocessor and an analog-to-digital converter of the designer's choice. The measured voltage is to be displayed on two seven-segment LEDs.

0.1.3 Operation

Figure 10.1 shows a block diagram of the digital voltmeter. It is composed of the microprocessor, 2K bytes of EPROM, 256 bytes of RAM with I/O, the A/D converter, and the display section.

The Intel 8085 microprocessor provides control over all address, data, and control information involved in program execution. It also provides for manipulation of data as taken from the A/D and sent to the display section.

The EPROM is a memory unit which stores the instructions necessary for system operation. The RAM and I/O section is a memory unit which provides for data storage as well as data transfers to and from the A/D and display. The A/D converter takes the voltage measured across Vin(+) and Vin(−) pins and converts it to an 8-bit binary value. The binary information is taken into the microprocessor via the I/O and converted to its decimal equivalent. The display section takes the converted binary information from the processor so that it may be read on two seven-segment LED displays. The leftmost display provides the integer portion of the measured voltage, while the rightmost provides the decimal portion.

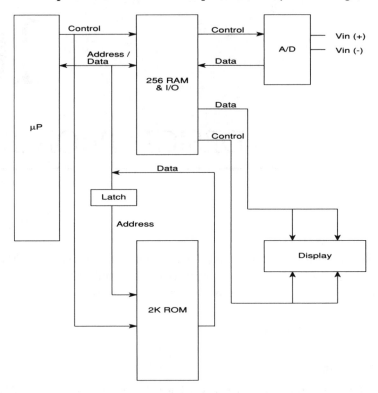

FIGURE 10.1   Voltmeter block diagram.

10.1.4 Hardware

Figure 10.2 shows the detailed hardware schematic. The system uses standard I/O with RAM memory mapped at 0800H-08FFH. The I/O ports of the 8155 are all used and are I/O mapped starting at 08H. Configured as an input port, port B is connected to the output of the 0804 A/D chip. Configured as an output port, port C is connected to the TIL311 displays. Bits 0–3 are the data outputs while bits 4 and 5 are connected to the latches of the TIL311. Only three bits of port A are used, configured as an output port, to control the select, read, and write lines of the A/D chip.

Using the fully decoded memory addressing, the 74LS318 decoder is used to select either the RAM or the EPROM. Also, a 74LS373 is used to latch the address lines for the EPROM. The RAM does not require such a chip because the 8155 RAM has its own internal latches. The ALE line of the 8085 microprocessor controls the latches as seen in the schematic.

The EPROM contains the instructions for converting the binary representation of the analog voltage (applied to the A/D converter) back to decimal representation. The instructions are used to control the system operation. The algorithm uses repeated subtraction to obtain the correct voltage in decimal form. The left display is the integer part and the right display is the decimal part.

The displays, as stated before, are TIL311 hexadecimal displays. In addition, the displays have their own latches which are active low. In the 8085 microprocessor, the interrupt RST 6.5 is used to jump to the address with the algorithm to convert and display the voltage. The INTR

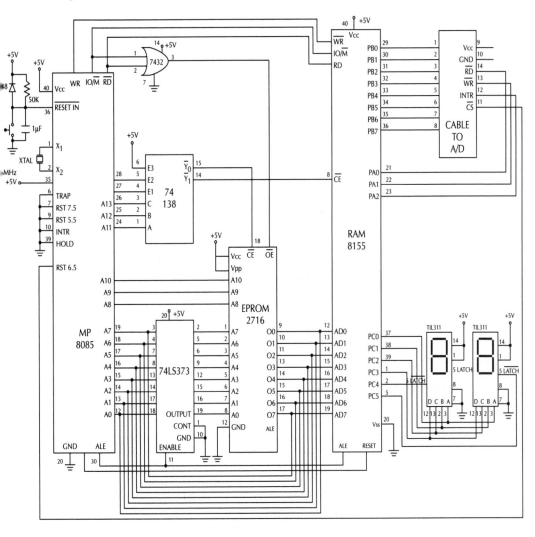

FIGURE 10.2 Digital voltmeter detailed schematic.

pin of the A/D is connected directly to the RST 6.5 pin of the microprocessor. First, an active low is sent to the chip select pin, and then the write pin of the A/D converter is toggled.

Upon completion of the A/D conversion, the 8085 is interrupted. The service routine outputs an active low onto the read pin of the A/D, which latches the data. After inputting the data via the port, the read pin is toggled which then tristates the A/D output.

0.1.5 Software

An important part of the software is to convert the A/D's 8-bit binary data into its decimal equivalent for the display. The decimal data will have two digits: one integer part and one fractional part. Two approaches can be used to accomplish this as follows.

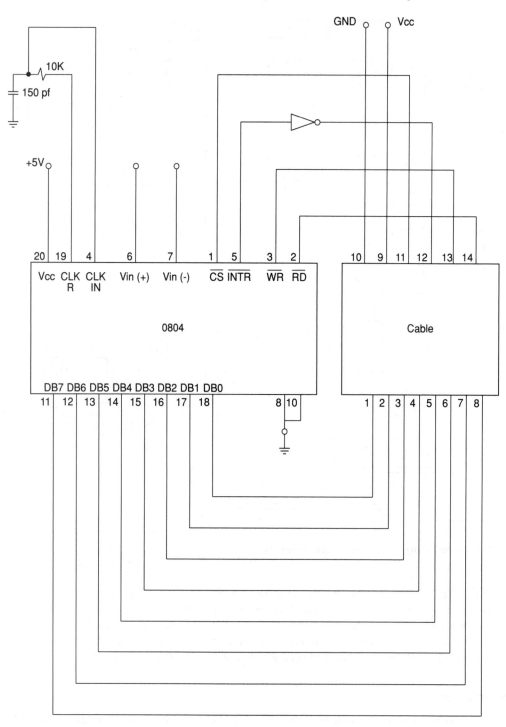

FIGURE 10.2   *Continued.*

Approach 1

Since the maximum decimal value that can be accommodated in 8 bits is 255_{10} (FF_{16}), the maximum voltage of 5 V will be equivalent to 255_{10}. This means that the display in decimal is given by:

$$D = 5*(Input/255)$$
$$= Input/51$$
$$= Quotient + (Remainder/51)$$
$$\uparrow$$
$$\text{Integer part}$$

The fractional part in decimal is

$$F = (Remainder/51)*10$$
$$\cong Remainder/5$$

Approach 2

In the second approach, the equivalent of 1 V ($255/5$ V $= 51_{10} = 33_{16}$) is subtracted from the input data. If the input data are greater than 1 V, a counter initially cleared to zero is incremented by one. This process continues until the data are less than 1 V. The register keeps count of how many subtractions take place with a remainder greater than 1 V and thus contains the integer portion of the measured voltage in decimal.

The decimal portion of the fractional part is obtained in the same way except that if the input data are less than 1 V, then they are compared with the decimal equivalent ($51/10 \cong 5$) of 0.1 V. If the measured data are greater than 0.1 V, a counter initially cleared to zero is incremented by one and the process continues until the input data are less than 0.1 V. The counter contains the fractional part of the display.

Approach 2 is used as a solution to this problem. A complete listing of the 8085 assembly language program to control the digital voltmeter is given below. The program is used to begin and end the A/D conversion process as well as to manipulate the binary data into their decimal form so that they can be displayed in an easily readable format.

```
FILE:  LIST1;RAT001  HEWLETT-PACKARD:  8085 Assembler

LOCATION        OBJECT
CODE            LINE                SOURCE LINE

                1  "8085"
<09A0>          2  STACKP   EQU    09A0H
<0009>          3  PORTA    EQU    0009H
<000A>          4  PORTB    EQU    000AH
<000B>          5  PORTC    EQU    000BH
<0008>          6  CSR      EQU    0008H
<0034>          7  INTR     EQU    0034H
<0000>          8  PROG     EQU    0000H
                9
                10          ORG    PROG
0000 3109A0     11          LXI    SP,STACKP    ;   INIT, STACK
0003 3E0D       12  START   MVI    A,0DH        ;   SET INTERRUPT MASK
0005 30         13          SIM                 ;   SET INTERRUPT MASK 6.5
0006 FB         14          EI                  ;   ENABLE INTERRUPT
0007 D308       15          OUT    CSR          ;   DEFINE PORTS A,B,C
0009 3E30       16          MVI    A,30H        ;   SET DISPLAY ENABLES
000B D30B       17          OUT    PORTC        ;
000D 3EFF       18          MVI    A,OFFH       ;
000F D309       19          OUT    PORTA        ;   SET /CS,/WR/RD HIGH
0011 3EFE       20          MVI    A,OFEH       ;
0013 D309       21          OUT    PORTA        ;   SEND /CS LOW
0015 3EFC       22          MVI    A,OFCH       ;
0017 D309       23          OUT    PORTA        ;
```

FILE: LIST1;RAT001 HEWLETT-PACKARD: 8085 Assembler (continued)

LOCATION CODE	OBJECT LINE					SOURCE LINE
0019 3EFE	24		MVI	A,0FEH	;	
001B D309	25		OUT	PORTA	;	TOGGLE /WR
001D 9B	26	WAIT	JMP	WAIT	;	WAIT FOR INTERRUPT
	27				;	
	28				;	
	29		ORG	INTR	;	INTERRUPT VECTOR ADDRESS
0034 210900	30		LXI	H,0900H	;	INIT, MEM. POINTER
0037 1600	31		MVI	D,00H	;	INIT, INTEGER COUNTER
0039 0E33	32		MVI	C,33H	;	
003B 3EFA	33		MVI	A,0FAH	;	SEND /RD LOW
003D D309	34		OUT	PORTA	;	
003F 00	35		NOP			
0040 DB0A	36		IN	PORTB	;	INPUT DATA
0042 47	37		MOV	B,A	;	MOVE DATA TEMP. TO B
0043 3EFE	38		MVI	A,0FEH	;	
0045 D309	39		OUT	PORTA	;	TOGGLE /RD
0047 70	40		MOV	M,B		
0048 78	41		MOV	A,B	;	MOVE DATA TO 0900H
0049 91	42	SUB1:	SUB	C	;	
004A DA0052	43		JC	CONT1	;	
004D 14	44		INR	D	;	
004E 77	45		MOV	M,A	;	
004F C30049	46		JMP	SUB1	;	
0052 7A	47	CONT1:	MOV	A,D	;	
0053 D30B	48		OUT	PORTC	;	
0055 E63F	49		ANI	3FH	;	
0057 D30B	50		OUT	PORTC	;	
0059 1600	51		MVI	D,00H	;	
005B 0E05	52		MVI	C,05H	;	
005D 7E	53		MOV	A,M	;	
005E 91	54	SUB2:	SUB	C	;	
005F DA0066	55		JC	CONT2	;	
0062 14	56		INR	D	;	
0063 C3005E	57		JMP	SUB2	;	
0066 JA	58	CONT2:	MOV	A,D	;	
0067 F610	59		ORI	10H	;	
0069 D30B	60		OUT	PORTC	;	
006B E63F	61		ANI	3FH	;	
006D D30B	62		OUT	PORTC	;	
006F 3EFC	63		MVI	A,0FCH	;	
0071 D309	64		OUT	PORTA	;	
0073 3EFE	65		MVI	A,0FEH	;	
0075 D309	66		OUT	PORTA	;	
0077 CB	67		RET		;	

Errors = 0

Lines 2–8 are assembler directives which equate a recognizable label with a hex value. This is useful for values which are to be used throughout the program.

Line 10 is another assembler directive which sets the beginning of the program at address 0000H.

Line 11 initializes the stack pointer at address 09A0H. This is necessary if we are to return to a current program after an interrupt has been serviced.

Lines 12–14 set the mask bits and enable interrupt RST6.5.

Line 15 defines port A as output, port B as input, and port C as output. Note that the data to configure the ports were already in the accumulator as per line 12.

Lines 16–17 send an active high to each display's data latch enable pin. This insures that the displays will output the correct data on the next high-to-low transition at the latch enable pins.

Lines 18–19 send a high to the chip's select ($\overline{CS}$), write ($\overline{WR}$), and read ($\overline{RD}$) pins of the A/D converter. This insures proper start-up of the converter.

Lines 20–21 first send an active low to the converter's $\overline{CS}$ pin. Next, lines 22–25 toggle the $\overline{WR}$ pin so that conversion starts. The combination of $\overline{CS}$ and $\overline{WR}$ active low resets the A/D internally and sets it up for the start of the conversion. By sending $\overline{WR}$ high, the conversion starts. Figure 10.3 shows the timing diagram for the A/D.

Line 26 is a "WAIT LOOP" which is provided as a delay to wait for the interrupt request. This is necessary since it may take as long as 127 µs for the interrupt to be asserted. This is equivalent to approximately 380 clock cycles for the 8085 operating at 3 MHz.

Line 29 continues the program at the interrupt vector for interrupt RST6.5.

Line 30 loads the HL register pair with a memory address to be used later in the program.

Lines 31–32 initialize the D and C registers. D register is to hold the integer portion of the measured voltage, while C register holds a hex value equivalent to 1 V for this system.

Lines 33–34 send an active low to the $\overline{RD}$ pin of the A/D converter so that the binary information corresponding to the measured voltage may be read by the microprocessor.

Lines 36–37 take the data from the A/D converter and store it into register B.

Lines 38–39 toggle the $\overline{RD}$ pin back to active high.

Lines 40–41 move the 8-bit data into memory location 0900H and then into the accumulator.

Lines 42–46 convert the binary data into their decimal equivalent so that the integer portion may be displayed. First the equivalent of 1 V is subtracted from the input data. If the measured voltage is less than 1 V, the program jumps to line 47. If the voltage is greater than one, the program continues at line 44 where register D is incremented by one. The remainder from the subtraction is temporarily stored in memory. The program then unconditionally jumps back to line 42 so that another subtraction takes place. This loop occurs until the remainder from the subtraction is less than 1 V. Register D keeps count of how many subtractions took place with a remainder greater than 1 V and thus counts the integer number of volts measured.

Lines 47–50 send the contents of register D to the leftmost display. The AND operation unlatches the data at the display.

Lines 51–52 again initialize registers D and C, but this time register D will be counting the fractional portion of the measured voltage and register C will hold the hex equivalent of 0.1 V.

Line 53 moves the last positive remainder from memory into the accumulator.

Lines 54–57 perform the same function as lines 42–46 but with the fractional portion of the measured data.

The remaining lines output the contents of register D into the rightmost display, toggle $\overline{WR}$, and returns to the main program.

10.2 Design Problem No. 2

10.2.1 Display Scroller Using the Intel 8086

10.2.1.a Introduction and Problem Statement

The objective of this project is to design and build an 8086-based system as shown in the block diagram of Figure 10.4. The system scans a 16-key keyboard and drives three seven-segment displays. The keyboard is scanned in a 4 × 4 X-Y matrix. The system will take each key pressed

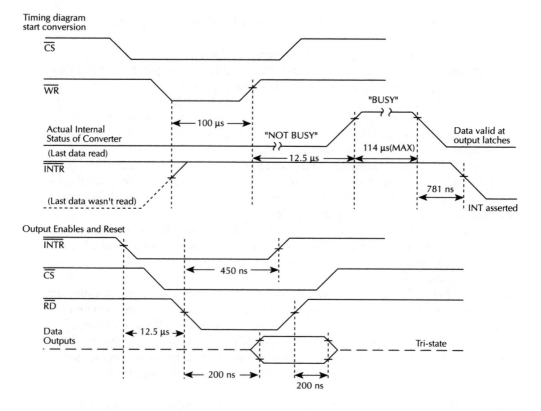

FIGURE 10.3 A/D timing diagram.

and scroll them in from the right side of the displays and keep scrolling as each key is pressed. The leftmost digit is discarded. The system continues indefinitely.

10.2.1.b Hardware Description

Figure 10.5 shows the hardware schematic. The microcomputer is designed using the 8086, 8255 I/O port chip and two 2716 EPROMs. The system does not contain any RAM since no stack is required.

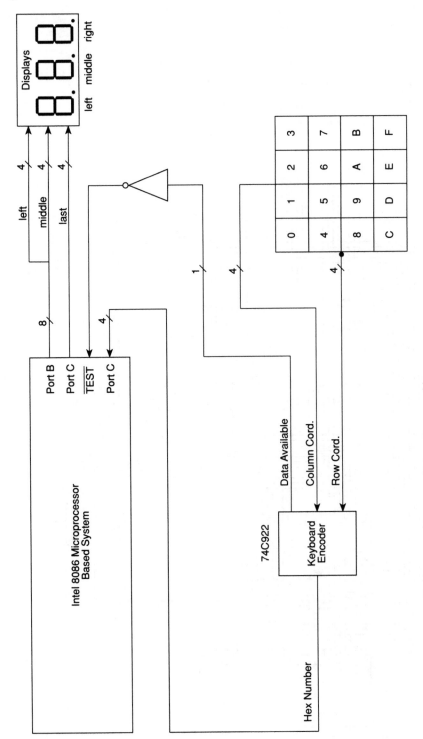

FIGURE 10.4   Keyboard scroller block diagram.

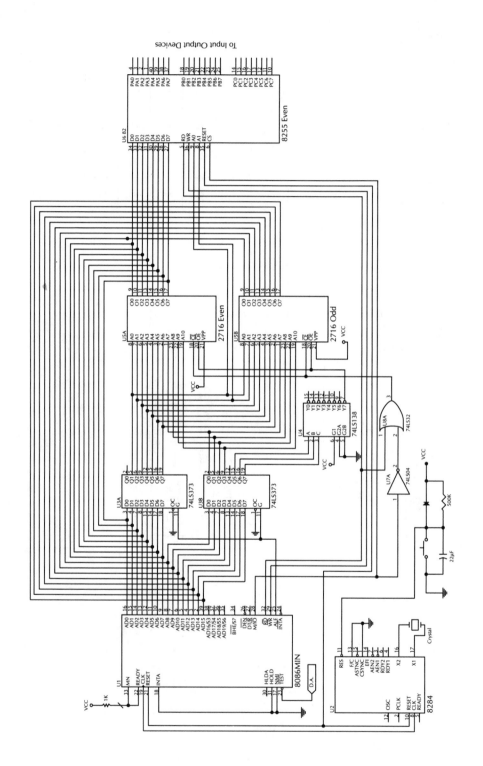

FIGURE 10.5   Keyboard scroller schematic diagram.

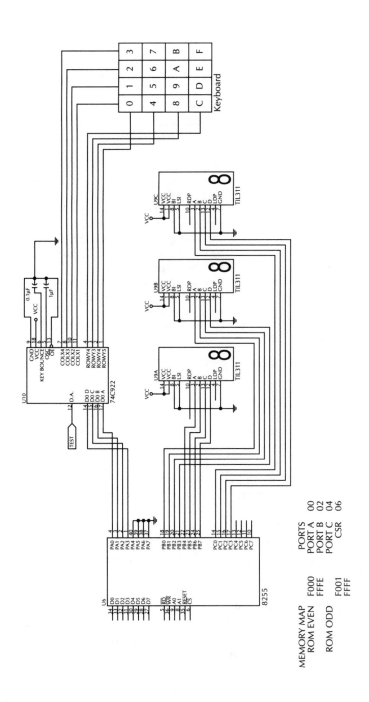

FIGURE 10.5 *Continued.*

```
PROG      SEGMENT
          ASSUME CS:PROG, DS:PROG
          .8086
PORTA     equ   00h
PORTB     equ   02h
PORTC     equ   04h
CSR       equ   06h
start:    mov CX, 0FFFh
LP1:      loop LP1          ; Delay test.
          mov AL, 90h       ; set ports A input, B & C output
          out CSR, AL
          xor BL, BL        ; clear BL
          xor AX, AX        ; clear AX
          out PORTB, AL     ; clear Displays
          out PORTC, AL
          mov CL, 04h       ; set CL to 04
over:     wait              ; wait for TEST pin to go low.
          in AL, PORTA      ; Get data
          out PORTC, AL     ; Out first number and
          shl AL, CL        ; rotate into place for
          shl AX, CL        ; 2nd and 3rd position.
          mov AL, BL        ; Move old inputed nibbles
          out PORTB, AL     ; to be outputed.
          mov BL, AH        ; Copy AH new to BL.
          mov CX, 0FFFh     ; Delay so D.A. signal
LP2:      loop LP2          ; can return to low.
          mov CL, 04h       ; Reset CL to 04h
          jmp over
PROG      ENDS
          END start
```

FIGURE 10.6a   8086 Assembly language program for the keyboard scroller.

Keyboard encoding is accomplished via hardware. The 74C922 chip is used for this purpose. This chip inputs a key pressed from a hexadecimal keypad and outputs the corresponding binary equivalent on its data lines. In order to indicate that a key has been pressed, the 74C922 sends a HIGH on its Data Available (DA) output pin. This signal is inverted and then connected to the TEST pin of the 8086. This means that the key actuation is indicated by a LOW on the 8086 TEST pin.

The displays are three TIL311s. The rightmost TIL311 is connected to bits 0–3 of port C. This display outputs the most recent key pressed. The middle and the leftmost displays are connected to port B. These two displays show the previous two keys pressed.

10.2.1.c Software Development

Figure 10.6a shows the 8086 assembly language program.

The program first initializes the ports and then waits in a loop for a key to be pressed. In this loop, the 8086 WAIT instruction checks the TEST pin for a LOW. As soon as a key is pressed, the DA pin of the 74C922 goes to a HIGH. This, in turn, drives the TEST pin of the 8086 to a LOW indicating that the data is available.

The 4-bit equivalent of the hex key pressed is input into the 8086 AL register and output to port C. The last two keys pressed are saved in BL. This data is moved to AL and then output to port B. The program loops back to the WAIT instruction and waits for the next key.

Figure 10.6b shows how the contents of the 8086 registers change as the keys are pressed on the keyboard.

Loop 1

					Operation
Ax	0	0	0	0	0 → Bx
Bx	0	0	0	0	
Ax	0	0	0	1	1 (keyboard) → AI
Bx	0	0	0	0	AI → Display #1
Ax	0	0	1	0	SHL AI
Bx	0	0	0	0	*4
Ax	0	1	0	0	SHL Ax
Bx	0	0	0	0	*4
Ax	0	1	0	0	BI → AI; Ah → BI
Bx	0	0	0	1	AI → Display #2 & 3

Loop 2

					Operation
Ax	0	1	0	2	2 (keyboard) → AI
Bx	0	0	0	1	AI → Display #1
Ax	0	1	2	0	SHL AI
Bx	0	0	0	1	*4
Ax	1	2	0	0	SHL Ax
Bx	0	0	0	1	*4
Ax	1	2	0	1	BI → AI; Ah → BI
Bx	0	0	1	2	AI → Display #2 & 3

Loop 3

					Operation
Ax	1	2	0	3	3 (keyboard) → AI
Bx	0	0	1	2	AI → Display #2 & 3
Ax	1	2	3	0	SHL AI
Bx	0	0	1	2	*4
Ax	2	3	0	0	SHL Ax
Bx	0	0	1	2	*4
Ax	2	3	1	2	BI → AI; Ah → BI
Bx	0	0	2	3	AI → Display #2 & 3

Loop 4

					Operation
Ax	2	3	0	4	4 (keyboard) → AI
Bx	0	0	2	3	AI → Display #1
Ax	2	3	4	0	SHL AI
Bx	0	0	2	3	*4
Ax	3	4	0	0	SHL Ax
Bx	0	0	2	3	*4
Ax	3	4	2	3	BI → AI; Ah → BI
Bx	0	0	3	4	AI → Display #2 & 3

Loop 4 →

Display registers:

	D1SP3	D1SP2	D1SP1
Loop (0)	0	0	0
Loop 1	0	0	1
Loop 2	0	1	2
Loop 3	1	2	3
Loop 4	2	3	4

FIGURE 10.6b Contents of program registers.

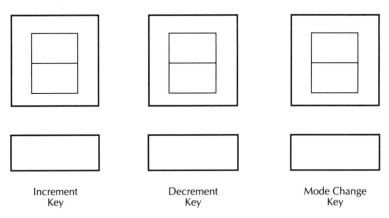

Increment
Key

Decrement
Key

Mode Change
Key

FIGURE 10.7   Block diagram for design problem no. 3.

10.3 Design Problem No. 3

10.3.1 Problem Statement

A 68000-based system is designed to drive three seven-segment displays and monitor three key switches. The system starts by displaying 000. If the increment key is pressed, it will increment the display by one. Similarly, if the decrement key is pressed, it will decrement the display by one. The display goes from 00-FF in the hex mode and from 000-255 in the BCD mode. The system will count correctly in either mode. The change mode key will cause the display to change from hex to decimal or vice versa, depending on its present mode. Figure 10.7 depicts the block diagram.

Two solutions are provided for this problem. Solution one uses programmed I/O with no interrupts, while solution two utilizes interrupt I/O but no programmed I/O.

10.3.2 Solution No. 1

The simplest and the most straightforward system possible is built to obtain the required results. This means that there will be no RAM in the system; therefore, no subroutine will be used in the software and only programmed I/O (no interrupt) is used.

10.3.2.a Hardware

Figure 10.8 shows the detailed hardware schematic. The circuit is divided into the following sections.

10.3.2.a.i Reset Circuit.   The reset circuit for the system is basically the same as the one used for the 8085. The circuit has a 0.1-μF capacitor and 1K resistor to provide an RC time constant of 10^{-4} s for power on reset. The $\overline{RESET}$ and $\overline{HALT}$ pins of the 68000 and the $\overline{RESET}$ pin of the 6821 are tied together for complete and total reset of the system.

10.3.2.a.ii Clock Signal.   An external pulse-generator is used to generate the clock signal for the system. The system is driven up to 3 MHz, the limit of the generator, without any problems.

10.3.2.a.iii Address Mapping.   The system has two 2K EPROM (2716s) and one 6800's peripheral I/O chip (6821). The 68000 address lines A1 through A11 are needed to address the EPROMs. So

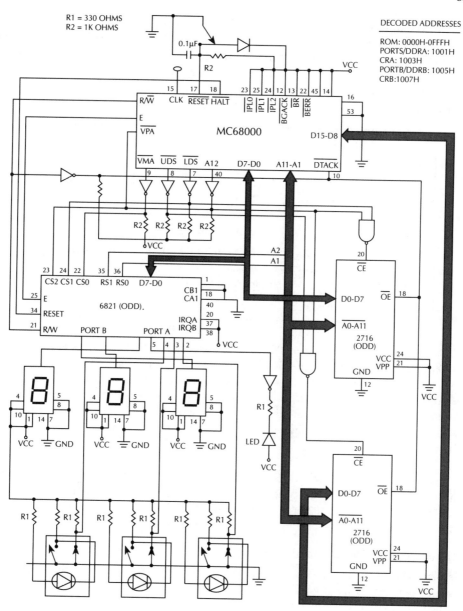

FIGURE 10.8 Detailed hardware schematic.

A12 is used to select between the 2716s and the 6821 (0 for 2716s and 1 for 6821). Memory access for the EPROMs is asynchronous, while the 6821 is synchronized with the E-clock. A12 is inverted, through the buffer, so the output of the inverter goes to $\overline{CS2}$ of the 6821 and also to $\overline{VPA}$ of the 68000 for synchronization. The 68000 $\overline{VMA}$ pin is also buffered and inverted and it goes to $\overline{CS0}$ of the 6821. The 6821 is chosen to be odd, so $\overline{CS1}$ is activated by the inverted $\overline{LDS}$ line. Finally, address lines A1 and A2 are connected to RS0 and RS1, respectively.

The $\overline{CE}$ for the two 2716s comes from two NAND gates. They are the results of the inverted A12 NANDed with the inverted $\overline{LDS}$ or the inverted $\overline{UDS}$, depending on whether the

EPROM is odd or even. The $\overline{\text{DTACK}}$ pin of the 68000 and the $\overline{\text{OE}}$ pins of the 2716s are activated by the signal of R/$\overline{\text{W}}$ inverted. When the 68000 wants to read the EPROMs this signal will be high, so its inverted signal will provide a low to $\overline{\text{DTACK}}$. This does not cause any problem because when the 68000 accesses the 6821, $\overline{\text{VPA}}$ is activated and so the 68000 will not look for $\overline{\text{DTACK}}$.

The configuration above causes the memory map to be as follows:

	A23...A13	A12	A11	A10	A9	A8	A7	A6	A5	A4	A3	A2	A1	HEX
4K of EPROM Memory	0 0	0	0	0	0	0	0	0	0	0	0	0	0	000000_{16} to $000FFF_{16}$
					THROUGH									
	0 0	0	1	1	1	1	1	1	1	1	1	1	1	
PA/DDRA	0 0	1	0	0	0	0	0	0	0	0	0	0	1	001001_{16}
CRA	0 0	1	0	0	0	0	0	0	0	0	0	1	1	001003_{16}
PB/DDRB	0 0	1	0	0	0	0	0	0	0	0	1	0	1	001005_{16}
CRB	0 0	1	0	0	0	0	0	0	0	0	1	1	1	001007_{16}

10.3.2.a.iv I/O.   There are 3 seven-segments displays in the system (TI311), an LED, and 3 switches. The 3 displays have internal latches and hex decoders. So the two least significant displays are connected directly to port B of the 6821 chip, and the most significant display is connected to the upper 4 bits of port A. The latches are tied to ground so as to enable the displays at all times. The LED, when ON, will indicate that the display is in the BCD mode. Each of the three switches, double-pole single throw type, with LED indicator are connected to the lowest 3 bits of port A.

10.3.2.a.v Unused-Input Pins Connection.   For the 68000 there are 6 unused, active-low, input pins which must be disabled by connecting them to 5 V. These are $\overline{\text{IPL}0}$, $\overline{\text{IPL}1}$, $\overline{\text{IPL}2}$, $\overline{\text{BERR}}$, $\overline{\text{BR}}$, and $\overline{\text{BGACK}}$. Two of the 6821 unused pins ($\overline{\text{IRQA}}$ and $\overline{\text{IRQB}}$) are also disabled this way, while CA1 and CB1 are disabled by connecting them to ground.

10.3.2.b Microcomputer Development System

Hewlett-Packard (HP) 64000 is used to design, develop, debug, and emulate the 68000-based system. Some details are given in this section.

The emulator is a very important part of the development of the software and hardware of the system. The 68000 emulator has most of the functions for emulation such as display memory or registers, modify memory or registers. But there is no single-step function. The HP 64000 emulator is divided into three modes of operations: initialization, emulation, and EPROM programming.

10.3.2.b.i Initialization

Edit.   The edit function is used to create the application program in mnemonic form. The first line of the program must be "68000" to indicate that the program is to be assembled by the 68000 assembler, and that a 68000 microprocessor is to be used for emulation.

Assemble.   When the application program has been completed and properly edited, the file is then assembled into a relocatable object code file. All errors indicated by the assembler should be corrected at this point.

In order to use the 68000 emulator special functions, a special monitor program is required. This can be copied as follows:

```
COPY Mon_68k:HP:source to MON_68k
```

Note upper and lower case. Once this monitor is copied, it must be assembled for no errors.

Linking. The two relocatable files must be linked together to create an absolute file for the emulation process. The files can be linked as follows:

```
Link <cr>
Object files? Mon_68k
Library file? <cr>
Prog, Data, Comm = 000100H, 000000H, 000000H
More files? yes
Object file? (name of application program file)
Library file? <cr>
Prog, Data, Comm = 001100H, 000000H, 000000H
More files? no
Absolute file? (name of absolute file)
```

There are reasons why the two files were linked this way. The monitor program must be stored at 0100H through 0FFFH, and since this I/O port is mapped starting at 1001H, the application program must be stored starting at a different address, so 1100H was used. Also, address 000H through 0FFF is used for exception vectors by the 68000 microprocessor.

10.3.2.b.ii Emulation. The emulator was used as a replacement for the actual 68000 chip to test the software logic and hardware before it was actually installed into the circuit.

To start the emulation process the following soft-key parameters were entered:

```
EMULATELOAD                  (absolute file name)
Processor clock?             external
Restrict to real time?       no
Memory block size?           256
Significant bits?            20
Break on write to ROM?       yes
Memory map:
0000H thru 0FFFH emulation RAM (monitor & exception vector)
1000H thru 10FFH user RAM (I/O PORT addresses)
1100H thru 1FFFH emulation ROM (application program)
Modify simulated I/O?        no
Reconfigure pod?             no
Command file name?           (name of emulation command file)
```

Note: Usually external clock requires external $\overline{\text{DTACK}}$; however, since the system only has EPROMs and for the purpose of emulation these EPROMs are not used, the external $\overline{\text{DTACK}}$ is not required.

Once the required files and memory maps are loaded, the system is ready for emulation. The monitor program must be running before the application program is executed. To run the monitor program, the following is used:

```
run from 0100H <cr>
```

Then the application program is run using:

```
run from 1100H <cr>
```

Another important part that the user should keep in mind is the processor status. There are three messages for the processor status which indicate that the emulator is not generating any bus cycles. They are

1. Reset — Indicates that the user's hardware is asserting the Reset input. The condition can only be terminated by releasing the user's hardware.
2. Wait — Indicates that the 68000 is waiting for a $\overline{\text{DTACK}}$ or other memory response. The condition can be terminated by asserting $\overline{\text{DTACK}}$, $\overline{\text{BERR}}$, $\overline{\text{VPA}}$, or entering "reset" from the keyboard.
3. No memory cycle — Indicates that the 68000 has executed a STOP instruction. The condition can be terminated by asserting "break" or "reset" from the keyboard.

10.3.2.b.iii EPROM Programming.   After the software and hardware have been emulated and they work properly, the final step is to program the EPROM and put the final circuit together. But before this, the program must be changed to include the addresses for the stack pointer and the initial PC. This is done by using the "ORG" and "DC" assembler directives. Then this new program is assembled and linked again. The EPROM is then programmed with the contents of this final "absolute file".

Programming EPROMs with the HP 64000 for 68000 is done by odd and even EPROMs. To program the lower 8 bits of data (odd ROM), the option bit 0 is selected, and bit 8 for the upper 8 bits (even ROM) is chosen as follows:

```
Prom_Prog          <cr>
2716               <cr>
Program from       (filename: absolute: bit 0 or 8)
```

Also, to check if the EPROM is clear, the command "check_sum" is used and if the result is F800H then the EPROM is clear.

10.3.2.c Software

The program consists of three major functions: initialize I/O ports and data registers, monitor and debounce key switches, and increment, decrement, or change mode. The program configures port B of the 6821 as an output port which will be used to display the two lower significant nibbles of data. The higher 4 bits of port A are configured as output to display the most significant nibble of the data. Bit 3 is also an output bit which turns ON and OFF the LED. The lowest 3 bits of port A are configured as inputs to detect the positions of three key switches. Register D3 is used to store the data in hex. Registers D4 and D7 are used to store the data in BCD mode with the low order byte in D4 and the high order byte in D7. Bit 3 of D0 contains a logic 1 representing BCD mode and logic 0 representing hex mode. Register D5 contains a 1 which will be used for incrementing BCD data, since ABCD doesn't have immediate mode. Register D6 contains 999 which is used for decrementing BCD.

The program monitors the three switches and stores the three input bits into register D0 if any of the keys is pressed. The processor then waits until the depressed key is released and then checks the input data one by one. The processor then branches to the increment, decrement, or mode change routine according to the depressed key. After execution, the processor will display the result on the three seven-segment displays.

Figure 10.9 shows the software flowchart. Note that the flowchart and the corresponding software does not include 'Mode LED' on/off feature.

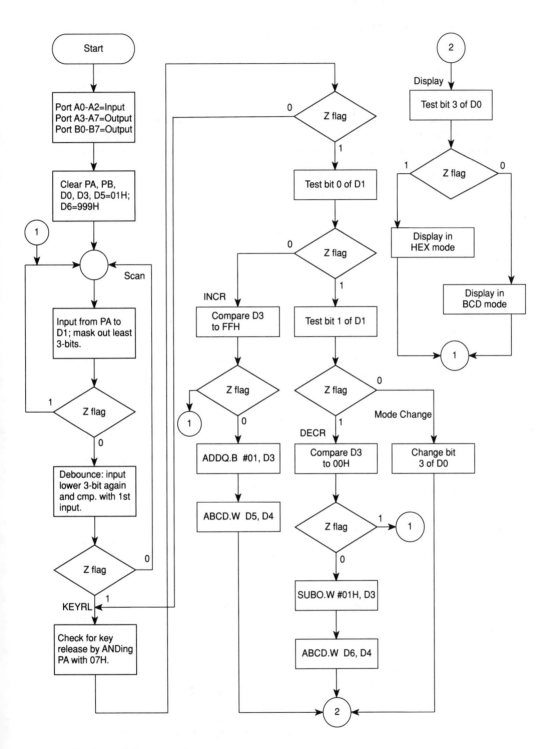

FIGURE 10.9 Software flowchart.

The assembly language program is listed below:

```
FILE: LAB2:KH0A22 HEWLETT-PACKARD: 68000 Assembler
```

```
1 "68000"
2 ************************************************************
3 *THIS PROGRAM STARTS DISPLAYING 000 AND MONITORS THREE KEY*
4 *SWITCHES THEN INCREMENT, DECREMENT, OR CONVERT HEX TO BCD*
5 *OR VICE VERSA, DEPENDING ON WHICH KEY IS DEPRESSED, THE  *
6 *DISPLAY GOES FROM 00-FF IN HEX MODE OR 00-255 IN BCD MODE*
7 ************************************************************
```

```
LOCATION OBJECT
CODE LINE                                    SOURCE LINE

<1001>               8 PA      EQU      001001H
<1001>               9 DDRA    EQU      001001H
<1003>              10 CRA     EQU      001003H
<1005>              11 PB      EQU      001005H
<1005>              12 DDRB    EQU      001005H
<1007>              13 CRB     EQU      001007H
                    14         ORG      00000000H
000000 FFFF FFFF    15         DC.L     0FFFFFFFFH
000004 0000 0008    16         DC.L     START
                    17 *
                    18 * CONFIGURE THE INPUT AND OUTPUT PORTS,
                    19 * DISPLAY 000 ON THE 7-SEGMENT DISPLAYS,
                    20 * AND INITIALIZE ALL THE DATA REGISTERS.
                    21 * THE HEX MODE IS STORED IN D3 AND THE
                    22 * BCD MODE IS STORED IN D7 AND D4.
000008 4238 1003    23 START   CLR.B    CRA
00000C 11FC 00FB    24         MOVE.B   #0F8H,DDRA  ;BIT 0-2 OF
                                                    ;PORT A AS
                                                    ; INPUT
000012 0BF8 0002    25         BSET.B   #02H,CRA    ;BIT 3-7 OF
                                                    ;PORT AS OUT-
                                                    ;PUT
000018 4238 1007    26         CLR.B    CRB
00001C 11FC 00FF    27         MOVE.B   #0FFH,DDRB  ;ALL 8 BITS OF
                                                    ;PORT B AS
                                                    ;OUTPUT
000022 08F8 0002    28         BSET.B   #02H,CRB
000028 4200         29         CLR.B    D0
00002A 11C0 1001    30         MOVE.B   D0,PA       ;DISPLAY 000
00002E 11C0 1005    31         MOVE.B   D0,PB
000032 4203         32         CLR.B    D3
000034 4244         33         CLR.W    D4
000036 7A01         34         MOVEQ.W  #01H,D5
000038 3C3C 0999    35         MOVE.W   #999H,D6
00003C 4207         36         CLR.B    D7
                    37 *
                    38 * DEBOUNCE THE KEY SWITCHES
                    39 *
00003E 1238 1001    40 SCAN    MOVE.B   PA,D1       ;MONITOR THE
                                                    ;KEYS
000042 0201 0007    41         ANDI.B   #07H,D1     ;MASK OUT THE
                                                    ;OUTPUT PINS
```

FILE: LAB2:KH0A22 HEWLETT-PACKARD: 68000 Assembler (*continued*)

LOCATION OBJECT
CODE LINE   SOURCE LINE

```
000046 67F6           42        BEQ      SCAN      ;IF NO KEY IS
                                                   ;DEPRESSED GO TO
                                                   ;SCAN
000048 1438 1001      43        MOVE.B   PA,D2     ;READ THE DATA
                                                    AGAIN
00004C 0202 0007      44        ANDI.B   #07H,D2
000050 B401           45        CMP.B    D1,D2     ;CHECK TO SEE IF
                                                   ;THE DATA REMAIN
                                                   ;UNCHANGED
000052 66EA           46        BNE      SCAN      ;IF IT CHANGES GO
                                                   ;TO SCAN
                      47 *
                      48 * CHECK TO MAKE SURE THAT THE KEY IS
                      49 * RELEASED BEFORE THE NEXT KEY CAN BE
                      50 * ENTERED.
000054 1438 1001      51 KEYRL  MOVE.B   PA,D2
000058 0202 0007      52        ANDI.B   #07H,D2
00005C 66F6           53        BNE      KEYRL
                      54 * CHECK TO SEE WHICH KEY HAS BEEN EN-
                      55 * TERED. BITS 0,1, AND 2 OF D1 REPRESENT
                      56 * INCREMENT, DECREMENT, AND MODE EX-
                      57 * CHANGE RESPECTIVELY.
00005E 0801 0000      58        BTST.B   #0H,D1
000062 6600 003C      59        BNE      INCR      ;IF BIT-0 OF D1
                                                   ;IS 1 GOTO INCR
000066 0801 0001      60        BTST.B   #1H,D1
00006A 6700 0044      61        BEQ      MODE      ;IF BIT-1 IS 1
                                                   ;DECREMENT, OTHER-
                                                   ;WISE GOTO MODE
                      62 *
                      63 * DECREMENT BOTH HEX AND BCD AT THE SAME
                      64 * TIME.
00006E 0C03 0000      65        CMPI.B   #00H,D3
000072 67CA           66        BEQ      SCAN      ;IF THE NUMBER IS
                                                   ;0 NO DECREMENT,
                                                   ;GOTO SCAN
000074 5303           67        SUBQ.B   #1H,D3    ;DECREMENT HEX BY 1
000076 C603           68        AND.B    D3,D3     ;CLEAR THE CARRY
000078 C906           69        ABCD.B   D6,D4     ;DECREMENT BCD BY
                                                   ;1 BY ADDING IT
                                                   ;WITH 999
00007A CF06           70        ABCD.B   D6,D7
                      71 *
                      72 * DISPLAY THE NUMBER IN HEX IF BIT-3 OF
                      73 * D0 IS 0, OTHERWISE DISPLAY IN BCD.
                      74 *
00007C 0800 0003      75 DISPLAY BTST.B  #3H,D0
000080 6700 0014      76        BEQ      HEX       ;IF BIT-3 OF D0
                                                   ;IS 0, GOTO HEX
000084 11C4 1005      77        MOVE.B   D4,PB     ;OUTPUT THE LSB
                                                   ;TO PORT B
000088 E94F           78        LSL.W    #4H,D7    ;SHIFT LEFT 4
                                                   ;TIMES
```

```
FILE: LAB2:KH0A22   HEWLETT-PACKARD: 68000 Assembler (continued)

LOCATION OBJECT
CODE LINE                                              SOURCE LINE

00008A 08C7 0003     79        BSET.B  #3H,D7    ;TURN OFF THE LED
00008E 11C7 1001     80        MOVE.B  D7,PA     ;OUTPUT THE MSB
                                                 ;TO UPPER 4 BITS
                                                 ;OF PORT A
000092 E84F          81        LSR.W   #4,D7
000094 60A8          82        BRA     SCAN
000096 11C0 1001     83 HEX    MOVE.B  D0,PA     ;OUTPUT 0 TO PORT A
00009A 11C3 1005     84        MOVE.B  D3,PB     ;OUTPUT THE HEX
                                                 ;NUMBER TO PORT B
00009E 609E          85        BRA     SCAN
                     86        *
                     87        * INCREMENT BOTH HEX AND BCD.
                     88        *
0000A0 0C03 00FF     89 INCR   CMPI.B  #0FFH,D3
0000A4 6798          90        BEQ     SCAN      ;IF THE NUMBER
                                                 ;IS FF NO
                                                 ;INCREMENT
0000A6 5203          91        ADDQ.B  #1H,D3    ;INCREMENT HEX
                                                 ;NUMBER BY 1
0000A8 4202          92        CLR.B   D2
0000AA C905          93        ABCD.B  D5,D4     ;INCREMENT LSB OF
                                                 ;BCD BY 1
0000AC CF02          94        ABCD    D2,D7     ;INCREMENT MSB OF
                                                 ;BCD BY 1 IF CARRY
                                                 ;IS 1
0000AE 60CC          95        BRA     DISPLAY
                     96 *
                     97 * EXCHANGE THE MODE THEN DISPLAY THE
                     98 * NUMBER.
0000B0 0840 0003     99 MODE   BCHG    #3H,D0    ;EXCHANGE MODE BY
                                                 ;CHANGING BIT-3
                                                 ;OF D0
0000B4 60C6          100       BRA     DISPLAY
Errors = 0

LINE#   SYMBOL     TYPE   REFERENCES

***     B          U      23,24,25,26,27,28,29,30,31,32,36,40,41,43,44,
                          45,51,52,58,60,65,67,68,69,70,75,77,79,80,
                          83,84,89,91,92
10      CRA        A      23,25
13      CRB        A      26,28
9       DDRA       A      24
12      DDRB       A      27
75      DISPLAY    A      95,100
83      HEX        A      76
89      INCR       A      59
51      KEYRL      A      53
***     L          U      15,16
99      MODE       A      61
8       PA         A      30,40,43,51,80,83
11      PB         A      31,77,84
40      SCAN       A      42,46,66,82,85,90
23      START      A      16
***     W          U      33,34,35,78,81
```

TABLE 10.1   Memory Map

$000000-$000FFF	EPROM
$003000-$003FFF	RAM
$005000-$005FFF	DISPLAYS
$009000-$009FFF	SWITCHES

10.3.3 Solution No. 2

The second solution approach uses interrupt I/O but no I/O ports.

10.3.3.a Hardware

The system includes a 3-digit display and three momentary function switches (increment, decrement, and mode select). In order to minimize the complexity of the project, no I/O chips are used. Instead, a buffer and some latches are used as the I/O ports. The buffer is used to input the status of the momentary switches and the latches are used to input the information coming from the data bus. To further the design, three TIL311 displays are used because they contain internal data latches. Because the 68000 has 23 address lines (not including A0), the memory is linearly decoded. The even and odd memory chips are enabled by decoding pins $\overline{\text{UDS}}$, $\overline{\text{LDS}}$, and $\overline{\text{AS}}$.

To display the three-digit number, the data lines are connected to the inputs of the three TIL311 displays (D0–D3 = LSD, D4–D7 = middle digit, D8–D11 = MSD). The address strobe ($\overline{\text{AS}}$) is NANDed with the address line A14 to latch the data onto the three displays. The memory map for the displays is given in Table 10.1. Because of linear decoding, the problem of foldback exists.

Two 2716s are used for the EPROM and two 6116s are used for the RAM. Both the RAM and the EPROM chips are divided into even and odd memory. The configuration enables the 68000 to access an even or an odd data bytes or a complete word in one bus cycle. The even and odd select lines are generated by ANDing the $\overline{\text{UDS}}$ and $\overline{\text{AS}}$ pins and the $\overline{\text{LDS}}$ and $\overline{\text{AS}}$ pins, respectively. To access a word, both the even and the odd enable signals are asserted. These signals are then NANDed with address lines A12 and A13 to select the EPROM and the RAM, respectively (see Figure 10.10). The odd memory chip data lines are connected to D0–D7 of the 68000. The even memory chip data lines are connected to D8–D15. Table 10.1 shows the memory map.

In the system, the interrupt pins are implemented by ANDing the status of the momentary switches and connecting the output of the gate to $\overline{\text{IPL}\,2}$. To achieve a level 5 interrupt, $\overline{\text{IPL1}}$ and $\overline{\text{IPL0}}$ pins are connected to Vcc and ground, respectively (see Figure 10.10). To reduce the number of components, the 68000 is instructed to generate an internal autovector to service the interrupt. This is accomplished by asserting $\overline{\text{VPA}}$ and $\overline{\text{IPL}\,2}$ at the same time. If an interrupt occurs (switch pressed), the 68000 will compute the autovector number $1D and the vector address $74. The processor will then go to a service routine that will find the switch that was pressed.

A 4-MHz crystal oscillator is used to clock the processor. Since the 68000 is operating at 4 MHz, $\overline{\text{AS}}$ is directly connected to $\overline{\text{DTACK}}$. This gave the EPROMs (450 ns access time) about 500 ns to provide valid data. A reset circuit similar to the one used in the 8085 system is used for the 68000-based microcomputer. However, on the 68000, both the $\overline{\text{RESET}}$ and $\overline{\text{HALT}}$ pins are tied together (see Figure 10.10). Figure 10.11 shows the board layout of all the chips.

10.3.3.b Software

The first major feature of the software is the inclusion of a start-up routine. The advantage of the start-up routine is to visibly verify the system performance. For example, if one of the displays malfunctions, the fault will not be known unless the user is able to see the display patterns. This requirement leads to the development of a start-up routine in which all three

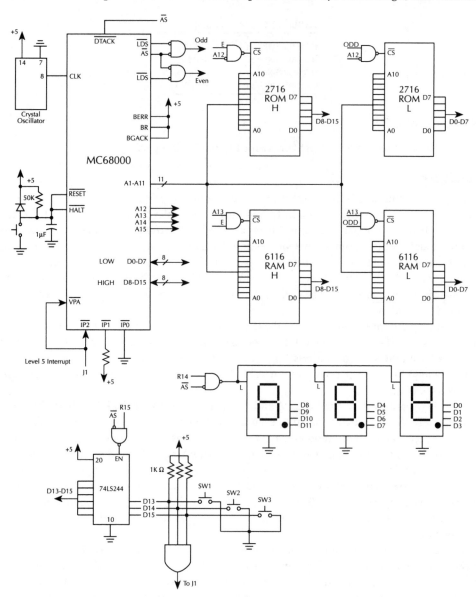

FIGURE 10.10 68000-based system for design problem no. 2.

displays count F down to 0 (in parallel). This routine uses a DBF loop in which the counter's value is duplicated to the two higher hex digits. The following is the actual start-up routine implemented in the program.

```
        MOVEQ    #0FH,D0        ; INITIALIZE LOOP COUNTER
                                ; TO $0000000F
LOOP    MOVE.W   D0,D1          ; COPY D0 TO D1
        ASL.W    #4,D1          ; SHIFT D1 LEFT 4 TIMES
        ADD      D0,D1          ; ADD D0 TO D1
        ASL.W    #4,D1          ; SHIFT D1 LEFT 4 TIMES
```

ROM Connections

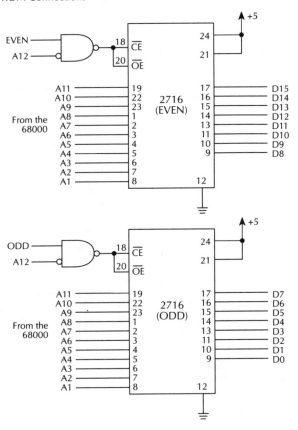

FIGURE 10.10   *continued.*

```
        ADD.W     D0,D1          ; ADD D0 TO D1
        MOVE.W    D1,DISPADDR    ; SEND RESULT TO DISPLAY
        MOVE.L    #VISIBLE,D6    ; LOAD DELAY TIME
        JSR       DELAY          ; CALL DELAY SUBROUTINE
        DBF.W     D0,LOOP        ; DEC BRANCH IF D0 ≠ -1, NOT
                                 ; TO THE LOOP
        CLR.L     D0             ; INITIALIZE COUNTER TO ZERO
        CLR.L     D7             ; INITIALIZE MODE TO DECIMAL
        MOVE.W    D0,DISPADDR    ; INITIALIZE DISPLAYS TO ZERO
        MOVE.W    #INTRMASK,SR   ; SET INTR AT LEVEL 5 AND
                                 ; SUPERVISOR MODE
WAIT    BRA.B     WAIT           ; WAIT FOR INTERRUPT
```

Upon successful completion of the start-up routine, the software directs the 68000 to enter an infinite wait loop. The wait loop serves to occupy the 68000 until a level 5 interrupt signals the processor. Upon interrupt, SR is pushed and the PC is also stacked. The 68000 accesses the long word located at address $74 and jumps to that service routine. The service routine exists at location $500. In response to the interrupt, the software directs the 68000 to move in the status switches to the low word of D0. A "C"-type priority case statement executes.

RAM Connections

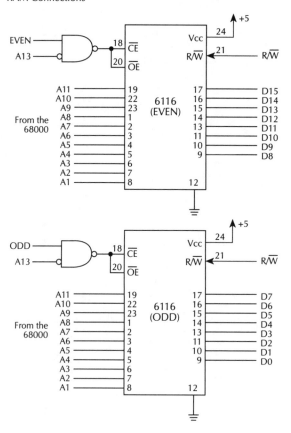

FIGURE 10.10   *continued.*

The case statement has the priority of up, down, then mode. Implementation of the case statement eliminates uncertainties when multiple keys are depressed. In the following, the case statement is shown.

```
RESPONSE    NOP                         ; ENTRY NO OPERATION
            MOVE.W   STATUS,D1          ; MOVE IN BUTTON STATUS WORD
            BTST.W   #UPBIT,D1          ; TEST INCREMENT BIT
            BEQ.W    INCREASE           ; IF UPBIT=0 BRANCH TO INCREASE
            BTST.W   #DOWNBIT,D1        ; TEST DECREMENT BIT
            BEQ.W    DECREASE           ; IF DOWNBIT=0 BRANCH TO
                                        ;   DECREASE
            BTST.W   #MODEBIT,D1        ; TEST MODEBIT
            BEQ.B    CHMODE,D1          ; IF MODEBIT=0 BRANCH TO CHANGE
                                        ;   MODE
            BRA.B    RESPONSE           ; NO RESPONSE, THEN SEARCH AGAIN
```

This segment utilizes the test bit facilities of the 68000. The algorithm first loads the switches. The switch word is then tested by the BTST instruction. The first bit test is the upbit. If the bit is found to be 0, the program branches to an increase-update routine. If the downbit

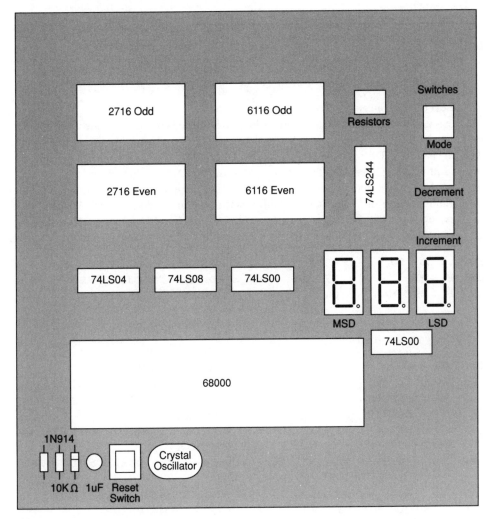

FIGURE 10.11 Board layout.

is found to be low, then the program reacts to decrement the displays. If the mode bit is found low, then the response is the base conversion of the displayed output.

The user may be tempted to indefinitely press a button or press multiple buttons. The habit is permissible. The program implements a 0.4-second wait loop at the end of any press of a key. This is a post-debounce. Without this feature, the 68000 will either count or change modes at speeds beyond recognition. The debounce routine also contains a priority. If the user constantly depresses multiple keys, the 68000 will service the input with the highest priority.

At this point, a deviation of the problem was made. The deviation was for ease of checking out the project. During checkout, when one wants to see a rollover, the increment or decrement key must be pressed 255 times. This is futile. At the end of the service routine, the software will not lock out a key entry, but rather the 68000 will immediately go to the wait state where the next interrupt may take place. To the user, it will appear that the 68000 is either autoincrementing, autodecrementing, or automatically changing modes. The post-debounce segment is displayed below.

```
VIEWER  NOP                      ; ENTRY NO OPERATION
        MOVE.1  #VISIBLE,D6  ; PLACE DELAY INTO D6
```

```
                JSR     DELAY           ; JUMP TO DELAY SUBROUTINE
                RTE                     ; RETURN FROM EXCEPTION
        DELAY   NOP                     ; ENTRY NO OPERATION
                DBF.W   D6,DELAY        ; DECREMENT FOR WAIT
                RTS                     ; RETURN FROM SUBROUTINE
```

The debounce routine implements a dummy loop that utilizes a large loop count. The routine is initialized by an immediate move long to D6. The debounce routine is called via the jump subroutine command. The delay loop contains a no-operation to increase loop time. After the NOP, the DBF.W will decrement D6 and branch if D6 is not equal to negative 1.

The software uses a hex base for counting; that is, all numbers whether decimal or hex will originate from a hex byte in data register 0 (D0). The display status exists in data register 7 (D7). If the contents of D7 are zero, this informs the program to display a decimal number on the next update. Otherwise, the program will send a hex value to the display. A typical decision-making segment (below) uses the 68000's ability to update flags on a move operation.

```
MOVE.   D7,D7       ; MOVE TO UPDATE FLAGS
BNE.B   HEX         ; IF Z=0 THEN SEND HEX TO DISPLAYS
BRA.B   DECIMAL     ; OTHERWISE, DECIMAL TO DISPLAYS
```

To convert a hex number to decimal format, the program uses the division/modulo algorithm shown in the following.

```
DECIMAL    NOP                     ; ENTRY NO OPERATION
           CLR.L   D2              ; INITIALIZE D2 TO ZERO
           CLR.L   D1              ; INITIALIZE D1 TO ZERO
           MOVE.B  D0,D1           ; COPY COUNT
           DIVU    #10,D1          ; DIVIDE D1 BY 10 MSD HEX →
                                   ; DECIMAL
           SWAP    D1              ; PLACE REMAINDER IN LOW WORD
                                   ; D1
           MOVE.W  D1,D2           ; MOVE REMAINDER TO D2
           CLR.W   D1              ; CLEAR REMAINDER
           SWAP    D1              ; REPLACE REMAINDER
           DIVU    #10,D1          ; DIVIDE D1 BY 10
           SWAP    D1              ; REMAINDER TO LOW WORD D1
           ASL.W   #4,D1           ; SHIFT REMAINDER UP ONE DIGIT
           ADD.W   D1,D2           ; ADD IN SECOND SIG FIG
           SWAP    D1              ; REPLACE QUOTIENT
           ASL.W   #4,D1           ; SHIFT QUOTIENT UP ONE DIGIT
           ASL.W   #4,D1           ; SHIFT QUOTIENT ANOTHER DIGIT
           ADD     D1,D2           ; ADD IN LSD HEX Æ DECIMAL
           MOVE.W  D2,DISPADDR     ; SEND DECIMAL RESULTS TO
                                   ; DISPLAY
           BRA.W   VIEWER          ; GO TO DISPLAY BRANCH
```

The algorithm exploits the DIVU (unsigned division) facilities of the 68000. The hex byte is moved to a long word register with zero-extend (assumed by CLR.L followed by a MOVE.B operation). The number is then divided by 10. The quotient remains in the low word of the destination register (D1); the remainder lies in the high word. With the use of SWAP, the remainder and quotient words are swapped. The remainder is moved (MOVE.W D1,D2) to another register D2 (initialized to zero). At this point, the remainder is cleared in D1, and swap is used to replace the quotient in the low word of D1. The next lower significant digit is extracted. Again, DIVU uses an immediate source of 10. The remainder in D1 is swapped into the low word, shifted up four times, then added to D2. The quotient is swapped back to the low word, shifted left eight times, then added to D2. The result of this routine is (at most) a three-digit BCD number which is suitable to send to the displays.

After the update of the displays, a time delay subroutine allows execution delays from the order of microseconds to the order of seconds. The time delay subroutine is shown below.

```
DELAY   NOP                 ; ENTRY NO OPERATION
        DBF.W   D6,DELAY    ; DECREMENT FOR WAIT
        RTS                 ; RETURN FROM SUBROUTINE
```

The NOP serves to increase the delay time of the loop. The NOP takes 4 clock cycles. DBF.W (decrement and branch on false) takes 10 clocks on a branch and 14 on a skip. JSR (jump subroutine) to the delay takes 23 cycles. The time analysis is simplified when consideration is taken only of the duration of the delay loop. A suitable delay for this project is about 0.4 s. This equates to (4-MHz clock) 1.6 million clock cycles! Because of the high number of cycles required, the "calls" and "returns" can be avoided because of their insignificance when compared to the massive number of clock delays required. A delay of about 0.4 s is used, which requires about 100,000 loops in the delay routine.

Some mention should be made of the mode features of the software. The change mode allows the user to liberally change the viewing format from hex to BCD or vice versa. The activation of this software feature simply complements D7 and then updates the displays via the previously mentioned methods. The increment facility increments D0 and updates the displays. Similarly, the decrement facility decrements D0 and updates the displays.

Expansion of the system is possible. Maybe for user entertainment, an uptone or downtone can be implemented. The tone can be generated through a variable delay routine. One of the address lines may be tied in series to a small speaker and every time the address is accessed, the speaker will "tick". Otherwise, the software, as it is, is suitable for the project. A listing of the assembly language program is provided below:

```
"68000"
;
; Microcomputer Applications
; M. Rafiquzzaman
; November 9, 1987
;
;
; This is the software routine for Design Problem
; No. 2
; THE BCD<>HEX COUNTER
; Language is 68000 MACRO
;
;
                NAME    BCD<>HEX_COUNTER
AUTO6           EQU     00000074H       ;SERVICE ROUTINE
                                        ;ADDRESS LOCATION
TSTACK          EQU     000037FCH       ;STACK INITIALIZE
RESPONSE        EQU     00000500H       ;INTERRUPT VECTOR
                                        ;ADDRESS
DISPADDR        EQU     00005000H       ;ADDRESS OF DISPLAY
PCINIT          EQU     00000400H       ;PC STARTUP ADDRESS
```

```
VISIBLE         EQU     00100000H       ;DELAY TIME APPROX. 0.4
                                        ;SECONDS
UPBIT           EQU     0FH             ;INCREMENT BIT LOCATION
DOWNBIT         EQU     0EH             ;DECREMENT BIT LOCATION
MODEBIT         EQU     0DH             ;MODE BIT LOCATION
STATUS          EQU     00009000H       ;STATUS WORD LOCATION
INSTRMSK        EQU     2500H           ;INTERRUPT MASK, LEVEL 7
;
; Top of the stack, program origin, and interrupt service
; location
;
                ORG     00000000H       ;STARTUP
                DC.L    TSTACK          ;INITIAL SUPERVISOR
                DC.L    PCINIT          ;FIRST PROGRAM INSTR
                                        ;LOC
                ORG     AUTO6           ;LOCATION OF AUTOVECTOR
                                        ;RESPONSE
                DC.L    RESPONSE        ;ADDRESS  OF  SERVICE
                                        ;ROUTINE 5
                DC.L    RESPONSE        ;ADDRESS  OF  SERVICE
                                        ;ROUTINE 6
                DC.L    RESPONSE        ;ADDRESS  OF  SERVICE
                                        ;ROUTINE NMI
;
; Startup routine
;
                ORG     PCINIT          ;BOOTUP AND TEST
                                        ;ROUTINE
                NOP                     ;ENTRY NO OPERATION
                MOVEQ   #0FH,D0         ;INITIALIZE LOOP
                                        ;COUNTER TO $0000000F
LOOP            MOVE.W  D0,D1           ;COPY D0 TO D1
                ASL.W   #4,D1           ;SHIFT D1 LEFT FOUR
                                        ;TIMES
                ADD.W   D0,D1           ;ADD D0 TO D1
                ASL.W   #4,D1           ;SHIFT D1 LEFT FOUR
                                        ;TIMES
                ADD.W   D0,D1           ;ADD D0 TO D1
                MOVE.W  D1,DISPADDR     ;SEND RESULT TO DISPLAY
                MOVE.L  #VISIBLE,D6     ;LOAD DELAY TIME
                JSR     DELAY           ;CALL DELAY SUBROUTINE
                DBF     D0,LOOP         ;DEC BRANCH IF D0 !=
                                        ;-1, NOT TO LOOP
                CLR.L   D0              ;INITIALIZE COUNTER TO
                                        ;ZERO
                CLR.L   D7              ;INITIALIZE MODE TO
                                        ;ZERO
                MOVE.W  D0,DISPADDR     ;INITIALIZE DISPLAYS TO
                                        ;ZERO
                MOVE.W  #INTRMASK, SR   ;SET INTR AT 5 AND
                                        ;SUPER MODE
WAIT            BRA.B   WAIT            ;WAIT FOR INTERRUPT
;
;  INTERRUPT ROUTINE
;
                ORG     RESPONSE
RESPONSE        NOP                     ;ENTRY NO OPERATION
                MOVE.W  STATUS,D1       ;MOVE IN BUTTON STATUS
                                        ;WORD
```

```
                BTST.W    #UPBIT,D1        ;TEST INCREMENT BIT
                BEQ.W     INCREASE         ;IF UPBIT=0 THEN BRANCH
                                           ;TO INCREASE
                BTST.W    #DOWNBIT,D1      ;TEST DECREMENT BIT
                BEQ.W     DECREASE         ;IF DOWNBIT=0 BRANCH TO
                                           ;DECREASE
                BTST.W    #MODEBIT,D1      ;TEST MODEBIT
                BEQ.B     CHMODE,D1        ;IF MODEBIT=0 THEN
                                           ;BRANCH TO CHANGE MODE
                BRA.B     RESPONSE         ;NO RESPONSE, THEN
                                           ; SEARCH AGAIN
;
; Change mode and update displays
;
CHMODE          NOP                        ;ENTRY NO OPERATION
                EORI.B    #0FFH,D7         ;COMPLEMENT MODE MASK
                BNE.B     HEX              ;IF EORI IS NOT ZERO
                                           ;THEN HEX OUT
                BRA.B     DECIMAL          ;EORI IS ZERO, THEN
                                           ;DECIMAL OUT
;
; Increment display count
;
INCREASE        NOP                        ;ENTRY NO OPERATION
                ADDQ.B    #1,D0            ;INCREMENT THE COUNT D0
                MOVE.B    D7,D7            ;MOVE TO UPDATE FLAGS
                BNE.B     HEX              ;IF Z=0 THEN SEND HEX
                                           ;TO DISPLAYS
                BRA.B     DECIMAL          ;OTHERWISE, DECIMAL TO
                                           ;DISPLAYS
;
; Decrement display count
;
DECREASE        NOP                        ;ENTRY NO OPERATION
                SUBQ.B    #1,D0            ;DECREMENT COUNT
                MOVE.B    D7,D7            ;MOVE TO UPDATE FLAGS
                BNE.B     HEX              ;IF Z=0 THEN SEND HEX
                                           ;TO DISPLAYS
                BRA.B     DECIMAL          ;OTHERWISE, DECIMAL
                                           ;DISPLAYS
; This routine sends hex contents of D0 to the displays
;
HEX             NOP                        ;ENTRY NO OPERATION
                MOVE.W    D0,DISPADDR      ;HEX DATA IS SENT TO
                                           ;DISPLAYS
                BRA.W     VIEWER           ;GO TO DELAY BRANCH
;
; HEX → Decimal converter
;
DECIMAL         NOP                        ;ENTRY NO OPERATION
                CLR.L     D2               ;INITIALIZE D2 TO ZERO
                CLR.L     D1               ;INITIALIZE D1 TO ZERO
                MOVE.B    D0,D1            ;COPY COUNT
                DIVU      #10,D1           ;DIVIDE D1 BY 10 MSD
                                           ;HEX → DECIMAL
                SWAP      D1               ;PLACE REMAINDER IN LOW
                                           ;WORD D1
                MOVE.W    D1,D2            ;MOVE REMAINDER TO D2
                CLR.W     D1               ;CLEAR REMAINDER
```

```
        SWAP     D1                      ;REPLACE REMAINDER
        DIVU     #10,D1                  ;DIVIDE D1 BY 10
        SWAP     D1                      ;REMAINDER TO LOW WORD
                                         ;D1
        ASL.W    #4,D1                   ;SHIFT REMAINDER UP ONE
                                         ;DIGIT
        ADD.W    D1,D2                   ;ADD IN SECOND SIG FIG
        SWAP     D1                      ;REPLACE QUOTIENT
        ASL.W    #4,D1                   ;SHIFT QUOTIENT UP ONE
                                         ;DIGIT
        ASL.W    #4,D1                   ;SHIFT QUOTIENT ANOTHER
                                         ;DIGIT
        ADD      D1,D2                   ;ADD IN LSD HEX →
                                         ;DECIMAL
        MOVE.W   D2,DISPADDR             ;SEND DECIMAL RESULTS
                                         ;TO DISPLAY
        BRA.W    VIEWER                  ;GO TO DISPLAY BRANCH
;
; This sends output to displays and implements delay of 0.7
; seconds
;
VIEWER          NOP                      ;ENTRY NO OPERATION
                MOVE.L   #VISIBLE,D6     ;PLACE ENTRY INTO D6
                JSR      DELAY           ;JUMP TO DELAY
                                         ;SUBROUTINE
                RTE                      ;RETURN FROM EXCEPTION

;
; Delay subroutine
;
DELAY           NOP                      ;ENTRY NO OPERATION
                DBF.B    D6,DELAY        ;DECREMENT FOR WAIT
                RTS                      ;RETURN FROM
                                         ;SUBROUTINE
```

The following shows 68000 Delay Analysis:

```
        MOVE.L   #VISIBLE,D6             ;MOVE IN DELAY LOOP
                                         ;COUNT
        JSR      DELAY                   ;CALL SUBROUTINE TO
                                         ;DELAY LOOP
DELAY   NOP                              ;TIME DELAY INCREASE
                                         ;LOOP
        DBF      D6,DELAY                ;DECREMENT BRANCH FALSE
        RTS                              ;RETURN SUBROUTINE
```

Execution Time:

```
        MOVE.L   #REG          12       CLOCKS
        JSR      ADDR          10       CLOCKS
        NOP                    4        CLOCKS
        DBF      REG,ADDR      14/10    CLOCKS
        RTS                    16       CLOCKS
```

First Pass: t1 = 12 + 20 + 4 + 14 = 50 CLOCKS

Middle Pass: t(n–2) = (n – 2) (4 + 14) CLOCKS

Last Pass: tn = 4 + 10 + 16 = 30 CLOCKS

General Pass: (for n > 3)

Typical Pass: (for large n)

For 0.45-s delay with a 4-MHz clock, 1.8 million clock cycles are required (0.45 s was chosen for ease of calculations).
Therefore,

$$18n = 1,800,000$$
$$n = 100,000$$

Figure 10.12 shows the flowchart for the start-up routine.

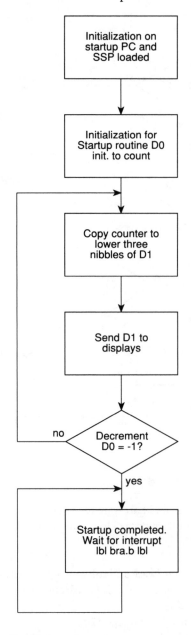

FIGURE 10.12 Start-up routine flowchart.

From Interrupt 6 Autovector

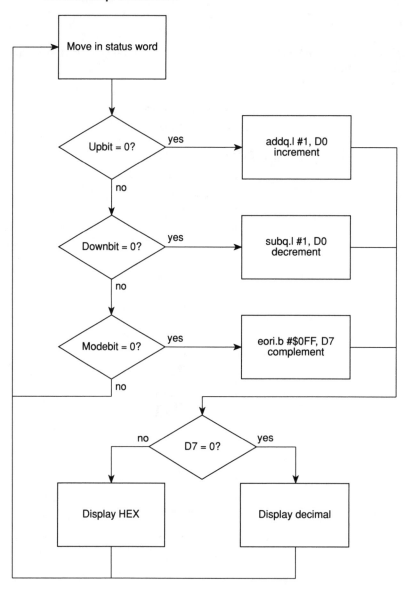

FIGURE 10.12   *continued.*

Questions and Problems

Design and develop the software and hardware for the following using a particular microprocessor (unless mentioned) and its support chips with a microcomputer development system of your choice.

10.1   Design and develop the hardware and software for a microprocessor-based system that would measure, compute, and display the Root-Mean-Square (RMS) value of a sinusoidal voltage. The system is required to:

    1.   Sample a 60-Hz sinusoidal voltage 128 times.

    2.   Digitize the sampled value through a microprocessor-controlled analog-to-digital converter.

3. Input the digitized value to the microprocessor using an interrupt.
4. Compute the RMS value of the waveform using the equation

$$\text{RMS value} = \sqrt{\frac{\sum \text{Xi}^2}{N}}$$

where Xi's are the samples and N is the total number of samples.
5. Display the RMS value using two digits.

10.2 Design a microcomputer-based capacitance meter using the following RC circuit:

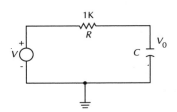 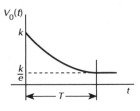

The voltage across the capacitor is V0(t) = ke$^{-t/RC}$. In one time constant RC, this voltage is discharged to the value k/e. For a specific value of R, the value of the capacitor C = t/R, where t is the time constant that can be counted by the microcomputer. Design the hardware and software for a microprocessor to charge a capacitor by using a pulse to a voltage of up to 10 V peak voltage via an amplifier. The microcomputer will then stop charging the capacitor, measure the discharge time for one time constant, and compute the capacitor value.

10.3 Design and develop the hardware and software for a microprocessor-based system to drive a four-digit seven-segment display for displaying a number from 0000H to FFFFH.

10.4 Design a microprocessor-based digital clock to display time in hours, minutes, and seconds on six-digit seven-segment displays in decimal.

10.5 Design a microcomputer-based temperature sensor. The microcomputer will measure the temperature of a thermistor. The thermistor controls the timing pulse duration of a monostable multivibrator. By using a counter to convert the timing pulse to a decimal count, the microcomputer will display the temperature in degrees Celsius.

10.6 Design a microprocessor-based system to test five different types of IC, namely, OR, NOR, AND, NAND, and XOR. The system will apply inputs to each chip and read the output. It will then compare the output with the truth table stored inside the memory. If the comparison passes, a red LED will be turned OFF. If the comparison fails, the red LED will be turned ON.

10.7 Design a microprocessor-based system that reads a thermistor via an A/D converter and then displays the temperature in degrees Celsius on three seven-segment displays.

10.8 Design a microprocessor-based system to measure the power absorbed by a 1K resistor. The system will input the voltage V across the 1K resistor and then compute the power using V²/R.

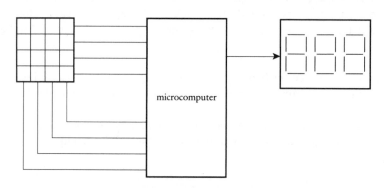

10.9   It is desired to design a priority vectored interrupt system using a daisy-chain structure for a microcomputer. Assume that the system includes four interrupt devices DEV0, DEV1, ..., DEV3, which, during the interrupt sequence, place the respective instructions RST0, RST1, ..., RST3 on the data bus. Also assume that DEV0, ..., DEV3 are Teledyne 8703 A/D converters (DEV3 highest, DEV0 lowest priority) or equivalent.
  i)   Flowchart the problem to provide service routines for inputting the A/D converters' outputs.
  ii) Design and develop the hardware and software.

10.10   It is desired to drive a six-digit display through six output lines of a microcomputer system. Use six Texas Instruments TIL311, 14-pin MSI hexadecimal displays or equivalent:
  i)   Design the interface with minimum hardware.
  ii)   Flowchart the software.
  iii) Convert the flowchart to the assembly language program.
  iv)   Implement the hardware and software.

10.11   Design a microcomputer-based combinational lock which has a combination of five digits. The five digits are entered from a hexadecimal keyboard and they are to be entered within 10 s. If the right combination is entered within the same limit, the lock will open. If after 10 s either all five digits are not entered or a wrong combination is entered, then the display will show an error signal by displaying "E". The system will then allow 5 s for the first digit to be entered the second time. If after this time the digit is not entered, the system will turn ON the alarm. If the second try fails, the alarm is also turned ON. When the alarm is ON, in order to reset the system, power has to be turned OFF.

10.12   Design a microcomputer-based stopwatch. The stopwatch will operate in the following way: the operator enters three digits (two digits for minutes and one digit for tenths of minutes) from a keyboard and then presses the GO key. The system counts down the remaining time on three seven-segment LED displays.

10.13   Design a microcomputer-based system as shown in the following diagram. The system scans a 16-key keyboard and drives three seven-segment displays. The keyboard is a 4 × 4 matrix. The system will take each key pressed and scroll them in from the right side of the displays and keep scrolling as each key is pressed. The leftmost digit is just discarded. The system continues indefinitely. Do not use any keyboard encoder chip. Use the 68000 microprocessor.

10.14 Design a microcomputer-based smart scale. The scale will measure the weight of an object in the range of 0–5 lb. The scale will use a load cell as a sensor such as the one manufactured by transducer, Inc. (Model # c462-10#-10pl strain-gage load cell). This load cell converts a weight in the range of 0–10 lb to an analog electrical voltage in the range 0–20 mV. The weight in lbs. and tenths should be displayed onto two BCD displays.

10.15 Design a microcomputer-based EPROM programmer to program a 2716.

10.16 Design a microcomputer-based system to control a stepper motor.

10.17 Design a microcomputer-based sprinkler control system.

10.18 Design a phone call controller. The controller will allow the user to pass only ten random phone numbers chosen by the user. The controller will use the touch-tone frequencies to encode the user information code numbers. A device will be used to decode the touch-tone signals and convert each into a seven-bit word. A microprocessor will then interpret this word and see if it is a match with one of the ten different numbers chosen by the user. The ten numbers are inputted by the user via the * button from the touch-tone system. The controller will have a manual override via the # button from the touch-tone system.

10.19 Design a microprocessor-based appointment reminder system with a clock. The system will alert the user before the present appointment time. The user has to set the appointments into fixed slots; for example: 9 AM or 2 PM. The system will deliver a voice message such as "Your next appointment is five minutes away" five minutes before the appointment time. A real time clock is to be included in the system to display the current time and will show the appointment time slots. You may use the Radio Shack SP0256 narrator speech processor.

10.20 Design a microcomputer-based autoranged ohmmeter with a range of 1 ohm to 999 kohm as follows: the microcomputer generates a pulse to charge a capacitor up to 10 V peak voltage through an amplifier and then stops charging the capacitor. The microcomputer measures the discharge time of the capacitor for one time constant and then computes the value of the resistor.

10.21 Interface two microcomputers to a pair 2K × 8 dual-ported RAMs (IDT7132) without using any bus locking mechanism. Two seven-segment displays will serve as an indicator. A program will be written to verify the dual-ported RAM contents. One processor will write some known data to the dual-ported RAM and the other processor will read and verify this data against the known data.

10.22 Design a microcomputer-based low frequency (1 Hz to 10 kHz) sine waveform generator. One cycle of a sine wave will be divided into a certain number of equal intervals. Each interval is defined as a phase increment. The precalculated sine values corresponding to the intervals are stored in ROM. The frequency of the signal will be set up by switches. When the system is started, the microprocessor will read the switches and will determine the time delay corresponding to the phase increment. The microprocessor will follow the time increments to send data to a D/A converter to convert the digital signal to an analog signal.

10.23 Design a microcomputer-based automobile alarm system. The purpose of this system is to prevent intruders from stealing a car or having enough time to steal a stereo or other valuable items in a car.

10.24 Design a microcomputer-based three-axis robot arm controller. The microcomputer will perform the calculations and the I/O to control movement of the arm. The microcomputer will receive destination data from an external source and perform coordinate transformations and boundary checking on the external data. It will then provide motor commands to the motor controllers to move the arm to the desired position.

10.25   Design a microcomputer-based home controller system. The system will simultaneously control six sprinkler stations, a heater, an air conditioner, and a burglar alarm. The system will contain a 12-hour clock and a temperature sensor. The user will program the system through a keypad. The time and temperature will be entered to control the sprinklers, the heater, and the A/C. The alarm will be armed or disarmed by entering a 4-digit code.

10.26   Design a microcomputer-based FM modulator. The microcomputer will read an analog input, convert the signal to digital, and perform several data manipulations to generate a digital representation of the FM signal. Finally, the microcomputer will convert the FM value to an analog signal.

10.27   Design and develop a microcomputer-based system for FFT (Fast Fourier Transform) computation. The microcomputer will sample eight data points using an A/D converter and compute the time-decimation FFT. After computation of the FFT, the result will be stored in system RAM where it can be used by another program for signal processing.

APPENDIX A

THE HEWLETT-PACKARD (HP) 64000

A.1 System Description

The HP 64000 Microprocessor Logic Development System is a universal development system which provides all of the necessary tools to create, develop, modify, and debug software for microprocessor-based systems. In-circuit emulation provides the capability of performing an in-depth analysis of hardware and software interfacing during the integration phase of the development process.

The HP 64000 Microprocessor Logic Development System is a multi-user development system, allowing up to as many as six users to operate on the system simultaneously. All users of the system share a line printer and a common data base in the form of a 12-megabyte Winchester Technology Disc Drive or a selection of one to eight Multi-Access Controller (MAC) disk drives connected to the system via the HP Interface Bus, commonly referred to as HP-IB. Eight disk drives can provide up to 960 metabytes of HP-formatted storage space.

A.2 Development Station Description

Figure A.1 shows the front view of the HP 64000 Development Station. The keyboard (Figure A.2) is divided into four areas: (1) an ASCII-encoded typewriter-type keyboard; (2) a group of edit keys, which facilitate movement of text or cursor when in the edit mode; (3) special function keys, for system reset or pause or to access a command recall buffer; (4) the all important system "soft keys," eight unobtrusive large key pads just beneath the bezel which surrounds the display.

The soft keys provide a quick and easy means to invoke system commands, virtually eliminating the typographical errors one usually has to contend with when having to enter commands character by character. The definition of each soft key is written on the display just above the bezel. The soft key syntax changes depending on the mode of operation and the

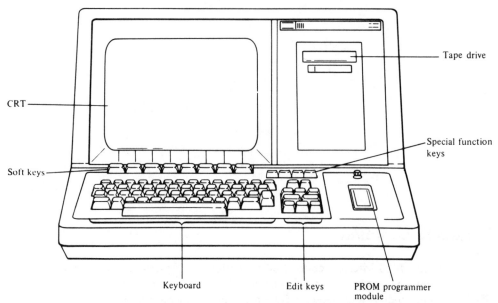

Figure A.1   Front view of the HP 64000 Development Station (Model 64100A). Front Panel: The seven major areas of the front panel are shown. Each area provides the interface necessary to operate and control the system. CRT Display: The CRT is a large-screen, raster-scan magnetic display. Screen capacity is 25 rows and 80 columns of characters. The standard 128-character (upper and lower case) ASCII set can be displayed. A blinking underline cursor is present as the prompt. Video enhancements are inverse video, blinking, and underline. Soft Keys: Just below the CRT are eight unlabeled keys. These keys are defined as the "soft keys." Each key ties to the soft key label line at the bottom of the CRT. During operation, the soft keys are labeled on the CRT screen. Source: Courtesy of Hewlett-Packard.

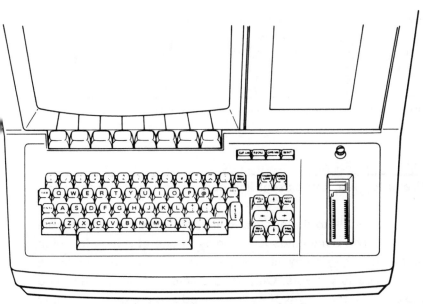

Figure A.2   Model 64100A keyboard. Source: Courtesy of Hewlett-Packard.

position of the cursor. This greatly enhances the ease of use of the system since it provides a list of alternatives available and guides the operator to use the system. In cases where the form of the input required is unknown, brackets surrounding a key word, a syntactical variable will prompt the user with the correct form of input the system expects.

The system display is a Raster Scan CRT which provides a display of 18 lines of text entry, a status line which always displays the system's status and date and time, three lines for command entry, and the soft key label line which indicates the function of each key. The display is 80 columns wide, but with the edit keys the display can be relocated to show text or data out to 240 columns. This is convenient for adding comments and really enhances the program documentation.

Other external station hardware includes RS232 ports for communication with either Data Communications Equipment (DCE) or Data Terminal Equipment (DTE). The RS232 port has a selectable baud rate up to 9,600 and uses the X-ON X-OFF convention for handshaking at baud rates 2,400 and above. There is a 20-mA current loop for TTY interfacing and two ports for triggering of external devices such as an oscilloscope during a logic trace. As system options, the front panel hosts a **PROM** programmer (Figure A.1) to the immediate right of the keyboard and a tape drive for file back-up. The tape drive performs a high-speed read and write and each cassette holds 250K bytes of data.

Figure A.3 shows special function keys and Table A.1 summarizes their functions.

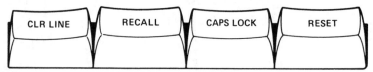

Figure A.3   Special function keys. Source: Courtesy of Hewlett-Packard.

Table A.1   Summary of Special Functions Keys

CLR LINE	Press to clear the current line containing cursor on the CRT.
RECALL	Used to recall, to the command line, previous commands from a stack. The commands are displayed one at a time for each time the RECALL key is pressed. The number of recallable commands is variable. Only valid commands are pushed into the stack. If the RECALL key is pressed and the buffer is empty the system responds with "Recall buffer is empty" message.
CAPS LOCK	Used to lock keyboard in all uppercase letters. A message is presented on the CRT indicating "CAPS LOCK on" or "CAPS LOCK off." At the next key stroke, the message is erased, but the mode remains in effect.
RESET	Pressing RESET once initiates a pause in system operation. A flashing "PAUSED" message, in inverse video, is presented on the status line. To continue operation, press any key except RESET .
	Pressing the RESET key the second time will clear the CRT and return the system to the system monitor.
SHIFT RESET	Holding the SHIFT key down and pressing RESET initiates a complete system reboot. This function should be regarded as a last resort when the system does not respond.
CNTL RESET	Holding CNTL key down and pressing RESET initiates system performance verification.

Source: Courtesy of Hewlett-Packard.

The following summarizes the HP 64000 soft keys, commands, assembler error codes, and other features.

System Monitor Soft Keys

The following provides a description of the system monitor soft keys:

userid
: The userid or user identification identifies each user as being unique within the system. This facilitates file management in that once the userid command is invoked all future references to files will be to files within that userid unless explicitly stated otherwise. The HP 64000 uses six characters and must begin with an uppercase alpha character.

time
: HH:MM. Allows the user to enter the correct time on the 24-hour clock displayed on the status line. This also facilitates file management since files can be referenced by time and date.

date
: DD/MM/YY. (Day/Month/Year) Allows the user to enter the correct date into the system. This aids the file management system since files can be referenced by date and time.

store
: This command will transfer files from the disk to the tape cartridge. The user specifies the file name and file type or all files. If all files are specified the system will store only the source files, linker command files, and emulator command files. Other file types may be stored but the file type must be specified. Other file types can readily be regenerated. This command will overwrite any previous contents of the tape cartridge.

append	Allows files to be appended to files previously stored on tape.
verify	Verify compares a file on the disk to a file resident on the tape cartridge. The user has the option of specifying a single file or all files on the tape assigned to the current userid.
restore	This command will transfer files from the tape cartridge to the disk. The user can specify file name or names and file type.
purge	This command will remove specified files from the active file list. A purged file can be recovered providing it has not been written over.
recover	Recover is used to recover files which have been purged. Files, if not written over, will be returned to the active file list.
rename	Allows the user to rename files. This is used to rename a file before recovering a previous file with the same name. This command also allows the user to transfer a file from one userid to another userid.
copy	Copy allows a disk file to be copied to, or from, the tape, display, or another file name or the RS232 port. The current display or a file may be copied to the printer.
directory	This command provides a listing of those files on the disk, the tape cartridge, and those recoverable files. The listing information consists of file name, file type, file size, last modified data, and last access date. Options: A directory can be made to include all userids, all types of files, before or after a specified date a file has been accessed or modified, and files on a specified disk unit.
library	This command is used to build libraries of relocatable files for use by the linker. These library files consist of relocatable files to be selectively loaded by the linker.
log	This creates a command file for all legal keystrokes. The log function is either toggled on or off by the "log" soft key.
(CMD_FILE)	This soft key represents a syntactical variable to be supplied by the user. This variable is a file name consisting of system commands which the development system will execute. A command file can be generated through the use of the editor or by using the log soft key.

Editor Commands

The 64000 editor commands are listed below:

revise
This mode is toggled ON and OFF and allows text to be modified. Modification may include character insertion or deletion. All appropriate command soft keys including "insert" are operational within the "revise" mode.

delete
This mode allows deletion of one line or a group of lines specified by the limit specified. The syntax "thru" includes deletion of the limit while the syntax "until" is not inclusive of the limit. The limit can be specified as a line #, string within a line, or as a start or end of text.

find
This command allows the user to search the text for the occurrence of the string. The find parameters include a (string) consisting of a single character or any combination of characters; (limit) allows the user to specify the boundaries of the search.

replace
This command allows text replacement of a string, a character, partial string with another character, string or partial string. There is an optional (limit) parameter that can specify boundaries of replacement.

(line #)
This command causes the line to become the current line of text.

end
This terminates the edit session and directs it to a specific destination. Usually this destination is a new file name. If no new file name is specified, the edit session terminates by purging the original file and replacing it with the edited file.

merge
Merge allows the user to merge an entire file or portions of it into the file being edited. Any text added to the file being edited will be added after the current line. Delimiters can be specified to determine the amount to be merged.

copy
Copy places specified text into a temporary storage buffer on disk for future use. The copy command will overwrite any text previously stored in the buffer. This is avoided by selecting the append option. The default value for (limit) is the current line only.

extract
This command removes the specified lines and places them into temporary storage space. If the append option is not selected, the extracted text will overwrite previously stored text. If (limit) is not specified, the current line will be extracted.

retrieve   This command retrieves the text from temporary storage and inserts into the program following the current line. The user has the option regarding the number of times the text is to be retrieved.

insert   This allows insertion of a combination of ASCII characters, after the current line of text. Insert is executable in the command mode, revise and insert mode.

list   This allows the user to list a file to another file or to a printer in numbered or unnumbered format. The listing will be exactly like the file text. There is also a (limit) option available.

renumber   This command renumbers the edited text starting from line one.

repeat   Repeat allows the user to duplicate the current line of text and add it immediately after the current line. The user can specify the number of times the repeat command is executed.

tabset   This command allows the user to set tabs in the desired column. The user has the choice of all 240 columns. Any character can be used to set tabs in any desired location.

range   Range restricts the columns to which find and replace commands are constrained. Columns 1 through 240 can be specified. The range function is toggled ON and OFF. When ON, the label range displays in inverse video.

autotab   This function provides an automatic tab function that is based on the first nonblank column of the previous line of text. Depressing the shift and the tab keys simultaneously allows tab back from autotab position.

Assembler Soft Key Definitions

The following provides the definitions of the 64000 assembler soft keys.

Key Label   **Definition**

(FILE)   This indicates the name of the source file that will be assembled.

listfile   This soft key specifies the destination of the assembler's output. The options available are listing the output to a specified file, to the display, to the printer, or to null (no generation of a list). If no list file option is specified, the assembler output listing defaults to the device previously specified by the user when the userid was declared.

options This soft key provides the user with a selection of five options specifying the type of output listing.

list This provides a listing of the source program excluding macro or data expansion. All no list pseudoinstructions in the source code are ignored.

nolist Selection of this soft key provides no listing except error messages. All list pseudoinstructions in the source code are ignored.

expand This soft key lists all source and macro generated codes. All list pseudoinstructions in the source program are ignored.

nocode This option causes the source program to be assembled without placing it in a relocatable file.

xref The selection of this option turns on the symbol cross-reference feature of the assembler and lists this table.

Assembler Pseudoinstructions

Pseudoinstructions are instructions used only by the assembler. They produce no executable code for the processor and normally do not take up any memory locations. They are used by the assembler to make programming easier. The following list contains those pseudo ops and their definitions supported by the HP 64000 assembler.

Op Code	Function
ASC	Stores data in memory in ASCII format.
BIN	Stores data in memory in binary format.
COMN	Assigns common block of data or code to a specific location in memory.
DATA	Assigns data to a specific location in memory.
DEC	Stores data in memory in decimal format.
END	Terminates the logical end of a program module. Operand field can be used to indicate starting address in memory for program execution.
EQU	Defines label field with operand field value. Symbol cannot be redefined.
EXPAND	Causes an output listing of all source and macro generated codes.
EXT	Indicates symbol defined in another program module.
GLB	Defines a global symbol that is used by other modules.
HEX	Stores data in memory in hexadecimal format.
LIST	Used to modify output listing of program.

MASK Performs and/or logical operations on designated ASCII string.

NAME Permits user to add comments for reference in the linker list.

NOLIST Suppresses output listings (except error messages).

ORG Sets program counter to specific memory address for absolute programming.

PROG Assigns source statements to a specific location in memory. Assembler default condition is "PROG" storage area.

REPT Enables user to repeat a source statement any given number of times.

SKIP Enables user to skip to a new page to continue program listing.

SPC Enables user to generate blank lines within program listing.

TITLE Enables user to create a text line at the top of each page listing for the source program.

The following pseudo ops are for the 8080 and 8085 assembler.

DB Stores data in consecutive memory locations starting with the current setting of the program counter.

DS Reserves the number of bytes of memory as indicated by the value in the operand field.

DW The define word pseudo stores each 16-bit value in the operand field as an address with the least significant byte stored at the current setting of the program counter. The most significant is byte stored at the next higher location.

Assembler Error Codes

The following provides a description of the 64000 assembler error codes.

Error Code	Definition

AS ASCII string; the length of the ASCII string was not valid or the string was not terminated properly.

CL Conditional label: Syntax of a conditional macro source statement requires a conditional label that is missing.

DE Definition error: Indicated symbol must be defined prior to its being referenced. Symbol may be defined later in the program sequence.

DS   Duplicate Symbol: Indicates that the noted symbol has been previously defined in the program. This occurs when the same symbol is equated to two values (using EQU directive) or when the same symbol labels two instructions.

DZ   Division by zero: Invalid mathematical operation resulting in the assembler trying to divide by zero.

EG   External Global: Externals cannot be defined as globals.

EO   External Overflow: Program module has too may external declarations (512 externals maximum).

ES   Expanded Source: Indicates insufficient input buffer area to perform macro expansion. It could be the result of too many arguments being specified for a parameter substitution, or too many symbols being entered into the macro definition.

ET   Expression Type: The resulting type of expression is invalid. Absolute expression was expected and not found or expression contains an illegal combination of relocatable types (refer to Chapter 2 of the *Assembler Manual* for rules and conventions).

IC   Illegal Constant: Indicates that the assembler encountered a constant that is not valid.

IE   Illegal Expression: Specified expression is either incomplete or an invalid term was found within the expression.

IO   Invalid Operand: Specified operand is either incomplete or inaccurately used for this operation. This occurs when an unexpected operand is encountered or the operand is missing. If the required operand is an expression, the error indicates that the first item in the operand field is illegal.

IP   Illegal Parameter: Illegal parameters in the macro header.

IS   Illegal Symbol: Syntax expected an identifier and encountered an illegal character or token.

LR   Legal Range: Address or displacement causes the location counter to exceed the maximum memory location of the instruction's addressing capability.

MC   Macro Condition: Relational (conditional) operator in macro is invalid.

MD   Macro Definition: Macro is called before being defined in the source file. Macro definition must precede the call.

ML   Macro Label: Label not found within the macro body.

MM   Missing Mend: Indicates that a macro definition with a missing mend directive was included in the program.

MO   Missing Operator: An arithmetic operator was expected but was not found.

MP   Mismatched Parenthesis: Missing right or left parenthesis.

MS   Macro Symbol: A local symbol within a macro body was not found.

NM   Nested Macro: A macro definition is not permitted within another macro.

PC   Parameter Call: Invalid parameter in macro header.

PE   Paremeter Error: An error has been detected in the macro parameter listed in the source statement.

RC   Repeat Call: Repeat cannot precede a macro call.

RM   Repeat Macro: The repeat pseudo operation code cannot precede a macro definition.

SE   Stack Error: Indicates that a statement or expression does not conform to the required syntax.

TR   Text Replacement: Indicates that the specified text replacement string is invalid.

UC   Undefined Conditional: Conditional operation code invalid.

UO   Undefined Operation code: Operation code encountered is not defined for the microprocessor, or the assembler disallows the operation to be processed in its current context. This occurs when the operation code is misspelled or an invalid delimiter follows the label field.

UP   Undefined Parameter: The parameter found in macro body was not included in the macro header.

US   Undefined Symbol: The indicated symbol is not defined as a label or declared an external.

Linker Commands

The 64000 linker commands are defined below:

Key Label	Definition
link	Initiates the link process.
(CMDFILE)	A syntactical variable supplied by the user. This would be the name of linker command file previously established.
listfile	Allows the user to select a destination other than the system default for the linker output listing.
display	Using this command designates the display as the output destination for the linker output listing.
(FILE)	Syntactical variable supplied by the user. This would be the name of a disk file to which the output of the linker would be directed.
null	Using this command suppresses the output listing. Error messages will still be output to the default destination as previously selected by the user.

printer	This designates the printer to be the destination of the linker output listing.
options	Soft key which precedes the selection of a linker option.
edit	Available linker option to edit a previously established linker command file.
nolist	Available linker option to suppress the generation of a linker load map.

Soft Key Definitions

The 64000 emulator soft key definitions are given below:

Label   **Description**

run	This starts program execution in the emulation processor. Execution begins at the location specified by "from" and ending under the conditions specified by "until." If no limits are specified, emulation will begin at the current address until halted by a "stop run" or by a boundary specified by "until."
	Syntax: run from (ADDRESS OR SYMBOL) until (ADDRESS OR SYMBOL)
step	This function causes the emulation processor to execute one instruction at a time. Once in the step mode, each depression of the return key will cause another instruction to be executed and displayed. The user can specify the number of steps to be executed each time the return key is pressed and the address from which stepping occurs. If these parameters are not specified, the system defaults to stepping from the current program counter location, executing one instruction each time the return key is pressed.
	Syntax: step # of (STATES) from (ADDRESS)
trace	This key is used to control the analysis function of the system, allowing the triggering and capturing of data of the emulation data bus.
	Syntax: trace in_sequence—permits tracing on a sequence of events.
	trace after—captures and displays data after the trigger qualifier word is satisfied.
	trace about—captures and displays data before and after the trigger qualifier.
	trace only—allows explicit definition of the information to be captured in the trace.

trace continuous—allows continuous monitoring of trace information without reentering the trace command.

display   This command causes the system to display a variety of data types on the development station's screen. Data types can be specified as global symbols, local symbols, and last active trace specification (valid only with the analysis card), the last active run specification, the trace buffer (valid only with analysis card), contents of proper emulation microprocessor registers, absolute or relative time display (valid only with analysis card), or contents of user or emulation memory.
Syntax: display trace
Syntax: display register (REGISTER NAME)
Syntax: display memory (ADDRESS)
Syntax: display trace specification
Syntax: display run specification
Syntax: display count
Syntax: display global symbols
Syntax: display local symbols

The mode option for the trace, register, and memory display provides the user with a choice of how the data will be presented on the screen. The following modes are defined:

static   The system will display the current conditions or contents one time only. No update will be shown.

dynamic   The system will continually update the display as data are changed in the emulation system.

absolute   The system displays data in absolute numeric code (i.e., hexadecimal or octal).

mnemonic   The system presents the data in the appropriate assembly language.

offset by   The system displays program modules so that the address values are offset by a specified value.

no offset   The system displays all addresses in program modules with those values assigned by the linking loader.

packed   The system displays opcodes and operands on the same line.

block   The system displays more data on the development station by displaying multiple columns of data.

modify   This command allows the user to change the contents of the emulation memory or processor registers to correspond to data entered from the console keyboard.
Syntax: modify (ADDRESS) to (VALUE)

	Syntax: modify memory (ADDRESS) thru (ADDRESS) to (VALUE)
	Syntax: modify register (REGISTER NAME) to (VALUE)
stop	This command halts the execution of either the run or trace commands. If stop-run is executed, it can be continued by a run command without skipping any of the intervening of the program code.
	Syntax: stop run
	Syntax: stop trace
end	Selecting this soft key changes the operating mode of the station, allowing other tasks to be performed. "end" does not stop the emulation process. Emulation continues even as other functions are performed on the system.
	Syntax: end_emulation
load	load_memory transfers abolute object files from the system's disk into emulation or user RAM memory.
	Syntax: load memory (FILE)
count	The count command is used in conjunction with a trace command. The count command is used to measure the elapsed time or the number of times certain user-specified events occurred between the start and end times specified by the trace.
	Syntax: count time
	count address = (ADDRESS)
copy	This command allows the user to transfer data from one location of emulation or user memory to the system's disk. The content of memory from which the data are taken remains unchanged.
	Syntax: copy (ADDRESS) thru (ADDRESS) to (FILENAME)
list_to	This command allows the user to make a permanent record of the contents of the stations display by writing it to a file on the disk or to the line printer.
	Syntax: list display to printer
	Syntax: list display to (FILE)
restart	Upon initializing the restart command, the microprocessor's program counter is reset to 0000H and the processor is reinitialized. It is important to execute the run command from the appropriate place in emulation memory.
edit_cnfg	(Edit-Configuration). This command recalls the series of queries which allows mapping of memory space and fault selection. When this command is invoked the previous responses can be modified by the user.
	Syntax: edit_cnfg

Following are the monitor level soft keys which will be in effect after December 1981:

```
edit compile assemble link emulate prom_prog run ---etc---
directory purge rename copy library recover log ---etc---
uerid date&time opt_test terminal (CMDFILE) --TAPE--- ---etc---
```

"--TAPE---" Soft Key

After --TAPE--- is returned the following soft keys are available.

```
store restore append verify tension directory ---etc---
```

"date&time" Soft Key

After date&time is depressed the following soft keys are available.

```
(DATE) (TIME)
```

"opt_test" Soft Key

This executes option test, which provides performance verification tests for options that are present.

Terminal Mode

"terminal Soft Key

This puts the station in an RS232 terminal mode which allows it to be a terminal to another system.

Passwords

The capability to have increased file security using passwords has been added. Following is the new syntax for userid.

"userid" Soft Key

After userid is depressed the following soft keys are available.

```
(USERID) listfile
```

After USERID is entered the following soft keys are available.

listfile password

After password is entered the following soft keys are available.

(PASSWD)

The user types in his password. This is nonprinting so he will not see on the display what he entered.

"HOST" PASCAL

"HOST" PASCAL consists of a compiler to allow users of the 64000 system to write programs that will execute on the internal host processor. In order to execute these programs the following syntax is used.

"run" Soft Key

After run is depressed the following soft key is available.

(FILE)

After a file is specified the following soft keys are available.

input output

After input is depressed the following soft keys are available.

(FILE) keyboard

After output is depressed the following soft keys are available.

(FILE) display display1 printer null

Summary of the HP 64000 Development System

Example A-1

This example shows how to create a new file and edit it. The file to be created is listed below:

```
"8085"
        ;   THIS PROGRAM STARTS AT 0 AND ADDS LOCATION 100H TO
        ;   LOCATION 101H AND STORES THE RESULT IN LOCATION 102H
            NAME    "ADD_WORKSHOP_1"
            ORG     2000H           ;   PROGRAM ORIGIN WILL BE AT
                                    ;   HEXADECIMAL 2000
START       LXI     H,100H          ;   LOAD HL PAIR WITH 100H
            MOV     A,M             ;   MOVE NUMBER IN LOCATION 100
                                    ;   INTO THE ACC.
            INX     H               ;   INCREMENT H AND L PAIR
            ADD     M               ;   ADD LOCATION 101 TO ACCUMULATOR
            INX     H               ;   INCREMENT HL PAIR
            MOV     M,A             ;   STORE ACC IN LOCATION 102H
            JMP     START
```

Procedure

Step 1: Press soft key "userid" and type in your USERID. The HP doesn't ask for the time and date after you enter your USERID. Rather, you have to press the soft key "Date&Time" to change it.

Step 2: To enter the edit mode, you have to create a new file. We'll call this new file "ADD".

edit into ADD (RETURN)

Step 2.5: It is a good idea to set up tabs so that if you want to you may jump to the opcode, operand or comment. You do this by pressing the softkey "Tabset". The editor will then display a tab row in which you can type "T" at the current cursor position. Then when you want to move faster, the "tab" key will jump to where you set your tabs. The way we did it was:

tabset 7 17 27 37

To save your tab sets, type the inverse softkey text "tabset" again.

Step 3: The first line of the program is the assembler directive, which lets the assembler know what microprocessor you wish to emulate.

"8085" (RETURN)

Step 4: To enter comments, type a "*" in column 1 and then start with your comments. If you want to start it anywhere else on that line, type a ";" and enter your comment.

* This is a comment (The "*" is at the leftmost edge of column 1)
; This is a comment (The ";" can be at any position).

Note: Your comments must come after you type "*" or ";".

Step 5: Enter "Name" and a brief explanation of the file. This lets you know what a particular file does in case you have to link many files. This is optional.

(TAB) NAME (TAB) "ADD_WORKSHOP_I" (RETURN)

Step 6: Enter "ORG" to let the assembler know the starting location of your program — in this case, at 2000H.

(TAB) ORG (TAB) 2000H (TAB) ; COMMENTS (RETURN)

Step 7: Enter the label "START" at column 1 of the next line along with the first instruction. This label helps the user to remember the English word rather than what the number was, if the user wants to loop or jump to that part of the program again.

START (TAB) LXI (TAB) H,100H (TAB) ; COMMENTS (RETURN)

Step 8: If you use only a few labels in your program, it is wise to use the softkey "AUTOTAB". This softkey jumps to the next line and moves the cursor right under the first word of the previous line. This saves time. Also note that the "TAB" key can also do this if you specified the tab sets.

Step 9: Enter the rest of the program. This is the program listing called "ADD".

Step 10: To list your file to the printer, make sure the printer is on-line (the light that's adjacent to the word should be on; if not, push the "on-line" button on the printer). Then type:

list printer all (RETURN)

Step 11: To save the file, type

end (RETURN)

Note: If your file wasn't named when you entered the edit mode, then type

end ADD (RETURN)

Step 12: The file is stored onto the hard drive or disk. To see your file on it, type

directory (RETURN)

You should then see the file "ADD" with the type "SOURCE". You will also see when you last modified it and accessed it. This information is important, so that you know how updated your file is. The directory listing will only show those files under your USERID.

Step 13: To re-edit the file, type

edit ADD (RETURN)

Notice that you don't type "edit into ADD". If you want to load your file from a disk, you have to specify the drive number. For disk drive X, type "ADD:X", where X is the disk drive number. If no drive number is specified, then the default is 0.

Step 14: To use the insert softkey, type a "NOP" after line 9 by

9 (RETURN)
insert (TAB) NOP (TAB) (TAB) ; NO OPERATION (RETURN)

Note: If you get an error, go to Step 15 and then back to 14. This may be because the editor was trying to find line 9, but your file has line numbers reading "NEW" instead.

Step 15: To renumber your file in order to give your editor a way to find what line to edit, type

renumber (RETURN)

Step 16: If your file is very large and you want to search your file for a particular word like "NOP", type

find "NOP" all (RETURN)

Note: Remember to enclose all strings with double quotes. If not, the editor will think it is a softkey command.

Step 17: To insert more text, type

insert (RETURN)

then move the cursor up, down, or sideways and begin typing the new line.

Step 18: To revise a line, enter

revise (RETURN)

This edits the line that the cursor is on. If that line isn't what you want, then move the cursor using the cursor keys.

Step 19: To move the display to allow for viewing all of the columns, depress the SHIFT and LEFT arrow keys simultaneously. Hitting SHIFT and

RIGHT keys will scroll the text right. To scroll the text up or down, hit the edit key ROLL UP or ROLL DOWN, respectively.

Step 20: To insert or delete character(s) when revising, hit the edit key INSERT CHAR or DELETE CHAR, respectively.

Step 21: To delete a line at the current cursor position, hit DELETE (RETURN). If you want to delete a line somewhere else, type

    extract (RETURN)

Then move the cursor to where you want to insert that line and type

    retrieve (RETURN)

Note: If you want to insert many copies of that line at the current cursor position, then type **retrieve** # (RETURN), where # is the number of copies.

Step 23: To abort the editor and not save your file, press the special function key RESET twice. Pressing it once will pause a running listing or program.

Step 24: To replace a word with another word type

    replace "word1" with "word2" all

This will replace all word1s and word2s. You can also specify where you want to stop replacing by using **thru** or **until** a certain line number.

Step 25: The "copy" command lets you copy a group of lines without erasing those lines, like "extract" does. First, place the cursor at the starting location line and then type

    copy thru line # (RETURN)

where # is the last of your lines to copy. Next, move the cursor to the place where you want it inserted, and type **retrieve**.

Step 26: The "merge" command lets you insert an entire file or copies a block of lines like the "copy" command does. To merge a file, type

    merge ADD (RETURN)

This lets you insert the file ADD at the current cursor position.

Example A-2

This example goes through the steps in assembling a file. Upon completion, it will create a "reloc" file to be later used for linking purposes.

Procedure

Step 1: Enter your userid, and optionally enter the time and date. The HP 64000 already has the current date and time, so updating isn't necessary.

userid USERID (RETURN)

Step 2: To assemble your source file called "ADD" and make a printout of it with the cross-reference table, type

assemble ADD listfile printer options xref (RETURN)

The printer will then output the assembled file that has both a source and object listing. It will then show the cross-reference table that lists any labels you put into that file and what lines accessed it or needed it to run.

Step 3: To show how the assembler command treats an error, we'll put one in by doing this

edit ADD (RETURN)
11 (RETURN)
insert (TAB) MVI (TAB) D,FFH (RETURN)
end (RETURN)

Step 4: Now assemble the file by typing the commands from Step 2. The assembler treated "FFH" as a symbol, not a hex value, so it generated an error. You will see the number of errors is one and that the error is an undefined symbol. On the xref table, FFH is a type U, which means it's undefined.

Step 5: To fix this error, all hex values beginning with an alpha character must be preceded with a "0". Do this by

edit ADD (RETURN)
12 (RETURN)
revise (TAB) (TAB) D,0FFH (RETURN)
end (RETURN)

Step 6: Now assemble the file again, and this time list it to the screen or display.

assemble ADD listfile display options xref (RETURN)

To pause the display from scrolling up so fast, type RESET. To resume scrolling, type any other key.

Step 7: Your assembled file is stored on your USERID directory. To see it type

directory (RETURN)

You will then see two more "ADD" files with the type "reloc" and "asmb_sym".

These files are useful for the assembler and linker programs. The "reloc" file is an object file containing the hex values of your program. It then must be made into an "absolute" file so it can run by itself.

Example A-3

This example goes through the steps in linking the "reloc" file to create an "absolute" file. This new file can then be run independently, emulated, or even used by the PROM programmer.

Procedure

Step 1: Initialize the linker and show the results to the display by

link listfile display (RETURN)
"Object files ?"

Step 2: This new message asks you what file(s) you want to be linked, so type

ADD (RETURN)
"Library files ?"

Step 3: There are no library routines in "ADD" so skip it by

(RETURN)
"Load addresses: PROG,DATA,COMN = 0000H,0000H,0000H"

Step 4: This command allows you to specify different memory areas for the program, data, and common modules. No memory assignment is needed because the "ADD" file already has an ORG statement, so skip it by

(RETURN)
"More files ?"

Step 5: Since there is only one file to be linked, respond by

no (RETURN)
"LIST,XREF, overlap_check,comp_db = on off on off"

Step 6: The linker is prompting the user to specify the output and declaring the default for the output listing. It then checks to see if your memory assignments overlap as well. Ignore this and type

(RETURN)
"Absolute file name ?"

Step 7: The linker wants you to enter the file name to be assigned to the absolute file.

ADD (RETURN)

Step 8: You will then see the linker examining your file

"STATUS:Linker : HP 640000S linker: Pass1"
"STATUS:Linker : HP 640000S linker: Pass2"
"STATUS:Linker : HP 640000S linker: End of link"

Note that the above display will be on a single line.

Step 9: The linker output will display the start and end location of your program, the current date and time, the assembler pseudo name "ADD_WORKSHOP_I", and extra data like XFER address and the total bytes loaded.

Step 10: To view your new files on the directory, type

directory (RETURN)

You should then see three new files with types "link_sym", "link_com", and "absolute". The absolute file is used for the emulator or the PROM programmer. The "link_com" is a command file that holds all of the data that you entered from Steps 1 to 7. This is good if you want to keep the same link configuration, but want to change or re-edit your source file(s).

If you want to save a linker listing to a file for later viewing type

link ADD listfile ADD_L (RETURN)

This will create a link listing similar to that of Step 9. Specifying "link ADD" will tell the linker to link it using its "link_com" file.

Example A-4

This example goes through the steps of emulating an absolute file without external hardware. This gives you a good idea of the importance of a development system like the HP 64000.

Procedure

Step 1: Before beginning, you must go through Examples 1 through 3. Also, the HP 64000 must have an 8085 emulator. If it is configured or has a 68000 emulator, emulation program will use only it, so be wary of this.

Step 2: To enter the emulation mode and load the absolute file do this

emulate load ADD (RETURN)
"Processor clock ?"

Step 3: This question asks if you want the source of the processor clock to be internal or external. Since there is no external hardware being used, type

internal (RETURN)
"Restrict processor to real-time runs ?"

Step 4: This question asks if you want to restrict the processor to real-time runs, which will limit the analysis functions that can be performed, such as debugging your program. An example of this would be "display registers blocked". So answer

no (RETURN)
"Stop processor on illegal opcodes ?"

Step 5: Specify "yes" so that the emulator will stop if an illegal opcode is detected.

yes (RETURN)

Step 6: The emulator will then want to specify the memory range for your emulation ram/rom and user ram/rom. Since all memory is internal (no user external hardware for this file), address 100H to 103H is used for storing the variables used in the program. Define this to be emulation RAM. To protect your program, define the memory to be emulation ROM. Thus, if something writes to your program, it will generate an error.

100H thru 103H emulation ram
2000H thru 20FFH emulation rom

Also notice that the ram and rom ranges are from 100H-3FFH and 2000H-20FFH. This may be because the emulator can only provide a range of memory area rather than a specified one.

Step 7: To keep your defined memory area, type

end (RETURN)
"Modify simulated I/O ?"

Step 8: Since we are not using any I/O ports, type

no (RETURN)
"Modify interactive measurement specification ?"

Step 8.5: This question is not in the book so type

(RETURN)
"Command file name ?"

Step 9: The emulator wants you to specify a file name to which to assign the emul_ com file configuration. Specify with

ADD (RETURN)

Once doing this, the emulator will then load your absolute file to the memory areas you specified. You should then see

"STATUS: 8085--Program loaded"

If you want to make any modifications, you have to start again, by re-editing, assembling, linking and then emulating. If you want the same emulating configuration that you specified in Steps 1 to 9, then type **emulate ADD load ADD**.

Step 11: To display your program with mnemonics, type

display memory 2000H mnemonic (RETURN)

You should then see the locations with their corresponding instructions of your program.

Step 12: To change the values that your program uses to add two numbers, you have to modify the emulation RAM by

modify memory 100H thru 102H to 02H (RETURN)
display memory 100H blocked (RETURN)

You should now see a display of your edited bytes. The memory block map will show you the address, data, and ASCII translation of each byte.

Step 13: To run your program, you can type either

run from 2000H (RETURN)

or

run from glob_sym START (RETURN)
"STATUS: 8085--Running"

Your program will keep running, so you can now modify the memory locations from Step 12 to something else, and then see the changes.

Step 14: You can also single-step the program to execute a single instruction at a time. This is a good debugging tool on the 64000.

```
break (RETURN)
"STATUS: 8085--break in background"
display registers (RETURN)
step from 2000H (RETURN)
```

or

```
step from glob_sym START (RETURN)
step (RETURN)
(RETURN)
```

Continuously pressing return will execute the "step" command again. Remember that pressing "return" will execute anything on the command prompt, no matter where the cursor is. You should then see a display of each instruction being executed with its corresponding register values. This is really good because you can trace any program, and scan the instructions, registers, flags, stack pointer, and the next IP.

Step 15: You can also set up breakpoints to stop the program when it reaches a certain argument. An example of this would be "run" from 2000H until address 2006H.

Step 16: To set up a breakpoint at address 2005H, type

```
run from 2000H until address 2005H (RETURN)
```

Step 17: Now say you want to halt the program after a memory write:

```
run from 2000H until status memory_write (RETURN)
```

Step 18: To end the emulation session, type

```
end (RETURN)
```

You can also press "RESET" twice too.

Step 19: If you want to get back to the emulation and keep the same emulator configuration, type

```
emulate ADD load ADD (RETURN)
```

If you want to change the emulator configuration type

```
modify_configuration (RETURN)
```

Note that the HP 64000 can perform in-circuit emulation with or without a large-system hardware.

Operation of the 68000 Emulator

In order to use the 68000 emulator, a monitor program must be included in the linking of the user's program. The purpose of the monitor is to provide special functions during emulation (including register display, software breakpoint setting, etc.).

The steps in the emulation process are as follows:

1. Create a program using the 64000's text edition.
    A. Make sure that "68000", including quotations, is the first line in the editor.
    B. Make the second line in the application program "PROG". (This will cause the monitor program to be successfully linked with the application.)
    C. Use "H" for Hex instead of "$" signs.
    D. Write your application program.
        Locate the program between 0FFH and 10000H-application size.
2. Assemble the program you write by typing:

    <Assemble> MON_68K

3. Make a copy of the assembly program "MON_68K" by typing:

    COPY Mon_68K:HP:source TO MON_68K

    (Note upper and lower case. Type it exactly as shown.)
4. Now assemble this program by typing:

    SOFTKEY
    <Assemble> MON_68K

5. Now, the application and monitor program must be linked together. Type:

    ```
    SOFTKEY
    <LINK>
    Object File?           MYFILE
    Library Files?         <CR>
    Prog,Data,Comn,A5=     000XXX,0H,0H,0H where XXX=100H to
                                10000H
    More Files?            <yes> Soft Key
    Object Files?          MON_68K
    Library Files?         <CR>
    Prog,Data,Comn,A5=     10000H,0H,0H,0H
    More Files?            <No> Soft Key
    Absolute File Name=    MYFILE
    ```

A. Notes: The monitor program is position independent. Since the TARGET SYSTEM has limited address space, it is suggested that you locate your program in **THAT** address range. Furthermore, this allows the user to specify the monitor program's address in upper-address space and in emulation RAM.

B. Care should be taken so that address ranges 000H→0FFH are reserved for vectors, and that user program addresses do not conflict with the monitor program whose size is about 1000_{16} bytes. I locate the monitor at address 10000H→10FFFH.

6. Finally, the emulator is entered by answering questions with their default values (unless the user wishes otherwise).

A. For the memory map, the following should be entered:

```
0000H-0FFFH     Emulation ROM (Vectors, Interrupts, User Prog)
10000H-10FFFH   Emulation RAM (Monitor Program)
0WWWH-0VVVH     User RAM (other space for ports, tables, etc.
                that exist in either software or the target
                system)
```

B. Some helpful commands during emulation (<XXXX> = Soft Key):

1. <Load> MYFILE	Loads both your file and the monitor program (linked)
2. <Modify><Config>	Lets you change the emulator configuration
3. <Modify><Softwre_bkpts>	Modifies software breakpoints so that you can stop program execution anywhere

NOTE: The 68000 emulator DOES NOT allow single-stepping.

4. <BREAK>	Enter the monitor program
5. <DISPLAY><MEMORY> <MNEMONIC>	Display disassembled code
6. <MODIFY><REGISTER> <MODIFY><MEMORY>	Modify address space and registers

DO NOT put ORG statements in your program except for the interrupt vectors.

APPENDIX B

MC68000L4
(4 MHz)
MC68000L6
(6 MHz)
MC68000L8
(8 MHz)
MC68000L10
(10 MHz)

HMOS
(HIGH-DENSITY, N-CHANNEL,
SILICON-GATE DEPLETION LOAD)

**16-BIT
MICROPROCESSOR**

Advance Information

16-BIT MICROPROCESSING UNIT

Advances in semiconductor technology have provided the capability to place on a single silicon chip a microprocessor at least an order of magnitude higher in performance and circuit complexity than has been previously available. The MC68000 is the first of a family of such VLSI microprocessors from Motorola. It combines state-of-the-art technology and advanced circuit design techniques with computer sciences to achieve an architecturally advanced 16-bit microprocessor.

The resources available to the MC68000 user consist of the following:

- 32-Bit Data and Address Registers
- 16 Megabyte Direct Addressing Range
- 56 Powerful Instruction Types
- Operations on Five Main Data Types
- Memory Mapped I/O
- 14 Addressing Modes

As shown in the programming model, the MC68000 offers seventeen 32-bit registers in addition to the 32-bit program counter and a 16-bit status register. The first eight registers (D0-D7) are used as data registers for byte (8-bit), word (16-bit), and long word (32-bit) data operations. The second set of seven registers (A0-A6) and the system stack pointer may be used as software stack pointers and base address registers. In addition, these registers may be used for word and long word address operations. All seventeen registers may be used as index registers.

L SUFFIX
CERAMIC PACKAGE
CASE 746

64-pin dual in-line package

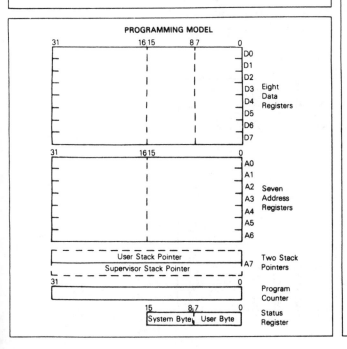

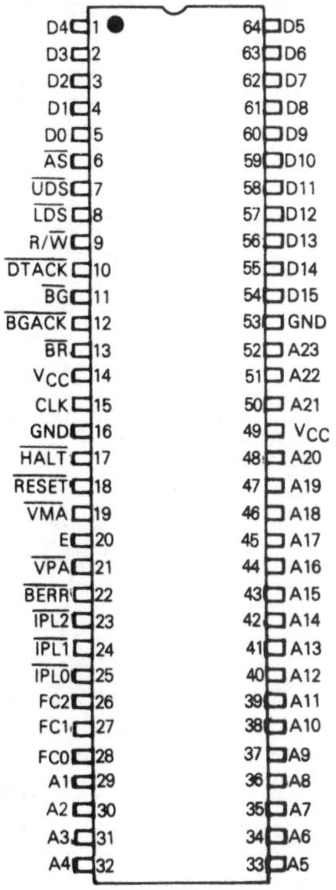

PROGRAMMING MODEL

D0–D7	Eight Data Registers
A0–A6	Seven Address Registers
User Stack Pointer / Supervisor Stack Pointer (A7)	Two Stack Pointers
Program Counter	
System Byte / User Byte	Status Register

64-pin pinout (left side, top to bottom):
D4 1, D3 2, D2 3, D1 4, D0 5, AS 6, UDS 7, LDS 8, R/W 9, DTACK 10, BG 11, BGACK 12, BR 13, Vcc 14, CLK 15, GND 16, HALT 17, RESET 18, VMA 19, E 20, VPA 21, BERR 22, IPL2 23, IPL1 24, IPL0 25, FC2 26, FC1 27, FC0 28, A1 29, A2 30, A3 31, A4 32

(right side, top to bottom):
64 D5, 63 D6, 62 D7, 61 D8, 60 D9, 59 D10, 58 D11, 57 D12, 56 D13, 55 D14, 54 D15, 53 GND, 52 A23, 51 A22, 50 A21, 49 Vcc, 48 A20, 47 A19, 46 A18, 45 A17, 44 A16, 43 A15, 42 A14, 41 A13, 40 A12, 39 A11, 38 A10, 37 A9, 36 A8, 35 A7, 34 A6, 33 A5

68-Terminal Chip Carrier

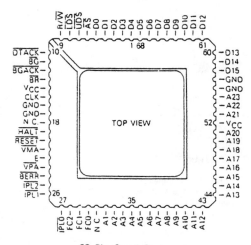

68-Pin Quad Pack

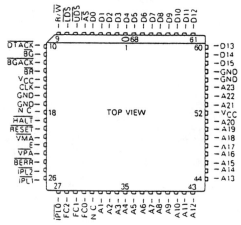

68-pin grid array.

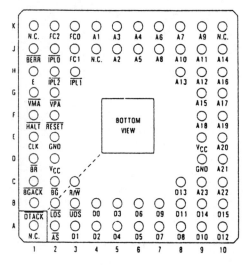

 **MOTOROLA**

Advance Information

MC68230 PARALLEL INTERFACE/TIMER

The MC68230 Parallel Interface/Timer provides versatile double buffered parallel interfaces and an operating system oriented timer to MC68000 systems. The parallel interfaces operate in unidirectional or bidirectional modes, either 8 or 16 bits wide. In the unidirectional modes, an associated data direction register determines whether the port pins are inputs or outputs. In the bidirectional modes the data direction registers are ignored and the direction is determined dynamically by the state of four handshake pins. These programmable handshake pins provide an interface flexible enough for connection to a wide variety of low, medium, or high speed peripherals or other computer systems. The PI/T ports allow use of vectored or autovectored interrupts, and also provide a DMA Request pin for connection to the MC68450 Direct Memory Access Controller or a similar circuit. The PI/T timer contains a 24-bit wide counter and a 5-bit prescaler. The timer may be clocked by the system clock (PI/T CLK pin) or by an external clock (TIN pin), and a 5-bit prescaler can be used. It can generate periodic interrupts, a square wave, or a single interrupt after a programmed time period. Also it can be used for elapsed time measurement or as a device watchdog.

- MC68000 Bus Compatible
- Port Modes Include:
  - Bit I/O
  - Unidirectonal 8-Bit and 16-Bit
  - Bidirectional 8-Bit and 16-Bit
- Selectable Handshaking Options
- 24-Bit Programmable Timer
- Software Programmable Timer Modes
- Contains Interrupt Vector Generation Logic
- Separate Port and Timer Interrupt Service Requests
- Registers are Read/Write and Directly Addressable
- Registers are Addressed for MOVEP (Move Peripheral) and DMAC Compatibility

MC68230L8
MC68230L10

HMOS
(HIGH-DENSITY N-CHANNEL SILICON-GATE)

PARALLEL INTERFACE/TIMER

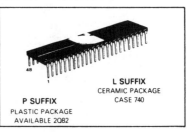

P SUFFIX
PLASTIC PACKAGE
AVAILABLE 2Q82

L SUFFIX
CERAMIC PACKAGE
CASE 740

PIN ASSIGNMENT

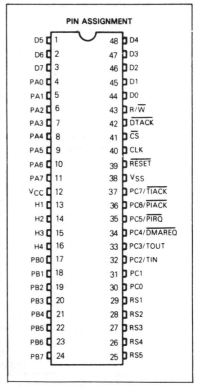

D5 — 1	48 — D4
D6 — 2	47 — D3
D7 — 3	46 — D2
PA0 — 4	45 — D1
PA1 — 5	44 — D0
PA2 — 6	43 — R/$\overline{W}$
PA3 — 7	42 — $\overline{DTACK}$
PA4 — 8	41 — $\overline{CS}$
PA5 — 9	40 — CLK
PA6 — 10	39 — $\overline{RESET}$
PA7 — 11	38 — VSS
VCC — 12	37 — PC7/$\overline{TIACK}$
H1 — 13	36 — PC6/$\overline{PIACK}$
H2 — 14	35 — PC5/$\overline{PIRQ}$
H3 — 15	34 — PC4/$\overline{DMAREQ}$
H4 — 16	33 — PC3/TOUT
PB0 — 17	32 — PC2/TIN
PB1 — 18	31 — PC1
PB2 — 19	30 — PC0
PB3 — 20	29 — RS1
PB4 — 21	28 — RS2
PB5 — 22	27 — RS3
PB6 — 23	26 — RS4
PB7 — 24	25 — RS5

 **MOTOROLA**

MC6821
(1.0 MHz)

MC68A21
(1.5 MHz)

MC68B21
(2.0 MHz)

PERIPHERAL INTERFACE ADAPTER (PIA)

The MC6821 Peripheral Interface Adapter provides the universal means of interfacing peripheral equipment to the M6800 family of microprocessors. This device is capable of interfacing the MPU to peripherals through two 8-bit bidirectional peripheral data buses and four control lines. No external logic is required for interfacing to most peripheral devices.

The functional configuration of the PIA is programmed by the MPU during system initialization. Each of the peripheral data lines can be programmed to act as an input or output, and each of the four control/interrupt lines may be programmed for one of several control modes. This allows a high degree of flexibility in the overall operation of the interface.

- 8-Bit Bidirectional Data Bus for Communication with the MPU
- Two Bidirectional 8-Bit Buses for Interface to Peripherals
- Two Programmable Control Registers
- Two Programmable Data Direction Registers
- Four Individually-Controlled Interrupt Input Lines; Two Usable as Peripheral Control Outputs
- Handshake Control Logic for Input and Output Peripheral Operation
- High-Impedance Three-State and Direct Transistor Drive Peripheral Lines
- Program Controlled Interrupt and Interrupt Disable Capability
- CMOS Drive Capability on Side A Peripheral Lines
- Two TTL Drive Capability on All A and B Side Buffers
- TTL-Compatible
- Static Operation

MOS
(N-CHANNEL, SILICON-GATE, DEPLETION LOAD)

PERIPHERAL INTERFACE ADAPTER

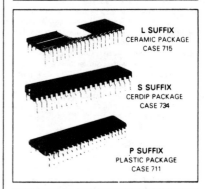

L SUFFIX
CERAMIC PACKAGE
CASE 715

S SUFFIX
CERDIP PACKAGE
CASE 734

P SUFFIX
PLASTIC PACKAGE
CASE 711

PIN ASSIGNMENT

V_{SS}	1 ●	40	CA1
PA0	2	39	CA2
PA1	3	38	$\overline{IRQA}$
PA2	4	37	$\overline{IRQB}$
PA3	5	36	RS0
PA4	6	35	RS1
PA5	7	34	$\overline{RESET}$
PA6	8	33	D0
PA7	9	32	D1
PB0	10	31	D2
PB1	11	30	D3
PB2	12	29	D4
PB3	13	28	D5
PB4	14	27	D6
PB5	15	26	D7
PB6	16	25	E
PB7	17	24	CS1
CB1	18	23	$\overline{CS2}$
CB2	19	22	CS0
V_{CC}	20	21	$R/\overline{W}$

MAXIMUM RATINGS

Characteristics	Symbol	Value	Unit
Supply Voltage	V_{CC}	−0.3 to +7.0	V
Input Voltage	V_{in}	−0.3 to +7.0	V
Operating Temperature Range MC6821, MC68A21, MC68B21 MC6821C, MC68A21C, MC68B21C	T_A	T_L to T_H 0 to 70 −40 to +85	°C
Storage Temperature Range	T_{stg}	−55 to +150	°C

THERMAL CHARACTERISTICS

Characteristic	Symbol	Value	Unit
Thermal Resistance Ceramic Plastic Cerdip	θ_{JA}	50 100 60	°C/W

This device contains circuitry to protect the inputs against damage due to high static voltages or electric fields; however, it is advised that normal precautions be taken to avoid application of any voltage higher than maximum-rated voltages to this high-impedance circuit. Reliability of operation is enhanced if unused inputs are tied to an appropriate logic voltage (i.e., either V_{SS} or V_{CC}).

The expanded block diagram of the MC6821 is shown in Figure B.1.

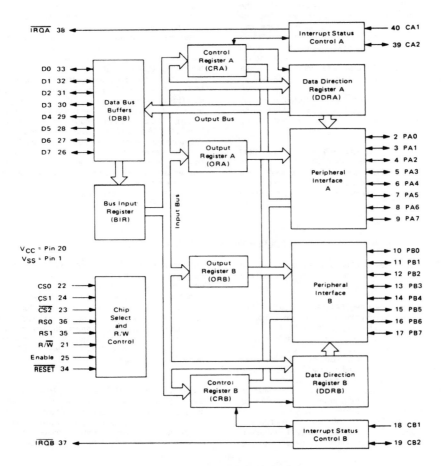

FIGURE B.1 EXPANDED BLOCK DIAGRAM

PIA INTERFACE SIGNALS FOR MPU

The PIA interfaces to the M6800 bus with an 8-bit bidirectional data bus, three chip select lines, two register select lines, two interrupt request lines, a read/write line, an enable line and a reset line. To ensure proper operation with the MC6800, MC6802, or MC6808 microprocessors, VMA should be used as an active part of the address decoding.

Bidirectional Data (D0-D7) — The bidirectional data lines (D0-D7) allow the transfer of data between the MPU and the PIA. The data bus output drivers are three-state devices that remain in the high-impedance (off) state except when the MPU performs a PIA read operation. The read/write line is in the read (high) state when the PIA is selected for a read operation.

Enable (E) — The enable pulse, E, is the only timing signal that is supplied to the PIA. Timing of all other signals is referenced to the leading and trailing edges of the E pulse.

Read/Write (R/W̄) — This signal is generated by the MPU to control the direction of data transfers on the data bus. A low state on the PIA read/write line enables the input buffers and data is transferred from the MPU to the PIA on the E signal if the device has been selected. A high on the read/write line sets up the PIA for a transfer of data to the bus. The PIA output buffers are enabled when the proper address and the enable pulse E are present.

RESET — The active low RESET line is used to reset all register bits in the PIA to a logical zero (low). This line can be used as a power-on reset and as a master reset during system operation.

Chip Selects (CS0, CS1, and CS2̄) — These three input signals are used to select the PIA. CS0 and CS1 must be high and CS2̄ must be low for selection of the device. Data transfers are then performed under the control of the enable and read/write signals. The chip select lines must be stable for the duration of the E pulse. The device is deselected when any of the chip selects are in the inactive state.

Register Selects (RS0 and RS1) — The two register select lines are used to select the various registers inside the PIA. These two lines are used in conjunction with internal Control Registers to select a particular register that is to be written or read.

The register and chip select lines should be stable for the duration of the E pulse while in the read or write cycle.

Interrupt Request (IRQA and IRQB) — The active low Interrupt Request lines (IRQA and IRQB) act to interrupt the MPU either directly or through interrupt priority circuitry. These lines are "open drain" (no load device on the chip). This permits all interrupt request lines to be tied together in a wire-OR configuration.

Each Interrupt Request line has two internal interrupt flag bits that can cause the Interrupt Request line to go low. Each flag bit is associated with a particular peripheral interrupt line. Also, four interrupt enable bits are provided in the PIA which may be used to inhibit a particular interrupt from a peripheral device.

Servicing an interrupt by the MPU may be accomplished by a software routine that, on a prioritized basis, sequentially reads and tests the two control registers in each PIA for interrupt flag bits that are set.

The interrupt flags are cleared (zeroed) as a result of an MPU Read Peripheral Data Operation of the corresponding data register. After being cleared, the interrupt flag bit cannot be enabled to be set until the PIA is deselected during an E pulse. The E pulse is used to condition the interrupt control lines (CA1, CA2, CB1, CB2). When these lines are used as interrupt inputs, at least one E pulse must occur from the inactive edge to the active edge of the interrupt input signal to condition the edge sense network. If the interrupt flag has been enabled and the edge sense circuit has been properly conditioned, the interrupt flag will be set on the next active transition of the interrupt input pin.

PIA PERIPHERAL INTERFACE LINES

The PIA provides two 8-bit bidirectional data buses and four interrupt/control lines for interfacing to peripheral devices.

Section A Peripheral Data (PA0-PA7) — Each of the peripheral data lines can be programmed to act as an input or output. This is accomplished by setting a "1" in the corresponding Data Direction Register bit for those lines which are to be outputs. A "0" in a bit of the Data Direction Register causes the corresponding peripheral data line to act as an input. During an MPU Read Peripheral Data Operation, the data on peripheral lines programmed to act as inputs appears directly on the corresponding MPU Data Bus lines. In the input mode, the internal pullup resistor on these lines represents a maximum of 1.5 standard TTL loads.

The data in Output Register A will appear on the data lines that are programmed to be outputs. A logical "1" written into the register will cause a "high" on the corresponding data line while a "0" results in a "low." Data in Output Register A may be read by an MPU "Read Peripheral Data A" operation when the corresponding lines are programmed as outputs. This data will be read property if the voltage on the peripheral data lines is greater than 2.0 volts for a logic "1" output and less than 0.8 volt for a logic "0" output. Loading the output lines such that the voltage on these lines does not reach full voltage causes the data transferred into the MPU on a Read operation to differ from that contained in the respective bit of Output Register A.

Section B Peripheral Data (PB0-PB7) — The peripheral data lines in the B Section of the PIA can be programmed to act as either inputs or outputs in a similar manner to PA0-PA7. They have three-state capabiity, allowing them to enter a high-impedance state when the peripheral data line is used as an input. In addition, data on the peripheral data lines

PB0-PB7 will be read properly from those lines programmed as outputs even if the voltages are below 2.0 volts for a "high" or above 0.8 V for a "low". As outputs, these lines are compatible with standard TTL and may also be used as a source of up to 1 milliampere at 1.5 volts to directly drive the base of a transistor switch.

Interrupt Input (CA1 and CB1) — Peripheral input lines CA1 and CB1 are input only lines that set the interrupt flags of the control registers. The active transition for these signals is also programmed by the two control registers.

Peripheral Control (CA2) — The peripheral control line CA2 can be programmed to act as an interrupt input or as a peripheral control output. As an output, this line is compatible with standard TTL; as an input the internal pullup resistor on this line represents 1.5 standard TTL loads. The function of this signal line is programmed with Control Register A.

Peripheral Control (CB2) — Peripheral Control line CB2 may also be programmed to act as an interrupt input or peripheral control output. As an input, this line has high input impedance and is compatible with standard TTL. As an output it is compatible with standard TTL and may also be used as a source of up to 1 milliampere at 1.5 volts to directly drive the base of a transistor switch. This line is programmed by Control Register B.

INTERNAL CONTROLS

INITIALIZATION

A $\overline{\text{RESET}}$ has the effect of zeroing all PIA registers. This will set PA0-PA7, PB0-PB7, CA2 and CB2 as inputs, and all interrupts disabled. The PIA must be configured during the restart program which follows the reset.

There are six locations within the PIA accessible to the MPU data bus: two Peripheral Registers, two Data Direction Registers, and two Control Registers. Selection of these locations is controlled by the RS0 and RS1 inputs together with bit 2 in the Control Register, as shown in Table B.1

Details of possible configurations of the Data Direction and Control Register are as follows:

TABLE B.1 INTERNAL ADDRESSING

RS1	RS0	Control Register Bit CRA-2	Control Register Bit CRB-2	Location Selected
0	0	1	X	Peripheral Register A
0	0	0	X	Data Direction Register A
0	1	X	X	Control Register A
1	0	X	1	Peripheral Register B
1	0	X	0	Data Direction Register B
1	1	X	X	Control Register B

X = Don't Care

PORT A-B HARDWARE CHARACTERISTICS

As shown in Figure .17, the MC6821 has a pair of I/O ports whose characteristics differ greatly. The A side is designed to drive CMOS logic to normal 30% to 70% levels, and incorporates an internal pullup device that remains connected even in the input mode. Because of this, the A side requires more drive current in the input mode than Port B. In contrast, the B side uses a normal three-state NMOS buffer which cannot pullup to CMOS levels without external resistors. The B side can drive extra loads such as Darlingtons without problem. When the PIA comes out of reset, the A port represents inputs with pullup resistors, whereas the B side (input mode also) will float high or low, depending upon the load connected to it.

Notice the differences between a Port A and Port B read operation when in the output mode. When reading Port A, the actual pin is read, whereas the B side read comes from an output latch, ahead of the actual pin.

CONTROL REGISTERS (CRA and CRB)

The two Control Registers (CRA and CRB) allow the MPU to control the operation of the four peripheral control lines CA1, CA2, CB1, and CB2. In addition they allow the MPU to enable the interrupt lines and monitor the status of the interrupt flags. Bits 0 through 5 of the two registers may be written or read by the MPU when the proper chip select and register select signals are applied. Bits 6 and 7 of the two registers are read only and are modified by external interrupts occurring on control lines CA1, CA2, CB1, or CB2. The format of the control words is shown in Figure B.3

DATA DIRECTION ACCESS CONTROL BIT (CRA-2 and CRB-2)

Bit 2, in each Control Register (CRA and CRB), determines selection of either a Peripheral Output Register or the corresponding Data Direction E Register when the proper register select signals are applied to RS0 and RS1. A "1" in bit 2 allows access of the Peripheral Interface Register, while a "0" causes the Data Direction Register to be addressed.

Interrupt Flags (CRA-6, CRA-7, CRB-6, and CRB-7) — The four interrupt flag bits are set by active transitions of signals on the four Interrupt and Peripheral Control lines when those lines are programmed to be inputs. These bits cannot be set directly from the MPU Data Bus and are reset indirectly by a Read Peripheral Data Operation on the appropriate section.

Control of CA2 and CB2 Peripheral Control Lines (CRA-3, CRA-4, CRA-5, CRB-3, CRB-4, and CRB-5) — Bits 3, 4, and 5 of the two control registers are used to control the CA2 and CB2 Peripheral Control lines. These bits determine if the control lines will be an interrupt input or an output control signal. If bit CRA-5 (CRB-5) is low, CA2 (CB2) is an interrupt input line similar to CA1 (CB1). When CRA-5 (CRB-5) is high, CA2 (CB2) becomes an output signal that may be used to control peripheral data transfers. When in the output mode, CA2 and CB2 have slightly different loading characteristics.

Control of CA1 and CB1 Interrupt Input Lines (CRA-0, CRB-1, CRA-1, and CRB-1) — The two lowest-order bits of the control registers are used to control the interrupt input lines CA1 and CB1. Bits CRA-0 and CRB-0 are used to enable the MPU interrupt signals $\overline{IRQA}$ and $\overline{IRQB}$, respectively. Bits CRA-1 and CRB-1 determine the active transition of the interrupt input signals CA1 and CB1.

FIGURE B.2 PORT A AND PORT B EQUIVALENT CIRCUITS

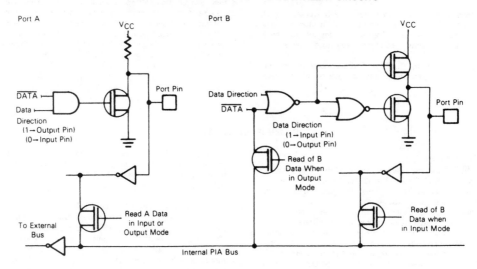

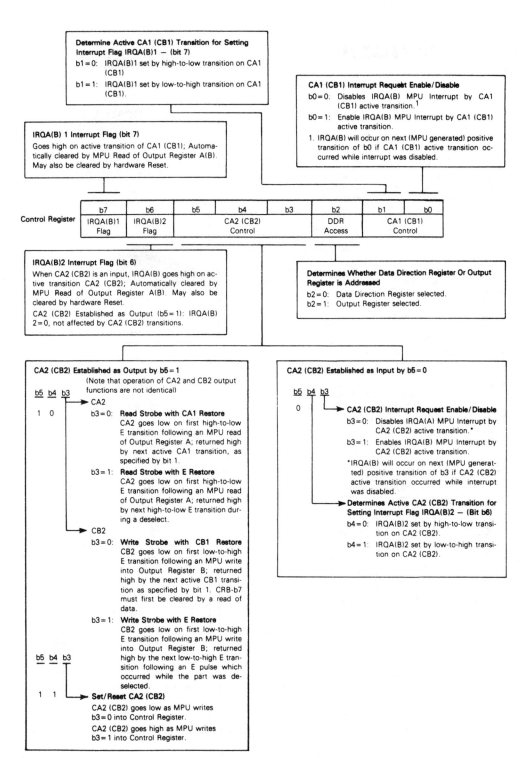

FIGURE B.3 CONTROL WORD FORMAT

 **MOTOROLA**

MCM6116

16K BIT STATIC RANDOM ACCESS MEMORY

The MCM6116 is a 16,384-bit Static Random Access Memory organized as 2048 words by 8 bits, fabricated using Motorola's high-performance silicon-gate CMOS (HCMOS) technology. It uses a design approach which provides the simple timing features associated with fully static memories and the reduced power associated with CMOS memories. This means low standby power without the need for clocks, nor reduced data rates due to cycle times that exceed access time.

Chip Enable ($\overline{E}$) controls the power-down feature. It is not a clock but rather a chip control that affects power consumption. In less than a cycle time after Chip Enable ($\overline{E}$) goes high, the part automatically reduces its power requirements and remains in this low-power standby as long as the Chip Enable ($\overline{E}$) remains high. The automatic power-down feature causes no performance degradation.

The MCM6116 is in a 24-pin dual-in-line package with the industry standard JEDEC approved pinout and is pinout compatible with the industry standard 16K EPROM/ROM.

- Single +5 V Supply
- 2048 Words by 8-Bit Operation
- HCMOS Technology
- Fully Static: No Clock or Timing Strobe Required
- Maximum Access Time: MCM6116-12 — 120 ns
  MCM6116-15 — 150 ns
  MCM6116-20 — 200 ns
- Power Dissipation: 70 mA Maximum (Active)
  15 mA Maximum (Standby-TTL Levels)
  2 mA Maximum (Standby)
- Low Power Version Also Available — MCM61L16
- Low Voltage Data Retention (MCM61L16 Only):
  50 µA Maximum

HCMOS
(COMPLEMENTARY MOS)

**2,048 × 8 BIT
STATIC RANDOM
ACCESS MEMORY**

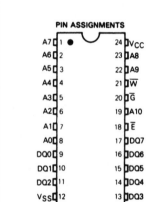

P SUFFIX
PLASTIC PACKAGE
CASE 709

PIN ASSIGNMENTS

A7	1 ●	24	V_CC
A6	2	23	A8
A5	3	22	A9
A4	4	21	$\overline{W}$
A3	5	20	$\overline{G}$
A2	6	19	A10
A1	7	18	$\overline{E}$
A0	8	17	DQ7
DQ0	9	16	DQ6
DQ1	10	15	DQ5
DQ2	11	14	DQ4
V_SS	12	13	DQ3

PIN NAMES

A0-A10	Address Input
DQ0-DQ7	Data Input/Output
$\overline{W}$	Write Enable
$\overline{G}$	Output Enable
$\overline{E}$	Chip Enable
V_CC	Power (+5 V)
V_SS	Ground

BLOCK DIAGRAM

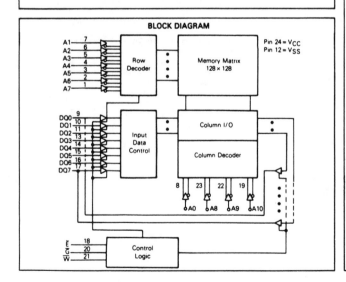

ABSOLUTE MAXIMUM RATINGS (See Note)

Rating	Value	Unit
Temperature Under Bias	− 10 to + 80	°C
Voltage on Any Pin With Respect to V$_{SS}$	− 1.0 to + 7.0	V
DC Output Current	20	mA
Power Dissipation	1.2	Watt
Operating Temperature Range	0 to + 70	°C
Storage Temperature Range	− 65 to + 150	°C

This device contains circuitry to protect the inputs against damage due to high static voltages or electric fields; however, it is advised that normal precautions be taken to avoid application of any voltage higher than maximum rated voltages to this high-impedance circuit.

NOTE: Permanent device damage may occur if ABSOLUTE MAXIMUM RATINGS are exceeded. Functional operation should be restricted to RECOMMENDED OPERATING CONDITIONS. Exposure to higher than recommended voltages for extended periods of time could affect device reliability.

DC OPERATING CONDITIONS AND CHARACTERISTICS
(Full operating voltage and temperature ranges unless otherwise noted.)

RECOMMENDED OPERATING CONDITIONS

Parameter	Symbol	Min	Typ	Max	Unit
Supply Voltage	V$_{CC}$	4.5	5.0	5.5	V
	V$_{SS}$	0	0	0	V
Input Voltage	V$_{IH}$	2.2	3.5	6.0	V
	V$_{IL}$	− 1.0*	−	0.8	V

*The device will withstand undershoots to the − 1.0 volt level with a maximum pulse width of 50 ns at the − 0.3 volt level. This is periodically sampled rather than 100% tested.

RECOMMENDED OPERATING CHARACTERISTICS

Parameter	Symbol	MCM6116 Min	MCM6116 Typ*	MCM6116 Max	MCM61L16 Min	MCM61L16 Typ*	MCM61L16 Max	Unit
Input Leakage Current (V$_{CC}$ = 5.5 V, V$_{in}$ = GND to V$_{CC}$)	|I$_{LI}$|	−	−	1	−	−	1	μA
Output Leakage Current ($\overline{E}$ = V$_{IH}$ or $\overline{G}$ = V$_{IH}$ V$_{I/O}$ = GND to V$_{CC}$)	|I$_{LO}$|	−	−	1	−	−	1	μA
Operating Power Supply Current ($\overline{E}$ = V$_{IL}$, I$_{I/O}$ = 0 mA)	I$_{CC}$	−	35	70	−	35	55	mA
Average Operating Current Minimum cycle, duty = 100%	I$_{CC2}$	−	35	70	−	35	55	mA
Standby Power ($\overline{E}$ = V$_{IH}$)	I$_{SB}$	−	5	15	−	5	12	mA
Supply Current ($\overline{E}$ ≥ V$_{CC}$ − 0.2 V, V$_{in}$ ≥ V$_{CC}$ − 0.2 V or V$_{in}$ ≤ 0.2 V)	I$_{SB1}$	−	20	2000	−	4	100	μA
Output Low Voltage (I$_{OL}$ = 2.1 mA)	V$_{OL}$	−	−	0.4	−	−	0.4	V
Output High Voltage (I$_{OH}$ = − 1.0 mA)**	V$_{OH}$	2.4	−	−	2.4	−	−	V

*V$_{CC}$ = 5 V, T$_A$ = 25°C
**Also, output voltages are compatible with Motorola's new high-speed CMOS logic family if the same power supply voltage is used.

CAPACITANCE (f = 1.0 MHz, T$_A$ = 25°C, periodically sampled rather than 100% tested.)

Characteristic	Symbol	Typ	Max	Unit
Input Capacitance except $\overline{E}$	C$_{in}$	3	5	pF
Input/Output Capacitance and $\overline{E}$ Input Capacitance	C$_{I/O}$	5	7	pF

MODE SELECTION

Mode	$\overline{E}$	$\overline{G}$	$\overline{W}$	V$_{CC}$ Current	DQ
Standby	H	X	X	I$_{SB}$, I$_{SB1}$	High Z
Read	L	L	H	I$_{CC}$	Q
Write Cycle (1)	L	H	L	I$_{CC}$	D
Write Cycle (2)	L	L	L	I$_{CC}$	D

AC OPERATING CONDITIONS AND CHARACTERISTICS

(Full operating voltage and temperature unless otherwise noted.)

Input Pulse Levels .... 0 Volt to 3.5 Volts Input and Output Timing Reference Levels .... 1.5 Volts

Input Rise and Fall Times .... 10 ns Output Load .... 1 TTL Gate and $C_L = 100$ pF

READ CYCLE

Parameter	Symbol	MCM6116-12 MCM61L16-12 Min	Max	MCM6116-15 MCM61L16-15 Min	Max	MCM6116-20 MCM61L16-20 Min	Max	Unit
Address Valid to Address Don't Care (Cycle Time when Chip Enable is Held Active)	t_{AVAX}	120	–	150	–	200	–	ns
Chip Enable Low to Chip Enable High	t_{ELEH}	120	–	150	–	200	–	ns
Address Valid to Output Valid (Access)	t_{AVQV}	–	120	–	150	–	200	ns
Chip Enable Low to Output Valid (Access)	t_{ELQV}	–	120	–	150	–	200	ns
Address Valid to Output Invalid	t_{AVQX}	10	–	15	–	15	–	ns
Chip Enable Low to Output Invalid	t_{ELQX}	10	–	15	–	15	–	ns
Chip Enable High to Output High Z	t_{EHQZ}	0	40	0	50	0	60	ns
Output Enable to Output Valid	t_{GLQV}	–	80	–	100	–	120	ns
Output Enable to Output Invalid	t_{GLQX}	10	–	15	–	15	–	ns
Output Enable to Output High Z	t_{GLQZ}	0	40	0	50	0	60	ns
Address Invalid to Output Invalid	t_{AXQX}	10	–	15	–	15	–	ns
Address Valid to Chip Enable Low (Address Setup)	t_{AVEL}	0	–	0	–	0	–	ns
Chip Enable to Power-Up Time	t_{PU}	0	–	0	–	0	–	ns
Chip Disable to Power-Down Time	t_{PD}	–	30	–	30	–	30	ns

WRITE CYCLE

Parameter	Symbol	MCM6116-12 MCM61L16-12 Min	Max	MCM6116-15 MCM61L16-15 Min	Max	MCM6116-20 MCM61L16-20 Min	Max	Unit
Chip Enable Low to Write High	t_{ELWH}	70	–	90	–	120	–	ns
Address Valid to Write High	t_{AVWH}	105	–	120	–	140	–	ns
Address Valid to Write Low (Address Setup)	t_{AVWL}	20	–	20	–	20	–	ns
Write Low to Write High (Write Pulse Width)	t_{WLWH}	70	–	90	–	120	–	ns
Write High to Address Don't Care	t_{WHAX}	5	–	10	–	10	–	ns
Data Valid to Write High	t_{DVWH}	35	–	40	–	60	–	ns
Write High to Data Don't Care (Data Hold)	t_{WHDX}	5	–	10	–	10	–	ns
Write Low to Output High Z	t_{WLQZ}	0	50	0	60	0	60	ns
Write High to Output Valid	t_{WHQV}	5	–	10	–	10	–	ns
Output Disable to Output High Z	t_{GHQZ}	0	40	0	50	0	60	ns

TIMING PARAMETER ABBREVIATIONS

```
                                   t X X X X
signal name from which interval is defined ──┘ | | |
       transition direction for first signal ────┘ | |
signal name to which interval is defined ──────────┘ |
  transition direction for second signal ────────────┘
```

The transition definitions used in this data sheet are:

H = transition to high
L = transition to low
V = transition to valid
X = transition to invalid or don't care
Z = transition to off (high impedance)

TIMING LIMITS

The table of timing values shows either a minimum or a maximum limit for each parameter. Input requirements are specified from the external system point of view. Thus, address setup time is shown as a minimum since the system must supply at least that much time (even though most devices do not require it). On the other hand, responses from the memory are specified from the device point of view. Thus, the access time is shown as a maximum since the device never provides data later than that time.

intel® 8085

X₁	1	40	V_cc

Pin layout:

```
X₁        1    40   V_CC
X₂        2    39   HOLD
RESET OUT 3    38   HLDA
SOD       4    37   CLK (OUT)
SID       5    36   RESET IN
TRAP      6    35   READY
RST 7.5   7    34   IO/M
RST 6.5   8    33   S₁
RST 5.5   9    32   RD
INTR     10    31   WR
INTA     11    30   ALE
AD₀      12    29   S₀
AD₁      13    28   A₁₅
AD₂      14    27   A₁₄
AD₃      15    26   A₁₃
AD₄      16    25   A₁₂
AD₅      17    24   A₁₁
AD₆      18    23   A₁₀
AD₇      19    22   A₉
V_SS     20    21   A₈
```

8085A

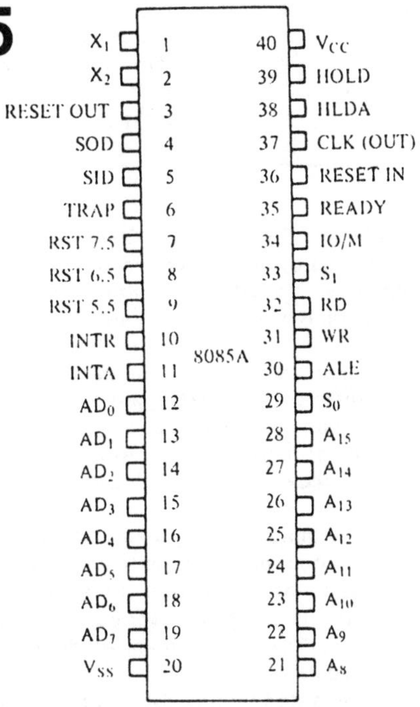

Figure C.1 8085 pinout

Figure C.1 shows 8085 pins and signals. The following table describes the function of each pin:

Symbol	Function
A_8–A_{15} (Output, three-state)	Address bus: The most significant 8 bits of the memory address or the 8 bits of the I/O address.
AD_{0-7} (Input / output, three-state)	Multiplexed address / data bus: Lower 8-bits of the memory address (or I/O address) appear on the bus during the first clock cycle (T state) of a machine cycle. It then becomes the data bus during the second and third clock cycles.
ALE (Output)	Address Latch Enable: It occurs during the first clock state of a machine cycle and enables the address to get latched into the on-chip latch.
S_0, S_1, and IO/$\overline{M}$ (Output)	Machine cycle status:

IO/$\overline{M}$	S_1	S_0	Status
0	0	1	Memory write
0	1	0	Memory read
1	0	1	I/O write

Symbol	Function			
	IO/$\overline{\text{M}}$	S$_1$	S$_0$	Status
	1	1	0	I/O read
	0	1	1	Op code fetch
	1	1	1	Interrupt acknowledge
	*	0	0	Halt
	*	X	X	Hold
	*	X	X	Reset

* = 3-state (high impedance)
X = unspecified

S$_1$ can be used as an advanced R/$\overline{\text{W}}$ status. IO/$\overline{\text{M}}$, S$_0$, and S$_1$ become valid at the beginning of a machine cycle and remain stable throughout the cycle. The falling edge of ALE may be used to latch the state of these lines.

$\overline{\text{RD}}$ (Output, three-state)	READ control: A low level on $\overline{\text{RD}}$ indicates the selected memory or I/O device is to be read.
$\overline{\text{WR}}$ (Output, three-state)	WRITE control: A low level on $\overline{\text{WR}}$ indicates the data on the data bus is to be written into the selected memory or I/O location.
READY (Input)	If READY is high during a read or write cycle, it indicates that the memory or peripheral is ready to send or receive data. If READY is low, the CPU will wait an integral number of clock cycles for READY to go high before completing the read or write cycle.
HOLD (Input)	HOLD indicates that another master is requesting the use of the address and data buses. The CPU, upon receiving the hold request, will relinquish the use of the bus as soon as the completion of the current bus transfer. Internal processing can continue. The processor can regain the bus only after the HOLD is removed. When the HOLD is acknowledged, the address, data, $\overline{\text{RD}}$, $\overline{\text{WR}}$, and IO/$\overline{\text{M}}$ lines are three-stated.
HLDA (Output)	HOLD ACKNOWLEDGE: Indicates that the CPU has received the HOLD request and that it will relinquish the bus in the next clock cycle. HLDA goes low after the HOLD request is removed. The CPU takes the bus one-half clock cycle after HLDA goes low.
INTR (Input)	INTERRUPT REQUEST: Is used as a general-purpose interrupt. It is sampled only during the next to the last clock cycle of an instruction and during HOLD and HALT states. If it is active, the PC will be inhibited from incrementing and an $\overline{\text{INTA}}$ will be issued. During this cycle a RESTART or CALL instruction can be inserted to jump to the interrupt service routine. The INTR is enabled and disabled by software. It is disabled by RESET and immediately after an interrupt is accepted.
$\overline{\text{INTA}}$ (Output)	INTERRUPT ACKNOWLEDGE: Is used instead of (and has the same timing as) $\overline{\text{RD}}$ during the instruction cycle after an INTR is accepted. It can be used to activate the 8259 interrupt chip or some other interrupt port.
RST5.5 RST6.5 RST7.5 (Inputs)	RESTART INTERRUPTS: These three inputs have the same timing as INTR except they cause an internal RESTART to be automatically inserted.
TRAP (Input)	Trap interrupt is a nonmaskable RESTART interrupt. It is recognized at the same time as INTR or RST5.5–7.5. It is unaffected by any mask or interrupt enable. It has the highest priority of any interrupt.
$\overline{\text{RESET IN}}$ (Input)	Sets the program counter to zero and resets the interrupt enable and HLDA flip-flops.
RESET OUT (Output)	Indicates CPU is being reset. Can be used as a system reset.

Symbol	Function
,X₂ (Input)	X₁ and X₂ are connected to a crystal, LC, or RC network to drive the internal clock generator. X₁ can also be an external clock input from a logic gate. The input frequency is divided by 2 to give the processor's internal operating frequency.
.K (Output)	Clock output for use as a system clock. The period of CLK is twice the X₁, X₂ input period.
D (Input)	Serial Input Data line. The data on this line is loaded into accumulator bit 7 whenever a RIM instruction is executed.
)D (Output)	Serial Output Data line. The output SOD is set or reset as specified by the SIM instruction.
c	+5 V supply.
s	'Ground reference.

8086/8086-2/8086-4
16-BIT HMOS MICROPROCESSOR

- Direct Addressing Capability to 1 MByte of Memory

- Assembly Language Compatible with 8080/8085

- 14 Word, By 16-Bit Register Set with Symmetrical Operations

- 24 Operand Addressing Modes

- Bit, Byte, Word, and Block Operations

- 8-and 16-Bit Signed and Unsigned Arithmetic in Binary or Decimal Including Multiply and Divide

- 5 MHz Clock Rate (8 MHz for 8086-2) (4 MHz for 8086-4)

- MULTIBUS™ System Compatible Interface

The Intel® 8086 is a new generation, high performance microprocessor implemented in N-channel, depletion load, silicon gate technology (HMOS), and packaged in a 40-pin CerDIP package. The processor has attributes of both 8- and 16-bit microprocessors. It addresses memory as a sequence of 8-bit bytes, but has a 16-bit wide physical path to memory for high performance.

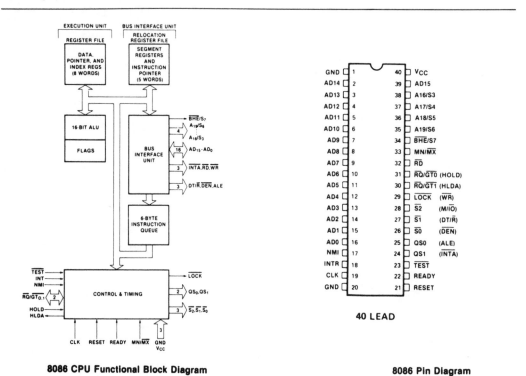

8086 CPU Functional Block Diagram

8086 Pin Diagram

I8284
CLOCK GENERATOR AND DRIVER
FOR 8086, 8088, 8089 PROCESSORS

- Generates the System Clock for the 8086, 8088 and 8089
- Uses a Crystal or a TTL Signal for Frequency Source
- Single +5V Power Supply
- 18-Pin Package

- Generates System Reset Output from Schmitt Trigger Input
- Provides Local Ready and MULTIBUS™ Ready Synchronization
- Capable of Clock Synchronization with other 8284's
- Industrial Temperature Range −40° to +85°C

The I8284 is a bipolar clock generator/driver designed to provide clock signals for the 8086, 8088 & 8089 and peripherals. It also contains READY logic for operation with two MULTIBUS™ systems and provides the processors required READY synchronization and timing. Reset logic with hysteresis and synchronization is also provided.

I8284 PIN CONFIGURATION

CYSNC	1	18	Vcc
PCLK	2	17	X1
AEN1	3	16	X2
RDY1	4	15	TNK
READY	5	14	EFI
RDY2	6	13	F/C̄
AEN2	7	12	OSC
CLK	8	11	RES
GND	9	10	RESET

I8284 BLOCK DIAGRAM

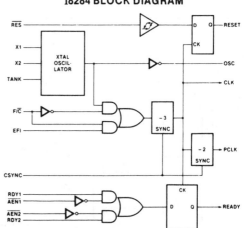

I8284 PIN NAMES

X1 X2	CONNECTIONS FOR CRYSTAL
TANK	USED WITH OVERTONE CRYSTAL
F/C̄	CLOCK SOURCE SELECT
EFI	EXTERNAL CLOCK INPUT
CSYNC	CLOCK SYNCHRONIZATION INPUT
RDY1 RDY2	READY SIGNAL FROM TWO MULTIBUS™ SYSTEMS
AEN1 AEN2	ADDRESS ENABLED QUALIFIERS FOR RDY1,2
RES	RESET INPUT
RESET	SYNCHRONIZED RESET OUTPUT
OSC	OSCILLATOR OUTPUT
CLK	MOS CLOCK FOR THE PROCESSOR
PCLK	TTL CLOCK FOR PERIPHERALS
READY	SYNCHRONIZED READY OUTPUT
Vcc	+5 VOLTS
GND	0 VOLTS

8288
BUS CONTROLLER
FOR 8086, 8088, 8089 PROCESSORS

- Bipolar Drive Capability

- Provides Advanced Commands

- Provides Wide Flexibility in System Configurations

- 3-State Command Output Drivers

- Configurable for Use with an I/O Bus

- Facilitates Interface to One or Two Multi-Master Busses

The Intel® 8288 Bus Controller is a 20-pin bipolar component for use with medium-to-large 8086 processing systems. The bus controller provides command and control timing generation as well as bipolar bus drive capability while optimizing system performance.

A strapping option on the bus controller configures it for use with a multi-master system bus and separate I/O bus.

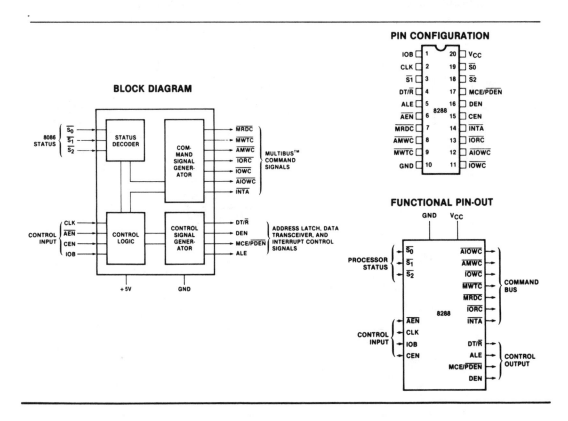

2716
16K (2K × 8) UV ERASABLE PROM

- **Fast Access Time**
  - **350 ns Max. 2716-1**
  - **390 ns Max. 2716-2**
  - **450 ns Max. 2716**
  - **650 ns Max. 2716-6**

- **Single +5V Power Supply**

- **Low Power Dissipation**
  - **525 mW Max. Active Power**
  - **132 mW Max. Standby Power**

- **Pin Compatible to Intel® 2732 EPROM**

- **Simple Programming Requirements**
  - **Single Location Programming**
  - **Programs with One 50 ms Pulse**

- **Inputs and Outputs TTL Compatible during Read and Program**

- **Completely Static**

The Intel® 2716 is a 16,384-bit ultraviolet erasable and electrically programmable read-only memory (EPROM). The 2716 operates from a single 5-volt power supply, has a static standby mode, and features fast single address location programming. It makes designing with EPROMs faster, easier and more economical.

The 2716, with its single 5-volt supply and with an access time up to 350 ns, is ideal for use with the newer high performance +5V microprocessors such as Intel's 8085 and 8086. The 2716 is also the first EPROM with a static standby mode which reduces the power dissipation without increasing access time. The maximum active power dissipation is 525 mW while the maximum standby power dissipation is only 132 mW, a 75% savings.

The 2716 has the simplest and fastest method yet devised for programming EPROMs — single pulse TTL level programming. No need for high voltage pulsing because all programming controls are handled by TTL signals. Program any location at any time—either individually, sequentially or at random, with the 2716's single address location programming. Total programming time for all 16,384 bits is only 100 seconds.

PIN CONFIGURATION

2716

A_7	1	24	V_{CC}
A_6	2	23	A_8
A_5	3	22	A_9
A_4	4	21	V_{PP}
A_3	5	20	$\overline{OE}$
A_2	6	19	A_{10}
A_1	7	18	$\overline{CE}$
A_0	8	17	O_7
O_0	9	16	O_6
O_1	10	15	O_5
O_2	11	14	O_4
GND	12	13	O_3

(16K)

2732†

A_7	1	24	V_{CC}
A_6	2	23	A_8
A_5	3	22	A_9
A_4	4	21	A_{11}
A_3	5	20	$\overline{OE}/V_{PP}$
A_2	6	19	A_{10}
A_1	7	18	$\overline{CE}$
A_0	8	17	O_7
O_0	9	16	O_6
O_1	10	15	O_5
O_2	11	14	O_4
GND	12	13	O_3

(32K)

†Refer to 2732 data sheet for specifications

PIN NAMES

$A_0 - A_{10}$	ADDRESSES
$\overline{CE}/PGM$	CHIP ENABLE/PROGRAM
$\overline{OE}$	OUTPUT ENABLE
$O_0 - O_7$	OUTPUTS

MODE SELECTION

PINS / MODE	$\overline{CE}/PGM$ (18)	$\overline{OE}$ (20)	V_{PP} (21)	V_{CC} (24)	OUTPUTS (9-11, 13-17)
Read	V_{IL}	V_{IL}	+5	+5	D_{OUT}
Standby	V_{IH}	Don't Care	+5	+5	High Z
Program	Pulsed V_{IL} to V_{IH}	V_{IH}	+25	+5	D_{IN}
Program Verify	V_{IL}	V_{IL}	+25	+5	D_{OUT}
Program Inhibit	V_{IL}	V_{IH}	+25	+5	High Z

BLOCK DIAGRAM

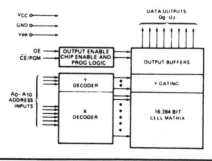

2732
32K (4K x 8) UV ERASABLE PROM

- **Fast Access Time:**
  - **450 ns Max. 2732**
  - **550 ns Max. 2732-6**

- **Single +5V ± 5% Power Supply**

- **Output Enable for MCS-85™ and MCS-86™ Compatibility**

- **Low Power Dissipation:**
  150mA Max. Active Current
  30mA Max. Standby Current

- **Pin Compatible to Intel® 2716 EPROM**

- **Completely Static**

- **Simple Programming Requirements**
  - **Single Location Programming**
  - **Programs with One 50ms Pulse**

- **Three-State Output for Direct Bus Interface**

The Intel® 2732 is a 32,768-bit ultraviolet erasable and electrically programmable read-only memory (EPROM). The 2732 operates from a single 5-volt power supply, has a standby mode, and features an output enable control. The total programming time for all bits is three and a half minutes. All these features make designing with the 2732 in microcomputer systems faster, easier, and more economical.

An important 2732 feature is the separate output control, Output Enable ($\overline{OE}$), from the Chip Enable control ($\overline{CE}$). The $\overline{OE}$ control eliminates bus contention in multiple bus microprocessor systems. Intel's Application Note AP-30 describes the microprocessor system implementation of the $\overline{OE}$ and $\overline{CE}$ controls on Intel's 2716 and 2732 EPROMs. AP-30 is available from Intel's Literature Department.

The 2732 has a standby mode which reduces the power dissipation without increasing access time. The maximum active current is 150mA, while the maximum standby current is only 30mA, an 80% savings. The standby mode is achieved by applying a TTL-high signal to the $\overline{CE}$ input.

PIN CONFIGURATION

A_7	1	24	V_{CC}
A_6	2	23	A_8
A_5	3	22	A_9
A_4	4	21	A_{11}
A_3	5	20	$\overline{OE}/V_{PP}$
A_2	6	19	A_{10}
A_1	7	18	$\overline{CE}$
A_0	8	17	O_7
O_0	9	16	O_6
O_1	10	15	O_5
O_2	11	14	O_4
GND	12	13	O_3

PIN NAMES

A_0-A_{11}	ADDRESSES
$\overline{CE}$	CHIP ENABLE
$\overline{OE}$	OUTPUT ENABLE
O_0-O_7	OUTPUTS

MODE SELECTION

PINS MODE	$\overline{CE}$ (18)	$\overline{OE}/V_{PP}$ (20)	V_{CC} (24)	OUTPUTS (9-11,13-17)
Read	V_{IL}	V_{IL}	+5	D_{OUT}
Standby	V_{IH}	Don't Care	+5	High Z
Program	V_{IL}	V_{PP}	+5	D_{IN}
Program Verify	V_{IL}	V_{IL}	+5	D_{OUT}
Program Inhibit	V_{IH}	V_{PP}	+5	High Z

BLOCK DIAGRAM

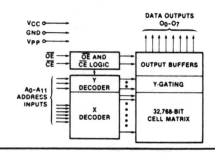

8355/8355-2
16,384-BIT ROM WITH I/O

- **2048 Words × 8 Bits**
- **Single +5V Power Supply**
- **Directly compatible with 8085A and 8088 Microprocessors**
- **2 General Purpose 8-Bit I/O Ports**

- **Each I/O Port Line Individually Programmable as Input or Output**
- **Multiplexed Address and Data Bus**
- **Internal Address Latch**
- **40-Pin DIP**

The Intel® 8355 is a ROM and I/O chip to be used in the 8085A and 8088 microprocessor systems. The ROM portion is organized as 2048 words by 8 bits. It has a maximum acess time of 400 ns to permit use with no wait states in the 8085A CPU.

The I/O portion consists of 2 general purpose I/O ports. Each I/O port has 8 port lines and each I/O port line is individually programmable as input or output.

The 8355-2 has a 300ns access time for compatibility with the 8085A-2 and full speed 5 MHz 8088 microprocessors.

PIN CONFIGURATION

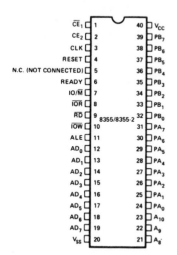

BLOCK DIAGRAM

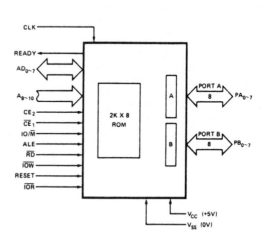

8355/8355-2

Symbol	Function	Symbol	Function
ALE (Input)	When ALE (Address Latch Enable is high, AD_{0-7}, $IO/\overline{M}$, A_{8-10}, CE, and $\overline{CE}$ enter address latched. The signals (AD, $IO/\overline{M}$, A_{8-10}, CE, $\overline{CE}$) are latched in at the trailing edge of ALE.	CLK (Input)	The CLK is used to force the READY into its high impedance state after it has been forced low by $\overline{CE}$ low, CE high and ALE high.
AD_{0-7} (Input)	Bidirectional Address/Data bus. The lower 8-bits of the ROM or I/O address are applied to the bus lines when ALE is high. During an I/O cycle, Port A or B are selected based on the latched value of AD_0. If $\overline{RD}$ or $\overline{IOR}$ is low when the latched chip enables are active, the output buffers present data on the bus.	READY (Output)	Ready is a 3-state output controlled by $\overline{CE_1}$, CE_2, ALE and CLK. READY is forced low when the Chip Enables are active during the time ALE is high, and remains low until the rising edge of the next CLK (see Figure 6).
A_{8-10} (Input)	These are the high order bits of the ROM address. They do not affect I/O operations.	PA_{0-7} (Input/ Output)	These are general purpose I/O pins. Their input/output direction is determined by the contents of Data Direction Register (DDR). Port A is selected for write operations when the Chip Enables are active and $\overline{IOW}$ is low and a 0 was previously latched from AD_0.
$\overline{CE_1}$ CE_2 (Input)	Chip Enable Inputs: $\overline{CE_1}$ is active low and CE_2 is active high. The 8355 can be accessed only when BOTH Chip Enables are active at the time the ALE signal latches them up. If either Chip Enable input is not active, the AD_{0-7} and READY outputs will be in a high impedance state.		Read operation is selected by either $\overline{IOR}$ low and active Chip Enables and AD_0 low, or $IO/\overline{M}$ high, $\overline{RD}$ low, active chip enables, and AD_0 low.
$IO/\overline{M}$ (Input)	If the latched $IO/\overline{M}$ is high when $\overline{RD}$ is low, the output data comes from an I/O port. If it is low the output data comes from the ROM.	PB_{0-7} (Input/ Output)	This general purpose I/O port is identical to Port A except that it is selected by a 1 latched from AD_0.
		RESET (Input)	An input high on RESET causes all pins in Port A and B to assume input mode.
$\overline{RD}$ (Input)	If the latched Chip Enables are active when $\overline{RD}$ goes low, the AD_{0-7} output buffers are enabled and output either the selected ROM location or I/O port. When both $\overline{RD}$ and $\overline{IOR}$ are high, the AD_{0-7} output buffers are 3-state.	$\overline{IOR}$ (Input)	When the Chip Enables are active, a low on $\overline{IOR}$ will output the selected I/O port onto the AD bus. $\overline{IOR}$ low performs the same function as the combination $IO/\overline{M}$ high and $\overline{RD}$ low. When $\overline{IOR}$ is not used in a system, $\overline{IOR}$ should be tied to V_{CC} ("1").
$\overline{IOW}$ (Input)	If the latched Chip Enables are active, a low on $\overline{IOW}$ causes the output port pointed to by the latched value of AD_0 to be written with the data on AD_{0-7}. The state of $IO/\overline{M}$ is ignored.	V_{CC}	+5 volt supply.
		V_{SS}	Ground Reference.

8755A/8755A-2
16,384-BIT EPROM WITH I/O

- 2048 Words × 8 Bits

- Single +5V Power Supply (V_{CC})

- Directly Compatible with 8085A and 8088 Microprocessors

- U.V. Erasable and Electrically Reprogrammable

- Internal Address Latch

- 2 General Purpose 8-Bit I/O Ports

- Each I/O Port Line Individually Programmable as Input or Output

- Multiplexed Address and Data Bus

- 40-Pin DIP

The Intel® 8755A is an erasable and electrically reprogrammable ROM (EPROM) and I/O chip to be used in the 8085A and 8088 microprocessor systems. The EPROM portion is organized as 2048 words by 8 bits. It has a maximum access time of 450 ns to permit use with no wait states in an 8085A CPU.

The I/O portion consists of 2 general purpose I/O ports. Each I/O port has 8 port lines, and each I/O port line is individually programmable as input or output.

The 8755A-2 is a high speed selected version of the 8755A compatible with the 5 MHz 8085A-2 and the full speed 5 MHz 8088.

PIN CONFIGURATION	BLOCK DIAGRAM

PIN CONFIGURATION:

PROG AND $\overline{CE}_1$	1	40	V_{CC}
CE_2	2	39	PB_7
CLK	3	38	PB_6
RESET	4	37	PB_5
V_{DD}	5	36	PB_4
READY	6	35	PB_3
$IO/\overline{M}$	7	34	PB_2
$\overline{IOR}$	8	33	PB_1
$\overline{RD}$	9	32	PB_0
$\overline{IOW}$	10	31	PA_7
ALE	11	30	PA_6
AD_0	12	29	PA_5
AD_1	13	28	PA_4
AD_2	14	27	PA_3
AD_3	15	26	PA_2
AD_4	16	25	PA_1
AD_5	17	24	PA_0
AD_6	18	23	A_{10}
AD_7	19	22	A_9
V_{SS}	20	21	A_8

8755A/8755A-2

BLOCK DIAGRAM:

CLK, READY, AD_{0-7}, A_{8-10}, CE_2, $IO/\overline{M}$, ALE, $\overline{RD}$, $\overline{IOW}$, RESET, $\overline{IOR}$ — 2K × 8 EPROM — A: PORT A 8 PA_{0-7} — B: PORT B 8 PB_{0-7}

$PROG/\overline{CE}_1$, V_{DD}, V_{CC} (+5V), V_{SS} (0V)

8755A/8755A-2

8755A FUNCTIONAL PIN DEFINITION

Symbol	Function
ALE (input)	When Address Latch Enable goes high, AD_{0-7}, IO/M, A_{8-10}, CE_2, and $\overline{CE_1}$ enter the address latches. The signals (AD, IO/M, A_{8-10}, CE) are latched in at the trailing edge of ALE.
AD_{0-7} (input/output)	Bidirectional Address/Data bus. The lower 8-bits of the PROM or I/O address are applied to the bus lines when ALE is high.
	During an I/O cycle, Port A or B are selected based on the latched value of AD_0. If $\overline{RD}$ or $\overline{IOR}$ is low when the latched Chip Enables are active, the output buffers present data on the bus.
A_{8-10} (input)	These are the high order bits of the PROM address. They do not affect I/O operations.
PROG/$\overline{CE_1}$ CE_2 (input)	Chip Enable Inputs: $\overline{CE_1}$ is active low and CE_2 is active high. The 8755A can be accessed only when *BOTH* Chip Enables are active at the time the ALE signal latches them up. If either Chip Enable input is not active, the AD_{0-7} and READY outputs will be in a high impedance state. $\overline{CE_1}$ is also used as a programming pin. (See section on programming.)
IO/M (input)	If the latched IO/M is high when $\overline{RD}$ is low, the output data comes from an I/O port. If it is low the output data comes from the PROM.
$\overline{RD}$ (input)	If the latched Chip Enables are active when $\overline{RD}$ goes low, the AD_{0-7} output buffers are enabled and output either the selected PROM location or I/O port. When both $\overline{RD}$ and $\overline{IOR}$ are high, the AD_{0-7} output buffers are 3-stated.
$\overline{IOW}$ (input)	If the latched Chip Enables are active, a low on $\overline{IOW}$ causes the output port pointed to by the latched value of AD_0 to be written with the data on AD_{0-7}. The state of IO/M is ignored.
CLK (input)	The CLK is used to force the READY into its high impedance state after it has been forced low by $\overline{CE_1}$ low, CE_2 high, and ALE high.

Symbol	Function
READY (output)	READY is a 3-state output controlled by CE_2, $\overline{CE_1}$, ALE and CLK. READY is forced low when the Chip Enables are active during the time ALE is high, and remains low until the rising edge of the next CLK. (See Figure 6.)
PA_{0-7} (input/output)	These are general purpose I/O pins. Their input/output direction is determined by the contents of Data Direction Register (DDR). Port A is selected for write operations when the Chip Enables are active and $\overline{IOW}$ is low and a 0 was previously latched from AD_0, AD_1.
	Read operation is selected by either $\overline{IOR}$ low and active Chip Enables and AD_0 and AD_1 low, *or* IO/M high, $\overline{RD}$ low, active Chip Enables, and AD_0 and AD_1 low.
PB_{0-7} (input/output)	This general purpose I/O port is identical to Port A except that it is selected by a 1 latched from AD_0 and a 0 from AD_1.
RESET (input)	In normal operation, an input high on RESET causes all pins in Ports A and B to assume input mode (clear DDR register).
$\overline{IOR}$ (input)	When the Chip Enables are active, a low on $\overline{IOR}$ will output the selected I/O port onto the AD bus. $\overline{IOR}$ low performs the same function as the combination of IO/M high and $\overline{RD}$ low. When $\overline{IOR}$ is not used in a system, $\overline{IOR}$ should be tied to V_{CC} ("1").
V_{CC}	+5 volt supply.
V_{SS}	Ground Reference.
V_{DD}	V_{DD} is a programming voltage, and must be tied to +5V when the 8755A is being read.
	For programming, a high voltage is supplied with $V_{DD} = 25V$, typical. (See section on programming.)

8155/8156/8155-2/8156-2
2048 BIT STATIC MOS RAM WITH I/O PORTS AND TIMER

- ■ **256 Word x 8 Bits**
- ■ **Single +5V Power Supply**
- ■ **Completely Static Operation**
- ■ **Internal Address Latch**
- ■ **2 Programmable 8 Bit I/O Ports**

- ■ **1 Programmable 6-Bit I/O Port**
- ■ **Programmable 14-Bit Binary Counter/ Timer**
- ■ **Compatible with 8085A and 8088 CPU**
- ■ **Multiplexed Address and Data Bus**
- ■ **40 Pin DIP**

The 8155 and 89156 are RAM and I/O chips to be used in the 8085A and 8088 microprocessor systems. The RAM portion is designed with 2048 static cells organized as 256 x 8. They have a maximum access time of 400 ns to permit use with no wait states in 8085A CPU. The 8155-2 and 8156-2 have maximum access times of 330 ns for use with the 8085A-2 and the full speed 5 MHz 8088 CPU.

The I/O portion consists of three general purpose I/O ports. One of the three ports can be programmed to be status pins, thus allowing the other two ports to operate in handshake mode.

A 14-bit programmable counter/timer is jalso included on chip to provide either a square wave or terminal count pulse for the CPU system depending on timer mode.

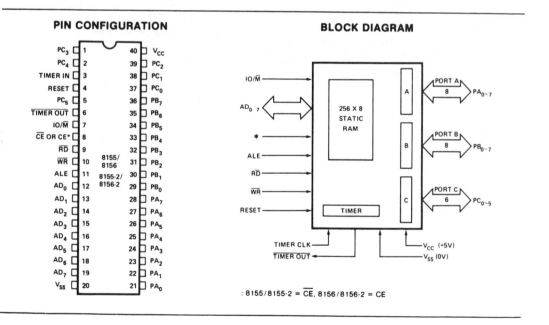

PIN CONFIGURATION   **BLOCK DIAGRAM**

: 8155/8155-2 = $\overline{CE}$, 8156/8156-2 = CE

8155/8156/8155-2/8156-2

8155/8156 PIN FUNCTIONS

Symbol	Function
RESET (input)	Pulse provided by the 8085A to initialize the system (connect to 8085A RESET OUT). Input high on this line resets the chip and initializes the three I/O ports to input mode. The width of RESET pulse should typically be two 8085A clock cycle times.
AD_{0-7} (input)	3-state Address/Data lines that interface with the CPU lower 8-bit Address/Data Bus. The 8-bit address is latched into the address latch inside the 8155/56 on the falling edge of ALE. The address can be either for the memory section or the I/O section depending on the $IO/\overline{M}$ input. The 8-bit data is either written into the chip or read from the chip, depending on the $\overline{WR}$ or $\overline{RD}$ input signal.
CE or $\overline{CE}$ (input)	Chip Enable: On the 8155, this pin is $\overline{CE}$ and is ACTIVE LOW. On the 8156, this pin is CE and is ACTIVE HIGH.
$\overline{RD}$ (input)	Read control: Input low on this line with the Chip Enable active enables and AD_{0-7} buffers. If $IO/\overline{M}$ pin is low, the RAM content will be read out to the AD bus. Otherwise the content of the selected I/O port or command/status registers will be read to the AD bus.
$\overline{WR}$ (input)	Write control: Input low on this line with the Chip Enable active causes the data on the Address/Data bus to be written to the RAM or I/O ports and command/status register depending on $IO/\overline{M}$.

Symbol	Function
ALE (input)	Address Latch Enable: This control signal latches both the address on the AD_{0-7} lines and the state of the Chip Enable and $IO/\overline{M}$ into the chip at the falling edge of ALE.
$IO/\overline{M}$ (input)	Selects memory if low and I/O and command/status registers if high.
$PA_{0-7}(8)$ (input/output)	These 8 pins are general purpose I/O pins. The in/out direction is selected by programming the command register.
$PB_{0-7}(8)$ (input/output)	These 8 pins are general purpose I/O pins. The in/out direction is selected by programming the command register.
$PC_{0-5}(6)$ (input/output)	These 6 pins can function as either input port, output port, or as control signals for PA and PB. Programming is done through the command register. When PC_{0-5} are used as control signals, they will provide the following: PC_0 — A INTR (Port A Interrupt) PC_1 — ABF (Port A Buffer Full) PC_2 — $\overline{A\ STB}$ (Port A Strobe) PC_3 — B INTR (Port B Interrupt) PC_4 — $\overline{B\ BF}$ (Port B Buffer Full) PC_5 — B STB (Port B Strobe)
TIMER IN (input)	Input to the counter-timer.
$\overline{\text{TIMER OUT}}$ (output)	Timer output. This output can be either a square wave or a pulse depending on the timer mode.
V_{CC}	+5 volt supply.
V_{SS}	Ground Reference.

8255A/8255A-5
PROGRAMMABLE PERIPHERAL INTERFACE

- MCS-85™ Compatible 8255A-5
- 24 Programmable I/O Pins
- Completely TTL Compatible
- Fully Compatible with Intel® Micro-processor Families
- Improved Timing Characteristics

- Direct Bit Set/Reset Capability Easing Control Application Interface
- 40-Pin Dual In-Line Package
- Reduces System Package Count
- Improved DC Driving Capability

The Intel® 8255A is a general purpose programmable I/O device designed for use with Intel® microprocessors. It has 24 I/O pins which may be individually programmed in 2 groups of 12 and used in 3 major modes of operation. In the first mode (MODE 0), each group of 12 I/O pins may be programmed in sets of 4 to be input or output. In MODE 1, the second mode, each group may be programmed to have 8 lines of input or output. Of the remaining 4 pins, 3 are used for hand-shaking and interrupt control signals. The third mode of operation (MODE 2) is a bidirectional bus mode which uses 8 lines for a bidirectional bus, and 5 lines, borrowing one from the other group, for handshaking.

PIN CONFIGURATION

8255A BLOCK DIAGRAM

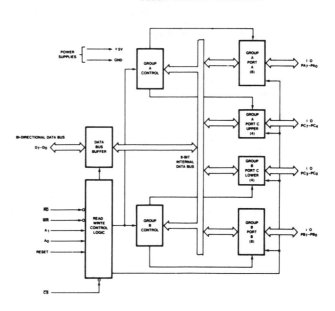

PIN NAMES

D₇–D₀	DATA BUS (BI DIRECTIONAL)
RESET	RESET INPUT
CS	CHIP SELECT
RD	READ INPUT
WR	WRITE INPUT
A0, A1	PORT ADDRESS
PA7-PA0	PORT A (BIT)
PB7-PB0	PORT B (BIT)
PC7-PC0	PORT C (BIT)
VCC	+5 VOLTS
GND	0 VOLTS

8279/8279-5
PROGRAMMABLE KEYBOARD/DISPLAY INTERFACE

- **Simultaneous Keyboard Display Operations**
- **Scanned Keyboard Mode**
- **Scanned Sensor Mode**
- **Strobed Input Entry Mode**
- **8-Character Keyboard FIFO**
- **2-Key Lockout or N-Key Rollover with Contact Debounce**
- **Dual 8- or 16-Numerical Display**

- **Single 16-Character Display**
- **Right or Left Entry 16-Byte Display RAM**
- **Mode Programmable from CPU**
- **Programmable Scan Timing**
- **Interrupt Output on Key Entry**
- **Available in EXPRESS**
  — Standard Temperature Range
  — Extended Temperature Range

The Intel® 8279 is a general purpose programmable keyboard and display I/O interface device designed for use with Intel® microprocessors. The keyboard portion can provide a scanned interface to a 64-contact key matrix. The keyboard portion will also interface to an array of sensors or a strobed interface keyboard, such as the hall effect and ferrite variety. Key depressions can be 2-key lockout or N-key rollover. Keyboard entries are debounced and strobed in an 8-character FIFO. If more than 8 characters are entered, overrun status is set. Key entries set the interrupt output line to the CPU.

The display portion provides a scanned display interface for LED, incandescent, and other popular display technologies. Both numeric and alphanumeric segment displays may be used as well as simple indicators. The 8279 has 16x8 display RAM which can be organized into dual 16x4. The RAM can be loaded or interrogated by the CPU. Both right entry, calculator and left entry typewriter display formats are possible. Both read and write of the display RAM can be done with auto-increment of the display RAM address.

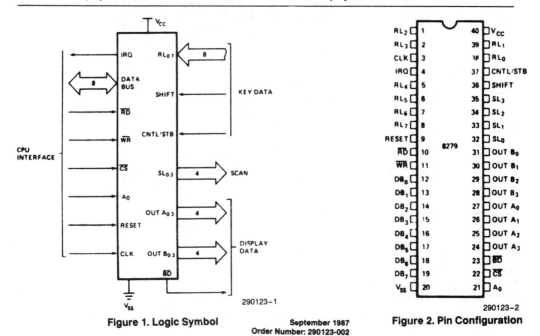

Figure 1. Logic Symbol

Figure 2. Pin Configuration

September 1987
Order Number: 290123-002

8279/8279-5

HARDWARE DESCRIPTION

The 8279 is packaged in a 40 pin DIP. The following is a functional description of each pin.

Table 1. Pin Description

Symbol	Pin No.	Name and Function
DB$_0$–DB$_7$	19–12	**BI-DIRECTIONAL DATA BUS:** All data and commands between the CPU and the 8279 are transmitted on these lines.
CLK	3	**CLOCK:** Clock from system used to generate internal timing.
RESET	9	**RESET:** A high signal on this pin resets the 8279. After being reset the 8279 is placed in the following mode: 1) 16 8-bit character display—left entry. 2) Encoded scan keyboard—2 key lockout. Along with this the program clock prescaler is set to 31.
CS	22	**CHIP SELECT:** A low on this pin enables the interface functions to receive or transmit.
A$_0$	21	**BUFFER ADDRESS:** A high on this line indicates the signals in or out are interpreted as a command or status. A low indicates that they are data.
$\overline{RD}$, $\overline{WR}$	10–11	**INPUT/OUTPUT READ AND WRITE:** These signals enable the data buffers to either send data to the external bus or receive it from the external bus.
IRQ	4	**INTERRUPT REQUEST:** In a keyboard mode, the interrupt line is high when there is data in the FIFO/Sensor RAM. The interrupt line goes low with each FIFO/Sensor RAM read and returns high if there is still information in the RAM. In a sensor mode, the interrupt line goes high whenever a change in a sensor is detected.
V$_{SS}$, V$_{CC}$	20, 40	**GROUND AND POWER SUPPLY PINS.**
SL$_0$–SL$_3$	32–35	**SCAN LINES:** Scan lines which are used to scan the key switch or sensor matrix and the display digits. These lines can be either encoded (1 of 16) or decoded (1 of 4).
RL$_0$–RL$_7$	38, 39, 1, 2, 5–8	**RETURN LINE:** Return line inputs which are connected to the scan lines through the keys or sensor switches. They have active internal pullups to keep them high until a switch closure pulls one low. They also serve as an 8-bit input in the Strobed Input mode.
SHIFT	36	**SHIFT:** The shift input status is stored along with the key position on key closure in the Scanned Keyboard modes. It has an active internal pullup to keep it high until a switch closure pulls it low.
CNTL/STB	37	**CONTROL/STROBED INPUT MODE:** For keyboard modes this line is used as a control input and stored like status on a key closure. The line is also the strobe line that enters the data into the FIFO in the Strobed Input mode. (Rising Edge). It has an active internal pullup to keep it high until a switch closure pulls it low.
OUT A$_0$–OUT A$_3$ OUT B$_0$–OUT B$_3$	27-24 31–28	**OUTPUTS:** These two ports are the outputs for the 16 x 4 display refresh registers. The data from these outputs is synchronized to the scan lines (SL$_0$–SL$_3$) for multiplexed digit displays. The two 4 bit ports may be blanked independently. These two ports may also be considered as one 8-bit port.
$\overline{BD}$	23	**BLANK DISPLAY:** This output is used to blank the display during digit switching or by a display blanking command.

APPENDIX D
MC68000 INSTRUCTION EXECUTION TIMES

D.1 INTRODUCTION

This Appendix contains listings of the instruction execution times in terms of external clock (CLK) periods. In this data, it is assumed that both memory read and write cycle times are four clock periods. A longer memory cycle will cause the generation of wait states which must be added to the total instruction time.

The number of bus read and write cycles for each instruction is also included with the timing data. This data is enclosed in parenthesis following the number of clock periods and is shown as: (r/w) where r is the number of read cycles and w is the number of write cycles included in the clock period number. Recalling that either a read or write cycle requires four clock periods, a timing number given as 18(3/1) relates to 12 clock periods for the three read cycles, plus 4 clock periods for the one write cycle, plus 2 cycles required for some internal function of the processor.

NOTE

The number of periods includes instruction fetch and all applicable operand fetches and stores.

D.2 OPERAND EFFECTIVE ADDRESS CALCULATION TIMING

Table D-1 lists the number of clock periods required to compute an instruction's effective address. It includes fetching of any extension words, the address computation, and fetching of the memory operand. The number of bus read and write cycles is shown in parenthesis as (r/w). Note there are no write cycles involved in processing the effective address.

Table D-1. Effective Address Calculation Times

Addressing Mode		Byte, Word	Long
Register			
Dn	Data Register Direct	0(0/0)	0(0/0)
An	Address Register Direct	0(0/0)	0(0/0)
Memory			
(An)	Address Register Indirect	4(1/0)	8(2/0)
(An) +	Address Register Indirect with Postincrement	4(1/0)	8(2/0)
− (An)	Address Register Indirect with Predecrement	6(1/0)	10(2/0)
d(An)	Address Register Indirect with Displacement	8(2/0)	12(3/0)
d(An, ix)*	Address Register Indirect with Index	10(2/0)	14(3/0)
xxx.W	Absolute Short	8(2/0)	12(3/0)
xxx.L	Absolute Long	12(3/0)	16(4/0)
d(PC)	Program Counter with Displacement	8(2/0)	12(3/0)
d(PC, ix)*	Program Counter with Index	10(2/0)	14(3/0)
#xxx	Immediate	4(1/0)	8(2/0)

*The size of the index register (ix) does not affect execution time.

D.3 MOVE INSTRUCTION EXECUTION TIMES

Tables D-2 and D-3 indicate the number of clock periods for the move instruction. This data includes instruction fetch, operand reads, and operand writes. The number of bus read and write cycles is shown in parenthesis as (r/w).

Table D-2. Move Byte and Word Instruction Execution Times

Source	Destination								
	Dn	An	(An)	(An) +	– (An)	d(An)	d(An, ix)*	xxx.W	xxx.L
Dn	4(1/0)	4(1/0)	8(1/1)	8(1/1)	8(1/1)	12(2/1)	14(2/1)	12(2/1)	16(3/1)
An	4(1/0)	4(1/0)	8(1/1)	8(1/1)	8(1/1)	12(2/1)	14(2/1)	12(2/1)	16(3/1)
(An)	8(2/0)	8(2/0)	12(2/1)	12(2/1)	12(2/1)	16(3/1)	18(3/1)	16(3/1)	20(4/1)
(An) +	8(2/0)	8(2/0)	12(2/1)	12(2/1)	12(2/1)	16(3/1)	18(3/1)	16(3/1)	20(4/1)
– (An)	10(2/0)	10(2/0)	14(2/1)	14(2/1)	14(2/1)	18(3/1)	20(3/1)	18(3/1)	22(4/1)
d(An)	12(3/0)	12(3/0)	16(3/1)	16(3/1)	16(3/1)	20(4/1)	22(4/1)	20(4/1)	24(5/1)
d(An, ix)*	14(3/0)	14(3/0)	18(3/1)	18(3/1)	18(3/1)	22(4/1)	24(4/1)	22(4/1)	26(5/1)
xxx.W	12(3/0)	12(3/0)	16(3/1)	16(3/1)	16(3/1)	20(4/1)	22(4/1)	20(4/1)	24(5/1)
xxx.L	16(4/0)	16(4/0)	20(4/1)	20(4/1)	20(4/1)	24(5/1)	26(5/1)	24(5/1)	28(6/1)
d(PC)	12(3/0)	12(3/0)	16(3/1)	16(3/1)	16(3/1)	20(4/1)	22(4/1)	20(4/1)	24(5/1)
d(PC, ix)*	14(3/0)	14(3/0)	18(3/1)	18(3/1)	18(3/1)	22(4/1)	24(4/1)	22(4/1)	26(5/1)
#xxx	8(2/0)	8(2/0)	12(2/1)	12(2/1)	12(2/1)	16(3/1)	18(3/1)	16(3/1)	20(4/1)

*The size of the index register (ix) does not affect execution time.

Table D-3. Move Long Instruction Execution Times

Source	Destination								
	Dn	An	(An)	(An) +	– (An)	d(An)	d(An, ix)*	xxx.W	xxx.L
Dn	4(1/0)	4(1/0)	12(1/2)	12(1/2)	12(1/2)	16(2/2)	18(2/2)	16(2/2)	20(3/2)
An	4(1/0)	4(1/0)	12(1/2)	12(1/2)	12(1/2)	16(2/2)	18(2/2)	16(2/2)	20(3/2)
(An)	12(3/0)	12(3/0)	20(3/2)	20(3/2)	20(3/2)	24(4/2)	26(4/2)	24(4/2)	28(5/2)
(An) +	12(3/0)	12(3/0)	20(3/2)	20(3/2)	20(3/2)	24(4/2)	26(4/2)	24(4/2)	28(5/2)
– (An)	14(3/0)	14(3/0)	22(3/2)	22(3/2)	22(3/2)	26(4/2)	28(4/2)	26(4/2)	30(5/2)
d(An)	16(4/0)	16(4/0)	24(4/2)	24(4/2)	24(4/2)	28(5/2)	30(5/2)	28(5/2)	32(6/2)
d(An, ix)*	18(4/0)	18(4/0)	26(4/2)	26(4/2)	26(4/2)	30(5/2)	32(5/2)	30(5/2)	34(6/2)
xxx.W	16(4/0)	16(4/0)	24(4/2)	24(4/2)	24(4/2)	28(5/2)	30(5/2)	28(5/2)	32(6/2)
xxx.L	20(5/0)	20(5/0)	28(5/2)	28(5/2)	28(5/2)	32(6/2)	34(6/2)	32(6/2)	36(7/2)
d(PC)	16(4/0)	16(4/0)	24(4/2)	24(4/2)	24(4/2)	28(5/2)	30(5/2)	28(5/2)	32(5/2)
d(PC, ix)*	18(4/0)	18(4/0)	26(4/2)	26(4/2)	26(4/2)	30(5/2)	32(5/2)	30(5/2)	34(6/2)
#xxx	12(3/0)	12(3/0)	20(3/2)	20(3/2)	20(3/2)	24(4/2)	26(4/2)	24(4/2)	28(5/2)

*The size of the index register (ix) does not affect execution time.

D.4 STANDARD INSTRUCTION EXECUTION TIMES

The number of clock periods shown in Table D-4 indicates the time required to perform the operations, store the results, and read the next instruction. The number of bus read and write cycles is shown in parenthesis as (r/w). The number of clock periods and the number of read and write cycles must be added respectively to those of the effective address calculation where indicated.

In Table D-4 the headings have the following meanings: An = address register operand, Dn = data register operand, ea = an operand specified by an effective address, and M = memory effective address operand.

Table D-4. Standard Instruction Execution Times

Instruction	Size	op<ea>, An†	op<ea>, Dn	op Dn, <M>
ADD	Byte, Word	8(1/0) +	4(1/0) +	8(1/1) +
	Long	6(1/0) + * *	6(1/0) + * *	12(1/2) +
AND	Byte, Word	—	4(1/0) +	8(1/1) +
	Long	—	6(1/0) + * *	12(1/2) +
CMP	Byte, Word	6(1/0) +	4(1/0) +	—
	Long	6(1/0) +	6(1/0) +	—
DIVS	—	—	158(1/0) + *	—
DIVU	—	—	140(1/0) + *	—
EOR	Byte, Word	—	4(1/0) * * *	8(1/1) +
	Long	—	8(1/0) * * *	12(1/2) +
MULS	—	—	70(1/0) + *	—
MULU	—	—	70(1/0) + *	— .
OR	Byte, Word	—	4(1/0) +	8(1/1) +
	Long	—	6(1/0) + * *	12(1/2) +
SUB	Byte, Word	8(1/0) +	4(1/0) +	8(1/1) +
	Long	6(1/0) + * *	6(1/0) + * *	12(1/2) +

NOTES:
+ add effective address calculation time
† word or long only
* indicates maximum value
* * The base time of six clock periods is increased to eight if the effective address mode is register direct or immediate (effective address time should also be added).
* * * Only available effective address mode is data register direct.
DIVS, DIVU — The divide algorithm used by the MC68000 provides less than 10% difference between the best and worst case timings.
MULS, MULU — The multiply algorithm requires 38 + 2n clocks where n is defined as:
MULU: n = the number of ones in the <ea>
MULS: n = concatanate the <ea> with a zero as the LSB; n is the resultant number of 10 or 01 patterns in the 17-bit source; i.e., worst case happens when the source is $5555.

D.5 IMMEDIATE INSTRUCTION EXECUTION TIMES

The number of clock periods shown in Table D-5 includes the time to fetch immediate operands, perform the operations, store the results, and read the next operation. The number of bus read and write cycles is shown in parenthesis as (r/w). The number of clock periods and the number of read and write cycles must be added respectively to those of the effective address calculation where indicated.

In Table D-5, the headings have the following meanings: # = immediate operand, Dn = data register operand, An = address register operand, and M = memory operand. SR = status register.

Table D-5. Immediate Instruction Execution Times

Instruction	Size	op #, Dn	op #, An	op #, M
ADDI	Byte, Word	8(2/0)	—	12(2/1) +
	Long	16(3/0)	—	20(3/2) +
ADDQ	Byte, Word	4(1/0)	8(1/0) *	8(1/1) +
	Long	8(1/0)	8(1/0)	12(1/2) +
ANDI	Byte, Word	8(2/0)	—	12(2/1) +
	Long	16(3/0)	—	20(3/1) +
CMPI	Byte, Word	8(2/0)	—	8(2/0) +
	Long	14(3/0)	—	12(3/0) +
EORI	Byte, Word	8(2/0)	—	12(2/1) +
	Long	16(3/0)	—	20(3/2) +
MOVEQ	Long	4(1/0)	—	—
ORI	Byte, Word	8(2/0)	—	12(2/1) +
	Long	16(3/0)	—	20(3/2) +
SUBI	Byte, Word	8(2/0)	—	12(2/1) +
	Long	16(3/0)	—	20(3/2) +
SUBQ	Byte, Word	4(1/0)	8(1/0) *	8(1/1) +
	Long	8(1/0)	8(1/0)	12(1/2) +

\+ add effective address calculation time
* word only

D.6 SINGLE OPERAND INSTRUCTION EXECUTION TIMES

Table D-6 indicates the number of clock periods for the single operand instructions. The number of bus read and write cycles is shown in parenthesis as (r/w). The number of clock periods and the number of read and write cycles must be added respectively to those of the effective address calculation where indicated.

Table D-6. Single Operand Instruction Execution Times

Instruction	Size	Register	Memory
CLR	Byte, Word	4(1/0)	8(1/1) +
	Long	6(1/0)	12(1/2) +
NBCD	Byte	6(1/0)	8(1/1) +
NEG	Byte, Word	4(1/0)	8(1/1) +
	Long	6(1/0)	12(1/2) +
NEGX	Byte, Word	4(1/0)	8(1/1) +
	Long	6(1/0)	12(1/2) +
NOT	Byte, Word	4(1/0)	8(1/1) +
	Long	6(1/0)	12(1/2) +
S$_{CC}$	Byte, False	4(1/0)	8(1/1) +
	Byte, True	6(1/0)	8(1/1) +
TAS	Byte	4(1/0)	10(1/1) +
TST	Byte, Word	4(1/0)	4(1/0) +
	Long	4(1/0)	4(1/0) +

+ add effective address calculation time

D.7 SHIFT/ROTATE INSTRUCTION EXECUTION TIMES

Table D-7 indicates the number of clock periods for the shift and rotate instructions. The number of bus read and write cycles is shown in parenthesis as (r/w). The number of clock periods and the number of read and write cycles must be added respectively to those of the effective address calculation where indicated.

Table D-7. Shift/Rotate Instruction Execution Times

Instruction	Size	Register	Memory
ASR, ASL	Byte, Word	6 + 2n(1/0)	8(1/1) +
	Long	8 + 2n(1/0)	−
LSR, LSL	Byte, Word	6 + 2n(1/0)	8(1/1) +
	Long	8 + 2n(1/0)	−
ROR, ROL	Byte, Word	6 + 2n(1/0)	8(1/1) +
	Long	8 + 2n(1/0)	−
ROXR, ROXL	Byte, Word	6 + 2n(1/0)	8(1/1) +
	Long	8 + 2n(1/0)	−

+ add effective address calculation time
n is the shift count

D.8 BIT MANIPULATION INSTRUCTION EXECUTION TIMES

Table D-8 indicates the number of clock periods required for the bit manipulation instructions. The number of bus read and write cycles is shown in parenthesis as (r/w). The number of clock periods and the number of read and write cycles must be added respectively to those of the effective address calculation where indicated.

Table D-8. Bit Manipulation Instruction Execution Times

Instruction	Size	Dynamic		Static	
		Register	Memory	Register	Memory
BCHG	Byte	–	8(1/1) +	–	12(2/1) +
	Long	8(1/0) *	–	12(2/0) *	–
BCLR	Byte	–	8(1/1) +	–	12(2/1) +
	Long	10(1/0) *	–	14(2/0) *	–
BSET	Byte	–	8(1/1) +	–	12(2/1) +
	Long	8(1/0) *	–	12(2/0) *	–
BTST	Byte	–	4(1/0) +	–	8(2/0) +
	Long	6(1/0)	–	10(2/0)	–

+ add effective address calculation time
* indicates maximum value

D.9 CONDITIONAL INSTRUCTION EXECUTION TIMES

Table D-9 indicates the number of clock periods required for the conditional instructions. The number of bus read and write cycles is indicated in parenthesis as (r/w). The number of clock periods and the number of read and write cycles must be added respectively to those of the effective address calculation where indicated.

Table D-9. Conditional Instruction Execution Times

Instruction	Displacement	Branch Taken	Branch Not Taken
B$_{CC}$	Byte	10(2/0)	8(1/0)
	Word	10(2/0)	12(2/0)
BRA	Byte	10(2/0)	–
	Word	10(2/0)	–
BSR	Byte	18(2/2)	–
	Word	18(2/2)	–
DB$_{CC}$	CC true	–	12(2/0)
	CC false	10(2/0)	14(3/0)

+ add effective address calculation time
* indicates maximum value

D.10 JMP, JSR, LEA, PEA, AND MOVEM INSTRUCTION EXECUTION TIMES

Table D-10 indicates the number of clock periods required for the jump, jump-to-subroutine, load effective address, push effective address, and move multiple registers instructions. The number of bus read and write cycles is shown in parenthesis as (r/w).

Table D-10. JMP, JSR, LEA, PEA, and MOVEM Instruction Execution Times

Instr	Size	(An)	(An) +	– (An)	d(An)	d(An, ix) +	xxx.W	xxx.L	d(PC)	d(PC, ix) *
JMP	–	8(2/0)	–	–	10(2/0)	14(3/0)	10(2/0)	12(3/0)	10(2/0)	14(3/0)
JSR	–	16(2/2)	–	–	18(2/2)	22(2/2)	18(2/2)	20(3/2)	18(2/2)	22(2/2)
LEA	–	4(1/0)	–	–	8(2/0)	12(2/0)	8(2/0)	12(3/0)	8(2/0)	12(2/0)
PEA	–	12(1/2)	–	–	16(2/2)	20(2/2)	16(2/2)	20(3/2)	16(2/2)	20(2/2)
MOVEM M → R	Word	12 + 4n (3 + n/0)	12 + 4n (3 + n/0)	–	16 + 4n (4 + n/0)	18 + 4n (4 + n/0)	16 + 4n (4 + n/0)	20 + 4n (5 + n/0)	16 + 4n (4 + n/0)	18 + 4n (4 + n/0)
	Long	12 + 8n (3 + 2n/0)	12 + 8n (3 + 2n/0)	–	16 + 8n (4 + 2n/0)	18 + 8n (4 + 2n/0)	16 + 8n (4 + 2n/0)	20 + 8n (5 + 2n/0)	16 + 8n (4 + 2n/0)	18 + 8n (4 + 2n/0)
MOVEM R → M	Word	8 + 4n (2/n)	–	8 + 4n (2/n)	12 + 4n (3/n)	14 + 4n (3/n)	12 + 4n (3/n)	16 + 4n (4/n)	–	–
	Long	8 + 8n (2/2n)	–	8 + 8n (2/2n)	12 + 8n (3/2n)	14 + 8n (3/2n)	12 + 8n (3/2n)	16 + 8n (4/2n)	–	–

n is the number of registers to move
* is the size of the index register (ix) does not affect the instruction's execution time

D 11 MULTI-PRECISION INSTRUCTION EXECUTION TIMES

Table D-11 indicates the number of clock periods for the multi-precision instructions. The number of clock periods includes the time to fetch both operands, peform the operations, store the results, and read the next instructions. The number of read and write cycles is shown in parenthesis as (r/w).

In Table D-11, the headings have the following meanings: Dn = data register operand and M = memory operand.

Table D-11. Multi-Precision Instruction Execution Times

Instruction	Size	op Dn, Dn	op M, M
ADDX	Byte, Word	4(1/0)	18(3/1)
	Long	8(1/0)	30(5/2)
CMPM	Byte, Word	–	12(3/0)
	Long	–	20(5/0)
SUBX	Byte, Word	4(1/0)	18(3/1)
	Long	8(1/0)	30(5/2)
ABCD	Byte	6(1/0)	18(3/1)
SBCD	Byte	6(1/0)	18(3/1)

D.12 MISCELLANEOUS INSTRUCTION EXECUTION TIMES

Tables D-12 and D-13 indicate the number of clock periods for the following miscellaneous instructions. The number of bus read and write cycles is shown in parenthesis as (r/w). The number of clock periods plus the number of read and write cycles must be added to those of the effective address calculation where indicated.

Table D-12. Miscellaneous Instruction Execution Times

Instruction	Size	Register	Memory
ANDI to CCR	Byte	20(3/0)	—
ANDI to SR	Word	20(3/0)	—
CHK	—	10(1/0)+	—
EORI to CCR	Byte	20(3/0)	—
EORI to SR	Word	20(3/0)	—
ORI to CCR	Byte	20(3/0)	—
ORI to SR	Word	20(3/0)	—
MOVE from SR	—	6(1/0)	8(1/1)+
MOVE to CCR	—	12(2/0)	12(2/0)+
MOVE to SR	—	12(2/0)	12(2/0)+
EXG	—	6(1/0)	—
EXT	Word	4(1/0)	—
EXT	Long	4(1/0)	—
LINK	—	16(2/2)	—
MOVE from USP	—	4(1/0)	—
MOVE to USP	—	4(1/0)	—
NOP	—	4(1/0)	—
RESET	—	132(1/0)	—
RTE	—	20(5/0)	—
RTR	—	20(5/0)	—
RTS	—	16(4/0)	—
STOP	—	4(0/0)	—
SWAP	—	4(1/0)	—
TRAPV	—	4(1/0)	—
UNLK	—	12(3/0)	—

+ add effective address calculation time

Table D-13. Move Peripheral Instruction Execution Times

Instruction	Size	Register → Memory	Memory → Register
MOVEP	Word	16(2/2)	16(4/0)
MOVEP	Long	24(2/4)	24(6/0)

D.13 EXCEPTION PROCESSING EXECUTION TIMES

Table D-14 indicates the number of clock periods for exception processing. The number of clock periods includes the time for all stacking, the vector fetch, and the fetch of the first two instruction words of the handler routine. The number of bus read and write cycles is shown in parenthesis as (r/w).

Table D-14. Exception Processing Execution Times

Exception	Periods
Address Error	50(4/7)
Bus Error	50(4/7)
CHK Instruction	44(5/4) +
Divide by Zero	42(5/4)
Illegal Instruction	34(4/3)
Interrupt	44(5/3) *
Privilege Violation	34(4/3)
RESET**	40(6/0)
Trace	34(4/3)
TRAP Instruction	38(4/4)
TRAPV Instruction	34(4/3)

+ add effective address calculation time

* The interrupt acknowledge cycle is assumed to take four clock periods.

** Indicates the time from when $\overline{\text{RESET}}$ and $\overline{\text{HALT}}$ are first sampled as negated to when instruction execution starts.

APPENDIX E

8086 INSTRUCTION SET REFERENCE DATA

AAA	**AAA** (no operands) ASCII adjust for addition			**Flags**	O D I T S Z A P C U U U X U X
Operands	**Clocks**	**Transfers***	**Bytes**	**Coding Example**	
(no operands)	4	—	1	AAA	

AAD	**AAD** (no operands) ASCII adjust for division			**Flags**	O D I T S Z A P C U X X U X U
Operands	**Clocks**	**Transfers***	**Bytes**	**Coding Example**	
(no operands)	60	—	2	AAD	

AAM	**AAM** (no operands) ASCII adjust for multiply			**Flags**	O D I T S Z A P C U X X U X U
Operands	**Clocks**	**Transfers***	**Bytes**	**Coding Example**	
(no operands)	83	—	1	AAM	

AAS	**AAS** (no operands) ASCII adjust for subtraction			**Flags**	O D I T S Z A P C U U U X U X
Operands	**Clocks**	**Transfers***	**Bytes**	**Coding Example**	
(no operands)	4	—	1	AAS	

*For the 8086, add four clocks for each 16-bit word transfer with an odd address. For the 8088, add four clocks for each 16-bit word transfer.

ADC	ADC destination,source Add with carry			Flags	O D I T S Z A P C X X X X X X
Operands	**Clocks**	**Transfers***	**Bytes**	**Coding Example**	
register, register	3	—	2	ADC AX, SI	
register, memory	9 + EA	1	2-4	ADC DX, BETA [SI]	
memory, register	16 + EA	2	2-4	ADC ALPHA [BX] [SI], DI	
register, immediate	4	—	3-4	ADC BX, 256	
memory, immediate	17 + EA	2	3-6	ADC GAMMA, 30H	
accumulator, immediate	4	—	2-3	ADC AL, 5	

ADD	ADD destination,source Addition			Flags	O D I T S Z A P C X X X X X X
Operands	**Clocks**	**Transfers***	**Bytes**	**Coding Example**	
register, register	3	—	2	ADD CX, DX	
register, memory	9 + EA	1	2-4	ADD DI, [BX].ALPHA	
memory, register	16 + EA	2	2-4	ADD TEMP, CL	
register, immediate	4	—	3-4	ADD CL, 2	
memory, immediate	17 + EA	2	3-6	ADD ALPHA, 2	
accumulator, immediate	4	—	2-3	ADD AX, 200	

AND	AND destination,source Logical and			Flags	O D I T S Z A P C 0 X X U X 0
Operands	**Clocks**	**Transfers***	**Bytes**	**Coding Example**	
register, register	3	—	2	AND AL,BL	
register, memory	9 + EA	1	2-4	AND CX,FLAG__WORD	
memory, register	16 + EA	2	2-4	AND ASCII [DI],AL	
register, immediate	4	—	3-4	AND CX,0F0H	
memory, immediate	17 + EA	2	3-6	AND BETA, 01H	
accumulator, immediate	4	—	2-3	AND AX, 01010000B	

CALL	CALL target Call a procedure			Flags	O D I T S Z A P C
Operands	**Clocks**	**Transfers***	**Bytes**	**Coding Examples**	
near-proc	19	1	3	CALL NEAR__PROC	
far-proc	28	2	5	CALL FAR__PROC	
memptr 16	21 + EA	2	2-4	CALL PROC__TABLE [SI]	
regptr 16	16	1	2	CALL AX	
memptr 32	37 + EA	4	2-4	CALL [BX].TASK [SI]	

CBW	CBW (no operands) Convert byte to word			Flags	O D I T S Z A P C
Operands	**Clocks**	**Transfers***	**Bytes**	**Coding Example**	
(no operands)	2	—	1	CBW	

*For the 8086, add four clocks for each 16-bit word transfer with an odd address. For the 8088, add four clocks for each 16-bit word transfer.

CLC	CLC (no operands) Clear carry flag				Flags O D I T S Z A P C 0
Operands	**Clocks**	**Transfers***	**Bytes**	**Coding Example**	
(no operands)	2	—	1	CLC	

CLD	CLD (no operands) Clear direction flag				Flags O D I T S Z A P C 0
Operands	**Clocks**	**Transfers***	**Bytes**	**Coding Example**	
(no operands)	2	—	1	CLD	

CLI	CLI (no operands) Clear interrupt flag				Flags O D I T S Z A P C 0
Operands	**Clocks**	**Transfers***	**Bytes**	**Coding Example**	
(no operands)	2	—	1	CLI	

CMC	CMC (no operands) Complement carry flag				Flags O D I T S Z A P C X
Operands	**Clocks**	**Transfers***	**Bytes**	**Coding Example**	
(no operands)	2	—	1	CMC	

CMP	CMP destination,source Compare destination to source				Flags O D I T S Z A P C X X X X X X
Operands	**Clocks**	**Transfers***	**Bytes**	**Coding Example**	
register, register	3	—	2	CMP BX, CX	
register, memory	9 + EA	1	2-4	CMP DH, ALPHA	
memory, register	9 + EA	1	2-4	CMP [BP + 2], SI	
register, immediate	4	—	3-4	CMP BL, 02H	
memory, immediate	10 + EA	1	3-6	CMP [BX].RADAR [DI], 3420H	
accumulator, immediate	4	—	2-3	CMP AL, 00010000B	

CMPS	CMPS dest-string,source-string Compare string				Flags O D I T S Z A P C X X X X X X
Operands	**Clocks**	**Transfers***	**Bytes**	**Coding Example**	
dest-string, source-string	22	2	1	CMPS BUFF1, BUFF2	
(repeat) dest-string, source-string	9 + 22/rep	2/rep	1	REPE CMPS ID, KEY	

*For the 8086, add four clocks for each 16-bit word transfer with an odd address. For the 8088, add four clocks for each 16-bit word transfer.

CWD	**CWD** (no operands) Convert word to doubleword			**Flags**	O D I T S Z A P C
Operands	**Clocks**	**Transfers***	**Bytes**	**Coding Example**	
(no operands)	5	—	1	CWD	

DAA	**DAA** (no operands) Decimal adjust for addition			**Flags**	O D I T S Z A P C X X X X X X
Operands	**Clocks**	**Transfers***	**Bytes**	**Coding Example**	
(no operands)	4	—	1	DAA	

DAS	**DAS** (no operands) Decimal adjust for subtraction			**Flags**	O D I T S Z A P C U X X X X X
Operands	**Clocks**	**Transfers***	**Bytes**	**Coding Example**	
(no operands)	4	—	1	DAS	

DEC	**DEC** destination Decrement by 1			**Flags**	O D I T S Z A P C X X X X X
Operands	**Clocks**	**Transfers***	**Bytes**	**Coding Example**	
reg16	2	—	1	DEC AX	
reg8	3	—	2	DEC AL	
memory	15 + EA	2	2-4	DEC ARRAY [SI]	

DIV	**DIV** source Division, unsigned			**Flags**	O D I T S Z A P C U U U U U U
Operands	**Clocks**	**Transfers***	**Bytes**	**Coding Example**	
reg8	80-90	—	2	DIV CL	
reg16	144-162	—	2	DIV BX	
mem8	(86-96) + EA	1	2-4	DIV ALPHA	
mem16	(150-168) + EA	1	2-4	DIV TABLE [SI]	

ESC	**ESC** external-opcode,source Escape			**Flags**	O D I T S Z A P C
Operands	**Clocks**	**Transfers***	**Bytes**	**Coding Example**	
immediate, memory	8 + EA	1	2-4	ESC 6,ARRAY [SI]	
immediate, register	2	—	2	ESC 20,AL	

*For the 8086, add four clocks for each 16-bit word transfer with an odd address. For the 8088, add four clocks for each 16-bit word transfer.

HLT	HLT (no operands) Halt			Flags	O D I T S Z A P C
Operands	**Clocks**	**Transfers***	**Bytes**	**Coding Example**	
(no operands)	2	—	1	HLT	

IDIV	IDIV source Integer division			Flags	O D I T S Z A P C U U U U U
Operands	**Clocks**	**Transfers***	**Bytes**	**Coding Example**	
reg8	101-112	—	2	IDIV BL	
reg16	165-184	—	2	IDIV CX	
mem8	(107-118) +EA	1	2-4	IDIV DIVISOR__BYTE [SI]	
mem16	(171-190) +EA	1	2-4	IDIV [BX].DIVISOR__WORD	

IMUL	IMUL source Integer multiplication			Flags	O D I T S Z A P C X U U U U X
Operands	**Clocks**	**Transfers***	**Bytes**	**Coding Example**	
reg8	80-98	—	2	IMUL CL	
reg16	128-154	—	2	IMUL BX	
mem8	(86-104) +EA	1	2-4	IMUL RATE__BYTE	
mem16	(134-160) +EA	1	2-4	IMUL RATE__WORD [BP] [DI]	

IN	IN accumulator,port Input byte or word			Flags	O D I T S Z A P C
Operands	**Clocks**	**Transfers***	**Bytes**	**Coding Example**	
accumulator, immed8	10	1	2	IN AL, 0FFEAH	
accumulator, DX	8	1	1	IN AX, DX	

INC	INC destination Increment by 1			Flags	O D I T S Z A P C X X X X X
Operands	**Clocks**	**Transfers***	**Bytes**	**Coding Example**	
reg16	2	—	1	INC CX	
reg8	3	—	2	INC BL	
memory	15+EA	2	2-4	INC ALPHA [DI] [BX]	

*For the 8086, add four clocks for each 16-bit word transfer with an odd address. For the 8088, add four clocks for each 16-bit word transfer.

INT	INT interrupt-type Interrupt			Flags	O D I T S Z A P C 0 0
Operands		**Clocks**	**Transfers***	**Bytes**	**Coding Example**
immed8 (type = 3)		52	5	1	INT 3
immed8 (type ≠ 3)		51	5	2	INT 67

INTR†	INTR (external maskable interrupt) Interrupt if INTR and IF=1			Flags	O D I T S Z A P C 0 0
Operands		**Clocks**	**Transfers***	**Bytes**	**Coding Example**
(no operands)		61	7	N/A	N/A

INTO	INTO (no operands) Interrupt if overflow			Flags	O D I T S Z A P C 0 0
Operands		**Clocks**	**Transfers***	**Bytes**	**Coding Example**
(no operands)		53 or 4	5	1	INTO

IRET	IRET (no operands) Interrupt Return			Flags	O D I T S Z A P C R R R R R R R R
Operands		**Clocks**	**Transfers***	**Bytes**	**Coding Example**
(no operands)		24	3	1	IRET

JA/JNBE	JA/JNBE short-label Jump if above / Jump if not below nor equal			Flags	O D I T S Z A P C
Operands		**Clocks**	**Transfers***	**Bytes**	**Coding Example**
short-label		16 or 4	—	2	JA ABOVE

JAE/JNB	JAE/JNB short-label Jump if above or equal / Jump if not below			Flags	O D I T S Z A P C
Operands		**Clocks**	**Transfers***	**Bytes**	**Coding Example**
short-label		16 or 4	—	2	JAE ABOVE__EQUAL

JB/JNAE	JB/JNAE short-label Jump if below / Jump if not above nor equal			Flags	O D I T S Z A P C
Operands		**Clocks**	**Transfers***	**Bytes**	**Coding Example**
short-label		16 or 4	—	2	JB BELOW

*For the 8086, add four clocks for each 16-bit word transfer with an odd address. For the 8088, add four clocks for each 16-bit word transfer.

†INTR is not an instruction; it is included in table 2-21 only for timing information.

JBE/JNA	**JBE/JNA** short-label Jump if below or equal/Jump if not above			**Flags** O D I T S Z A P C
Operands	Clocks	Transfers*	Bytes	Coding Example
short-label	16 or 4	—	2	JNA NOT__ABOVE

JC	**JC** short-label Jump if carry			**Flags** O D I T S Z A P C
Operands	Clocks	Transfers*	Bytes	Coding Example
short-label	16 or 4	—	2	JC CARRY__SET

JCXZ	**JCXZ** short-label Jump if CX is zero			**Flags** O D I T S Z A P C
Operands	Clocks	Transfers*	Bytes	Coding Example
short-label	18 or 6	—	2	JCXZ COUNT__DONE

JE/JZ	**JE/JZ** short-label Jump if equal/Jump if zero			**Flags** O D I T S Z A P C
Operands	Clocks	Transfers*	Bytes	Coding Example
short-label	16 or 4	—	2	JZ ZERO

JG/JNLE	**JG/JNLE** short-label Jump if greater/Jump if not less nor equal			**Flags** O D I T S Z A P C
Operands	Clocks	Transfers*	Bytes	Coding Example
short-label	16 or 4	—	2	JG GREATER

JGE/JNL	**JGE/JNL** short-label Jump if greater or equal/Jump if not less			**Flags** O D I T S Z A P C
Operands	Clocks	Transfers*	Bytes	Coding Example
short-label	16 or 4	—	2	JGE GREATER__EQUAL

JL/JNGE	**JL/JNGE** short-label Jump if less/Jump if not greater nor equal			**Flags** O D I T S Z A P C
Operands	Clocks	Transfers*	Bytes	Coding Example
short-label	16 or 4	—	2	JL LESS

*For the 8086, add four clocks for each 16-bit word transfer with an odd address. For the 8088, add four clocks for each 16-bit word transfer.

JLE/JNG	JLE/JNG short-label Jump if less or equal/Jump if not greater			Flags	O D I T S Z A P C
Operands		**Clocks**	**Transfers***	**Bytes**	**Coding Example**
short-label		16 or 4	—	2	JNG NOT__GREATER

JMP	JMP target Jump			Flags	O D I T S Z A P C
Operands		**Clocks**	**Transfers***	**Bytes**	**Coding Example**
short-label		15	—	2	JMP SHORT
near-label		15	—	3	JMP WITHIN__SEGMENT
far-label		15	—	5	JMP FAR__LABEL
memptr16		18 + EA	1	2-4	JMP [BX].TARGET
regptr16		11	—	2	JMP CX
memptr32		24 + EA	2	2-4	JMP OTHER.SEG [SI]

JNC	JNC short-label Jump if not carry			Flags	O D I T S Z A P C
Operands		**Clocks**	**Transfers***	**Bytes**	**Coding Example**
short-label		16 or 4	—	2	JNC NOT__CARRY

JNE/JNZ	JNE/JNZ short-label Jump if not equal/Jump if not zero			Flags	O D I T S Z A P C
Operands		**Clocks**	**Transfers***	**Bytes**	**Coding Example**
short-label		16 or 4	—	2	JNE NOT__EQUAL

JNO	JNO short-label Jump if not overflow			Flags	O D I T S Z A P C
Operands		**Clocks**	**Transfers***	**Bytes**	**Coding Example**
short-label		16 or 4	—	2	JNO NO__OVERFLOW

JNP/JPO	JNP/JPO short-label Jump if not parity/Jump if parity odd			Flags	O D I T S Z A P C
Operands		**Clocks**	**Transfers***	**Bytes**	**Coding Example**
short-label		16 or 4	—	2	JPO ODD__PARITY

JNS	JNS short-label Jump if not sign			Flags	O D I T S Z A P C
Operands		**Clocks**	**Transfers***	**Bytes**	**Coding Example**
short-label		16 or 4	—	2	JNS POSITIVE

*For the 8086, add four clocks for each 16-bit word transfer with an odd address. For the 8088, add four clocks for each 16-bit word transfer.

JO	JO short-label Jump if overflow			Flags	O D I T S Z A P C
Operands		**Clocks**	**Transfers***	**Bytes**	**Coding Example**
short-label		16 or 4	—	2	JO SIGNED__OVRFLW

JP/JPE	JP/JPE short-label Jump if parity / Jump if parity even			Flags	O D I T S Z A P C
Operands		**Clocks**	**Transfers***	**Bytes**	**Coding Example**
short-label		16 or 4	—	2	JPE EVEN__PARITY

JS	JS short-label Jump if sign			Flags	O D I T S Z A P C
Operands		**Clocks**	**Transfers***	**Bytes**	**Coding Example**
short-label		16 or 4	—	2	JS NEGATIVE

LAHF	LAHF (no operands) Load AH from flags			Flags	O D I T S Z A P C
Operands		**Clocks**	**Transfers***	**Bytes**	**Coding Example**
(no operands)		4	—	1	LAHF

LDS	LDS destination,source Load pointer using DS			Flags	O D I T S Z A P C
Operands		**Clocks**	**Transfers**	**Bytes**	**Coding Example**
reg16, mem32		16 + EA	2	2-4	LDS SI,DATA.SEG [DI]

LEA	LEA destination,source Load effective address			Flags	O D I T S Z A P C
Operands		**Clocks**	**Transfers***	**Bytes**	**Coding Example**
reg16, mem16		2 + EA	—	2-4	LEA BX, [BP] [DI]

LES	LES destination,source Load pointer using ES			Flags	O D I T S Z A P C
Operands		**Clocks**	**Transfers***	**Bytes**	**Coding Example**
reg16, mem32		16 + EA	2	2-4	LES DI, [BX].TEXT__BUFF

*For the 8086, add four clocks for each 16-bit word transfer with an odd address. For the 8088, add four clocks for each 16-bit word transfer.

LOCK	LOCK (no operands) Lock bus			Flags	O D I T S Z A P C
Operands	**Clocks**	**Transfers***	**Bytes**	**Coding Example**	
(no operands)	2	—	1	LOCK XCHG FLAG,AL	

LODS	LODS source-string Load string			Flags	O D I T S Z A P C
Operands	**Clocks**	**Transfers***	**Bytes**	**Coding Example**	
source-string (repeat) source-string	12 9+13/rep	1 1/rep	1 1	LODS CUSTOMER__NAME REP LODS NAME	

LOOP	LOOP short-label Loop			Flags	O D I T S Z A P C
Operands	**Clocks**	**Transfers***	**Bytes**	**Coding Example**	
short-label	17/5	—	2	LOOP AGAIN	

LOOPE/LOOPZ	LOOPE/LOOPZ short-label Loop if equal/Loop if zero			Flags	O D I T S Z A P C
Operands	**Clocks**	**Transfers***	**Bytes**	**Coding Example**	
short-label	18 or 6	—	2	LOOPE AGAIN	

LOOPNE/LOOPNZ	LOOPNE/LOOPNZ short-label Loop if not equal/Loop if not zero			Flags	O D I T S Z A P C
Operands	**Clocks**	**Transfers***	**Bytes**	**Coding Example**	
short-label	19 or 5	—	2	LOOPNE AGAIN	

NMI†	NMI (external nonmaskable interrupt) Interrupt if NMI = 1			Flags	O S I T S Z A P C 0 0
Operands	**Clocks**	**Transfers***	**Bytes**	**Coding Example**	
(no operands)	50˙	5	N/A	N/A	

*For the 8086, add four clocks for each 16-bit word transfer with an odd address. For the 8088, add four clocks for each 16-bit word transfer.

†NMI is not an instruction; it is included in table 2-21 only for timing information.

MOV	**MOV** destination, source Move			**Flags** O D I T S Z A P C
Operands	**Clocks**	**Transfers***	**Bytes**	**Coding Example**
memory, accumulator	10	1	3	MOV ARRAY [SI]; AL
accumulator, memory	10	1	3	MOV AX, TEMP__RESULT
register, register	2	—	2	MOV AX,CX
register, memory	8+EA	1	2-4	MOV BP, STACK__TOP
memory, register	9+EA	1	2-4	MOV COUNT [DI], CX
register, immediate	4	—	2-3	MOV CL, 2
memory, immediate	10+EA	1	3-6	MOV MASK [BX] [SI], 2CH
seg-reg, reg16	2	—	2	MOV ES, CX
seg-reg, mem16	8+EA	1	2-4	MOV DS, SEGMENT__BASE
reg16, seg-reg	2	—	2	MOV BP, SS
memory, seg-reg	9+EA	1	2-4	MOV [BX].SEG__SAVE, CS

MOVS	**MOVS** dest-string, source-string Move string			**Flags** O D I T S Z A P C
Operands	**Clocks**	**Transfers***	**Bytes**	**Coding Example**
dest-string, source-string	18	2	1	MOVS LINE EDIT__DATA
(repeat) dest-string, source-string	9+17/rep	2/rep	1	REP MOVS SCREEN, BUFFER

MOVSB/MOVSW	**MOVSB/MOVSW** (no operands) Move string (byte/word)			**Flags** O D I T S Z A P C
Operands	**Clocks**	**Transfers***	**Bytes**	**Coding Example**
(no operands)	18	2	1	MOVSB
(repeat) (no operands)	9+17/rep	2/rep	1	REP MOVSW

MUL	**MUL** source Multiplication, unsigned			**Flags** O D I T S Z A P C X U U U U X
Operands	**Clocks**	**Transfers***	**Bytes**	**Coding Example**
reg8	70-77	—	2	MUL BL
reg16	118-133	—	2	MUL CX
mem8	(76-83) +EA	1	2-4	MUL MONTH [SI]
mem16	(124-139) +EA	1	2-4	MUL BAUD__RATE

*For the 8086, add four clocks for each 16-bit word transfer with an odd address. For the 8088, add four clocks for each 16-bit word transfer.

NEG	NEG destination Negate			Flags	O D I T S Z A P C X X X X X 1*
Operands	**Clocks**	**Transfers***	**Bytes**	**Coding Example**	
register memory	3 16 + EA	— 2	2 2-4	NEG AL NEG MULTIPLIER	

*0 if destination = 0

NOP	NOP (no operands) No Operation			Flags	O D I T S Z A P C
Operands	**Clocks**	**Transfers***	**Bytes**	**Coding Example**	
(no operands)	3	—	1	NOP	

NOT	NOT destination Logical not			Flags	O D I T S Z A P C
Operands	**Clocks**	**Transfers***	**Bytes**	**Coding Example**	
register memory	3 16 + EA	— 2	2 2-4	NOT AX NOT CHARACTER	

OR	OR destination, source Logical inclusive or			Flags	O D I T S Z A P C 0 X X U X 0
Operands	**Clocks**	**Transfers***	**Bytes**	**Coding Example**	
register, register register, memory memory, register accumulator, immediate register, immediate memory, immediate	3 9 + EA 16 + EA 4 4 17 + EA	— 1 2 — — 2	2 2-4 2-4 2-3 3-4 3-6	OR AL, BL OR DX, PORT__ID [DI] OR FLAG__BYTE, CL OR AL, 01101100B OR CX,01H OR [BX].CMD__WORD,0CFH	

OUT	OUT port, accumulator Output byte or word			Flags	O D I T S Z A P C
Operands	**Clocks**	**Transfers***	**Bytes**	**Coding Example**	
immed8, accumulator DX, accumulator	10 8	1 1	2 1	OUT 44, AX OUT DX, AL	

POP	POP destination Pop word off stack			Flags	O D I T S Z A P C
Operands	**Clocks**	**Transfers***	**Bytes**	**Coding Example**	
register seg-reg (CS illegal) memory	8 8 17 + EA	1 1 2	1 1 2-4	POP DX POP DS POP PARAMETER	

*For the 8086, add four clocks for each 16-bit word transfer with an odd address. For the 8088, add four clocks for each 16-bit word transfer.

POPF	POPF (no operands) Pop flags off stack			Flags	O D I T S Z A P C R R R R R R R R
Operands	**Clocks**	**Transfers***	**Bytes**	**Coding Example**	
(no operands)	8	1	1	POPF	

PUSH	PUSH source Push word onto stack			Flags	O D I T S Z A P C
Operands	**Clocks**	**Transfers***	**Bytes**	**Coding Example**	
register	11	1	1	PUSH SI	
seg-reg (CS legal)	10	1	1	PUSH ES	
memory	16 + EA	2	2-4	PUSH RETURN_CODE [SI]	

PUSHF	PUSHF (no operands) Push flags onto stack			Flags	O D I T S Z A P C
Operands	**Clocks**	**Transfers***	**Bytes**	**Coding Example**	
(no operands)	10	1	1	PUSHF	

RCL	RCL destination,count Rotate left through carry			Flags	O D I T S Z A P C X X
Operands	**Clocks**	**Transfers***	**Bytes**	**Coding Example**	
register, 1	2	—	2	RCL CX, 1	
register, CL	8 + 4/bit	—	2	RCL AL, CL	
memory, 1	15 + EA	2	2-4	RCL ALPHA, 1	
memory, CL	20 + EA + 4/bit	2	2-4	RCL [BP].PARM, CL	

RCR	RCR designation,count Rotate right through carry			Flags	O D I T S Z A P C X X
Operands	**Clocks**	**Transfers***	**Bytes**	**Coding Example**	
register, 1	2	—	2	RCR BX, 1	
register, CL	8 + 4/bit	—	2	RCR BL, CL	
memory, 1	15 + EA	2	2-4	RCR [BX].STATUS, 1	
memory, CL	20 + EA + 4/bit	2	2-4	RCR ARRAY [DI], CL	

REP	REP (no operands) Repeat string operation			Flags	O D I T S Z A P C
Operands	**Clocks**	**Transfers***	**Bytes**	**Coding Example**	
(no operands)	2	—	1	REP MOVS DEST, SRCE	

*For the 8086, add four clocks for each 16-bit word transfer with an odd address. For the 8088, add four clocks for each 16-bit word transfer.

REPE/REPZ	REPE/REPZ (no operands) Repeat string operation while equal/while zero			Flags	O D I T S Z A P C
Operands	**Clocks**	**Transfers***	**Bytes**	**Coding Example**	
(no operands)	2	—	1	REPE CMPS DATA, KEY	

REPNE/REPNZ	REPNE/REPNZ (no operands) Repeat string operation while not equal/not zero			Flags	O D I T S Z A P C
Operands	**Clocks**	**Transfers***	**Bytes**	**Coding Example**	
(no operands)	2	—	1	REPNE SCAS INPUT__LINE	

RET	RET optional-pop-value Return from procedure			Flags	O D I T S Z A P C
Operands	**Clocks**	**Transfers***	**Bytes**	**Coding Example**	
(intra-segment, no pop)	8	1	1	RET	
(intra-segment, pop)	12	1	3	RET 4	
(inter-segment, no pop)	18	2	1	RET	
(inter-segment, pop)	17	2	3	RET 2	

ROL	ROL destination,count Rotate left			Flags	O D I T S Z A P C X X
Operands	**Clocks**	**Transfers**	**Bytes**	**Coding Examples**	
register, 1	2	—	2	ROL BX, 1	
register, CL	8 + 4/bit	—	2	ROL DI, CL	
memory, 1	15 + EA	2	2-4	ROL FLAG__BYTE [DI],1	
memory, CL	20 + EA + 4/bit	2	2-4	ROL ALPHA , CL	

ROR	ROR destination,count Rotate right			Flags	O D I T S Z A P C X X
Operand	**Clocks**	**Transfers***	**Bytes**	**Coding Example**	
register, 1	2	—	2	ROR AL, 1	
register, CL	8 + 4/bit	—	2	ROR BX, CL	
memory, 1	15 + EA	2	2-4	ROR PORT__STATUS, 1	
memory, CL	20 + EA + 4/bit	2	2-4	ROR CMD__WORD, CL	

SAHF	SAHF (no operands) Store AH into flags			Flags	O D I T S Z A P C R R R R R
Operands	**Clocks**	**Transfers***	**Bytes**	**Coding Example**	
(no operands)	4	—	1	SAHF	

*For the 8086, add four clocks for each 16-bit word transfer with an odd address. For the 8088, add four clocks for each 16-bit word transfer.

SAL/SHL	**SAL/SHL** destination,count Shift arithmetic left/Shift logical left			**Flags**	O D I T S Z A P C X X
Operands	Clocks	Transfers*	Bytes		Coding Examples
register,1	2	—	2		SAL AL,1
register, CL	8 + 4/bit	—	2		SHL DI, CL
memory,1	15 + EA	2	2-4		SHL [BX].OVERDRAW, 1
memory, CL	20 + EA + 4/bit	2	2-4		SAL STORE__COUNT, CL

SAR	**SAR** destination,source Shift arithmetic right			**Flags**	O D I T S Z A P C X X X U X X
Operands	Clocks	Transfers*	Bytes		Coding Example
register, 1	2	—	2		SAR DX, 1
register, CL	8 + 4/bit	—	2		SAR DI, CL
memory, 1	15 + EA	2	2-4		SAR N__BLOCKS, 1
memory, CL	20 + EA + 4/bit	2	2-4		SAR N__BLOCKS, CL

SBB	**SBB** destination,source Subtract with borrow			**Flags**	O D I T S Z A P C X X X X X X
Operands	Clocks	Transfers*	Bytes		Coding Example
register, register	3	—	2		SBB BX, CX
register, memory	9 + EA	1	2-4		SBB DI, [BX].PAYMENT
memory, register	16 + EA	2	2-4		SBB BALANCE, AX
accumulator, immediate	4	—	2-3		SBB AX, 2
register, immediate	4	—	3-4		SBB CL, 1
memory, immediate	17 + EA	2	3-6		SBB COUNT [SI], 10

SCAS	**SCAS** dest-string Scan string			**Flags**	O D I T S Z A P C X X X X X X
Operands	Clocks	Transfers*	Bytes		Coding Example
dest-string	15	1	1		SCAS INPUT__LINE
(repeat) dest-string	9 + 15/rep	1/rep	1		REPNE SCAS BUFFER

SEGMENT†	**SEGMENT** override prefix Override to specified segment			**Flags**	O D I T S Z A P C
Operands	Clocks	Transfers*	Bytes		Coding Example
(no operands)	2	—	1		MOV SS:PARAMETER, AX

*For the 8086, add four clocks for each 16-bit word transfer with an odd address. For the 8088, add four clocks for each 16-bit word transfer.

†ASM-86 incorporates the segment override prefix into the operand specification and not as a separate instruction. SEGMENT is included in table 2-21 only for timing information.

SHR	**SHR** destination,count Shift logical right			**Flags**	O D I T S Z A P C X X
Operands	**Clocks**	**Transfers***	**Bytes**		**Coding Example**
register, 1 register, CL memory, 1 memory, CL	2 8 + 4/bit 15 + EA 20 + EA + 4/bit	— — 2 2	2 2 2-4 2-4		SHR SI, 1 SHR SI, CL SHR ID__BYTE [SI] [BX], 1 SHR INPUT__WORD, CL

SINGLE STEP†	**SINGLE STEP** (Trap flag interrupt) Interrupt if TF = 1			**Flags**	O D I T S Z A P C 0 0
Operands	**Clocks**	**Transfers***	**Bytes**		**Coding Example**
(no operands)	50	5	N/A		N/A

STC	**STC** (no operands) Set carry flag			**Flags**	O D I T S Z A P C 1
Operands	**Clocks**	**Transfers***	**Bytes**		**Coding Example**
(no operands)	2	—	1		STC

STD	**STD** (no operands) Set direction flag			**Flags**	O D I T S Z A P C 1
Operands	**Clocks**	**Transfers***	**Bytes**		**Coding Example**
(no operands)	2	—	1		STD

STI	**STI** (no operands) Set interrupt enable flag			**Flags**	O D I T S Z A P C 1
Operands	**Clocks**	**Transfers***	**Bytes**		**Coding Example**
(no operands)	2	—	1		STI

STOS	**STOS** dest-string Store byte or word string			**Flags**	O D I T S Z A P C
Operands	**Clocks**	**Transfers***	**Bytes**		**Coding Example**
dest-string (repeat) dest-string	11 9 + 10/rep	1 1/rep	1 1		STOS PRINT__LINE REP STOS DISPLAY

*For the 8086, add four clocks for each 16-bit word transfer with an odd address. For the 8088, add four clocks for each 16-bit word transfer.

†SINGLE STEP is not an instruction; it is included in table 2-21 only for timing information.

SUB	**SUB** destination,source Subtraction				Flags O D I T S Z A P C X X X X X X
Operands		**Clocks**	**Transfers***	**Bytes**	**Coding Example**
register, register		3	—	2	SUB CX, BX
register, memory		9 + EA	1	2-4	SUB DX, MATH__TOTAL [SI]
memory, register		16 + EA	2	2-4	SUB [BP + 2], CL
accumulator, immediate		4	—	2-3	SUB AL, 10
register, immediate		4	—	3-4	SUB SI, 5280
memory, immediate		17 + EA	2	3-6	SUB [BP].BALANCE, 1000

TEST	**TEST** destination,source Test or non-destructive logical and				Flags O D I T S Z A P C 0 X X U X 0
Operands		**Clocks**	**Transfers***	**Bytes**	**Coding Example**
register, register		3	—	2	TEST SI, DI
register, memory		9 + EA	1	2-4	TEST SI, END__COUNT
accumulator, immediate		4	—	2-3	TEST AL, 00100000B
register, immediate		5	—	3-4	TEST BX, 0CC4H
memory, immediate		11 + EA	—	3-6	TEST RETURN__CODE, 01H

WAIT	**WAIT** (no operands) Wait while $\overline{\text{TEST}}$ pin not asserted				Flags O D I T S Z A P C
Operands		**Clocks**	**Transfers***	**Bytes**	**Coding Example**
(no operands)		3 + 5n	—	1	WAIT

XCHG	**XCHG** destination,source Exchange				Flags O D I T S Z A P C
Operands		**Clocks**	**Transfers***	**Bytes**	**Coding Example**
accumulator, reg16		3	—	1	XCHG AX, BX
memory, register		17 + EA	2	2-4	XCHG SEMAPHORE, AX
register, register		4	—	2	XCHG AL, BL

XLAT	**XLAT** source-table Translate				Flags O D I T S Z A P C
Operands		**Clocks**	**Transfers***	**Bytes**	**Coding Example**
source-table		11	1	1	XLAT ASCII__TAB

*For the 8086, add four clocks for each 16-bit word transfer with an odd address. For the 8088, add four clocks for each 16-bit word transfer.

XOR	XOR destination,source Logical exclusive or			Flags	O D I T S Z A P C 0 X X U X 0
Operands	**Clocks**	**Transfers***	**Bytes**	**Coding Example**	
register, register	3	—	2	XOR CX, BX	
register, memory	9 + EA	1	2-4	XOR CL, MASK__BYTE	
memory, register	16 + EA	2	2-4	XOR ALPHA [SI], DX	
accumulator, immediate	4	—	2-3	XOR AL, 01000010B	
register, immediate	4	—	3-4	XOR SI, 00C2H	
memory, immediate	17 + EA	2	3-6	XOR RETURN__CODE, 0D2H	

*For the 8086, add four clocks for each 16-bit word transfer with an odd address. For the 8088, add four clocks for each 16-bit word transfer.

APPENDIX F
GLOSSARY

Absolute Addressing: The specific identification number (address) permanently assigned to a storage location, device, or register by the machine designer. Used to locate information and assist in circuit fault diagnosis.

Accumulator: Used for storing the result after most ALU operations; 8 bits long for an 8-bit microprocessor.

Address: A unique identification number (or locator) of some source or destination of data. That part of an instruction which specifies the register or memory location of an operand involved in the instruction.

Addressing Mode: The manner in which a microprocessor determines the operand and destination addresses in an instruction cycle.

Address Register: A register used to store the address (label for a memory location) of data being fetched or stored, a sequence of instructions to be executed, or the location to which control will be transferred.

Address Space: The number of storage locations in a microcomputer's memory that can be directly addressed by the microprocessor. The addressing range is determined by the number of address pins provided with the microprocessor chip.

American Standard Code for Information Interchange (ASCII): An 8-bit code commonly used with microprocessors for representing alphanumeric codes.

Analog-to-Digital (A/D) Converter: Transforms an analog voltage into its digital equivalent.

Architecture: The organizational structure or hardware configuration of a computer system.

Arithmetic and Logic Unit (ALU): A digital circuit which performs arithmetic and logic operations on two n-bit numbers.

Assembler: A program that translates an assembly language program into a machine language program.

Assembly Language: A type of microprocessor programming language that uses a semi-English-language statement.

Asynchronous Operation: The execution of a sequence of steps such that each step is initiated upon completion of the previous step. For bus structures, this implies a timing protocol that uses no clock and has no period; hence system operation proceeds at a rate governed by the time-constants of the enabled circuitry.

Asynchronous Serial Data Transmission: The transmitting device does not need to be synchronized with the receiving device.

Autodecrement Addressing Mode: The contents of the specified microprocessor register are first decremented by K (1 for byte, 2 for 16-bit, and 4 for 32-bit) and then the resulting value is used as the address of the operand.

Autoincrement Addressing Mode: The contents of a specified microprocessor register are used as the address of the operand first and then the register contents are automatically incremented by K (1 for byte, 2 for 16-bit, and 4 for 32-bit).

Bandwidth: Bandwidth of a bus or memory is a measure of communications throughput and can be represented as the product of the maximum number of transactions per second and number of data bits per transaction.

Barrel Shifter: A specially configured shift register that is normally included in 32-bit microprocessors for fast shift operations.

Base Address: An address that is used to convert all relative addresses in a program to absolute (machine) addresses.

Base Page Addressing: This instruction typically uses two bytes: the first byte is the op code, and the second byte is the low-order address byte. The high-order address byte is assumed to be the base-page number.

Baud Rate: Rate of data transmission in bits per second.

Binary-Coded Decimal (BCD): The representation of 10 decimal digits, 0 through 9, by their corresponding 4-bit binary numbers.

Bit: An abbreviation for a binary digit. A unit of information equal to one binary decision or one of two possible states (one or zero, on or off, true or false) and represents the smallest piece of information in a binary notation system.

Bit-Slice Microprocessor: Divides the elements of a central processing unit (ALU, registers, and control unit) among several ICs. The registers and ALU are usually contained in a single chip. These microprocessors can be cascaded to produce microprocessors of variable word lengths such as 8, 12, 16, 32. The control unit of a bit-slice microprocessor is typically microprogrammed.

Block Transfer DMA: A peripheral device requests the DMA transfer via the DMA request line, which is connected directly or through a DMA controller chip to the microprocessor. The DMA controller chip completes the DMA transfer and transfers the control of the bus to the microprocessor.

Branch: The branch instruction allows the computer to skip or jump out of program sequence to a designated instruction either unconditionally or conditionally (based on conditions such as carry or sign).

Breakpoint: Allows the user to execute the section of a program until one of the breakpoint conditions is met. It is then halted. The designer may then single step or examine memory and registers. Typically breakpoint conditions are program counter address or data references. Breakpoints are used in debugging assembly language programs.

Buffer: A temporary memory storage device designed to compensate for the different data rates between a transmitting device and a receiving device (for example, between a CPU and a peripheral). Current amplifiers are also sometimes referred to as buffers.

Bus: A collection of parallel unbroken electrical signal lines that interconnect or link computer modules. The typical microcomputer interface includes separate buses for address, data, control, and power functions.

Bus Arbitration: Bus operation protocols that guarantee conflict-free access to a bus. Arbitration is the process of selecting one respondent from a collection of several candidates that concurrently request service.

Bus Cycle: The period of time in which a microprocessor carries out all the necessary bus communications to implement a standard operation.

Byte: An 8-bit word.

Cache Memory: An ultra-high speed, directly accessible, relatively small semiconductor memory block used to store data/instructions that the microcomputer may need in the immediate future. Increases system bandwidth by reducing the number of external memory fetches required by the processor. Typical 32-bit microprocessors are normally provided with on-chip cache memory.

Cathode Ray Tube (CRT): Evacuated glass tube with a fluorescent coating on the inner side of the screen.

Central Processing Unit (CPU): The portion of a computer containing the ALU, register section, and control unit.

Clock: Timing signals providing synchronization among the various components in a microcomputer system.

Code: A system of symbols or sets of rules for the representation of data in a digital computer. Some examples include binary, BCD, and ASCII.

Compiler: A software program which translates the source code written in a high-level programming language into machine language that is understandable to the processor.

Complementary Metal Oxide Semiconductor (CMOS): Provides low power density and high noise immunity.

Concurrency: The occurrence of one or more operations at a time (see Parallel Operation).

Conditional Branching: Conditional branch instructions are used to change the order of execution of a program based on the conditions set by the status flags.

Condition Code Register: Contains information such as carry, sign, zero, and overflow based on ALU operations.

Control Store: Used to contain microcode (usually in ROM) in order to provide for microprogrammed "firmware" control functions. An integral part of a microprogrammed system controller.

Control Unit: Part of the microprocessor; its purpose is to read and decode instructions from the memory.

Controller/Sequencer: The hardware circuits which provide signals to carry out selection and retrieval of instructions from storage in sequence, interpret them, and initiate the required operation. The system functions may be implemented by hardware control, firmware control, or software control.

Coprocessor: A companion microprocessor that performs specific functions such as floating-point operations independently from the microprocessor to speed up overall operations.

CPU Space: Protected memory space addressable only by the microprocessor itself; it is used for a processor's internal functions or vectored exception processing.

CRT Controller: Provides all logic functions for interfacing the microprocessor to a CRT.

Cycle Stealing DMA: The DMA controller transfers a byte of data between the microcomputer's memory and a peripheral device such as the disk by stealing a clock cycle of the microprocessor.

Data: Basic elements of information represented in binary form (that is, digits consisting of bits) that can be processed or produced by a microcomputer. Data represents any group of operands made up of numbers, letters, or symbols denoting any condition, value, or state. Current typical microcomputer operand sizes include: a word, which typically contains 2 bytes or 16 bits; a long word, which contains 4 bytes or 32 bits; a quad word, which contains 8 bytes or 64 bits.

Data Counter (DC): Also known as Memory Address Register (MAR). Stores the address of data; typically, 16 bits long for 8-bit microprocessors.

Data Register: A register used to temporarily hold operational data being sent to and from a peripheral device.

Debugger: A program that executes and debugs the object program generated by the assembler or compiler. The debugger provides a single stepping, breakpoints, and program tracing.

Decoder: A device capable of generating 2n output lines based on n inputs.

Direct Memory Access (DMA): A type of input/output technique in which data can be transferred between the microcomputer memory and external devices without the microprocessor's involvement.

Directly Addressable Memory: The memory address space in which the microprocessor can directly execute programs. The maximum directly addressable memory is determined by the number of the microprocessor's address pins.

Dynamic RAM: Stores data in capacitors and, therefore, must be refreshed; uses refresh circuitry.

EAROM (Electrically Alterable Read-Only Memory): Can be programmed without removing the memory from its sockets. This memory is also called read-mostly memory since it has much slower write times than read times.

Editor: A program that produces an error-free source program, written in assembly or high-level languages.

Effective Address: The final address used to carry out an instruction. Determined by the addressing mode.

Emulator: A hardware device that allows a microcomputer system to emulate (that is, mimic the procedures or protocols) another microcomputer system.

Encode: To apply the rules governing a specific code. For example, the selection of which hardware devices to enable during an operation can occur automatically by encoding individual device identifications into the instructions themselves. Hence, to encode is to convert data from its natural form into a machine-readable code usable to the computer.

EPROM (Erasable Programmable Read-Only Memory): Can be programmed and erased using ultraviolet light. The chip must be removed from the microcomputer system for programming.

Exception Processing: The CPU processing state associated with interrupts, trap instructions, tracing, and other exceptional conditions, whether they are initiated internally or externally.

Extended Binary-Coded Decimal Interchange Code (EBCDIC): An 8-bit code commonly used with microprocessors for representing character codes.

Firmware: Permanently stored, unalterable program instructions contained in the ROM section of a computer's memory (see Control Store).

Flag(s): An indicator, often a single bit, to indicate some conditions such as trace, carry, zero, and overflow.

Flash Memory: Nonvolatile and reprogrammable memory. Fabricated by using ETOXII (EPROM tunnel oxide) technology which is a combination of EPROM and EEPROM technologies. Can be reprogrammed while embedded in the board. However, one can only change a sector or block (consisting of multiple bytes) at a time.

Flowchart: Representation of a program in a schematic form. It is convenient to flowchart a problem before writing the actual programs.

Global Bus: A computer bus system that is available to and shared by a number of processors connected together in a multiprocessor system environment.

Handshaking: Data transfer via exchange of control signals between the microprocessor and an external device.

Hardware: The physical electronic circuits (chips) that make up the microcomputer system.

HCMOS: Low-power HMOS.

Hexadecimal Number System: Base-16 number system.

Hierarchical Memory: A memory organization or informational structure in which functional relationships are associated with different levels.

High-Level Language: A type of programming language that uses a more understandable human-oriented language such as Pascal.

HMOS: High-performance MOS reduces the channel length of the NMOS transistor and provides increased density and speed in LSI and VLSI circuits.

Immediate Address: An address that is used as an operand by the instruction itself.

Implied Address: An address not specified, but contained implicitly in the instruction.

In-Circuit Emulation: The most powerful hardware debugging technique; especially valuable when hardware and software are being debugged simultaneously.

Index: A symbol used to identify or place a particular quantity in an array (list) of similar quantities. Also, an ordered list of references to the contents of a larger body of data such as a file or record.

Indexed Addressing: Typically uses 3 bytes: the first byte for the op code and the next 2 bytes for the 16-bit address. The effective address of the instruction is determined by the sum of the 16-bit address and the contents of the index register.

Index Register: A register used to hold a value used in indexing data, such as when a value is used in indexed addressing to increment a base address contained within an instruction.

Indirect Address: A register holding a memory address to be accessed.

Instruction: A program statement (step) that causes the microcomputer to carry out an operation, and specifies the values or locations of all operands.

Instruction Cycle: The sequence of operations that a microprocessor has to carry out while executing an instruction.

Instruction Register (IR): A register storing instructions; typically 8 bits long for an 8-bit microprocessor.

Instruction Set: Lists all the instructions (available in machine code) that the microcomputer can execute.

Interleaved DMA: Using this technique, the DMA controller takes over the system bus when the microprocessor is not using it.

Internal Interrupt: Activated internally by exceptionally conditions such as overflow and division by zero.

Interpreter: A program that executes a set of machine language instructions in response to each high-level statement in order to carry out the function.

Interrupt I/O: An external device can force the microcomputer system to stop executing the current program temporarily so that it can execute another program known as the interrupt service routine.

Interrupts: A temporary break in a sequence of a program, initiated externally, causing control to pass to a routine, which performs some action while the program is stopped.

I/O (Input/Output): Describes that portion of a microcomputer system that exchanges data between the microcomputer system and the external world, or the data itself.

I/O Port: A module that contains control logic and data storage used to connect a microcomputer to external peripherals.

Keyboard: Has a number of pushbutton-type switches configured in a matrix form (rows × columns).

Keybounce: When a mechanical switch opens or closes, it bounces (vibrates) for a small period of time (about 10–20 ms) before settling down.

Large-Scale Integration (LSI): An LSI chip contains more than 100 gates.

Linkage Editors: Connect the individual programs together which are assembled or compiled independently.

Linked Programming: The process of joining a subprogram with a main program or joining two separate programs together to form a single program.

Local Area Network: A collection of devices and communication channels that connect a group of computers and peripherals devices together so that they can communicate with each other.

Logic Analyzer: A hardware development aid for microprocessor-based design; gathers data on the fly and displays it.

Logical Address Space: All storage locations with a programmer's addressing range.

Loops: A programming control structure where a sequence of microcomputer instructions are executed repeatedly (looped) until a terminating condition (result) is satisfied.

Machine Code: A binary code (composed of bit patterns) that a microcomputer can sense, read, interpret, recognize, and manipulate.

Machine Language: A type of microprocessor programming language that uses binary or hexadecimal numbers.

Macroinstruction: Commonly known as an instruction; initiates execution of a complete microprogram.

Macroprogram: The assembly language program.

Mask: A pattern of bits used to specify (or mask) which bit parts of another bit pattern are to be operated on and which bits are to be ignored or "masked" out.

Mask ROM: Programmed by a masking operation performed on the chip during the manufacturing process; its contents cannot be changed by the user.

Maskable Interrupt: Can be enabled or disabled by executing typically the instructions such as EI and DI, respectively. If the microprocessor's interrupt is disabled, the microprocessor ignores the interrupt.

Memory: Any storage device which can accept, retain, and read back data. Usually refers to a computer subsystem of internal RAM- or ROM-based storage devices.

Memory Access Time: Average time taken to read a unit of information from the memory.

Memory Address Register (MAR): Also known as the Data Counter (DC). Stores the address of the data; typically 16 bits long for 8-bit microprocessors.

Memory Cycle Time: Average time lapse between two successive read operations.

Memory Management Unit (MMU): Performs address translation and protection functions.

Memory Map: A representation of the physical location of software within a microcomputer's addressable main storage.

Memory-Mapped I/O: A microprocessor communications methodology (addressing scheme) where the data, address, and control buses extend throughout the system, with every connected device treated as if it were a memory location with a specific address. Manipulation of I/O data occurs in "interface registers" (as opposed to memory locations); hence there are no input (read) or output (write) instructions used in memory-mapped I/O.

Microcode: A set of "subcommands" or "pseudocommands" built into the hardware (usually stored in ROM) of a microcomputer (that is, firmware) to handle the decoding the execution of higher-level instructions such as arithmetic operation.

Microcomputer: Consists of a microprocessor, a memory unit, and an input/output unit.

Microcontroller: Typically includes a microcomputer, timer, A/D (Analog to Digital) and D/A (Digital to Analog) converters in the same chip.

Microinstruction: Most microprocessors have an internal memory called control memory. This memory is used to store a number of codes called microinstructions. These microinstructions are combined to design the instruction set of the microprocessor.

Microprocessor: The Central Processing Unit (CPU) of a microcomputer.

Microprocessor Development System: A tool for designing and debugging both hardware and software for microcomputer-based systems.

Microprocessor-Halt DMA: Data transfer is performed between the microcomputer's memory and a peripheral device either by completely stopping the microprocessor or by a technique called cycle stealing.

Microprogramming: The microprocessor can use microprogramming to design the instruction set. Each instruction in the instruction register initiates execution of a microprogram in the control unit to perform the operation required by the instruction.

Module: (1) Any single hardware arrangement (device or component) within a microcomputer system. (2) Any software, routine, or subroutine.

Monitor: Consists of a number of subroutines grouped together to provide "intelligence" to a microcomputer system. This intelligence gives the microcomputer system the capabilities for debugging a user program, system design, and displays.

Multiplexer: A hardware device which allows a microprocessor to be physically connected to a number of communication channels to receive or transmit data.

Multiprocessing: The process of executing two or more programs in parallel, handled by multiple processors all under common control. Typically each processor will be assigned specific processing tasks.

Multitasking: Operating system software that permits more than one program to run on a single microprocessor. Even though each program is given a small time slice in which to execute, the user has the impression that all tasks (different programs) are executing at the same time.

Multiuser: Describes a computer operating system that permits a number of users to access the system on a time-sharing basis.

Nested Subroutine: A commonly used programming technique that includes one subroutine entirely embedded within the "scope" of another subroutine.

Nibble: A 4-bit word.

NMOS: Denser and faster in comparison to PMOS. Most 8-bit microprocessors and some 16-bit microprocessors are fabricated using this technology.

Noncontiguous: Noncontiguous in nature. Refers to breaks in the linear sequential flow of any information structure.

Nonmaskable Interrupt: Occurrence of this type of interrupt cannot be ignored by the microprocessor, even though the interrupt capability of the microprocessor is disabled. Its effect cannot be disabled by instruction.

Non-Multiplexed: A non-multiplexed system indicates a direct single communication channel (that is, electrical wires) connection to the microprocessor.

Object Code: The binary (machine) code into which a source program is translated by a compiler, assembler, or interpreter.

Octal Number System: Base-8 number system.

One-Pass Assembler: This assembler goes through the assembly language program once and translates the assembly language program into a machine language program. This assembler has the problem of defining forward references. See Two-Pass Assembler.

Op Code (Operation Code): The instruction represented in binary form.

Operand: A datum or information item involved in an operation from which the result is obtained as a consequence of defined actions (that is, data which is operated on by an instruction). Various operand types contain information, such as source address, destination address, or immediate data.

Operating System: Consists of a number of program modules to provide resource management. Typical resources include microprocessors, disks, and printers.

Operation: (1) Means by which a result is obtained from an operand(s). (2) An action defined by a single instruction or single logical element.

Page: Some microprocessors, such as the Motorola 6800 and the MOS 6502, divide the 65,536 memory locations into 256 blocks. Each of these blocks is called a page and contains 256 addresses.

Parallel Operation: Any operation carried out simultaneously with a related operation.

Parallel Transmission: Each bit of binary data is transmitted over a separate wire.

Parity: The number of 1's in a word is odd for odd parity and even for even parity.

Peripherals: An I/O device capable of being operated under the control of a CPU through communication channels. Examples include disk drives, keyboards, CRTs, printers, modems, etc.

Personal Computer: Low-cost, affordable microcomputer used by an individual or a small group for video games, daily schedules, and industrial applications.

Physical Address Space: Address space is defined by the address pins of the microprocessor.

Pipeline: A technique that allows a microcomputer processing operation to be broken down into several steps (dictated by the number of pipeline levels or stages) so that the individual step outputs can be handled by the computer in parallel. Often used to fetch the processor's next instruction while executing the current instruction, which considerably speeds up the overall operation of the microcomputer.

Pointer: A storage location (usually a register within a microprocessor) that contains the address of (or points to) a required item of data or subroutine.

Polled Interrupt: A software approach for determining the source of interrupt in a multiple interrupt system.

POP Operation: Reading from the top or bottom of the stack.

Port: An access point (a register) for a microcomputer through which communication data may be passed to peripheral devices.

Primary Memory Store: That memory storage which is considered main, integral, or internal to the computing system. It is that storage which is physically most closely associated with the microprocessor and is directly controlled by it.

Primitives: A basic or fundamental unit; often refers to the lowest level of machine instruction or the lowest unit of programming language instruction.

Privileged Instructions: An instruction which is reserved for use by a computer's operating system, which will determine the range of system resources that the user is allowed to exploit.

Processor Memory: A set of microprocessor registers for holding temporary results when a computation is in progress.

Program: A self-contained sequence of computer software instructions (source code) that, when converted into machine code, directs the computer to perform specific operations for the purpose of accomplishing some processing task.

Program Array Logic (PAL): Similar to a ROM in concept except that it does not provide full decoding of the input lines. PAL's are used with 32-bit microprocessors for performing the memory decode function.

Program Counter (PC): A register that normally contains the address of the next instruction in the sequence of operations.

Programmed I/O: The microprocessor executes a program to perform all data transfers between the microcomputer system and external devices.

PROM (Programmable Read-Only Memory): Can be programmed by the user by using proper equipment. Once programmed, its contents cannot be altered.

Protocol: A list of data transmission conventions or procedures that encompass the timing, control, formatting, and data representations by which two devices are to communicate. Also known as hardware "handshaking", which is used to permit asynchronous communication.

PUSH Operation: Writing to the top or bottom of the stack.

Random Access Memory (RAM): A read/write memory. RAMs (static or dynamic) are volatile in nature (in other words, information is lost when power is removed).

Read-Only-Memory (ROM): A memory in which any addressable operand can be read from, but not written to, after initial programming. It is an asynchronous device whose access time is dictated by its internal circuit time delays. ROM storage is non-volatile (information is not lost when power is removed).

Real-Time Software: Computer code that allows processes to be performed during the actual time that a related physical I/O action takes place.

Reduced Instruction Set Computer (RISC): A necessary and sufficient instruction set is included. The RISC architecture maximizes speed by reducing clock cycles per instruction. Performs infrequent operations in software and frequent functions in hardware.

Register: A one-word, high-speed memory device usually constructed from flip-flops (electronic switches) that are directly accessible to the processor. It can also refer to a specific location in memory that contains word(s) used during arithmetic, logic, and transfer operations.

Register Indirect: Uses a register pair which contains the address of data.

Relative Address: An address used to designate the position of a memory location in a routine or program.

Rollover: Occurs when more than one key is pushed simultaneously.

Routine: A group of instructions for carrying out a specific processing operation. Usually refers to part of a larger program. A routine and subroutine have essentially the same meaning, but a subroutine could be interpreted as a self-contained routine nested within a routine or program.

Sample and Hold Circuit: When connected to the input of an A/D converter, it keeps a rapidly varying analog signal fixed during the A/D conversion process by storing it in a capacitor.

Scalar Microprocessor: Provided with one pipeline. Can execute one instruction per clock cycle. The 80486 is a scalar microprocessor.

Scaling: To adjust values or bring them into a range that is acceptable to a microcomputer.

Secondary Memory Storage: An auxiliary data storing device that supplements the main (primary) internal memory of a microcomputer. It is used to hold programs and data that would otherwise exceed the capacity of the main memory. Although it has a much slower access time, secondary storage is less expensive, Common devices include magnetic disk (floppy and hard), cassette tape, and videodisk.

Serial Transmission: Only one line is used to transmit the complete binary data bit by bit.

Single-Chip Microcomputer: Microcomputer (CPU, memory, and input/output) on a chip.

Single-Chip Microprocessor: Microcomputer CPU (microprocessor) on a chip.

Single Step: Allows the user to execute a program one instruction at a time and examine memory and registers.

Software: Programs in a microcomputer.

Source Code: The high-level language code used by a programmer to write computer instructions. This code must be translated to the object (machine) code to be usable to the microcomputer.

Stack: An area of read/write memory reserved to hold information about the status of a microcomputer the instant an interrupt occurs so that the microcomputer can continue processing after the interrupt has been handled. Another common use is in handling the accessing sequence of "nested" subroutines. The stacks are the last in/first out (LIFO) devices that are manipulated by using PUSH or POP instructions.

Stack Pointer: An address or register used to keep track of the storage and retrieval of each byte or word of information in the system stack.

Standard I/O: Utilizes a control pin on the microprocessor chip called the IO/M pin, in order to distinguish between input/output and memory; typically, IN and OUT instructions are used for performing input/output operations.

Static RAM: Stores data in flip-flops; does not need to be refreshed. Information is lost upon power failure unless backed up by battery.

Status Register: A register which contains information concerning the activity within the microprocessor or about the condition of a functional unit or peripheral device.

Subroutine: A program carrying out a particular function and which can be called by another program known as the main program. A subroutine needs to be placed only once in memory and can be called by the main program as many times as the programmer wants.

Superscalar Microprocessor: Provided with dual pipelining and executes more than one instruction per clock cycle. The Pentium is a superscalar microprocessor.

Supervisor: Provides the procedures or instructions for coordinating the use of system resources and maintaining the flow of operations through a microprocessor to perform I/O operations.

Supervisor State: When internal microprocessor system processing operations are conducted at a higher privilege level, it is usually in the supervisor state. An operating system typically executes in the supervisor state to protect the integrity of "basic" system operations from user influences.

Synchronous Operation: Operations that occur at intervals directly related to a clock period. Also, a bus protocol in such data transactions is controlled by a master clock and is completed within a fixed clock period.

Synchronous Serial Data Transmission: Data is transmitted or received based on a clock signal.

Tracing: A dynamic diagnostic technique in which a record of internal counter events is made to permit analysis (debugging) of the program's execution.

Tristate Buffer: Has three output states: logic 0, 1, and a high-impedance state. It is typically enabled by a control signal to provide logic 0 or 1 outputs. This type of buffer can also be disabled by the control signal to place it in a high-impedance state.

2's Complement: The 2's complement of a binary number is obtained by replacing each 0 with a 1 and each 1 with a 0 and adding one to the resulting number.

Two-Pass Assembler: This assembler goes through the assembly language program twice. In the first pass, the assembler defines the labels with the addresses. In the second pass, the assembler translates the assembly language program to the machine language. See One-Pass Assembler.

UART (Universal Asynchronous Receiver Transmitter): A chip that provides all the interface functions when a microprocessor transmits or receives data to or from a serial device. Converts serial data to parallel and vice versa.

User State: Typical microprocessor operations processing conducted at the user level. The user state is usually at lower privilege level than the supervisor state. This protects basic system operation resources (the operating system).

Vectored Interrupts: A device identification technique in which the highest priority device with a pending interrupt request forces program execution to branch to an interrupt routine to handle exception processing for the device.

Very Large Scale Integration (VLSI): A VLSI chip contains more than 1000 gates.

Virtual Machine: A microcomputer whose hardware and software architecture is specifically designed to support virtual storage techniques. The virtual machine concept is widely used within multiprogramming environments.

Virtual Memory: A memory management operating system technique that allows programs or data to exceed the physical size of the main, internal, directly accessed memory. Program or data segments/pages are swapped from external disk storage as needed. The swapping is invisible (transparent) to the programmer. Therefore the programmer need not be concerned with the actual physical size of internal memory while writing the code.

Word: The bit size of a microprocessor refers to the number of bits that can be processed simultaneously by the basic arithmetic circuits of the microprocessor. A number of bits taken as a group in this manner is called a word.

Table for American Standard Code for Information Interchange (ASCII), Standard No. X3.4—1968 of the American National Standards Institute.

$b_3b_2b_1b_0$	Row (hex)	$b_6b_5b_4$ (column)							
		000 0	001 1	010 2	011 3	100 4	101 5	110 6	111 7
0000	0	NUL	DLE	SP	0	@	P	`	p
0001	1	SOH	DC1	!	1	A	Q	a	q
0010	2	STX	DC2	"	2	B	R	b	r
0011	3	ETX	DC3	#	3	C	S	c	s
0100	4	EOT	DC4	$	4	D	T	d	t
0101	5	ENQ	NAK	%	5	E	U	e	u
0110	6	ACK	SYN	&	6	F	V	f	v
0111	7	BEL	ETB	'	7	G	W	g	w
1000	8	BS	CAN	(	8	H	X	h	x
1001	9	HT	EM	)	9	I	Y	i	y
1010	A	LF	SUB	*	:	J	Z	j	z
1011	B	VT	ESC	+	;	K	[	k	{
1100	C	FF	FS	,	<	L	\	l		
1101	D	CR	GS	–	=	M	]	m	}
1110	E	SO	RS	.	>	N	^	n	~
1111	F	SI	US	/	?	O	_	o	DEL

Control codes

NUL	Null	DLE	Data link escape
SOH	Start of heading	DC1	Device control 1
STX	Start of Text	DC2	Device control 2
ETX	End of text	DC3	Device control 3
EOT	End of transmission	DC4	Device control 4
ENQ	Enquiry	NAK	Negative acknowledge
ACK	Acknowledge	SYN	Synchronize
BEL	Bell	ETB	End transmitted block
BS	Backspace	CAN	Cancel
HT	Horizontal tab	EM	End of medium
LF	Line feed	SUB	Substitute
VT	Vertical tab	EXC	Escape
FF	Form feed	FS	File separator
CR	Carriage return	GS	Group separator
SO	Shift out	RS	Record separator
SI	Shift in	US	Unit separator
SP	Space	DEL	Delete or rubout

Bibliography

Allison, D. R., "A Design Philosophy for Microcomputer Architectures", *IEEE Trans. Computers.*

Artwick, B. A., *Microcomputer Interfacing,* Prentice-Hall, 1980.

Baer, J.-L., *Computer Systems Architecture,* Computer Science Press, 1980.

Boyce, J. C., *Microprocessor and Microcomputer Basics,* Prentice-Hall, 1979.

Breeding, K., *Microprocessor System Design Fundamentals,* Prentice-Hall, 1995.

Brey, B., *The Motorola Microprocessor Family: 68000, 68008, 68010, 68020, 68030, and 68040,* Saunders College Publishing, 1992.

Burns, J., "Within the 68020," *Electronics and Wireless Word,* pp 209–212, February 1985; pp 103–106, March 1985.

Chi, C. S., "Advances in Mass Storage Technology," *IEEE Computer,* Vol. 15, no. 5, pp 60–74, May 1982.

Chow, C. K., "On Optimization of Storage Hierarchies," *IBM Journal of Research and Development,* pp 194–203, May 1974.

Cohn, D. L. and Melsa, J. L., *A Step by Step Introduction to 8080 Microprocessor Systems,* Dilithium Press, 1977.

Cramer, W. and Kane, G., *68000 Microprocessor Handbook,* 2nd ed., Osborne/McGraw-Hill, 1986.

Danhor, K. J. and Smith, C. L., *Computing System Fundamentals: An Approach Based on Microcomputers,* Addison-Wesley, 1981.

Denning, P. J., "Virtual Memory," *ACM Computing Surveys,* Vol. 2, no. 3, pp 153–159, September 1970.

Electronic Industries Association, Washington, D.C., EIA Standard RS-232-C Interface, Electronic Industries Association, 1969.

Faggin, F., "How VLSI Impacts Computer Architecture," *IEEE Spectrum,* pp 28–31, May 1978.

Feibus, M. and Slater, M., "Pentium Power," *PC Magazine,* April 27, 1993.

Fisher, E. and Jensen, C. W., *Pet and the IEEE 488 Bus (BPIB),* Osborne/McGraw-Hill, 1979.

Friedman, A. D., *Logical Design of Digital Systems,* Computer Science Press, 1975.

Garland, H., *Introduction to Microprocessor System Design,* McGraw-Hill, 1979.

Gay, "6800 Family Memory Management — Part 1," *Electronic Engineering,* pp 39–48, June 1986.

Gibson, G. A. and Liu, Y., *Microprocessors for Engineers and Scientists,* Prentice-Hall, 1980.

Gill, A., *Machine and Assembly Language Programming of the PDP-11,* 2nd ed., Prentice-Hall, 1983.

Girsline, G., *16-Bit Modern Microcomputers, The Intel 8086 Family,* Prentice-Hall, 1985.

Gladstone, B. E., "Comparing Microcomputer Development System Capabilities," *Computer Design,* pp 83–90, February 1979.

Goody, R. W., *Intelligent Microcomputer,* SRA, 1982.

Goody, R., *The Versatile Microcomputer, The Motorola Family,* SRA, 1984.

Greenfield, J. D., *Practical Digital Design Using IC's,* John Wiley & Sons, 1977.

Greenfield, J. D. and Wray, W. C., *Using Microprocessors and Microcomputers: The 6800 Family,* John Wiley & Sons, 1983.

Greenfield, J. D., *Practical Digital Design Using IC's,* John Wiley & Sons, 1983.

Grinich, V. H. and Jackson, H. G., *Introduction to Integrated Circuits,* McGraw-Hill, 1975.

Hall, D. V., *Microprocessors and Digital Systems,* McGraw-Hill, 1980.

Hamacher, V. C., Vranesic, Z. G., and Zaky, S. G., *Computer Organization,* McGraw-Hill, 1978.

Hamacher, V. C., Vranesic, Z. G., and Zaky, S. G., *Computer Organization,* McGraw-Hill, 1984.

Harman, T. L. and Lawson, B., *The Motorola MC68000 Microprocessor Family,* Prentice-Hall, 1984.

Hartman, B., "16-Bit 68000 Microprocessor Concepts on 32-Bit Frontier," MC 68000 Article Reprints, Motorola, pp 50–57, March 1981.

Hayes, J. P., *Computer Architecture and Organization,* McGraw-Hill, 1978.

Hayes, J. P., *Digital System Design and Microprocessors,* McGraw-Hill, 1984.

Haynes, J. L., "Circuit Design with Lotus 1-2-3," *BYTE,* Vol. 10, no. 11, pp 143–156, 1985.

Hewlett-Packard, "HP 64000," *Hewlett-Packard Journal,* 1980.

Hnatek, E. R., *A User's Handbook of Semiconductor Memories,* John Wiley & Sons, 1977.

Holt, C. A., *Electronic Circuits — Digital and Analog,* John Wiley & Sons, 1978.

Horden, I., "Microcontrollers Offer Realtime Robotics Control," *Computer Design,* pp 98–101, October 15, 1985.

IEEE, "Technology 1994" — *The Spectrum,* January 1994.

IEEE, "Technology 1995" — *The Spectrum,* January 1995.

Intel, *Microprocessors and Peripheral Handbook,* Vol. 1, Microprocessors, Intel Corporation, 1988.

Intel, *Microprocessors and Peripheral Handbook,* Vol. 2, Peripheral, Intel Corporation, 1988.

Intel, *80386 Programmer's Reference Manual,* Intel Corporation, 1986.

Intel, *80386 Hardware Reference Manual,* Intel Corporation, 1986.

Intel, 80386 Advance Information, Intel Corporation, 1985.

Intel, *80387 Programmer's Reference Manual,* 1987.

Intel, *Intel 486 Microprocessor Family Programmer's Reference Manual,* 1992.

Intel, *Intel 486 Microprocessor Hardware Reference Manual,* 1992.

Intel, *i960 SA/SB Microprocessor,* 1991.

Intel, *Pentium Processor User's Manual,* 1993.

Intel, *8080 and 8085 Assembly Language Programming Manual,* Intel Corporation, 1978.

Intel, *The 8086 Family User's Family,* Intel Corporation, 1979.

Intel, *Intel Component Data Catalog,* Intel Corporation, 1979.

Intel, *MCS-85 User's Manual,* Intel Corporation, 1978.

Intel, *MCS-86 User's Manual,* Intel Corporation, 1982.

Intel, *Memory Components Handbook,* Intel Corporation, 1982.

Intel, *SDK-85 User's Manual,* Intel Corporation, 1978.

Intel, "Marketing Communications," *The Semiconductor Memory Book,* John Wiley & Sons, 1978.

Isaacson, R. et al., "The Oregon Report — Personal Computing," selected reprints from *IEEE Computer,* pp 226–237.

Johnson, "A Comparison of Mc68000 Family Processors," *BYTE,* pp 205–218, September 1986.

Johnson, C. D., *Process Control Instrumentation Technology,* John Wiley & Sons, 1977.

Johnson, R. C., "Microsystems Exploit Mainframe Methods," *Electronics,* 1981.

Kane, G., *CRT Controller Handbook,* Osborne/McGraw-Hill, 1980.

Kane, G., Hawkins, D., and Leventhal, L., *68000 Assembly Language Programming,* Osborne/McGraw-Hill, 1981.

King, T. and Knight, B., *Programming the MC68000,* Addison-Wesley, 1983.

Krutz, R. L., *Microprocessors and Logic Design,* John Wiley & Sons, 1980.

Krutz, R. L., *Microprocessors and Logic Design,* John Wiley & Sons, 1977.

Lesea, A. and Zaks, R., *Microprocessor Interfacing Techniques,* Sybex, 1978.

Leventhal, L. A., *8080A/8085 Assembly Language Programming,* Osborne/McGraw-Hill, 1978.

Leventhal, L. A., *Introduction to Microprocessors: Software, Hardware Programming,* Prentice-Hall, 1978.

Leventhal, L. and Walsh, C., *Microcomputer Experimentation with the Intel SDK-85,* Prentice-Hall, 1980.

Lewin, M., *Logic Design and Computer Organization*, Addison-Wesley, 1983.

Lipschutz, S., *Essential Computer Mathematics*, Schaum Outline Series, McGraw-Hill, 1982.

MacGregor, Mothersole, Meyer, *The Motorola MC68020*," IEEE MICRO, pp 101–116, August 1984.

MacGregor, "Diverse Applications Put Spotlight on 68020's Improvements," *Electronic Design*, pp 155–164, February 7, 1985.

MacGregor, "Hardware and Software Strategies for the MC68020," *EDN*, pp 163–168, June 20, 1985.

Mano, M., *Computer System Architecture*, Prentice-Hall, 1983.

McCartney, Groepler, "The 32-Bit 68020's Power Flows Fully Through a Versatile Interface," *Electronic Design*, pp 335–343, January 10, 1985.

Miller, M., Raskin, R., and Rupley, S., "The Pentium that Stole Christmas," *PC Magazine*, February 27, 1995.

MITS-ALTAIR, *S-100 Bus*, MITS, Inc., Albuquerque, NM.

Morse, S., *The 8086/8088 Primer*, 2nd ed., Hayden, 1982.

Motorola, *6809 Applications Notes*, Motorola Corporation, 1978.

Motorola, *MC68000 User's Manual*, Motorola Corporation, 1979.

Motorola, *16-Bit Microprocessor — MC68000 User's Manual*, 4th ed., Prentice-Hall, 1984.

Motorola, *MC68000 16-Bit Microprocessor User's Manual*, Motorola Corporation, 1982.

Motorola, *MC68000 Supplement Material (Technical Training)*, Motorola Corporation, 1982.

Motorola, *Microprocessor Data Material*, Motorola Corporation, 1981.

Motorola, *MC68020 User's Manual*, Motorola Corporation, 1985.

Motorola, "MC68020 Course Notes," MTTA20 REV 2, July 1987.

Motorola, "MC68020/68030/88100 Audio Course Notes," 1988.

Motorola, *MC88100 Data Sheets*, Motorola Corporation, 1988.

Motorola, *MC68020 User's Manual*, 2nd ed., MC68020 UM/AD Rev. 1, Prentice-Hall, 1984.

Motorola, *Programmer's Reference Manual* (Includes CPU 32 Instructions), 1989.

Motorola, *MC68040 User's Manual*, 1989.

Motorola, *Power PC 601, RISC Microprocessor User's Manual*, 1993.

Motorola Technical Summary, *32-Bit Virtual Memory Microprocessor*, MC68020 BR243/D. Rev. 2, Motorola Corporation, 1987.

Myers, G. and Budde, D., *The 80960 Microprocessor Architecture*, John Wiley & Sons, 1988.

Osborne, A., *An Introduction to Microprocessors*, Vol. 1, Basic Concepts, rev. ed., Osborne/McGraw-Hill, 1980; 2nd ed., 1982.

Osborne, A. and Kane, G., *The Osborne Four- and Eight-Bit Microprocessor Handbook*, Osborne/McGraw-Hill, 1980.

Osborne, A. and Kane, G., *The Osborne 16-Bit Microprocessor Handbook*, Osborne/McGraw-Hill, 1981.

Rafiquzzaman, M., *Microprocessors and Microcomputer Development Systems — Designing Microprocessor-Based Systems*, Harper and Row, 1984.

Rafiquzzaman, M., *Microprocessors and Microcomputer Development Systems*, John Wiley & Sons, 1984.

Rafiquzzaman, M., *Microcomputer Theory and Applications with the INTEL SDK-85*, 2nd ed., John Wiley & Sons, 1987.

Rafiquzzaman, M., *Microprocessors — Theory and Applications — Intel and Motorola*, Prentice-Hall, 1992.

Rafiquzzaman, M. and Chandra, *Modern Computer Architecture*, West, 1988.

RCA, *Evaluation Kit Manual for the RCA CDP1802 COSMAC Microprocessor*, RCA Solid State Division, Somerville, NJ.

Rector, R. and Alexy, G., *The 8086 Book*, Osborne/McGraw-Hill, 1980.

Reichborn-Kjennerud, G., "Novel Methods of Integer Multiplication and Division," *BYTE*, Vol. 8, no. 6, pp 364–374, June 1983.

Ripps, Mushinsky, "32-Bit Up Speeds Code Design and Execution," *EDN*, pp 163–168, June 27, 1985.

Rockwell International, *Microelectronic Devices Data Catalog,* 1979.

Short, K. L., *Microprocessors and Programmed Logic,* Prentice-Hall, 1981.

Sloan, M. E., *Introduction to Minicomputers and Microcomputers,* Addison-Wesley, 1980.

Smith, J. and Weiss, S., "Power PC 601 and Alpha 21064: A Tale of Two RISCs," *IEEE Computer,* June 1994.

Solomon, "Motorola's Muscular 68020," *Computers & Electronics,* pp 74–79, October 1984.

Sowell, E. F., *Programming in Assembly Language, MACRO II,* Addison-Wesley, 1984.

Starnes, T. W., "Compact Instruction Set Gives the MC68000 Power While Simplifying Its Operation," MC68000 Article Reprints, Motorola, pp 43–47, March 1981.

Strauss, E., *The Waite Group, Inside the 80286,* A Brady Book published by Prentice-Hall, 1986.

Stone, H. S., *Introduction to Computer Architecture,* SRA, 1980.

Stone, H. S., *Microcomputer Interfacing,* Addison-Wesley, 1982.

Streitmatter, G. A. and Fiore, V., *Microprocessors, Theory and Applications,* Reston Publishing, 1979.

Stritter, E. and Gunter, T., "A Microprocessor Architecture for a Changing World: The Motorola 68000," *IEEE Computers,* Vol. 12, no. 2, pp 43–52, February 1970.

Tanenbaum, A. S., *Structured Computer Organization,* Prentice-Hall, 1984.

Teledyne, *Teledyne Semiconductor Catalog,* 1977.

Texas Instruments, *The TTL Data Book,* Vol. 1, 1984.

Texas Instruments, *The TTL Data Book for Design Engineers,* 2nd ed., 1976.

Tocci, R. J. and Laskowski, L. P., *Microprocessors and Microcomputers: Hardware and Software,* Prentice-Hall, 1979.

Triebel, W., *The 80386 DX Microprocessor,* Prentice-Hall, 1992.

Twaddel, "32-Bit Extension to the 68000 Family Addresses 7 GBytes, Runs at 3 MIPS," *EDN,* pp 75–77, July 12, 1984.

Wakerly, J. F., *Microcomputer Architecture and Programming,* John Wiley & Sons, 1981.

Zilog, *Z8000 Advance Specification,* Zilog, Inc., 1978.

Zoch, B., "68020 Dynamically Adjusts Its Data Transfers to Match Peripheral Ports," *Electronic Design,* pp 219–225, January 10, 1985.

Zorpette, G., "Microprocessors — The Beauty of 32-Bits," *IEEE Spectrum,* Vol. 22, no. 9, pp 65–71, September 1985.

Credits

The following material was reprinted by permission of the sources indicated below:

Motorola Corporation, Inc.: **Chapter 1**: Figures 1.10, 1.11, 1.12; **Chapter 5**: Figures 5.1 to 5.5, 5.7, 5.10 to 5.12, 5.14, 5.15, 5.17, 5.21 to 5.24, Tables 5.1, 5.2, 5.14 to 5.16, table on page 332; **Chapter 6**: Examples 6.1 and 6.8, all figures, tables, and graphics; **Chapter 7**: Figures 7.1 to 7.29, Tables 7.1 to 7.7; **Chapter 8**: Figures 8.5 to 8.18, Tables 8.7 to 8.18; **Chapter 9**: Figures 9.7, 9.8; **Appendix B**: data sheets; **Appendix D**: 68000 Instructions.

Intel Corporation: **Chapter 2**: Figures 2.3, 2.4; 2.6 to 2.8, 2.10 to 2.15, 2.17, 2.23 to 2.25, Tables 2.9, 2.12, 2.13; **Chapter 3**: Figures 3.1 to 3.3, 3.5 to 3.13, 3.24, 3.26a and b, Table 3.2, structure on page 165; **Chapter 4**: Figures 4.1, 4.2, 4.6 to 4.28, Tables 4.1 to 4.3, 4.5 to 4.11, table on page 244; **Chapter 7**: Figure 7.30; **Chapter 8**: Figures 8.1 to 8.4, Tables 8.1 to 8.3; **Chapter 9**: Figures 9.15 , 9.16; **Appendix C**: figures and data sheets; **Appendix E**: 88086 Instructions.

All mnemonics in Tables 2.1 and 3.1 are courtesy of Intel Corporation.

The 80386 microprocessor referred to in text as the i386™, the 80486 as the i486™ and the Pentium as the Pentium™, trademarks of Intel Corporation.

Rafiquzzaman, M., *Microcomputer Theory and Applications with the Intel SDK-85*, John Wiley & Sons, Inc., New York, New York, 1987, reprinted by permission of John Wiley & Sons, Inc.: **Chapter 2**: Tables 2.1 to 2.8, Section 2.6; **Chapter 5**: Tables 5.6 to 5.13, Examples 5.1 to 5.5.

Rafiquzzaman, M., *Microcomputer Theory and Applications with the Intel SDK-85*, John Wiley & Sons, Inc., New York, New York, 1987, reprinted by permission of Prentice Hall, Inc., Englewood Cliffs, New Jersey: **Chapter 2**: Figures 2.18 to 2.20, 2.24, 2.25, 2.29a and b, 2.30, figure on page 92, Sections 2.5 and 2.9.2, pages 53 to 56, 77 to 89, 93 to 99; **Chapter 5**: Figures 5.6, 5.7, 5.11, 5.14, 5.24, 5.31, 5.32, Tables 5.2, 5.4, 5.5, text on pages 280 to 303, 305 to 307, 309 to 312, 314 to 319, 323, 328, 334 to 335, 341 to 346.

Rafiquzzaman, M., *Microprocessors and Microcomputer Development Systems*, copyright John Wiley & Sons, Inc., New York, New York, ©1984, reprinted by permission of John Wiley & Sons, Inc.: **Chapter 9**: Figures 9.1, 9.2, 9.8 to 9.13, Section 9.1.2; **Chapter 10**: Problems 10.2 to 10.13; **Appendix A**; **Appendix C** (excluding figures and data sheets).

Morse, S. amd Albert, D., *The 80286 Architecture*, John Wiley & Sons, Inc., New York, New York, ©1986, reprinted by permission of John Wiley & Sons, Inc.: **Chapter 4**: Figures 4.3 and 4.4.

Practical Microprocessors — Hardware, Software and Troubleshooting, Hewlett Packard, Palo Alto, California: **Chapter 2**: Figure 2.1.

Burns, D. and Jones, D., Within the 68020, *Electronics and Wireless World*, Surrey, United Kingdom, ©1987: **Chapter 7**: Figures 7.3, 7.27, Example 7.1.

Gay, C., MC68000 Family Memory Management, *Electronic Engineering*, 58(714), June 1986, reprinted by permission of Electronic Engineering, London, United Kingdom, ©1986: **Chapter 7**: Figure 7.23.

Index